Plant Nutrition

Food security and sustainability of agro-ecosystems through basic and applied research

Edited by

W.J. Horst
University of Hannover,
Hannover, Germany

M.K. Schenk
University of Hannover,
Hannover, Germany

A. Bürkert
University of Kassel,
Witzenhausen, Germany

N. Claassen
University of Göttingen,
Göttingen, Germany

H. Flessa
University of Göttingen,
Göttingen, Germany

W.B. Frommer
University of Tübingen,
Tübingen, Germany

H. Goldbach
University of Bonn,
Bonn, Germany

H.-W. Olfs
Pflanzenernährung und Umweltforschung,
Hydro Agri,
Dülmen, Germany

V. Römheld
University of Hohenheim,
Stuttgart, Germany

B. Sattelmacher
University of Kiel,
Kiel, Germany

U. Schmidhalter
TU München,
Freising-Weihenstephan, Germany

S. Schubert
University of Gießen,
Gießen, Germany

N. v. Wirén
University of Tübingen,
Tübingen, Germany

L. Wittenmayer
University of Halle-Wittenberg,
Halle, Germany

Part 3

KLUWER ACADEMIC PUBLISHERS
DORDRECHT / BOSTON / LONDON

W. J. Horst et al.(Eds.), Plant nutrition – Food security and sustainability of agro-ecosystems. 676~677, 2001

Nitrogen release of wheat roots grown under semi-sterile in comparison to unsterile conditions

A. Deubel, M. Richert, and W. Merbach
Institute of Soil Science and Plant Nutrition, Martin Luther-University of Halle-Wittenberg, A.-Kuckhoff-Str. 17b, Halle (Saale), D 06108, Germany, deubel@landw.uni-halle.de

Key words: microbial colonization, ^{15}N labelling, net-N release, root exudation, wheat

Abstract

It is often postulated that microorganisms have a promotional effect on root exudation. To determine the influence on microbial colonisation in the rhizosphere on the release of nitrogen compounds spring wheat (*Triticum aestivum*) was grown for four weeks in pots containing a 5:1 mixture of soil and quartz sand under semi-sterile and unsterile conditions. ^{15}N was applied three times to the shoots as $^{15}NH_3$ for four hours. Plants were harvested after three days (tillering stage) and 28 days (shooting stage) respectively after the last labelling period. Total N and ^{15}N were measured in shoots, roots, and soil compartments using an elemental analyser coupled with an emission spectrometer.
Approximately 30 % of the N present in the plant tissues three days after the last labelling period were derived from atmospheric ammonia. Approx. 90 % of the ^{15}N incorporated by the plants were found in the shoots, 8 % in the roots and 2-3 % were released in the soil. ^{15}N-root portion and net-N release of unsterile growing wheat significantly increased up to the second harvest (13 % and 7 % respectively) while shoot portion declined to 80 %. Under semi-sterile conditions, the nitrogen release decreased to 40 % in comparison to unsterile plants and also a significantly higher percentage of ^{15}N in the rhizosphere was water soluble.

Introduction

An understanding of net-N release by living plant roots is of great importance in the assessment of nitrogen balances in agrarian ecosystems and for calculation of N fertilisation. Interactions between plant roots, microorganisms and soil particles in the rooted soil zones and especially in the root–soil contact zone (rhizosphere) are of high ecological importance. Root-borne organic substances influence the solubility, sorption and transport of nutrients and pollutants, the decomposition of organic materials as well as the activity and the turnover of microorganisms (Curl and Truelove, 1986). A promotive effect of microorganisms on root exudation is often postulated. Detailed studies with plants in soil substrates, however, have shown both higher (Meharg and Killham, 1991) and lower (Whipps and Lynch, 1983) exudation rates. Most results are based on the determination of the carbon distribution or the release of specific compounds, e.g. amino acids. To determine the influence of microbial colonisation in the rhizosphere on the release of nitrogen compounds by roots, wheat plants were grown under semi-sterile and unsterile conditions. In order to distinguish between root-borne and soil-derived organic substances, it is necessary to label the plant material with isotopes. The results of Janzen and Bruinsma (1989) and Merbach *et al.* (2000) indicate that foliar application of gaseous $^{15}NH_3$ is well suitable to achieve sufficient labelling of the plants with ^{15}N and to quantify the ^{15}N distribution in plant-soil systems also under sterile conditions.

Materials and methods

Plants were grown in 750 ml glass pots with a gas-tight separation between the shoot compartment and the sterilisable root compartment. A 5:1 (m/m) mixture of quartz sand and soil (Haplic Phaeozem) was used as substrate. Following fertilisation was given per 1 kg substrate: 110 mg N as NH_4NO_3, 95 mg P as $CaHPO_4 \cdot 2 H_2O$, 240 mg K as K_2SO_4, 54 mg Mg as $MgSO_4 \cdot 7 H_2O$, 3 mg Fe as $FeCl_3$ solution (5 %), 0,18 mg A-Z solution of Hoagland (a+b).

The substrate was autoclaved four times with intermittent incubation (semi-sterile treatment) and recontaminated for the unsterile treatment by mixing with 10 % fresh substrate. Seeds were surface-sterilised under vacuum with 5 % NaClO. After four weeks (tillering stage) ^{15}N was applied to the shoots as $^{15}NH_3$ for four hours (100 ppm NH_3 with 95 at.-% $^{15}N_{exc.}$). This procedure was repeated three times. Three days after the last labelling period the first plant samples were harvested. The second harvest took place four weeks after the last labelling (shooting stage). Plant roots were carefully removed from the bulk soil and dipped into water for two minutes to obtain water-soluble root exudates and rhizosphere soil. Total N and ^{15}N were measured in shoots, roots, soil compartments and in water using an elemental analyser coupled with an ^{15}N emission spectrometer (NOI 7, FAN Leipzig).

Results

Wheat plants assimilated more than 50 % of the ^{15}N supplied. The ^{15}N abundance in shoots and roots reached approximately 30 and 20 at.-% $^{15}N_{exc.}$ respectively three days after labelling. The ^{15}N abundance in shoots was comparable in both treatments and at both times. After 28 days however, the roots of unsterile grown wheat plants had a higher ^{15}N abundance in comparison with plants under semi-sterile conditions (Fig. 1). In addition the total

N content of the unsterile grown roots was higher (3.1 %) than that of the semi-sterile grown roots (2.5 %).

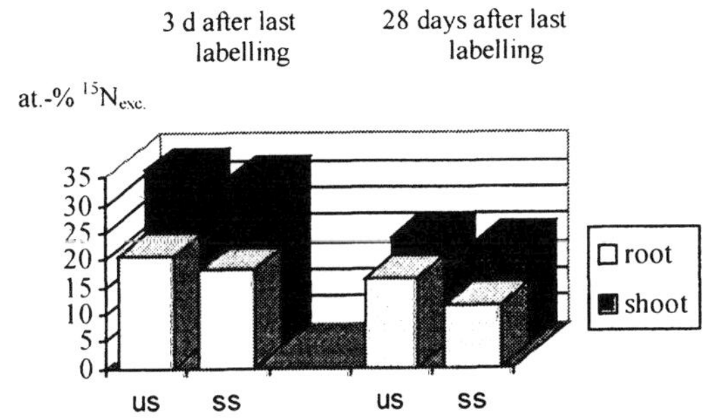

Figure 1. ^{15}N abundance in roots and shoots of wheat plants grown under unsterile (us) or semi-sterile (ss) conditions.

Most of the ^{15}N in the entire system was found in the plant shoots (Fig. 2). The net-N release of wheat plants under unsterile conditions was in the range of 3 % three days after the last labelling and 7 % after 28 days. Under semi-sterile conditions, the nitrogen release decreased to 40 % in comparison with unsterile plants. Also the ^{15}N amount in the root system of unsterile growing plants increased from first to second harvest while it remained constant under semi-sterile conditions.

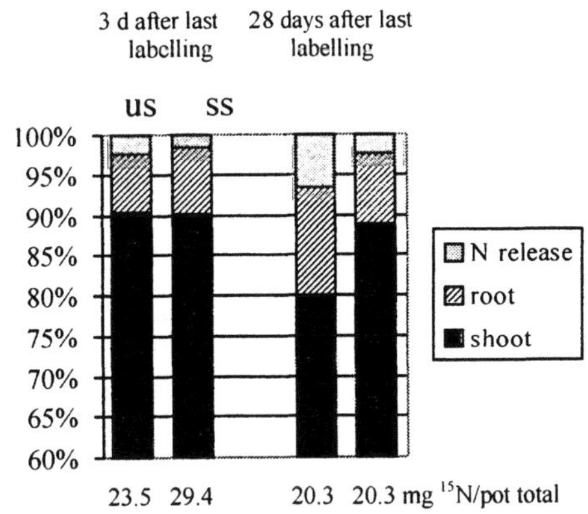

Figure 2. ^{15}N distribution in roots, shoots and N release of wheat plants grown under unsterile (us) or semisterile (ss) conditions

Approximately 90 % of ^{15}N released by roots was found in the bulk soil. While under semi-sterile conditions the biggest portion of labelled nitrogen compounds in the rhizosphere was water-soluble, under unsterile conditions nearly the half of ^{15}N in the rhizosphere was insoluble in water (Fig. 3).

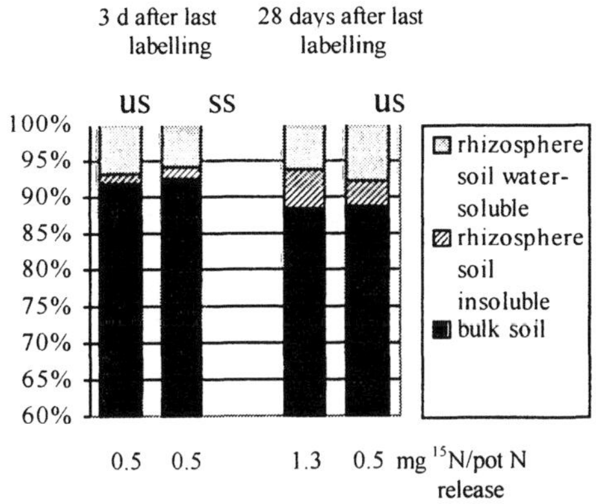

Figure 3. ^{15}N distribution in the rhizo-deposition of wheat plants grown under unsterile (us) or semi-sterile (ss) conditions.

Discussion

Approximately 30 % of the N present in the plant tissues three days after the last labelling period was derived from atmospheric ammonia. According to the results of Janzen and Bruinsma (1989), a high labelling of nitrogen compounds with this method is possible. Up to 7 % (28 days after the last labelling, shooting stage) of the ^{15}N incorporated in the plants was released into the soil under unsterile conditions. This value is in a similar range to those observed by Janzen and Bruinsma (1993) and Reining *et al.* (1995) although different methods of ^{15}N labelling were used. It can be concluded that microbial colonization resulted in a considerable increase of nitrogen translocation to roots and a significant higher net-N release. Rhizosphere flora seems to be a sink for nitrogen. Similar effects are known for the release of carbon compounds (Merbach and Ruppel, 1992, Biondini *et al.*, 1988). We found visible differences in root morphology between semi-sterile and unsterile plants. Also microbial autolysis of root cortex cells is conceivable. The decrease in water solubility of root deposits under unsterile conditions could be a proof for secondary metabolisation of primary root-borne compounds by microorganisms.

References

Biondini M, Klein DA and Redente EF 1988 Soil Biol. Biochem. 20, 477-482.

Curl EA and Truelove B 1986 The rhizosphere. Springer, Berlin, Heidelberg, New York, Tokyo.

Janzen HH and Bruinsma Y 1989 Soil Biol. Biochem. 21, 189-196.

Janzen HH and Bruinsma Y 1993 Soil Biol. Biochem. 25, 631-632.

Meharg AA and Killham K 1991 Plant Soil 113, 111-116.

Merbach W and Ruppel S 1992 Photosynth. 26, 551-554.

Merbach W, Schulze J, Richert M, Rrocco E and Mengel K 2000 J. Plant Nutr. Soil Sci. 163, 375-379.

Reining E, Merbach W and Knof G 1995 Isotopes Environ. Health Stud. 31, 345-349.

Whipps SM and Lynch JM 1983 New Phytol. 95, 605-623.

W. J. Horst et al.(Eds.), Plant nutrition – Food security and sustainability of agro-ecosystems. 678~679, 2001

Exudation of nitrogenous compounds by clover and ryegrass in sterile and non-sterile micro-lysimeters

F. Paynel, C. Clement, J. Bigot and J. B. Cliquet

UMR INRA 950 de Physiologie et Biochimie Végétales, IRBA, Université de Caen, 14032 CAEN Cedex, France, cliquet@ibba.unicaen.

Key words: amino acid, ammonium, *Lolium perenne* L., micro-organisms, root exudates, *Trifolium repens* L.

Abstract

Exudation of ammonium and amino acids from white clover and perennial ryegrass roots was investigated in a micro-lysimeter system.

A higher proportion of ammonium than amino acids was observed in the exudate. The most abundant amino acids in both clover and ryegrass exudates were serine and glycine. There was no correlation between amino acid profiles of exudates and root extracts: asparagine was the major amino acid in clover roots and glutamine in ryegrass roots. The presence of micro-organisms in the growth medium had no effect on the amounts of ammonium, serine and glycine measured in exudates, while amounts of glutamate and aspartate were increased in non-sterile clover exudates. It is proposed that bacteria have no substantial effect on the N-exudation in sand by young clover and ryegrass plants.

Introduction

Legumes/grasses mixtures and particularly mixtures of white clover (*Trifolium repens* L.) and perennial ryegrass (*Lolium perenne* L.) are commonly used in agriculture. Indeed, it has been shown that part of the grass nitrogen is derived from N-fixation by the clover allowing benefit for production. The N-transfer between plants can occur by an indirect pathway involving senescence and degradation of donor plant tissues or exudation of N-compounds into the rhizosphere by the plant followed by uptake of these compounds by the receiver plant. To study the first step of this second pathway, the quantity and quality of N-compounds in exudates obtained from white clover and perennial ryegrass were measured.

Many studies indicate that mineral nitrogen is released mainly as ammonium and organic N as amino acids (Ta *et al.* 1986). The proportions of ammonium and amino acids measured in root exudates depend on plant species and plant age (Shepherd and Davies, 1994). In the present study, these compounds were measured in the exudates of young clover and ryegrass plants. In addition, the effect of bacteria on this exudation was also studied in a micro-lysimeter experiment to test their competitive ability for acquisition of N released by plants.

Materials and methods

Surface sterilised seeds were germinated in micro-lysimeters that are each composed of a 60 ml syringe containing sterilised sand as the growth medium and allowing collection of exudates.

The plants were supplied with a nutrient solution containing 0.5 mM KNO_3.

Three levels of sterility were compared:

1- micro-lysimeters, sand, nutrient solution were sterilised. The root system was maintained under sterile conditions, isolated from the atmosphere with cotton and the nutrient solution was added through a tube across the cotton (noted "sterile" in the results),

2- same as 1, but roots were not protected with cotton (noted "semi-sterile" in the results),

3- nothing, except the seeds and sand were sterilised (noted "non-sterile" in the results).

An aliquot of the exudate was used to count the bacterial population present in the micro-lysimeters. No bacteria were found in the sterile exudate, while the number of Colony Forming Units reached 10^5 ml^{-1} in the exudate from the two other conditions.

Exudates of 39-day-old ryegrass and 67-day-old clover were collected and analysed for ammonium and amino acids contents. The plants were harvested and amino acids were extracted and analysed.

Results

Both species released large amounts of ammonium (135 $nmol.plant^{-1}.day^{-1}$ for clover and 150 for ryegrass). The level of sterility had no effect on this compound exudation (Figure 1). Amino acids, were released in smaller amounts (between 4 and 6.5 nmol $plant^{-1}$ day^{-1}). Glycine and serine were the major amino acids exuded by both plants. Aspartate and glutamate were also released in significant amounts. The same amounts and profiles of amino acids were recovered from non-sterile and sterile micro-lysimeters containing ryegrass. In contrast, slightly more amino acids were analysed in exudates obtained from clover grown in non-sterile conditions compared to sterile clover (6.3 and 4.3 nmol $plant^{-1}$ day^{-1} respectively). This was due to higher amounts of glutamate and aspartate in the non-sterile experiment.

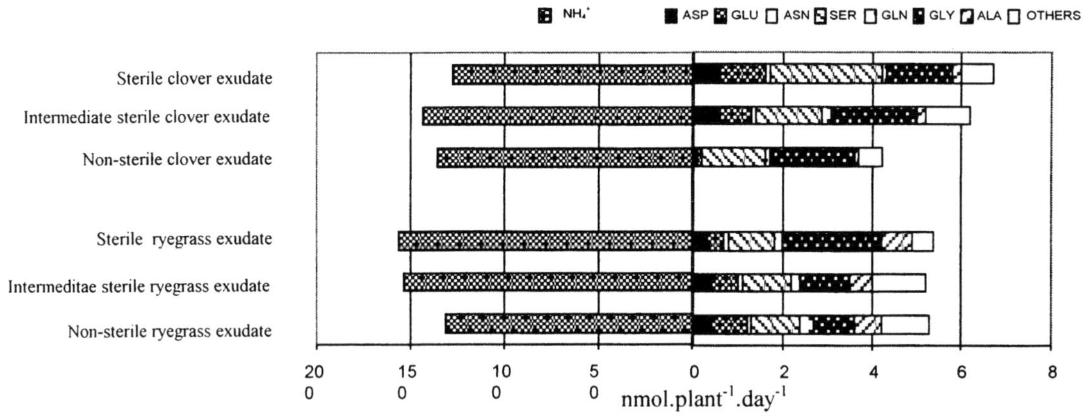

Figure 1. Comparison of ammonium and amino acids from exudates of clover and ryegrass grown under 3 levels of sterility (sterile, intermediate and non-sterile : *cf.* "Materials and methods").

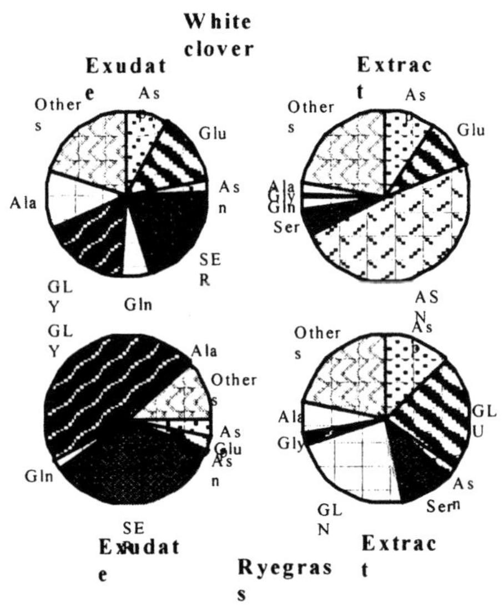

Figure 2. Amino Acid profiles of root extracts and sterile exudates of white clover and perennial ryegrass. Exudates and extracts were analysed from 39-day-old ryegrass and 67-day-old clover

The comparison between amino acids profiles of root extracts and exudates of clover and ryegrass shows that there was no correlation between both profiles (Figure 2). Asparagine was the major amino acid in clover roots, glutamine and glutamate were the major ones in ryegrass roots, while in exudates these amino acids were present in in very low amounts .

Discussion

Ammonium is the major N-compound released by both clover and ryegrass. The small proportion of amino acid exudation confirms previous results with soybean (Ofosu-Budu *et al.*, 1990) and alfalfa (Brophy and Heichel, 1989).

Serine and glycine were the major amino acids released by clover and ryegrass as it is shown in sterile forage rape (Shepherd and Davies, 1994) and sterile alfalfa (Richter *et al.*, 1968). The fact that these two amino acids were found in the exudate while asparagine and glutamine, which were the major ones in clover and ryegrass roots, have never been measured in significant amounts in exudates, could be explained by a selective root exudation as proposed by Ta *et al.* (1986).

The similar amounts of ammonium observed in sterile and non-sterile exudates show that this compound originates directly from young plants. However, concerning amino acids, the presence of micro-organisms increased the amounts of glutamate and aspartate released only for clover, but did not modify the exudation of the other amino acids. The higher release of these two compounds in the non-sterile exudates does not seem to result from a deamination of glutamine and asparagine by the micro-organisms as suggested by Murray *et al.* (1996). In conclusion, bacteria present in the sand have no substantial effect on the N-exudation by young clover and ryegrass plants.

References

Brophy LS and Heichel GH 1989 Plant Soil 116, 77-84.

Murray PJ, Hatch DJ and Cliquet JB 1996 Can. J. Bot. 74, 1591-1595.

Ofosu-Budu KG, Fujita K and Ogata S 1990 Plant Soil 128, 135-142.

Richter M, Wilms W and Scheffer F 1968 Plant Physiol. 43, 1747-1754.

Shepherd T and Davies HV 1994 Plant Soil 158, 99-109.

Ta TC, MacDowall DH and Faris MA 1986 Can. J. Bot. 64, 2063-2067.

W. J. Horst et al.(Eds.), Plant nutrition – Food security and sustainability of agro-ecosystems. 680–681, 2001

Uptake of organic labelled sulphur by oilseed rape and barley in the rhizosphere and in corresponding non-rhizosphere soil

P. C. Vong, F. Lasserre-Joulin and A. Guckert

Laboratoire Agronomie et Environnement ENSAIA-INRA, 2, avenue de la Forêt de Haye, Vandoeuvre les Nancy, F 54505, France, phuy-chhoy.vong@ensaia.inpl-nancy.fr

Key words: barley, hot water sulphur, labile sulphur, rape, rhizosphere, sulphur nutrition

Abstract

Spring rape and barley were grown in "rhizobag", in a calcareous soil previously labelled with ^{35}S. Before sowing, the soil had received two levels of N and five levels of S. Three pot destructive samplings were made at 20, 42 and 56 days after sowing. Despite the dilutions made by levels of added $S\text{-}SO_4^{2}$, the results averaged over the three sampling dates showed a significant increase ($p<0.05$) in ^{35}S uptake by rape, in contrary to barley that showed a trend to decrease with increasing ^{35}S dilutions. Comparatively, rape rhizosphere was able to mobilise more ^{35}S organic extracted by hot water at 70°C as well as ^{35}S-labile (^{35}S extracted by 0.01 M $CaCl_2$ + ^{35}S-hot water) than barley rhizosphere, especially at high levels of S addition. Significant linear correlation coefficients between ^{35}S-uptake and ^{35}S-hot water ($r^2 = 0,72$, $p<0.01$), and ^{35}S-labile ($r^2 = 0.75$, $p<0.01$) were found in rape but not in barley rhizosphere. It was concluded that rape rhizosphere was endowed with specific microbial biomass that could regulate the uptake of organic ^{35}S when compared with barley. Thus, a high S demanding plant, such as rape, needs S adding to maintain its normal development.

Introduction

Rape rhizosphere was found to be able to harbour predominantly sulphur oxidising bacteria as well as fungi which can oxidise S to SO_4^{2-} (Grayston and Germida, 1991). This oxidising mechanisms release protons that acidify the microenvironment around the roots and solubilise S, P, Fe and other nutrients (Hoffland, 1992). Moreover, the imbalanced nutrition between anions and cations (Hinsinger, 1998), the ATPase mediated net H^+ extrusion (Crawford and Glass, 1998) may contribute to the main acidity in the rhizosphere as well. This study aimed to compare the uptake of ^{35}S by rape and barley and the subsequent mobilisations of organic ^{35}S in their respective rhizosphere and corresponding non-rhizosphere soil.

Materials and methods

A pot experiment with spring rape (cultivar '*Tanto*') and barley (cultivar '*Scarlett*') on a calcareous soil (rendzina) was conducted in a growth chamber using the "rhizobag" technique. The soil was previously labelled with ^{35}S carrier-free sodium sulphate [^{35}S] for three months prior to planting. Before sowing, the soil had received two levels of N: 0 and single dose of 76.9 mg N kg^{-1} soil as ammonium nitrate and five levels of S: 0, 15.4, 20.4, 30.8 and 61.5 mg S kg^{-1} soil as $MgSO_4$. Each treatment had 4 replicates. Considering the mineral N (65.5 mg kg^{-1} soil) and sulphate-S (10.3 mg kg^{-1} soil) present in the soil solution at the start of the experiment, the levels of added fertilisers brought the available N/S ratio values to 6.4, 5.6, 4.6, 3.5 and 2.0 respectively. Three pot destructive samplings were made at 20, 42 and 56 days after sowing (DAS). At each sampling, shoots, roots were processed and rhizosphere, non-rhizosphere soil successively extracted by 0.01 M $CaCl_2$ and hot water at 70°C (Sparling *et al.*, 1998) and was analysed for ^{35}S.

Results

With increased levels of added $S\text{-}SO_4^{2-}$ of 0, 15.4, 20.4, 30.8 and 61.5 mg kg^{-1} soil, the average results (Table 1) showed that rape took up ^{35}S significantly higher than barley (13.6, 20.0, 19.6, 19.2 and 20.3% vs. 15.9, 15.5, 16.4, 15.2 and 11.8% of soil ^{35}S respectively).

Table 1. Mean percentage values of ^{35}S uptake by rape and barley. Letters refer to significant differences within plant types and N/S ratios (Newman-Keuls test at $p<0.05$)

Plant	N/S ratio				
	6.4	5.6	4.6	3.5	2.0
Rape	13.6cd	20.0a	19.6a	19.2a	20.3a
Barley	15.9bc	15.5bc	16.4b	15.2bc	11.8d

Besides, except the treatment S = 0 (N/S = 6.4), the mean dry matter (shoots + roots) of both plants was not significantly different (data not shown). These results suggest that, despite the dilutions brought about by SO_4^{2-}, rape was able to mobilise more endogenous ^{35}S compared to barley.

As expected, the high uptake by rape was revealed by the significantly larger decrease in mean percentage values of ^{35}S found in hot water as well as in labile fraction ($^{35}S\text{-}SO_4^{2-}$ extracted with 0.01 M $CaCl_2$ and hot water) and this especially at high levels of added $S\text{-}SO_4^{2-}$ (Fig. 1). However, without S addition (N/S = 6.4), the values of both fractions were lower in barley rhizosphere.

In rape rhizosphere soil significant linear correlation coefficients were observed between ^{35}S uptake and ^{35}S-hot water ($r^2 = 0.72$**, $p<0.01$), as well as ^{35}S-labile ($r^2 = 0.75$**, $p<0.01$). This explains the higher ability of rape rhizosphere to regulate ^{35}S uptake compared to barley.

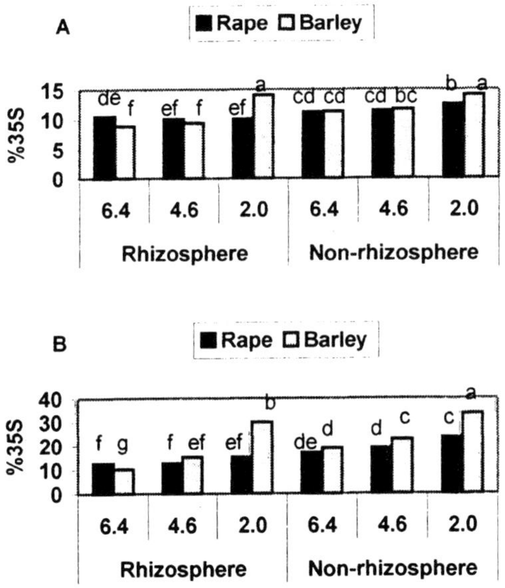

Figure 1. Percentage values of ^{35}S present in hot water **(A)** and labile ($^{35}S\text{-}SO_4^{2-}$ + ^{35}S-hot water) fraction **(B)** of rape and barley. For practical reasons, the values related to N/S ratios 5.6 and 3.5 were not presented. Letters refer to significant differences within soil compartments, N/S ratios and plant types (Newman-Keuls test at $p<0.05$)

As to the non-rhizosphere soil, significant coefficients found for both plants and for all studied parameters (Table 2) suggested a common mechanism of mobilisation of ^{35}S which would operate mainly by mass flow.

Table 2. Coefficients of linear correlation between ^{35}S uptake and ^{35}S-labile ($^{35}S\text{-}SO_4^{2-}$ + ^{35}S-hot water), $^{35}S\text{-}SO_4^{2-}$ and ^{35}S-hot water obtained over the three samplings. Significance level ($p^{**}<0.01$, $n = 15$). NS is not significant

Soil compartment	Plant	^{35}S-labile	$^{35}S\text{-}SO_4$	^{35}S-hot water
Rhizosphere	Rape	0.75**	NS	0.72**
	Barley	NS	NS	NS
Non-rhizosphere	Rape	0.75**	0.73**	0.76**
	Barley	0.90**	0.86**	0.95**

Discussion

In agreement with Grayston and Germida (1991), rape rhizosphere is endowed with specific microbial biomass that could regulate the release of organic ^{35}S (^{35}S-hot water). This fact would be linked to the coarse root system of rape which presents low microsite refuge and therefore harbours quite homogeneous microbial biomass specific for S immobilisation/mobilisation. Whereas the root architecture of barley is more dense and more fibrous, which implies a greater microsite refuge for diversity of microorganisms non specific in sulphur cycling. The existence of a significant correlation coefficient between ^{35}S uptake and ^{35}S-hot water observed only with rape, together with the higher uptake of ^{35}S by rape than barley (Tables 1-2), confirms the above observation. Also, this correlation suggests that rape but not barley can directly use ^{35}S- organic components that are extracted by hot water. In addition, significant coefficient found between ^{35}S-uptake and ^{35}S-labile observed in rape only suggests that this pool contains some of the organic components released by rape rhizosphere. In consequence, when S is not limiting, rape can regulate and use, via its rhizosphere microorganisms, both freshly immobilised organic and inorganic ^{35}S. This underlines the characteristic feature of microbial biomass of rape rhizosphere that is known to be acidic. Nevertheless, under S limiting conditions (without S addition, N/S = 6.4), ^{35}S previously added would be rather stored by microorganisms of rape rhizosphere than used for growth as evidenced by the larger percentage value of ^{35}S-hot water when compared with barley (Fig. 1).

The significant correlation coefficients found in the rhizosphere of rape is very interesting, because these data may be used as indices for a sound fertilisation. ^{35}S-hot water and $^{35}S\text{-}CaCl_2$ of rape rhizosphere are not single pools. For a full understanding, further study is needed to identify the S organic components of these pools that are actually involved in plant uptake.

References

Crawford NM and Glass ADM 1998 Trends in Plant Sci. 3, 389-395.

Hinsinger P 1998 Adv. Agron. 64, 225-265.

Hoffland E 1992 Plant Soil 140, 279-289.

Grayston SJ and Germida JJ 1991 Can. J. Microbiol. 37, 521-529.

Sparling G, Vojvodic-Vukovic M and Shipper LA 1998 Soil Biol. Biochem. 30, 1469-1472.

W. J. Horst et al.(Eds.), Plant nutrition – Food security and sustainability of agro-ecosystems. 682~683, 2001

Allelopathy versus competition between plants and microbes

E. Wiegard[1)] and S. Jutzi[2)]

[1)] Institute of Agricultural Chemistry, University of Göttingen, 37075 Göttingen, Germany, ewiegar@gwdg.de
[2)] Food and Agriculture Organisation of the UN (FAO), Rome, Italy, Samuel.Jutzi@fao.org

Key words: allelopathy, *Eucalyptus globulus, Juglans regia, Leucaena leucocephala*, N mineralisation

Abstract

Leaf material from *Leucaena leucocephala*, *Juglans regia* and *Eucalyptus globulus* was incubated in a sand/nutrient solution and tested for its effects on microbial activity and N mineralisation as affected by incubation time. In parallel, the influence of incubated plant material was tested on *Sorghum bicolor* seedlings at seven-day intervals. The results of this study illustrate the relationship between substrate respiration and plant growth. *Eucalyptus* and *Leucaena* had significant inhibitory effects on *Sorghum* on days 7, 14 and 21 after incubation of the plant material. This growth inhibition was negatively correlated with the microbial activity of the substrate. A negative effect of mineral immobilisation through microorganisms on the growth of test plants was observed with *Eucalyptus* and *Juglans* material. The growth reduction of test plants by *Leucaena* on days 7, 14 and 21 could not be attributed exclusively to mineral N immobilisation effects.
The study confirmed the substantial difficulties in quantifying the allelopathic potential of tree materials through mulch trials. A differentiation between decomposition effects through mineral immobilisation on the one hand, and microbial metabolism on the other hand from the allelopathic potential of the suspected plant material studied cannot be reliably established in such trials.

Introduction

Plant–plant interference may imply on the one hand competition for light, water and nutrients, and on the other hand allelopathic interaction through organic compounds released into the environment. In order to evaluate the allelopathic potential of a plant species various tests are in use, especially mulch trials. This is done in a special substrate which is prepared according to practical considerations (Chou, 1986; Nakahisa *et al.*, 1993): in comparison to a control (substrate without mulch material) reduced growth may be observed in test plants on substrate with mulch, which is often interpreted as the result of allelopathic effects. Such growth reduction may, however, just as well be attributed to mineral nutrient immobilisation by the stimulation of microbial activity. The objective of the research reported here was to assess the suitability of mulch trials for allelopathic assessment, taking into consideration possible mulch decomposition effects on test plant growth.

Material and methods

The research was conducted at the Institute of Crop Science, University of Kassel, Germany. Plant material from *Leucaena leucocephala, Juglans regia* and *Eucalyptus globulus* was used; it was analysed for total phenol, total N, P, Ca, Mg and sugar content. The plant material was mixed with sand and nutrient solution and incubated at 25 °C. At weekly intervals, samples were taken to run growth tests with *Sorghum bicolor* seedlings for four weeks and to analyse mineral N, P, C, sugar, total phenol content and respiration of the substrate.

Results

The results show a clear relationship between substrate respiration, N immobilisation and test plant growth. N immobilisation was related to the activity of micro-organisms. The plant material with the highest N and P content and the highest soluble carbohydrate content clearly produced a higher stimulation of the micro-organisms (results not shown). *Eucalyptus* had significant inhibitory effects on *Sorghum* in the course of the whole incubation period and, simultaneously, it caused N immobilisation up to the end of the experiment (Figures 1 and 2).

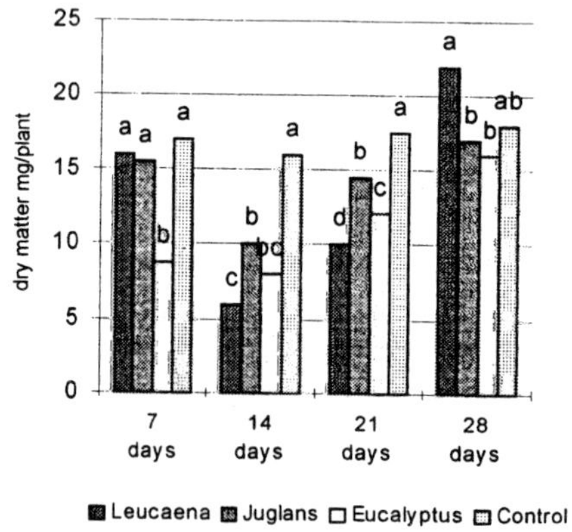

Figure 1. Shoot dry matter of *Sorghum bicolor* after one week grown in fertile sand 7, 14, 21, and 28 days after incubation of plant material (P < 0.05 SNK-test)

This growth inhibition was negatively correlated with the microbial activity in the substrate. *Juglans* induced *Sorghum* growth inhibition up to 28 days after the incorporation of plant material, and this growth inhibition was correlated with N immobilisation.

Table 1. CO_2-production rate ($\mu g\ CO_2\ g^{-1}\ FM^{-1} \cdot d^{-1}$) in the substrate as related to mulch treatment, ($p<0{,}05$ SNK-test).

	7 days	14 days	21 days	28 days
Leucaena	183 b	393 a	218 a	142 a
Juglans	114 c	180 b	121 b	152 a
Eucalyptus	568 a	163 b	87 b	86 b

Table 2. Correlation coefficients relating the CO_2-production rate to initial chemical properties of mulch material during incubation with substrate (across mulch sources).

Days	N/CO_2	P/CO_2	C/CO_2	$C{:}N/CO_2$	$sugar/CO_2$
7	–0.2	0.51	0.81	0.91	0.925
14	0.87	–0.21	–0.27	–0.63	–0.149
21	–0.01	–0.97	–0.93	–0.80	–0.778
28	–0.45	–0.90	–0.94	–0.64	–0.906

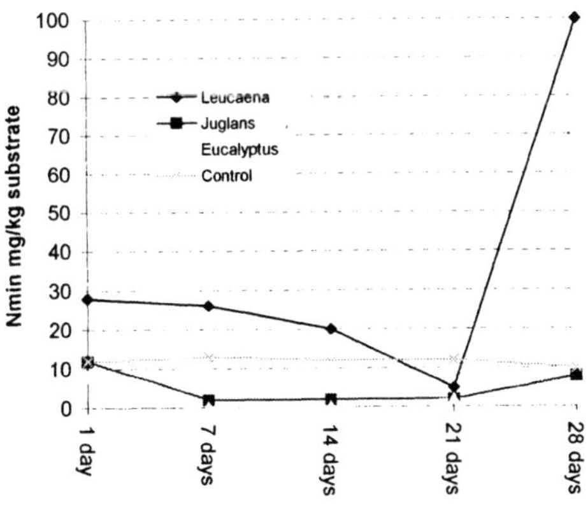

Figure 2. Substrate mineral N content after incorporation of mulch material.

The growth inhibition of *Sorghum* by *Leucaena*, particularly on day 14, cannot be explained exclusively by N immobilisation effects since, at that time, enough mineral N was available in the substrate. In comparison, *Leucaena* produced only a short-term N-immobilisation up to day 21 after the incorporation in the plant; it had, however, a significant inhibitory effect on *Sorghum* up to that stage (Figures 1 and 2). The microbial degradation of phenol derivatives took place in *Leucaena* within 14 days of incubation. In *Eucalyptus*, only half of the phenol derivatives were metabolised (Fig. 3).

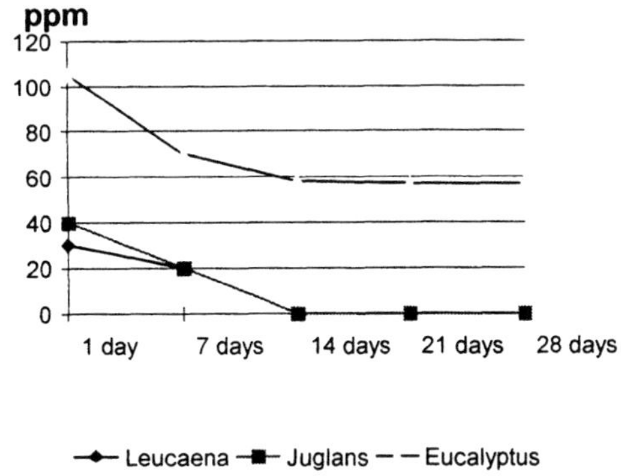

Figure 3. Effect of incubation time on the total phenol content (ppm) of mulch substrates.

Conclusions

The results demonstrate the difficulties in quantifying the allelopathic potential using mulch trials. The separation of the decomposition effects (mineral immobilisation) from the allelopathic effects of the plant material can only be obtained by controlling the nutrient supply to the test plants.

References

Chou CH and Kuo YL 1986 J. Chem. Ecol. 12, 1431-1448.

Nakahisa K, Tsuzuki E and Mitsumizo 1993 Jpn. J. Crop Sci. 62, 294-299.

W. J. Horst et al.(Eds.), Plant nutrition – Food security and sustainability of agro-ecosystems. 684–685, 2001

Effect of rhizosphere indole-3-acetic acid on growth and phosphorus nutrition of maize plants (*Zea mays* L.)

L. Wittenmayer
Institute of Soil Science and Plant Nutrition, Martin Luther-University of Halle-Wittenberg, A.-Kuckhoff-Str. 17b, Halle (Saale), D 06108, Germany, wittenmayer@landw.uni-halle.de

Key words: IAA, phosphorus availability, phytohormones, rhizosphere, root exudates

Abstract

The influence of rhizosphere indole-3-acetic acid (IAA) on the growth of maize plants (*Zea mays* L. cv. 'Boss') supplied with P sources of different solubility was investigated under greenhouse conditions. For this purpose, P-sufficient plants (+P) were supplied with adequate amounts of $Ca(H_2PO_4)_2$ whereas the P source of P-deficient plants (–P) was sparingly soluble Ca_3PO_4. In addition to one half of each P treatment, IAA was applied to the rhizosphere solution every day.
After four weeks of treatment, a greater dry matter accumulation and P concentration was observed in +P plants. IAA had no significant influence on growth parameters and P concentration of P sufficient plants. In –P plants, IAA application caused a significant decrease in acid phosphatase activity of fully developed leaves indicating an improved P supply, which may be partly attributed to morphological changes of the roots.

Introduction

Plants affect rhizospheric processes by root exudation. The exuded compounds may directly influence nutrient availability or may serve as a substrate for various nutrient solubilizing soil microorganisms (Schilling *et al.*, 1998). Furthermore, plant growth promoting effects of rhizosphere bacteria include the production of a vast range of metabolites (biological active substances) that may affect plant growth directly after being taken up by the plant, or indirectly by modifying the soil environment (Frankenberger and Arshad, 1995). One of these biologically active substances is the phytohormone indole-3-acetic acid (Kumar and Narula, 1999).

It is well documented that many plant growth-promoting bacteria are known to produce this phytohormone in considerable amounts. But there are only few reports available as to the effect of this compound on rhizosphere processes of intact plants due its fast decomposition in soil (Raczkowska-Blach *et al.*, 1995)

Materials and methods

Maize plants (cultivar 'Boss') were grown under greenhouse conditions in pots with 1.4 kg coarse sand (1 to 2 mm diameter) as substrate. Control plants (+P) were supplied with nutrients by daily watering with a nutrient solution according to Johnson *et al.* (1994) but modified by replacing Fe-EDTA by $FeCl_3$. P-deficient plants (–P) were cultivated in a substrate containing one mmole sparingly soluble $Ca_3(PO_4)_2$. Other nutrients were added by watering with an adequate P-free nutrient solution. In addition, the rhizosphere IAA concentration was maintained by daily application of 0.1 mM IAA to the nutrient solution supplied to one half of each P treatment (+H).

Plants were harvested four weeks after establishing the hormone treatment (emergence of the sixth leaf) and the following parameters were measured: dry matter accumulation and distribution, root volume, evapotranspiration, total P content of shoots and roots, concentration of water-soluble P in the rhizosphere solution, acidic phosphates activity of fully developed leaves (indicator of P deficiency), and their pigment content (chlorophyll a and chlorophyll b, carotenoids).

Results

At harvest, water-soluble P concentration of the rhizosphere solution in the +P treatment was 0.3 mM irrespective of hormone application. In the –P treatment, its concentration decreased significantly to 0.13 mM and 0.10 mM for plants with and without IAA application respectively (Table 1).

Phosphorus availability significantly influenced plant growth and P concentration of roots and shoots. Significantly higher dry matter yields, P concentrations and P contents were observed in +P plants. Hormone treatment did not affect dry matter accumulation and total P concentration, but substantially increased the total P content of roots in P deficient plants (+31 %) and tended to increase that of shoots (+ 13%).

The acid phosphatase activity in fully developed leaves (AP) was lowest in P sufficient plants. In the +P treatments, IAA application had no influence on this parameter. Under conditions of low P availability, activity of AP significantly increased (+84 %). The highest activity was observed in plants without IAA application. Addition of indole-3-acetic acid to –P plants significantly decreased the enzyme activity by 26 % indicating a lower P stress in these plants.

The pigment content was slightly influenced by hormone treatment and P availability. In contrast, transpiration and root volume were increased by both P availability and IAA.

Discussion

Maize plants used both of the P sources differently. The original concentration of water-soluble $Ca(H_2PO_4)_2$ in the nutrient solution was 0.5 mM (1 mM P). After 24 hours, its concentration decreased to 0.3 mM. That means that about 70 % of the added P were utilized by plants. In the –P treatment, one mmole $Ca_3(PO_4)_2$ (62 mg P) was added to the substrate. Only 1.86 mg P were found in the plants without hormone treatment. Consequently, maize plants were not able to use this P source efficiently. The addition of IAA to the nutrient solution increased the P content by 18 % to 2.20 mg per maize plant.

In addition, the acid phosphatase activity significantly decreased when IAA was applied to P-deficient plants. Simultaneously the root volume and the evapotranspiration rate were increased indicating that the effects of rhizosphere IAA are at least partly based on morphological changes, especially in the roots. In ongoing analyses, the influence of rhizosphere indole-3-acetic acid on quantitative and qualitative characteristics of root exudates as a further possible strategy of hormone-induced improvement of nutrient solubilization are being investigated.

References

Frankenberger WT and Arshad M 1995 Phytohormones in Soils. Marcel Dekker Inc., New York, Basel, Hong Kong.

Johnson JF, Allan DL and Vance CP 1994 Plant Physiol. 104, 657-665.

Kumar V and Narula N 1999 Biol. Fert. Soils 28, 301-305.

Raczkowska-Blach E, Rozycki H, Strelczyk E and Pokojska A 1995 Microbiol. Res. 150, 265-270.

Schilling G, Gransee A, Deubel A, Lezovic G and Ruppel S 1998 J. Plant Nutr. Soil Sci. 161, 465-478.

Table 1. Effect of rhizosphere indole-3-acetic acid on selected growth parameters of maize plants (cv. 'Boss') depending on P source. Pot experiment, six replicates. Harvested six weeks after emergence and four weeks after beginning of hormone application.

	$Ca(H_2PO_4)_2$ –IAA	+IAA	$Ca_3(PO_4)_2$ –IAA	+IAA	GD Tukey (P = 0.05)
P concentration of the rhizosphere solution, mM	0.31	0.31	0.10	0.13	0.09
P concentration in dry matter, mg g^{-1}					
shoots	2.16	2.18	0.92	1.03	0.13
roots	1.43	1.41	0.94	1.12	0.21
P content per plant, mg					
shoots	4.03	4.10	1.31	1.48	0.33
roots	0.97	1.07	0.55	0.72	0.25
plant	5.00	5.17	1.86	2.20	0.34
acidic phosphatase activity, nmol s^{-1} g^{-1} dry matter	278	234	537	406	109
Evapotranspiration per plant, ml h^{-1}	2.35	2.52	1.67	1.79	0.27
Root volume per plant, ml	11.1	12.6	9.3	10.8	2.3

Effect of long-term fertilisation and cropping systems on selected soil enzyme activities

F. Eivazi[1)] and M. R. Bayan[2)]

1) *Cooperative Research, Lincoln University, 212 Foster Hall, Jefferson City, Missouri 65102-0029, United States, eivazif@lincolnu.edu*

2) *The E-TEC Group, LLC., 818 Harvest Dr., Jefferson City, Missouri 65109, United States, raybayan@aol.com*

Key words: Long-term fertilisation, residue, rotation, soil enzymes, tillage

Abstract

Studies in recent years have shown that soil enzyme activities are sensitive in discriminating among soil management effects. Enzymes are important components involved in the dynamics of soil nutrient cycling. Enzyme activity in the soil environment is considered to be a major factor contributing to overall soil microbial activity and, more recently, to soil quality. This study was designed to examine the effects of long-term fertilisation, tillage and crop rotations on the activities of acid and alkaline phosphatases, □-glucosidase, and arylsulfatase. Soil samples were collected from the historical Sanborn Field. The Sanborn Field has been under various cropping and management practices since 1888. In general, enzymatic activities were reflective of the cropping and management histories and/or soil properties observed on the Sanborn Field.

Introduction

Sanborn Field was initiated in late 1888 to demonstrate the value of crop rotations and manure in grain production. The field has undergone several changes since 1888. The changes were made in 1914 to introduce commercial fertilisers and reduce the number of plots receiving manure. Limestone was introduced in 1928 along with increased rates of nitrogen. Some minor changes were made in 1940.

Long-term cropping systems and N-fertilisation can influence important soil properties and nutrient cycles within soil profile (Simek *et al.*, 1999). Acid phosphatase and arylsulfatase play role in cycling of important plant nutrients (P, S). Glucosidases are widely distributed in nature and their hydrolysis products are important source of energy for soil microorganisms (Tabatabai, 1994). Acid phosphatase mediates the release of inorganic phosphorus (P) from organically bound P returned to the soil as litter and other organic debris. Arylsulfatase is the enzyme that catalyzes the hydrolysis of an arylsulfatase anion by fission on the oxygen-sulphur (O-S) bond. □ -Glucosidase catalyses the hydrolysis of □ -D-glucopyranoside and is involved in the hydrolysis of maltose. Investigation on limited number of enzymes show that agricultural management practices affect their activity (Dick, 1992). A wide range of enzymes has not systematically investigated for their potential to reflect short- and long-term soil management effects in relation to soil quality. The objective of this study was to study the effects of long-term management and fertilisation on the activities of above mentioned enzymes.

Materials and methods

The soil within the study area is a Mexico silt loam (fine, montmorillonitic, mesic Udollic Ochraqualf) which is representative of gently rolling, more erodible soils of the Midwest claypan area. Soil samples from plots no. 6, 7, 17, 18, (continuous corn); 22 and 23 (continuous timothy); 2, 5, 9, 10 (continuous wheat); 25, 26, 27, 28 (rotation C-W-red clover) and plot 39 (full fertility, continuous soybean) were collected before planting in April 1999. Soil samples were taken from 0-10, 10-20, and 20-30 cm depth (after removing the O_1 horizon) and processed. Enzyme activity determinations were made using the appropriate procedures described by Tabatabai (1994). Full fertility plots were fertilized based on soil test recommendation and manure was applied at 13.4 Mg/ha. Measurements of temperature and precipitation have been done in Sanborn Field since 1890.

Results

The activities of all four enzymes studied decreased with depth and was correlated with decrease in organic matter content in the soil profile. In general, differences were detected in enzyme activities as related to cropping systems. For example, acid phosphatase activity was significantly higher for continuous corn under no-till (plot 7) than for continuous corn under conventional tillage (plot 6) that had received full fertility. Acid and alkaline phosphatases activities were higher for continuous corn with manure fertility (plot 18) compared with no fertility treatment (plot 17). Significant effects on enzyme activities were observed in relation to rotation and N-fertilisation treatment (plots 25, 26, 27, 28). However, the highest values for organic matter and enzyme activities were found in plot 25 which has been under the 3-year corn-wheat-clover rotation since 1888 with manure as only source of fertiliser. The trends were similar for all the four enzymes. Plot 39 has been under soybean monoculture with full fertility and conventional tillage since 1990 and showed the lowest organic matter content and enzyme activities.

Discussion

Significant and positive correlation between enzyme activities and organic C that was found in this study is in agreement with findings of other investigators. Eivazi and

Tabatabai (1990) reported glucosidase and galactosidase activities to be significantly and positively correlated with organic C content. Bandick and Dick (1999) reported that glucosidases, galactosidases, arylsulfatase and urease activities were significantly correlated with total-C in a residue utilisation plot established in 1931. Our results on effect of tillage on four enzymes support previous suggestions that accumulation of organic C and N in surface soils under a reduced tillage system and a greater accumulation of inorganic nutrients under no-till affects enzyme activities (Deng and Tabatabai, 1996). In a study conducted by Jordan *et al.* (1995) in Sanborn Field, significantly higher soil-microbial biomass C was found in continuous corn, full fertility plot under no-till (plot 7) compared with conventional till (plot 6). Soils under crop rotations with a high input and diversity of organic materials generally contain higher concentrations of microbial biomass and enzymes compared with monocropping systems (Klose and Tabatabai, 2000). Our findings from the Sanborn Field confirm other results on effect of inorganic N fertilisation on soil microorganisms and enzyme activities. Dick *et al.* (1988) showed that amidase and urease activity decreased with increasing application of ammonium-based N-fertiliser. In general inorganic N fertilisation can have significant effect on soil microbial and enzyme activities through higher plant yields and impact on soil pH.

References

Bandick AK and Dick RP 1999 Soil Biol. Biochem. 31, 1472-1479.

Deng SP and Tabatabai MA 1996 Biol. Fertil. Soils 22, 202-207.

Dick RP, Rasmussen PE and Kerle EA 1988 Biol. Fertil. Soils 6, 159-164.

Dick RP 1992 Agricult. Ecosys. Environ. 40, 25-36.

Eivazi F and Tabatabai MA 1990 Soil Biol. Biochem. 22, 891-897.

Klose S and Tabatabai MA 2000 Biol. Fertil. Soils 31, 191-199.

Jordan D, Kremer RJ, Bergfield WA, Kim KY and CacnioVN 1995 Biol. Fertil. Soils 19, 297-302.

Simek M, Hopkins DW, Kalcik J, Picek T, Santruckova J and Travnik K 1999 Biol. Fertil. Soils 29, 300-308.

Tabatabai MA 1994 *In* Methods of soil analysis. Eds. Weaver RW, Angel GS and Bottomley PS. pp. 775-833. Soil Sci Soc Am, Madison, Wis.

W. J. Horst et al.(Eds.), Plant nutrition – Food security and sustainability of agro-ecosystems. 688~689, 2001

Root production and turnover in *Populus* grown under elevated CO_2 using a free air enrichment system POPFACE

M. Lukac[1], C. Calfapietra[2] and D. L. Godbold[1]

[1] *School of Agricultural and Forest Sciences, University of Wales Bangor, Deiniol road, Bangor, LL57 2UW, United Kingdom, afp450@bangor.ac.uk*

[2] *Di.S.A.F.Ri., Università degli Studi della Tuscia, Viterbo, Via S.Camillo de Lellis, s.n.c.,, Viterbo, 01100, Italy, ecophys@unitus.it*

Key words: auger, elevated CO_2, ingrowth core, POPFACE, *Populus*, root

Abstract

Rising levels of atmospheric CO_2 may change root production, root turnover and ecosystem carbon balance. In order to quantify these possible alterations over longer time period and at an ecosystem level a novel technique of air CO_2 enrichment was applied to a plantation of three fast growing *Populus* species. Root production and turnover were measured both by auger and ingrowth coring methods. In two years of this experiment, both methods applied show increased production of fine and coarse roots under elevated CO_2 conditions for all three clones utilised. These results suggest that future levels of atmospheric CO_2 will have to be taken into account when constructing a carbon budget of an ecosystem.

Introduction

With rising level of atmospheric CO_2 and its potentially large effect on green plants (Tingley *et al.*, 2000) many questions concerning ecosystems' response remain unanswered. When considering the capacity of plants to sequester carbon the key components determining the fate of assimilated carbon may remain unseen – the belowground systems of roots (Norby and Jackson, 2000). A number of studies have evaluated the effect of elevated CO_2 on root growth (Ceulemans and Mousseau, 1994; Taylor *et al.* 1994; Godbold *et al.*, 1997; Rogers *et al.* 1999). Their findings indicate that elevated atmospheric CO_2 leads to increased biomass production accompanied by greater proportion of C allocated belowground. However, majority of published studies report on experiments carried out with seedling in greenhouses or growth chambers. Extrapolating these finding to an ecosystem level is of limited value.

POPFACE experiment (for more detail see Miglietta *et al.* 2001) aims at assessing the response of whole ecosystem to elevated CO_2. Commercial *Populus* plantation was exposed to ambient and elevated atmospheric CO_2 conditions for three years.

The objective of this paper is to report on the belowground response of *Populus* trees in the first two years of the experiment. In particular, findings on root production and root turnover are discussed.

Materials and methods

Three *Populus* species: *P. alba, P. nigra* and *P. euramericana* ('I-214') were chosen for this experiment because of their fast growth facilitating early canopy closure. Sampling was carried out in 6 plots evenly spaced within 9 ha of 'I-214' plantation located in Central Italy (42°37' N, 11°51' E). Each circular plot of 20 m in diameter was divided into three sections occupied by plants of single species planted at 1 m × 1 m spacing. Elevated (550 ppm) CO_2 concentration in three of the plots was provided by a novel Free Air Carbon Enrichment (FACE) technique. Remaining three plots were used as controls with ambient air CO_2 concentration.

All trees were planted in early June 1999, roots sampling campaigns were carried out in November 1999, March, July and November 2000. Standing root biomass was measured by auger method with 7 cm core diameter at two depths: 0–20 cm and 20–40 cm, respectively Fine root production estimate was obtained by use of ingrowth cores (4 cm diameter., 40 cm depth) wrapped in nylon mesh with 1 mm openings. Both types of cores were manually washed on site shortly after extraction, roots were collected in paper bags, dried at 60 °C for seven days and weighed.

Results and discussion

Along with other studies (Rogers *et al.* 1994; Berntson and Bazzaz, 1996) we observed an enhancement of root growth under elevated CO_2 conditions. All three clones showed greater total root biomass in FACE plots in top 40 cm of soil. Increase of root biomass was 28 % for *P.alba*, 89 % for *P.nigra* and 80 % for *P.euramericana* on average over all sampling points.

Enhanced growth of *Populus* root biomass was already detectable when the first sampling was carried out in Nov 99 as Figure 1 illustrates for *P. euramericana.* However, statistically significant difference between ambient and elevated CO2 treatments was observed only for *P. nigra* in Jul 00 (**) and Nov 00 (*) and for *P. euramericana* in Nov 00 (*).

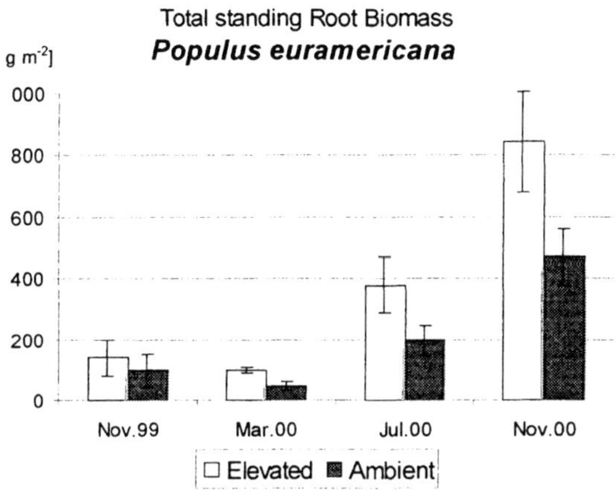

Figure 1. Total root biomass of *P. euramericana*

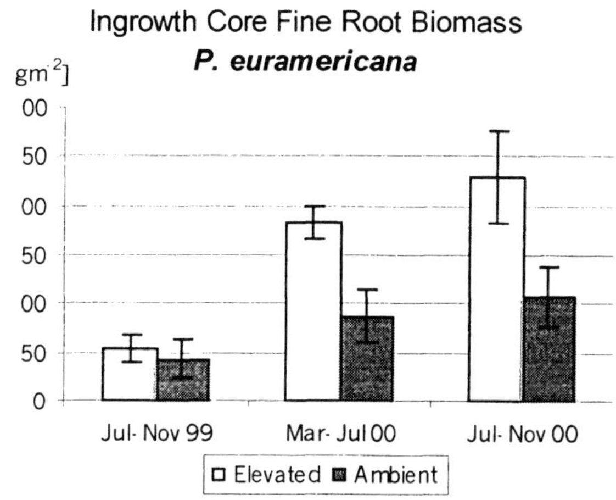

Figure 2. Ingrowth core fine root biomass

Analysis of vertical distribution of roots showed increased allocation of biomass into deeper soil horizon (20-40cm) under elevated CO_2 conditions for *P. alba* and *P. nigra*, from 23 to 36 % and from 20 to 39 % respectively. Ratio of deeper horizon biomass over total measured (0–40cm) for *P.euramericana* (37 %) did not change with CO_2 treatment.

Ingrowth core measurement showed a corresponding increase of fine root production. Elevated CO_2 treatment enhanced fine root growth of *P.alba* by 56 %, *P.nigra* by 97 % and *P.euramericana* by 73 %. Similarly to total root biomass, statistically significant difference between the treatments was observed only for some of the measurements. Pattern of relatively greater investment into roots grown in FACE plots is again visible from the onset of the experiment (Figure 2 for *P.euramericana*).

Annual fine root production data can be utilised for estimating root turnover when compared to maximum standing crop (Gill and Jackson, 2000). Calculated root turnover of more than 2 yr^{-1} for all three clones is probably an overestimation caused by ingrowth coring method utilised. Even though, a relative comparison of FACE versus control is still valid. This indicates faster fine root turnover under elevated CO_2 conditions by 53 %, 23 % and 84 % respectively for *P.alba, P.nigra* and *P.euramericana*.

Given such an early stage in the development of this plantation, uneven spatial distribution of roots might explain high variability between the measurements. Despite this drawback, results presented in this report are in line with widely accepted fact that elevated air CO_2 conditions lead to greater belowground biomass investment.

Acknowledgement

We thank EC for supporting this research through its Fourth Framework Programme as research contract No. ENV4-CT97-0657.

References

Bertson GM and Bazzaz FA 1996 Plant Soil 187, 119-131.

Ceulemans R and Mousseau M 1994 New Phytol. 127, 425-446.

Gill RA and Jackson RB 2000 New Phytol. 147, 13-31.

Godbold DL, Berntson GM and Bazzaz FA 1997 New Phytol. 137, 433-440.

Miglietta F, Zaldei A and Peressotti A 2001 New Phytol., (in press).

Norby RJ and Jackson RB 2000 New Phytol. 147, 3-12.

Rogers HH, Runion GB and Krupa SV 1994 Environ. Pollution 83, 155-189.

Rogers HH, Runion GB, Prior SA and Tolbert HA 1999 *In* Carbon Dioxide and Environmental Stress. Eds. Y Luo and HA Mooney. pp. 215-244. Academic Press, New York.

Taylor G, Ranasinghe S, Bosac C, Gardner SDL and Ferris R 1994 J. Exp. Bot. 45, 1761-1774.

Tingley DT, Phillips DL and Johnson MG 2000 New Phytol. 147, 87-103.

Symposium 10

Fertiliser use with regard to optimum yield and environment

Convener:

H. Kuhlmann

Editor:

Hans-Werner Olfs

W. J. Horst et al., Plant nutrition – Food security and sustainability of agro-ecosystems. 692~693, 2001

Precision agriculture: A challenge for crop nutrition management

P. C. Robert
Precision agriculture Center, University of Minnesota. 1991 Upper Buford Circle, St. Paul, MN 55108, USA. probert@umn.edu

Key words: corn, crop nutrition, precision agriculture, site-specific management, soybean, wheat

Abstract

Precision agriculture was initiated in the early 1980's using newly available technologies to improve the application of fertilizers by varying rates and blends as needed within fields. Presently, its adoption varies significantly by crops, regions, and countries but it is progressively introduced or evaluated around the world. Several kinds of challenges limit a broader adoption: socio-economical, agronomical, and technological. Socio-economical barriers are principally costs and lack of skills. Agronomical challenges are lack of basic information, inadequate sampling and scouting procedures, absence of site-specific fertilizer recommendations, misuse of information, and lack of qualified agronomic services. Technological barriers are multiple. They relate to machinery, sensor, GPS, software, and remote sensing. However, these barriers will be progressively lifted and precision agriculture will be a significant element of the agricultural system of the future.

Introduction

A better awareness of soil and crop condition variability within fields and the development of new technologies such as microcomputer, geographic information system (GIS), and global positioning system (GPS) initiated the concept of site-specific management (SSM) for variable rate applications of fertilizers to grain crops. Presently, the SSM concept has been considered for most common cropping systems and some specialty crops, worldwide. However, the adoption of the concept is still in infancy and has been slower than initially thought because of several kinds of significant challenges: socio-economical, agronomical, and technological.

Challenge: Socio-economical barriers

Costs.

The implementation of precision agriculture (PA) practices generates additional costs. Most common expenditures include new services such as creating soil nutrient status map, data processing for the development of site-specific application maps, and variable rate applications of fertilizers; acquisition of new technologies such as sensors, GPS receiver, GIS software; new or modified farm equipment; and a high-quality computer.

Skills.

PA adoption surveys have found that age, attitude, and education of producers are significant barriers. PA requires new skills and only a limited number of farmers have them or willing to get them.

Challenge: Agronomical barriers

Early adopters that have applied carefully and efficiently some PA practices such as yield mapping, selected site-specific practices, including crop nutrition, and precise farm information systems, have in most cases benefited financially from improved soil and input management, crop yield and quality, and marketing. They have progressively modified some of their practices using site-specific information. However they have found that present PA agronomics have major barriers.

Basic soil information.

Present country soil surveys commonly have 1:20000 or smaller scales. They were designed for general planning. They cannot or should not be used for PA management, particularly for crop nutrition.

Basic terrain information.

PA research results show the value of the landscape or Digital Elevation Model (DEM) for site-specific management, particularly the development of management zones (Robert *et al.*, 2000). But topographic maps at an appropriate scale are not readily available.

Soil sampling.

Presently, the most common practice is grid sampling. A recent US survey on the adoption of PA (Akridge and Whipker, 1999) indicates that 36% of dealers/consultants used a sampling grid size greater than 1 hectare to lower costs. Research indicates that such resulting nutrient status maps are of inconsistent values. Another common practice in the Midwest for selecting soil sampling sites, 31 % of dealers (Akridge and Whipker, 1999), is the use of the standard soil survey at the 1:20,000 scale. This also will generally result in a poor characterization of soil nutrient status because of soil type variability within map units and past field management. Presently, a preferred method is directed or smart sampling based on management zones. However, this method requires basic field and crop information.

Crop monitoring.

Monitoring of field crops during the growing season for problem identification varies greatly in intensity and quality depending on the crop, farm, and region. Sampling techniques to map distribution and intensity of a crop problem are rarely adequate and remote sensing techniques are seldom used.

Site-specific recommendations.

The utilization of conventional fertilization rate recommendations do not optimize site-specific nutrient management. But nutrient recommendations corresponding to within field site-specific characteristics are rarely available.

Inefficient management and use of agronomic information.

PA is an information rich based management. This is creating major problems to farmers used to make decisions from limited records and rules. Deductions are often to quickly made without careful study of site-specific natural resources conditions, sensor calibration, software bias, field management history, machinery performance, season special events, weather status, and previous yield maps.

Agroconsultants.

PA requires new skills most often missing on farms. Presently, most farmers depend on dealers, agroconsultants, cooperative staff, and extension agents to use site-specific fertilizer management and spatial information systems. If these services are often available in the Midwest, they are generally rare or absent in other regions and countries.

Challenge: Technological barriers

The concept of PA was always linked to technology development. However, there is still a very significant need to improve present PA technologies and develop new ones for machinery, sensor, GPS, GIS, software, and remote sensing.

Farm equipment.

PA machinery is not always available for local fertilization practices, crops, and local natural and socio-economical conditions and may require greater accuracy.

Sensors.

Sensors are an essential part of PA. However, sensors may not exist for local needs and lack accuracy.

GPS.

GPS correction signal may require a costly annual subscription.

Software.

Software for data acquisition, processing, and analysis in a GIS environment has made significant progress in ease of use, openness, spatial analysis and display (3D) but greater enhancement are still needed. Expert systems or decision tool systems will have to be developed to optimize the use of multi-layers and multi-years spatial information.

Remote sensing.

Remote sensing use on the farm is still rare and faces multiple barriers: lack of understanding of its applications and benefits, costs, availability of imagery (cloud cover), quality and resolution of imagery, delay in getting the imagery, and lack of skills in rural areas for the interpretation of the imagery.

Conclusions

Recent US surveys seem to indicate a continuing increase in the adoption of PA by dealers and farmers with a significant variability by regions and crops. Several challenges of socio-economical, agronomical, and technological nature limit the use of PA. Significant research and development must remove these barriers before PA adoption expands from progressive farmers to a majority of farmers. This is not specific to PA but common to the adoption of new technologies. A broad adoption of the tractor took more than 30 years. When barriers are lifted, site-specific practices, including crop nutrition, will be an essential aspect of the agricultural system of the 21st century because it offers a variety of potential benefits in profitability, productivity, sustainability, crop quality, food safety, environmental protection, on-farm quality of life, and rural economic development.

References

Akridge JT and Whipker LD 1999 Precision agricultural services and enhanced seed dealership survey results. Center for Agricultural Business, Purdue University. West Lafayette. In Staff Paper No. 99-6.

Robert PC, Rust RH and Larson WE (eds.) 2000 Proceedings of the Fifth International Conference on Precision Agriculture. 16-19 July, 2000. Minneapolis, MN. ASA, CSSA, SSSA: Madison, WI.

W. J. Horst et al., Plant nutrition – Food security and sustainability of agro-ecosystems. 694–695, 2001

Tractor based remote sensing for variable nitrogen fertilizer application

J. Lammel, J. Wollring† and S. Reusch

Centre for Plant Nutrition and Environmental Research Hanninghof, Hydro Agri, Hanninghof 35, D-48249 Duelmen, Germany. joachim.lammel@hydro.com, stefan.reusch@hydro.com, † deceased January 2001

Key words: grain yield, nitrogen, precision farming, remote sensing, variable N application

Abstract

Precision farming requires accurate and efficient tools to determine the N fertilizer requirement not only as an average for the field, but also to enable variable application of N within a field. A tractor based remote sensing tool has been developed to determine the N status of the crop and deduce its future N fertiliser demand. Connected to a variable rate spreader, this technique enables an on-line variable N application according to the actual N demand of the crop. In 1999 the performance of the sensor system was tested on cereals (winter wheat, winter barley) in 96 field trials. Variable nitrogen application increased yield (on average by 1.6 dt ha^{-1}) and protein content, compared to the uniform application of the same total N amount.

Introduction

It is well known that the crop N demand vary within single fields according to the spatial differences in soil conditions (McBratney and Pringle, 1997). Existing methods of soil and plant analysis have proven to be costly and time consuming in delivering information on the actual N demand at the required high spatial resolution (Wollring *et al.*, 1998). However, previous research has shown that the N requirement of a crop can be determined by crop analysis. Spectral indices derived from the reflectance spectra have shown to be indirectly related to the N status of crops (Reusch, 1997). Therefore the potential of remote sensing to determine N demand of crops within a field during the growing season has been investigated.

Materials and methods

A tractor mounted multispectral scanner has been developed to measure the reflectance properties of crop canopies. In the visible wavelength range (400 – 700 nm) the reflectance is indicative of leaf chlorophyll content. In the near-infrared region (700 – 1000 nm) higher reflectance is caused by increasing biomass (Fig. 1).

The unit consists of two diode array spectrometers: one spectrometer to collect the crop reflectance data on each side of the tractor, the other to measure the irradiance in order to correct the reflectance signal. A terminal, which logs crop and GPS data on a chip card and displays the status of the crop controls the system.

Field trials with winter wheat on traditional small plots were established to determine the relationship between the sensor values and the N status of the crop (1996 –1998). N was applied at different rates in order to obtain the N response curve to determine the optimum N fertilizer rate.

In 1999, trials with cereals on 96 large fields were carried out in several countries (CZ, D, DK, F, GB, H, I, S, and USA) to test the performance of the obtained calibration for variable N application according to the sensor readings. Treatments with a uniform N application rate were established as a reference.

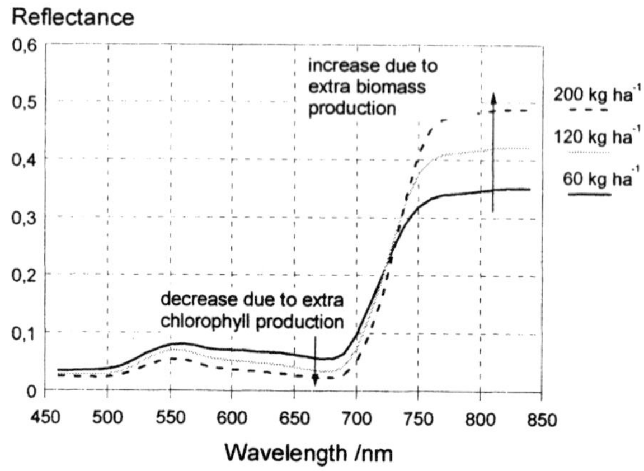

Figure 1. Typical reflectance spectra of crops: effect of different N status due to N rates from 60–200 kg N ha^{-1}

Results

The reflectance spectra of the crop at different N treatments contained information about the actual N status (Fig. 1). It was possible to convert this information into a sensor value. From the results of the N response trials a linear relationship between the sensor value and the N fertilizer demand has been calculated (Fig. 2): the higher the sensor value, the lower the optimum N application rate. However, this relationship has proven to be dependent on crop variety. Therefore, a reference value was built into the calibration in order to compensate for differences between varieties.

At high sensor values the N demand is low because of the good plant N and biomass status of the crop and *vice versa*. However, thin or badly developed crops with low biomass and chlorophyll contents would result in an inappropriately high N application rate. Therefore a biomass cut-off value has been established (Fig. 3): if the sensor value is below this value, the application rate is decreased. In addition, a minimum and a maximum rate have been defined.

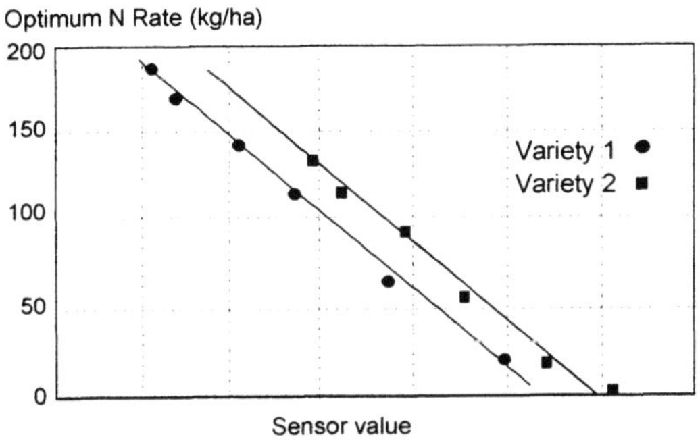

Figure 2. Relationship between sensor values and the optimum N rate for two winter wheat varieties

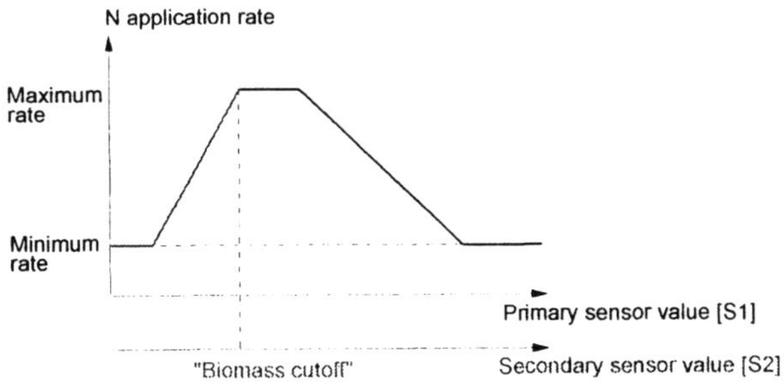

Figure 3. Principle of the agronomic calibration (S1 and S2 sensor values are obtained by different interpretation of the reflectance data)

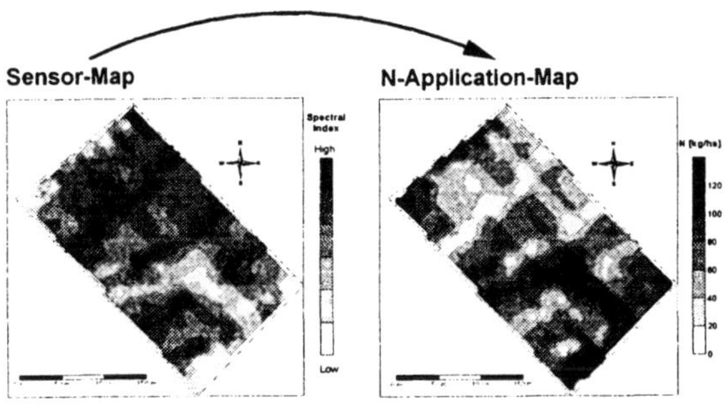

Figure 4. Calculated N application rate based on reflectance measurement

Maps of the sensor values before the 2nd and 3rd N application across a typical field show characteristic heterogeneity. Accordingly the variable N application rate varied between 0 to 100 kg N ha^{-1} for this field (Fig. 4).

Variable N application increased grain yield on average by 1.6 dt ha^{-1} for the 96 trials, compared to the uniform application of the same total N amount (Fig. 5). In addition grain protein content was increased and its variability over the field has been reduced (data not shown).

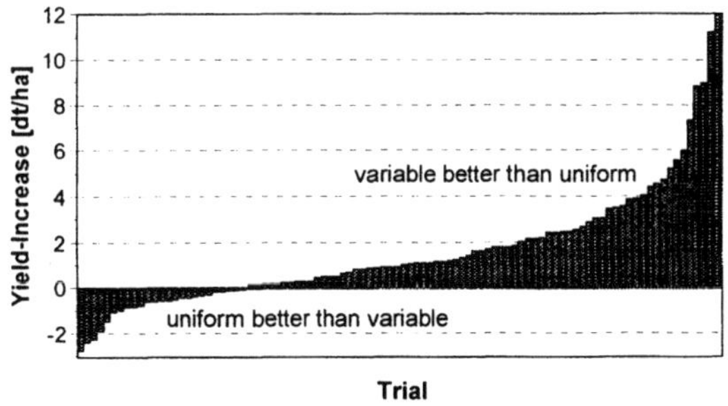

Figure 5. Yield increase due to variable N application rates compared to a uniform N application rate

Conclusion

It is possible to detect differences in the crop N status within a field by a tractor based remote sensing system and to apply N fertilizer accordingly. Varying N application rates according to spatially variable crop N requirements has positive effects: over- or under-fertilization within a field can be avoided. This will reduce lodging and increase yield. Other positive effects are a more homogeneous ripening of the crop, which reduces harvesting costs and grain losses.

References

McBratney AB and Pringle MJ 1997 *In* Precision Agriculture 1997, Vol 1. Ed. IV Stafford. pp. 3-31. BIOS Scientific Publishers Ltd., Oxford, UK.

Reusch S 1997 Entwicklung eines reflexionsoptischen Sensors zur Erfassung der Stickstoffversorgung landwirtschaftlicher Kulturpflanzen. Forschungsbericht Agrartechnik Band 303.

Wollring J, Reusch S and Karlsson C 1998 The International Fertilizer Society Proceedings No 423, 3-27.

W. J. Horst et al., Plant nutrition – Food security and sustainability of agro-ecosystems. 696~697, 2001

Bioassays to improve the determination of P and K availability in soil for fertilization recommendations

D. J. Bonfil, I. Mufradi, S. Asido and B. Dolgin
Agricultural Research Organization, Department of Field Crops and Natural Resources, Gilat Experimental Station, MP Negev, Israel, 85280, Israel, bonfil@netvision.net.il

Key words: bioassay, fertilization, phosphorus determination, potassium determination, soil test

Abstract

Precise fertilization is necessary to ensure grain yield, increase profitability and reduce environmental pollution. Pre-sowing soil tests of N and P are usually made to determine the recommended fertilizer applications. An accurate bioassay was developed to determine soil N availability for the Northern Negev area almost 20 years ago, but soil P was extracted by $NaCHO_3$, the common Olsen's test. Soil K status was never determined. We tested the use of the same N bioassay for P and K availability for wheat crops in a low rainfall area. A comparison between P-Olsen, K-$CaCl_2$, K-NH_4OAc versus P-Bio and K-Bio was studied on commercial and experimental fields during the 1998-2000 seasons. Total P uptake by field grown wheat correlated with the P-Bioassay and with P-Olsen, both with r =0.42. Potassium uptake correlated with the K-Bioassay (r=0.38), but was not correlated with the soil extraction methods (r=0.03-0.20). For fields where P or K was applied, correlation with the biological assays was decreased. The P-Bioassay might be used to determine P availability, and for fertilization recommendations. The K-Bioassay estimated the K availability much better than the common soil tests, however in wheat, K applications do not normally affect grain yield.

Introduction

Precise fertilization is needed to ensure grain yield, increase profitability and reduce environmental pollution. Pre-sowing soil tests of N and P are usually made to determine fertilizer applications. Water is the main limiting factor in the Negev semi-desert region. An accurate bioassay was developed to determine soil N availability for the Northern Negev area (Amir and Ephrat, 1971; Amir *et al*., 1994). This practical N-Bioassay has been in use for almost 20 years. Olsen-P is commonly used as an index of soil-available P (Olsen et al, 1959), whilst $CaCl_2$ extraction provides a good estimate of plant-available K. The neutral NH_4OAc extraction method is the most widely used procedure to extract exchangeable cations. However, due to the low response of wheat to additions of K on the loess soils of the region, farmers do not analyse soil K, and do not apply potassium fertilizers to field crops. Other crops, however, have responded to K fertilization.

In the present work we tested the same bioassay used to determine soil N availability for estimation of P and K availability for wheat.

Materials and methods

A comparison between P-Olsen and P-Bio, and K-$CaCl_2$, K-NH_4OAc and K-Bio was tested on soils sampled in commercial and experimental fields during the 1998-2000 seasons. Most soils in the region are sandy loam loess soils, although a small part of the area is covered by sandy or clay soils. Soils were sampled during the summer, by a mechanical auger with 15 cm diameter at 6-10 points per 30-100 ha. The samples were taken to 0-30, 0-45 or 0-60 cm depth according to the average rainfall in the fields. From each sample, 500 g of soil was taken for P and K chemical extraction, while the remaining 8 kg of soil was used for growing the maize in the bioassay tests. One month after emergence, the bioassay plants were harvested and analysed for total NPK content. Available P and K were calculated with the same calibration that was used for N (Amir *et al*., 1994). Actual uptake of P and K by wheat plants was determined by taking plants from each field (at least 4 replications of 0.33 m^2) at heading and analysing for NPK content. The correlation between the estimation of plant available P or K as determined by all the bioassays and extraction procedures and the actual uptake of P and K by wheat was calculated.

Results

Very low precipitation and high temperatures during the years 1998-2000 prevented most rain-fed grain yield production in most fields. Due to severe drought in 1999, only 30% of the experimental plots and fields reached heading, and only 10% produced grain yield. Therefore, all fields were divided into 3 categories: 1- Farms, represent commercial wheat fields all over the Negev, 2- Gilat Dryland, represent non irrigated experimental plots in Gilat permanent wheat experiment, 3- Gilat Irrigation, represent experimental plots in Gilat that received supplemental irrigation. A wide range of soil phosphorus and potassium was found in these soils (Table 1).

Total P uptake at heading was correlated with the P-Bioassay and with P-Olsen (Table 2). When fertilizer P had been applied, the correlation between the soil test and P uptake was not significant. However, in other cases significant correlations were detected, except in Gilat Irrigation plots where high variation was observed. Higher correlation was found between potassium uptake and the K-Bioassay than with either K-NH_4OAc or K-$CaCl_2$ extractions (Table 3). Moreover, combining the results of $CaCl_2$ and NH_4OAc extracts (Thomas, 1967) did not increase the correlation with K uptake.

Discussion

Different extractants remove quite different amounts of nutrients from soils. Sodium bicarbonate extractable P is commonly used as an index of soil-available P. $CaCl_2$ extracts the water-soluble K and about 20% of the exchangeable K, while NH_4OAc extracts the exchangeable fraction (McLean and Watson, 1985; Thomas, 1967). These methods are in use to assess the amount of plant available P and K. Although widely used, all the extraction methods require strict extraction conditions of constant shaking time and speed, pH and temperature. These methods require many small soil samples to be taken from the field. The bioassay method on the other hand is more complicated and takes a long time from field sampling to obtaining the results. In the bioassay only a few sampling points, each of about 10 kg are taken from each field, but it results in good forecasting of field N available to wheat. Since this method is already in field practice to determine nitrogen, determining P and K content in the plants as a bioassay involves relatively little extra work.

The main factors limiting wheat yield in the Negev are water and nitrogen. Despite the relatively low correlations between the respective bioassays and actual P or K uptake, the correlations showed that P-Bioassay could be used to determine P availability. During the last two years, about 70% of the fields that were commercially analysed for N have also been tested for P by the same bioassay. The K-Bioassay also estimated K availability much better than the common soil extraction tests. Due to the low response of wheat to K fertilization, we will test K-Bio availability for other crops that respond to K application. The time and labour required for the bioassay procedure can be reduced by using near-infra-red technique. This will be tested in further studies.

References

Amir J and Ephrat J 1971 Agron. J. 63, 385-388.

Amir J, Mufradi I, Klitman S and Asido S 1994 Plant Soil 158, 223-231.

McLean E O and Watson M E 1985 *In* Potassium in agriculture. Ed. Munson RD. pp 277-308. ASA Madison, Wis.

Olsen SR, Cole CV, Watanabe FS and Dean LA 1954 US Dept. Agric. Circ. 939.

Thomas GW 1967 *In* Soil testing and plant analysis. pp 37-54. SSSA Madison, Wis.

Table 1. Available phosphorus and potassium content in soils

Extraction method	Extracted mineral	Farm plots[a]		Gilat Dryland experimental plot		Gilat Irrigation experimental plot	
		mean	range	mean	range	mean	range
Bioassay	P (kg/ha)	11.1	1.9-54.1	6.9	2.7-14.4	9.8	4.9-22.4
Olsen	P (μg/g)	10.7	1.8-50.2	7.9	3.9-22.5	5.3	2.7-18.8
Bioassay	K (kg/ha)	258	86-846	192	94-375	201	110-525
NH_4OAc	K (μg/g)	124	55-296	144	74-207	121	69-199
$CaCl_2$	K (μg/g)	50	18-222	75	24-156	58	24-168
Number of plots tested		596		139		80	

a) Farms, represent commercial wheat fields all over the Negev Gilat, Dryland, represent non irrigated experimental plots in Gilat permanent wheat experiment, Gilat Irrigation, represent experimental plots in Gilat that received supplemental irrigation.

Table 2. Correlation coefficients between soil test for P and phosphorus uptake by wheat plants at heading.

Field	P Fertilization	P-Bioassay r	P-Bioassay p(F)[a]	P-Olsen R	P-Olsen p(F)
Total		0.415	***[b]	0.426	***
Farm	None	0.503	***	0.296	*
	Yes	0.290	ns	0.428	Ns
Gilat Dryland	None	0.382	*	0.417	*
	Yes	0.245	ns	0.336	Ns
Gilat Irrigation	None	0.348	ns	0.385	Ns
	Yes	0.313	ns	0.256	Ns

a) p(F): probability of F, b) significant level: ns- 0.05<pF, *- 0.01<pF<0.05, **- 0.001<pF<0.01, **- pF<0.001

Table 3. Correlation coefficients between soil test for K and potassium uptake by wheat plants at heading.

Field	K Fertilization	K-Bioassay r	K-Bioassay p(F)[a]	K-NH_4OAc r	K-NH_4OAc p(F)	K-$CaCl_2$ r	K-$CaCl_2$ p(F)
Total		0.376	***[b]	0.033	ns	0.205	*
Farm	None	0.328	*	0.352	*	0.075	ns
Gilat Dryland	None	0.068	ns	0.396	*	0.224	ns
	Yes	0.266	ns	0.095	ns	0.087	ns
Gilat Irrigation	None	0.566	***	0.041	ns	0.035	ns
	Yes	0.639	**	0.446	ns	0.374	ns

a) p(F): probability of F, b) significant level: ns- 0.05<pF, *- 0.01<pF<0.05, **- 0.001<pF<0.01, **- pF<0.001

W. J. Horst et al., Plant nutrition – Food security and sustainability of agro-ecosystems. 698~699, 2001

Uptake of fresh and residual P fertilisers by *Lolium perenne* in two Swiss agricultural soils

A. Gallet[1)], S. Sinaj[1)], R. Flisch.[2)], J. P. Ryser[3)] and E. Frossard[1)]

[1)] *Institute of Plant Sciences, Swiss Federal Institute of Technology (ETH), Eschikon 33, CH-8315 Lindau, Switzerland, anne.gallet@ipw.agrl.ethz.ch*

[2)] *Swiss Federal Research Station for Agroecology FAL, Reckenholzstrasse 191, CH-8046 Zürich, Switzerland, rene.flisch@admin.fal.ch*

[3)] *Swiss Federal Research Station for Plant Production RAC, CH-1260 Nyon, Switzerland, jean-pierre.ryser@rac.admin.ch*

Key words: fresh fertilisation, isotopic methods, *Lolium perenne*, phosphorus, residual fertilisation

Abstract

Residual phosphorus may significantly contribute to plant nutrition. To test this hypothesis, an experiment was conducted with *Lolium perenne* (*cv* Bastion) growing on two Swiss agricultural soils under controlled conditions. Treatments with or without the application of a fresh soluble P-fertiliser on soils with or without residual P allowed the estimation of the quantities of P taken up by plants coming from different sources of fertilisers. Addition of P to soils with residual P had no effect on the production of dry matter. Residual P contributed to more than 40% of the P taken up by to *Lolium perenne*, and significantly decreased the uptake of recently added P. The coefficient of P utilisation from recently applied P fertiliser was quite high for both soils (more than 24%), and decreased significantly (by almost a factor 2) in presence of residual P for the sandy Cadenazzo soil. The improved understanding of the contribution of residual P to P supply of rye grass will assist in the design of more appropriate and environmentally sensitive P fertilisation strategies.

Introduction

Most European soils have been overfertilised for many years. Swiss agricultural soils contain large reserves of P present in their upper horizons (Flisch, 1987) that can be transferred to surface waters. It is of agronomic and environmental interest, therefore, to know whether a fraction of the P fertiliser added in the past can contribute to the P nutrition of present crops.

Isotopic techniques have commonly been used to evaluate the agronomic efficiency of P fertilisers (Fardeau, 1996). These techniques allow above all, to distinguish P taken up by the crop that is derived from different sources (soil-P, residual and recently applied P).

The objective of this work was to determine, using isotopic techniques, the uptake of P from recently applied fertiliser versus uptake from residual P by *Lolium perenne* in two Swiss agricultural soils.

Materials and methods

Soil samples were collected from the 0-20cm layer of two treatments [0F: no P applied since the beginning of the trial, and F: P applied as superphosphate in quantities equivalent to removals by the crop (*i.e.* off take) since the beginning of the trial] of two Swiss long-term field experiments (Cadenazzo, canton of Tessin, and Changins, canton of Geneva: Tab.1).

A further treatment factors was added: $(NH_4)_2.HPO_4$ was added at the rate of 15 mg P/kg soil, to give a factorial combination of low (0F) and high (F) residual P +/- DAP (0F+DAP, F+DAP). The soils (500g per pot) were then watered with a P-free nutrient solution, sown with 0.5 g seeds of *Lolium perenne* (*cv* Bastion) and placed in a growth chamber for 4 months under controlled conditions. There were four replicates of each treatment. In order to distinguish the different sources of P (P derived from the soil, residual P and P from recently applied fertiliser) in the plant either the soil available P or the fresh fertiliser was labelled with carrier free ^{33}P in the form of orthophosphate. Dry matter (DM) production of the aerial parts and P uptake (^{33}P and ^{31}P) was determined for four successive cuts. Phosphorus in the crop derived from fresh (PDFff%) and residual (PDFrf%) fertiliser as well as the coefficient of P utilisation from fresh (CUff%) and residual (CUrf%) fertiliser were calculated according to Morel and Fardeau (1989).

Table 1. Some characteristics of the two soils. Parameters C_P (concentration of phosphate ions in solution), R/r1 (P-fixing capacity) and E1min (quantity of isotopically exchanged P within 1min) were determined with the isotopic exchange kinetics method (Fardeau, 1996). Total phosphorus Pt was determined with the Saunders and Williams method (1955).

	pH H_2O	clay g kg^{-1}	C_P mg L^{-1}	R/r1	E_{1min} mg kg^{-1}	Pt mg kg^{-1}
Cadenazzo						
0F	6.5	80	0.20	2.5	5.0	1159
F			0.44	2.9	12.9	1302
Changins						
0F	6.7	540	0.06	5.0	3.0	553
F			0.27	4.0	10.6	750

Results and discussion

Dry matter yield was similar across the four treatments in the Cadenazzo soil, whereas in the Changins soil the addition of DAP increased DM in the 0F treatment (Fig.1).

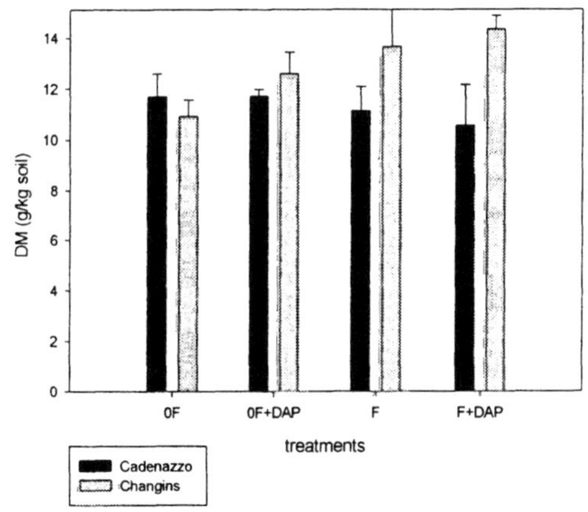

Figure 1. Effect of previous P fertiliser history and DAP on perennial rye grass above ground dry matter production for the Cadenazzo and Changins soils. Error bars represent standard deviation of 4 replicates

The absence of an effect of DAP application on the Cadenazzo 0F soil and the F soils from both locations suggests that P is not a limiting factor for ryegrass on these soils with a high total P content (i.e. above ca. 700 mg/kg) (Tab.1). The increase of DM associated with DAP addition to the 0F Changins soil could be explained by the very low C_P and E1min and values observed for this soil without residual P.

In both soils and treatments, the PDFrf was higher than 40% (Fig.2). The CUrf increased for the F treatment from 8.5 to 11.5% and 6.5 to 9%, respectively for Cadenazzo and Changins soils when DAP was applied. The lower values observed for Changins could be explained by the higher P-fixing capacity of this soil (Tab.1).

The PDFff was approximately halved (from 16 to 7% at Cadenazzo and from 26 to 13% at Changins) when rye grass was grown on the treatment with residual fertiliser (Fig.2). This decrease was in agreement with other observations (Morel and Fardeau, 1989; Sharpley, 1987).

The CUff decreased for the Cadenazzo soil from 26% (0F) to 14 % (F) whereas for the Changins soils the CUff increased from 24% (0F) to 29% (F). The increase in CUff observed on the fertilised Changins soil (F) compared to the non fertilised Changins soil (0F) is consistent with the decrease in soil P-fixing capacity (Tab.1) and the increase in DM production associated with past P applications. At Cadenazzo neither the soil fixing capacity nor the DM production was affected by DAP application.

Figure 2. Phosphorus in the crop derived from different

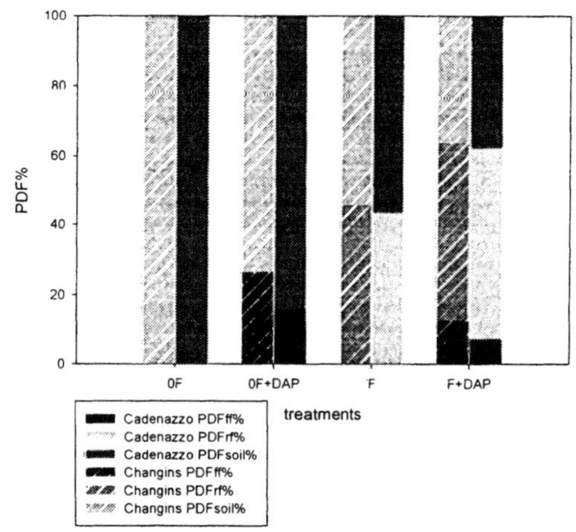

sources of P at Cadenazzo and Changins soils

Conclusion

Under controlled conditions residual P contributed greatly to rye grass nutrition for two Swiss agricultural soils with a high total P content. The proportion of P in the plant derived from the fresh fertiliser decreased with increasing soil P availability caused by past P fertiliser applications. Such studies using isotopic techniques allows an improved understanding of soil P dynamics and P fertiliser transfer in soil-plant systems. They will be useful for the better agronomic and environmental management of P.

References

Fardeau JC 1996 Fert. Res. 45, 101-109.

Flisch R 1987 Mitteilungen für die Schweizerische Landwirdschaft 11, 281-294.

Morel C and Fardeau JC 1989 Appl. Radiat. Isot. 40, No.4, 273-278.

Morel C and Fardeau JC 1989 Plant Soil 115, 123-128.

Sharpley AN 1987 Soil Sci. Soc. Am. J. 51, 1531-1535.

Saunders WMH, Williams EG 1955 J. Soil Sci. 6, 247-267.

W. J. Horst et al., Plant nutrition – Food security and sustainability of agro-ecosystems. 700–701, 2001

Fluid fertilisers - A better solution for growing wheat in alkaline soils?

I. Bertrand[1)], R. E. Holloway[2)] and M. J. McLaughlin[3)]

[1)] *CSIRO Land & Water, Adelaide, Australia, PMB 2 Glen Osmond, South Australia, 5064, Australia, Isabelle.bertrand@adl.clw.csiro.au*

[2)] *SARDI, Minnipa Research Centre, Minnipa Research Centre, Minnipa, SA, 5654, Australia, Holloway.bob@saugov.sa.gov.au*

[3)] *CSIRO Land & Water, PMB 2 Glen Osmond, SA 5064, Adelaide, 5064 SA, Australia, Mike.Mclaughlin@adl.clw.csiro.au*

Key words: fertilisers, fluids, granular, phosphorus, soils

Abstract

The major constraints for agricultural productivity associated with alkaline and particularly calcareous soils under the dryland climate of southern Australia, include low availability of P and micronutrient deficiencies. For 50 years granular fertilisers were applied to cope with these problems, but the responses of wheat to increasing rates of granular fertilisers were very low and not economically viable.

To address this issue, experiments in field and laboratory were performed to test the efficiency of fluid fertilisers directly injected into the root zone compared to the efficiency of granular products such as MAP, DAP or TSP. Results showed that in the case of fluid fertilisers applied in calcareous soils, grain yields of wheat were increased by 20 to 40% compared to the granular products. Experiments in the laboratory were performed to explain the higher efficiency of fluid fertilisers which was partially due to a chemical effect.

Introduction

Alkaline soils represent an important fraction of the soils throughout the world which are associated with problems such as sodicity and nutrient and micronutrient deficiencies. These soils represent an important agronomic area of southern Australia. The major constraints for agricultural productivity associated with alkaline and particularly highly calcareous soils under the dryland climate of southern Australia include low availability of P and micronutrient (Zn, Cu, Mn, Fe) deficiencies. To cope with these fertility problems, granular P fertilisers have been used for more than 50 years in southern Australia – originally superphosphate, then double superphosphate (DSP), triple superphosphate (TSP) and later monammonium phosphate (MAP) and diammonium phosphate (DAP). However, on highly calcareous soils, responses of wheat to increasing rates of granular P fertiliser are very low and not economically viable. To improve the efficiency of P fertilisers in alkaline soils, new fertilisation strategies were assessed using fluid formulations, injected directly into the root zone at sowing. Laboratory experiments were also performed to: i) test if the reported increased efficiency of fluid compared to granular fertilisers on some alkaline soils of southern Australia was due to a placement effect rather than a chemical effect; and ii) understand the chemical interactions between fertilisers (fluid and granular) and soils.

Materials and methods

Field experiments were conducted on 2 soil types on the upper Eyre Peninsula in South Australia. Soil A was a red-brown calcareous sandy loam (5% CaCO3) while soil B was a grey highly calcareous sandy loam (67% CaCO3). Fertilisers applied in the field were MAP (10:22:0) or TSP (0:20:0) as granular products and Technical Grade MAP (TGMAP)(12:26:0), ZIP® (0:18:3, 14%Zn w/v) or orthophosphoric acid (H3PO4)(0:26:0 w/w) as fluids. Nitrogen and zinc were applied so that all plots received 15 kg N ha-1 and 2.5 kg Zn ha-1. The amount of water applied was adjusted to give 400 L ha-1 for all treatments and both forms of fertiliser (when N and P were applied as granular product, Zn was applied in solution). Controls consisted of nil P or nil fertiliser (nil N and Zn) and all treatments were replicated four times and randomised.

In controlled environment conditions, pot experiments were performed with soils A and B and also with two non-calcareous soils, C and D, which were classified as a Vertosol and a Sodosol, respectively. In this experiment fertilisers were added as MAP, DAP (18:20:0) and TSP as granular products and as TGMAP, ammonium polyphosphate (13:20:0, 1% Zn) and H3PO4 as fluids. Rates of P, N and Zn applied were 60 kg P ha-1, 60 kg N ha-1 and 12 kg Zn ha-1 for all treatments. Fertilisers were uniformly mixed with the soil and the amount of water added was adjusted to 90% of the water holding capacity for each soil. Treatments were replicated four times. Soils with and without fertilisers were then labelled with ^{32}P and ^{65}Zn. Plant-available P and Zn pools were assessed using isotope dilution procedures (L value). Residual effects of fluid and granular fertilisers and effects of time on isotopically exchangeable P (E value) after the addition of fertilisers were also measured.

Results and discussion

The use of fluid fertilisers in the field increased wheat yields by up to 31% on highly calcareous soils in a soil moisture limiting area (Figure 1) and the relative effectiveness of fluid formulations was up to 15-fold greater than similar products applied in granular form in these conditions. This contrasts with results from studies in North America and Canada which concluded that fluid fertilisers did not provide any significant advantages over granular products. Part of the improved efficiency of fluids is hypothesised to occur due to placement effects of fluids vs. granules in the soil, as well as changes to solution chemistry in the fertiliser reaction zone.

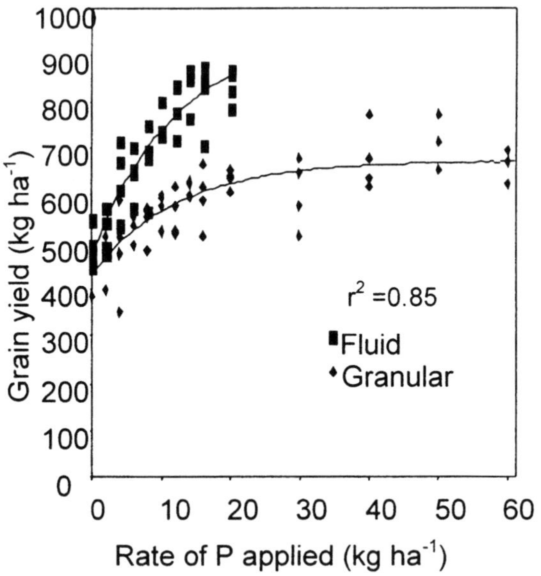

Figure 1. Relationship between wheat grain yield, and amount and form of P applied in a grey highly calcareous soil.

In the pot experiment wheat dry matter yields were again significantly greater using fluid fertilisers compared to similar granular products in both the highly calcareous grey soil and in the Vertosol (data not shown). In the controlled environment pot experiments, fluid and granular fertilisers were uniformly mixed with soils, consequently the higher efficiency of fluid fertilisers in these conditions suggest that the greater efficiency was not only related to a placement effect but also due to differences in fertiliser reaction products in the soil. Phosphorus from fluid fertilisers was also more easily taken up by plants than P from granular fertilisers (Figure 2), except in the Sodosol. Sorption experiments and determination of E values confirmed increased P availability using fluid fertilisers, and that the differences in pH between the fluid and granular products could not explain the large increases in yield using fluid fertilisers. The effect of soil moisture on the efficiency of fluid fertilisers also appears to be important and further investigations are underway to elucidate the mechanisms involved.

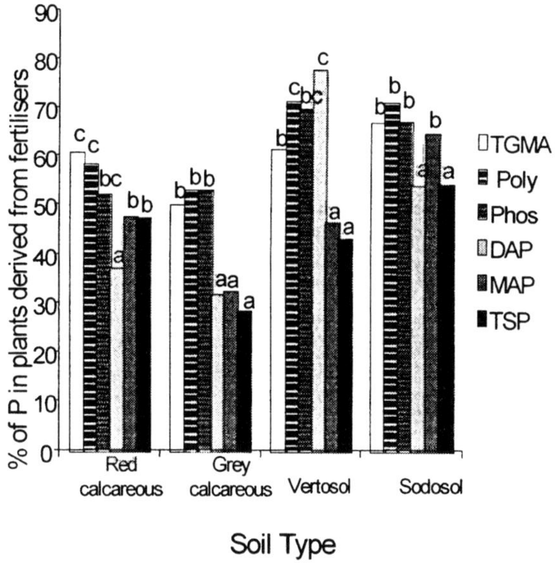

Figure 2. Proportion of P in plants derived from the fertiliser and expressed as the percent of total P in shoot, determined by isotopic dilution analyses.

Conclusions

In conclusion, fluid fertilisers appear to offer significant advantages in providing N, P and micronutrients for wheat production in alkaline soils of southern Australia. Furthermore, fluid fertilisers have the advantage of offering a simple means of providing several nutrients simultaneously, in variable proportions, compared to granulated products. In field conditions, the effects of water dilution and of different product formulations have been tested (Holloway *et al.*,2001).

References

Holloway RE, Bertrand I, Frischke AJ, Brace DM, McLaughlin MJ, Shepperd W and Graham RD 2001 IPNC Proceedings, (this volume).

W. J. Horst et al., Plant nutrition – Food security and sustainability of agro-ecosystems. 702~703, 2001

The new nitrification inhibitor DMPP (ENTEC®) – Comparisons with DCD in model studies and field applications

A. H. Wissemeier[1)], W. Linzmeier[2)], R. Gutser[2)], W. Weigelt[1)] and U. Schmidhalter[2)]

[1)] BASF AG, Agricultural Center Limburgerhof, D-67114 Limburgerhof, Germany, alexander.wissemeier@basf-ag.de

[2)] Chair of Plant Nutrition, Department of Plant Science, Technical University of Munich, Am Hochanger 2, D-85350 Freising, Germany

Key words: dicyandiamide, DCD, 3,4-dimethylpyrazole phosphate, DMPP, nitrification inhibitor

Abstract

DMPP (3,4-dimethylpyrazole phosphate (ENTEC®) is a new nitrification inhibitor (NI), the use of which is likely to advantage the environment and farmers. Comparisons with DCD (dicyandiamide) show that more than an order of magnitude less DMPP than DCD is needed to inhibit nitrification. Further features of DMPP are a soil – specific longer time period of efficacy and a substantial higher plant compatibility than DCD. Due to sorption DMPP is less mobile than DCD in soils and inhibits nitrification in the top soil even after simulation of heavy rainfall. Hence, NO_3^- leaching is more efficiently restricted by DMPP than by DCD under leaching conditions. Field experiments indicate reduced N losses due to N_2O volatilisation after DMPP or DCD application and no increase of NH_3 volatilisation from ammonium containing fertilisers stabilised with DMPP .

Introduction

The new nitrification inhibitor (NI) 3,4-dimethylpyrazole phosphate (DMPP) has been registered in Germany under the Chemical Act, and under fertiliser legislation in Austria, Belgium, France, Germany, Italy, Spain and The Netherlands. The present paper compares DMPP with the well established NI dicyandiamide (DCD), both in model studies and for selected aspects of field application. Crop yield responses following use of DMPP are reported elsewhere (Pasda *et al.*, 2001)

Materials and methods

Soil incubation experiments were conducted at 20°C at about 60% water holding capacity. 10 mg N as ammonium sulphate nitrate (ASN) was added to 100 g soil. Limburgerhof soil is a loamy sand and Ruchheim soil a loam with pH($CaCl_2$) 6.8 and 7.3, respectively. If not otherwise stated, application rate of DMPP and DCD were 1% and 10% (w/w) from NH_4-N, respectively (standard rate). 1% (w/v) K_2SO_4 solution was used to extract NH_4^+ and NO_3^- for analysis. The field experiments were conducted near Munich in 1998 (for details see text).

Results and discussion

The efficiency of DMPP in comparison to DCD in inhibiting nitrification was tested in an incubation experiment with slurry using different application rates of the 2 NI (Fig. 1). After 3 weeks of incubation, more than an order of magnitude less DMPP than DCD is needed to recover fertilised NH_4^+ in Ruchheim soil. In Limburgerhof soil, DCD even failed to stabilise NH_4^+ for 3 weeks, while small amounts of DMPP were highly effective. Comparing different soils and periods of time DMPP is 15 to 30 times more efficient than DCD (data not shown).

To test the longevity of DMPP in comparison to DCD soil was sampled from a pot experiment with mustard grown either on Ruchheim or Limburgerhof soils for 37 d during summer (June – July). In an incubation test with these soils the residual effect of the NI applied with ASN at the start of the pot trial was in focus. Therefore additional ASN was applied and after 14 d the recovery of ammonium was determined. DCD had lost any effectiveness in preventing NH_4^+ oxidation in these soils, while for the DMPP treated soils about 25% of the ammonium was recovered (Tab. 1).

Table 1. Residual effect of DMPP and DCD in a 14 d incubation experiment at 20°C using soils where ASN had been applied without extra NI (n = 4, ±SD, controls without ASN subtracted)

Fertiliser history	% NH_4^+ recovery Ruchheim soil	Limburgerhof soil
ASN	0 ± 0	0 ± 0
ASN/DCD	-0.6 ± 1.7	0 ± 2.5
ASN/DMPP	22.8 ± 6.5	25.4 ± 3.1

Plant compatibility is another feature where DMPP outperforms DCD. Hardly any leaf edge burn on lettuce leaves after DMPP application at 2.6 times the standard application rate could be observed, while DCD caused massive leaf burn at doses as low as 1.1 times the standard application rate (Tab. 2).

A major task of a NI is to keep NH_4^+ in the rooting zone and to prevent N leaching. Model studies in soil columns show indirectly that also in this important respect DMPP outperforms DCD. After leaching with 48 mm water, no residual DCD effect in preventing nitrification could be observed, while DMPP still inhibited nitrification as high NH_4^+ and low NO_3^- contents indicate (Tab. 3). Furthermore the low mobility of DMPP in soil has been shown in a lysimeter study by Fettweis *et al.* (1999), where no DMPP could be detected in the leaching water (limit of detection of 0.05 µg L^{-1}).

Table 2. Leaf edge burn of mature lettuce heads (cv 'Nadine') in pot experiments fertilised with ASN and extra doses of DMPP or DCD (n = 3, ±SD)

NI compared to standard rate	Percentages of affected leaf margin per leaf summed up over individual leaves per head	
Control (without NI)	24	± 9
1.1 x DCP	440	±178
2.2 x DCP	753	± 51
5.6 x DCD	1198	±120
2.6 x DMPP	96	± 41
5.2 x DMPP	149	± 37

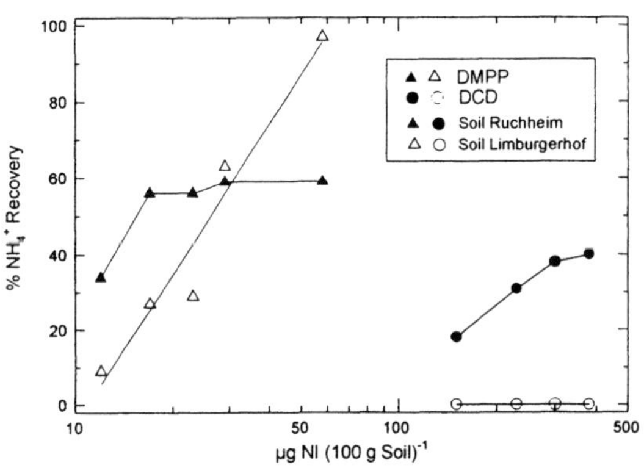

Figure 1. NH_4^+ recovery from liquid manure in two soils incubated for 3 weeks at 20°C either with DMPP or DCD at different concentrations

Table 3. N_{min} of 100 g Limburgerhof soil in columns (∅ 4 cm) together with 3 granules of ASN, ASN/DCD or ASN/DMPP embedded into the soil surface before and after leaching with 48 mm water and a subsequent incubation (14 d, 20°C, at 60% water holding capacity (n = 6, ±SD, controls without fertiliser subtracted)

	N_{min} in 100 g soil	
	mg NH_4-N	mg NO_3-N
Before leaching:		
ASN, ASN/DCD, ASN/DMPP	~ 16	~ 5
After leaching and subsequent incubation for 14 d:		
ASN	0.2 ± 0.03	6.9 ± 0.7
ASN/DCD	0.2 ± 0.02	10.0 ± 0.7
ASN/DMPP	7.5 ± 1.10	2.0 ± 0.8

An environmental concern is the emission of both N_2O and NH_3 after nitrogen fertiliser application. Field studies with the closed chamber technique on a brown earth with a silty loam texture, [pH($CaCl_2$) 6.4], show that N_2O emission derived from both soil-N as well as fertiliser-N is reduced by stabilised fertilisers as compared to ASN alone. This effect is particularly pronounced shortly after fertiliser application (Table 4). ^{15}N studies showed that the fertiliser derived N_2O emission (about 10-40% of total N_2O) was markedly reduced by stabilised fertilisers and further reduced by the granular form as compared to liquid application (data not shown). A significantly higher reduction of N_2O emission after fertilisation with ASN/DMPP compared to ASN/DCD or ASN without NI in a field experiment over three years is reported by Weiske *et al.* (2001).

Table 4. N_2O emission during 2 sampling periods in the field after soil surface application of liquid ASN (160 kg N ha^{-1}) on 27.03.98 (mean over 5 and 15 measurement dates, respectively, with n = 2–8 per treatment and date)

	N_2O [g ha^{-1} d^{-1}]	
	02.04.-17.04.98	22.04.–14.07.98
no N	6.0 (28)	5.2 (48)
ASN	21.8 (100)	10.8 (100)
ASN/DCD	17.3 (79)	10.7 (99)
ASN/DMPP	15.6 (72)	10.2 (94)

To estimate volatile NH_3 losses from fertilisers containing NI, a wind tunnel experiment was conducted. Calcium ammonium nitrate (CAN) was used as control (Table 5). After 25 d cumulativ NH_4-N losses from the ASN/DCD treatment was slightly higher compared to the control, while the losses from the ASN/DMPP treatment were even lower than from CAN.

Table 5. Loss of N as NH_3 after fertilisation with 160 kg N ha^{-1} to wheat (81 and 117 kg NH_4-N ha^{-1} as CAN and ASN, respectively) in a wind tunnel experiment for 25 d with one simulated rainfall event

	NH_3-N loss [kg ha^{-1}] cumulative for 25 d
no N (17 d only)	3.6
CAN	15.4 (100)
ASN/DCD	18.0 (117)
ASN/DMPP	14.2 (92)

In summary DMPP proved to be a highly efficient NI with several advantages compared to DCD.

References

Fettweis U, Mittelstaedt W, Führ F, and Schimansky C 1999. VDLUFA-Schriftenreihe 52 (Kongressband 1999), 457-460.

Pasda G, Hähndel, R, and Zerulla W 2001. IPNC Proceed., (this volume).

Weiske A, Benckiser G, and Ottow JCG 2001. IPNC Proceed., (this volume).

W. J. Horst et al., Plant nutrition – Food security and sustainability of agro-ecosystems. 704–705, 2001

Efficiency of foliar Zn fertilizers in coffee and citrus

C. A. Rosolem and L. V. S. Sacramento

College of Agricultural Sciences, São Paulo State University, Department of Crop Science, CP 237, 18603-970 Botucatu SP. Rosolem@fca.unesp.br

Key words: chelate, citrus, coffee, EDTA, foliar fertilization, lignosulphonate, micronutrient; zinc

Abstract

Zinc deficiency is widespread in coffee and citrus plantations in Brazil. Foliar fertilisation has been recommended to overcome the deficiencies. Laboratory studies showed that salt-Zn is absorbed in larger proportions than chelate-Zn, but chelates have a higher translocation rate within the plant. Four experiments were conducted in São Paulo State, Brazil, comparing coffee, orange, and lemon responses to foliar Zn applied as EDTA, lignosulphonate and sulphate. Zinc was applied in rates ranging from 0.0 to 4.9 kg ha^{-1} in coffee, from 0.0 to 5.5 kg ha^{-1} in lemon and from 0.0 to 4.6 kg ha^{-1} in orange trees, in three applications. Leaves were sampled periodically up to 180 days after the first application in coffee and up to 270 days in citrus. There was no effect of treatments on yields. All of the treatments were effective in maintaining Zn levels in the leaves above the threshold and above those observed in the check plots during most of the season. When applied at the same rate, in three out of four experiments there was no difference among the Zn sources. In just one experiment, with coffee, zinc sulphate was more effective in maintaining leaf Zn concentrations above threshold values than the chelates. In two out of four experiments it would be safe to reduce the chelate rates to 1/3 of the salt rate. There was evidence of Zn translocation to new growth irrespective of Zn source.

Introduction

Zinc deficiency is widespread throughout coffee and citrus plantations in Brazil. In acidic clay soils that require liming, Zn may be adsorbed to clay particles, complexed by the organic matter, or be fixed due to a higher pH and, therefore by unavailable to plants (Franco, 1982). Foliar applications of Zn are often recommended in regular fertilisation programs in coffee and citrus plantations as a tool to overcome these soil constraints.

Zinc absorption by leaves, as well as its translocation within the plant, depends on the source of the nutrient. Part of the applied element may be retained in the cuticle and cell walls but the retention is lower when Zn is applied as EDTA (Ferrandon and Chamel, 1988). Garcia and Salgado (1981) observed no significant differences when using zinc sulphate or zinc-EDTA in coffee but Malavolta *et al.* (1995) concluded that there was a lower retention of Zn in coffee leaves when it was applied as EDTA or lignosulphonate. Rosolem and Sacramento (2000) observed that, irrespective of Zn source, there was greater translocation of the nutrient to growing points when the plants were not Zn-deficient.

This experiment was designed to study the responses of coffee and citrus plantations to Zn when applied as a salt, a week chelate and a strong chelate.

Materials and methods

Four experiments were conducted on oxisols with 80 to 133 g kg^{-1} of clay. Zinc as $ZnSO_4$, lignosulphonate (weak chelate), or EDTA (strong chelate) was applied to coffee plantations at São Manuel and Pederneiras, a lemon orchard in Botucatu, and an orange orchard in Tabatinga, all in the State of São Paulo.

Zinc sulphate was applied at a concentration of 6 g L^{-1}, as recommended for coffee, and 3.3 g L^{-1}, as recommended for citrus. Urea and boron were added to the solutions. The chelates were applied at the full rate and the sulphate at 2/3 and 1/3 of the recommended rate. Plants were sprayed three times, 60 days apart, from December to April, during fruit development, using 250 ml of solution per plant in coffee, 7,800 ml of solution per plant in lemon, and 8,000 ml of solution per plant in orange. The amount of Zn applied (full rate) was 1,63 and 1,74 kg ha^{-1} in coffee and citrus, respectively. Coffee leaves were sampled regularly up to 180 days after the first application and citrus leaves were regularly sampled up to 270 days after the first application. The samples were collected from fruiting and non-fruiting. By sampling leaves from branches not bearing fruits, at each sampling new leaves are taken. So, it was possible to estimate the translocation of the nutrient to new leaves.

Results

During the first 60 to 90 days of the experiment, Zn levels of coffee leaves were below the threshold level of 10 mg kg^{-1} (Bataglia *et al.*, 1992) but increased to 16 mg kg^{-1} in Pederneiras and 18 mg kg^{-1} in São Manuel afterwards. Coffee yields averaged 5,449 kg ha^{-1} in Pederneiras and 4,500 kg ha^{-1} in São Manuel. There was no effect of the treatments. In Botucatu, lemon leaves of the check plots showed Zn levels above the threshold level of 25 mg kg^{-1} (Bataglia *et al.*, 1992) during the first 180 days after the first application. In Tabatinga, Zn levels in orange leaves of the check plots were above the critical level for only 45 days. Even so, there were no significant differences in yields of orange or lemon due the treatments. The yields averaged 31,200 kg ha^{-1} in Tabatinga and 35,000 kg ha^{-1} in Botucatu.

From the second application until 150 days after the first application there was no significant differences in Zn levels in coffee leaves in Pederneiras when the nutrient was applied as sulphate or lignosulphonate. The same was true for EDTAapplied at full rate or 2/3 of the full rate. In São Manuel, natrel and microfert were able to maintain Zn levels in the leaves above the threshold but lower than those obtained in plots sprayed with sulphate.

In lemon leaves, from 30 to 180 days after the first application the chelates performed like the sulphate when applied at the full rate or 2/3 of the full rate. Afterwards there was a decrease in foliar Zn levels, with no differences among fertiliser sources. In orange, for most of the time Zn concentrations in the leaves were no different when the nutrient was applied as lignosulphonate, EDTA, or sulphate. Only in samples taken around 180 days after the first application leaf Zn concentrations were greater with the sulphate source than other treatments.

Zinc concentrations were greater in leaves on branches not bearing fruits of sprayed plants than in the check plants of coffee around 150 days after the first application irrespective of Zn source. There was no difference in Zn levels when Zn was applied at full rate. For the 2/3 or 1/3 rate treatments, Zn concentrations in leaves were proportional to the rate applied. Very similar results were obtained in lemon and orange 30, 150 and 240 days after the first application.

Results obtained from leaves taken from branches bearing fruits do not support the conclusion of Ferrandon and Chamel (1982). We found no evidence of greater cuticular Zn retention when it was applied as sulphate. Furthermore, Zn translocation to regions of new growth did not differ when the nutrient was applied as a salt (sulphate), a weak chelate (lignosulphonate), or strong chelate (EDTA). Sacramento and Rosolem noticed greater translocation of Zn to growing points of coffee trees when the plants were not Zn deficient but there was little difference when the nutrient was applied as sulphate, lignosulphonate or EDTA.

In summary, all of the treatments were effective in maintaining leaf Zn levels above the threshold and above those observed in the check plots during most of the season. There was no difference in leaf Zn concentrations or yields among the Zn sources when they were applied at the same rate, in three out of four experiments,. In just one experiment, with coffee, zinc sulphate performed better than the chelates. In two of four experiments, it would have been safe to reduce the chelate rates to 1/3 of the salt rate and still retain efficacy. There was evidence of Zn translocation to new growth irrespective of Zn source.

References

Bataglia OC, Dechen AR and Santos WR 1992 *In* Adubação produtividade e ecologia. Eds. Dechen AR Boaretto AE and Verdade FC. pp 369-393. Fundação Cargill, Campinas, SP, Brazil.

Ferrandon M and Chamel AR 1988 J. Plant Nutrition 11, 247-263.

Franco CM 1982 *In* Micronutrientes, Fundação Cargill, Campinas, SP, Brazil.

Garcia AWR and Salgado AR 1981 *In* Congresso Brasileiro de Pesquisas Cafeeiras, 9, pp 39-47, São Lourenço, MG, Brazil.

Malavolta E, Malavolta M, Stepanutti R. Lima Filho OF, Nascimento Filho V and Cabral CP *In* Congresso Brasileiro de Pesquisas Cafeeiras, 21, pp 114-115, Caxambu, MG, Brazil.

Sacramento LVS and Rosolem CA 2000 *In* Colloquium for the Optimization of Plant Nutrition, 10, pp 232-233, Cairo.

W. J. Horst et al.(Eds.), Plant nutrition – Food security and sustainability of agro-ecosystems. 706~707, 2001

Determining P and N status of a tropical timber species (teak): Assessment of 'quick' chemical tests and a root phosphatase assay

M. J. Webb[1], P. Reddell[2], S. Nath[3] and R. J. Srivastava[4]

[1] *CSIRO Land and Water, PMB, Aitkenvale, Queensland, 4814, Australia, michael.webb@.tvl.clw.csiro.au*

[2] *CSIRO Land and Water, Tropical Forest Research Centre, PO Box 780, Atherton, Queensland, 4883, Australia*

[3] *Indian Council of Forestry Research and Education, Midnapore, West Bengal, 721 101, India*

[4] *Forest Research Institute, 18-G.T. Rd, Kanpur, Uttar Pradesh, 208024, India*

Key words: diagnostics, nitrogen deficiency, phosphatase, phosphorus deficiency, 'quick' tests, *Tectona grandis*, test strips, teak

Abstract

Although P and N deficiencies are known to be widespread in plantation grown trees in the humid tropics of the south-western Pacific, forest managers have difficulties in developing appropriate fertiliser strategies because there are currently no tests available that allow rapid assessment of the nutrient status (sufficiency or deficiency) of these trees. Thus we have assessed the suitability of 'quick' chemical test strips and an assay of root phosphatase activity for assessing the P and N status of seedlings of *Tectona grandis* (teak). In the 'quick' tests, phosphate or nitrate were not detected in the sap under the bark until these nutrients were in adequate supply for maximum growth. Consequently, these tests confirmed the adequacy of nutrient status (rather than determining the extent of severity of deficiency). The activity of root phosphatase was extremely sensitive to P status; decreasing rapidly with increasing P status. Phosphatase activity can thus be used to determine the severity of deficiency.

Introduction

Many of the soils commonly available for plantation forestry in the humid tropics of the south-western Pacific have inadequate P and N to maintain maximum tree growth rates and productivity (Webb *et al.*, 2000).

Traditional methods of nutrient diagnosis based on determining tissue elemental concentrations are problematic in such situations as (i) sampling protocols and diagnostic criteria for determining nutrient status have not been determined for important species, and (ii) laboratory facilities for elemental analysis are not readily accessible. Simple and robust alternatives to laboratory-based analysis of nutrient elements are desirable.

In agriculture, 'quick' tests, such as commercially available 'test strips' have been used successfully for nutrient diagnosis in a number of species (eg Williams and Maier, 1990), and phosphatase activity has been used for diagnosis of P status (Dracup *et al.*, 1984). This paper describes the potential of these tests for diagnosing P and N status in teak.

Materials and methods

Teak (*Tectona grandis*) seedlings were grown for two months in a glasshouse (19-36°C) in a randomised block design of three replicates. Soil had been mixed with basal nutrients and a range of rates of phosphorus or nitrogen.

P rate experiment

Soil, P-deficient soil (Galmara series – see Webb *et al.*, 1997); basal nutrient salts (mg kg^{-1}soil), NH_4NO_3 280, K_2SO_4 202, $CaCl_2{\cdot}2H_2O$ 300, $MgSO_4{\cdot}7H_2O$ 160, FeNaEDTA 20, $MnSO_4{\cdot}4H_2O$ 15, $ZnSO_4{\cdot}7H_2O$ 40, $CuSO_4{\cdot}5H_2O$ 10, H_3BO_3 1.4, $CoSO_4{\cdot}5H_2O$ 0.5, $NaMoO_4{\cdot}2H_2O$ 0.4, $NiSO_4{\cdot}6H_2O$ 0.2; P rates 0, 25, 50, 100, 150, 200, 250, 500, 1000 mg P kg^{-1} soil as $CaH_2(PO_4){\cdot}H_2O$.

N rate experiment

Soil, N-deficient soil (Pin Gin series – see Webb *et al.*, 1997); basal nutrients as for P rate experiment except that the NH_4NO_3 was replaced with KH_2PO_4 (88 mg kg^{-1}); N rates 0, 11, 23, 47, 94, 188, 281, 375, 563 mg N kg^{-1} soil as NH_4NO_3.

After two months, plants were separated into parts (bark, roots, and other parts). Phosphate test strips and nitrate test strips (Merckoquant PO_4^{3-}; NO_3^-, Merck) were pressed against the moist inner surface of freshly peeled bark to determine nutrient concentrations.

Root tips (3cm) were excised and assayed intact for phosphatase activity (Dracup *et al.*, 1984).

Results

Growth and PO_4 response to P supply

Increasing P supply substantially increased dry matter with maximum growth at 500 mg kg^{-1} (Fig. 1a). Over this same range, increasing P also increased PO_4 concentration in the sap under the bark (Fig. 1b). However, the response was different to that of dry matter; PO_4 concentration remained near zero at the lowest levels of P supply even though dry matter was increasing, but at high levels of P supply PO_4 continued to increase even when maximum dry matter had been reached.

Growth and NO_3 response to N supply

By contrast, increasing N supply to young teak seedlings had little or no effect on dry matter after two months (Fig. 1e) but doubled the concentration of NO_3 in the sap under the bark (Fig. 1g), increasing rapidly with small increases in N supply to a maximum detectable level of 500 mg kg^{-1}. At the second harvest, several months later, there was a substantial dry matter response to N and the NO_3 response was similar to that of PO_4 in that there was little increase in NO_3 until dry matter had reached a maximum (data not shown)

Root phosphatase response to P supply

Root phosphatase activity decreased with increasing P supply reaching near minimum activity at 500 mg P kg^{-1} soil, at about the same level which produced maximum growth (Fig. 1d). Indeed, the phosphatase activity relationship is almost a mirror image of the dry matter relationship; changing rapidly with small increases in P supply and then more slowly at higher rates of P supply. In many other tropical timber species that we have studied (data not shown), the root phosphatase response to increasing P supply was even more sensitive and less variable than that for young teak seedlings.

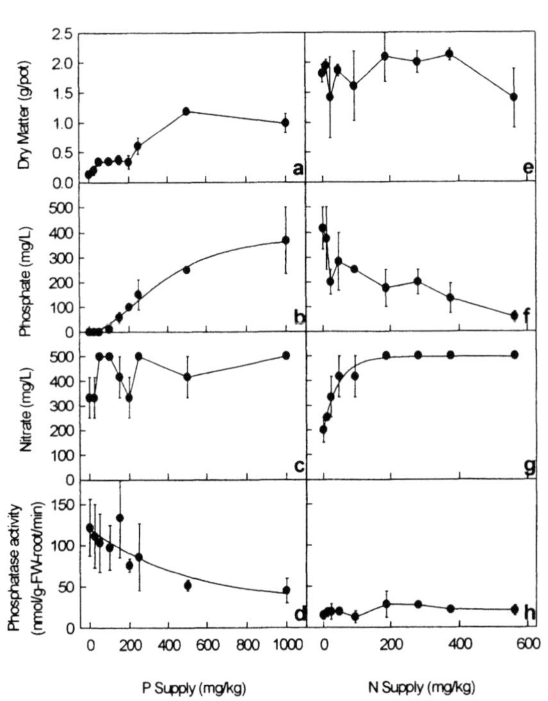

Figure 1. Effect of P and N supply on growth, nutrient concentration and phosphatase activity in *Tectona grandis*

Specificity of responses to nutrient supply

Each of the tests showed some degree of specificity to the nutrient status they were being used to assess.

For example, although NO_3 was highly responsive to N supply (Fig. 1g) it showed little response to P supply (Fig. 1c) even though there was a 10-fold response in dry matter. Similarly, root phosphatase activity was largely unaffected by N supply (Fig. 1h; *cf.* Fig. 1d).

By contrast, PO_4 was affected slightly by N supply (Fig. 1f). However, at all but the very highest levels of N supply, the PO_4 concentration was above 200 mg kg^{-1}; a level which would suggest P sufficiency (*cf.* Fig. 1b).

Discussion

While the NO_3 concentration in young teak increased rapidly with increasing N supply, in older teak plants and with other species we have studied, the NO_3 response is similar to that of PO_4; that is, remains low until the dry matter response approaches its maximum. This is likely to indicate that the PO_4 and NO_3 being measured is excess to the immediate requirements for growth.

If it is true that PO_4 and NO_3 only appear in easily detectable amounts when trees are well satisfied with these nutrients, then these tests will only be useful in establishing sufficiency. That is, they will not be able to discriminate between trees that are just adequate and those that are grossly deficient.

By contrast, phosphatase activity, does appear to offer a means of discriminating between trees that are just adequate and those that are grossly deficient.

Nitrate test strips in practice – a field example

In a commercial forest nursery in the south-western Pacific, *Eucalyptus deglupta* seedlings had symptoms appearing to be N deficiency. Using a 'nitrate test strip' it was immediately clear that N deficiency was not the problem. Further investigation determined that it was Fe deficiency that was the problem. Reliance on a laboratory based analysis would not have been feasible or allowed such rapid 'problem' solving.

Acknowledgement

We are pleased to acknowledge the assistance provided by Sue Joyce and Grace Baker. We are also grateful for the financial support for this work from the Australian Centre for International Agricultural Research (Forestry) and FAO/World Bank Project "Forestry Research Education and Extension Project" and for a travel grant from the Plant Nutrition Trust (Australia).

References

Dracup MNH, Barrett-Lennard EG, Greenway H and Robson AD 1984 J. Exp. Bot. 35, 466-480.

Webb MJ, Reddell P, Hambleton A and Mazza GM 1997 Aust. For. 60, 46-52.

Webb MJ, Reddell P, Singh K, Bevu T, Iakopo M and Pouli T 2000 *In* Opportunities for the New Millenium. Australian Forest Growers Conference. Eds. A Snell and S Vize. pp 256-260. Australian Forest Growers, Kingston, Australia.

Williams CMJ and Maier NA 1990 J. Plant Nutr. 13, 985-993.

W. J. Horst et al.(Eds.), Plant nutrition – Food security and sustainability of agro-ecosystems. 708–709, 2001

Corn stalk nitrate concentration profile: Implications for the end-of-season stalk nitrate test

W. W. Wilhelm, G. E. Varvel, and J. S. Schepers
Soil and Water Conservation Research Unit, Agricultural Research Service, United States Dep. of Agriculture, 120 Kiem Hall, Univeristy of Nebraska, Lincoln, Nebraska, 68583-0934, United States, wwilhelm1@unl.edu

Key words: corn, end-of-season stalk nitrate test, N nutrition, nitrate concentration, *Zea mays* L.

Abstract

The end-of-season corn (*Zea mays* L.) stalk nitrate-N test was developed as a post-mortem to determine if excessive or insufficient N was available to the corn crop during the latter part of the season. The stalk section specified for the test was very specific, the 20 cm-long section between 15 and 35 cm above the soil. Under production conditions, it may not always be possible to collect this precise stalk section. The objective of this study was to determine how nitrate concentration varied within the stalk from the soil level to the ear node, and how this variation could affect interpretations of the stalk nitrate test. Field grown (140 kg N ha^{-1}) corn stalks were collected and separated into phytomers (the node plus leaf, internode, and bud developing from it). Phytomers were further divided into six segments; the node and five equal length segments of the internode. All samples were analysed for NO_3-N with a nitrate-ion specific electrode after extraction with 0.04 *M* $(NH_4)_2SO_4$. Nitrate concentrations of individual samples varied from less than 100 to greater than 8000 mg NO_3-N kg^{-1} dry weight, and increased down the stalk from ear to soil. Generally, the nitrate concentrations of segments within a phytomer were similar. These results indicated new critical values, approximately 35% greater than the original ones, may be needed to determine if limiting or excessive amounts of N were available to the crop, i.e. 950 vs. 700 and 2700 vs. 2000 mg NO_3-N kg^{-1} for insufficient and excessive levels, respectively. However, the general interpretation of test would remain unchanged because stalk nitrate concentrations vary so widely under field conditions from less than 100 to greater than 5000 mg NO_3-N kg^{-1}.

Introduction

The end-of-season corn (*Zea mays* L.) stalk nitrate-N test (hereafter referred to as stalk nitrate test) was proposed by Binford *et al.* (1990) as a method of determining if excessive or insufficient N was available to the corn crop during the latter part of the season. In the test, 20-cm sections of corn stalks (between 15 and 35 cm above the soil) are collected from several plants, dried, ground, and analysed for nitrate-N. Nitrate-N concentrations less than 700 mg kg^{-1} dry stalk tissue indicated that available N limited grain yield whereas concentrations greater than 2,000 mg kg^{-1} indicated excessive N was available to the crop (Binford *et al.*, 1992). Other researchers have evaluated the test and concurred that stalk nitrate concentrations greater than 2000 mg kg^{-1} indicated that excessive N was available to the crop (Varvel *et al.*, 1997). These studies suggested that the stalk nitrate assay could be used as a post-mortem to determine if yield-limiting or excessive N was present to guide future fertilizer-N management, thereby improving profitability and reducing environmental degradation.

To be a useful crop management tool, methods for collecting and analysing samples must be straightforward and reasonably flexible so producers and consultants can adapt to field conditions as they gather samples. It is unrealistic to assume that samples will always be exactly 20 cm long and can always be collected between 15 and 35 cm above the ground under field situations. The objective of this study was to determine how stalk nitrate test results and interpretations would be affected if samples were collected from a portion of the stalk different from that specified by Binford *et al.* (1990).

Materials and methods

Stalks were collected, after physiological maturity, from a rain fed field of corn gown with 140 kg N ha^{-1}. Two plants were collected from each of 10 replications. Ears and leaf blades were removed from the stalks in the field and the portion of the stalk above the ear node was discarded. Stalks were cut at the soil surface, labelled and dried at 60° C. Leaf sheaths were removed from the stalks. Dried stalks were divided and numbered by phytomer unit (internode and subtending node, see Figure 1). Phytomer units were further divided into nodes and internodes, and internodes were subsequently divided into five equal length segments. Segments from the same relative position on the plant (i.e., same phytomer and same segment within the phytomer) from the two plants in each replication were combined and ground to pass a 2-mm screen. Nitrate-N concentration, in a 0.25 g samples, was determined with a nitrate-ion specific electrode after extraction with 50 mL of 0.04 *M* $(NH_4)_2SO_4$ (Wilhelm *et al.*, 2000). Analysis of variance (Proc MIXED, Littell *et al.*, 1996) was used to determine if phytomer, tissue type (node vs. internode), and segment within internode differed in nitrate-N concentration.

Results

Across all plant parts, nitrate concentrations varied from less than 100 mg NO_3-N kg^{-1} dry weight to greater than 8000 mg NO_3-N kg^{-1} dry weight. Analyses showed that tissue types (node vs. internode), phytomers, and segments were significant sources of variation for nitrate-N concentration. Their interaction was not significant suggesting that although nitrate concentration differed with phytomer and segment, the pattern was similar among segments within all phytomers (Table 1).

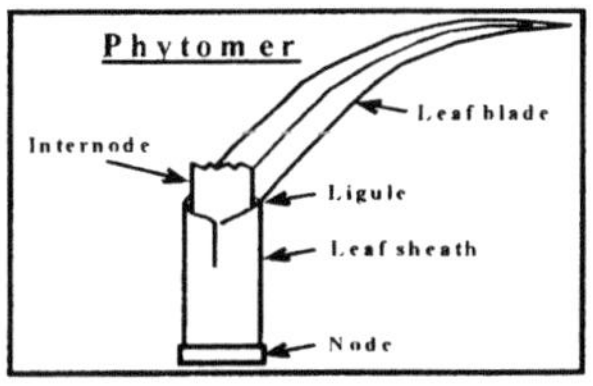

Figure 1. The phytomer is the basic repeating building block of the grass shoot and consists of four tissues, the node and leaf, internode, and axillary bud derived (above) from that node. The axillary bud in this diagram of the phytomer is hidden from view by the leaf sheath.

Table 1. Analysis of variance for nitrate-N concentration in several phytomers and segments of the corn stalks.

Source of variation	Degrees of freedom	F-value	Probability of a greater F
Phytomer	3	361	0.0001
Segment	5	4	0.0039
Phytomer x Segment	15	1	0.4751

Table 2. Mean nitrate-N concentration in several phytomers, divided into several segments, of corn stalks.

Segment[1]	Phytomer[2]				Mean
	S+1	S+2	E-2	E-1	
	Nitrate-N concentration (mg NO_3-N kg^{-1})				
5	3906	2261	126	146	1610
4	4456	2820	155	93	1881
3	4796	3304	189	105	2098
2	5026	3870	332	124	2338
1	4699	3628	423	163	2228
Node	4657	3643	382	175	2214
Mean	4590	3254	268	134	2061

[1] Stalk portions of each phytomer were divided into six segments, the node and five equal length segments of the internode. Segments were numbered upward from the node.
[2] Phytomer S+1 was the first phytomer above the soil surface, Phytomer S+2 was the second above the soil, Phytomer E-2 was second below the ear, and Phytomer E-1 was immediately below the ear.

Since not all plants within a sample had the same number of phytomers between the soil surface and ear, a continuous description of NO_3-N concentration for the entire stalk was difficult to produce. For comparison purposes, we report NO_3-N concentrations for only four phytomers, the two closest to the soil and the two immediately subtending the ear. Depending on the plant, zero to two phytomers may have existed between the phytomers near the soil and ear. As a result, nitrate profiles reported here (Table 2) appear to have a discontinuity that may not exist if data from all phytomers were available. Following the sampling procedures defined by Binford *et al.* (1990) for the stalk nitrate test, the second phytomer above the soil (S+2) would most likely be sampled.

The first two phytomers below the ear (E-1 and E-2) did not differ in nitrate concentration, but phytomers S+1 and S+2 differed from the former two and also differed from each other (Table 2). Nitrate concentration increased gradually moving down the stalk from the ear to the soil. Nitrate concentration in node tissue was similar to that of internode tissue in phytomers E-1, E-2, and S+1. However in phytomer S+2, the nitrate-N concentration in the upper two segments (4 and 5) was less than in the node tissue ($p < 0.05$). Nitrate-N concentration increased linearly as phytomers were further from the ear ($p < 0.0001$). Likewise, NO_3-N concentration of segments, within phytomers, increased linearly as they neared the node ($p < 0.001$).

Discussion

These results indicated that reasonable care must be taken in collecting field samples for the stalk nitrate test. If samples are collected closer to the soil than described in the original procedure (< 15 cm above the soil), the resulting analyses will likely produce greater nitrate-N concentrations. As a result, the critical values suggested by Binford *et al.* (1990, 1992) may have to be adjusted upward. Data from this initial study with an adequate to excessive N rate (N application, 140 kg ha^{-1}; yield 6.35 Mg ha^{-1}; stalk nitrate-N, 3254 mg NO_3-N kg^{-1}) suggests critical values may need to be increased by about 35% (comparison of S+2 to S+1 in Table 2). New critical values would be about 950 and 2700 mg NO_3-N kg^{-1} for yield-limiting and excessive levels of N [compared to 700 and 2000 mg NO_3-N kg^{-1}, respectively, as reported by Binford *et al.* (1990)]. Further research, with an array of N rates from yield limiting through excessive, will be required to accurately establish new values. However, interpretation and management decisions based on samples collected from stalk segments nearer the soil and new critical values would be similar to interpretations and decisions based on samples collected as described by Binford *et al.* (1990).

In fact, collecting samples from segments nearer the soil surface may offer a level of assurance because change in nitrate concentration was greater as samples were collected closer to the ear. Also because nitrate-N concentration declined rapidly in phytomers nearer the ear, field modification of collection procedures to acquire stalk sections more than 30 cm above the ground may result in inappropriate future N management procedures from a yield/profitability or environmental standpoint.

References

Binford GD, Blackmer AM and El-Hout NM 1990 Agron. J. 82, 124-129.
Binford GD, Blackmer AM and Meese BG 1992 Agron. J. 84, 881-887.
Littell RC, Milliken GA, Stroup WW and Wolfinger RW 1996 SAS System for Mixed Models. SAS Institute, Inc., Cray, NC (USA).
Varvel GE, Schepers JS and Francis DD 1997 J. Agric. Prod. 10, 147-151.
Wilhelm WW, Arnold SL and Schepers JS 2000 Agron. J. 92, 186-189.

W. J. Horst et al.(Eds.), Plant nutrition – Food security and sustainability of agro-ecosystems. 710–711, 2001

Diagnosing nitrogen requirements of potatoes in farmers' crops

I. Papastylianou and S. Gregoriou

Agricultural Research Institute, Ministry of Agriculture, Natural Resources and Environment, 1516 Nicosia, P.O.Box 22016, Cyprus. e-mail: papastyl@arinet.ari.gov.cy

Key words: critical nitrate range, nitrogen, potato petiole nitrate

Abstract

Diagnosis of the nitrogen nutrition status of the potato plant could be based on nitrate concentration in petioles. The applicability and diagnostic precision of this method were tested in fourteen experiments with different nitrogen fertilizer rates, established in farmers' crops. Nitrate concentration in petiole sap was monitored throughout the growing season and it was related to tuber yield. In thirteen out of fourteen cases the nitrate concentration diagnosed the nitrogen nutrition status of the crop and predicted the necessity of nitrogen fertilizer application during the growing season. A practical, simple test for analysing petiole sap for nitrate was developed. The test could be practised in situ by the farmer, and the necessity of nitrogen fertilizer application could be decided immediately.

Introduction

The petiole nitrate concentration (PNC) is a useful criterion for the N nutrition status of the potato *(Solanum tuberosum L.)* crop (Williams and Maier, 1990a). For this purpose, Williams and Maier (1990 b) established critical nitrate ranges (CNR) on which the decision for N fertilization could be based.

The present study aimed to examine: (a) the applicability of CNR in farmer fields, (b) the accuracy of CNR in determing N nutrition status, and (c) the efficiency of the method in diagnosing the necessity for N fertilization.

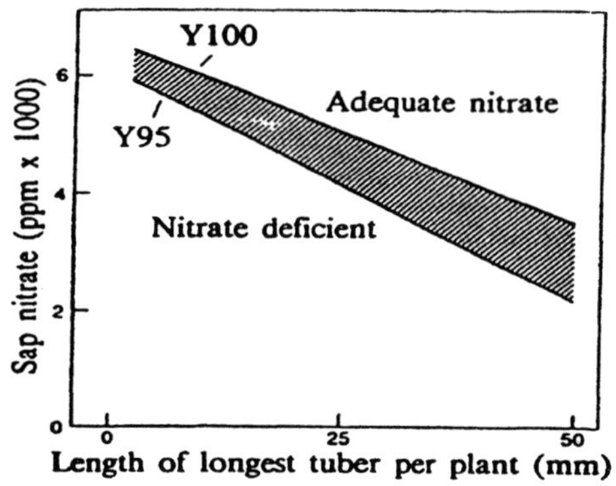

Figure 1. The relationships between sap nitrate for 95 and 100% of maximum yield (termed Y95 and Y100, respectively) and length of the longest tuber per plant (Williams and Maier, 1990b).

Materials and methods

During 1999 fourteen trials were established in farmers' crops at Kokkinochoria area in Cyprus.

The experiments were established on a part of the crop where the farmer had not applied any nitrogen fertilizer. At planting, nitrogen fertilizer in the form of ammonium sulphate was applied at rates zero and 150 kg N/ha in 8 x 20 m plots, in three replicates. Immediately after the first plant sampling see below, each plot was divided in two subplots (4 x 20 m) one of which received ammonium nitrate at 150 kg N/ha.

Nitrate concentration in petioles of the youngest fully expanded leaf (YFEL) was measured. Sampling was conducted in the morning, 1 to 2 hours after sunrise and twenty YFEL were collected. The first sampling started with the appearance of the first YFEL and continued every 15 days throughout the growing season.

Leaf blades were separated from the petioles. Petiole sap was used to measure nitrate concentration after it was diluted 1:39 in distilled water. The "RQflex" meter using "Reflectoquant" strips (MERCK, Darmstadt), were used to determine PNC.

Response of tuber yield to nitrogen application was determined by comparing plots that received fertilizer with plots that received no fertilizer.

Depending on the tuber size the CNR level at the first sampling was determined based on figure 1.

Results

Using the CNR criterion determind at the first sampling, PNC predicted the status of nitrogen nutrition and whether or not there was a necessity for nitrogen top-dressing in 12 out of the 14 potato crops tested. Only in the case of crop No 11, PNC indicated sufficient nitrogen nutrition, even though, there was a response to nitrogen application. In one case (crop No 1) PNC at the first sampling indicated deficiency of nitrogen nutrition, even though there was a negative response to nitrogen application (Table 1).

Discussion

All farmers applied nitrogen fertilizer, although in 70% of the cases it was not needed. The CNR method was shown to be useful for deciding the necessity of nitrogen application during the growing season (Williams and Maier, 1990b). A critical assessment of the CNR method by Mackerron *et al.* (1995) indicated that variability of PNC through the season, is a weakness of the method.

In the present study the PNC criterion was successful in 85 percent of the cases. However, there were two cases of discrepancy regarding accuracy of the CNR method, and this indicates that further research should elaborate on the factors involved in these faulty predictions. Considering that the farmers tend to fertilise their crops without examining the actual need for fertilization, it is important that the PNC at its present accuracy level is used in practice. Further improvement of the test should go along with the implementation of a better strategy of N fertilization.

References

Mackerron DKL, Young MW and Davies HV 1995 Plant Soil, 172, 247-260.

Williams CMJ and Maier NA 1990a J. Plant Nutr. 13, 985-993.

Williams CMJ and Maier NA 1990b J. Plant Nutr. 13, 971-984.

Table 1. Nitrogen nutrition status of the 14 farmers' crops on zero N rate, in relation to Critical Nitrate Range (CNR), and response to N fertilization.

Crop No.	Sampling*			Response to fertilization**
	First	Second	Third	
1	b	a	a	D
2	b	b	w	I
3	b	b	w	n.r.
4	w	a	w	n.r.
5	a	a	a	n.r.
6	a	a	a	n.r.
7	a	a	a	n.r.
8	a	a	a	n.r.
9	a	w	a	n.r.
10	w	a	w	n.r.
11	w	a	a	I
12	b	b	w	I
13	w	a	w	n.r
14	b	b	b	I

* Concentration of NO_3 in petioles in relation to the Critical Nitrate Range: b = below, a = above, w = within

** Tuber yield response to N fertilization: i=yield increase, d = yield decrease, n.r. = no response

W. J. Horst et al.(Eds.), Plant nutrition – Food security and sustainability of agro-ecosystems. 712~713, 2001

Nutrition diagnosis of potato (*Solanum tuberosum* L.) using leaf petiole sap

M. Takebe, Y. Kasahara and T. Karasawa

Hokkaido National Agricultural Experiment Station, Hitsujigaoka 1, Toyohira-ku, 062-8555, Japan, mtakebe@cryo.affrc.go.jp

Key words: inorganic P, leaf petiole sap, nitrate nitrogen, plant nutrition diagnosis, potato (*Solanum tuberosum* L.)

Abstract

Sap analysis is an effective method for diagnosing the nutritional status of field crops in order to determine appropriate fertilizer application for stable yield, high quality and low load to the environment. We conducted field experiments on potatoes to determine the sap properties of leaf petioles and to determine the optimum values of nitrogen and phosphorus for diagnosis in potato. The weight ratios of extracted sap to leaf petiole were 0.22 - 0.37. The NO_3-N concentration in the petiole sap was nearly the same as that in the leaf petiole, but the inorganic P concentration in the petiole sap was lower than that in the leaf petiole. NO_3-N and inorganic P concentrations in the sap were a good reflection of the plant nutritional status of nitrogen and phosphorus at the bud sprouting stage. For sufficient yield and quality, the optimum value of NO_3-N concentration in the sap of leaf petiole at the bud sprouting stage was estimated to be 1.3 - 1.5 g L^{-1}, and that of inorganic P was estimated to be about 100 mg L^{-1}.

Introduction

Nutrition diagnosis using leaf petiole sap is an effective method that provides real-time information on crops (Roppongi, 1998). The nitrate-nitrogen (NO_3-N) concentration in leaf petiole sap serves as an indicator of the nitrogen nutrition status. For fruit vegetables (e.g., cucumber, tomato, strawberry) and flowers (e.g., rose, cyclamen, chrysanthemum), the optimum concentrations of NO_3-N in leaf petiole sap have been determined to control the amount of top-dressing (Takebe, 1994). Sap analysis is also considered to be an effective method for cultivating field crops that have a stable yield, high quality and low load to the environment. In this study we conducted field experiments on potatoes (*Solanum tuberosum* L.) to know sap properties of leaf petiole and to determine the optimum value of nitrogen. In these experiments, we also considered the phosphorus nutrition status and tried to determine the optimum value of phosphorus.

Materials and methods

In 1997 – 1999, potatoes (Danshakuimo) were grown in Ando soil in Sapporo with 4 levels of nitrogen (0, 4, 8 and 16 g N m^{-2}) and 3 levels of phosphorous (0, 7.9 and 15.7 g P m^{-2}). Sap was collected from the leaf petioles at the bud sprouting stage and at the flowering stage. Freeze-dried samples were also prepared at same time. The leaf petiole sap was pressed by a garlic squeezer after cutting the petiole into several millimeter pieces. NO_3-N concentration was measured by an ion chromatograph (Dionex QIC analyzer; column, IonPac AS4A) using 50 to 1000-times diluted sap. Inorganic P (Pi) concentration was measured by the colorimetric method (molybdenum blue assay) using 5 to 10-times diluted sap. To determine the NO_3-N and Pi concentrations in the leaf petiole, the freeze-dried sample of leaf petiole was extracted by water. NO_3-N and Pi in the extract were measured, and their concentrations per water in petiole were calculated. Dried samples of leaf blade and leaf petiole were digested by the Kjeldahl method, and total N and total P concentrations were measured by the flow-injection method (Tecator FIAstar 5010) and colorimetric method, respectively. Starch value was measured with the gravimeter (improved Reiman's balance).

Results and discussion

The weight ratios of pressed sap to leaf petiole were 22 – 30 % in 1998 and 24 – 37 % in 1999. The NO_3-N concentration in the sap was nearly the same as that in the leaf petiole, but the Pi concentration in the sap was lower than that in the leaf petiole (Table 1). The NO_3-N concentration in the sap was a good reflection of the plant nutritional status of nitrogen, because it was strongly correlated with the total N concentrations in the leaf petiole and the plant top (Fig. 1). The Pi concentration was also correlated with total P concentration in the leaf petiole and the plant top (data not shown). The NO_3-N concentration in the sap reflected the amount of applied N at the bud sprouting stage and at the flowering stage, and inorganic P concentration in the sap reflected the amount of applied P only at the bud sprouting stage (Table 1). Thus, NO_3-N and Pi concentrations in leaf petiole sap are useful indicators of the nutrition status of potatoes.

The level of tuber yield was low in 1997 and high in 1998 and 1999. In 1997 and 1998, The NO_3-N concentration in the leaf petiole sap required for a high yield was 1.3 – 1.5 mg L^{-1} at the bud sprouting stage (Fig. 2). In 1999, NO_3-N concentration for a high yield was higher than 1.5 mg L^{-1} (Fig. 2), but the starch value became lower than 14 when the NO_3-N concentration exceeded 1.5 mg L^{-1} (Fig. 3). Therefore, for sufficient yield and quality, the optimum value of NO_3-N concentration in the sap of leaf petiole at the bud sprouting stage is 1.3 - 1.5 g L^{-1}. This value was similar to the value at the early sampling time of former study of potato (Williams and Maier, 1990), though the variety and culturing method were different.

The optimum value of Pi in the leaf petiole sap at the bud sprouting stage was estimated to be about 100 mg L^{-1}, considering the relationship of Pi in the sap and tuber yield (Fig. 4). There was no relationship between the Pi concentration in the sap and starch content.

References

Roppongi K 1998 Jpn. J. Soil Sci. Plant Nutr. 69, 235-238.
Takebe M 1994 Farming Japan 28(6), 22-27.
Williams CMJ and Maier NA 1990 J. Plant Nutr. 13, 985-993.

Table 1. NO_3-N and inorganic P concentrations in leaf petiole sap and leaf petiole

	treatment	bud sprouting stage			flowering stage		
	g m^{-2}	sap (a)	petiole(b)	a/b	sap (a)	petiole(b)	a/b
	N 0	0.30	0.30	1.02	0.01	0.00	-
NO_3-N	N 4	0.78	0.83	0.94	0.14	0.23	0.60
g L^{-1}	N 8	1.27	1.25	1.01	0.83	0.79	1.05
	N 16	1.53	1.53	1.00	1.58	1.59	0.99
Inorganic P	P 0	80	94	0.85	36	51	0.70
mg L^{-1}	P 7.9	96	120	0.80	41	49	0.83
	P 15.7	113	138	0.82	41	59	0.69

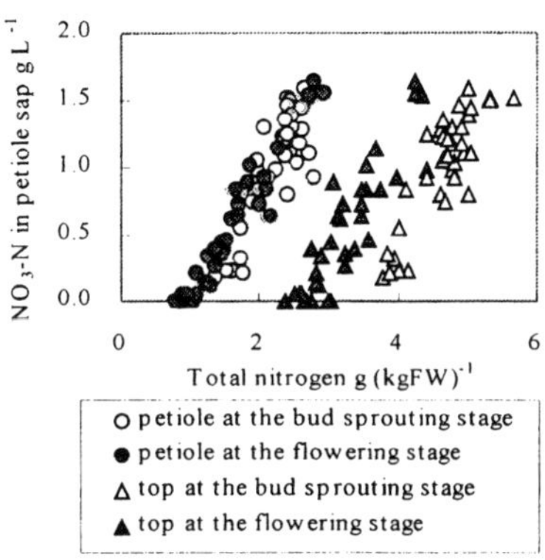

Fig. 1. Relationship between NO_3-N concentration in petiole sap and total nirtrogen content in petiole and top.

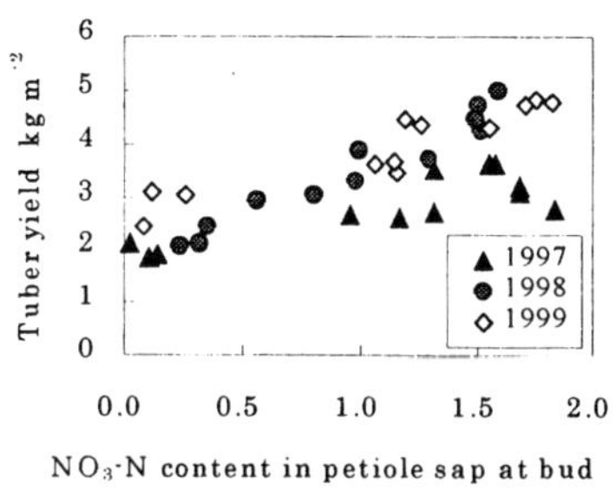

Fig. 2. Relationship between tuber yield and NO_3-N concentration in petiole sap at the bud sprouting stage.

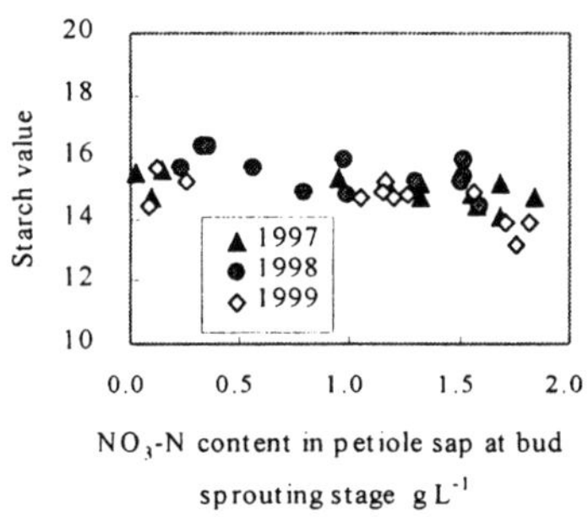

Fig. 3. Relationship between starch value and NO_3-N concentration in petiole sap at the bud sprouting stage.

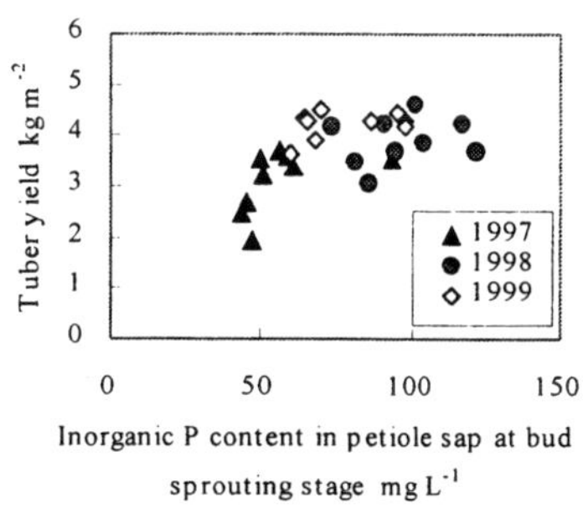

Fig. 4. Relationship between tuber yield and inorganic P in petiole sap at the bud sprouting stage.

W. J. Horst et al.(Eds.), Plant nutrition – Food security and sustainability of agro-ecosystems. 714~715, 2001

Critical phosphate (P) and inorganic phosphate (Pi) concentrations for the growth of winter oilseed rape (*Brassica napus*)

B. J. Major and P. B. Barraclough
Institute of Arable Crops Research - Rothamsted, Harpenden, Hertfordshire, AL5 5BX, United Kingdom, bernard.major@bbsrc.ac.uk

Key words: *Brassica napus*, critical %P, critical Pi, growth, phosphate, phosphorus

Abstract

Oilseed rape (*Brassica napus*) (OSR) has a high phosphorus (P) requirement for growth. Plant testing procedures give a measure of the actual physiological P status of OSR and indicate whether the plants requirements are being met by soil P supplies. This paper considers the effect of soil P supply on the amount of P and inorganic Pi ($H_2PO_4^-$) within tissues of OSR and relates these results to effects on growth. OSR (cv. Apex) was grown in a field experiment, consisting of plots with soil Olsen-P values ranging between 4-69 ppm P-PO_4. The crop was sampled at the rosette growth stage and tissue analysed for Pi mM concentration and total %P content in dry matter. The dry weight of the shoot was measured and related to soil Olsen-P, tissue Pi mM and %P concentration by fitting Mitscherlich curves to the data. Shoot dry weight varied from 144-306 g m^{-2}, with a critical soil Olsen-P of 21 ppm PO_4-P (R^2=0.58) giving 95% of the maximum shoot dry weight. A critical shoot Pi of 13.8 mM Pi (R^2=0.77) or 0.67% P, and a critical largest-leaf blade Pi of 5.9 mM (R^2=0.78) or 0.53% P and a largest-leaf petiole Pi of 1.8 mM (R^2=0.72) or 0.48% P were calculated for 95% maximum shoot dry weight. Preliminary plant diagnostic values for assessing whether OSR plants are sufficient in Pi for growth have been established.

Introduction

There are increasing economic pressures to use P fertilisers more sparingly. Additionally the inefficient use of P fertilisers is blamed for pollution and eutrophication in some water courses. On the basis of soil testing only 8% of UK fields growing oilseed rape (*Brassica napus*) (OSR) crops are believed to be P deficient (Skinner and Todd, 1998).

Plant tissue testing could provide information on whether OSR plants are actually sufficient or deficient in P, giving growers a guide to the successfulness of their long-term P fertiliser application strategy which is commonly based on soil tests taken every 3-4 years. Traditionally P tissue testing has focused on measuring total nutrient concentration in dry matter (%P). Measuring physiologically important inorganic ions within the plant may however give better relationships to plant growth (Barraclough, 1993).

This paper considers the effect of P supply on the concentration of inorganic $H_2PO_4^-$ (Pi) within tissues of OSR and relates these results to early plant growth.

Materials and methods

The experiment was undertaken at Rothamsted (51° 49' N, 0° 22' W) in a free-draining silty clay loam, sited 128 m above sea level. Soil samples were taken on the 17/08/99 and soil Olsen-P values determined (Olsen, 1954). The plots were not replicated, but the wide range of soil-P values obtained allowed a curve fitting approach to be applied to the data. Conventional winter OSR (cv Apex) was drilled on the 31/08/99. Establishment was good (72 plants m^{-2}) and the shoots were sampled at the rosette growth stage (Sylvester-Bradley, 1985) on the 22/12/99.

Shoot samples and the largest-leaf were split into leaf blades and petioles. Samples were microwave-oven dried to prevent conversion of organic-P compounds to Pi and the dry weight determined (Bollons and Barraclough, 1997). The molybdate-blue method (Murphy and Riley, 1962) was used to measure Pi in dried plant samples using 2% acetic acid as an extractant. Pi concentration was then calculated using the sample moisture content. Total P (%P in dry matter) was measured by inductively-coupled plasma (ICP) atomic emission spectrometry after the samples had been digested with concentrated perchloric and nitric acid (Zarcinas *et al.*, 1987).

The dry weight (g m^{-2}) of the shoot (y) was related to soil Olsen-P, tissue Pi mM and % P concentration (x) by fitting Mitscherlich curves to the data (Ware *et al.*, 1982). Critical values were calculated on a 95% threshold basis or by interpolating maximum growth from the maximum %P or Pi concentration observed.

Results

Soil Olsen-P values, at sowing, ranged between 4-69 ppm P-PO_4, giving shoot dry weights varying between 145-306 g m^{-2} at the rosette growth stage. A critical soil Olsen-P of 21 ppm PO_4-P was calculated to give 95% of the maximum shoot dry weight (R^2= 0.58).

Shoot Pi values ranged from 2.9-13.9 mM Pi and total P in shoots varied from 0.28-0.66% P. Pi values ranged from 3.7-16.9 mM Pi in leaf blades and 1.1-12.3 mM Pi in petioles. Similarly total P in leaf blades varied from 0.31-0.80% P, with lower values in petioles ranging from 0.18-0.60% P. ICP analysis indicated all other nutrients were at adequate concentrations (Reuter and Robinson, 1997).

Critical shoot Pi mM and total P concentrations for 95% maximum shoot dry weight are shown in Table 1. Shoot Pi values showed slightly better relationships to shoot dry weight than % P values. The youngest fully emerged leaf blade is usually used as an indicator tissue for plant testing, however identifying this equivocally in rape was problematical, hence the largest-leaf blade was selected instead. Good relationships between shoot dry weight and mM Pi concentrations in the largest-leaf blade (Figure 1) and petiole were identified (Table 1).

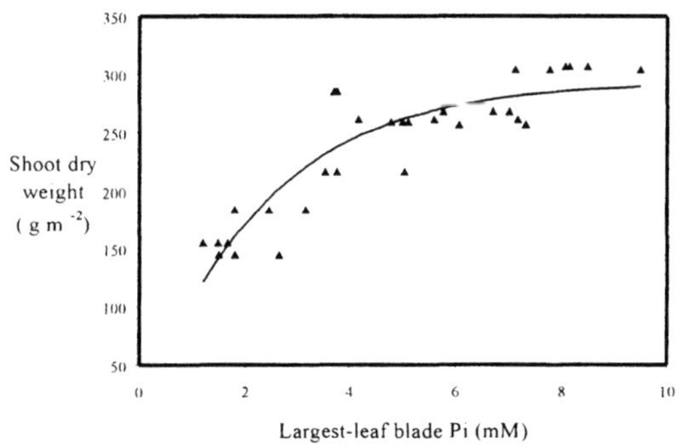

Figure 1. Largest-leaf blade Pi concentration and shoot dry weight of oilseed rape at the rosette growth stage.

Discussion

A critical Olsen-P of 21 ppm was calculated to give maximum shoot dry weight. For an OSR crop yielding 5t/ha seed, soil Olsen-P values of 21-23 ppm on a similar soil type have been stated to be required to maintain adequate P uptake (Barraclough, 1989). It is thus encouraging that the calculated critical Olsen-P value for growth at the rosette stage falls in this range.

There was little difference in the closeness of fit of the relationships between shoot dry weight and Pi mM and % P, suggesting either of the variables could be related to shoot dry weight. No published references could be found to compare the critical Pi mM values obtained. The largest-leaf blade critical % P value lies within the range of 0.35-0.75% P quoted by Bergmann (1993) as adequate for growth. This value is however considerable higher than the 90% critical values of 0.26% P in the youngest fully expanded leaf quoted by Pinkerton *et al.* (1989). They however included both the leaf blade and the petiole in their sample, which also had a lower range of % P contents than in other studies (Holmes, 1980), thus leading to their lower values.

It should be remembered that early autumn growth does not necessarily relate directly to final seed yield (Mendham *et al.*, 1981). Nonetheless, growth during this period sets the potential for seed yield, with growth and development following flowering determining the fulfilment of this yield potential.

The establishment of these critical plant Pi mM values allows the diagnosis of whether P deficiency is affecting the growth of OSR to be made independent of soil type. Measuring Pi mM in tissue-water is undergoing investigation as a possible *in situ* field test for detecting P sufficiency.

Acknowledgement

This work is funded by the Home-Grown Cereals Authority.

References

Barraclough PB 1989 Plant Soil 119, 59-70.

Barraclough PB 1993 Plant Soil 155/156, 175-178.

Bergmann B 1993 Ernährungsstörungen bei Kulturpflanzen. Gustav Fischer, Germany. 835pp.

Bollons HM and Barraclough PB 1997 J. Plant Nutr. 20, 641-655.

Holmes MRJ 1980 Nutrition of the oilseed rape crop. Applied Science Publishers Ltd, London, UK. 68-81.

Mendham NJ, Shipway PA and Scott RK 1981 J. Agric. Sci., Camb. 96, 389-416.

Murphy J and Riley J 1962 Anal. Chim. Acta 27, 31-36.

Olsen RA, Cole CV, Watanabe FS and Dean LA 1954 USDA Circular 939.

Pinkerton A, Spencer K and Govaars AG 1989 Aust. J. Exp. Agric. 29, 861-865.

Reuter DJ and Robinson JB 1997 Plant Anal. CSIRO Publishing, Australia, 97-105.

Skinner RJ and Todd AD 1998 Soil Use and Management 14, 162-169.

Sylvester-Bradley R 1985 Aspects of Applied Biology 10, 395-400.

Ware GO, Ohki K and Moon LC 1982 Agron. J. 74, 88-91.

Zarcinas BA, Cartwright B and Spouncer LR 1987 Comm. Soil Sci. Plant Anal. 18, 131-146.

Table 1. Critical %P and Pi mM concentrations for 95% maximum shoot dry weight (y) in oilseed rape at the rosette stage.

Element (x)	Tissue	Mitscherlich parameters[1] a	b	R^2	Critical value for 95% shoot dry weight (g m $^{-2}$)
%P	Whole shoot	527	1.19	0.80	0.67 %P [2]
	Large leaf-blade	673	1.06	0.79	0.53 %P [2]
	Large petiole	324	4.71	0.73	0.48 %P [2]
Pi (mM)	Whole shoot	305	0.217	0.77	13.8 mM
	Large-leaf blade	292	0.507	0.78	5.9 mM
	Large-leaf petiole	272	1.629	0.72	1.8 mM

[1] Mitscherlich curve $y = a[1-e^{(-bx)}]$, [2] Calculated on a maximum observed y basis

W. J. Horst et al.(Eds.), Plant nutrition – Food security and sustainability of agro-ecosystems. 716–717, 2001

Suitability of the SPAD meter and the petiole nitrate test for nitrogen management in nursery potatoes

J. Gerendás and I. Pieper
Institute for Plant Nutrition and Soil Science, University of Kiel, Olshausenstr. 40, Kiel, D 24098, Germany, jgerendas@plantnutrition.uni-kiel.de.

Key words: N supply, nitrate petiole test, seed potatoes, SPAD meter

Abstract

The analysis of plant constituents is an important management tool for monitoring and evaluating the nutrient status of crops. Conventional dried tissue analysis is rather time-consuming and for evaluating the N status of a crop, the SPAD chlorophyll meter and the nitrate petiole test are gaining importance. In collaboration with SaKa-Ragis (Germany) the suitability of these techniques to optimise the N management in tissue culture explants for seed potato production was evaluated. Initially the SPAD readings remained quite stable and only declined gradually except for the lowest N level. The nitrate petiole test is based on the determination of nitrate in expressed sap. After diluting the petiole sap appropriately (10-100 fold), the nitrate concentration can be determined either by test strips or ion sensitive electrodes. Reasonable agreement between the standard procedure and the simpler tests was obtained. In all cultivars the nitrate concentration declined earlier and more pronounced than the SPAD values. It is suggested that the nitrate petiole test provides a reliable and simple tool for monitoring the N status of nursery potatoes.

Introduction

To optimise N management for nursery potato production several approaches can be used. While conventional total N determinations are time consuming, alternative rapid tests are available as indices of tissue N status. The SPAD meter estimates the N status based on its correlation with the chlorophyll content (Martin et. al., 1995), while the nitrate petiole test determines the concentration of nitrate as the inorganic N reserve (Westcott *et al.*, 1993). Both methods have been used for field-grown potato crops, but little information is available regarding the production of nursery potatoes from tissue culture explants. The suitability of these techniques to optimise the N management was evaluated.

Materials and methods

The experiments were carried out with tissue culture explants of three cultivars (Arnika, Baltica, Secura) transplanted into 30 l boxes (12 explants per box) filled with a peat-based potting mix supplemented with all nutrients. N was supplied as $NH_4NO_3 \times (NH_4)_2SO_4$ at 4 levels (50, 100, 150 and 200%) with respect to the standard dose of 4.3 g N per box. The experimental layout was a completely randomised design consisting of 3 varieties × 4 N levels × 4 replications. The growth period was 100 days. The 150% and 200% N treatments were top-dressed 50 days after planting (DAP). SPAD readings and nitrate petiole tests were conducted on a weekly basis. The petiole sap of the first fully expanded leaf was pressed out and the nitrate concentration determined using three methods: 1. Reflectoquant test strips after appropriate dilution (RQflex, Merck, Germany); 2. Nitrate sensitive electrode (Ionplus Nitrate Combination Plastic Membrane Electrode, Orion Research) after appropriate dilution in the presence of 2% ionic strength adjustor (ISA, 26% $(NH_4)_2SO_4$); 3. Continuous flow analysis (CFA) with online dialysis (Zhu *et al.*, 1997). SPAD meter (Minolta) readings were taken from the leaf blades of the sampled petioles, and analysed for chlorophyll according to Bruinsma (1963). Tubers greater than 11 mm in diameter were weighed and counted. The N content of tubers, plant residues and leaf blades was determined according to Novozamski *et al.* (1983).

Results and discussion

Tuber Yield

The yield parameters of the 3 varieties showed a differential response to increasing N supply (Tab. 1). The tuber number of Arnika was always higher and more responsive than those of Baltica and Secura and reached its maximum at a N level of 150%. The tuber number of Baltica was very stable. Arnika and Secura generated the maximum yield (FW) at 150% N, while Baltica showed no yield decline at the highest N level. Two alternative rapid tests were evaluated regarding their suitability to optimise tuber yield.

SPAD measurements

As in other crops limiting N supply ultimately results in reduced protein and hence chlorophyll contents, which can be estimated by the Minolta SPAD meter. The exponential calibration curve of the SPAD readings vs. chlorophyll content was excellent (r^2 >90%), but high SPAD values did not correlate well with the total N content (data not shown). This is presumably due to the accumulation of nitrate and free amino acids. The SPAD readings themselves were quite different for the three varieties, reflecting typical genotypic differences in chlorophyll content (data not shown).

Table 1. Influence of N supply on tuber yield of three cultivars

N	N = 50%	N = 100%	N = 150%	N = 200%
Tuber number				
Arnika	85.3±10.2	98.3± 6.7	99.8± 7.4	95.0±13.4
Baltica	72.5±24.9	70.0±15.2	71.5± 5.2	73.8± 6.7
Secura	61.0± 2.9	57.3± 4.2	65.8± 8.5	58.3±10.8
Tuber FW [g]				
Arnika	816±40	1177±18	1397±24	1404±39
Baltica	785±38	1099±28	1261±51	1297±73
Secura	824± 9	1071±34	1269±20	1218±57

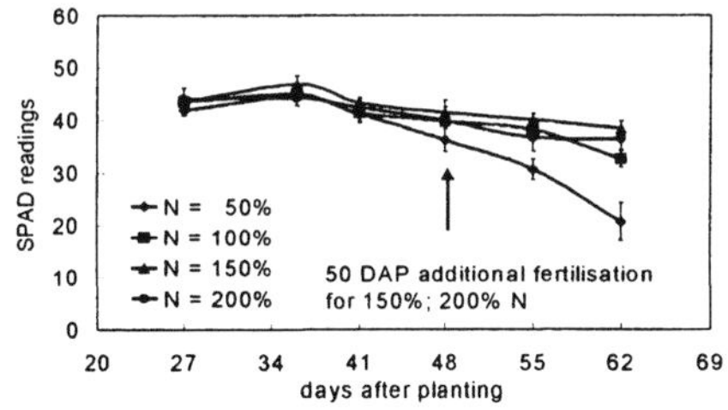

Figure 1. Influence of N supply on SPAD readings of the cultivar Secura.

The SPAD readings remained quite stable initially and, except for the lowest N level, only declined gradually (Fig. 1). The higher N levels did not show any differences between the treatments. The lowest N level, which was characterised by lower yield, showed a very late reaction, while the 100% level did not differ from higher N levels within the time window suitable for top-dressing. Therefore, a rapid test is needed that will detect the decline in the N status earlier then the SPAD meter does.

Nitrate petiole test

The petiole sap nitrate concentration (PSNC) can be determined using test strips after appropriate dilution (10-100-fold). Reasonable agreement between the standard photometric procedure (CFA) and the simpler tests was obtained. In all cultivars the nitrate concentration declined earlier and the decline was more pronounced than the decline in the SPAD values (Fig. 1). It is suggested that the petiole nitrate test provides a more reliable and simple tool for monitoring the N status of nursery potatoes in comparison to the SPAD meter. The lowest N level showed a decline in the PNSC from the 2nd set of measurements onwards. PNSC of the other N treatments remained stable until at least 50 DAP. PSNC of the 100% treatment declined from that point. A decline in the 150% and 200% treatments also decreased, but to a lesser extent. The range of responses seen over the range of N supply treatments suggest that petiole nitrate test allows a more sensitive description of the N status than the SPAD meter. Early diagnosis of the N availability is a prerequisite for the precise N management.

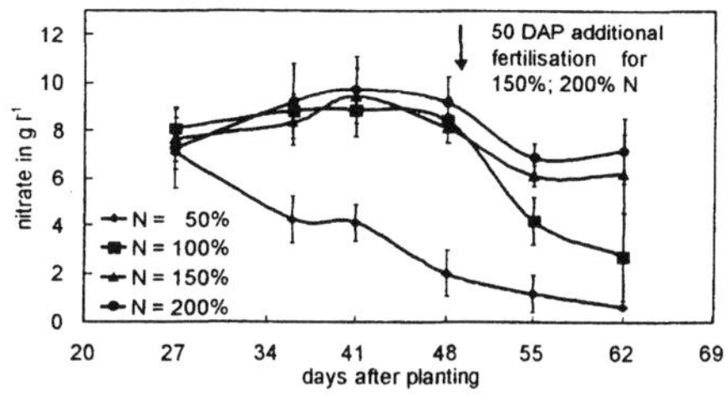

Figure 2. Petiole sap nitrate concentration of the cultivar Secura (tested reflectometrically using the RQflex).

The PSNC may also be determined by ion sensitive electrodes after appropriate dilution (e.g. 100-fold) in the presence of 2% ISA. The electrode measurements were in excellent agreement (r^2 > 95%) with the reference procedure (CFA, Fig. 3), although nitrate by CFA is analysed photometrically after diazotation, while the electrode determines the nitrate activity potentiometrically. In the petiole sap matrix effects during electrode measurements were not significant, which might be related to the strong dilution. This was additionally tested by standard addition. Chloride, which shows strong interferences in potentiometric nitrate determinations, is unlikely to cause major problems since fertilisers low in chloride are used for nursery potato production from tissue culture explants. Both rapid nitrate tests proved equally efficient and easy to use.

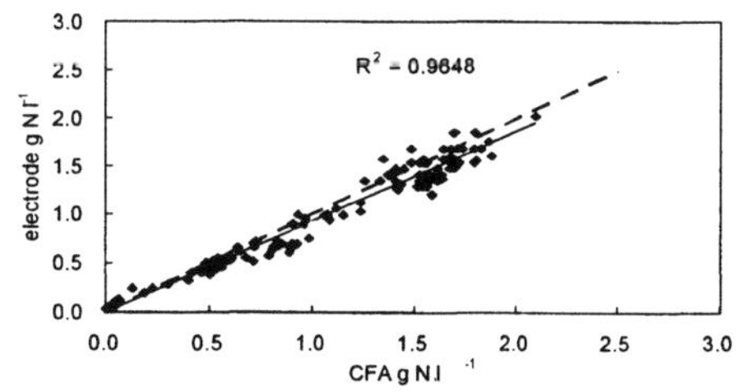

Figure 3. Relationship between NO_3-N concentration measured by continuous flow analysis and by nitrate sensitive electrode.

References

Bruinsma J 1963 Photochem. Photobiol. 2, 241-249.

Martin RJ 1995 Proceedings Agronomy Society of N.Z. 25, 91-95.

Novozamsky I, Houba VJG, van Eck R and van Vark W 1983 Communications in Soil Science and Plant Analysis 14, 239-248.

Westcott M 1993 *In* Fertilizer Facts No 5 June 1993.

Zhu Z, Gerendás J and Sattelmacher B 1997 *In* Plant Nutrition - for sustainable food production and environment. Eds. Ando *et al.*. pp. 963-964. Kluwer Academic Publishers.

W. J. Horst et al.(Eds.), Plant nutrition – Food security and sustainability of agro-ecosystems. 718~719, 2001

Nitrogen levels and its correlation with the chlorophyll content in three coffee cultivars

E. Furlani Jr.[1)], E. Lazarini[1)], E. M. Paulo[2)] and R. C. M. Furlani[1)]

[1)] *Department of Crop Sciences, Sao Paulo State University, Av. Brasil 56, Ilha Solteira-SP, !5385-000, Brazil. enes@agr.feis.unesp.br*

[2)] *Instituto Agronômico, Av. Barao de Itapura 1481, Campinas-SP, 13810-710, Brazil.*

Key words: chlorophyll, coffee, cultivars, nitrogen levels

Abstract

Three trials were conducted under greenhouse conditions using seedlings of 3 coffee cultivars Mundo Novo and Catuaí (both *Coffea arabica* L.) and Apoatã (*Coffea canephora* Pierre)and a range of N supply treatments from 0 to 98 mg/seedling. The chlorophyll contents were estimated using the SPAD 501 chlorophyll meter. Nitrogen supply and chlorophyll content were poorly related up to 20 days after N application. Subsequent to this, chlorophyll content of *arabica* cultivars correlated well with N supply, presumably as the seedlings became more dependent on soil N. Chlorophyll contents were maximal from 30 days after N addition at a N supply of 74 mg/seedling for the arabica cultivars and 84 mg/seedling for the Apoatã cultivar. As N supply increased above these levels, chlorophyll content decreased.

Introduction

Nitrogen deficiency can be observed throughout the life cycle of coffee plants, but quantifying the extent of the deficiency is difficult. Advances in other crops (e.g. common bean, potato and wheat) using the SPAD 501 chlorophyll meter prompted an evaluation of the instrument for quantifying N and chlorophyll levels in coffee.

Materials and methods

This work was carried out at the Experimental Station of the São Paulo State University (Unesp) located in Ilha Solteira-SP with the financial support of the Fundação de Amparo à Pesquisa do Estado de São Paulo (Fapesp). Three trials were conducted under greenhouse conditions with the seedlings of the cultivars Mundo Novo and Catuaí (both *Coffea arabica* L.) and Apoatã (*Coffea canephora* Pierre) in pots. N was supplied at 0; 14; 28; 42; 56; 70; 84 and 98 mg/seedling in a completely randomized design with 10 replications. The SPAD 501 chlorophyll meter readings were converted into chlophyll content using the following equiation: chlorophyll content 0.0966 Spad units – 0.125 (Furlani Jr *et al.*, 1996). Eight sets of readings were collected at approximately 7 days intervals.

Results and discussion

N supply did not affect chlorophyll content of the seedlings from the cultivar Mundo Novo until the 3rd set of readings were collected (Fig.1). This probably reflects the depletion of seed N reserves and the increasing reliance of the seedlings on soil mineral N. Thereafter increasing N supply up to 70 mg/seedling was associated with increasing chlorophyll content, and N supply above 70 mg/seedling was associated with lower chlorophyll content (Fig.1).

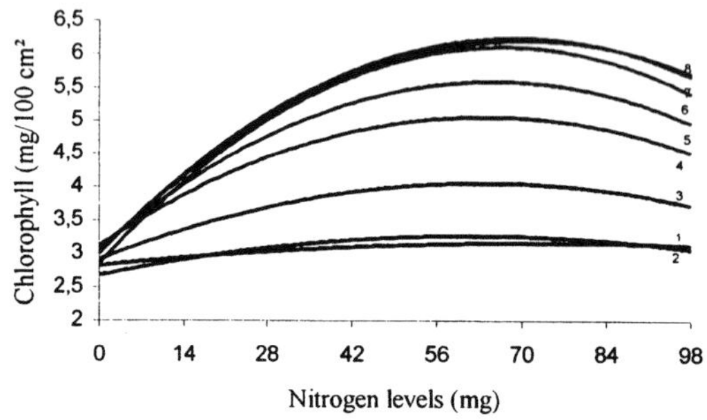

Figure 1. Effect of N supply on chlorophyll content of seedlings of coffee (*Coffea arabica*) cv Mundo Novo determined at 8 samplings at 7 days intervals

The seedling of cultivar Apoatã had a similar response curve, but the maximum chlorophyll concentration was observed at a N supply of 84 mg/seedling (Fig.2).

Correlation between N supply and chlorophyll levels was not as strong as for the two arabica cultivars (Tab.1).

These data suggest that the SPAD 501 meter may be useful for some coffee cultivars for estimating chlorophyll content as affected by N supply, but that plant age is a significant source of error.

Acknowledgement

Financial support of the Fundaçao de Amparo à Pesquisa do Estado de Sao Paulo.

References

Furlani Jr E, Bulhões LJ, Moreira JAA and Nakagawa J 1996 Bragantia Campinas 55, 171-175.

Table 1. Correlation coefficient for the realation between nitrogen levels and chlorophyll content in seedlings of 3 coffee cultivars (Catuaí, Mundo Novo and Apoatã) at 8 sampling dates.

	Catuai	Mundo Novo	Apoatã
	R	R	R
1 - 13/04/2000	0.14	0.36	-
2 - 20/04/2000	0.0	0.51	0.17
3 - 28/04/2000	0.77	0.71	0.16
4 - 05/05/2000	0.89	0.81	0.57
5 - 12/05/2000	0.90	0.85	0.65
6 - 19/05/2000	0.92	0.88	0.73
7 - 28/05/2000	0.92	0.89	0.79
8 - 02/06/2000	0.92	0.89	0.75

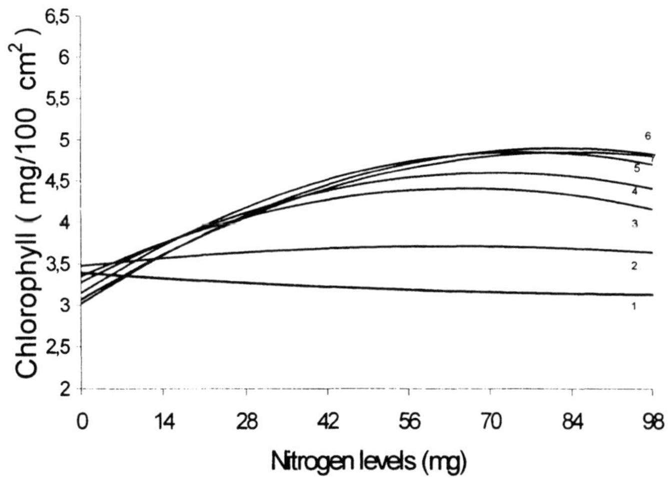

Figure 2. Effect of N supply on chlorophyll content of seedlings of coffee (*Coffea canephora* Pierre) c.v. Apoatã determined at 8 samplings at 7 days intervals

W. J. Horst et al.(Eds.), Plant nutrition – Food security and sustainability of agro-ecosystems. 720–721, 2001

Chlorophyll content distribution in leaves, stems, and ears in winter wheat

I. Karele

Department of Crop Production, Faculty of Agriculture, Latvia University of Agriculture, Liela Iela 2, Jelgava, LV-3001, Latvia; ilze.karele@llkc.lv

Key words: chlorophyll content, fertilizer, plant parts, winter wheat

Abstract

To obtain high crop yields different plant requirements such as demand for nitrogen have to be full filled. Estimation of the nitrogen status in the plant may be done with leaf diagnostics based on the relationship between nitrogen content and chlorophyll content. This study aimed to determine chlorophyll contents in individual plant parts and relationships between nitrogen and chlorophyll content of winter wheat varieties. Field trials were carried out at the experimental field trial station of the Latvia University of Agriculture during 1998-1999. Four winter wheat varieties were tested: early varieties ('Donskaja polukarlikovaja' and 'Sirvintas-1') and medium to late varieties ('Otto' and 'Bussard'). Chlorophyll concentration in the whole plant depends on the fertilizer rate. Chlorophyll concentration reaches a maximum during the flowering stage. A good correlation exists between leaf chlorophyll content and nitrogen content at beginning of stem elongation (Zadoks Growth Stage 32).

Introduction

Addition of nitrogen fertilization in accordance with the plant requirement is necessary to obtain high yields of winter wheat. Many authors (e.g. Shadchina and Dmitrieva, 1995; Fox *et al.*, 1994) have shown that nitrogen content correlates closely with chlorophyll content in leaves. Thus, estimation of the nitrogen status in plants may be possible by measuring chlorophyll content (Castelli *et al.*, 1996) as chlorophyll content depends on plant nitrogen status (Ruza, 1996). This study aimed to determine chlorophyll content during vegetative growth in individual plant parts and to determine the relationship between nitrogen and chlorophyll content in winter wheat.

Materials and methods

Field trial studies with winter wheat were done at the Experimental Station of Latvia University of Agriculture for two years. The soil can be characterized as a sod-calcareous medium loam. Before sowing 200 kg ha^{-1} NPK (6:24:30) were applied. There were four winter wheat varieties with different nitrogen applications: early varieties 'Donskaja polukarlikovaja' with one nitrogen regime of N-60+60 (i.e. 60 kg N ha^{-1} at 1st N dressing + 60 kg ha^{-1} at 2nd N dressing), as well as 'Sirvintas-1' with two fertilizer regimes (N-0, and N-60+60) and late varieties 'Otto' with three nitrogen regimes (N-0, N-60+60, and N-60+70+40), and 'Bussard' with one nitrogen regime (N-60+70+40). Split nitrogen dressing was done according to the following pattern: 1st N rate applied at the start of the growing period, 2nd N rate applied at the end of shooting and 3rd N rate applied at beginning of inflorescence emergence. The samples for chlorophyll analyses was collected at the Zadoks Growth Stage (ZGS; Zadoks *et al.* 1974) 32 (beginning of stem elongation), at ZGS 51 (beginning of inflorescence emergence), at ZGS 69 (end of anthesis). Chlorophyll content in leaves, stems, and ears was extracted in ethyl alcohol and estimated spectrophotometrically (Pavule, 1978). Chlorophyll content was calculated in mg per g fresh weight of plant material.

Results

The results of the experiment show that chlorophyll content is depending on the time of sampling in the vegetation period, as well as on the level of nitrogen fertilization. At the beginning of stem elongation (ZGS 32) leaf chlorophyll content of early varieties was found to be in the range of 1.40-1.83 mg per g fresh weight (Fig. 1), while in late varieties chlorophyll content was lower (1.09-1.64 mg per g fresh weight). Chlorophyll content in leaves and stems increased during the vegetation period (Fig. 2). The stems of variety Donskaja showed the highest chlorophyll content at each growth stage. The ears of early varieties reach the maximum chlorophyll content at ZGS (Fig. 3). Depending on the fertilizer regime the values ranged from 0.31-0.54 mg g^{-1}. Until ZGS 69 ears chlorophyll content decreased by 33 to 64 %.

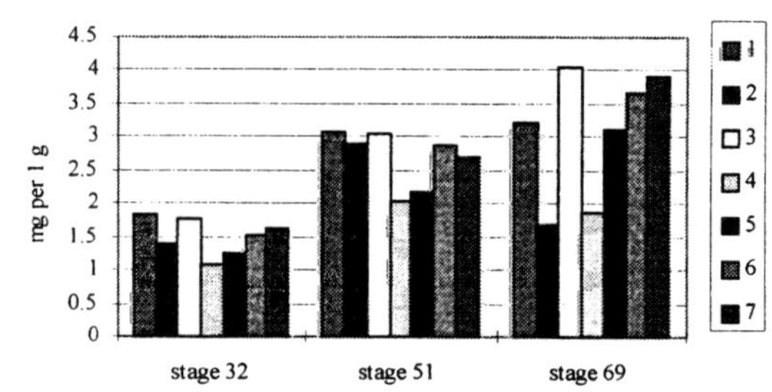

Figure 1. Chlorophyll content of winter wheat leaves (treatments: 1 - 'Donskaja polukarlikovaja' N-60+60, 2 - 'Sirvintas-1' N-0, 3 - 'Sirvintas-1' N-60+60, 4 - 'Otto' N-0, 5 - 'Otto' N-60+60, 6 - 'Otto' N-70+60+40, 7 - 'Bussard' N-60+70+40)

For the late variety 'Otto' in the N-0 regime and in the N-60+60 nitrogen regime a decrease by 13 and 28 %, respectively, was found. For varieties 'Otto' and 'Bussard' with late nitrogen application chlorophyll content increased approximately 1.3 times.

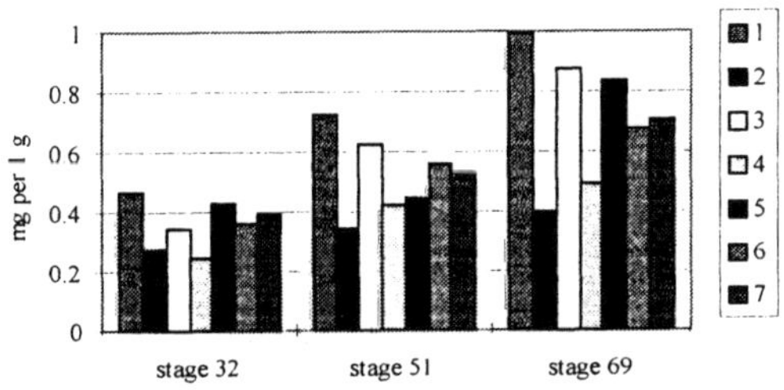

Figure 2. Chlorophyll content of winter wheat stems. (Legend details in Fig. 1)

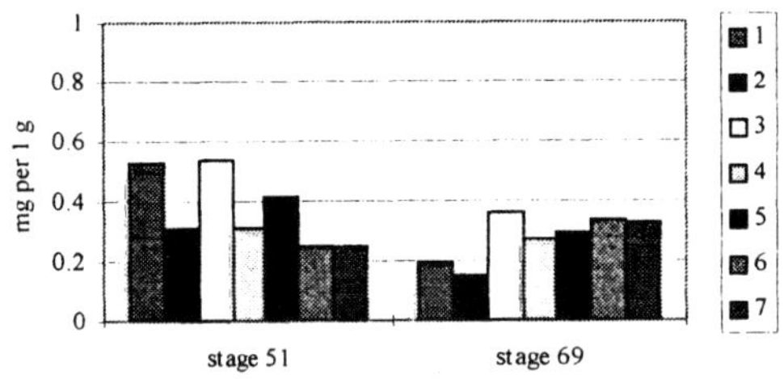

Figure 3. Chlorophyll content of winter wheat ears. (Legend details in Fig. 1)

Relationship between chlorophyll content and nitrogen content in individual plant parts were calculated. A significant correlation exists between chlorophyll content and nitrogen content in leaves at ZGS 32 (Table 1).

Table 1. Correlation coefficients between chlorophyll content (mg g^{-1}) and nitrogen content in leaves (%)

Plant put	ZGS 32	ZGS 51	ZGS 69
Leaves	0.784	0.543	0.577
Stems	0.428	0.697	0.686
Ears	-	-0.186	0.148

$LSD_{0.05}$=0.754

Discussion

Chlorophyll content in individual plant parts is dependent on the nitrogen fertilizer regime. Chlorophyll content in the leaves and stems increased until anthesis. A close correlation exists between chlorophyll and nitrogen content in leaves at ZGS 32. It seems to be possible to deduce N fertilizer recommendations for winter wheat based on this relationship. However, such N recommendations have to be adopted to different varieties.

References

Castelli F, Contillo R and Miceli F 1996 Agron. Crop Sci. 177, 275-285.

Fox RH, Piekielek WP and Macneal KM 1994 Comm. Soil Sci. Plant Anal. 25, 171-181.

Pavule A 1978 Manual of Agrochemistry. 290-298. Riga, Lv.

Ruza A 1996 Agroecological substantiation of Wheat and Barley Yields and its Quality. LLU, Jelgava, LV.

Shadchina TM and Dmitrieva VV 1995 J. Plant Nutr. 18, 1427-1437.

Zadoks IC, Chang TT and Konzak CF 1974 Weed Research 14, 415-421.

W. J. Horst et al.(Eds.), Plant nutrition – Food security and sustainability of agro-ecosystems. 722~723, 2001

Effect of water stress on chlorophyll meter readings in winter wheat

P. B. Barraclough and J. Kyte

IACR-Rothamsted, Crop and Disease Management Department, Harpenden, Herts, AL5 2JQ, United Kingdom, peter.barraclough@bbsrc.ac.uk

Key words: chlorophyll, critical, diagnostics, growth, nitrogen, SPAD, water, wheat

Abstract

Chlorophyll and nitrogen (N) are intimately linked in plants, and there is much interest in using chlorophyll as a surrogate for diagnosing crop N status. This is mainly due to the ease with which chlorophyll can be estimated with devices such as the Minolta SPAD meter. However, factors other than N can influence growth/chlorophyll/N relationships, and hence the interpretation of SPAD meter readings. The aim of the present study was to see how water stress affected critical SPAD meter readings for maximum growth of winter wheat. Winter wheat was grown under controlled conditions in sand culture to the 6-leaf stage, with 7 rates of N (deficient to luxury) and 2 water regimes (stressed and non-stressed). Water stress was achieved by adding polyethylene glycol (PEG) to the nutrient solution. Nitrogen and water stress reduced plant growth, but maximum growth was attained at the same N supply irrespective of water supply. SPAD readings were highly correlated with leaf chlorophyll and N. The best and most universal calibration, independent of water supply, was with N concentration expressed on a leaf area basis. Critical SPAD values for maximum shoot growth were 53 and 44 at low and high water supply, respectively.

Introduction

Accurate diagnosis of crop nitrogen (N) status is important for optimal fertiliser application. Established practice uses %N in plant dry matter (DM). However, N is tedious to measure and the effects of interacting growth factors on critical N concentrations are largely unknown. Chlorophyll and N are intimately linked in plants (Evans, 1983), and mainly as a result of the ease with which chlorophyll can now be measured, there is much interest in using it as a surrogate for diagnosing crop N status. Chlorophyll can be estimated, non-destructively, in single leaves with a hand-held meter such as the Minolta SPAD meter (Watanabe *et al.*, 1980) or in crop canopies by remote sensing (reflectance radiometers). However, factors other than N can influence growth/chlorophyll/N relationships, and hence the interpretation of SPAD meter readings. The aim of the present study was to see how water stress affected SPAD/chlorophyll/N relationships in winter wheat and to determine critical leaf %N and SPAD for maximum growth.

Materials and methods

Winter wheat (cv. Hereward) was grown in sand culture in a growth room (300 µE m^{-2} s^{-1} , 20/16 °C, 16 h day). There were 7 N rates (1, 3, 5, 7, 10, 15 and 20 mM) spanning deficient to luxury combined with 2 rates of water supply (low and high). Treatments were replicated 4 times. Pots were filled with 2 kg of sand and sown with 12 plants. The pots were watered with excess nutrient solution, which contained adequate amounts of all nutrients except N, on alternate days until leaf(4) was fully expanded (30 DAS), and daily thereafter. At this stage, the low water regime was introduced by adding PEG-4000 (100 g/l) to the nutrient solution (Izzo *et al.*, 1989).

Plants were harvested at the 6-leaf stage (45 DAS) and split into main stem leaf(5) and (6), and remaining whole shoots. Fresh and dry weights, areas, total-N ('Leco' N-Analyser) and total-chlorophyll (acetone extracts) were measured. Minolta SPAD-502 readings were taken mid-blade on 12 leaves per pot. Data were analysed statistically by ANOVA.

Results

Growth Shoot dry weight increased 3-fold with N, and was reduced by up to 40% by water stress (Fig. 1). The responses were caused mainly by differences in tiller growth. Leaves were smaller under N and water stress. Leaf(6) was bigger and had greater DM than leaf(5), but only at high water. Maximum shoot growth was attained at the same external N supply irrespective of water supply.

SPAD as N meter SPAD readings increased in the growth response range and were significantly greater at low than at high water (Fig. 2). There were no differences between leaf(5) and (6). SPAD was linearly related to leaf %N up to *ca.* 5%, but the relationship was significantly affected by water supply (Fig. 3). When N concentrations were expressed on a leaf area (or fresh weight) basis, a single relationship for all leaves and treatments sufficed (Fig. 4). Similar relationships were obtained between SPAD and chlorophyll concentrations, but the correlation was stronger for N and SPAD (r^2 = 0.92) than for chlorophyll and SPAD (r^2 = 0.80).

SPAD as diagnostic Shoot DM was linearly related to SPAD. Critical SPAD in leaf(6) for maximum shoot growth was 53.0 and 43.7 at low and high water, respectively. The relationship between shoot DM and leaf %N was asymptotic. Critical leaf %N was 4.85% irrespective of leaf or water supply. Critical shoot %N was 4.50% and 4.14% at high and low water, respectively.

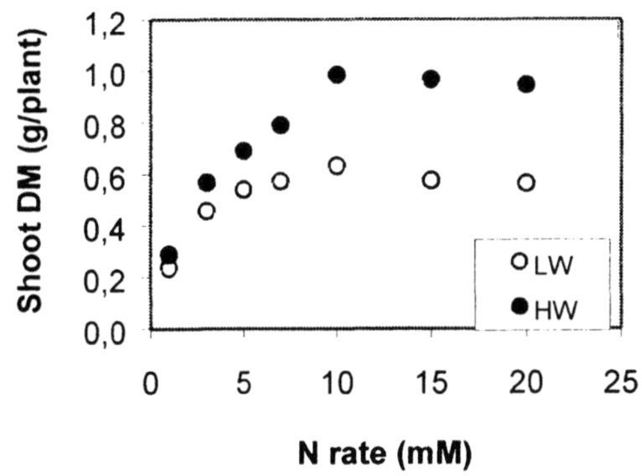

Figure 1. Effect of N and water supply (LW - low water HW - high water) on shoot DM (LSD (5%) NxW - 0.11).

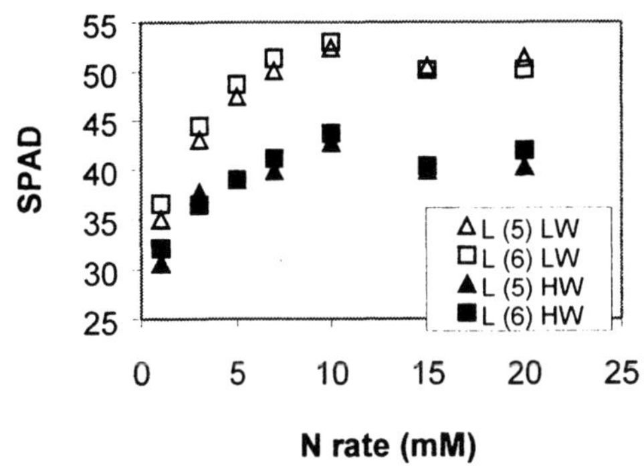

Figure 2. Effect of N and water supply on SPAD readings in leaf(5) and (6) (LSD (5%) NxWxL - 2.0)

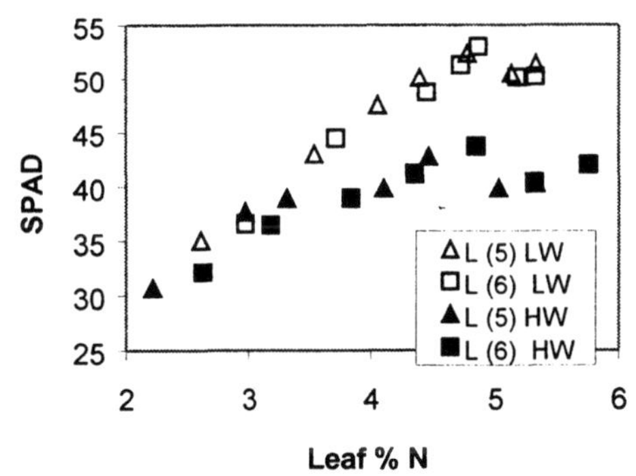

Figure 3. Relationship between SPAD readings and leaf N (leaf dry weight basis)

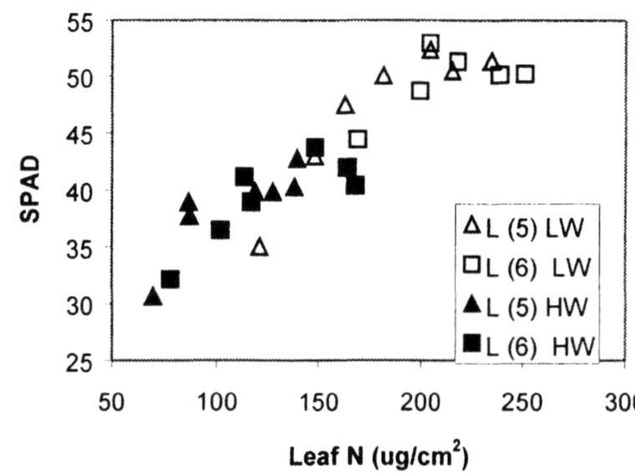

Figure 4. Relationship between SPAD readings and leaf N (leaf area basis)

Discussion

Correlations between SPAD, chlorophyll and nitrogen concentrations in leaves have been reported for many species (Wood *et al.*, 1993). In the present study, SPAD readings were well correlated with leaf %N, but the relationship was affected by water supply. Expressing N concentrations on a leaf fresh weight or area basis gave a more universal calibration. This was also reported by Peng *et al.* (1995) for rice. The conventional diagnostic, leaf %N, appeared to be the superior method in having a critical value unaffected by water supply, unlike shoot %N and SPAD. However, under a given water regime, SPAD and shoot DM correlated well, and SPAD offers logistical advantages.

It is difficult to extrapolate the significance of the present results to field crops. Firstly, the relative severity of the water stress is unknown. Secondly, SPAD and leaf N have been used diagnostically, *i.e.* to indicate the adequacy of N supply for growth up to the time of testing. In the field, most grain growers are interested in using SPAD prognostically, *i.e.* to predict future N requirements for desired grain yield and quality. This carries a greater risk than diagnosis. It would be inadvisable to act on SPAD readings taken under dry conditions unless the meter has been specifically calibrated for such conditions.

Acknowledgement

This work was supported by Hydro Agri. IACR receives grant-aided support from the Biotechnology and Biological Sciences Research Council (BBSRC) of the UK.

References

Evans JR 1983 Plant Physiol. 72, 297-302.

Izzo R, Navari-Izzo F and Quartacci MF 1989 J. Plant Nutr. 12, 1175-1193.

Peng S, Laza RC, Garcia FV and Cassman KG 1995 Commun. Soil Sci. Plant Anal. 26, 927-935.

Watanabe S, Hatanaka Y and Inada K 1980 Jpn. J. Crop Sci. 49, 89-90.

Wood CW, Reeves DW and Himelrick DG 1993 Proc. Agron. Soc. N.Z. 23, 1-9.

W. J. Horst et al.(Eds.), Plant nutrition – Food security and sustainability of agro-ecosystems. 724~725, 2001

Modelling conversion factors for leaf analyses from fruiting and nonfruiting terminals of citrus to asses the nutrient status of trees

I. A. Khan[1)], A. Srikandakumar[1)] and T. W. Embleton[2)]

[1)] *College of Agriculture, Sultan Qaboos University, P.O. Box 34, PC-123, Muscat, Sultanate of Oman*

[2)] *Department of Botany and Plant Sciences, University of California, Riverside, CA 92521, U.S.A*

Key words: fertilizer applications, least square means, nutrient requirements, prediction, regression

Abstract

Leaf analysis has been widely used to formulate fertilizer applications for citrus trees. Results of leaf analysis vary with age and position of the leaves sampled. The prevalent methods are to sample 4 to 7 months old spring flush leaves from the nonfruiting terminals and leaves from behind the fruit (fruiting terminal). We have reported significant differences between the leaves taken from two positions for 10 of the 12 elements analysed. Our long-term studies were aimed at developing a model conversion factor for leaf analysis based on samples taken from fruiting and nonfruiting terminals. A regression analysis was performed on a four years data set for 5 elements showing significant differences between two categories of leaves. Highly significant ($P<0.0001$) correlations were found between the nutrient contents of two categories of leaves. The regression analysis provided the basis for developing model conversion factors for different elements. A ratio of LSM (least square means) for the two categories of leaves gave highly reliable conversion rates.

Introduction

Fertilizer application guidelines for citrus are based on leaf analysis (DeVilliers and Beyers, 1961). The methods of sampling leaves for reliable assessment of nutrient status of trees have been controversial issues. Two important considerations are the age and position of the leaves to be sampled. The prevalent methods are to sample 4 to 7 months old spring flush leaves from the nonfruiting shoot terminals as advocated by the California standards (Embleton *et al.*, 1996) and leaves from behind the fruit (fruiting shoot terminals) as practised in South Africa (du Plessis and Koen, 1996). Both of the methods have following in different countries. The sampling from nonfruiting terminals has been considered to reflect the nutrient status of the trees more closer to reality because the leaves taken from behind the fruit are competing with an active 'sink' i.e. the fruit. However, sampling from the fruiting terminals is much easier and unambiguous. Model conversion factors are needed to assist the interpretation of leaf analysis data gathered from two leaf sampling methods.

We have reported significant differences between the leaves taken from two positions for 10 of the 12 elements analysed. N, P, K, Fe, Cu, Mn and B were significantly higher in nonfruiting leaves and Ca, Mg and Cl were significantly higher in fruiting leaves. There were no significant differences between two categories of leaves for Na and Zn (Khan *et al.*, 1992).

Our goal was to develop a model conversion between the two types of leaf analysis to facilitate the exchange of information and help improve fertilizer use.

Materials and methods

A long-term experiment was conducted in California to understand the interrelation of nutrient status of two kinds of leaves and to develop conversion factors. An earlier report (Khan *et al.*, 1992) presented data collected for seven years from two field experiments comprising 20 and 23 single tree replicates. The experimental trees were subjected to 5 levels of N-applications (0 to 1.82kg N per tree).

A subset of data was used in this report. It includes N, K, Cl, Mn and Fe for two experimental sites each for fruiting (F) as well as nonfruiting (NF) leaves. The combined data from each of the five N applications at both sites were subjected to regression analysis followed by LSM ratios, which were used to compute conversion factors. LSM are the estimates to minimize the sum of squares of differences between the actual response values and the values predicted by the model (SAS, 1996).

Results and discussion

Regression Analysis

The regression of data for fruiting and nonfruiting leaves for five elements under five applications of nitrogen at two experimental sites are presented in Fig. 1. The effect of five N application levels were along the same regression line despite significant differences among N application levels. The R-square values were found to be highly significant ($P<0.0001$). The regression equations for all elements were tested to predict the values of X or Y axis. The predictions were generally found on the higher side of actual values. To further improve the analysis, analysis of covariance was used and the predicted values for two procedures were comparable.

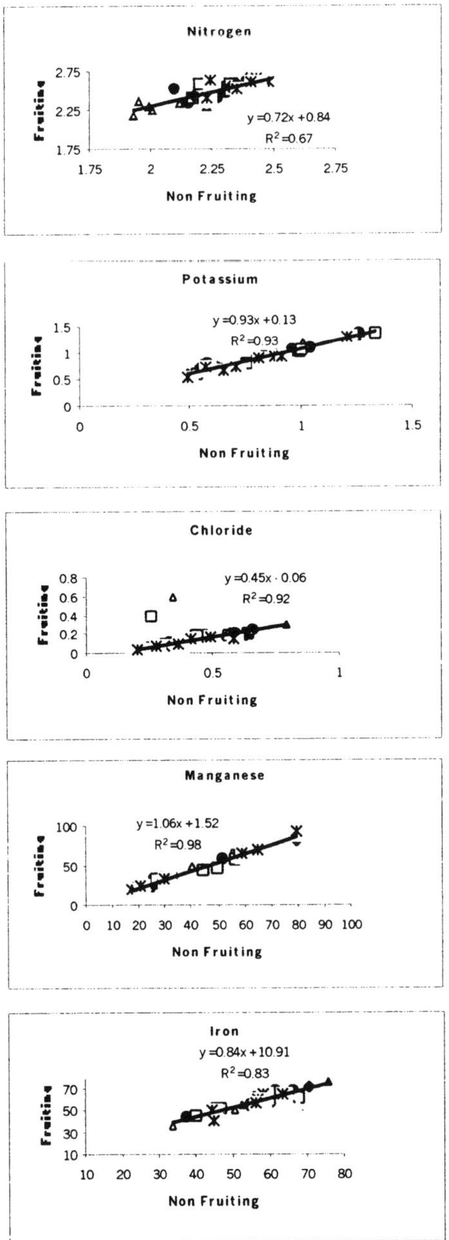

Figure 1. Regression analysis of five elements from leaves sampled from fruiting (y-axis) and nonfruiting (x-axis) shoot terminals. The symbols i.e. triangles, circles, squares, diamonds and asterisks are five levels of N applications i.e. 0, 0.45, 0.91, 1.36 and 1.82 kg per tree, respectively.

LS Means Conversion Factors

As an alternative to the regression equation, LSM procedure was followed to calculate representative values for five elements from fruiting and nonfruiting leaves (Table1). The LSM were used to calculate ratios of F/NF and NF/F to be used as conversion factors (CF). The CF values given in Table 1 were also tested randomly to predict the values for given fruiting or nonfruiting leaf data. The converted values approximated within a range of 95-99% of the actual values. However, the prediction value of CF's presented in this paper must be viewed as representative for the limits set on the LSM by the range of values of analysis for each element (Table 2).

Table 1. Conversion factors (CF) derived from least square means (LSM) analysis of leaves from fruiting (F) and nonfruiting (NF) shoot terminals.

Elements	LSM (F)	LSM (NF)	SEM	N-effect	CF: F to NF	CF: NF to F
N	2.29	2.52	0.013	P<0.001	1.100	0.908
K	0.82	0.91	0.013	P<0.001	1.104	0.905
Cl	0.42	0.13	0.009	P<0.001	0.303	3.302
Mn	35.85	39.79	1.669	P<0.001	1.109	0.901
Fe	55.66	58.15	0.618	P=0.684	1.045	0.957

Table 2. Range of values for leaf analysis for five elements in leaves taken from fruiting (F) and nonfruiting (NF) shoot terminals.

Leaf Type	N (%)	K (%)	Cl (%)	Mn (ppm)	Fe (ppm)
F	1.93-2.49	0.49-1.33	0.20-0.79	15.44-79.49	44.14-75.88
NF	2.18-2.67	0.54-1.37	0.03-0.30	18.34-92.20	35.29-76.91

The LSM approach for determining CF values for five elements followed in this report could be applied to the remaining elements. This information will facilitate more efficient use of leaf analysis data for the purpose of predicting fertilizer requirements of citrus trees.

References

DeVilliers JI and Beyers CJ 1961 *In* Plant Analysis and Fertilizer Problems. Ed. W Reuther. pp. 107-119. Amer. Inst. Biol. Sci. Publ. 8, Washington, D.C., USA

Du Plessis SF and Koen TJ 1996 South Africa. Proc. Int. Soc. Citriculture, 1283-1285.

Embleton TW, Coggins CW Jr and Witney GW 1996 Proc. Int. Soc. Citriculture, 1261-1264.

Khan IA, Embleton TW Matsumura M and Atkin DR 1992 Proc. Int. Soc. Citriculture, 564-569.

SAS 1996 Statistical Analysis System, SAS Institute Inc., Cany, NC, USA.

W. J. Horst et al.(Eds.), Plant nutrition – Food security and sustainability of agro-ecosystems. 726–727, 2001

Laser-induced chlorophyll fluorescence to determine the nitrogen status of plants

C. Bredemeier and U. Schmidhalter

Department of Plant Sciences, Chair of Plant Nutrition, Technical University of Munich, Am Hochanger 2, Freising-Weihenstephan, D 85350, Germany, bredemei@weihenstephan.de

Key words: chlorophyll fluorescence, chlorophyll meter, nitrogen, wheat

Abstract

The primary roles played by nitrogen in chlorophyll synthesis and photosynthesis suggest that deficiencies in this nutrient in the plant could be detected on the basis of changes in the plant's fluorescence spectra. The relationship between laser-induced fluorescence intensity at 680 and 740 nm, fluorescence ratio F680/F740 and the nitrogen nutritional status of wheat plants was characterized in two growth chamber experiments. The results showed that fluorescence intensity is related to the leaf chlorophyll content, especially to the intensity of the 680 nm fluorescence band. The fluorescence ratio F680/F740 is negatively correlated with the chlorophyll meter reading (r^2=0,62; p<0,01), as well as with the chlorophyll a content (r^2=0,62; p<0,01). The results showed that N-fertilization levels can be differentiated by means of fluorescence ratio F680/F740 measurements. Furthermore, the fluorescence ratio appears to be insensitive to light and air temperature conditions.

Introduction

The role played by N in chlorophyll synthesis suggests that a deficiency in this nutrient in the plant could be detected on the basis of changes in the plant's fluorescence spectra. Nitrogen is the major constituent of the tetrapyrole nucleus of chlorophyll. In consequence, the chlorophyll content of the leaf is positively correlated with the N concentration.

The main advantage of fluorescence detection compared to reflection measurements is the greater sensitivity achievable because the fluorescence signal has a very low background, since the signal comes only from the green plant parts (Lichtenthaler & Rinderle, 1988).

The chlorophyll fluorescence spectra of the upper side of the leaf exhibit two fluorescence maxima: one near 680 nm and a second one around 735 nm (Lichtenthaler & Rinderle, 1988). With increasing chlorophyll content, the 690 nm fluorescence band is decreased by a preferential reabsorption of the emitted fluorescence at 690 nm by chlorophylls. It occurs due to the partial overlapping of the absorption spectrum of the chlorophylls with the fluorescence emission spectrum between 640 nm to around 710 nm (Lichtenthaler & Rinderle, 1988).

The experiments had the objective to determine the relationship between the fluorescence intensity at 680 nm and 740 nm and the fluorescence ratio F680/F740 and the chlorophyll content of wheat leaves, as well as to investigate the influence of nitrogen supply and environmental conditions on the fluorescence yield and on the fluorescence ratio F680/F740.

Materials and methods

Two experiments were carried out under controlled environmental conditions. Wheat plants (*Triticum aestivum* cv. Thassos) were grown at a density of 12 plants per pot in 8 L pots (20 cm diameter) filled with a loamy soil. The average air temperature for day/night was 18/14°C, the relative air humidity was 60/75% day/night, and the maximum photosynthetically active radiation was 400 μmol m^{-2} s^{-1} during the day. The light period was 12 h. Every afternoon, the plants received an adequate amount of water.

The one-week-old wheat plants were divided in five groups, which received different amounts of N-fertilizer (NH_4NO_3 form) once a week. The final amount of N applied varied from 0,15 g N to 1,25 g N per pot.

The chlorophyll fluorescence intensity was measured at 18°C air temperature (if not otherwise specified) under steady-state conditions of fluorescence, i.e., in light-adapted plants. The excitation and the sensing of the fluorescence was performed at the adaxial leaf surface of the youngest fully expanded leaves. The leaves were cut from the plants and the fluorescence yield was quickly measured. The leaf area illuminated by the laser beam is about 7 mm^2 at about 14 cm distance. Before the measurements were done, the chlorophyll content of the leaf was estimated by using the chlorophyll meter (Minolta SPAD-502®).

The excitation wavelength of the laser beam was 640 nm, with an energy of 15 mW. Chlorophyll fluorescence was detected at two wavelengths (680 nm and 740 nm).

The photosynthetic pigments (chlorophylls) were extracted with acetone 100% and determined spectrophotometrically using the extinction coefficients and equations described by Schopfer (1989).

Results and discussion

The results of the experiments showed that the fluorescence intensity at 680 and 740 nm is affected by the chlorophyll content of wheat leaves, as estimated by the chlorophyll meter (Fig. 1). The 680 nm band decreased as the chlorophyll meter reading increased (r^2=0,20, p<0,05), while the fluorescence band at 740 nm slightly increased as the reading of the chlorophyll meter increased (r^2=0,05 not significant). With increasing chlorophyll content, the 680 nm fluorescence band is decreased by a reabsorption of the 680 nm band by the photosynthetic pigments. It occurs due to the partial overlapping of the absorption spectrum of the chlorophylls with the fluorescence emission spectrum. However, the correlation between the fluorescence intensity of single bands and the chlorophyll meter readings is low.

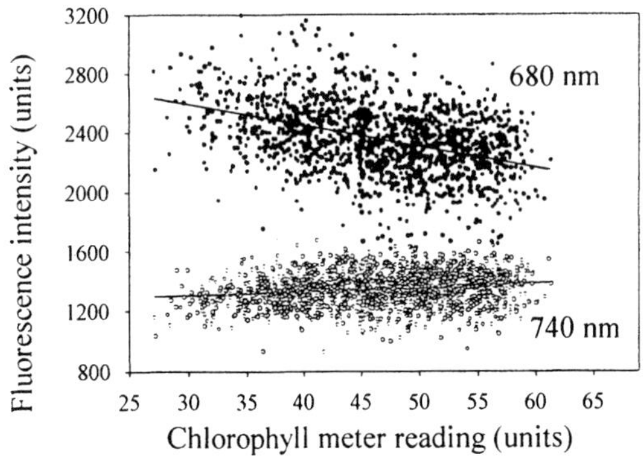

Figure 1. Fluorescence intensity at 680 and 740 nm in wheat leaves as related to the chlorophyll meter reading.

However, the correlation between fluorescence measurements and the chlorophyll meter readings was improved by using the ratio F680/F740 (Fig. 2). The ratio F680/F740 was better correlated with the chlorophyll meter readings (r^2=0,62, p<0,01) than were the values of the single fluorescence bands (Fig. 1 and 2).

The fluorescence ratio varies much less from leaf to leaf than the absolute fluorescence yield. For this reason, the ratio F680/F740 is more sensitive to estimate the chlorophyll content than are fluorescence values alone.

The fluorescence ratio F680/F740 was also negatively correlated with the chlorophyll *a* content (r^2=0,62, p<0,01) and with the total chlorophyll (*a+b*) content (r^2=0,60, p<0,01). Other authors also have shown that there is a negative correlation between the ratio F680/F740 and the chlorophyll content of leaves. Günther *et al.* (1991) showed that the fluorescence ratio of maize leaves decreased from 0,91 to 0,7 with increasing chlorophyll content from 24 to 36 µg cm^{-2}. In our work, we found a decrease in the fluorescence ratio from about 1,9 to 1,55 with increasing total chlorophyll content from about 25 to 65 µg cm^{-2}.

The results showed that N-fertilization levels can be differentiated by means of fluorescence measurements, as shown in Table 1. The mean fluorescence ratio in wheat leaves varied from 1,88 (plants without N-fertilization) to 1,55 (plants receiving 1,25 g N/pot up to this moment). The results also showed that 40 measurements per treatment were enough to get a reasonably reliable estimate of the chlorophyll status of the plant.

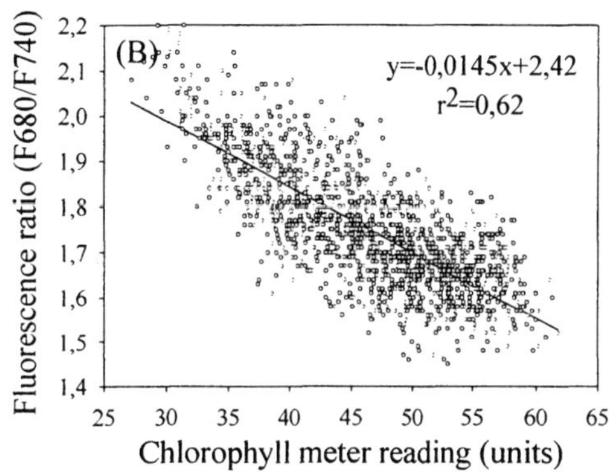

Figure 2. Fluorescence ratio F680/F740 in wheat leaves as related to the chlorophyll meter reading.

Table 1. Chlorophyll meter reading and the fluorescence ratio F680/F740 at the 8-leaf-stage as affected by the nitrogen supply.

Nitrogen applied	Chlorophyll meter Reading	Fluorescence Ratio[2]
g N/pot	units	F680/F740
without N	36,9 c[1]	1,88 c
0,15	39,4 c	1,83 c
0,40	45,8 b	1,66 b
0,85	52,5 a	1,58 a
1,25	54,5 a	1,55 a

[1] In a column, means followed by a common letter are not significantly different by Tukey's Multiple Range Test (α=0.01).
[2] Mean of 40 measurements.

Conclusions

The ratio F680/F740 is inversely related to the chlorophyll content. Taking into account that a positive correlation between leaf chlorophyll and N content exists, the use of this fluorescence method in remote sensing systems will permit monitoring on-line the spatial variation of the N-status of plants. The results of this project may lead to the development of a laser-induced fluorescence sensor, which allows farmers site-specific N fertilization.

References

Günther KP, Lüdeker W and Dahn HG 1991 Proceedings of the International Colloquium - Physical Measurements and Signatures in Remote Sensing. 723-726.

Lichtenthaler HK and Rinderle U 1988 Critical Reviews in Analytical Chemistry 19, 29-85.

W. J. Horst et al.(Eds.), Plant nutrition – Food security and sustainability of agro-ecosystems. 728~729, 2001

Reference populations for evaluation of the nutritional status of coffee by DRIS

O. C. Bataglia, W. R. Santos and J. A. Quaggio
Instituto Agronômico, Centro de Solos e Recursos Agroambientais, Av. Barão de Itapura, 1481, Cx. Postal 28, Campinas/SP, 13001-970, Brazil, ondino@iac.br

Key words: *Coffea arabica*, DRIS

Abstract

The Diagnosis and Recommendation Integrated System (DRIS) was tested on *Coffea arabica* with the objective of defining an efficient reference population. The data bank contained results on leaf samples taken from about 800 production fields during the summer in the states of São Paulo, Paraná, and Minas Gerais, Brazil. From the original population, a second one, called normal, was defined by excluding boron, zinc, and copper data when these elements were considered as contaminants. Three populations were selected and tested as references: a) the original medium yield, b) the normal high yield, and c) the normal medium yield with balanced nutrient concentrations. DRIS indices for macro and micronutrients from 15 selected samples were tested by comparison with the sufficiency range approach. For the original population there were a low number of correct diagnoses indicating the need to exclude contaminated samples. Use of the high yield population resulted in a poor assessment of N and K nutrition due to nutrient accumulation or removal in years of low and high yields, respectively. The high percentage of correct diagnostics with the normal medium yield nutrient equilibrated population suggested it was the best option to calculate DRIS indices for coffee.

Introduction

Monitoring the nutrient status by leaf analysis is a mandatory practice for sustainable coffee production in Brazil. Although critical level and sufficiency range have been the normal diagnostic approaches, the DRIS has recently received considerable attention. DRIS indices used for nutrient diagnosis are calculated by comparing nutrient relations of leaf samples with a reference population. The nutrient balance index (NBI) is the average of the absolute values of DRIS indices for a sample. Some accepted concepts in the DRIS literature, such as, the negative regression between yield and NBI, have failed for coffee yielding above 3,000 kg ha^{-1} due to the effect of the normal biennial cycling of high and low yields in this crop. The outcome of this situation is erroneous nutrient indices when the reference population is selected only by separating populations into high and low yield groups. In this study, the approach for the selecting reference populations was based not only on yield, but also on the nutrient equilibrium, as originally suggested by Beaufils (1973).

Materials and methods

Leaf analysis from about 800 *Coffea arabica* production fields were used to establish DRIS norms. Standardized samples, third leaf of yielding branches, were harvested during the summer on plantations from the main production regions in Brazil. Most of the samples were collected along the decades of eighties and nineties, while others came from nutritional surveys published by Lott *et al.* (1961) and Gallo *et al.* (1967).

The procedure was based on the original suggestion of Beaufils (1973) which was to select a normal, nutrient equilibrated population. Three types of populations derived from the original data were tested: a) the original medium yield (OP MY); b) the normal high yield (NP HY, as traditionally used on DRIS studies), and c) the normal medium yield nutrient equilibrated (NP MY NE).

DRIS indices were calculated using the formula defined by Jones (1981). To evaluate the efficiency of the reference populations, DRIS indices were calculated for 15 selected samples. The nutrient diagnostic by DRIS was compared to the diagnosis by the sufficiency range approach.

Figure 1. Frequency distribution for the original (OP MY), medium (NP MY NE) and high (NP HY) yield coffee populations.

Results

Figure 1 shows the yield distribution of original, medium and high yield populations. Very high standard deviations were observed for micronutrients on the original population.

On normal populations samples with excessive contents of B, Cu and Zn were discarded, making their average much closer to natural occurrence on coffee leaves.

The comparison of DRIS nutrient diagnostics of 15 selected samples with the diagnosis using the sufficiency range approach is shown in Table 1. The percentage of correct interpretation increased from the original medium yield to the normal nutrient equilibrated populations.

Table 1. Percentage of correct diagnoses relative to the sufficiency range approach using three different reference populations for coffee.

	Reference populations		
Nutrient	OP MY	NP HY	NP MY NE
N	60	40	80
P	100	80	87
K	47	60	73
Ca	60	80	80
Mg	60	67	67
S	86	79	79
B	57	35	43
Cu	60	93	87
Fe	60	93	87
Mn	100	73	80
Zn	67	80	87

Discussion

The usual procedure to select a reference population for DRIS studies is to separate high and low yield groups of crop leaf analysis. The nutrient relations to calculate DRIS indices are chosen according to the most significant variances of rates between the groups of high and low yields (Letzsch, 1985). The present results however, indicate that this method of selecting reference population for Coffea arabica fails (Table 1), mainly due to the effect of biennial cycle of high and low yields. On the other hand, this may not be true for other coffee species (Wadt *et al.*, 1999).

The original concept of Beaufils (1973) selecting plants of medium productivity and nutrient equilibrated proved to be the most efficient for coffee (Table 1).

Additional support may come from observation of figure 2. The usual negative relation between yield and NBI (Walworth and Sumner, 1987) is not followed for coffee yields above 3,000 kg.ha^{-1}. The nutrient inbalance at high yields is mainly a consequence of high nutrient drainage from the leaves to the fruits.

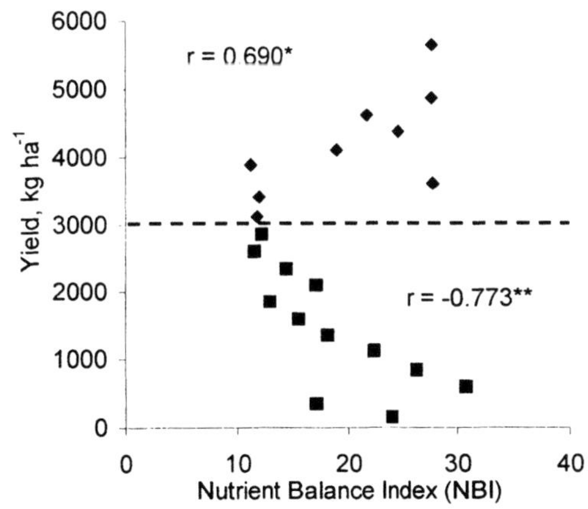

Figure 2. Relations between yield and NBI for 21 yield classis of the original population. Correlation coeficients calculated for yield and NBI relations bellow and above 3,000 Kg ha^{-1}.

Acknowledgement

Research was supported by CNP&D-Café and CNPq.

References

Beaufils ER 1973 Diagnosis and recommendation Integrated System (DRIS). 132. University of Natal, Pietermaritzburg.

Gallo JR, Hiroce R, Coelho FAS and Toledo SV 1967 Bragantia 26, 103-118.

Jones CA 1981 Commun. Soil Sci. Plant Anal. 12, 785-794.

Letzsch WS 1985 Commun. Soil Sci. Plant Anal. 16, 339-347.

Lott WL, McClung AC, Vita R and Gallo JR 1961 72. IBEC Research Institute, São Paulo.

Wadt PGS, Novais RF, Venegas VHA and Braganca SM 1999 Scientia Agricola 56, 83-92.

Walworth JL and Sumner ME 1987 Advanced Soil Science. 149-188. Springer-Verlag, New York.

W. J. Horst et al.(Eds.), Plant nutrition – Food security and sustainability of agro-ecosystems. 730~731, 2001

Towards a simple, generic modelling approach for developing site-specific nitrogen management strategies in irrigated rice

C. Witt[1)], J. M. Arah[2)] and P. Siband[1)]

[1)] *International Rice Research Institute, Crop, Soil and Water Sciences Division, DAPO Box 777, Metro Manila, Philippines, C.Witt@cgiar.org*

[2)] *AAT Consultants, 15 Clerk Street, Edinburgh, EH8 9JH, United Kingdom, JonArah@aol.com*

Key words: critical N, models, nitrogen, nutrient requirements, QUEFTS, rice

Abstract

To manage the large spatial and temporal variability in N requires a combination of both predictive and intervening management strategies to achieve the best congruence between crop N demand and soil N supply at all growth stages. An integrated approach featuring predictive models combined with temporally dense, real-time information will be required to properly manage N site-specifically in order to maximize plant N use efficiency and minimize N losses into the environment. In this paper, we attempt to integrate the critical N concept by Greenwood *et al.* (1990) into an existing modelling framework for site-specific nutrient management in irrigated rice (Dobermann and White, 1999) that is based on the model QUEFTS (Janssen *et al.*, 1990). QUEFTS allows the calculation of total fertilizer rates for targeted yields, but does not consider plant nutrient requirements during crop growth.

Introduction

Two strategies have been widely used to improve N management in irrigated rice. One is based on estimating total fertilizer rates and splitting patterns required to achieve a desired yield target or an optimal yield response. The other aims at managing N on a real time basis targeted towards maintaining an optimal plant (leaf) N concentration during crop growth using decision tools such as the chlorophyll meter. The underlying concepts of these strategies can be associated with two simple models: The model QUEFTS (Janssen *et al.*, 1990) can be used to generate total fertilizer N (P and K) rates for desired yield targets, while the theoretical basis for real time N management could be seen in the concept of critical N concentration described by Greenwood *et al.* (1990). The two models are not commonly linked, because one refers to the nutrient status during crop growth (Greenwood), the other at harvest (QUEFTS). We attempt to integrate the two modelling concepts into one framework for developing and evaluating N management options in irrigated rice.

Theory

The QUEFTS model

We used QUEFTS to describe a hypothetical relationship between above-ground plant dry-matter yield (Y) and plant N, P and K accumulation at physiological maturity as a function of nutrient supply and potential yield (Ymax). The minimum and maximum nutrient efficiencies ε (kg Y kg^{-1} plant nutrient) required for model calibration were estimated from the data set reported by Witt et al (1999) and are given in Table 1.

The nutrient efficiencies εXmax and εXmin are usually referred to as constants aX and dX in QUEFTS, describing the envelope functions depicted in Fig. 1. The line of optimal N dilution refers to the target situation of balanced nutrition at physiological maturity, where nutrients are neither limiting nor excessively supplied. For N, QUEFTS predicts an optimal or target εN of 136 kg kg^{-1}, until Y approaches Ymax leading to a decrease in εN.

Table 1. Nutrient efficiencies εX (kg plant dry-matter Y kg^{-1} plant nutrient X) and corresponding nutrient concentrations ρX (kg nutrient kg^{-1} Y) in rice, where X is either N, P or K.

Nutrient	Nutrient efficiencies ($kg\ kg^{-1}$)		Nutrient concentrations ($g\ kg^{-1}$)	
	εXmax	εXmin	ρXmax	ρXmin
N	84	192	11.9	5.2
P	412	1244	2.4	0.8
K	72	230	13.8	4.3

The Greenwood curve

Greenwood et al (1990) described a relationship between the potential plant dry-matter Y' (kg ha^{-1}) of an optimally fertilized crop during the cropping season and the corresponding critical ρN defined as the minimum ρN required for maximum growth rate at any time. The relationship appears to be independent of climatic zones, and is usually presented in terms of the critical above-ground N concentration ρN':

$$\rho N' = aY'^{-b}, \quad (Y' \geq Y_{min}) \qquad [1]$$

The parameter Y_{min} is a threshold plant dry-matter level below which equation [1] does not hold, and *a* and *b* are supposedly universal at least within cultivars. We used *a* = 5.18 and *b* = 0.52 (Sheehy et al, 1998).

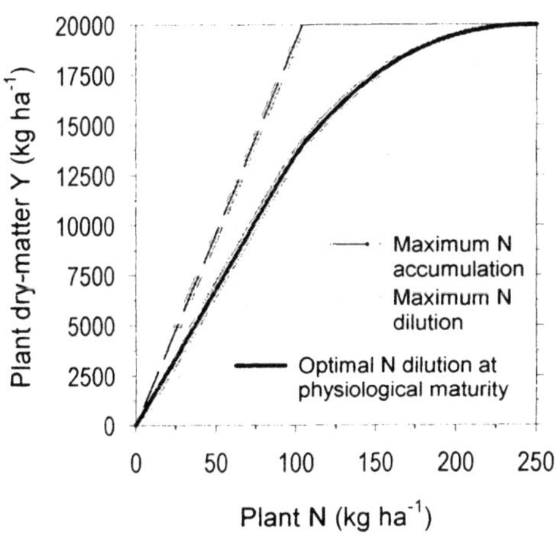

Figure 1. The QUEFTS envelope of maximum N accumulation and dilution (εNmax, εNmin), and the model predicted line of optimal N dilution at Ymax = 20 t ha^{-1}.

The N status index

Based on Greenwood et al (1986), the N status index ΦN was calculated as the ratio between the actual ρN at any point in time during crop growth and the corresponding critical concentration ρN' considering the minimum plant N concentration ρN_{min} (Table 1) required for crop growth:

$$\Phi N = (\rho N - \rho N_{min}) / (\rho N' - \rho N_{min}) \quad [2]$$

Results and Discussion

Fig. 2 depicts the typical Greenwood curve and fitted data from two fertilizer N treatments in the Philippines. All three curves show a decrease in ρN during crop growth and end within the QUEFTS boundaries given by ρNmax and ρNmin at harvest. The Greenwood curve intersects at ρNmax, which corresponds well with the predicted plant N requirement of 238 kg N at Ymax of 20 t ha^{-1} (Fig. 1). However, the two lines would not intersect at lower yield potentials. A major assumption of the Greenwood concept is that the line of critical ρN' is universal and holds for any Ymax, i.e. ρN' is greater at Ymax of less than 20 t ha^{-1}. Instead, QUEFTS assumes a constant ρNmax for any given Ymax. Adjustments are obviously necessary to successfully link QUEFTS and Greenwood curve.

However, we further explored the link between the two concepts based on the observation that the N dilution index ΦN seemed to be relatively constant for the two fertilizer N treatments depicted in Fig. 2.

A family of Greenwood curves was derived for different N dilution indices describing optimal dilution curves of plant N concentrations during crop growth to intersect with the QUEFTS line of plant N accumulation at harvest for an optimally N fertilized crop (Fig. 3). This clearly demonstrates the potential of generating plant nutrient requirements during crop growth based on the Greenwood concept - for target yields and corresponding plant N requirements at harvest generated by QUEFTS.

This framework could be used to further investigate the relationship between indices describing the plant N status during crop growth and indicators of leaf N concentration such as chlorophyll meter or leaf color chart for real time N management.

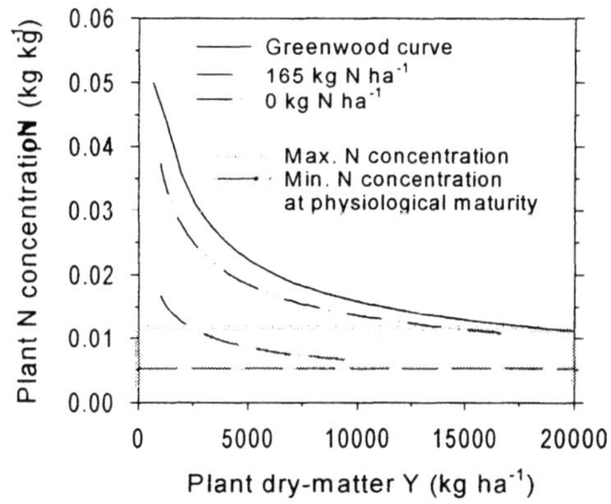

Figure 2. Nitrogen dilution curves of critical N (Greenwood) and of two rice crops grown at different fertilizer N levels at IRRI (Sheehy et al, 1998). Maximum and minimum N concentrations refer to ρNmax and ρNmin given in Table 1.

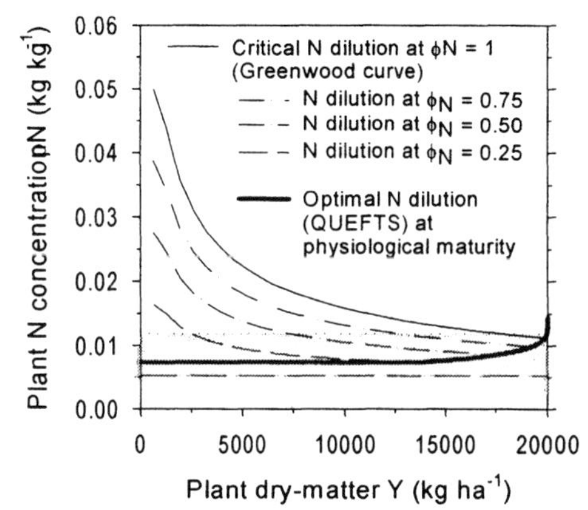

Figure 3. Nitrogen dilution curves during crop growth for four crops with different N status indices ranging from ΦN 0.25 to 1 (ΦN = 1 is the critical N dilution according to Greenwood). Lines intersect with the QUEFTS curve describing the optimal N dilution at physiological maturity.

References

Dobermann A and White PF 1999 Nutr. Cycl. Agroecosyst. 53, 1-18.

Greenwood DJ, Neeteson JJ and Draycott A 1986 Plant Soil 91, 281-301.

Greenwood DJ, Lemaire G, Gosse G, Cruz P, Draycott A and Neeteson JJ 1990 Ann. Bot. 66, 425-436.

Janssen BH, Guiking FCT, van der Eijk D, Smaling EMA, Wolf J and van Reuler H 1990 Geoderma 46, 299-318.

Sheehy JE, Dionora MJA, Mitchell PL, Peng S, Cassman KG, Lemaire G and Williams R L 1998 Field Crops Res. 59, 31-41.

Witt C, Dobermann A, Abdulrachman S, Gines HC, Wang G, Nagarajan R, Satawatananont S, Tran Thuc Son, Pham Sy Tan, Le Van Tiem, Simbahan G, and Olk DC 1999 Field Crops Res. 63, 113-138.

W. J. Horst et al.(Eds.), Plant nutrition – Food security and sustainability of agro-ecosystems. 732~733, 2001

Field scale variability of N_2-fixation in legume and grass mixtures

F. Locher, H. Heuwinkel, R. Gutser, U. Schmidhalter
Chair of Plant Nutrition, Department of Plant Sciences, Technical University of Munich, Am Hochanger 2, D 85350 Freising, Germany, locher@weihenstephan.de

Key words: legume content, legume-grass-mixture, N_2-fixation, NIRS, spatial variability

Abstract

Spatial variability of N_2-fixation of a legume-grass mixture has not been previously described. Legume content is a main factor to describe the amount of N_2-fixed in a multi-species legume-grass-mixture. In this study, Fourier-Transform Near Infrared Reflectance Spectroscopy (FT-NIRS) was used to estimate the legume content of oven dried (60°C) ground (1.5 mm) samples to get information about the spatial variability of N_2-fixation. The FT-NIRS was calibrated with 411 calibration samples. The cross validation of the calibration model showed a high coefficient of determination (R^2 = 0.99) and a small prediction error (RMSECV = 3%). In a second step this calibration model was applied to field samples (n = 90, field size 2.25 ha). The obtained legume content data were used as a factor in the calculation of N_2-fixation. For one harvest N_2-fixation varied in a wide range from 3 to more than 11 g N m^{-2} within one field.

Introduction

The amount of N_2 fixed in legume-grass mixtures is mainly influenced by yield and legume content (Weißbach *et al.* 1995, Boller 1988). N_{fixed} can be calculated as shown in Equation [1].

Equation [1]

$$N_{fixed}\ [g\ m^{-2}] = Yield\ [g\ m^{-2}] * Leg\ [\%] * N_{leg}\ [\%] * N_{dfA}\ [\%]$$

where:-

Yield	is the dry matter (DM) yield of the legume grass mixture
Leg	is the legume content of the mixture on a DM basis
N_{leg}	is the nitrogen content of the legume fraction
N_{dfA}	is the proportion of nitrogen derived from the atmosphere of the total legume-N

The spatial variability of N_2-fixation has not been previously described, although it is supposed to be an important factor for calculating accurate N-balances. This is due to several problems when working on field scale. First, there is no adequate procedure for the measurement of yield with a sufficient accuracy and spatial resolution apart from laborious manual weighing. Secondly, the other supposed main factor, legume content, can only be estimated by eye. Visual estimates are subjective and involve high errors and low repeatability. On the other hand highly accurate but time consuming botanical analyses are not suitable for spatial studies. Therefore a fast and accurate assessment of the legume content is essential for the description of the spatial variability of N_2-fixation. NIRS has been described to be a promising tool for the fast determination of the botanical composition of legume-grass mixtures (Coleman *et al.* 1985, Shaffer *et al.* 1990).

In our project a multivariate calibration was developed, that is suitable for the measurement of legume content of multi-species-mixtures. This allows a more accurate description of the spatial variability of N_2-fixation than by a visual estimation of the legume content. Additionally it was tested whether N_{leg} and N_{dfA} can be used as constant parameters for describing the spatial variability of N_2-fixation.

Materials and methods

In 1999 experiments were run with a multi-species legume-grass mixture *(Medicago sativa, Trifolium pratense, Trifolium repens, Dactylis glomerata, Festuca pratense, Arrhenaterum elatior)* on field A09 (2.25 ha) of the FAM Research Station Scheyern, Germany. Spatial variability of N_2-fixation was described by measurements of yield and legume content. Yield was measured on 90 plots each 12 m^2 in size. At each harvest (1st: 18th May 1999; 2nd: 16th July 1999; 3rd: 26.th August 1999; 4th: 7 th October 1999) dry matter yield (60°C, 48h) was assessed by a representative subsample. The dried subsamples were ground in a mill (BRABENDER, Duisburg, Germany) to pass a 1.5 mm screen and the legume content was measured with FT-NIRS (Vector 22/N, BRUKER, Ettlingen, Germany). The FT-NIRS was calibrated with hand-sorted pure grass and legume samples taken at each harvest. Additionally artificial mixtures of grass and legume samples, varying in legume contents of 5% increments up to 95% were used in calibration (411 calibration samples). N_{leg} and N_{dfA} were measured on three sub-plots (at least six replicates) which represented the heterogeneity of the field. N_{dfA} was measured with ^{15}N isotope dilution method (Warembourg 1993). For the calculation of N_2-fixation by Equation [1] NIRS based legume content was used. All data were interpolated (inverse distant weighting) with ArcView 3.1 GIS (ESRI, New York, USA) in order to describe the spatial variability of N_2-fixation.

Results and discussion

In 1999 yield and legume content were assessed four times (Table 1). DM-yield was highest at harvest 2 whereas mean legume content increased until harvest 3 from 55% to 73%.

Table 1. Yield parameters and calculated N_2-fixation of a clover-grass mixture of field A09 at four harvests in 1999.

Harvest	Yield		Legume content		N_2-fixation	
No.	M	s	M	s	M	s
	--- [g m^{-2}] ---		----- [%] ------		---[g N m^{-2}] ---	
1	368	77	55	15	6.2	2.3
2	452	105	65	12	6.6	2.1
3	348	70	73	11	7.9	2.0
4	103	38	72	15	2.9	1.3

M= mean, s = standard deviation
The data were assessed for the whole field (n=90)

N_{leg} and N_{dfA} varied from harvest to harvest, but both showed negligible variation in the field except for N_{dfA} at harvest 4 (Table 2).

Table 2. N content (N_{leg}) and N derived from the atmosphere (N_{dfA}) of the legume fraction at four harvests in 1999

Harvest	N_{leg}		N_{dfA}	
No.	M	s	M	s
	--------------- [%] ---------------			
1	3.2	0.2	95	1.8
2	2.4	0.2	95	2.5
3	3.4	0.1	91	4.9
4	4.1	0.3	86	10

M= mean, s = standard deviation
Data were assessed in three plots with at least six replicates

This observation justified the use of them as constant factors in Equation [1]. Therefore, only yield and legume content contributed to the spatial variability of N_2-fixation within one field at one harvest. Both varied in a wide range. High values of yield did not necessarily reflect high percentages of legumes in the mixture and vice versa (Figure 1). It was concluded that the pattern of N_2-fixation was only described precisely, if both, total DM-yield and legume content were included in the calculation. The variability caused a wide range of N_2-fixation at each harvest. Although both parameters varied strongly at each harvest on the field scale, the pattern was quite similar to the harvest shown in Figure 1.

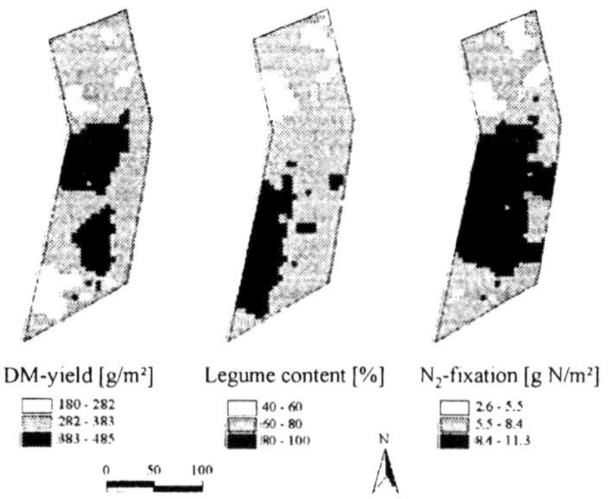

Figure 1. Yield, legume content predicted by NIRS and calculated N_2-fixation of the legume-grass mixture at harvest 3 of field A09 in 1999.

Conclusions

It could be shown for the first time, that N_2-fixation of a legume-grass-mixture varies strongly within one field. This differentiation was caused by the different patterns of both DM-yield and legume content. Therefore, the assessment of DM-yield is not sufficient to describe variation of N_2-fixation, legume content has to be taken into account as well. NIRS is a promising tool for easily and precisely measuring legume content in multi-species mixtures. The general applicability of the calibration has to be examined by broadening the variation of the included samples.

Acknowledgement

This research was supported by the Federal Ministry of Education and Research (BMBF), Berlin, Germany (Project No. 0339370I).

References

Boller BC 1988 Landwirtsch. Schweiz, 1, 251-253.

Coleman SW, Barton FE and Meyer RD 1985 Crop Sci. 25, 834-837.

Shaffer JA, Jung GA, Shenk JS and Abrams SM 1990 Agron. J. 82, 669-673.

Warembourg FR 1993 *In* Nitrogen isotope techniques. Eds. Knowles R and Blackburn TH. pp 127-156. Academic Press, San Diego.

Weißbach F 1995 Landbauforschung Völkenrode, 45, 67-74.

W. J. Horst et al.(Eds.), Plant nutrition – Food security and sustainability of agro-ecosystems. 734~735, 2001

Use of xylem sap to determine differences in the availability of mineral elements in soils

A. Noguchi, I. Hasegawa and J. Yazaki

Department of Agricultural and Biological Chemistry, College of Bioresource Sciences, Nihon University, 1866 Kameino, Fujisawa City, Kanagawa, 252-8510, Japan, noguchi@brs.nihon-u.ac.jp

Key words: availability, extraction, *Luffa cylindrica*, mineral elements, xylem sap

Abstract

Changes of concentration of mineral elements (P, K, Ca, Mg, Mn, Zn and Cu) in xylem sap obtained from cut hypocotyls of plants reflected the differences in the behaviour of elements in soil. A high correlation coefficient was obtained between concentrations of the elements in plant tops and those in xylem sap. Therefore, "xylem sap method" was proposed as a new method in evaluating the difference in availability of elements in soils, instead of the usual "extraction method".

Introduction

Availability of mineral elements in soil can be determined using extraction solutions of diluted acids, neutral salts, and chelating reagents (Leschber, Davis and L'Hermite, 1985) ("extraction method"). With this method, the extractants have to be changed every time depending on the kind and concentration of elements and soil properties, which may lead to inefficiency in the analysis and difficulty in the interpretation of data. As yet, no extractant has been identified for which extractability sufficiently corresponds to actual availability for any kind of soil (Kuboi and Noguchi, 1987). It is also unclear whether the data obtained using the "extraction methods" actually reflect how elements are obtained by the plant from the soil. We examined whether element concentrations in xylem sap, obtained at the cut end of the hypocotyl of plants, were applicable in estimating the availability of mineral elements in soil ("xylem sap method").

Materials and methods

Seedlings of sponge gourd (*Luffa cylindrica* Roem.) with newly developed cotyledons were transplanted into plastic pots (500mL) containing cultivation media as follows 1) 400g of quartz sand in which 1/4, 2/4, 3/4 and full strength of Hoagland's solution was added until 60% of maximum water holding capacity (MWHC) of quartz sand, 2) 400g of a mixture of alluvial soil and quartz sand at the ratio of 10:0, 8:2, 6:4, 5:5, 4:6 and 2:8, and 3) a mixture of 200g of alluvial soil and 200g of quartz sand containing synthesized zeolite and/or humic acid to make CEC of the media 20 and 45 cmol kg^{-1}. Cultivation media were irrigated with Hoagland's solution for 1) and water for 2) and 3) to 60% of MWHC of the quartz sand and the soil, respectively.

Upon emergence of the fourth leaf, each plant was cut off at the hypocotyl area and xylem sap was collected for 24 hours through vinyl tubes connected at the cut portions. Concentrations of P in xylem sap were measured by colorimetry and that of K, Ca, Mg, Mn, Zn and Cu were determined by atomic absorption spectrometry. Plant tops were then digested to analyse for theses elements. The elements in the soil were extracted by water, 1M CH_3COONH_4, 1M NH_4NO_3 (Symeonides and McRae, 1977), 0.1M HCl, and 0.05M EDTA (Mitchell, 1964) after cultivation in the case of 3).

Results and discussion

Relationship between concentrations of elements in xylem sap and element levels in media

To avoid complexation by soil and maintain a constant supply rate of mineral elements to plants, sand culture was used and sponge gourd was grown on quartz sand containing several concentrations of Hoagland's solution. Concentrations of all analyzed elements in xylem sap were proportional to that in the nutrient solution added to quartz sand, and correlation coefficients were more than 0.8 (Fig. 1). Therefore, concentrations of mineral elements present in xylem sap are assumed to reflect the soluble mineral elements present in the rhizosphere.

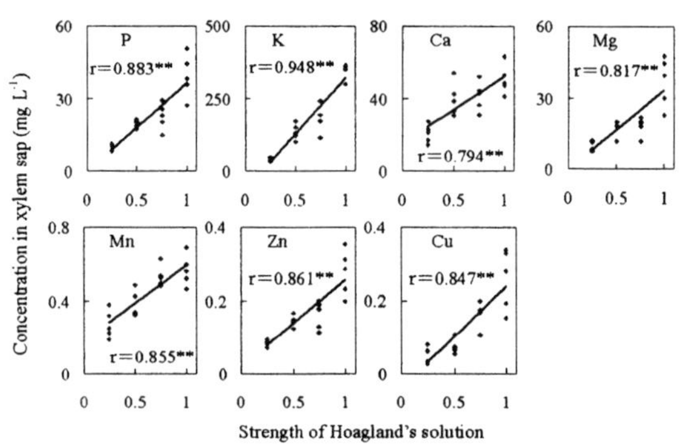

Figure 1. Relationship between concentrations of elements in xylem sap and element level in media. (*,** Significant at 5% and 1% level, respectively.)

Effect of mixing ratio of soil to quartz sand in medium on concentrations of elements in xylem sap

Although weight of roots and concentrations of Ca and Mg in xylem sap increased as the ratio of soil to quartz sand increased, concentrations of P and micronutrient cations in xylem sap were not proportional to the ratio as shown in Fig.2 (medium 2). Generally, concentrations of nutrient elements in soil solution might be constant, even if plant roots absorb them, because they are supplied from soil colloids and absorption of most of mineral elements by plant roots is dependent on mass flow and diffusion in soil solution. On the other hand, Ca and Mg are, mainly, absorbed by contact exchange with plant roots. Therefore, the results suggested that changes of concentrations of mineral elements in xylem sap may reflect differences in the behaviour of elements in soil as well as the manner of their absorption by plant roots.

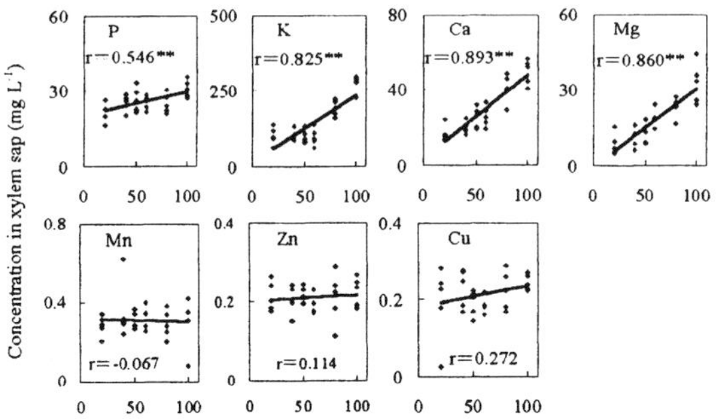

Figure 2. Effect of mixing ratio of soil to quartz sand in medium on concentrations of elements in xylem sap. (** Significant at 1% level.)

Comparing "extraction method" with "xylem sap method"

The extraction method was compared with the xylem sap method to evaluate contents of available elements in soil (medium 3). The relationship between concentrations of mineral elements in tops and those in xylem sap and concentrations of solution-extractable elements in soil is shown in Fig. 3 for Ca. There was linearity between the concentrations in tops and the concentration in xylem sap (r= 0.95), whereas lower correlation coefficient was obtained between those and the concentrations of extractable Ca except for EDTA-extractable Ca in soil. Correlation coefficients between concentrations of elements in plant tops and xylem sap and those of solution-extractable elements in soil are shown in Table 1. High correlations between concentrations in the tops and both those in xylem sap and soil solution were obtained for K. Concentrations of the other extractable elements analyzed in soil and plant tops did not show a high correlation. However, a high correlation coefficient was obtained between concentrations of elements in tops and those in xylem sap.

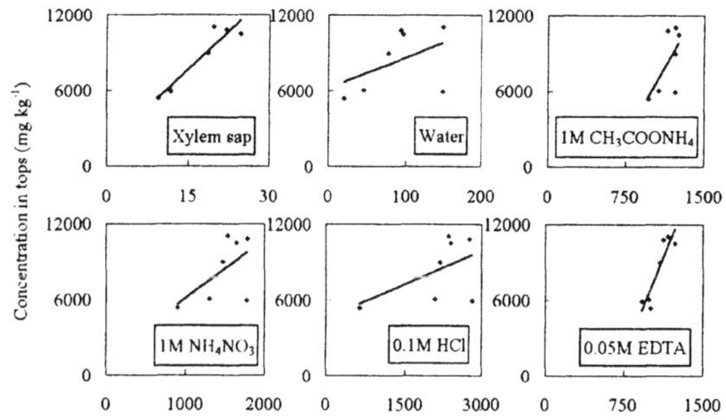

Figure 3. Relationship between Ca concentration in tops and the concentration in xylem sap and of solution-extractable Ca.

Table 1. Correlation coefficients between concentrations of elements in plant tops and xylem sap and those of solution-extractable elements in soil.

	P	K	Ca	Mg	Mn	Zn	Cu
Xylem sap	0.958**	0.925**	0.951**	0.982**	0.985**	0.987**	0.903**
Water	-0.147	0.939**	0.466	0.565	0.887**	0.812**	0.605*
1M CH_3COONH_4		0.922**	0.631	0.671	0.900**	0.965**	0.897**
1M NH_4NO_3	-0.701	0.950**	0.570	0.556	0.948**	0.881**	0.801**
0.1M HCl	-0.623	0.984**	0.508	0.491	0.805**	0.517	0.948**
0.05M EDTA	-0.433	0.915**	0.898**	0.575	0.334	0.848**	0.824**

*,** Significant at 5% and 1% level, respectively.

Therefore, availability of mineral elements in soil can be determined using the concentrations in xylem sap obtained from the hypocotyl area of cut plants.

References

Kuboi T and Noguchi A 1987 Commun. Soil Sci. Plant Anal. 18, 445-456.

Leschber R, Davis RD and L'Hermite P 1985 Chemical methods for assessing bio-available metals in sludges and soils. 96. Elsevier Scientific Pub. Inc., Essex, UK.

Mitchell R L 1964 *In* Chemistry of the soil. Ed. FE Bear. pp. 320-368. Am. Chem. Soc. no.160, Reinhold, New York.

Symeonides C and McRae SG 1977 J. Environ. Qual., 120-123.

W. J. Horst et al.(Eds.), Plant nutrition – Food security and sustainability of agro-ecosystems. 736~737, 2001

Soil properties and plant elemental compositions of faba bean at the flowering stage

S. Z. Labuda
Department of Agricultural Chemistry, Agricultural University of Lublin, Akademicka 15, 20-950 Lublin 1, P.O. Box 158, Poland, labuda@agros.ar.lublin.pl

Key words: elements ratio, flowering stage, *Vicia faba var. minor* L.

Abstract

The aim of the research was to define important soil properties in faba bean cultivation, and to define element ratios in faba bean plants, which may have important practical usefulness in faba bean fertilization. The investigations were carried out at 16 field sites in the Lublin region of Poland. 25 different soil properties were determined in the soil samples, and in the dry matter of the tops (whole aboveground parts) and roots of faba bean plants at the flowering stage were investigated. In the plant samples from all sites the 8 elements: N, P, K, Ca, Mg, S, Na and Cl were determined, and 28 element ratios in tops and roots were also calculated. On the basis of the relationships between soil indices and plant element ratios significant correlations were chosen, and then the relationships with the dry matter of tops and roots were evaluated. It was found that important soil properties of the agrosystems in faba bean cultivation were organic-C, available-P, available-K, and cation exchange capacity. The most important element ratios for faba bean fertilization were N/Ca, P/Ca and K/Ca in the tops, and Ca/Cl and S/Cl in the roots.

Introduction

It is commonly considered that balanced fertilization can bring excellent results when it is based on both very good chemical and agricultural recognition of agrosystems and natural environments (Irshewood, 1998; Lang 1997).

The flowering stage is an important plant development stage for the determination of elemental composition. In this stage, all elements taken up by the plants are mainly accumulated in the green vegetative parts, and the flowering stage is in fact the most comparable across all crops (Labuda, 1991). The basis of elemental composition at the flowering stage would also be useful for calculating the nutrient requirements of crops (Labuda, 2000).

The aim of the present studies was to define important indices for soil properties and ratios of elements in plants at the flowering stage, which would be practically useful in faba bean fertilization.

Materials and methods

Five whole plants of faba bean *Vicia faba var. minor* L. cv. Nadwislanski in the flowering stage were collected from each of 16 field sites in the Lublin region of Poland. At the same time soil samples were also collected from where the plants were grown. The dry matter of tops, as the entire aboveground parts and roots of faba bean plants were studied. The general chemical indices in soil samples were recorded: composition of soil particles, pH (H_2O), pH (KCl), organic-C, total-N, available-P, available-K, available-Mg, exchangeable cations (K, Ca, Mg, Na), TEB (Total Exchangeable Bases), Hydrolytic acidity, CEC (Cation Exchange Capacity) and BS (Base Saturation). Nitrogen was determined in the whole aboveground parts and roots using Kjeldahl's method. Phosphorus was analyzed by means of the vanado-molybdate method after dry digestion; K, Ca and Mg using AAS technique; sulfur was determined by nephelometry, sodium and chlorine were determined after extraction with acetic acid: Na by means of AAS, Cl colorimetrically with silver nitrate.

The following molar ratios, established before for elements in aboveground parts and roots of plants were calculated (28 ratios): N/P, N/K, N/Ca, N/Mg, N/S, N/Na, N/Cl, P/K, P/Ca, P/Mg, P/S, P/Na, P/Cl, K/Ca, K/Mg, K/S, K/Na, K/Cl, Ca/Mg, Ca/S, Ca/Na, Ca/Cl, Mg/S, Mg/Na, Mg/Cl, S/Na, S/Cl.

Correlation and causation by means of linear regression method ($y = bx + c$) were calculated on the basis of soil and plant analysis results.

Results

All chemical indices determined in the 16 soil samples were significantly differentiated, and included a very wide range of values in the experiments performed. This significantly affected the variation of both element contents and the calculated elemental ratios in the faba bean plants.

The wide range of soil and plant characteristics provided the basis for the calculation of dependencies between characteristics studied and evaluation of characteristic chemical indices useful for faba bean fertilization. On the basis of the relationships between all plants and soil characteristics, significant correlations were chosen. It was found that important soil indices for faba bean cultivation were organic C, available P and K, and cation exchange capacity. The most important element ratios for faba bean fertilization were N/Ca, P/Ca and K/Ca in tops, and Ca/Cl and S/Cl in roots in the Lublin region (Tab. 1).

Discussion

The balanced fertilization should comprise N, P, K, Mg and S supply, and it could be based exclusively on the recommended ratios of the components. However, such a thesis would require proving that it would be better to fertilize using low rates of five components (N, P, K, Mg, S) in proper ratios than optimum rates of only three fertilizing elements, e.g. N, P and K (Labuda, 1996). However, the present investigation indicated that six fertilizing elements N, P, K, Ca, Mg and S should be considered in faba bean fertilization. Element ratios can be considered in fertilizers, soils and plants as useful indices for the fertilization of agrosystems and evaluation of environmental hazards in ecosystems (Labuda, 1994, 1997, 1998).

Contents of organic-C, available-P, available-K, and total cation exchange capacity were important chemical soil indices in faba bean fertilization. Ratios of N/Ca, P/Ca and K/Ca in whole aboveground parts, and Ca/Cl and S/Cl in the roots of faba bean plant at the flowering stage could be considered as the most useful indices in faba bean fertilization.

References

Isherwood KF 1998 Bibl. Fragm. Agron., 157-170.

Labuda S 1994 Ann. Univ. Mariae Curie-Sklod. 49 Supl., 133-147.

Labuda S 1996 Proc. IXth International Colloquium for the Optimization of Plant Nutrition, 585-587. Prague, Czech Republic.

Labuda S 1997 Proc. 11th World Fertilizer Congress of CIEC, 325-331. Gent, Belgium.

Labuda S 1998 Fol. Univ. Agric. Stetin. 190 Agricultura 72, 191-197.

Labuda S 2000 Bibl. Fragm. Agron. 8, 181-190.

Lang I 1997 Proc. 11th World Fertilizer Congress of CIEC, 439-448. Gent, Belgium.

Labuda S 1991 Commun. Soil Sci. Plant Anal. 22, 15/16, 1591-1595.

Table 1. Relationships between soil properties and plant indices of faba bean at the flowering stage

Variable		Coefficient	
X	Y	r_{xy}	b_{yx}
Roots dry matter g	Tops dry matter g	0.8618**	5.5061
Tops dry matter g	Roots dry matter g	0.8618**	0.1349
Soil C %	Tops N/Ca	-0.7020**	-3.0589
Soil C %	Tops P/Ca	-0.6425**	-0.2306
Soil C %	Tops K/Ca	-0.6033*	-1.8756
Soil C %	Roots P/Ca	-0.4915*	-0.6925
Soil C %	Roots K/Ca	-0.6009*	-4.6066
Soil P mg/kg	Roots S/Cl	0.5020*	0.0010
Soil K mg/kg	Roots Ca/Cl	-0.3172	
Soil K mg/kg	Roots S/Cl	-0.1286	
Soil CEC mmol (+)/kg	Tops P/Ca	-0.5193*	-0.0011
Soil CEC mmol (+)/kg	Roots P/Ca	-0.6291**	-0.0053
Soil CEC mmol (+)/kg	Tops K/Ca	-0.5464*	-0.0102
Soil CEC mmol (+)/kg	Roots K/Ca	-0.7087**	-0.0328
Tops N/Ca	Roots dry matter g	0.5154*	0.4801
Tops P/Ca	Roots dry matter g	0.5124*	5.7952
Tops K/Ca	Roots dry matter g	0.5546*	0.7242
Roots Ca/Cl	Tops dry matter g	-0.5181*	-12.4791
Roots S/Cl	Tops dry matter g	-0.5026*	-24.3805

*, **, significant at the 0.05, 0.01 probability levels, respectively

W. J. Horst et al.(Eds.), Plant nutrition – Food security and sustainability of agro-ecosystems. 738~739, 2001

Using resin-extraction of soil nutrients for genotype selection and fertility management

A. Olness

USDA-ARS, North Central Soil Conservation Research Laboratory, 803 Iowa Avenue, Morris, Minnesota, 56267, USA

Key words: fertility management, maize, soybean

Abstract

Efficient use of resources requires matching plant and soil potential. This is aided by assessments of the soil resource and correlation with crop yield. Two hybrids of maize (*Zea mays* L.), three varieties of soybean (*Glycine max* L.) and two varieties of wheat (*Triticum aestivum* L.) were grown in rotational succession annually on a 3.2 ha field site. Soil samples (0 to 15 cm) were extracted with cation- and anion-exchange resins. Extracted chemical concentrations were regressed against seed yield. Each genotype was sensitive to a unique suite of chemistries. Soybean variety 9091 and maize hybrid 2292 both showed marked responses to the Mg:(Mg+Ca) ratio, which suggests that Mg may be a reasonable fertilizer. While most genotypes were sensitive to the presence of vanadium (V) and/or to the V:(V+P) ratio, one variety each of soybean and wheat were insensitive to this ratio. Wheat yields increased with Fe and showed a curvi-linear response to K. Losses of yield potentials due to the V:(V+P) ratio ranged from nil to as much as 25 % as the ratio ranged from 0 to about 0.15 for all crops (about 700 kg for Soybean, 2000 kg for maize and 600 kg for wheat ha^{-1}).

Introduction

Most studies of crop yield have focused on nitrogen, phosphorus, and potassium as factors in grain yield. Little attention is given to many of the remaining elements. Using a seldom applied resin-extraction (RE) technique, Olness and Rinke (1994) were able to obtain suites of chemical elements for a 3.2 ha site in western Minnesota which had soybean (*Glycine max* L.), maize, wheat (*Triticum aestivum* L.) grown in rotation. Evaluation of soil samples collected during soybean production showed a strong correlation between the V:(V+P) ratio and seed yield for soybean (Olness *et al.*, 2000). A distinct varietal difference in sensitivity to the V:(V+P) ratio was also noted.

Internal cellular reduction of $H_2VO_4^-$ to VO^{2+} led Olness *et al.* (2001) to speculate on the possibility of interactions with Ca^{2+} and Mg^{2+} in a set of observations from soybean. An analysis of the data showed a strong Mg^{2+}:(Mg^{2+} + Ca^{2+}) molar ratio effect for an Fe-chlorosis resistant variety.

The fact that V seems toxic at μmolar levels to a wide range of plants led to a review of the data for all three crops. This review included Ca^{2+} and Mg^{2+} as well as V and P.

Materials and methods

The field portion of the experiment was conducted on the Barnes-Aastad Association's Swan Lake Research Farm. The site consisted of about 3 ha measuring 108 m by 300 m.

The field was divided into 3.2 equal portions and planted successively to a soybean, maize, and wheat rotation. Soil samples (0 to 15 cm) had been taken from plots planted to soybean. Resin extractions were conducted by using the procedure described by Olness and Rinke (1994) with modifications described in Olness *et al.* (2000). Extracts were analyzed using ICP-AES methods.

Grain yields were plotted against various extracted chemical species and reviewed for evidence of curvi-linearity before statistical analyses were conducted. Standard regression analyses were applied using SAS PROC REG with forward, backward and max R^2 and PROC GLM programs (SAS 1989).

Results and discussion

Each genotype showed a unique response to the intensities of the resin-extracted elements (Table 1). Five of the genotypes showed a strong negative relationship with V:(V+P). Yield decreases of as much as 20 % are noted in most cases as V:(V+P) approaches 0.2. Both maize hybrids and two of the soybean varieties show a rather strong negative relationship with resin extractable V. Soybean variety 9091 showed no evidence of V toxicity from the field data but this variety shows a strong positive relationship with Mg:(Mg+Ca). The yield increase for 9091 exceeds 20% over the range of ratios found in the field. This is a rather large response but not surprising if V is interfering with a Mg function within the plant. The ease of reduction of anionic vanadate, a probable form in arable soils, to cationic vanadyl within the plant cell would suggest the possibility of such an effect. A similar Mg:(Mg+Ca) effect is observed with maize hybrid 2295. The complexity of the effects is shown in the usually positive coefficients for Ca and the negative coefficients for Mg. The relative extracted amounts of both elements are rather large.

Like both maize and soybean, wheat is also sensitive to V:(V+P). However, the wheat varieties showed little effect of Ca and no apparent complex relationships. However, wheat showed a significant response to extractable Fe (linear, not shown) and K (Fig. 1)

References

Olness AE and Rinke JL 1994 Can. J. Soil Sci. 74,461-463.

Olness A, Nelsen T, Rinke JL and Voorhees WB 2000 J. Agron. Crop Sci. 185:145-152.

Olness AE, Gesch RW Barbour NW and Rinke JL 2001 5th Internl. Conf .Precision Agric., Richfield, MN. July 14-19, 2000 (in Press).

SAS Institute Inc. 1989. SAS/STAT User's Guide, Version 6, Fourth Edn. Vol. 2 SAS Inst Cary, North Carolina. 846 pp.

Table 1. Resin extractable elements and ratios.

Statistic	P	V	Ca	Mg	V/(V+P)	Mg/(Mg+Ca)	Ca/(Ca+1000V)	Mg/(Mg+1000V)
	--- nmoles g^{-1} ---		--- µmoles g^{-1} ---		mole $mole^{-1}$			
Average	443	1.65	8.65	4.44	0.022	0.394	0.815	0.394
Minimum	nd[†]	nd	0.152	0.166	0.000	0.092	0.031	0.034
Maximum	3010	18.3	28.3	24.2	0.405	0.725	1.00	1.00

[†] nd = non detectable.

Table 2. Summary of 8 variable coefficients affecting soybean, maize, and wheat yields.

Variable	Regression Coefficients[†]						
	Soybean			Maize		Wheat	
	9091	704	9061	2295	3895	86	2375
Intercept	2.71	3.69	3.25	7.07	9.58	3.35	3.64
V	ns[‡]	**-1.63×10^{-1}**[§]	**-4.37×10^{-2}**	**-1.91×10^{-1}**	**-1.12×10^{-1}**	ns	ns
P	ns	**8.82×10^{-4}**	ns	ns	ns	ns	ns
Mg	ns	ns	ns	**-2.87×10^{-1}**	**-1.50×10^{-1}**	ns	ns
Ca	ns	**9.31×10^{-3}**	ns	**1.86×10^{-1}**	**1.07×10^{-1}**	**2.04×10^{-2}**	**-3.70×10^{-1}**
V:(V+P)	ns	**-1.09**	**-4.07**	ns	**-4.84**	**-2.98**	**-3.16**
Mg:(Mg+Ca)	**1.28**	ns	ns	**7.90**	ns	ns	ns
Mg:(Mg+1000V)	ns	**-8.81×10^{-1}**	ns	ns	**20.7**	ns	ns
Ca:(Ca+1000V)	ns	ns	ns	ns	**-20.6**	ns	ns
R^2	0.215	0.485	0.276	0.314	0.281	0.152	0.251

[†] Bold-faced coefficients have $p < 0.10$.
[‡] ns = non-significant.
[§] All chemical concentrations in micro- or nano-molar g^{-1} soil; intercept values are Megagrams ha^{-1}.

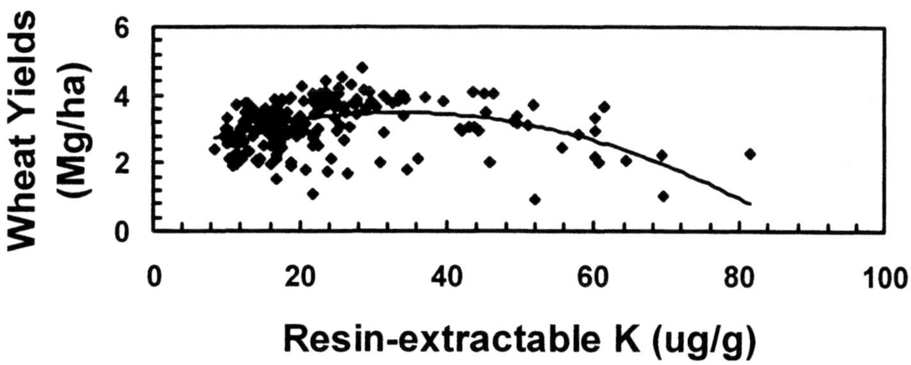

Figure 1. The relationship between resin-extractable K and wheat yield.

W. J. Horst et al.(Eds.), Plant nutrition – Food security and sustainability of agro-ecosystems. 740~741, 2001

The attempt to utilize chemical composition of soil solution in fertilization diagnostics

J. Labętowicz and B. Rutkowska

Faculty of Science of Soil Environment, Department of Agricultural Chemistry, Warsaw Agricultural University, Rakowiecka 26/30, Warsaw, 02 - 528, Poland, rol_kcr@delta.sggw.waw.pl

Key words: fertilization diagnostics, macronutrients, micronutrients, soil solution

Abstract

Soil solution is a significant link in the process of plant nutrition. Analysis of the soil solution as a background of fertilization diagnostics may be very promising. Concentration of ions, being the nutrients in the soil solution is relatively low. Therefore, the questions arises: to what degree - in quantity aspect - the components of the soil solution cover the nutritional requirements of plants. The degree of quantitative coverage of nutritional requirements of the selected plants by ions (P, K, Ca, Mg, B, Cu, Mn, Mo, Zn), contained in the soil solution of arable layer of cultivated soils from Poland, was evaluated. To illustrate this relationship the the synthetic index, which has been called "solution recovery coefficient" (SRC), was calculated.

For phosphorus, potassium and magnesium the nutritional requirements of the plants exceeded the quantities of these components in the soil solution of arable layer of the soil. It indicates a small meaning of the soil solution as a direct source of these nutrients. The possibilities of utilizing the measurements of their concentration, are therefore, smaller in fertilizing diagnostics. In case of calcium and micronutrients the nutritional requirements are similar and often lower than the quantities of these components in the soil solution arable layer of the soil. It indicates the possibility of diagnosing the fertilizing requirements with the use of these components from the information on their concentration in the soil solution.

Introduction

Soil solution is a significant link in the processes of plant nutrition. Its chemical composition and concentration of the particular ions are decisive for the rate of nutrients uptake by the plants (Marshner, 1993). The measurements of chemical composition of the soil solution provide the important data concerning the state of the nutrients in the soil and may be a major element of fertilization diagnostics (Adams and Odom, 1985). At present, we may observe the increased world interest in the methods for determination of the assimilable components, using the-so called mild extractors, which may become a joint extractive for determination of several elements. Analysis of the soil solution as a background of fertilization diagnostics may be, due to this reason, very promising (Jones and Kalra 1992).

Concentration of ions, being the nutrients in the soil solution is relatively low. Therefore, the question arises: to what degree the components of the soil solution cover the nutritional requirements of plants.

Materials and methods

The research material consisted of the soil samples, collected from the territory of Poland. The soil samples were collected from arable layer of soil. To obtain the soil solution, the vacuum method was employed. In the obtained soil solution, the concentration of particular ions (P, K, Ca, Mg, B, Cu, Mo, Mn, Zn) was determined by the plasma emission spectrometry (ICP). To illustrate the relationship between the quantity of nutrients in the soil solution and the nutritional requirements of the plants, the "solution recovery coefficient" (SRC) was calculated for the particular components.

$$SRC = \frac{\text{uptake of component in plant yield } [kg \cdot ha^{-1}]}{\text{quantity of component in solution of arable layer } [kg \cdot ha^{-1}]}$$

For calculation of SRC, it was adopted:

- mean uptake of components by the plants (based on literature),

- mean concentration of components in the soil solution of the arable layer of the soil from which the mean quantity of the components in the arable layer of the soil was calculated.

SRC value equal 1 indicates that the quantity of a given components, as uptaken by the plants from the arable layer from 1 ha area is equal to the quantity of this component in a form, dissolved in the soil solution. Values below and above 1 indicate, respectively that the quantity of the components, uptaken by the plants is lower or higher than the amount being found in the soil solution.

Results

The nutritional requirements of the plants in relation to phosphorus, potassium and magnesium exceeded the quantity of these components in the solution of arable layer of soil by few or even by several times (Tab. 1).

The highest value of SRC was reached for phosphorus. In order to satisfy the nutritional requirements of the plants in respect of phosphorus, the soil solution must be supplemented, to a high degree, with the reserves of this component from the solid phase of the soil. The quantity of phosphorus, uptaken by the plants was several times higher than the content of this component in arable layer solution (Tab. 1). In relation to potassium and magnesium, SRC values are lower what shows that their quantities in the soil solution satisfy the nutritional requirements of the plants in greater degree than in case of phosphorus. The quantities of potassium and magnesium, uptaken by the plants are several times higher than their content in the soil solution (Tab. 1). The SRC values for macronutrients (P, K, Mg) are higher than 1. It indicates the small possibilities of utilizing the measurements of their concentration in the soil solution, in determination of fertilizing requirements.

Table 1. Ratio of macroelements uptaken by the plants and their quantity in the soil solution (SRC)

Plant	P	K	Ca	Mg
Rye	13,60	3,23	0,28	2,01
Winter wheat	11,68	1,87	0,18	1,51
Spring barley	11,58	2,05	0,24	1,60
Potatoes	13,07	5,08	0,22	2,24
Sugar beets	18,92	7,62	4,61	5,54
Winter rape	17,12	4,17	1,40	2,56
Red clever	9,66	4,10	0,64	1,93

Table 2. Ratio of microelements uptaken by the plants and their quantity in the soil solution (SRC)

Plant	B	Cu	Mn	Mo	Zn
Rye	0,06	0,77	0,11	0,91	1,32
Winter wheat	0,04	0,76	0,06	0,62	1,06
Spring barley	0,04	0,79	0,05	0,58	1,05
Potatoes	0,16	1,37	0,04	0,97	0,74
Sugar beets	0,64	2,01	0,13	1,39	2,76
Winter rape	0,30	0,61	0,05	0,60	0,78
Red clever	0,28	1,00	0,05	0,66	1,31

In case of calcium and micronutrients, SRC values are often lower than 1 what indicates the possibility of satisfying the nutritional requirements of the plants from the soil solution (Tab. 1, Tab. 2).

From SRC values for calcium and micronutrients, it may be however concluded that the nutritional requirements of the plants in relation to these elements may be completely satisfied directly from the soil solution. It indicates the necessity of diagnosing the requirements in respect of fertilizing with these components from the knowledge of their concentration level in the soil solution.

Discussion

To illustrate the relationship between the quantity of nutrients in the solution of arable layer of the soil and the nutritional requirements of the plants, the "solution recovery coefficient" was calculated for the particular ions (Łabętowicz, 1995).

Analysis of SRC coefficient indicates that the role of the soil solution as a direct source of nutritional components is different for particular elements. For phosphorus, potassium and magnesium for which SRC values exceed considerably 1, the soil solution has a minor meaning as a direct source of components (Tab. 1). It seems that for the discussed components, and especially for phosphorus, nutrition of plants will be decisively dependent on the soil's capacity of supplementing quickly the soil solution, gradually with their uptake by the plants (Khasawneh, 1971).

In case of calcium and micronutrients the nutritional requirements of plants may be largely or even completely covered from the soil solution (Tab.1, Tab.2). Buffering capacities of the soil have, in this case, relatively smaller meaning. Only in case of zinc, SRC values exceed slightly 1 (Tab. 2). It has been probably connected with the decreased intensity of zinc dissolving in analyzed soils. The studies of Curtin and Smillie (1983) showed that liming of acidic soils caused often incidence of zinc deficits in the plants, even on the soils, being relatively rich in this element.

We should, however, stress the tentative character of "solution recovery coefficient" (SRC). Its calculations consider only ions, contained in the arable layer of the soil and assume the state of maximum capillary water capacity, which occurs only directly after rains.

References

Adams F and Odom J W 1985 Soil Sci. 140, 701-705.

Curtin D and Smillie G W 1983 Soil Sci. Soc. Am. J. 47, 701-707.

Jones J B and Kalra J P 1992 Commun. Soil Sci. Plant Anal. 23, 2015-2028.

Khasawneh F E 1971 Soil Sci. Soc. Am. Proc. 35, 426-436.

Łabętowicz J 1995 Skład chemiczny roztworu glebowego w zróżnicowanych warunkach glebowych i nawozowych. SGGW, Warszawa.

Marschner H 1995 Mineral nutrition of higher plants, 2nd edn. Academic Press, San Diego, CA.

W. J. Horst et al.(Eds.), Plant nutrition – Food security and sustainability of agro-ecosystems. 742~743, 2001

Potassium availability indices and plant response

M. Shenker and X. Huang

Department of Soil and Water Sciences, Faculty of Agricultural, Food and Environmental Quality Sciences, The Hebrew University of Jerusalem, P. O. Box 12, Rehovot, 76-100, Israel, shenker@agri.huji.ac.il

Key words: activity ratio, availability, buffer capacity, intensity, K, plant response, potassium, quantity, resin-extraction, soil-test

Abstract

Potassium adsorption by cation-exchange resin membranes was correlated to standard soil-K availability indices in a wide range of soil types and K status of the soils, and to plant responses in these soils, in order to ascertain whether it reflected quantity (Q) or intensity (I) factors of soils. The effects of soil-water content and the time of soil-resin contact on the adsorption of potassium and competing cations were tested to study kinetic effects that control K-adsorption by the membrane. Thus we aimed to check whether this method reflected Q-factors, I-factors, or kinetic-factors that control K availability for plants. The amount of membrane-adsorbed-K (K_{memb}) was sensitive to water content that control K movement towards the membrane-sink. As long as the membrane was dominated by pre-sorbed Na it accumulated K, however when Ca was adsorbed to a level that dominates the membrane, K was gradually desorbed. Modification of plant responses by multiplying by log potential buffer capacity (PBC) of the soils had resulted in highly significant correlation between K_{memb} and plant response. Correlation with Q and I-indices of the soils and with plant response, indicate that K_{memb} at a water tension of 0.2 bar and 24-h of soil-membrane contact, reflect the I-factors rather than the Q-factors of the soils.

Introduction

Two main factors are considered traditionally to control the capacity of a soil to provide potassium for plant uptake: a quantity (Q) factor and an intensity (I) factor. A set of well established soil tests were developed to quantify these factors. Thus ammonium acetate (1-*M*) is often used to quantify the total amount of exchangeable K (K_{Exch}) as a Q-factor measure and $CaCl_2$ (0.01-*M*) is used to quantify the "easily exchangeable K" ($K_{E\text{-}exch}$). The ratio $(K)/[(Ca)+(Mg)]^{1/2}$ (AR) in soil solution, or its equivalent energy term ΔF ($=R \cdot T \cdot \ln AR$) (Woodruff, 1955) are common measures of the I-factor. However, even though these measures were calibrated as soil tests and critical values are available as guidelines for the need in fertilisation, none of them alone provides a good tool for fertilisation recommendation. Thus plants grown in a clay soil with a high Q-value may suffer from K deficiency if the I-value is low, while a sandy soil with a high I-value may yet have too low Q-value to supply the needed amount of K for crop production. Else, a low I-value that indicates fertilisation need does not provide any clue for the needed amount of K-fertiliser. The potential buffer capacity (PBC) of the soil to maintain its I-value as was described by Beckett (1964) may bridge this gap, however a third factor, the kinetics of K release and movement in the soil towards the uptake sites of the root system, is yet not resolved. This last factor is becoming important as intensive agriculture is practiced. To address this complication a new approach of plant simulation by a resin sink was suggested (Yang *et al.*, 1991). Various soil-resin systems were used to simplify separation of the resin from the soil. Most studies were carried in suspension, however in this case the method is not sensitive for kinetics of K-diffusion in the soil. In other studies the extraction step was carried out in more realistic water content to reflect field conditions. It was assumed that by using this approach, K accumulation might reflect all factors (i.e. Q, I, and the kinetics of K release and movement) that regulate K supply to the root-sink, and if a long period of resin-soil contact is allowed, the resin will integrate these factors over the tested period (Skogley and Dobermann, 1996). The aim of our study was to correlate resin-extracted K to the common soil tests and to plant response for K-fertilisation in a wide range of soil types and K status, in order to resolve the correlation of the resin approach to the Q, I, and kinetic-factors of the soils and to plant responses.

Materials and methods

A Vertisol, Loess, and sandy soil (for selected soil properties see Table 1) were pre-treated to give a wide range of K status. Resin (BDH cation-exchange resin membrane, 55165 2U) extracted K at various water content levels of the soils and time of soil-membrane contact was correlated to Q and I K-measures of the soils. Corn (*Zea mays* L.) plants were cultured in the soils in a greenhouse experiment with or without K fertilisation and dry weight accumulation and K content in the shoots were correlated with all soil tests. All together 24 K-levels were employed for the soil-test comparison: four levels for the sandy soil, eight for the loess, and 12 for the vertisol, and 12 levels, (four for each soil), were used for plant culture.

Results

Wide ranges of K-status indices were the result of pre-treatment of the three soils. The ranges of exchangeable-K (K_{Exch}, extracted by 1-*M* ammonium acetate), easily exchangeable K ($K_{E\text{-}exch}$, extracted by 10-*mM* $CaCl_2$), AR, PBC, and membrane-extracted K (K_{memb}) at water tension of 0.2 bar for 24 h, are given in table 2.

Table 1. Selected properties of the soils

	Vertisol	Loess	Sandy soil
Soil texture			
sand (%)	37.5	66.2	94.9
silt (%)	13.5	7.8	0.0
clay (%)	49.0	26.0	5.1
pH (H_2O)	7.7	7.5	7.1
$CaCO_3$ (%)	26.5	14.3	0.1
CEC[i] (cmol kg^{-1})	40.6	24.2	7.5
SSA[ii] ($m^2 g^{-1}$)	274.2	177.6	25.7
Organic matter (%)	4.8	2.3	0.6

i) cation exchange capacity
ii) specific surface area

Table 2. Ranges of K-status indices

K index \ Soil	Vertisol	Loess	Sandy soil
K_{Exch}[i] (cmol kg^{-1})	0.4-1.5	0.3-0.6	0.1-0.3
$K_{E\text{-}exch}$[ii] (mmol kg^{-1})	0.7-3.6	0.4-1.5	0.3-2.1
AR[iii] ($M^{1/2}$)×1000	0.9-6.5	1.0-3.1	2.2-18.0
PBC (cmol kg^{-1} $M^{-1/2}$)	103-233	83-177	11-16
K_{memb}[iv] (nM cm^{-2})	13-85	13-38	42-335

i) exchangeable-K by 1-*M* ammonium acetate
ii) easily exchangeable K by 10-*mM* $CaCl_2$
iii) $(K)/[(Ca)+(Mg)]^{1/2}$ in 10 mM $CaCl_2$ extract
iv) membrane adsorbed K, 0.2 bar, 24 h

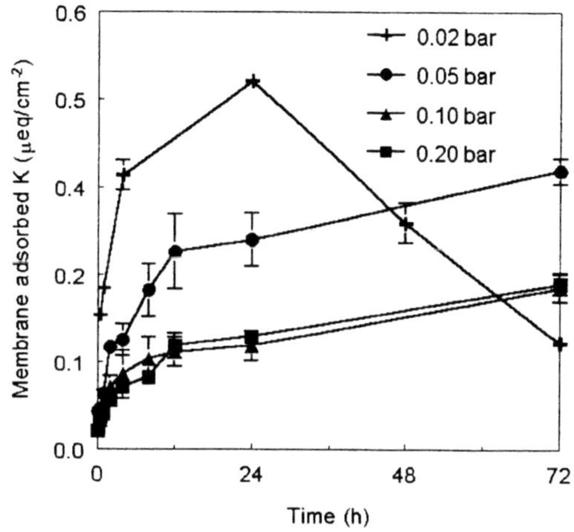

Figure 1. Effect of soil water tension and period of membrane-soil contact on the adsorption of potassium from sandy soil.

Potassium adsorption to the membrane was highly affected by water content of the soils and by the period of membrane-soil contact (Fig. 1, data for sandy soil). Much of the variation is explained by kinetics of K-diffusion towards the membrane-sink. At lower water tension K-accumulation by the membrane increased, however at certain point (e.g., about 20 h in sandy soil at 0.02 bar, Fig. 1) K was desorbed due to Ca competition. Correlation of K_{memb} at various water tension levels (0.02 to 0.2 bar) and time of membrane-soil contact (1 to 72 h) to Q and I K-indices of the soils was found to be highly significant for each soil separately (usually at $\alpha<0.01$). However, when data of all soils were analyzed together K_{memb} was significantly correlated only to the I-indices and not the Q-indices of potassium.

Correlation between soil test indices of potassium and plant response indicates that the Q-index K_{Exch}, but not the I-index AR, was significantly correlated to shoot dry weight and shoot K-content of plants grown without K-fertilization. However, when K-fertilization was applied the shoot dry weight was highly correlated with AR but not with K_{Exch}, yet shoot K-content was better correlated to K_{Exch}. Highly significant linear correlations of plant responses to K-fertilization were found with K_{Exch} but not with AR or ΔF: relative dry weight (RDW, fertilized as percentage of not fertilized shoot dry weight) was highly correlated ($R^2>0.9$, $\alpha<0.001$) with $1/K_{Exch}$, and relative K uptake (RKU, fertilized as percentage of not fertilized K uptake to shoot) was highly correlated ($R^2>0.9$, $\alpha<0.001$) with $\log(K_{Exch})$. Such high correlations of the I-indices were achieved only by modifying RDW and RKU by multiplying with log(PBC). By this modification correlations with AR were as high ($R^2>0.9$, $\alpha<0.001$) as the above correlations with K_{Exch}. Similarly, K_{memb} was significantly correlated to RDW and RKU only after plant response was modified according to PBC. Thus it was shown that Q-factors can be distinguished from I-factors according to their relations with plant responses. Under our conditions K_{memb} represents the I rather than the Q-factors of potassium in soils.

References

Beckett PHT 1964 J. Soil Sci. 15, 9-23

Skogley EO and Dobermann A 1996 J. Environ. Qual. 25, 13-24.

Yang JE, Skogley EO, Georgitis SJ, Schaff BE and Ferguson AH 1991 SSSA J 55, 1358-1365.

Woodruff CM 1955 Proc. Soil Sci. Soc. Am. 19, 36-40.

W. J. Horst et al.(Eds.), Plant nutrition – Food security and sustainability of agro-ecosystems. 744~745, 2001

Soil potassium determination in pasture soils in the basque country

A. Aizpurua, S. Alava, A. Castellón, A. Alonso and G. Besga
NEIKER, Departamento de Agricultura y Produción Animal, C/ Berreaga, 1, 48.160 Derio (Bizkaia), Spain, aaizpurua@neiker.net

Key words: ammonium acetate, Italian ryegrass, soil potassium, sodium tetraphenyl boron

Abstract

Practical experience has indicated a discordance between soil K content (extracted by ammonium acetate) and responses to fertilization as recommended using the nutrient balance method (based on K inputs and outputs) for pasture systems. The aim of this work was to compare different methods for determining the soil available potassium in order to improve the K use in pastures. For this purpose, eight soils with three clay levels and for each clay level two different soil K contents were selected. These soils were studied in a greenhouse pot trial with Italian ryegrass (Lolium multiflorum Lam.) where two doses of K (equivalent to 0 and 200 kg K/ha) were applied. Dry matter yield, and K herbage concentration were measured in six cuts. Soil K content was determined by the ammonium acetate and sodium tetraphenyl boron methods before sowing, and after the third and sixth cuts. K addition improved the yield only when the initial soil K content was low. Considering the initial and final soil K contents extracted by ammonium acetate and the K uptake by the plant, the results showed that when the initial soil K content was low a great percentage of the potassium absorbed by the plant was from non-exchangeable sources. The soil K content determined by the sodium tetraphenyl boron method was better correlated with the potassium extracted by the plant than the soil potassium extracted by the ammonium acetate method. The ammonium acetate method, one of the most widely used, could be improved by using an extractant which can extract part of the non exchangeable K that plants are evidently able to take up.

Introduction

There are difficulties in establishing the relationship between soil and plant factors and K availability (Beringer, 1987). The ammonium acetate method measures only solution-K and exchangeable-K in soil. It fails to account for "fixed" K that can become available to crops during the growing season. Plants deplete solution K by removing it from solution during active uptake (Joern, 2000). Practical experience in the Basque Country has shown a discordance between soil K content (as measured by ammonium acetate) and responses to K fertilization recommended by the nutrient balance method used in pastures (Oyanarte *et al.*, 1995).

Materials and methods

Table 1. Soil K and clay contents of the eight soils used in the greenhouse pot-experiment.

Soil K content	mg/kg (amm. acetate)	n*	Soil clay content	%	n*
High	<130	5	Low	15-20	3
Low	>130	3	Medium	22-25	3
			High	>30	2

* number of soils in each class

A greenhouse pot experiment with completely randomized design and four replicates was established. The size of the pots was 19.6 cm diameter and 21 cm in height. The pot were filled with 3.75 kg of soil. Italian ryegrass (*Lolium multiflorum* Lam.) was grown in eight different soils with two initial soil K levels and three clay contents (Tab. 1) and two K doses were applied (equivalent to 0 and 200 kg K/ha). Dry matter yield and K concentrations in the plant material were measured in six cuts. Soil K content was determined by ammonium acetate and sodium tetraphenyl boron methods before sowing, and after the third and sixth cuts. Released K was calculated by the ecuation: Released K=K uptaked by the plant-(final soil K-initial soil K). Differences between the soils were tested using the Duncan test at the 5% level, and means not followed by the same letter are significantly different at the 5% level (Table 2, 3, 4 and 5).

Results

There was a significant interaction of K soil content with clay content and K addition, so data analysis was separated by soil K content. The K addition only improved yield when the initial soil K content was low (Tab. 3). There was a negative effect of the clay content on total yield (Tab. 2 and Tab. 3) probably caused by lower root activity due to soil compaction.

The initial soil K content influenced the K uptake by the plant, because of the effect on yield and K herbage concentrations. The maximum K uptake by Italian ryegrass occurred when the soil K content was high (189-195 mg/soil kg) for soils with medium and low clay contents. The soils with initially low K content, medium clay content and no addition of K gave the lowest uptake (Tab. 4 and Tab. 5).

Considering the initial and final soil K contents extracted by ammonium acetate and K uptake by the plants in the pots with no addition of K, the results showed that when the initial soil K content was low a great percentage of the K absorbed by the plant was relesed from soil reserves not determined with the ammonium acetate method (Fig.1).

The soil K content determined by the sodium tetraphenyl boron method after the third cut was better correlated with the K extracted by the plant in the following cuts (4th, 5th and 6th) than the soil K extracted by ammonium acetate (Fig. 2).

Table 2. Italian ryegrass yield of six cuts with regard to soil clay contents and K supply in soils with high K content

Soil clay content	K supply (kg K/ha) 0	200	Mean
	Yield (DM kg/ha)		
High	4877	4851	4864 c
Medium	6690	6728	6708 b
Low	7832	7216	7523 a
Mean	6466 a	6264 a	

Table 3. Italian ryegrass yield of six cuts with regard to soil clay contents and K supply in soils with low K content

Soil clay content	K supply (kg K/ha) 0	200	Mean
	Yield (DM kg/ha)		
High	4472	4816	4645 c
Medium	4878	5634	5341 b
Low	5788	6799	6293 a
Mean	4943 b	5563 a	

Table 4. Italian ryegrass K uptake of six cuts with regard to soil clay content and K supply in soils with high K content

Soil clay content	K supply (kg/ha) 0	200	Mean
	Uptake (mg K/kg of soil)		
High	137	145	141 b
Medium	190	192	191 a
Low	189	195	192 a
Mean	172 a	177 a	

Table 5. Italian ryegrass K uptake of six cuts with regard to soil clay content and K supply in soils with low K content

Soil clay content	K supply (kg/ha) 0	200	Mean
	Uptake (mg K/kg of soil)		
High	102	122	112 a
Medium	68	117	92 b
Low	88	142	115 a
Mean	86 b	124 a	

Discussion

When the initial soil K content was low the plant used so-called "fixed K" to a greater extend, as the K release process by diffusion is accelerated (Sparks *et al.*, 1980; Sparks, 1986). This capability of the plant to absorb K forms initially not classified as "plant available" is related to the high density of the root system in gramineae (Fergus and Martin, 1974). In agreement with Cox *et al.* (1999) soil K extraction with sodium tetraphenyl boron was a better predictor of plant available K than ammonium acetate.

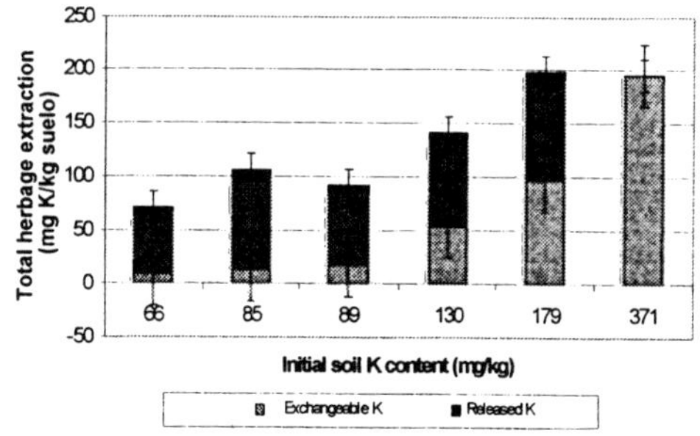

Figure 1. Contribution of the exchangeable K and the initially fixed K (released) to K plant uptake with regard to the initial soil K content extracted with ammonium acetate (Error bars indicate +-1 SE).

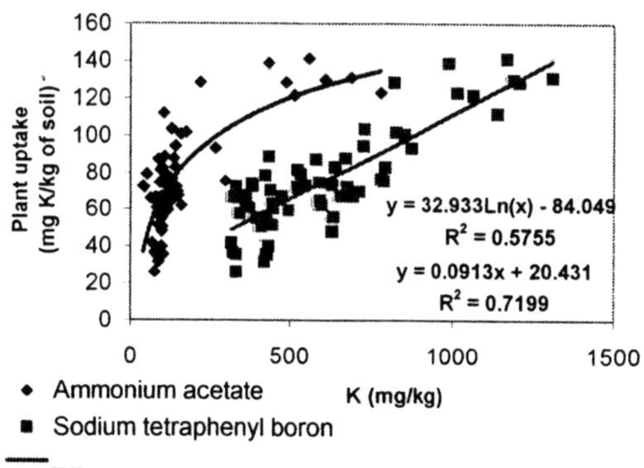

Figure 2. Relation between soil K content extracted by ammonium acetate and sodium tetraphenyl boron and plant K uptake.

References

Beringer H. 1987. In: Methodology in soil K research. Potash Review 1:15-16.

Cox A E, Joern B C, Brouder S M, Gao D. 1999. Soil Sci. Soc. Am. J. 63, 902-911.

Fergus I F, Martin A E. 1974. Aust. J. Soil Res. 12, 147-158.

Joern B. 2000. The Soil-Plant Analyst. Summer 2000, 3-5.

Oyanarte M, Rodríguez M, Besga G, Echeandia A. 1995. Sustrai 38, 30-33.

Sparks D L. 1986. In: Nutrient balances and the need for potassium. 13th Congress of the Inter.Potash Inst. pp 83-95. Rheims.

Sparks D L, Zelazny L W, Martens D C. 1980. Soil Sci.Soc. Am. J. 44, 1205-1208.

W. J. Horst et al.(Eds.), Plant nutrition – Food security and sustainability of agro-ecosystems. 746~747, 2001

Nitrogen fertilizer recommendation for sugar beet according to the EUF soil testing system

D. Horn[1)] and F. Fürstenfeld[2)]

[1)] *EUF-Arbeitsgemeinschaft für Bodenfruchtbarkeit und Bodengesundheit, Marktbreiter Straße 74, Ochsenfurt, D 97199, Germany, Dietmar.Horn@Suedzucker.de*

[2)] *BGD-Bodengesundheitsdienst GmbH, Marktbreiter Straße 74, Ochsenfurt, D 97199, Germany, Fred.Fuerstenfeld@Bodengesundheitsdienst.de*

Key words: EUF, fertilizer recommendation, nitrogen, sugar beet, residual nitrate

Abstract

Poor sugar concentration and juice quality of sugar beets required an adaptation of N fertilization. After testing of different N fertilizer recommendation systems, the EUF method for soil analysis was introduced in Austria and Germany in 1974 and 1982, respectively. Since the introduction, more than 1.2 million soil samples have been analysed by EUF. Today, EUF fertilizer recommendations are given each year to different crops on 150.000 ha in southern Germany and 50.000 ha in Austria. About 70 % of the EUF fertilizer recommendations are given for sugar beets, while the rest is done for cereals, potatoes, maize, vine and other crops.

Field trials with sugar beets and other crops were performed to calibrate the EUF fertilizer recommendation system. The extracted inorganic N (mainly nitrate) and organic N are the basis for the N fertilizer recommendation. Based on the EUF soil testing an average of about 80-90 kg N ha^{-1} is recommended for sugar beets. Adjusted N doses resulted in better quality due to decreased amino-N concentration of sugar beets by about 30% and less residual nitrate after harvest of sugar beets. To further improve the nitrogen recommendation system, soluble organic carbon is tested in EUF research as an additional analytical parameter to be used for the calculation of N fertilizer requirement.

Introduction

The cultivation of a crop needs an economical use of applied fertilizer. Excess of nitrogen fertilizer is an environmental hazard. A reliable soil test method on which the nitrogen fertilizer rate can be based is the EUF method (Mengel, 1991). The principle of the Electro Ultrafiltration (EUF) was described by Németh (1979). A soil suspension is exposed to an electrical field in which anions migrate to the anode and cations to the cathode, where they are recovered. Extracted nitrogen comprises nitrate, ammonium, and organic nitrogen. About 50 % of this organic nitrogen consists of amino N, which is easily mineralised (Mengel *et al.*, 1999). Therefore, the EUF extractable organic N is an indicator for the amount of organic N mineralised during a growth period. The EUF method is applied to N fertilizer calculation of various field crops (Fürstenfeld *et al.*, 1994). This paper reviews the results of the adjustment of N fertilization to N requirement of sugar beet and the impact for sugar beet growers, sugar industry, and the environment are presented.

Materials and methods

Soil samples were taken from the top soil of farmers' fields in the year before cropping when nutrient uptake of the precrop has finished. The soils were dried at 40°C, crushed and sieved (1mm mesh). The EUF extraction follows the method of Németh (1979, 1982): the first fraction is obtained at 20°C for 30 min and 200 V, while the second fraction is extracted at 80°C for 5 min and 400 V. Then the extracts were analysed for nitrate and organic nitrogen which are used as the basis of the N fertilizer recommendation. The estimation of the N fractions were elaborated in field experiments where 1 mg NO_3-N kg^{-1} soil and 1 mg organic N kg^{-1} soil is equal to 4.4 kg ha^{-1} (Németh *et al.*, 1991). In addition to EUF data, precrop, intercrop and slurry application are taken into account for calculating the N fertilizer requirement.

Results and discussion

Since the introduction of the EUF system in Austria and Germany in 1973 and 1982, respectively, more than 1.2 million soil samples have been analysed by EUF. Today, EUF fertilizer recommendations are given each year to different crops on 150.000 ha in southern Germany and 50.000 ha in Austria. About 70 % of the fertilizer recommendations are given for sugar beets, while the rest is done for cereals, potatoes, maize, vine and other crops.

The EUF extractable nitrate ranges in 98 % of all soils analysed from 2 to 50 mg kg^{-1} (mean: 14 mg kg^{-1}) while organic N ranges from 10 to 35 mg kg^{-1} soil (mean: 22 mg kg^{-1}). This results in N fertilizer recommendation from 30 to 160 kg N ha^{-1} with an average of about 80-90 kg N ha^{-1} for sugar beets in the year 2001. A data collection from 218 sugar beet growers using EUF showed a substantial decrease of N fertilization for sugar beets of more than 200 kg N ha^{-1} in 1980 to an average of 120 kg N ha^{-1} in 1993 (Altweck, 1994). In 2000 the average N fertilizer is below 100 kg N ha^{-1}. The reduction of N fertilization resulted in better quality and increased yield of sugar beets. The introduction of the EUF system with adjusted N fertilization decreased the amino-N concentration of sugar beets by about 30% (Fig. 1), which resulted in less N freight into sugar factories (Fig. 2) (Köhler, 1997). With regard to the costs for N stripping, this is of much economical and ecological importance for the sugar factory. At the same time the costs of sugar beet growers are decreased.

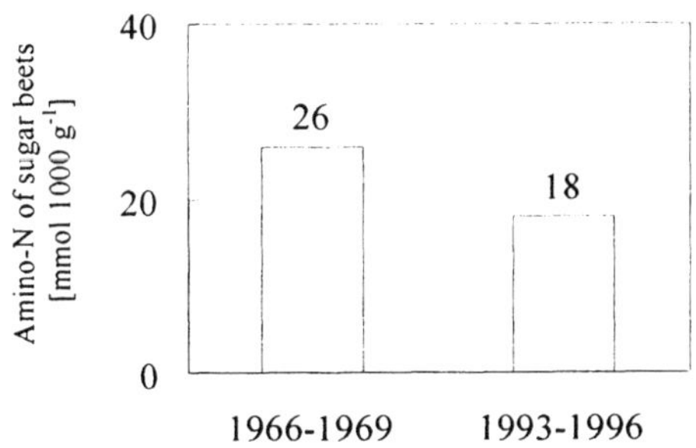

Figure 1. Amino-N concentration of sugar beets of Südzucker in Rain, Regensburg and Plattling in the period 1966-1969 and 1993-1996

Lowering the N fertilizer rates according to the EUF recommendation decreases the residual nitrate in soils at harvest of sugar beets from an average of 77 to 34 kg N ha^{-1} (Fig. 3). Similar results of low residual NO_3^- content after sugar beet harvest were found by Timmermann *et al.* (1995) in soil samples from fields in water protection areas in Baden-Württemberg. Therefore, the reduction of N fertilization is of ecological relevance.

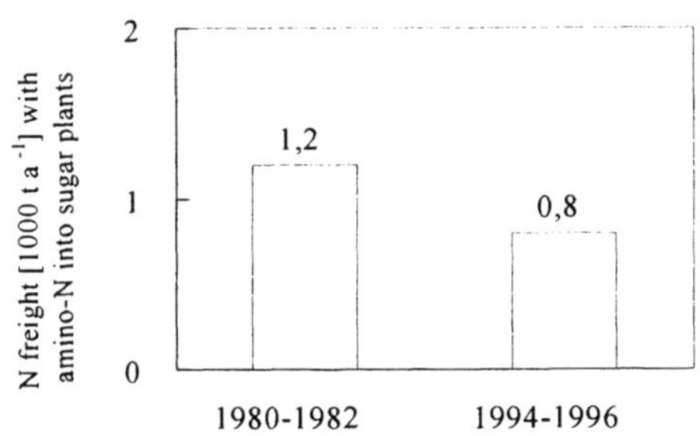

Figure 2. Nitrogen freight per year with amino-N into sugar factories of Südzucker in Rain, Regensburg and Plattling in the period 1980-1982 and 1994-1996

These results show that the use of a fertilizer recommendation system based on EUF soil tests is highly successful in its target to improve the parameters of technical quality and lowering the residual nitrate after harvest.

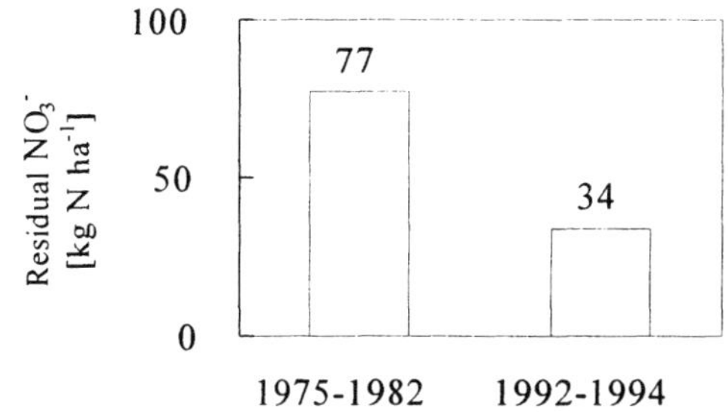

Figure 3. Residual nitrate in soil profile (0-90cm) after harvest of sugar beets in 1975-1982 (average of 29 trials of "Kuratorium für Versuchswesen im Zucker-rübenanbau") and 1992-1994 (average of 24 trials of "BGD-Bodengesundheitsdienst GmbH"), respectively

EUF with its advantages in respect to soil sampling depth and sampling time has been accepted by farmers on voluntary basis. Comparisons of EUF and different N recommendation systems by official advisory bodies showed that EUF gives reliable recommendations for different crops (Fürstenfeld *et al.*, 1994). To further improve the nitrogen recommendation system, soluble organic carbon is tested in EUF research as an additional analytical parameter to be used for the calculation of N fertilizer requirement (Kempl *et al.*, 2001).

References

Altweck A 1994 DZZ 6, 16.

Fürstenfeld F, Deller B and Schulz R 1994 VDLUFA-Schriftenreihe, Kongressband 38, 215-218.

Kempl F, Steffens D, Horn D and Schubert S 2001 Proceedings of XIV International Plant Nutrition Colloquim Eds Kluwer Academic Publishers

Köhler R 1997 *In* Düngemittelmärkte und Düngeberatung in Deutschland. Eds. Bundesarbeitskreis Düngung pp. 97-111.

Németh K 1979 Adv. Agron. 31, 155-188.

Németh K 1982 Plant Soil 64, 7-23.

Németh K, Recke R, Heuer C 1991 Zuckerindustrie 116, 991-996.

Mengel K 1991 Fert. Res. 28, 251-262.

Mengel K, Schneider B, Kosegarten H 1999 J. Plant Nutr. Soil Sci. 162, 139-148.

Timmermann F, Übelhör W, Klinger S, Schmidt G, Lindörfer G, Dühne J 1995 *In* Nitrat-Bericht 1991-1994. Eds. Ministerium für Ländlichen Raum, Ernährung, Landwirtschaft und Forsten, Baden-Württemberg.

W. J. Horst et al.(Eds.), Plant nutrition – Food security and sustainability of agro-ecosystems. 748~749, 2001

Net mineralization of organic N in soils related to EUF (Electro-ultrafiltration) extractable organic N, organic C, nitrate and growth

A. Maci[1)] and K. Mengel[2)]

[1)] *Department of Agronomy, Agricultural University of Tirana, Albania*
[2)] *Institute of Plant Nutrition, Justus Liebig University, 35392 Gießen, Germany*

Key words: EUF extractable carbon, EUF extractable nitrogen, immobilization, interlayer NH_4^+, mineralization, nitrate.

Abstract

The objective of the investigation was to study the relationship between EUF (Electro-ultrafiltration) extractable N, EUF extractable carbon and interlayer NH_4^+ on net N mineralization and N uptake of ryegrass (*Lolium multiflorum* cv. Aubade) cultivated in Mitscherlich pots on 20 different arable soils. Grass of the first harvest (first cut) exclusively fed from nitrate and N uptake of the first cut was significantly correlated with the EUF extracted nitrate in the soil at seeding (r= 0.96***). In average from this nitrate about half was taken up by the grass and half was assimilated (immobilized) by soil microorganisms. This immobilized N was correlated with the organic C extracted by EUF at seeding (r = 0.50*). The grass of the second and third harvest exclusively drew N from the organic soil N pool and N uptake of grass was correlated with EUF-Norg (r = 0.66**, second cut and r = 0.55*, third cut). Interlayer ammonium significantly decreased during the growth period, but this decrease was not correlated with the N uptake of the grass. The experiment shows that the N sources for grass differed during the experimental period with high nitrate concentrations at the beginning of the growth period causing N immobilization and later plants feeding from organic N. From the results it is concluded that the microbial assimilation (immobilization) of inorganic N depends on the presence of inorganic N, mainly nitrate, and on the supply of microbes with metabolizable organic carbon predominantly provided by plant roots. Therefore, it is suggested that the nitrate immobilization mainly occurs in the rhizosphere. EUF extractable soil N is an indicator for the mineralizable organic soil N; EUF Corg was correlated with the immobilization of inorganic soil N. Since, however, the correlation was only weakly significant further efforts are neccessary for elaborating a more precise extraction technique for the organic C which provides the energy for microbial mineral N assimilation.

Introduction

Efficient use of nitrogen fertilizer requires a knowledge about the potential of net N mineralization in soils. Net N mineralization results from the difference between the potential of mineralizable N and the potential for the immobilization of mineral N. Mengel *et al.* (1999) found in pot experiments with ryegrass that net N mineralization differed enormously for the same soils depending on whether the soils were fallow or grown with grass, being highest in the fallow soils. The authors concluded from their results that the N immobilization depended on the nitrate and on the available organic carbon, which in non-fallow soils may originate to a large extent by root excretion and root debris.

We therefore tested in our investigation whether the organic carbon extracted by EUF was related to nitrate immobilization. In addition the relationship between the EUF extractable N (mineral N, organic N = EUF-Norg) *versus* N uptake of the grass was studied. The research comprised also interlayer NH_4^+ because this may function as source or a sink for NH_4^+ (Scherer, 1993).

Materials and methods

Twenty soils were taken from the upper layer of arable fields. Their most important characteristics are shown in Table 1. Soils were dried at air, crushed, sieved and filled into Mitscherlich pots, each soil 4 replicates.

Table 1. Means and range of main characteristics of soils and means and range of EUF extractable nitrogen and carbon at the beginnning of the experiment.

Soil characteristics	Mean	Range
pH, $CaCl_2$ solution	7.4	6.9 - 7.8
Sand, g kg^{-1}	146	29 - 864
Silt, g kg^{-1}	669	90 - 857
Clay, g kg^{-1}	185	45 - 258
Total organic C, g kg^{-1}	10.7	5.4 – 17.9
Total organic N, g kg^{-1}	1.0	0.4 – 1.4
C/N ratio	10.6	8.6 – 13.9
EUF-NH_4^+, mg kg^{-1}	1.78	1.2 – 3.1
EUF-NO_3^-, mg kg^{-1}	27.9	9.0 – 64.4
EUF-organic N, mg kg^{-1}	17.9	12.4 – 24.4
EUF-organic C, mg kg^{-1}	304.2	167.6 – 544.7

For a normal growth of the plants mineral N should amount at least to 20 mg N kg^{-1} soil. This was the case for 19 soils; only one soil (sandy soil) received a N fertilizer dressing as $Ca(NO_3)_2$. Soils were incubated in pots for 14 days at 60% of the maximum water capacity (MWC). Soil samples of 40 g moist soil were taken from each pot on 05.05.1998 (day 0) and analyzed for EUF extractable N and C and interlayer NH_4^+. Soils were sown with 0.5 g seed/ pot of *Lolium multiflorum* (cv Aubade) at day 48. The grass was daily watered with demineralized water to 60% of MWC. Grass was harvested by cutting it directly at the surface of the soil at day 76, 106 and 155. After each cut soil samples were taken and analyzed for EUF extractable N and C. The dry matter of grass was ground and then analyzed for total N with a Macro-N analyzer. At the third cut also roots were carefully taken out of the soil and washed. N was determined in the dry root matter. Details of N analysis in the EUF fractions are described by Mengel *et al.* (1999). The organic carbon in the EUF extracts was colorimetrically determined by using an autoanalyzer. All EUF data presented below are pooled for two extractions. Interlayer NH_4^+ in soils was also determined at seeding and after third cut.

Results and discussion

From day zero (beginning of soil incubation in pots) to the day 48 (seeding) a significant increase in nitrate was found, followed by a steep decrease during the period of grass growth of the first cut (Fig. 1). Grass of the first cut (day 76) exclusively fed from nitrate and N uptake of the first cut was significantly correlated with the EUF extracted nitrate in the soil at seeding (r= 0.96***). On an average from this nitrate about half was taken up by the grass and half was assimilated (immobilized) by soil microorganisms. This immobilized N was correlated with the organic C extracted by EUF at seeding (r = 0.50*). At the first cut EUF-NH_4^+ and EUF-NO_3^- concentrations in soils were extremely low and remained at this low level throughout the experimental period. Therefore the grass of the second and third cut exclusively drew N from the organic soil N pool and N uptake of grass was correlated with EUF-Norg (r = 0.66**, second cut and r = 0.55*, third cut). The EUF extractable Norg significantly declined from seeding until the second cut. Interlayer ammonium significantly decreased during the growth period, but this decrease was not correlated with the N uptake of the grass. The contribution of interlayer NH_4^+ to total N uptake of grass was only about 7 %. Significant correlations were found between the cumulative N uptake of the grass *versus* EUF nitrate (r = 0.86***), EUF Norg (r= 0.72***), total EUF extractable N (ammonium + nitrate + Norg; r = 0.81***). The experiment shows that the N sources for grass differed during the experimental period with high nitrate concentrations at the beginning of the growth period causing N immobilization and later plants feeding from organic N. From the results it is concluded that the microbial assimilation (immobilization) of inorganic N depends on the presence of inorganic N, mainly nitrate, and on the supply of microbes with metabolizable organic carbon predominantly provided by plant roots. Therefore it is suggested that the nitrate immobilization mainly occurs in the rhizosphere.

References

Mengel K, Schneider B and Kosegarten H 1999 Z. Pflanzenernähr. Bodenk. 162, 139-148.

Scherer HW 1993 Europ. J. Agron. 2, 149-160.

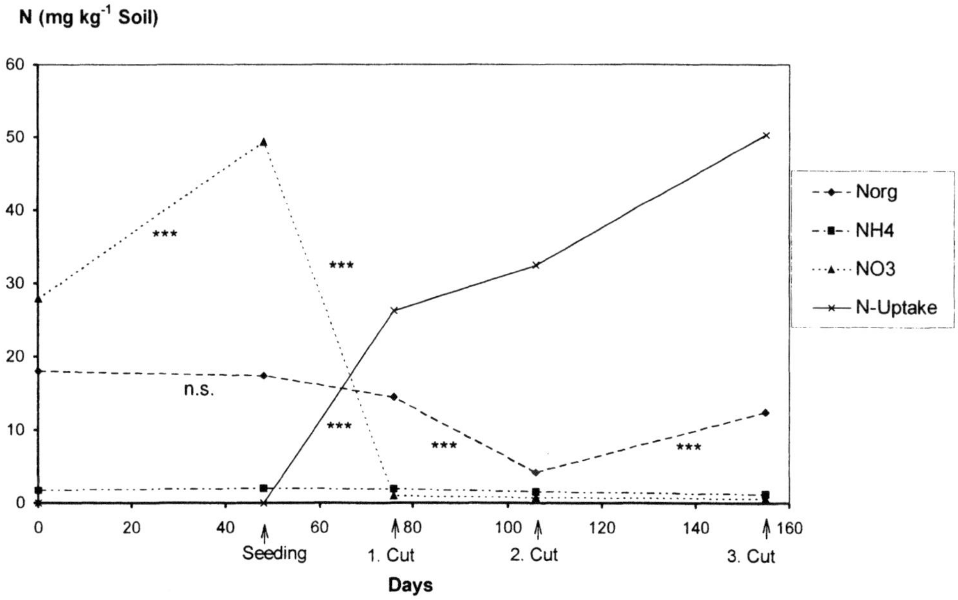

Figure 1. Course of EUF extractable N throughout the experimental period and cumulative N depletion of soils by three cuts of ryegrass. *** significant difference between cuts at $p < 0.001$. Means of 20 soils.

W. J. Horst et al.(Eds.), Plant nutrition – Food security and sustainability of agro-ecosystems. 750~751, 2001

Importance of EUF-extractable organic C for soil N mineralization and for N management of sugar beet

F. Kempl[1)], D. Steffens[1)], D. Horn[2)] and S. Schubert[1)]

[1)] *Institute of Plant Nutrition, Justus Liebig University, Heinrich-Buff-Ring 26-32, D 35392 Giessen, Germany*
[2)] *EUF-Arbeitsgemeinschaft, Marktbreiter Str. 74, D 97199 Ochsenfurt, Germany*

Key words: *Beta vulgaris*, carbon, EUF, N management, nitrogen, quality, sugar beet

Abstract

The quality of sugar beet (*Beta vulgaris*) is mainly influenced by N nutrition. Nitrogen mineralization is not only influenced by the amount of organic soil N, but also by organic soil C. The objective of our investigations was to determine the importance of organic C extracted by means of EUF for N mineralization and for N fertilizer recommendation of sugar beet. The importance of EUF-extractable organic C was investigated at 21 different sugar beet sites which were fertilized with 0 (=control treatment), 40, 80, 120, 160, and 200 kg N/ha. The experimental sites were located in southern Germany and in Austria.

The results of the field experiments showed that EUF-extractable organic C was a parameter which improved the estimation of the N fertilization requirement of sugar beet. The N fertilization requirement of sugar beet could be predicted with a fitted equation containing EUF-nitrate and EUF-N_{org} with $r^2 = 0.556^{***}$. The N fertilization requirement of sugar beet was better predicted with a fitted equation containing the additional parameter EUF-C_{org} with $r^2 = 0.830^{***}$. This indicated that EUF-extractable organic C is an additional parameter for improvement of N fertilizer recommendation of sugar beet.

Introduction

Nitrogen nutrition is one of the most important factors in sugar beet production. Therefore the estimation of N mineralization in soil should be considered in N fertilization.

N fertilization is either recommended according to the N_{min}-method (Wehrmann and Scharpf 1986) or to the electro-ultrafiltration (EUF) technique (Wiklicky and Németh 1981, Németh *et al.* 1991). However, the potential of N mineralization depends not only on soil organic-N, but also on soil organic-C (Körschens *et al.*, 1998). The objective of our research was to investigate the influence of EUF-extractable organic C in soil on N mineralization and for the estimation of sugar beet N requirements.

Materials and methods

In November 1998, soil samples were taken (0 - 30 cm depth) from sites in southern Germany and Austria. The soil samples were dried at 37 °C, ground to pass a 1 mm sieve and were extracted by means of EUF (Németh and Wiklicky 1982). In addition to N fractions (NO_3-N, N_{org}-N) the organic C in the EUF extracts was determined using a continuous flow autoanalyzer (Bran and Luebbe Co., Germany).

Field experiments were conducted at 21 sites in 1999. The design of the N fertilization experiments was either a randomized block design (Germany) or a latin square design (Austria). The N fertilizer application varied from 0 to 200 kg N/ha with increments of 40 kg N. Nitrogen fertilizer was applied as nitrochalk after emergence of the sugar beet. Application of more than 120 kg N/ha was split and the second dose was applied at the six-leaf stage. At the sites in Germany four replications per treatment were used, while in Austria 6 application were done.

Representative samples of beet roots and leaves were harvested by hand. The sugar concentration of the beets was determined by means of the polarization technique. Beet and leaf samples were dried at 105 °C to determine dry matter. After grinding, the N-concentration of the plant samples was determined by means of a makro-N analyzer (Foss Heraeus Co., Germany).

Sugar yield was calculated as the product of sugar concentration and beet fresh weight. Sugar yield was plotted against N fertilization using a yield function of Boguslawski and Schneider (1964) to determine the maximum yield of sugar. N-uptake was calculated as the product of beet and leaf fresh weight and dry matter content and N-content.

Data and the regression analysis for the relations between EUF-data and N requirement for maximum sugar yield were calculated using the statistics program SPSS.

Results and discussion

The mean values for EUF-NO_3 and EUF-N_{org} were 0.63 and 2.47 mg N/100 g soil and for EUF-C_{org} 33.34 mg C/100 g soil, respectively (Tab. 1). The EUF-NO_3 values in this investigation were comparable to those in previous experiments in November 1989 (Steffens *et al.*, 1990). However, the EUF-N_{org} concentrations were higher than at the sites analyzed by Steffens *et al.* (1990).

The mean sugar yield of the 21 sites increased with increasing N fertilization from 0 to 80 kg N/ha. Mean sugar yields were 135 and 152 dt/ha in the control and 80 kg N/ha treatments, respectively (Fig. 1).

Table 1. Mean values of EUF-NO_3-N, N_{org}-N and C_{org} (Σ of the 1. and 2. EUF-fraction) in Nov. 1998

n = 21	EUF-NO_3-N	EUF-N_{org}	EUF-C_{org}
	mg / 100 g soil		
Mean	0.63	2.47	33.34
Standarddev.	0.28	0.61	5.30

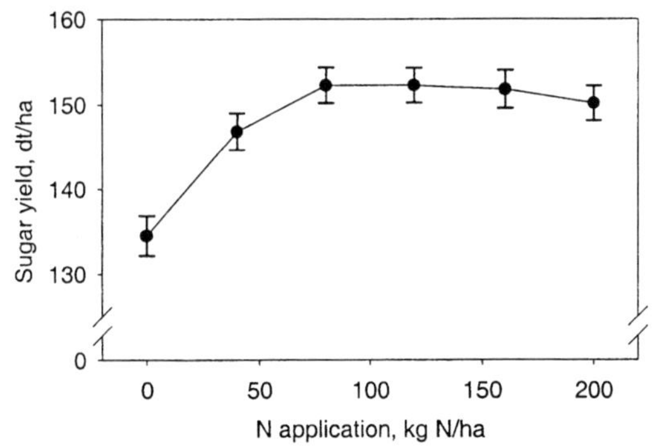

Figure 1. Effect of N fertilization on the mean sugar yield of sugar beet at 21 sites (± standard error).

The individual correlations between the N fertilization requirements for maximum sugar yield and the results of soil analysis were well fitted for EUF-C_{org} ($r^2 = 0.773$***), followed by EUF-N_{org} ($r^2 = 0.448$***) and EUF-NO_3 ($r^2 = 0.200$*).

In further calculations we used multiple regression analysis for the prediction of N fertilizer demand for maximum sugar yield. The results of this multiple regression analysis are shown in Tab. 2. The coefficient of determination was $r^2 = 0.556$*** for the relation between the N fertilization for maximum sugar yield and the EUF-N fractions (NO_3-N and N_{org}-N). The coefficient of determination increased to $r^2 = 0.830$*** if the EUF-extractable C_{org} was incorporated in the multiple regression analysis for prediction of N fertilization demand for maximum sugar yield.

Table 2. Relationship between predicted N fertilization for maximum sugar yield and measured EUF fractions. 21 field sites in 1998/99

$y = 257.57 - 54.47*NO_3\text{-}N - 44.41*N_{org}$ $r^2 = 0.556$***
$y = 368.62 - 14.26*NO_3\text{-}N - 20.18*N_{org} - 5.57*C_{org}$ $r^2 = 0.830$***

In further calculations, multiple regression analysis was employed for the prediction of the difference between N-uptake at maximum sugar yield and N-uptake of control. The coefficient of determination increased from $r^2 = 0.504$*** to $r^2 = 0.798$*** (Fig 2).

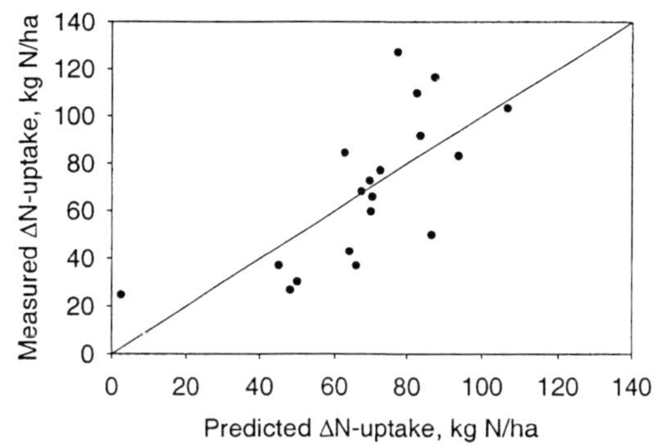

$y = 160.33 - 30.90*NO_3\text{-}N - 22.08*N_{org}$
$r^2 = 0.504$***

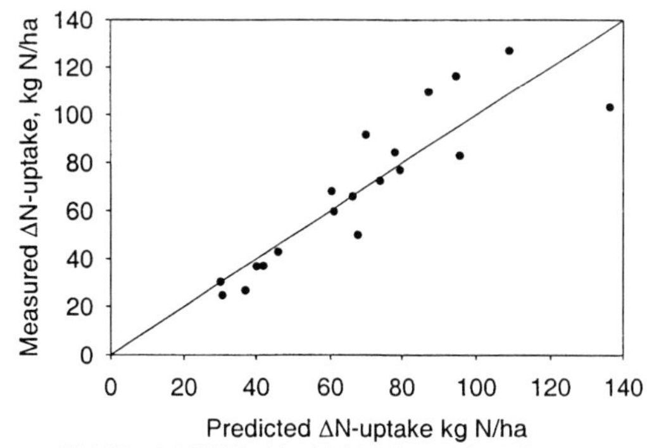

$y = 236.50 - 3.17*NO_3\text{-}N - 12.44*N_{org} - 4.04 * C_{org}$
$r^2 = 0.798$***

Figure 2. Relationship between predicted and measured ΔN-uptake of sugar beet (ΔN-uptake: Difference between N-uptake at maximum sugar yield and N-uptake of control treatment)

Our results indicated that organic C extracted by means of EUF can be an additional parameter for improvement of N management in sugar beet production. However, data presented here are from one growing season only and have to be evaluated in further growing seasons.

References

Boguslawsi E and Schneider B 1964 Z. Acker- und Pflanzenbau 119, 1-28.

Körschens M, Weigel A and Schulz E 1998 Z. Pflanzenernähr. Bodenk. 161, 409-424.

Németh K and Wiklicky L 1982 Zuckerind. 107, 958-962.

Németh K, Recke H and Heuer C 1991 Zuckerind. 116, 991-996.

Steffens D, Barekzai A, Bohring J and Poos F 1990 Agribiol. Res. 43, 319-329.

Wehrmann J and Scharpf HC 1986 Z. Pflanzenernähr. Bodenk. 149, 428-440.

Wiklicky L and Németh K 1981 Zuckerind. 106, 982-988.

W. J. Horst et al.(Eds.), Plant nutrition – Food security and sustainability of agro-ecosystems. 752~753, 2001

NIR-Spectroscopy to estimate soil nitrogen supply

B. Wagner, R. Gutser and U. Schmidhalter
Chair of Plantnutrition, Department of Plant Science, Technical University of Munich, Am Hochanger 2, Freising, D 85350, Germany, bwagner@weihenstephan.de

Key words: N-potential, NIR-spectroscopy, soil parameters

Abstract

The aim of this study is to establish the applicability of NIR-spectroscopy as a rapid method for an accurate estimation of N-potentials in agricultural soils. Furthermore, the usefulness of this method to determine other important soil parameters was tested. Three subsequent N-uptakes of by oat in rye grass in pot trials conducted with 106 soils were used to define short-term or long-term N-potentials. Using NIR-spectra (3850-10000 cm^{-1}), models were calculated to predict N-potentials and other soil-parameters (i.e. clay). The results demonstrate that the prediction of short-term N-potentials by NIRS-models is inadequate (R^2=0.33), whereas models for long-term N-potentials are good (R^2=0.77). In addition, valuable information concerning the N-turnover as N_t, C_{org} or soil texture (i.e. clay, silt) (R^2=0.91) can be obtained. NIR-spectroscopy is a useful tool to estimate long-term soil nitrogen supply.

Introduction

Owing to the rapid progress in data-processing during the last decade, the use of NIR-spectroscopy in chemical, biological or agricultural sciences has been enhanced. Referring to agriculture, NIR-spectroscopy is used for the rapid analysis of the quality of forages, rape seed or cereals. The subject of this study was to prove the applicability of NIR-spectroscopy for the fast and accurate determination of N-potentials in agricultural soils.

Materials and methods

Pot trial

From 1997 to 1999 four subsequent pot experiments were conducted with 106 soils representing a wide range of different composition and origin (Table 1).

Table 1. Soil characteristics

Parameter	Range	Parameter	Range
clay (%)	6 – 42	C_{org} (%)	0.59 – 4.24
silt (%)	9 – 79	N_t (%)	0.07 – 0.46
sand (%)	3 – 84	C/N	7.7 – 12.7
pH ($CaCl_2$)	4.7 – 7.5		

Owing to the broad origin (fields and long-term fertilising experiments) of the samples the NO_3-N-content of the soils differed considerably. Its influence on the N-potential was excluded with the first pot trial (08/97-10/97) (Figure 1) (results not shown). On soil samples, taken after the 1st trial (10/97), NIR-spectra were taken. The total N-uptake was defined as the criteria for the total N-potential, divided into short-term N-potential (i.e. N-uptake 3/98 – 6/98) and long-term N-potential (i.e. N-uptake 7/98 – 6/99), respectively (Figure 1).

Methods

NIR-spectra were obtained by using the Bruker/Vector 22N system (Bruker©, Ettlingen, Germany) equipped with a PbS detector. For the NIR-spectroscopy, soils samples (10/97) were sieved (2 mm) and dried (40°C, 24 h). Scanning was performed from 3850 – 10000 cm^{-1} (λ = 1000 – 2600 nm), 64 scans per sample, data were collected every 8 cm^{-1}. Five measurements were averaged to calculate a mean NIR-spectra. From all NIR-spectra, models for the N-potentials and other soil characteristics were developed based on the Partial Least Square regression and cross validation using Opus/Quant2-Software (Bruker©). Further, several standard methods as EUF-extraction (Nemeth, 1982) or determination of microbial biomass (Anderson and Domsch, 1978) were carried out and the data were also included with the N-potentials in regression equation to describe N-potentials.

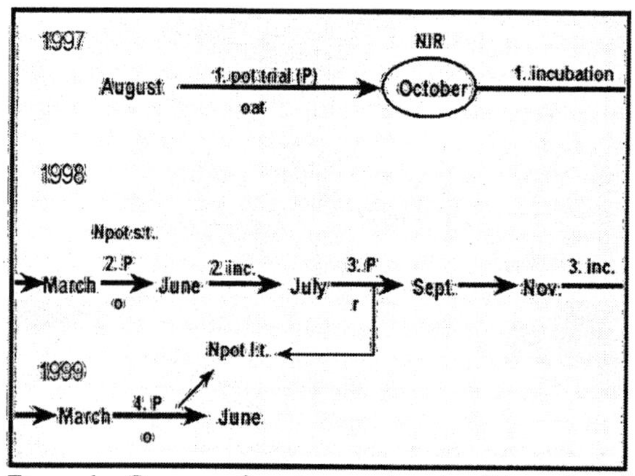

Figure 1. Course of pot experiments (P) and soil incubation periods (inc.), 1997 to 1999 (o = oat, r = rye grass), Npot s.t. = short-term N-potential, Npot l.t. =long-term N-potential

Results

The long-term N-potential was accurately predicted by the NIR-models (R^2=77 %), where as the prediction for the short-term N-potential was inadequate (Table 2). For the long-term N-potential the proportional mean prediction error was approx. 12 %. The total N-uptake was also well represented by the NIR-spectra.

Table 2. Results of cross-validation of the NIRS-models for the N-potentials (n=106).

	Range	R^2 (%)	PE (%)
Npot s.t.	0.4-4.2 mg N/100g d.s.	34	30
Npot l.t.	1.1-7.1 mg N/100g d.s.	77	12
Npot total	2.5-11.1 mg N/100g d.s.	63	15

(PE = ∅ prediction error)

Compared to the other tested methods for the evaluation of long-term and total N-potentials, the NIR-models showed the highest regression coefficient together with the lowest prediction error (Table 3).

Table 3. Regression coefficients and mean prediction errors of NIRS-models compared to models calculated with results of conventional methods (n=106).

		R^2 (%)	PE (%)
Npot l.t.	• NIRS	77	12
	• Norg (EUF)	66	14
	• N_t	71	13
	• microb.biomass	58	18
Npot total	• NIRS	63	15
	• Norg (EUF)	55	17
	• N_t	53	17
	• microb.biomass	50	18

Beside the development of prediction-models for N-potentials by the means of NIR-spectroscopy, the NIR-spectra were also computed for important soil parameters such as C_{org}, N_t and soil texture (Table 4). Especially for C_{org}, N_t, clay, and silt the prediction is very accurate. For sand, the regression coefficient is also very high, but the model has an inadequate prediction error.

Table 4. Results of cross-validation of the NIRS-models for soil texture and organic matter content.

Parameter	R^2 (%)	PE (%)
Clay (%)	91	11
silt (%)	91	15
Sand (%)	89	35
C_{org} (%)	90	12
N_t(%)	90	11

The presented NIR-models are based on cross-validations, i.e. all 106 samples were entered into the regression. The general applicability of NIR-prediction for C_{org} and N_t has been established by a validation with test samples (n=42, R^2=80 %, PE =9 %) (data not presented). The relation between N-potentials and N_t is quite close (Table 3), therefore NIR-spectroscopy is likely to work on the prediction of N-potentials. A conclusive evidence will only be possible by the conduction of further pot experiments with unknown test-samples.

By the means of adequate NIRS-models, information on important soil characteristics and on details of the N-cycle can be obtained by NIR-spectroscopy as an easy and rapid method for analysis.

Discussion

Only few information is available on the use of NIR-spectroscopy for the prediction of N-potentials. Börjesson *et al.* (1999) described the successful use of NIRS to predict the N-uptake on unfertilised plots, whereas Reeves *et al.* (1999) found no relationship between NIR-spectra and potential N-mineralisation in agricultural soils. Both studies dealt with soils taken from plots only at few locations. In our study, the locations were scattered over a broad area. This indicates, that NIR-spectroscopy can be generally used to predict long-term N-potentials in soils. The presented results possibly enable the classification of arable soils into different levels of N-potentials. However, the N-potential was measured through pot trials, the realisation of the potentials in field depends on environmental factors as climate or soil cultivation, for example, and has yet not been taken into consideration. To apply our results to agricultural practice, further field experiments on different soils conducted over several years are required.

The NIR-spectroscopy has already been demonstrated as an accurate method to obtain valuable information on soil texture and organic matter (Stenberg *et al.*, 1995), especially of soils of homogeneous origin (Ben-Dor *et al.*, 1995). These results have been confirmed by our studies with soils of very different origin and composition.

The stability of NIR-models over several years has not yet been demonstrated (Börjesson *et al.*, 1999). To establish a general applicability of NIR-spectroscopy for the determination of soil characteristics and N-potentials in agricultural soils further investigations are necessary.

References

Anderson JPE and Domsch KH 1978 Soil Biol. Biochem. 10, 215-221.

Ben-Dor E and Banin A 1995 Soil Sci.Am. 59, 364-372.

Börjesson T, Stenberg B, Linden B and Jonsson A 1999 Plant Soil 214, 75-83.

Nemeth K 1982 Plant Soil 64, 7-23.

Reeves JB III, McCarthy GW and Meisinger JJ 1999 J. Near Infrared Spectrosc. 7, 179-193.

Stenberg B, Nordkvist E and Salomonsson L 1995 Soil Sci. 159, 109-114.

W. J. Horst et al.(Eds.), Plant nutrition – Food security and sustainability of agro-ecosystems. 754~755, 2001

The new nitrification inhibitor DMPP (ENTEC®) for use in agricultural and horticultural crops – an overview

W. Zerulla, G. Pasda, R. Hähndel and A. H. Wissemeier
BASF AG, Agricultural Center Limburgerhof, D-67114 Limburgerhof, Germany, wolfram.zerulla@basf-ag.de

Key words: 3,4-dimethylpyrazole phosphate, DMPP, N losses, N efficiency, nitrification inhibitor

Abstract

Nitrification inhibitors constitute an efficient means to reduce the environmental problems caused by nitrogen in the nitrate form (nitrate leaching and denitrification losses). These effects are obtained by curbing the activity of the soil bacteria *Nitrosomonas spp.* responsible for the transformation of ammonium into nitrite and by thus retarding this process over a certain period of time.

DMPP (3,4-dimethypyrazole phosphate (ENTEC®) is a new and highly efficient nitrification inhibitor, which is effective at low application rates and can be added to fertiliser that contain ammonium or urea.

The addition of DMPP to fertilisers results in a number of advantages for farmers and vegetable growers: They can increase the efficiency of N fertiliser use by either reduced nitrogen rates or increased yield. Other studies have shown advantages for crop quality and the percentage of marketable produce.

Introduction

Nitrification inhibitors retard the natural transformation of ammonium into nitrate, by inhibiting the activity of *Nitrosomonas spp.* the bacteria responsible for the first step of the nitrification process (conversion of ammonium to nitrite).

By adding nitrification inhibitors to a fertiliser, the ammonium contributed by the fertiliser is "stabilised" in the soil for a certain period of time after its application. In consequence, as long as the nitrification is effectively inhibited, the soil ammonium content is increased and the nitrate content is reduced.

Since nitrate is seen as the major cause of the environmental problems which are related to soil N dynamics (nitrate leaching, denitrification), nitrification inhibitors can significantly contribute to the reduction of these problems.

Although several chemicals are known to reduce the rate of nitrification, only a few comply with the requirements for commercial products, for example N-serve [Nitrapyrine = 2-chloro-6-(trichloromethyl)pyridine] in North America or Didin [DCD = dicyandiamide] in Western Europe (Trenkel, 1997).

This paper describes the properties of the new nitrification inhibitor DMPP (3,4-dimethylpyrazole phosphate = ENTEC ®) by presenting a short number of representative experiments evaluating the effects of DMPP on soil ammonium content, nitrate leaching and N-efficiency.

Materials and methods

Four field trials were conducted to test the effect of DMPP on the soil NH_4-content under winter rape seed. 80 kg/ha N were applied at onset of vegetation. In a pot trial the effect of DMPP on N leaching after irrigation was tested. The effect of DMPP on yield and N-productivity had been tested in 50 field trials with different cereals under various soil and climatic conditions around Europe. The application rates for DMPP were 0.5 to 1.5 kg/ha. More details are given in the respective table.

Results

DMPP is a highly efficient nitrification inhibitor that delayed the transformation of ammonium to nitrite by *Nitrosomonas spp.* for 4 weeks or more in trials in Germany (Table 1). Similar results have been obtained in other parts of Europe, including Mediterranean regions (data not shown).

In model experiments in Mitscherlich pots DMPP significantly reduced nitrate leaching (Table 2), and this was confirmed in lysimeter experiments conducted under adverse environmental conditions in Spain (Serna *et al.*, 2000).

The use of DMPP-containing fertilisers improves fertiliser N-efficiency. A number of field trials of DMPP showed that the inhibitor increased fertiliser N efficiency by 7.2 to 16.7%, depending on the crop (Table 3).

Discussion

As a result of the above described and other trials (data not shown) mineral fertilisers only need to contain 1% active ingredient, based on their ammonium-N or urea-N content to inhibit the first step of the nitrification process in soil for 4 to 10 weeks. It can be used either on solid or liquid fertilisers or in slurry. The duration of effectiveness depends on climatic conditions, site characteristics and probably cultivated crop (Barth *et al.*, 2001; Pasda *et al.*, 2001).

The use of DMPP promises several advantages for farmers and vegetable growers who are likely to benefit from its use by either reduced nitrogen rates or increased yield (Table 3), by a lower number of fertiliser applications (Table 3) and - often – improved quality and increased proportion of marketable yield (Pasda *et al.*, 2001). Leaf vegetables show a darker green and thus look more attractive, several vegetable species have a lower nitrate content at harvest and starch content is often improved (Pasda *et al.*, 2001).

Besides the positive effects on the reduction of nitrate leaching (Table 2) the use of DMPP has other environmental benefits like the reduction of N_2O emission (Linzmeier *et al.*, 2001), without a negative effect on the methane sorption-capacity of the soil (Weiske *et al.*, 2001).

DMPP is not translocated within the soil profile. In a 3-year lysimeter-study at the Jülich Research Center using a slightly pseudogleyic brown earth from aeolic sand, Fettweis *et al.* (1999), could find no evidence of leaching of DMPP. Thus there is no spatial separation of the nitrification inhibitor from the fertiliser ammonium. The risk of DMPP being leached into the groundwater can thus also be excluded. Only mechanical incorporation into the ploughed soil layer by soil tillage could be possible.

Table 1. Effect of DMPP on soil NH_4-content (0 – 30 cm) after application of ammonium sulphate nitrate (ASN) and ASN + DMPP (average of 4 field trials, Germany)

Fertiliser	DMPP	NH_4-N (kg ha^{-1}) weeks after application			
		2	4	6	8
ASN	-	39	36	14	8
ASN	+	61	65	28	26

According to European legislation, DMPP is a new substance, and as such has been submitted to extensive toxicology and ecotoxicology tests. None of these assays indicates any problematic characteristics of DMPP, so that neither DMPP-containing fertilisers nor liquid DMPP formulations as additives for UAN (urea ammonium nitrate solution) or slurry need to be marked as hazardous substances. DMPP has both been declared under the Chemicals Act and been internationally registered according to fertiliser laws.

Table 2. Nitrate leaching under spinach after application of ASN and ASN + DMPP (soil: Ruchheim, L, pH ($CaCl_2$): 7.6; CEC (meq $100g^{-1}$ soil): 14; C_t (%): 1.2; N_t (%): 0.1; N-application: 0.75 g N pot^{-1}

DAT[1)]	ASN	ASN + DMPP
	(% of fertilised N in leachate)	
7[2)]	10.9	6.2
18[3)]	8.2	1.2
22[4)]	2.6	0.2
Σ	21.7	7.6

[1)] days after treatment
[2, 3, 4] Mitscherlich pots were watered with 20 mm[(2)], 10 mm[(3)], and 10 mm[(4)] of distilled water on the indicated DAT

References

Barth G, von Tucher S and Schmidhalter U 2001 IPNC Proceed., (this volume).

Fettweis U, Mittelstaedt W, Führ F and Schimansky C 1999 VDLUFA-Schriftenr. 52, 457-460.

Linzmeier W, Gutser R and Schmidhalter U 2001 IPNC Proceed., (this volume).

Pasda G, Hähndel R and Zerulla W 2001 IPNC Proceed., (this volume).

Serna MD, Banuls J, Quinones A, Primo-Millo E and Legaz F 2000 Biol Fertil Soils 32, 41-46.

Trenkel ME 1997 International Fertilizer Industry Association (Publ.), Paris.

Weiske A, Benckiser G and Ottow JCG 2001 IPNC Proceed., (this volume).

Table 3. Effect of different DMPP fertiliser application systems on yield and N-productivity of different crops (data from field-trials in Germany, Italy and Spain, 1997-1999)

Crop	N fertiliser	N rate (kg ha^{-1}) [1)]	Yield (t ha^{-1})	N-productivity (kg grain kg^{-1} N)	Improvement of N-productivity (%)
winter wheat	without N	-	4.47 a[2)]	-	
n=28	ASN	180	7.96 b	19.39	
	ASN + DMPP	180	8.21 c	20.78	7.17
wetland rice	without N	-	5.47 a	-	
n=8	Urea	120	7.20 b	14.42	
	urea + DMPP	120	7.49 c	16.83	16.71
grain maize	without N	-	6.69 a	-	
n=14	ASN	140	9.10 b	17.21	
	ASN + DMPP	140	9.34 c	18.93	9.99

[1)]timing of application: wheat: ASN: 80 kg/ha N onset of vegetation, 40 kg/ha N GS 30/32, 60 kg/ha N GS 49/51
ASN + DMPP: 100 kg/ha N at onset of vegetation, 80 kg/ha N GS 32/37
rice: ASN, ASN+DMPP: 120 kg/ha N at seeding
maize: ASN, ASN+DMPP: 140 kg/ha N at seeding
[2)] values with the same letter within a crop do not differ significantly (Duncan-Test, 5%)

W. J. Horst et al.(Eds.), Plant nutrition – Food security and sustainability of agro-ecosystems. 756~757, 2001

Influence of soil parameters on the efficiency of the new nitrification inhibitor DMPP (ENTEC®)

G. Barth, S. v. Tucher, and U. Schmidhalter

Department of Plant Sciences, Chair of Plant Nutrition, Technical University of Munich, Am Hochanger 2, 85350 Freising, Germany, tucher@weihenstephan.de.

Key words: adsorption, 3,4-dimethylpyrazole phosphate, nitrification inhibitor

Abstract

The influence of soil properties on the efficiency of 3,4-dimethylpyrazole phosphate (DMPP) as nitrification inhibitor was investigated in long term (25d) and short (5h) incubation experiments. The efficacy of DMPP was explained by a multiple regression analysis with the parameters soil clay and silt content, supplemented by additional factors (i.e. soil catalase activity and potential nitrification). Clay and silt content were related to the DMPP adsorption in soil ($R^2 = 0.64$). The adsorption of DMPP to inorganic soil surfaces is supposed to play a major role in DMPP efficiency.

Introduction

Nitrification inhibitors retard the first step of the oxidation of NH_4^+ to NO_3^-. The application of these inhibitors, like the recently developed 3,4-dimethylpyrazole phosphate (DMPP), represents a tool to increase N-fertiliser use efficiency, reducing N losses by NO_3^- leaching and by gaseous emissions from nitrification and de-nitrification (Linzmeier *et al.*, 1999; Weiske *et al.*, 1999).

Field studies gave evidence, that the efficacy of DMPP to inhibit the nitrification of ammonium fertilisers is soil influenced by conditions. Stabilisation of NH_4^+ (Linzmeier *et al.*, 1999) as well as positive yield effects (Pasda *et al.*, 2001) were more pronounced in light textured soils.

This study aims to investigate the influence of soil parameters on the nitrification inhibitory effect of DMPP.

Materials and methods

Long term (25 days) and short term (5 h) incubation experiments were conducted to study the influence of soil properties on the effect of the nitrification inhibitor DMPP on ammonium conversion.

Forth long term incubation experiment soil sampled from a loamy sand and a silty loam were equilibrated (1 week) to a soil matric potential of -50 kPa (non limiting for nitrifier activity), equivalent to water contents of 0.123 kg kg^{-1} (loamy sand) or 0.255 kg kg^{-1} (silty loam). Thereafter, fertiliser granules of ammonium-sulfate-nitrate with or without DMPP (Entec®) were added. Samples were incubated for 25 days at 25°C. NH_4^+ was determined colorimetrically by an indophenol blue method after extraction with 1 M KCl (1:2; w:v).

Short term incubation experiments were conducted with 22 soils covering a wide range of soil types (soil parameters see Barth *et al.*, 2001). The potential nitrification was determined according to Barth *et al.* (2001). The efficiency of the nitrification inhibitor DMPP was calculated by the formation of NO_2^-, (stabilised by $NaClO_4$) from added NH_4^+ (at a DMPP concentration of 5 mg DMPP kg^{-1} soil) compared with the NO_2^- formation without DMPP (termed "relative NO_2^- formation" in Figure 2).

DMPP adsorption to soil components was determined by shaking soil samples (20 g dry weight) with 100 ml 0.01 M $CaCl_2$ containing 0.2, 2.0, or 20.0 mg DMPP (Barth *et al.*, 2001) and was calculated as difference between initial and equilibrium DMPP concentrations, and related to soil dry weight (mean of three concentrations).

Coefficients of multiple regressions between relative NO_2^- formation or DMPP sorption and selected soil parameters (see Figure 3 and 4) were calculated with SAS version 8.01 (SAS Institute Inc. Cary, NC, USA).

Results and discussion

Ammonium fertiliser without DMPP was degraded within 25 days in both soils, the loamy sand and the silty loam (Figure 1). DMPP retarded the NH_4^+ oxidation in both soils, but the inhibition effect was much stronger for the loamy sand. This soil dependent higher efficiency of DMPP in light textured soils is in agreement with result from Pasda *et al.* (2001). In long term incubation experiments, this performance might be associated

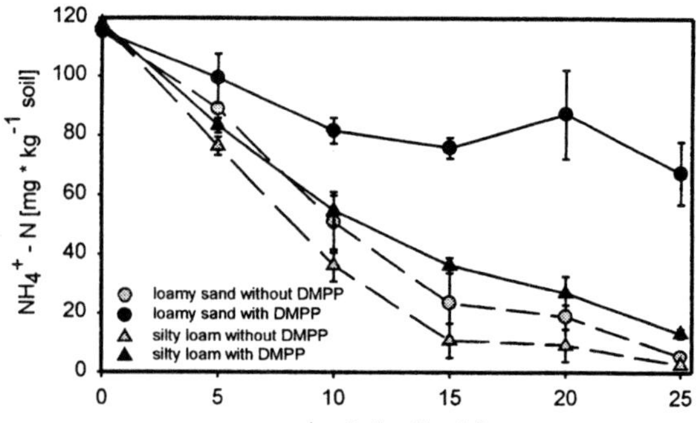

Figure 1. Effect of DMPP on ammonium degradation in two different soils at a soil matric potential of – 50 kPa. Error bars represent standard deviation.

with several interdependent factors, like differences in 1) the activity of the nitrifiers, 2) the degradation of

DMPP, 3) the availability of the nitrification inhibitor to microorganisms.

In agreement with theses results from the long term study, also in short term experiments, where DMPP degradation can be disregarded, the DMPP inhibition effect on the formation of NO_2^- was related to the soil type. The relative NO_2^- formation was more reduced in the *coarse textured soils (Figure 2).*

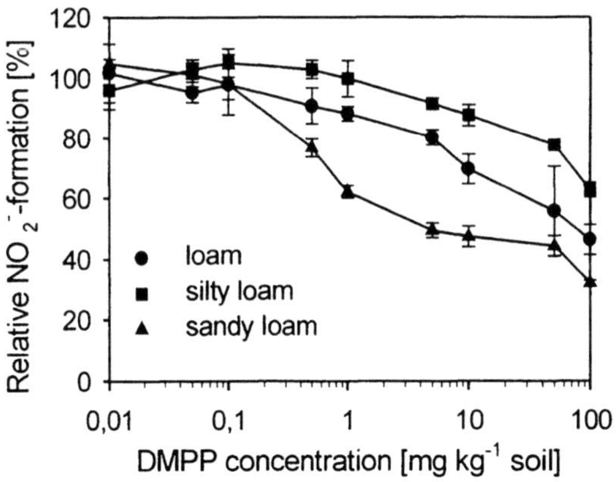

Figure 2. Influence of DMPP in short term incubation experiments on the relative NO_2^- formation. Three examples selected from 22 soils. Error bars represent standard deviation.

Soil texture proved to be a major factor to describe the inhibitory effect of DMPP in short term experiments, but NO_2^- formation was only slightly related to individual soil parameters (Barth *et al.*, 2001). In multiple regression analysis the efficiency of DMPP was best explained by supplementing textural parameters with additional factors, i.e. catalase activity and soil pH value (Barth *et al.* 2001) or catalase activity and potential nitrification, representing the potential activity of the nitrifiers (Figure 3).

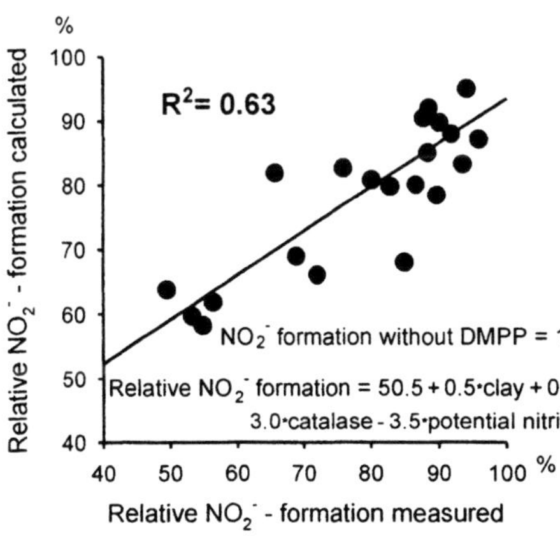

Figure 3. Influence of soil parameters on the relative NO_2^- formation in short term incubation experiments. Measured versus calculated values (DMPP concentration 5 mg kg^{-1} soil). Signifi-cance of the regression coefficient $p < 0.05$.

Soil components might act as binding partners for the adsorption of DMPP to soil surfaces and hence determine the availability of DMPP to microorganisms. Adsorption studies demonstrated, that about 8 to 30 % of the added DMPP were adsorbed to the soil. Regression analysis gave evidence, that 64 % of the DMPP adsorption was described by soil clay and silt content (Figure 4).

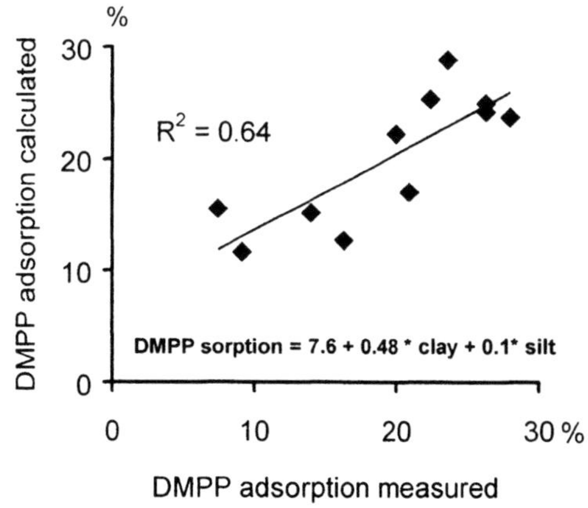

Figure 4. Adsorption of DMPP to soil. Relation to soil clay and silt content. Measured versus calculated values. Significance of the regression coefficient $p < 0.05$.

Adsorption of DMPP to soil surfaces is in agreement with results from Fettweis *et al.* (2001). In this study no DMPP was found in leachates collected from lysimeters.

Conclusions

Soil dependent differences in the efficiency of DMPP are related to soil type and the adsorption of DMPP to inorganic soil surfaces. Hence, textural properties, supplemented by additional factors as soil catalase activity, pH value and nitrifier activity, will play an important role in the prediction of the effect of DMPP on the inhibition of NH_4^+ oxidation. Further research is neccessary to evaluate the role of these parameters for the long term inhibition effect of DMPP.

Acknowledgement

This research was supported by Bundesministerium für Bildung und Forschung (BMBF), Bonn, Germany (project No. 02 WT 9659/0).

References

Barth G, von Tucher S and Schmidhalter U 2001 Biol. Fertil. Soils, (in press).

Fettweis U, Mittelstaedt W, Schimansky Ch and Führ F 2001 Biol. Fertil. Soils, in press.

Linzmeier W, Schmidhalter U and Gutser R 1999 VDLUFA Schriftenreihe 52, 485-488.

Pasda G, Hähndel R and Zerulla W 2001 Biol. Fertil. Soils, (in press).

Weiske A, Benckiser G and Ottow J C G 1999 VDLUFA Schriftenreihe 52, 517-520.

W. J. Horst et al.(Eds.), Plant nutrition – Food security and sustainability of agro-ecosystems. 758~759, 2001

The new nitrification inhibitor DMPP (ENTEC®) – Effects on yield and quality of agricultural and horticultural crops

G. Pasda, R. Hähndel, W. Zerulla
BASF AG, BASF Agricultural Center Limburgerhof, D-67114 Limburgerhof, Germany, gregor.pasda@basf-ag.de, reinhard.haehndel@basf-ag.de, wolfram.zerulla@basf-ag.de

Key words: 3,4-dimethylpyrazole phosphate, crop yield, crop quality, DMPP, N-fertiliser, nitrification inhibitor

Abstract

In the years 1997 to 2000, 161 field trials were conducted under various soil-climatic conditions in west and south Europe, in order to assess the effects of N-fertilisers with the new nitrification inhibitor DMPP (3,4-dimethylpyrazole phosphate) on yield and quality of various agricultural and horticultural crops. Results show that DMPP may increase the mean crop yield (grain yield: winter wheat + 0.24 t ha^{-1}, wetland rice + 0.43 t ha^{-1}, grain maize + 0.24 t ha^{-1}; tuber yield: potatoes + 1.9 t ha^{-1}, corrected sugar yield: sugar beets + 0.24 t ha^{-1}; biomass: carrots + 1.9 t ha^{-1}, lambs' lettuce + 2.6 t ha^{-1}, onions + 1.0 t ha^{-1}, radish + 4.6 t ha^{-1}, lettuce + 1.7 t ha^{-1}, cauliflower + 2.3 t ha^{-1}, leek + 3.1 t ha^{-1}, celeriac + 1.9 t ha^{-1}).

Introduction

The use of nitrification inhibitors (NI) in combination with nitrogenous fertilisers offers several environmental advantages, compared to the application of conventional N fertilisers. Especially, N losses by nitrate leaching and the volatilization of N compounds are reduced (Trenkel, 1997). The new nitrification inhibitor DMPP (= 3,4-dimethylpyrazole phosphate) which has been developed by BASF Aktiengesellschaft, Germany, seems to be especially effective in this respect (Serna *et al.*, 2000; Linzmeier *et al.*, 2001; Weiske *et al.*, 2001; Wissemeier *et al.*, 2001; Zerulla *et al.*, 2001). Therefore, the use of N-fertilisers with NI in intensive agriculture and horticulture should be beneficial for the environment. The objective of this study was to establish whether the new nitrification inhibitor DMPP also offers agronomic advantages when added to N-fertilisers. As it could be expected that agro-climatic factors would have a significant influence on the trial results (Trenkel, 1997), a great number of trials with different important agricultural and horticultural crops was conducted both in western and southern Europe.

Materials and methods

Various fertilisers containing ammonium nitrate (N, NPK) or urea together with a DMPP (= ENTEC®, registered trademark of COMPO GmbH & Co. KG, Münster, Germany) content of ~1 % relative to the content of ammonium-N or urea-N, were tested, compared to conventional fertilisers without DMPP and to a control without nitrogen. Amounts of N (except for the control without N), P, K, Mg and S were equal in all treatments. With the exception of nitrogen fertiliser application and harvest, all other crop management operations in the trial fields, such as choice of crop variety, soil tillage, basal fertilization (P, K, Mg), liming, irrigation, plant protection, growth regulator use, were carried out by the respective farmer, uniform over all treatments, according to Good Agricultural Practice so as to achieve the site-specific optimum. For all trials a randomized block design with four replications was used. Most of the trials were laid out in farmers' fields, some trials on experimental sites of BASF Aktiengesellschaft, Germany.

Results and discussion

In Table 1 the observed yield effects are summarized. On average, DMPP increased the mean yield substantially, especially in lambs lettuce, radish, cauliflower, leek, cereals and potatoes. In some crops, the same yield level as obtained with the control (fertiliser without DMPP) was achieved with one less N application, or with a reduced N application rate (data not shown). The positive effect of DMPP on crop yield was especially pronounced at sites with high precipitation rate or intensive irrigation, and/or light sandy soil (data not shown, compare also Barth *et al.*, 2001). DMPP had a negative effect on the biomass yield of nitrophilic crops with a relatively short growing period such as spinach. The application of DMPP containing fertilisers had a slight negative effect on the crude protein content of winter wheat (compare Linzmeier *et al.*, 2001). This can be explained by the advanced late topdressing in the case of DMPP containing fertilisers, and by a dilution effect due to the above mentioned yield increase. To produce high winter wheat yields with a high crude protein content, it is therefore advisable to use a combination of fertilisers with and without DMPP for the late topdressing (Linzmeier *et al.*, 2001). In vegetable crops and in potatoes, a lower nitrate content was frequently found after DMPP application, compared to fertilization without DMPP (data not shown). Vegetable crops, which had received DMPP, showed a clearly darker green leaf coloration than the control without DMPP. Leaf color is an important quality characteristic of directly marketed vegetables.

Table 1 Yield results with fertiliser with DMPP for different agricultural and horticultural crops in comparison to the control (fertiliser without DMPP) and an unfertilized treatment (1997-2000)

Crop	No. of trials	Yield	Yield (t/ha) No fertiliser	Fertiliser without DMPP	Fertiliser with DMPP	Additional yield with DMPP
carrots	6	root	65.7	82.5	84.4	+ 1.9
lambs lettuce	10	leaf	2.4	6.4	9.0	+ 2.6*
onions	7	bulb	55.6	74.4	75.4	+ 1.0
radish	3	root	11.4	25.6	30.2	+ 4.6*
lettuce	18	head	12.1	36.7	38.4	+ 1.7
cauliflower	9	curl	19.3	34.5	36.8	+ 2.3*
leek	8	plant	36.3	44.8	47.9	+ 3.1*
celeriac	6	tuber	41.3	69.8	71.7	+ 1.9
winter wheat	35	grain	4.53	7.95	8.19	+ 0.24*
grain maize	14	grain	6.69	9.10	9.34	+ 0.24*
wetland rice	9	grain	5.35	6.93	7.36	+ 0.43*
potatoes	21	tuber	34.0	45.7	47.6	+ 1.9*
sugar beets	15	corrected sugar	8.45	10.46	10.70	+ 0.24*

*significant differences between without and with DMPP (Duncan-Test, $\alpha = 5$ %)

When looking at the absolute amounts of N which are lost by leaching and volatilization under field conditions, it becomes obvious that their reduction alone by application of an NI (Serna *et al.*, 2000; Linzmeier *et al.*, 2001; Weiske *et al.*, 2001) is frequently not sufficient to explain the observed yield differences between the treatments with and without an NI. An additional explanation for the observed NI effects could be the effects of partial NH_4^+ nutrition of the plants. As NH_4^+-N is supplied to the crop at a higher rate and for a longer period of time when fertilized with NI, a substantial amount of NH_4^+-N rather than NO_3-N will be taken up by the plants. This partial ammonium nutrition offers several advantages: Firstly, plants have to spend less energy for ammonium uptake than for that of nitrate, especially at high NH_4^+ supply. Secondly, ammonium can be used directly for protein metabolism. Thirdly, it has been established in various studies that ammonium has a positive effect on the synthesis of the polyamines, cytokinins and gibberellins (e.g. Wang and Below, 1992; Gerendás and Sattelmacher, 1995; Ali *et al.*, 1998). Higher contents of these two phytohormones, which retard senescence and thus chlorophyll degradation, could also give an explanation why plants fertilized with DMPP have a deeper green color than plants without DMPP. Further, NI induced partial ammonium nutrition may decrease soil pH in the rhizosphere. This low pH should improve the availability of other nutrients, especially micronutrients, for plant uptake. Another reason for the positive yield effect of NI is the improved N supply caused by different fertiliser application schedules (Linzmeier *et al.*, 2001).

References

Ali IA, Kafkafi U, Ymaguchi I, Sugimoto Y and Inanaga S 1998 J. Plant Nutr. 21, 1463-1481.

Barth G, von Tucher S and Schmidhalter U 2001 IPNC Proceed., (this volume).

Gerendás J and Sattelmacher B 1995 Z. Pflanzenernaehr. Bodenk. 158, 299-305.

Linzmeier W, Gutser R and Schmidhalter U 2001 IPNC Proceed., (this volume).

Serna MD, Banuls J, Quinones A, Primo-Millo E and Legaz F 2000 Biol. Fertil. Soils 32, 41-46.

Trenkel M 1997 International Fertilizer Industry Association (Publ.), Paris.

Wang X and Below FE 1998 Crop Sci. 36, 121-126.

Weiske A, Benckiser G and Ottow JCG 2001 IPNC Proceed., (this volume).

Wissemeier AH, Linzmeier W, Gutser R, Weigelt W and Schmidhalter U 2001 IPNC Proceed., (this volume).

Zerulla W, Pasda G, Hähndel R and Wissemeier AH 2001 IPNC Proceed., (this volume).

W. J. Horst et al.(Eds.), Plant nutrition – Food security and sustainability of agro-ecosystems. 760–761, 2001

The new nitrification inhibitor DMPP (ENTEC ®) allows increased N-efficiency with simplified fertilizing strategies

W. Linzmeier, R. Gutser and U. Schmidhalter
Chair of Plant Nutrition, Department of Plant Sciences, Technical University of Munich, Am Hochanger 2, Freising, D-85350, Germany, schmidhalter@weihenstephan.de

Key words: 3,4-dimethylpyrazole phosphate (DMPP), gaseous N-losses, nitrate leaching, nitrogen fertilization

Abstract

In short-term field trials and a long-term lysimeter experiment simplified fertilizing strategies with stabilized fertilizers were tested and yield and N-losses (leaching, N_2O-emissions) were measured. Compared to conventional fertilizer practice, strategies with stabilized fertilizers resulted in higher yields together with lower gaseous N-losses in field trials. In the long-term lysimeter study, stabilized fertilizers showed also higher yields and a reduction of NO_3^--leaching. Simplified fertilization strategies with stabilized fertilizers (fewer applications, enhanced fertilization in earlier growth stage) lead to the dual benefits of increased yield and a reduced risk of N-losses.

Introduction

An optimized nitrogen fertilization strategy aims to combine high yield and good quality of product with minimized risks of nitrate leaching and gaseous N-losses. The form of applied mineral N-fertilizer (NH_4^+ or NO_3^-) influences nitrogen utilization and efficiency (Trenkel, 1997). With 3,4-dimethylpyrazole phosphate (DMPP) a new nitrification inhibitor is available - ENTEC®. Stabilized fertilizers enable the development of new specific application strategies with fewer applications and an enhanced N-fertilization in earlier growth stages. These strategies have the benefits of improved yield and reduced N-losses by nitrate leaching, compared to conventional fertilization practice (Gutser, 1999; Linzmeier *et al.*, 1999). In this paper, we compare simplified fertilizing strategies with stabilized fertilizers with a conventional strategy (i.e. non-stabilized fertilizer in three split applications) regarding yield, N-uptake, and N-losses determined in short-term field experiments and long-term lysimeter studies. Differences are due to the fertilizer system and not to the fertilizer alone.

Materials and methods

The field trials were conducted on two locations near Freising, Germany, brown earth soils with silty loam (10 % clay, 26 % silt) or loamy sand (24 % clay, 51 % silt) (annual precipitation 800 mm, mean annual air temperature 7.4°C).

For winter wheat (Triticum aestivum cv. Astron) 3 simplified fertilizing strategies with stabilized ammonium sulfate nitrate (ASN; NH_4-N 75 %, NO_3-N 25% of total N , addition of DMPP 0,3 %) surface broadcast as granules were compared with a conventional fertilizing strategy with calcium ammonium nitrate (CAN, NH_4-N 50 %; NO_3-N 50%). All treatments received a total of 160 kg N ha^{-1} (Table 1), applied on the soil surface. In 1999, N_2O-emissions (closed-chamber technique) were determined for the treatments II and IV with ^{15}N-labelled fertilizers applied in liquid form (Linzmeier *et al.*, 2001). NH_3-losses after N-fertilization were studied in a wind tunnel system. The data are presented by Linzmeier *et al.* (1999) and in this issue by Wissemeier *et al.* (2001).

Table 1. N-fertilization and splitting of tested strategies.

		Growth stages[1]			
		early spring	30/31	32	49/51
	Treatment	kg N ha^{-1}			
I	no N	-	-	-	-
II	CAN	70	40	-	50
III	ASN/DMPP	160	-	-	-
IV	ASN/DMPP	110	-	50	-
V	ASN/DMPP	70	90	-	-

[1] according to Federal Biological Research Center for Agriculture and Forestry (1997)

In a long-term lysimeter study in Freising (1982 – 1998, silty loam, pH 6.4) CAN was compared to the stabilized fertilizer ASN/DCD (3 % Dicyandiamide) at two N-levels. The number of applications for ASN/DCD was fewer than for CAN. Yield, N-removal of plants, and nitrate leaching were measured. Crop rotation was sugar beet, winter wheat, winter barley, from 1994 with catch crops after winter barley (Gutser, 1999).

Results and discussion

Yield and quality

At both locations, the stabilized ASN resulted in higher yields compared to CAN (1997: treatment V, high N-contribution during shooting; 1998: treatments III and IV, high N-contribution at early spring) (Table 2). These results were influenced by the occurrence and quantity of rain in both years (data not shown).

However, crude protein contents in grain were reduced by stabilized fertilizers (Table 2). In order to achieve high grain crude protein levels a late N application with a conventional fertilizer (e.g. CAN) is recommended.

Table 2. Grain yield (Y) and crude protein content (P)[1] of winter wheat receiving DMPP-fertilizers compared to CAN (=100) on two locations and three years. Significant differences are indicated by an asterisk.

		Treatment					
		II	I	III	IV	V	LSD 5%
fertilization		CAN	no N	ASN/DMPP			
silty loam		% of treatment II					
1997	Y	69,3[2]	46	106	106	113*	13
	P	11,2[2]	80	89	92	92	-
1998	Y	78,7	40	110	102	96	12
	P	12,5	78	88	93	96	-
1999	Y	78,4	48	99	110*	104	8
	P	12,3	82	86	90	93	-
loamy sand							
1997	Y	69,2	55	101	112*	118*	11
	P	13,3	88	88	84	88	-
1998	Y	74,7	64	115*	111*	95	11
	P	14,6	70	95	97	102	-
1999	Y	82,1	50	101	105*	103	5
	P	12,9	86	87	86	85	-

[1] N content in grain x 5,7
[2] absolute: yield (dt/ha) and crude protein (% of dry matter), resp.

Gaseous losses

The short term N_2O-emission after N-application was determined for a simplified (ASN/DMPP, 2 applications) and a conventional strategy (CAN, 3 applications) (Fig.1).

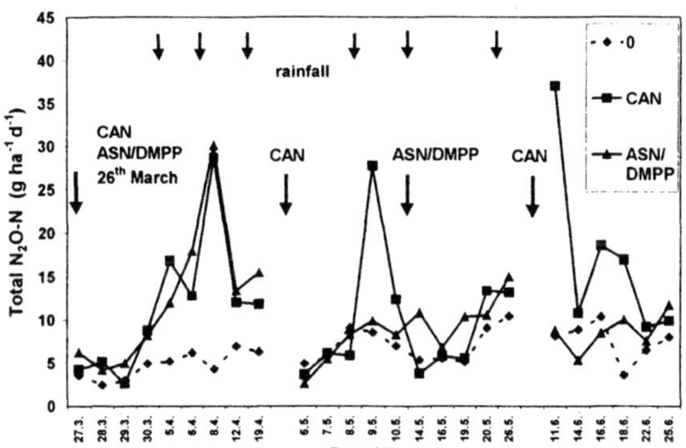

Figure 1. Total N_2O emission of treatments I, II, and IV on a silty loam soil in 1999.

For the stabilized fertilizer, the N_2O-emission was decreased by 20 % (Figure 1). ^{15}N labelling of the N-fertilizers showed that 10-40 % of the total N_2O-emission was attributed to fertilizer-N and 60-90% originated from soil-N (Linzmeier *et al.*, 2001). The N-pool of the soil represents an important source for N_2O-losses. The influence of fertilizers on the microbial N-turnover is likely to depend on the form of application (liquid or granules) (Gutser, 1999) and the chemical form of applied N (NO_3^- or NH_4^+) (Jenkinson *et al.*, 1985). Due to high mobility of nitrate, a larger area in the soil is probably influenced by nitrate-fertilization than by ammonium, which is easily adsorbed to clay minerals (Linzmeier *et al.*, 2001). For fertilizers with nitrification inhibitors, the reduced N_2O-emissions (see Figure 1) are probably mainly due to their lower NO_3^--N content, the inhibition of nitrification leading to longer persistence of ammonium and the reduced number of applications.

Long-term lysimeter studies

In the long-term lysimeter study particularly at high N-application stabilized fertilizers increased yield compared to the conventional fertilizer CAN (Table 3). Furthermore, for stabilized fertilizers nitrate-leaching was reduced. Yield increases were dependent on the crop (winter barley 12%; winter wheat 7%, sugar beet 3% (n.s.)) (Gutser, 1999). No differences in N-removal were determined at equal N-levels (Table 3). Comparable N-removals are due to lower N-contents in grain for the stabilized fertilizer treatment. Furthermore, this fertilizer shows a slightly higher N-immobilisation than conventional NH_4-fertilizer.

Table 3. Efficiency of NH_4^+-stabilized N-fertilizer in a lysimeter experiment from 1982 to 1998 at 0, 117, and 168 kg N ha^{-1}. Different letters within rows indicate significant differences.

N-fertilization		yield (grain/ beet)	removal	leaching	leaching water
			kg N $ha^{-1}a^{-1}$		mg NO_3^- l^{-1}
N_0	Control	61	44	30	37
N_{117}	CAN	=100 a[2]	97	29 a[2]	47
	ASN/NI[1]	101 a	97	27 a	46
N_{168}	CAN	106 b	121	35 b	59
	ASN/NI[1]	111 c	121	29 a	52

[1] NI = DCD   [2] homogeneous groups (Duncan-test)

According to our results, simplified strategies of stabilized fertilizer (fewer applications, enhanced fertilization in earlier growth stage) decrease N-losses and increase yield compared to conventional fertilizers.

References

Federal Biological Research Center for agriculture and Forestry 1997 Growth stages of mono- and dicotyledonous plants BBCH-Monograph. Blackwell Wissenschaftsverlag, Berlin, Wien.

Gutser R 1999 VDLUFA Schriftenr. 52, 465-468.

Jenkinson DS, Fox RH and Rayner JH 1985 J. Soil Sci. 36, 425-466.

Linzmeier W, Gutser R and Schmidhalter U 2001 Biol. Fert. Soils, (in press).

Linzmeier W, Schmidhalter U and Gutser R 1999 VDLUFA-Schriftenr. 52, 485-488.

Trenkel M 1997 *In* Improving fertilizer use efficiency. Ed International Fertilizer Industry Association. Paris.

Wissemeier A, Linzmeier W, Gutser R, Weigelt W and Schmidhalter U 2001 Proceed. IPNC Colloquium.

W. J. Horst et al.(Eds.), Plant nutrition – Food security and sustainability of agro-ecosystems. 762~763, 2001

A new nitrification inhibitor (DMPP) improves the nitrogen fertilizer efficiency in citrus-growing systems

J. Bañuls, A Quiñones, E. Primo-Millo and F Legaz

Instituto Valenciano de Investigaciones Agrarias, Departamento de Citricultura y Otros Frutales, Apartado Oficial 46113 Moncada, Valencia, Spain, mjbanuls@ivia.es

Key words: citrus, 3,4-dimethylpyrazole phosphate, nitrate leaching, nitrification inhibitor, nitrogen fertilizer efficiency

Abstract

The objectives of this work were to evaluate the inhibitory action on nitrification of 3,4-dimethylpyrazole phosphate (DMPP) added to ammonium sulphate (AS) applied as a fertilizer to *Citrus* plants, and its effects on plant growth, nitrogen uptake and mineral-N in the soil. In a greenhouse experiment, 2 g N as AS either alone or with 1% DMPP (relative to NH_4^+-N) was applied six times at 20-day intervals to plants grown in 14-l pots. The addition of DMPP to AS resulted in higher levels of NH_4^+-N and lower levels of NO_3^--N in the soil during the whole experimental period. Nitrogen concentration in leaves and roots was higher with the application of AS+DMPP than with AS alone. The NO_3^--N concentration in drainage water was lower in the AS plus DMPP- treated pots. The results indicate that the DMPP nitrification inhibitor improved N fertilizer efficiency and reduced nitrate-leaching losses by retaining the applied N in the ammoniacal form.

Introduction

Ammonium-based nitrogen (N) fertilizers are frequently used in *Citrus* production. The NH_4^+ released by solubilization of these N fertilizers is rapidly transformed into NO_3^-. Part of this NO_3^- can be lost through leaching. This loss contributes to NO_3^- pollution of groundwater and also accounts for a low efficiency of N fertilization in this crop (Feigenbaum *et al.*, 1987). The use of nitrification inhibitors may retain N in the NH_4^+ form that is less susceptible to leaching and thus minimize N loss (McCarty and Bremner, 1990). In a previous set of experiments in *Citrus* cultivated under glasshouse conditions, it was shown that the nitrification inhibitor dicyandiamide (DCD) improved N-fertilizer efficiency (Serna *et al.*, 1994). The purpose of this work was to evaluate the performance of DMPP added to ammonium sulphate to inhibit nitrification and to reduce N loss in *Citrus* plants.

Materials and methods

Two-year-old Valencia late orange plants (*Citrus sinensis* (L.) Osbeck) grafted on citrange Carrizo rootstock (*C. Sinensis x Poncirus trifoliata*) were grown under glasshouse conditions by using a drip irrigation system. Individual plants were grown in 14-l pots filled with a sandy-loam soil (76.4% sand, 16.6% silt and 7% clay), pH=7.8, low levels of $CaCO_3$ (<1%) and organic matter (<0.5%). Fertilization was carried out with 2 g N as ammonium sulphate (AS) either alone or with 1% DMPP (BASF, Germany) with respect to NH_4^+-N. N fertilizer was applied 6 times (12 g N/pot in total) at 20 days intervals between July and October. Unfertilized control plants were also included. Four uniform plants were used for each treatment. Soil samples were taken from 0 to 0.20-m-deep-layer, every 20 days (4 days after each N application and 4, 24 and 44 days after the last N application) and NH_4^+-N and NO_3^--N concentrations measured. The drainage water of each pot was collected and ammonium and nitrate concentrations were determined. After 140 days, plants were harvested and separated into leaves, stems and roots and total N in the tissue was measured.

Results and discussion

Growth and nitrogen content in plants

Dry matter of leaves, stem and roots increased in response to treatment with AS and the inclusion of DMPP to AS led to a further increase in root weight.

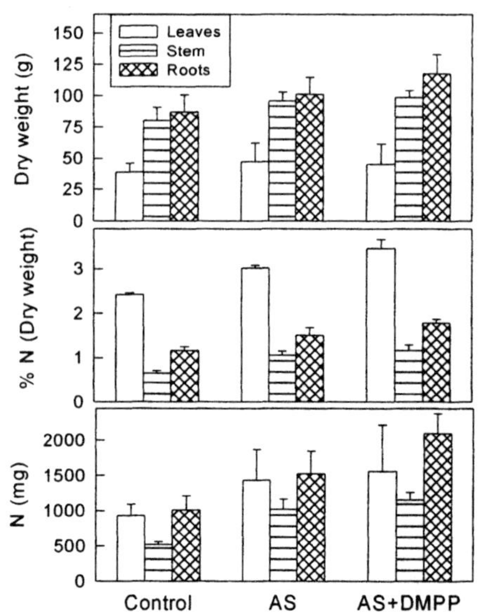

Figure.1. Dry weight and N content in different fractions of the plants. Values are means of four replicates and bars show SD.

However, none of these effects reached significance in the 140 days of the experiment (Figure 1). %N followed a similar trend, although the inclusion of DMPP led to a further increase in roots and also in leaves. N content in roots increased significantly with the addition of DMPP to AS. N balance calculations indicated that uptake of fertilizer N accounted for 19.5% and 12.7% of N applied as AS+DMPP and AS treatments, respectively.

Changes in NH_4^+-N and NO_3^--N concentrations in soil

Four days after the first application, the concentration of NH_4^+ was slightly higher in the soil treated with DMPP than in soil that had received AS alone, the opposite occurred with NO_3^- levels (Figure 2). In the determinations performed 20, 40, 60, 80 and 100 days later, after the successive N applications, ammonium levels increased progressively in the soil treated with DMPP. NH_4^+ concentrations fell rapidly after the last application (at day 100), reaching the level of the unfertilised soil 40 days later. In both AS and AS+DMPP treatment, NO_3^- concentrations in soil increased progressively during the experimental period, although NO_3^- levels were lower in the DMPP-treated soil.

The mineral-N (NO_3^--N plus NH_4^+-N) contents in the soil after the last irrigation (at day 140) were 291, 4066 and 4592 mg N per pot for the control, AS and AS+DMPP treatments respectively. Calculating the mineral N derived from the Fertilizer by subtraction of that supplied in the control, 31.5 % and 35.8% of N applied as and AS+DMPP, respectively, remained in the soil at the end of the experiment.

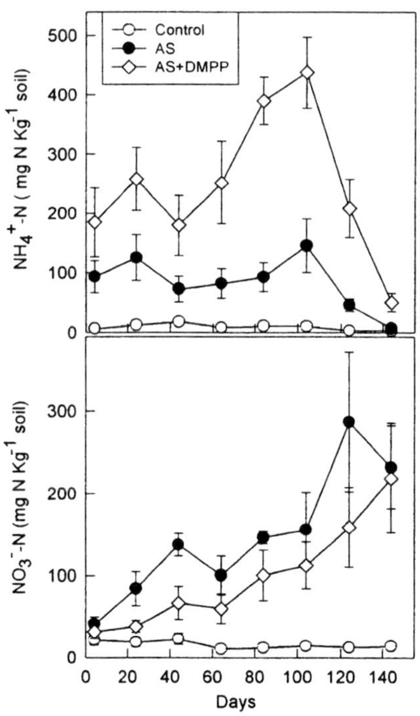

Figure. 2. Changes in NH_4^+-N and NO_3^--N concentrations in the soil after successive N fertilizer applications

Nitrogen losses in the drainage water

Drainage water recovered from fertilized pots contained elevated nitrate levels and very low levels of ammonium (Figure 3). The total amount of NO_3^- leached from the AS+DMPP treatment was much lower than that found in the AS treatment. Leaching resulted in accumulative total N losses of 756, 5190 and 2730 mg N per pot for control, AS and AS+DMPP treatments respectively. Thus, subtracting the loss in the control, N leached from AS and AS+DMP represented 37.0% and 16.5% of the N applied respectively.

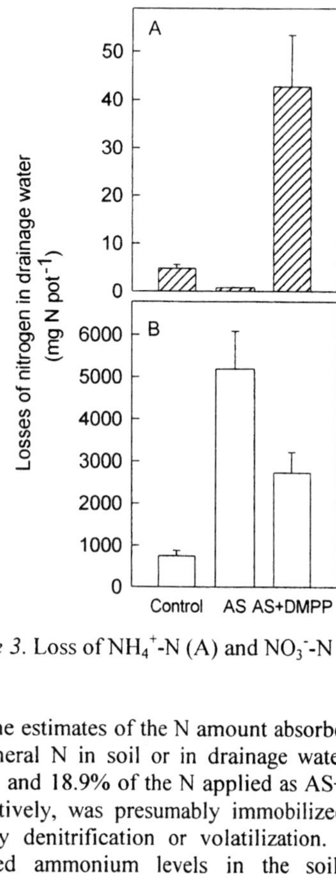

Figure 3. Loss of NH_4^+-N (A) and NO_3^--N (B) in drainage water.

The estimates of the N amount absorbed by plants and of mineral N in soil or in drainage water indicated that 28.2% and 18.9% of the N applied as AS+DMPP and AS respectively, was presumably immobilized in the soil or lost by denitrification or volatilization. It appears that elevated ammonium levels in the soil may increase immobilization of N because the micro-organisms that prefer NH_4^+ rather than NO_3^-.

The results of the experiment indicate that the DMPP reduced nitrate leaching and improved the efficiency of the N fertilizer uptake by *Citrus* plants.

References

Feigenbaum S, Bielora IH, Erner Y and Dasberg S 1987 Plant Soil 97, 179-187.

McCarty GW and Bremner JM 1990 Biol. Fertil. Soils 9, 252-256.

Serna MD, Legaz F and Primo-Millo E 1994 Soil Sci. Soc. Am. J. 58, 1817-1824.

W. J. Horst et al.(Eds.), Plant nutrition – Food security and sustainability of agro-ecosystems. 764~765, 2001

Field evaluation of DMPP as a nitrification inhibitor in the area irrigated by the Canal d'Urgell (Northeast Spain)

I. Carrasco[1,2] and J. M. Villar [1]

[1] *Environment and Soil Science Department, University of Lleida, Av. Rovira Roure 177, Lleida, 25198, Spain*

[2] *COMPO Agricultura S.L. Paseo de Gracia 99, Barcelona, 08008, Spain, israel.carrasco@compo.es*

Key words: DMPP, N use efficiency, nitrification inhibitor, nitrogen fertilisation, pig slurry, soil mineral N

Abstract

Nitrification inhibitors (NIs) can play an important role in maintaining soil N in the form of ammonium, as they inhibit activity of the *Nitrosomonas* bacteria. A study was conducted in the area irrigated by the Canal d'Urgell (Northeast Spain) to assess whether the use of DMPP with mineral and organic fertilisers presented environmental, productive and economic advantages for irrigated maize and wheat. The results of this study have demonstrated that DMPP enhanced the presence of NH_4^+-N in the soil, but was more important in winter conditions, where four months after DMPP application, 30% of soil mineral N still remained in this form. Treatments without DMPP had higher NO_3^--N in the soil profile for most of the growing season, indicating a greater risk of NO_3^- leaching. Therefore, DMPP could be an interesting N management practice for reducing N loss in so-called "Nitrate Pollution Vulnerable Zones".

Introduction

In the area irrigated by the Canal d'Urgell (Northeast Spain), the N applications using mineral N and pig slurry often exceed the crop requirements. As a consequence, there is clear experimental evidence of N loss (leaching, denitrification, etc.) to the environment (Teira, 1998; Ferrer, 1999). The first way to reduce these losses is by splitting the application of N fertiliser, but this is not easy to achieve for several crops, such as maize. Another way of reducing the risk of N losses involves preserving soil mineral N (SMN) as ammonium (Ball-Coelho and Roy, 1999), which is less susceptible to losses. It has been suggested that nitrification inhibitors (NIs) can play an important role in maintaining soil N in the ammonia form, inhibiting activity of the *Nitrosomonas* bacteria (Prasad and Power, 1995; Trenkel, 1997). A large number of chemicals have been reported to be NIs, but of these, only Nitrapiryn (NP), Dicyandiamide (DCD) and 3,4-dimetilpirazolphosphate (DMPP) are widely commercialised. Compared to other NIs, DMPP has the best inhibition effect (Wissemeier and Weigelt, 1999). Therefore, smaller doses of DMPP are required to obtain the same inhibition effect, which results in a reduced risk to the environment and in lower costs. The objective of this study was to assess how the use of DMPP with mineral and organic fertilisers affected soil mineral N profiles and yields, N uptake and the efficiency of N use, and whether it presented environmental, productive and economic advantages for irrigated maize and wheat.

Materials and methods

The study was conducted in the area irrigated by the Canal d'Urgell (Northeast Spain) on fields with winter wheat and maize. Maize experiments have been carried out at two different sites. In 1998, five fertiliser treatments were applied before planting: 100 and 200 kg N ha^{-1} either as ammonium nitrate-sulphate (ANS) + DMPP (ENTEC 26®) or as urea. In addition there was also a control without any N fertilisation. Three new fertiliser treatments were added in 1999: 100 and 200 kg N ha^{-1} as urea + DMPP applied before planting and 200 kg N ha^{-1} as ANS (50 % applied before planting and the rest at 6 leaves stage). In the 1999-2000 wheat growing season, a field experiment was conducted with the following treatments: a control, 30 m^3 pig slurry ha^{-1}, 30 m^3 pig slurry +1 L DMPP ha^{-1}, 30 m^3 pig slurry ha^{-1}+sidedress ANS, 30 m^3 pig slurry ha^{-1} + sidedress ANS+1 L DMPP ha^{-1}. In all experiments, soil samples were collected at different depths before sowing and during the growing season in order to determine N-NO_3^- and N-NH_4^+. Plants were also sampled during the growing season to determine dry matter production, plant N content and plant N uptake. A randomised block design with three replications was used in all of experiments.

Results

Maize experiments with inorganic fertiliser

The results of this study have demonstrated that DMPP enhanced the presence of NH_4^+-N in the soil (Figure 1). However, this effect was only detectable during the first 40 days after application. Apparent recovery of N fertiliser was low in all treatments (Table 1). Also, highest agronomic efficiencies (higher than economic threshold) were obtained with ANS + DMPP (200 kg N ha^{-1}) in 1998 and with UREA + DMPP (200 kg N ha^{-1}) and ANS + DMPP (200 kg N ha^{-1}) in 1999.

Winter wheat experiment with pig slurry (PS)

The maintenance of SMN as ammonium was more important during winter, where four months after DMPP application, 30% of SMN still remained in this form (Figure 2). Results presented in Table 2 show that sidedress mineral N applications did not result in significant wheat yield increases. However, in order to reach a similar N content for grain, applying only PS+

DMPP at preplanting had the same effect as applying extra N at the end of the winter. These results also show that the application of DMPP did not result in clear yield increase, although it resulted in a more efficient crop N uptake from the soil, as reflected by uptake efficiency. Total N uptake in PS+DMPP treatment was 11 % higher than using only pig slurry (Table 2).

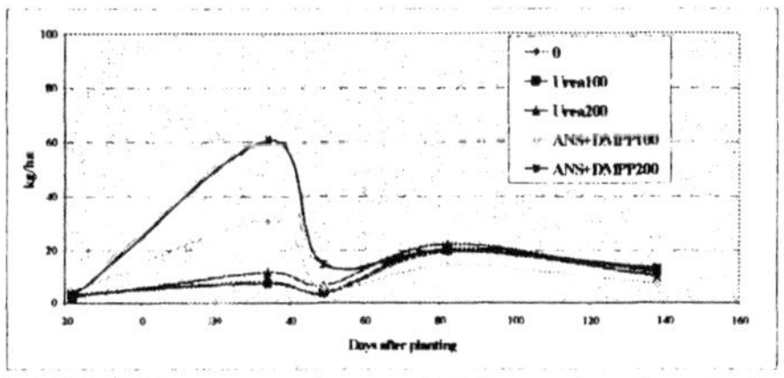

Figure 1. NH_4^+-N content in the 0-30 cm depth soil during 1998 maize experiment.

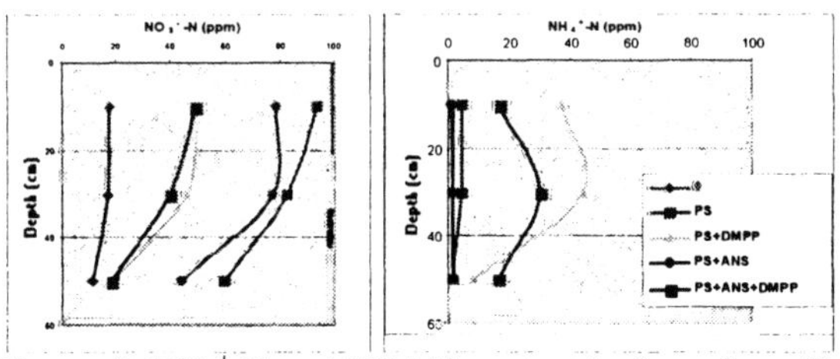

Figure. 2. NH_4^+-N and NO_3^--N content in the soil profile before sidedress N applications in the winter wheat experiment

Discussion

The experiments with mineral fertilisers applied to maize confirmed that even though the inhibition effect of DMPP was restricted to a period of 40 days, it affected the soil NO_3^--N and NH_4^+-N profiles and prevented accumulation of NO_3^--N in deeper soil layers during the crop season. The observed NUE indexes supported the theory that using DMPP enhances N recovery and profitability with each unit of fertiliser applied. However, more field experiments should be conducted to confirm these promising environmental, productive and economic results. The experiment with PS applied to winter wheat demonstrated that when DMPP was added, SMN could be maintained as ammonium for up to 4 months after its application. Lower soil temperatures prolonged the efficiency of DMPP. Treatments without DMPP had higher NO_3^--N in the profile for most of the growing season, indicating a greater risk of NO_3^- leaching. A general improvement in NUE was observed following the addition of DMPP to PS, suggesting that this could be an interesting N management practice for controlling N loss in so-called "Nitrate Vulnerable Zones".

References

Ball-Coelho BR and Roy RC 1999. Nutrient Cycling in Agroecosystems 54, 73-80.

Ferrer F 1999. Ph D Thesis. University of Lleida, Spain.

Prasad R and Power JF1995 Adv. Agron. 54, 233-281.

Teira-Esmatges MR, Van Cleemput O, and Porta J 1998 J. Environ. Qual. 27, 687-697.

Trenkel ME 1997 Controlled-Release and Stabilized Fertilisers in Agriculture. IFA, Paris.

Wissemeier A and Weigelt W 1999 Düngen mit einer neuen Technologie. BASF Aktiengesellschaft, Limburgerhof.

Table 1. Dry matter of maize, total N uptake and N use efficiencies for field experiments in 1998 and 1999. Values followed by the same letter within a column do not differ significantly (P=0,05).

	Treatment (kg N applied ha^{-1})	Corn yield (kg ha^{-1})	Total N uptake (kg ha^{-1})	Apparent N Rec. (%)	Agronomic Efficiency (kg kg^{-1})	Economic Threshold (kg kg^{-1})	NUE[c] (kg kg^{-1})	NUE[d] (kg kg^{-1})	NUE[e] (kg kg^{-1})
1988 Exp.	0	12880 b	236 b	-	-	-	-	-	54,9
	UREA 100	12630 b	263 ab	28	< 0	3	126,3	2,63	48,3
	UREA 200	13372 ab	266 ab	16	2,5	3	66,9	1,33	50,2
	ANS+DMPP 100	13414 ab	246 b	11	5,3	7	134,1	2,46	54,5
	ANS+DMPP 200	14530 a	286 a	26	8,2	7	72,6	1,43	50,8
1999 Experiment	0	8564 b	169 b	-	-	-	-	-	50,9
	UREA 100	9144 ab	183 ab	14	5,8	3	91,4	1,83	50,9
	UREA 200	10215 ab	184 ab	7	9,1	3	51,1	0,92	56,2
	ANS+DMPP 100	8167 b	166 b	< 0	< 0	7	81,7	1,66	49,0
	ANS+DMPP 200	10527 a	198 ab	15	10,6	7	52,6	0,99	53,2
	UREA+DMPP 100	8973 ab	189 ab	20	4,1	4	89,7	1,89	48,0
	UREA+DMPP 200	10829 a	221 a	26	12,2	4	54,1	1,10	49,2
	ANS (100+100)	10419 a	209 ab	20	10,1	5,3	52,1	1,04	49,9

Table 2. Dry matter of wheat, total N uptake, N in the grain and N use efficiencies for field experiment in 1999-2000. Values followed by the same letter within a column do not differ significantly (P=0,05)

Treatment	Wheat yield (kg ha^{-1})	Total N uptake (kg ha^{-1})	N in the grain (%)	Apparent N Recovery[a] (%)	Agronomic Efficiency[b] (kg kg^{-1})	NUE[c] (kg kg^{-1})	NUE[d] (kg kg^{-1})	NUE[e] (kg kg^{-1})
0	4293 b	76 d	1,47 c	-	-	-	-	56,5
Pig slurry (PS)	6717 a	153 c	1,73 b	28,7	9,0	24,9	0,57	43,9
PS+DMPP	7116 a	170 b	1,81 ab	34,9	10,5	26,4	0,63	41,8
PS+ANS	7289 a	185 a	1,82 ab	34,2	9,4	22,8	0,58	39,3
PS+DMPP+ANS	7390 a	191 a	1,85 a	36,0	9,7	23,1	0,60	38,7

[a] ($Nuptake_T$ - $Nuptake_0$) / Fertiliser N applied [b] (Grain $yield_T$ – Grain $yield_0$) / Fertiliser N applied
[c] Grain yield / Fertiliser N applied [d] N uptake / Fertiliser N applied [e] Grain yield / N uptake

W. J. Horst et al.(Eds.), Plant nutrition – Food security and sustainability of agro-ecosystems. 766~767, 2001

The new nitrification inhibitor DMPP – effects on gaseous emissions (N_2O, CO_2, CH_4) from soil under field conditions

A. Weiske, G. Benckiser and J. C. G. Ottow
Institute for Applied Microbiology, Justus-Liebig University, Heinrich-Buff-Ring 26-32, D-35392 Giessen, Germany, Johannes.C.Ottow@agrar.uni-giessen.de

Key words: ammonium oxidation, carbon dioxide fluxes, ClMP, DCD, DMPP, methane oxidation, nitrification inhibitor, nitrous oxide

Abstract

From an ecological viewpoint using a nitrification inhibitor (NI) with ammonium based fertilizers may be one potential management strategy to lower the fluxes of trace gases thus mitigating their undesirable effects. During 3 years of repeated application in the field, the effects of the new NI's DMPP (3,4-dimethylpyrazole phosphate) and ClMP (4-chloro-3-methylpyrazole) on net fluxes of N_2O, CH_4 and CO_2 from soil were investigated in comparison to DCD (dicyandiamide). Compared to the control plots without NI, the average N_2O emissions of DMPP, ClMP and DCD treated plots were 49, 26 and 25 % lower. Furthermore, DMPP apparently stimulated CH_4 oxidation throughout the 3 growing seasons by 28 % in comparison to control. Surprisingly, our data suggest that all 3 NI's affect C-mineralisation as quantified by CO_2 emissions. The NI's decreased the release of CO_2 on average of the 3 years by 7 % (DCD), 6 % (ClMP) and by 28 % (DMPP). Altogether, DCD and ClMP reduced the global warming potential (GWP) by 7 % and DMPP by 30 %, respectively. Results are site and experiment specific.

Introduction

In recent years, the use of specific nitrification inhibitors (NI) in agriculture and horticulture has increased again. NI's are of interest because certain compounds are highly effective in blocking N_2O emissions from nitrification (nitritation) at extremely low concentrations when applied to soils in conjunction with N fertilizers. Most of the fertilizer N applied to arable soils in form of ammonium or ammonium-producing compounds is usually oxidized quite rapidly to nitrate by nitrifying microorganisms. New fertilizer strategies such as ammonium granules formulated with specific NI are being developed to optimize the N-use-efficiency by crops and to decrease nitrate losses by leaching and N_2O emissions through nitrification and denitrification (Trenkel, 1997). Substantial progress has been made towards understanding the biochemistry of autotrophic NH_3 oxidation and the possible modes of action observed with the various specific inhibitors (McCarty, 1999). Among numerous chemical substances, several of these potent inhibitors of nitritation (such as nitrapyrin and etridiazole) may affect both ammonium monooxygenase (AMO) as well as methane monooxygenase (MMO), presumably because of the close structural relationship between AMO and MMO. The new heterocyclic NI's 3,4-dimethylpyrazole phosphate (DMPP) and 4-chloro-3-methylpyrazole (ClMP) are promising NI's in developement by the BASF AG, Agricultural Center Limburgerhof, D-67114 Limburgerhof. In a 3 year field experiments (1997-1999) the effect of the new NI's DMPP and ClMP on N-transformations, on the release of N_2O and CO_2 as well as on CH_4 oxidation was examined in comparison to dicyandiamide (DCD).

Materials and methods

The experiments were carried out on an allochtone brown earth (Fluvisol) at the Experimental Station "Weilburger Grenze" of the Justus-Liebig University in Giessen. The topsoil (0-35 cm) was a clayey loam (clay 31 %, silt 60 % and sand 5 %) with a pH (H_2O) 6.0-6.8. For each experimental site four microplots were established inside larger experimental plots with 4 repetitions. The NI's were formulated on fertilizer granules (DMPP ~1.6 %, ClMP ~0.9 % and DCD ~16 % to NH_4^+-N; BASF Agricultural Center, Limburgerhof) of ammonium sulphate nitrate (ASN, 26 % N: 18.5 % NH_4^+-N and 7.5 % NO_3^--N) and manually applied in springtime to barley (*Hordeum vulgare* L.; 90 kg ASN-N ha^{-1}) in 1997, to maize (*Zea mays* L.; 160 kg ASN-N ha^{-1}) in 1998 and to winter wheat (*Triticum aestivum* L.; 180 kg ASN-N ha^{-1}) in 1999 in one dose. ASN without NI's was used as control. On all microplots the N_2O emissions, N_2O retained in soil, NH_4^+ decrease, NO_2^- and NO_3^- production, grain yields as well as degradation kinetics of DCD and DMPP were measured throughout the cropping periods in 1997-1999. The effect of DMPP, ClMP and DCD on non-target processes of the C-cycle was examined by measuring methane oxidation and CO_2 emissions. Gases (N_2O, CO_2, CH_4) were measured by use of a closed chamber method (Hutchinson and Mosier, 1981). At each experimental site (4 replicate plots) steel chambers (40 × 40 cm) were permanently implanted into the soil. The gas sampling plots were kept fallow to quantify gas emissions from soil but not from plants. Gas samples were collected two times weekly between 14^{00}-15^{00} h from the chamber atmosphere and analysed using a gas chromatography system designed for simultaneous analyses of N_2O and CO_2 (ECD) as well as CH_4 (FID) in the same sample.

Results and discussion

In the 3 year field experiment the treatments with either DMPP, ClMP or DCD had no significant effect on the grain yield of summer barley (1997) and maize (1998). However, yield of winter wheat was affected in 1999 (DMPP +0.24, DCD +0.30, ClMP –0.23 t ha^{-1}; statistically not significant). The inhibitory effect of DMPP on N_2O emissions of the N fertilized plots ranged from 41 ± 6.7 % in 1997, to 47 ± 3.6 % in 1998, and to 53 ± 2.6 % in 1999. On average over the 3 years, DMPP decreased the N_2O emissions by 49 % while the efficiency of ClMP and DCD was significantly lower: DCD 30 ± 7.5 % (1997), 22 ± 4.6 % (1998) or 29 ± 5.2 % (1999) (mean 26 %) and of ClMP 38 ± 10.2 % (1997), 14 ± 4.9 % (1998) or 30 ± 4.1 % (1999) (mean 25 %). With regard to methane, all NI treated and control plots acted as sinks rather than sources throughout the three cropping periods. In contrast to the treatments with ClMP and DCD, the plots which received DMPP even seemed to stimulate methane oxidation (ClMP –10 %, DCD +1 %, DMPP +28 %). Surprisingly, all 3 NI's seemed to decrease the CO_2 emissions. The release of CO_2 was reduced significantly on average of the 3 years by 7 % (DCD), 6 % (ClMP) and by 28 % (DMPP). From these data it seems as if ClMP and DCD but particularly DMPP may affect C-mineralisation in soil. So far, laboratory experiments on soil respiration with the same soil and NI's could not confirm the field experiments (unpublished results). Although the concentrations of CO_2 are considerably higher than those of N_2O or CH_4 it is necessary to take all relevant trace-gas fluxes into account because the latter gases have significant higher impact on the greenhouse effect than CO_2. The global warming potential (GWP) of methane and nitrous oxide is 21 and 310 times, respectively, the GWP of CO_2 when calculated over a 100-year period (IPCC, 1996). If expressed in terms of CO_2 equivalents by multiplying emission data and the GWP factors DCD as well as ClMP reduced the total release of trace gases (not significantly) by only 7 % (Table 1). DMPP, however, minimized the GWP significantly ($p < 0.001$) by 30 %. Further, DCD, DMPP and ClMP diminished the amount of N_2O retained in soil by 52, 61 and 66 %. The concentrations of ammonium remained unaffected by all NI's whereas the amounts of nitrite diminished in the plots treated with DCD by 25 %, with DMPP by 20 % and with ClMP by 38 %. Nitrate concentrations in soil were in all NI treatments about 23 % lower than in the control. Although DMPP was applied at rates approximately 10 times lower than DCD, DMPP was degraded in soil more slowly than DCD in all experiments. Throughout the growing seasons 1997 to 1999 DCD diminished rapidly within 3 months whereas DMPP persisted for 5 months. At the end of the growing seasons, DMPP was still detectable at concentrations of about 16 % (1997), 6 % (1998) and 15 % (1999), respectively, of the original amount. The higher persistence of DMPP may also account for its lasting inhibitory effect on the release of N_2O in comparison to DCD, particularly towards the end of the cropping period.

Table 1. Effect of DCD, ClMP and DMPP in comparison to the fertilized control without NI on total CO_2, CH_4 and N_2O emissions as well as on equivalent CO_2 concentrations calculated following the recommendations of the IPCC (1996)

	CO_2	CH_4	N_2O	CO_2 equivalent[1]
Control	5840594	-130	828	6094553 **(100 %)**
DCD	5456339	-130	614	5643788 **(93 %)**
DMPP	4185470	-166	425	4313752 **(70 %)**
ClMP	5476759	-117	625	5668147 **(93 %)**

[1] $= CO_2 + (CH_4 \times 21) + (N_2O \times 310)$

The results presented here are site as well as experiment specific and cannot be generalized so far. Final conclusions must await further field and laboratory experiments. Especially, the inhibitory effect of the NI's on C-mineralisation should be studied both in further model experiments as well as in field experiments under various soil conditions. However, if DMPP should reduce N_2O, CH_4 and CO_2 production from all fertilized arable soils this NI could be a useful and cost-effective strategy in mitigating trace gas fluxes.

References

Hutchinson GL and Mosier AR 1981 Soil Sci. Soc. Am. J. 45, 311-316.

IPCC 1996 *In* Climate Change 1995 Eds. JT Houghton, LG Meira Filho, BA Callander, N Harris, A Kattenberg and K Maskell. p. 22. Cambridge University Press, Cambridge.

McCarty A 1999 Biol. Fertil. Soils 29, 1-9.

Trenkel ME 1997 Improving fertilizer use efficiency – Controlled-release and stabilized fertilizers in agriculture. International Fertilizer Industry Association, Paris, France.

W. J. Horst et al.(Eds.), Plant nutrition – Food security and sustainability of agro-ecosystems. 768~769, 2001

Nitrous oxide emissions from injected ^{15}N-labelled cattle slurry into grassland soil as affected by DMPP nitrification inhibitor

K. Dittert[1)], R. Bol[2)], D. Chadwick[2)] and D. Hatch[2)]

[1)] *Institute of Plant Nutrition and Soil Science, Kiel University, Olshausenstr. 40, 24118 Kiel, Germany, kdittert@plantnutrition.uni-kiel.de*

[2)] *Institute of Grassland and Environmental Sciences (IGER) North Wyke, Okehampton, Devon, EX20 2SB, United Kingdom*

Key words: denitrification, DMPP, nitrification, nitrification inhibitor, nitrous oxide, slurry injection, stable isotope labelling

Abstract

On a grassland soil in Devon, South England, a field experiment on the effect of DMPP on nitrous oxide (N_2O) emissions from injected dairy slurry was carried out. ^{15}N enriched slurry was used and the isotopic label in the translocated soil nitrogen and in the emitted N_2O-N were studied.

Major N_2O emissions occurred during the first ten days after slurry injection in August 2000. N_2O emission rates and ^{15}N content of the emitted N_2O-N were highest over the slurry injection slots, showing a steep decrease towards the non-treated centre between slurry injection slots. DMPP proved to be very efficient in reducing N_2O emissions. At a rate of 2 kg DMPP ha^{-1} the total amount of N_2O emitted was reduced by 32 % as compared to the DMPP free slurry injection plots.

The ^{15}N label of the N_2O-N revealed that only slurry derived N_2O-N emissions were affected by DMPP, while soil nitrogen derived N_2O-N emissions were similar in +DMPP and control. The isotopic label furthermore revealed two distinct peaks of N_2O emissions within the first 10 days. There is evidence that the first peak is denitrification related, while the second peak seems to originate from nitrification, since it coincides with the first appearance of substantial amounts of nitrate in the soil.

Introduction

Among the trace gases responsible for the atmosphere's depletion in ozone, N_2O is one of the most important contributors to the greenhouse effect. Even though there are uncertainties about the release from fossil fuel or industrial processes, soils are seen as major sources. Particularly under agricultural land-use, N_2O emissions are increased, as a result of nitrogen fertiliser and livestock manure inputs, delivering either the substrate, or the substrate and energy source for nitrification and denitrification, respectively.

The use of nitrification inhibitors such as di-methyl pyrazole phosphate (DMPP) has been shown to successfully reduce N_2O emissions from mineral fertilisers (Linzmeier *et al.*, 1999). The aim of the present study was therefore, to evaluate DMPP as a tool to control N_2O emissions from slurry application in grassland systems. Stable isotope labelling (^{15}N) was used in order to study slurry N transformations in the soil and to differentiate between N_2O emissions emanating from the soil and/or from the slurry N pool.

Materials and methods

In August 2000, a field experiment was carried out on a humid grassland site at IGER North Wyke, UK. The soil type was a clay loam, characterised by a high carbon content of 5 % and moderate acidity of pH 6. The grassland sward was previously fertilised with 102 kg N ha^{-1} in March and 100 kg N ha^{-1} in June, and two weeks prior to the start of the experiment the sward was cut for silage. The mean annual rainfall is 1030 mm, and the actual climatic conditions were typical for the location and time of the year, such that during the first 10-day period of the experiment there were some warmer days, followed by a cooler weather period.

A ^{15}N labelled slurry was produced by spiking dairy cattle slurry with ^{15}N enriched ammonium sulphate (10 at% ^{15}N) and separated into two aliquots. To one aliquot, DMPP nitrification inhibitor was added at a rate of 66g DMPP m^{-3}, which is equivalent to 2 kg ha^{-1}. The KCl extractable ammonium nitrogen of the slurry N had an isotopic label of 4.29 at%^{15}N. On August 15, the slurry was injected into the grassland soil to a mean injection slot depth of 4 ± 2cm. Slots were 30cm apart.

Sampling of N_2O emissions was carried out in the early afternoon of every second day starting at day 4. Closed-chambers of 5cm width and 50cm length were used, which were centred above the slot, or at 5 cm, 10 or 15 cm distance from the slurry injection slot. Gas samples were taken into evacuated serum stopper bottles of 100 ml volume. Concentrations and ^{15}N contents of N_2O were determined on a PDZ Europa Trace Gas II system linked to a 20/20 isotope ratio mass spectrometer and calibrated with a reference gas of known N_2O concentration. On the basis of its isotopic enrichment (APE), the rate of N_2O-N derived from slurry was calculated according to equation 1 (Nason and Myrold, 1991). Soil nitrogen-derived N_2O-N emission rates were calculated as the difference between total and slurry nitrogen-derived N_2O-N emission rates.

$$(1) \qquad \%N_{dff} = \frac{ape_{sample}}{ape_{fertiliser(slurry)}} * 100$$

Results

DMPP, the new nitrification inhibitor, reduced the total N_2O emissions by 32 % from 1.6 to 1.1 kg N_2O-N ha^{-1}, and this effect was most pronounced in the first 10 days after slurry injection. N_2O emissions were concentrated on the injection slots, regardless of the addition of nitrification inhibitor. Also, in the period after day 10, emissions from the slurry injection slots significantly dominated, but the absolute rates were much lower.

Highest N_2O emissions were measured on day four after slurry injection, which was the first sampling day of the study (Figure 1). After 10 days, total emissions decreased rapidly, which might be attributed to falling temperature. Furthermore after day 5-6 of the experimental period, the soil nitrate content increased considerably (Bol, personal communic.), which can be seen as onset of nitrification, leading to an increase in soil nitrate concentration from 0 to 63 and 26 mg NO_3-N (kg dry soil)$^{-1}$ on day 9 for -DMPP and +DMPP, respectively. The nitrate peaks corresponded to the N_2O emission peaks measured at day 8, in both, the +DMPP and the control treatment. This second peak of N_2O emissions is seen as indicating that nitrification is responsible for the second N_2O peak, while the absence of noteworthy nitrification during the initial 4-5 days suggests that denitrification of the trace amounts of nitrate in the slurry was the dominant source of the first N_2O emission peak.

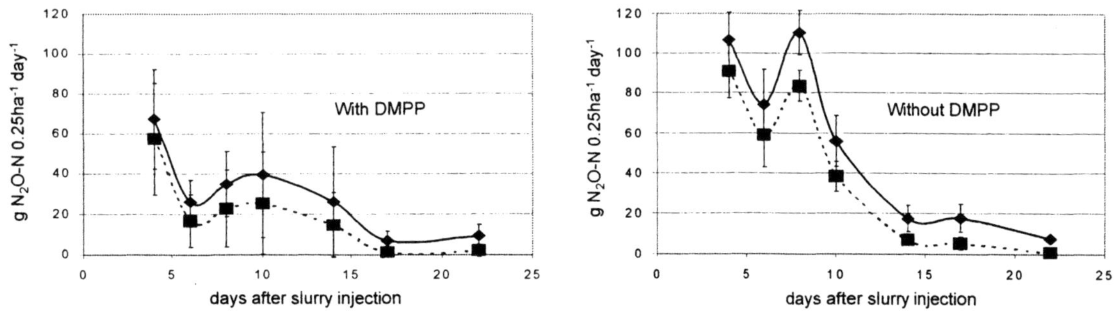

Figure 1. Nitrous oxide fluxes from grassland soil after injection of ^{15}N-labelled cattle slurry. Left: emission after injection of DMPP amended slurry; right: emission after slurry injection without DMPP addition. Solid lines total N_2O-N flux ± stddev, dashed lines slurry derived N_2O-N flux ± stddev.

Discussion

The mitigation potential of nitrification inhibitors for N_2O emissions from slurry using has seldom been studied. In a field experiment on the use of DCD with slurry, only a 5% reduction of N_2O emissions was observed (Schwarz *et al.*, 1991). In contrast, in a study of the combined effect of slurry acidification and DCD addition, a 50% reduction of total nitrification and denitrification loss (N_2O and N_2) was measured (Pain *et al.*, 1994). In the present study, a 32% reduction of total N_2O emission was obtained. In contrast to our experiment, slurry was broadcast applied in the H_2SO_4 + DCD study, and ammonia volatilisation stated in that contribution was clearly higher than volatilisation measured in our investigation (Chadwick, personal communication). Therefore in that experiment, the substrate pool for nitrification and denitrification was significantly reduced, which is likely to have increased the nitrification inhibitor efficiency.

Reductions in a similar range were reported with application of mineral fertiliser plus DMPP (Ottow *et al.*, 1999). Here DMPP led to a reduction of 40-47 % of the total N_2O emissions as compared to the control treatment without DMPP.

The question arises, whether for grassland systems nitrification inhibitors can reduce N_2O emissions from sources other than fertiliser or slurry. Particularly grassland soils with their high organic matter inputs from plant debris and/or organic manure are likely to show high rates of N_2O-N emission from turnover of their soil organic nitrogen pool. The great differences in nitrification inhibitor efficiency observed in grassland systems suggest highly variable contributions to N_2O-N emitted from fertiliser/manure N, or soil N. They require differentiation for mitigation strategies.

References

Linzmeier W and Schmidhalter U 1999 VDLUFA Congress Proceedings 52, 485-488.

Nason GE and Myrold DD 1981 Agriculture, Ecosystems and Environment 34, 427-441.

Ottow JCG, Weiske A and Benckiser G 1999 Scientific Colloquium Proceedings, Limburgerhof, May 17-18, 1999. 51-55. BASF AG., Limburgerhof, Germany.

Pain BF and Misselbrook TH 1994 Grass and Forage Science 49, 209-215.

Schwarz J, Kapp M, Benckiser G and Ottow JCG 1990 VDLUFA Congress Proceedings 32, 659-664.

W. J. Horst et al.(Eds.), Plant nutrition – Food security and sustainability of agro-ecosystems. 770~771, 2001

The effect of nitrogen supply by NH_4-beaker-deposits on tomatoes

K. R. Chang[1)] and Sommer K.[2)]

[1)] *Department of Food Resources, Division of Applied Biological Scienc, University of Sunmoon, Galsanri 100, Asan, Republik of Korea*

[2)] *Institute of Agriculturchemie, University of Bonn, Karlrobert Kreiten Str. 13, D-53115 Bonn, Germany*

Key words: ammonium deposit, ammonium nutrition, environment fertilizer, point placement

Abstract

This study was conducted to investigate the effect of nitrogen supply by NH_4-beaker deposit fertilizer on plant growth and nitrogen uptake of tomatoes. The NH_4-beaker deposit fertilizer was developed to supply ammonium continuously during the whole growing season for crops with a long cultivation period, using the principles of the "CULTAN" (Controlled Uptake Long Term Ammonium Nutrition) method. Tomato roots spread into the beaker, presumably due to attraction by ammonium ions. The experiment sought to establish the optimum amount of gypsum to be added to the ammonium fertilizer as a binder. It was found that a loam soil gave the best results as a cover for the fertilizer-gypsum mixture in the beaker to act as a regulator of diffusion of water and growth of roots into the beaker.

Introduction

Crops like tomato, capsicum and melon which produce yield over an extended period require a continuous supply of nitrogen (N) during the growing period. When such crops are grown in the glasshouse or in the field with fertigation their cultivation might result in high losses of nitrate by leaching (Geissler, 1991). To meet the demand for N and avoid losses, the "CULTAN" beaker deposit technique has been developed. The beakers contain urea and ammonium sulfate, a binder and a medium to regulate diffusion of water into the beaker. This paper describes a glasshouse experiment with tomato to investigate the use of NH_4-beaker deposit fertilizers.

Material and method

Urea [$CO(NH_2)_2$] and ammonium [$(NH_4)_2SO_4$] in a 2:1 ratio were placed in a PVC-beaker (6cm high and 5cm diameter) (Fig. 1). The N fertilizers were covered with gypsum as a binder and soil or compost (with different C/N ratios) to regulate diffusion of water and root growth into the beakers(Table 1). Tomato plants were grown in 39cm high by 45cm diameter containers. One deposit beaker was applied per plant and container. To minimize the leaching of N from the beakers they were placed in the soil with a distance of 5 cm to the nursery ball at transplanting with the open end down. Ammonium nitrate in the control treatment was applied on the soil surface 3 times during the growth period. 10g N per plant was applicated for all treatments. As base dressing the following nutrients were applied: 20g P_2O_5, 15g K_2O and 5.4g MgO per plant. The water was supplied by drip irrigation. Four replicates were used for each treatment. Data were tested by one-way analysis of variance using the Statistical Analysis System (SAS).

Results

None of the NH_4-beaker treatments gave tomato fruit yields significantly different from the NH_4NO_3 control treatment. Although, 7.5g gypsum per beaker with loam soil as regulator material for diffusion appeared to be the optimum rate (Table 1). The N uptake of tomatoes in the NH_4-beaker deposit fertilizer with 7.5g gypsum as binder showed statistical differences to the control and to the treatment with 3.8g gypsum as binder (Table 1). N uptake by ripe fruits was highest when the loam soil was used as the diffusion regulator (treatment No. 7), but it was not statistically different at the 5% level from the control. The exploration of the soil volume in the beakers by roots was particularly marked in the treatments with loam soil and compost at the C/N ratio of 16/1, where roots grew upwards into the beakers against geotropism (Fig. 2).

Discussion

Supplying N protected against leaching in an inverted beaker was equally as effective as surface application of ammonium nitrate for N uptake and yield of greenhouse-grown tomato plants. The inferior performance of beakers containing the lowest level of gypsum (3.8g) could be interpreted as a consequence of having not enough binder for nitrogen salts. The material used to cover the fertilizer and binder in the beaker and act to regulate the passage of water and roots into the beaker was important in determining N uptake. Loam soil and compost with a high C/N ratio used as materials to regulate diffusion appeared to give the better performance than other materials like sandy soil and compost of a C/N ratio 10/1. The reasons for this may be the higher clay and organic matter contents in those materials. Ammonium nitrogen adsorbed at clay surfaces (Blue and Eno, 1954; Page, 1975) may be spontaneously released due to changes in concentration gradients after irrigation and are than available for uptake. From this work we conclude that N fertilizer protected within a beaker can provide a continuous supply of N for plants over an extended period, with reduced risk of N loss. This could provide benefits for farmers as well as for the environment. In addition, improvement of the NH_4-beaker deposit fertilizer concept is expected by supplementation of micro-nutrients and pesticides to the beaker deposits in the future.

References

Blue WG and Eno CF 1954 Soil Sci. Soc. Proceedings 18, 420-424.

Geissler T 1991 Gemueseproduktion unter Glas und Plasten, Deutscher Landwirtschaftsverlag Berlin.

Page ER 1975 J. Agri. Sci. Cambridge 85, 65-74.

Table 1. Yield of tomato fruits and nitrogen uptake of tomato fruits and shoot by application of NH_4-beaker deposit fertilizers containing various amounts of gypsum as a binder and various regulator materials for diffusion regulation.

No	Treatment	Fresh weight (g /plant)		Nitrogen uptake (mg N/plant)		
		Ripe fruits	Green fruits[z]	Ripe fruits	Green fruits	Shoot
1	NH_4NO_3 (control)	4736	1129	6046	1170	2107
2	NC + 3.8g gypsum	3520	1328	5291	1662	2415
3	NC + 7.5g gypsum	4850	1495	7193	1804	2592
4	NC + 15.0g gypsum	4339	1310	6357	1483	2546
5	NC + 30.0g gypsum	4401	1218	6018	1466	3039
6	NC + 7.5g gypsum + sandy soil	4479	638	6650	963	1539
7	NC + 7.5g gypsum + loam soil	4878	974	7480	1166	2002
8	NC + 7.5g gypsum + compost[a]	3595	1005	5385	1070	1995
9	NC + 7.5g gypsum + compost[b]	4404	1374	6239	1559	2321
LSD 5%		1596	883	1067	1001	1104

Green fruit = non ripe tomato fruits at last harvest
NC = 6.6 g N as $(NH_4)_2SO_4$ + 3.4 g N as $CO(NH_2)_2$
[a]Compost = OM 15%, C/N ratio 10/1
[b]Compost = OM 29%, C/N ratio 16/1

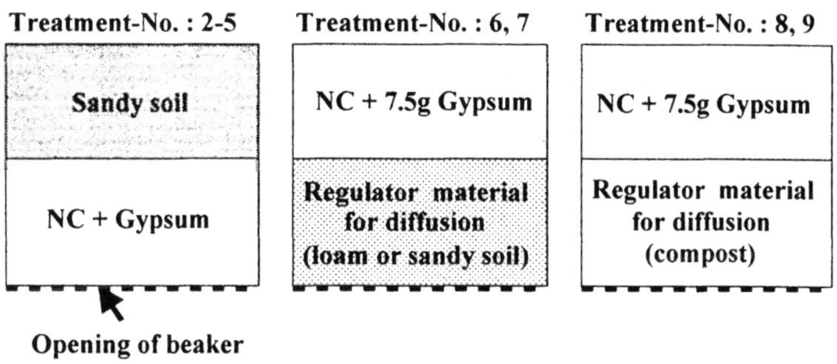

Figure 1. Distribution of fertilizer and diffusion regulator material in the NH_4-beaker deposits.

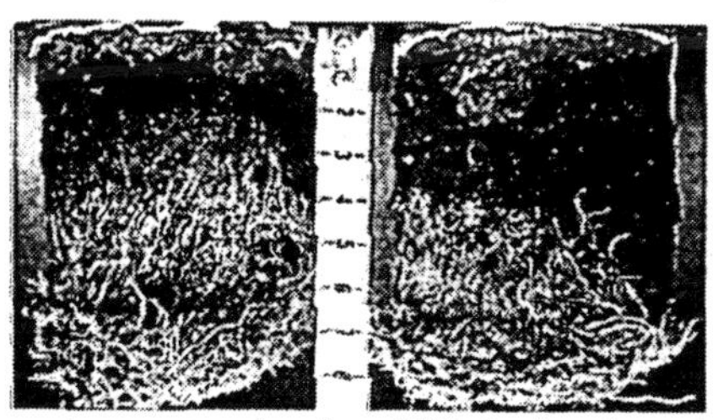

Figure 2. Root growth of tomato plants against geotropism in to NH_4-beaker deposits with open ends down.

W. J. Horst et al.(Eds.), Plant nutrition – Food security and sustainability of agro-ecosystems. 772~773, 2001

Effect of calcareous irrigation water and N-management on the Ca^{2+} and SO_4^{2-} distribution in sunflower (*Helianthus annuus L.*)

B. Jaeger, H. Goldbach and K. Sommer

[1] *Institute for Agricultural Chemestry, University of Bonn, Karlrobert-Kreiten Str.13, D 53115 Bonn, Germany, b.jaeger@uni-bonn.de*

Key words: bicarbonate, calcareous irrigation, CULTAN, sulphate, iron

Abstract

The effect of calcareous irrigation water and different N-management on the SO_4^{2-} and Ca^{2+} distribution in sunflower (*Helianthus annuus L.*) was investigated. Different N-management consisted of conventional NO_3^- and of CULTAN nutrition based on $(NH_4)_2SO_4$. In former experiment it was found out, that plants nourished by CULTAN contain less Ca^{2+} in the shoot than NO_3^--fed plants when grown on calcareous soil. Results show that the reason for this is an influence of SO_4^{2-} in the roots, being predominant in CULTAN treatments and also being strongly influenced by irrigation water containing $Ca(HCO_3)_2$. This might play a crucial role in maintaining a charge balance in plants grown on calcareous soils, for it has been shown in former experiments that NO_3^--fed plants show high surpluses of easily extractable and alkalinely active cations compared to CULTAN-fed plants, where the balance is more or less even. It is suggested that either Ca^{2+} is precipitated as $CaSO_4$ in the roots or the transport to the shoot is somehow hampered.

Introduction

Iron deficiency on calcareous soils is still a widespread problem throughout the world. There is evidence that HCO_3^- as well as NO_3^- induce iron chlorosis by raising the apoplastic pH (Kosegarten *et al.*, 1999). However under conventional fertiliser practice plants are exposed to NO_3^- nutrition and high HCO_3^- concentrations in the soil solution. These anions are accompanied by Ca^{2+} being the predominant cation (Kinzel 1982). However, with the CULTAN-cropping system it is possible to offer plants an exclusive NH_4^+ nutrition which is stabilised in a depot located in the root zone. By CULTAN, an appreviation for Controlled Uptake Long Term Ammonium Nutrition, plants are fed exclusively by NH_4^+. The NH_4^+ is located at one spot in the soil and is stabilised against mineralization by high NH_4^+ concentrations (Sommer, 1993). Plants are able to find the depot due to the attraction of NH_4^+ to roots (Sommer, 1989). In the case of applying the CULTAN-system to calcareous soils, the NH_4^+ is given as $(NH_4)_2SO_4$ in a depot acidified with H_2SO_4 to pH 2, thus providing the plants with extra amounts of SO_4^{2-}. After NH_4^+ is taken over into the protein metabolism, SO_4^{2-} is released and able to contribute to charge balance within the plants, which has been found to be unbalanced under conventional conditions (Jaeger *et al.*, 2000). Thus besides the change from NO_3^- to NH_4^+ nutrition it is likely that SO_4^{2-} coming from $(NH_4)_2SO_4$ and H_2SO_4 plays a significant role in compensating excessive cation uptake and thus can prevent increases in apoplastic pH. The influence of additional chlorides to a calcareous irrigation on the charge balance was also studied (data not shown).

Materials and methods

A split-root-design for nutrient solution cultures was developed in order to be able to harvest roots from CULTAN-fed and from NO_3-fed plants without soil contamination. The system divided the roots into three groups, one taking up the nitrogen, one growing in $Ca(HCO_3)_2$-solution and one growing in H_2O (Fig. 1).

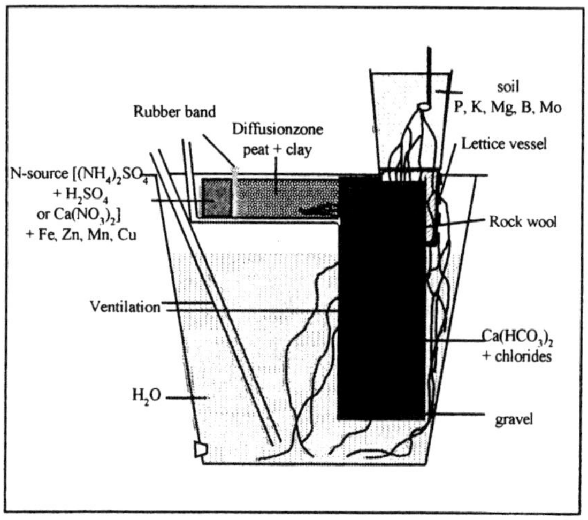

Figure 1. Split-root-design for CULTAN- and for NO_3-fed plants.

300 mg N was supplied either as $Ca(NO_3)_2$ or as $(NH_4)_2SO_4$ acidified with H_2SO_4 to pH 2. By this specific arrangement of the system cross-contamination between the solutions was avoided. However, to be safe solutions were changed every third day. 300 mg P, 300 mg K, 100 mg Mg, 0,5 mg B and 0,25 mg Mo (as KH_2PO_4, K_2SO_4, $MgSO_4$ x 7 H_2O, H_3BO_3, Na_2MoO_4x 2 H_2O) were applied to the soil in which three sunflowers (*Helianthus annuus* L. cv. Optisol) per pot were grown for five weeks. The soil was a parabrown soil with a pH of 6.8. Fe, Zn, Mn and Cu were given as sulphates together with the nitrogen. It is clear that CULTAN-treatments received extra SO_4

compared to NO_3-treatments, where SO_4 came from $MgSO_4$ and from the trace elements. However CULTAN is the only possible way to feed plants with $(NH_4)_2SO_4$ + H_2SO_4 on calcareous soils, without the formation of gypsum within the soil.
$Ca(HCO_3)_2$-solution was produced by adding $CaCO_3$ to water and pumping air through the mixture for at least 6 weeks in order to provide time for sufficient $Ca(HCO_3)_2$-formation. Roots were harvested from two sections: the one growing in water and the one growing in $Ca(HCO_3)_2$-enriched water. After harvest the roots were first washed with H_2O and afterwards with 1 N KCl, gaining solutions representing the nutrient concentration in the water free space (WFS) and the donnan free space (DFS). SO_4^{2-} analysis was conducted by ionchromotography (Dionex IC20) and Ca^{2+} by flame photometer (Eppendorf Elex 6361).

Results

Analysis of Ca^{2+} and SO_4^{2-} in the roots extracted with H_2O and 1 N KCl (representing the concentrations in the WFS and DFS respectively), showed significant differences between NO_3^-- and CULTAN-fed plants as well as between H_2O- and $Ca(HCO_3)_2$-containing water (*Figure 2*). $CaCl_2$ enhanced the Ca^{2+} content in the H_2O

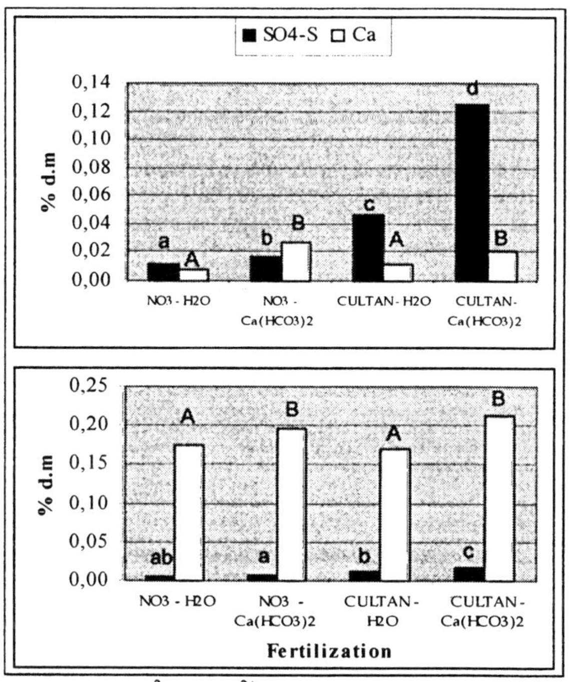

Figure 2. $SO_4^{2-}$and Ca^{2+} contents in different parts of sunflower roots according to the split-root-design in a fraction washed out with H_2O (upper part) and 1 N KCl (lower part). Significant differences are expressed with different letters at the 5% level.

fraction and in the 1 N KCl fraction (data not shown). For SO_4^{2-} the CULTAN-treatments showed higher contents than NO_3^- treatments. The most striking result is the enhancement of SO_4^{2-}-contents when roots grew in solution containing $Ca(HCO_3)_2$ compared to H_2O for both NO_3^- and CULTAN treatments.

Discussion

Because the uptake of SO_4^{2-} was restricted to an isolated region of the roots, SO_4^{2-} was transported from the place of uptake, which in this experimental set-up had no contact to the water, to the other regions of the roots. The Ca^{2+} content did not show a surprising behaviour regarding the enrichment with $Ca(HCO_3)_2$. The results clearly indicate, that SO_4^{2-} was preferentially transported to roots which grew in a solution containing $Ca(HCO_3)_2$. This was the case for NO_3^--fed plants as well as for CULTAN-fed plants, indicating the probable role of SO_4^{2-} in maintaining charge balance within the roots. This was not only true for the SO_4^{2-}-contents of the fractions representing the WFS and the DFS as shown in Fig. 2, but also for a hotwater-extractable fraction and for the total SO_4^{2-}-content of the roots (data not shown).
This effect was more striking for CULTAN-fed plants, obviously because of the extra SO_4 in the $(NH_4)_2SO_4$ + H_2SO_4 depots. However, it is important to notice, that SO_4^{2-} is only effective in charge balancing when it is supplied as $(NH_4)_2SO_4$ (or as H_2SO_4), because NH_4^+ is taken over into protein metabolism in the roots immediately after uptake. Excessive SO_4^{2-} can accumulate and act as a counter anion to balance excessive cations. In this context it is called "active SO_4^{2-}". It has to be pointed out that this is the aim of the CULTAN-system when used on calcareous soils and it is the only possible way to nourish plants with NH_4^+ and "active SO_4^{2-}" on these soils. Chlorides depressed SO_4^{2-} contents, indicating the probable role of Cl^- in maintaining a charge balance (data not shown). However, Cl^- may also enhance cation contents due to its role as accompanying anion for Ca^{2+} (Jaeger *et al.*, 2000).
It is likely that Ca^{2+} is precipitated as $CaSO_4$ within the roots due to the low solubility of this compound. This would explain why CULTAN-fed plants regularly show lower contents of Ca^{2+} in the shoot (Jaeger *et al.*, 2000).

References

Jaeger B, Goldbach H and Sommer K 2000 *In* Acta Horticulturae 531. Ed Blanke, M & Pohlan, J Eds. pp 107-113.

Kinzel H 1982 Pflanzenökologie und Mineralstoffwechsel. Verlag Eugen Ulmer, Stuttgart.

Kosegarten H, Grolig F, Esch A, Glüsenkamp K H and Mengel K 1999 Planta 209, 444-452.

Sommer K 1989 7th International Colloquium for the Optimation of Plant Nutrition, Nyborg, Denmark, 401, 1-9.

Sommer K 1993 International Worshop on Controlled/Slow Release Fertilizers, Haifa, Israel, paper no. 15, p. 13.

W. J. Horst et al.(Eds.), Plant nutrition – Food security and sustainability of agro-ecosystems. 774~775, 2001

Effect of acetylene generated from carbide on nitrification in soil, yield of irrigated maize and growth of maize seedlings.

P. J. Randall[1)], J. R.Freney[1)], J. Hodgkin[2)] and T. C. Morton[2)]

[1)] *CSIRO Plant Industry, PO Box 1600, Canberra, ACT, 2614, Australia, P.Randall@pi.csiro.au*

[2)] *CSIRO Molecular Science, Private Bag 10, Clayton South MDC, Vic, 3169, Australia*

Key words: carbide, maize, nitrification inhibitor, nitrogen use efficiency

Abstract

The inhibition of nitrification in soils has been pursued as a strategy to improve recovery of fertilizer nitrogen (N) by crops through lowering losses by leaching of nitrate or denitrification, in order to benefit yields and the environment. Acetylene is known to inhibit nitrification in soils and recently, we have described a novel matrix consisting of calcium carbide and polyethylene which delivers acetylene in soil over an extended period, resulting in delayed nitrification. In work with irrigated maize, the matrix delayed the disappearance of ammonium derived from urea fertilizer, but gave no benefit for crop yield in a soil that was highly responsive to N. One possibility is that the loss processes (nitrate leaching and/or denitrification) were not significant at the site. Also, it was shown in a pot experiment that seedling growth of maize was impaired by the matrix, and this may provide another explanation for the lack of yield response noted in the field. Further tests are required before the matrix can be recommended for practical use.

Introduction

Poor recovery of fertilizer nitrogen (N) by crops has been attributed in part to losses due to leaching of nitrate or denitrification. Inhibiting nitrification in soil and retaining N in the ammonium form may reduce N loss and increase the recovery of N and crop yield (Hauck, 1983). In addition to the agronomic benefits, there would be environmental benefits; the lowering of nitrate contamination of water bodies and lower emissions of nitrogen oxides to the atmosphere.

Acetylene is known to inhibit nitrification in soils and wax-coated calcium carbide has been tested as a slow release source of acetylene (formed when soil water penetrates the coating and reacts with the carbide) with promising results for improving recovery of N and crop yield (Zhang *et al.*, 1992; Freney *et al.*, 1993). However, the wax-coated material is difficult to handle and recently we have devised an improved slow release source of acetylene using a novel matrix consisting of calcium carbide and polyethylene. In laboratory studies this matrix delivered acetylene over an extended period resulting in delayed nitrification of ammonium added to soil (Freney *et al.*, 2000). The matrix has now been assessed in the field and in this preliminary report the effect on soil ammonium concentration, plant N and yield of irrigated maize are described. The paper also describes effects of the matrix on early seedling growth of maize in pots.

Materials and methods

The inhibitor matrix described by Freney et al (2000) was used in a factorial experiment with irrigated maize (*Zea mays* L cv 008 with high amylose) sown in a clay loam (pH 4.9) overlying a heavy clay subsoil in the Colleambally Irrigation Area of New South Wales. Basal fertilizer (20 kg P ha $^{-1}$ as Trifos), urea and matrix were mixed just prior to application and applied in the drill row ~5 cm below the seed. There were 5 rates of matrix (only data for 0 and one rate of matrix, 30 kg carbide ha^{-1}, are given here) arranged factorially across 5 rates of urea fertilizer (0, 30, 60, 150 and 300 kg N ha^{-1}) in each of 4 randomised blocks. The trial was furrow irrigated as required. Soil samples were taken with a 20 x 2 cm rectangular tube augur across the row to 20 cm depth, sieved moist, and spread out to air dry in a glasshouse. Ten grams of the air-dried soil was then extracted with 50 mL 2M KCl containing 5 mg L^{-1} phenyl mercuric acetate and ammonium was determined by distillation with magnesium oxide. Total N in plant samples was determined by a Kjeldahl method.

To examine the effect of the inhibitor on early seedling performance, maize seeds were sown in free draining pots of soil. The pots were partially filled with dry soil, the matrix was added followed by 3 cm of soil, the seed, and a further 2 cm of soil. Water was added to bring the soil to 85% of field capacity. Brown paper was laid over the top of the pots and after 4 days seedlings were recovered for root length measurements.

Results

In the absence of the matrix, soil ammonium declined steadily from the first sampling, 20 days after sowing. The matrix delayed the disappearance of ammonium (Fig. 1a). Data for the 60 kg N ha^{-1} only is presented because that was a suboptimal rate of N where inhibiting nitrification might be particularly important in terms of availability of N for crop growth.

Nitrogen concentrations in maize plants at 60 days after sowing tended to be elevated by inhibitor treatments at all N application rates, but the effect did not reach the 5% significance level (Fig. 1b).

Grain yields were increased almost 2-fold by the application of N fertilizer, but were not affected by the nitrification inhibitor (Table 1).

In the short-term pot experiment the matrix applied below the seed at sowing reduced growth of both roots and shoots (Fig. 2).

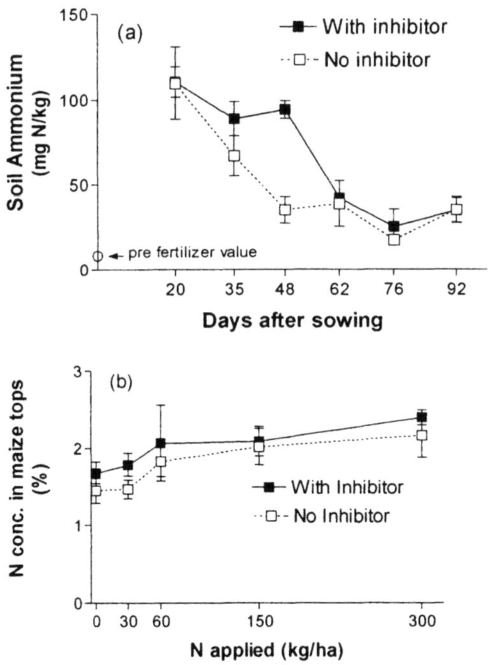

Figure 1. (a) Changes in soil ammonium with time after application of fertilizer N (60 kg ha $^{-1}$) with or without inhibitor matrix. (b) Concentration of N in maize tops 60 days after sowing as affected by the nitrification inhibitor. Error bars are ± S.E

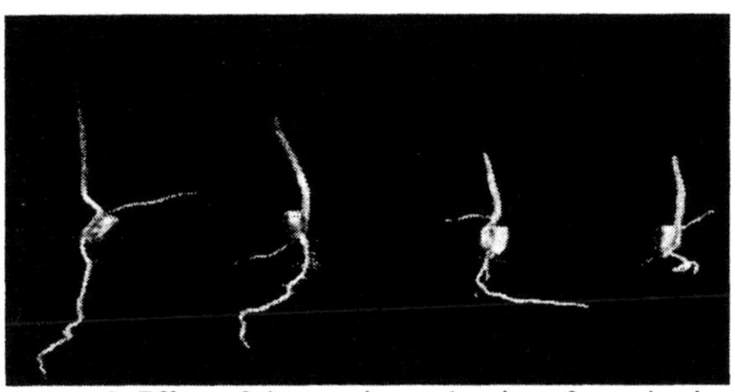

Figure 2. Effect of the matrix on the size of germinating seedlings of maize after 4 days. Matrix rates L to R: no inhibitor, 15, 150, 1500 kg carbide ha^{-1}

Table 1. Effect of urea fertilizer applied with or without a nitrification inhibitor on yield of maize grain

Urea applied kg N ha^{-1}	No inhibitor	With inhibitor
	Yield of grain (t ha^{-1} ± S.E.)	
0	3.6 ±0.27	3.1±0.81
30	4.5±0.84	4.3±0.69
60	3.9±0.21	3.7±0.59
150	5.7±0.19	5.3±0.53
300	6.4±0.56	4.9±0.37

Discussion

In a soil where maize responded to applied N fertilizer it would be expected that any increase in availability of soil N, particularly at sub-optimal rates of N application, would be reflected in yield. With the heavy clay subsoil and the furrow irrigation system used we anticipated large losses of N through denitrification, and no leaching of nitrate. The observation that there was no increase in yield in response to the matrix suggests that despite its effect in slowing nitrification, losses of N at this site may not have been significant by either process. It is possible that denitrification at this site was limited by the lack of available carbon for the denitrifying organisms. In addition, direct adverse effects of the matrix on seedling growth could have counteracted any benefits of increased nitrogen availability. Adverse effects of acetylene applied to plant roots have been previously reported (Daum and Schenk, 1997), and the use of acetylene-generating inhibitors of nitrification for application at the time of sowing may not prove feasible. In other tests in pots, when sowing was delayed for 10 days after the matrix was applied, seedling growth was not impaired (data not shown).

Before such inhibitors can be recommended in practice, more work is needed to predict situations where losses of fertilizer N by leaching or denitrification are likely to be serious and, consequently, where inhibition of nitrification may be of benefit. Further, the timing and placement of the matrix will have to be investigated in order to avoid damage to seedlings.

References

Daum D and Schenk MK 1997 Biol. Fertil. Soils. 24, 111-117.

Hauck RD 1983 *In* Gaseous loss of nitrogen from plant-soil systems. Eds. Freney JR and Simpson JR. pp. 285-312. Martinus Nijhoff Publishers, The Hague.

Freney JR, Chen DL, Mosier AR, Rochester IJ, Constable GA and Chalk PM 1993 Fert. Res. 34, 37-44.

Freney JR, Randall PJ, Smith JWB, Hodgkin J, Harrington KJ and Morton TC 2000 Nutr. Cycl. Agroecosystems. 56, 241-251.

Zhang SL, Cai GX, Wang XZ, Xu YH, Zhu ZL and Freney JR 1992 Pedosphere 2, 171-1.

W. J. Horst et al.(Eds.), Plant nutrition – Food security and sustainability of agro-ecosystems. 776~777, 2001

Release of nitrogen from two controlled release fertilizers

L. Andersen
Danish Institute of Agricultural Sciences, Department of Horticulture, Kirstinebjergvej 10, P.O. 102, DK-5792 Aarslev Denmark, e-mail: lillie.andersen@agrsci.dk

Key words: ammonium, container, controlled release fertilizer, nitrate, plants

Abstract

One-year-old seedlings (*Hebe x andersonii* 'Variegata' and *Aubrieta* 'Blaumeise') were transplanted into containers and grown outdoors from May to September (*Hebe*) and from June to November (*Aubrieta*). Osmocote Plus ® and Osmocote Exact Low ® (both classified as 5-6 month release types) was embedded in each container beneath the seedling at transplanting. From samples taken during the cultivation period the release of ammonium and nitrate was calculated as the difference between initial and final amount of ammonium and nitrate. Release of ammonium from the controlled release fertilizers was much faster than the release of nitrate in both fertilizers. The release of total nitrogen (ammonium and nitrate) from Osmocote Plus ® could be related to the sum of air temperature calculated from the time of transplanting, independently of growing period and cultivars.

Introduction

Controlled release fertilizers are widely used in Northern Europe in cultivation of container plants. Release of nutrients from controlled release fertilizers is known to depend on temperature (Lamont et al., 1987). However, the experiments concerning nutrient release are mainly studied in the laboratory. Hence, it is unclear whether the results from these experiments can be used to predict the release of nutrients during cultivation of container plants grown outdoors. The aim of the present experiment was to study the release of nitrogen from two controlled release fertilizers, Osmocote Plus ® and Osmocote Exact Low ® during cultivation of two cultivars, *Hebe x andersonii* 'Variegata' and *Aubrieta* 'Blaumeise' grown outdoors in a container area at two irrigation regimes.

Materials and methods

One-year-old seedlings (*Hebe x andersonii* 'Variegata' and *Aubrieta* 'Blaumeise') were transplanted into limed, but unfertilized peat in 0.67 litre polyethylene containers at 11 May (*Hebe*) and 15 June (*Aubrieta*). The controlled release fertilizers, Osmocote Plus ® and Osmocote Exact Low ® (both classified as 5-6 month release types), were placed in a polypropylene bag beneath the seedling in the container. The amount of controlled release fertilizer was 5 gram per plant (*Hebe*) and 3 gram per plant (*Aubrieta*). The plants were placed outdoors on trays topped with 1 cm sand. The experimental design was a randomised block design with 4 blocks and 30 (*Hebe*) and 80 (*Aubrieta*) plants in each block. Irrigation was initiated by a tensiometer (Steck-Tensio 100 KV1, Tensio-Technik, Geisenheim, Germany) at 5 kPa in the growing medium and supplied through a moving irrigation machine at an intensity of 2 mm or 4 mm with a leaching fraction (leached as fraction of added) of 0.05 and 0.5, respectively. Liquid fertilizers were added from 1 August in *Hebe* and from 15 August in *Aubrieta* at each irrigation. During the cultivation period, the polypropylene bag was removed from the container and samples were taken. Fertilizer granules in the polypropylene bag were ground and diluted with deionised water before analysing ammonium according to Crook & Simpson (1971) and nitrate using an nitrate sensitive electrode (Orion). Release of ammonium and nitrate was calculated as difference between initial and final amount.

Results and discussion

At low irrigation intensity (2 mm) no release of nitrate from Osmocote Exact Low ® could be detected in the containers grown with Hebe during the first month. From Osmocote Plus ® only 6% of initial nitrate content was released (Table 1). The release of nitrate at high irrigation intensity was 5% and 10% from Osmocote Exact Low ® and Osmocote Plus ®, respectively, but not significantly different to the nitrate release at low irrigation intensity. In Aubrieta the release of nitrate at low irrigation intensity was significantly lower than at high irrigation intensity.

Release of ammonium was much larger than release of nitrate from both fertilizers. Within 27 days 45% of the initial ammonium content was released from Osmocote Plus ® in Hebe at high irrigation intensity and 37% at low intensity. The release from Osmocote Exact Low ® was 29% and 27% at 4 and 2 mm intensity, respectivaley. However, the difference between irrigation intensity 4mm and 2 mm was not significant.

A relationship between the release of total nitrogen from Osmocote Plus ® and the air temperature sum (calculated as the sum of the mean daily air temperature from the time of transplanting) was found for both varieties and both irrigation intensities (Figure 1). However, such a relationship could not be found for the release of total nitrogen from Osmocote Exact Low ® (data not shown).

References

Crook WM and Simpson WE 1971 J. Sci. Fd. Agric. 22, 9-10.

Lamont GP, Worrall RJ and O'Connell MA 1987 Sci. Hort. 32, 265-273.

Table 1. Effect of irrigation intensity on nitrate [%NO_3-N] and ammonium [%NH_4-N] remaining in Osmocote Plus ® and Osmocote Exact low ® 27 days after transplanting of *Hebe* pot-plants and 63 days after transplantning of *Aubrieta* pot-plants.

	Osmocote Plus ®				Osmocote Exact low ®				$LSD_{0.95}$	
	4 mm		2 mm		4 mm		2 mm			
	NO3-N	NH_4-N	NO_3-N	NH_4-N	NO_3-N	NH_4-N	NO_3-N	NH_4-N	NO_3-N	NH_4-N
Hebe										
27 days after transplanting	90 b	55 b	94 ab	63 ab	95 ab	71 a	100 a	73a	8.27	13.62
Aubrieta										
63 days after transplanting	72 d	49 b	78 c	50 b	86 b	68 a	94 a	67 a	5.43	3.61

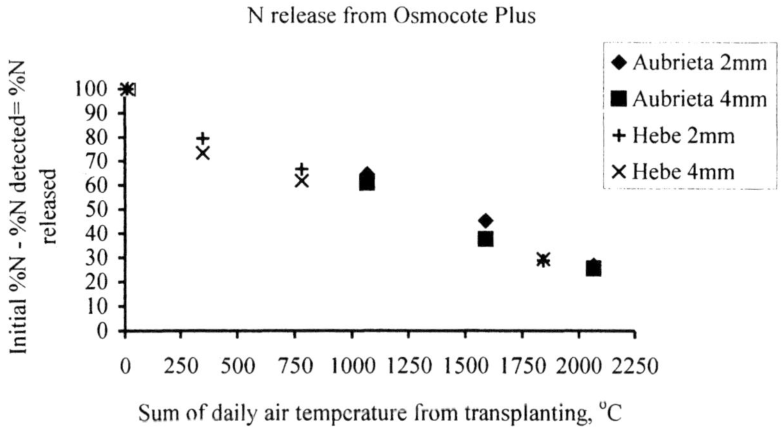

Figure 1. Percentage of total nitrogen (ammonium and nitrate) remaining in Osmocote Plus ® in relation to daily air temperature sum calculated from time of transplanting in *Hebe x andersonii* 'Variegata' and *Aubrieta* 'Blaumeise'

W. J. Horst et al.(Eds.), Plant nutrition – Food security and sustainability of agro-ecosystems. 778~779, 2001

Fertiliser and irrigation best management practices (BMPs) for optimal citrus production and water quality in sandy entisols of Florida

S. Paramasivam[1], A. K. Alva[2], K. S. Sajwan[3], J. P. Syvertsen[4], T. A. Wheaton[5] and D. P. H. Tucker[6]

[1] Center for Marine, Environmental Sciences and Biotechnology Research, Savannah State University, P.O. Box 20600, Savannah, Georgia 31404, United States, siva@savstate.edu

[2] USDA-ARS-PWA, 24106 N. Bunn Road, Prosser, Washington 99350, United States, aalva@tricity.wsu.edu

[3] Center for Marine, Environmental Sciences and Biotechnology Research, Savannah State University , P.O. Box 20600, Savannah, Georgia 31404, United States, sajwank@savstate.edu

[4] Citrus Research and Education Center, Institute of Food and Agricultural Sciences, University of Florida, 700 Experiment Station Road, Lake Alfred, Florida 33850, United States, jmsn@lal.ifas.ufl.edu

[5] Citrus Research and Education Center, Institute of Food and Agricultural Sciences, University of Florida, 700, Experiment Station Road, Lake Alfred, Florida 33850, United States, taw@lal.ifas.ufl.edu

[6] Citrus Research and Education Center, Institute of Food and Agricultural Sciences, University of Florida, 700 Experiment Station Road, Lake Alfred, Florida 33850, United States, dpht@lal.ifas.ufl.edu

Key words: best management practices, citrus, sandy entisols, water quality

Abstract

In order to increase efficient use of applied fertiliser by citrus, economically feasible best irrigation and fertiliser management practices (BMPs) are necessary. A six year field study was conducted on a well-drained Tavares fine sand in Highlands county, Florida, USA, to evaluate the effects of N source, placement, and rate (112 to 280 kg $ha^{-1}yr^{-1}$) with optimum irrigation management on (i) the fruit yield, quality and leaf nutrient contents of bearing mature (25 + year old) 'Hamlin' orange trees on Cleopatra mandarin rootstock and (ii) changes in NO_3-N status in soil solution collected from vadose zone below the rootzone and in surficial groundwater. Results of this study indicated that optimum fruit yield could be obtained at about 250 kg N ha^{-1}, which supported optimum range of spring flush N concentration. Leaching of NO_3-N accounted for only 22 to 34 kg $ha^{-1}yr^{-1}$ and the NO_3-N concentration in soil solution in the vadose zone and in surficial groundwater remained below 10 mg L^{-1} under optimal irrigation and N management conditions used in this study.

Introduction

Commercial citrus production in Florida occupies 342,200 ha with a production of 11.0 and 2.08 million mega gram (Mg) of oranges and grapefruit, respectively. This accounts for 75% of the total U.S. citrus production. Nearly 28% of Florida citrus acreage is on sandy Entisols along the Central Florida Ridge. Due to high sand content (>97%), low organic matter, lack of a confining soil horizon, poorly distributed high annual rainfall (mean annual rainfall of 1400 mm, with close to 60% of annual rainfall received during June to September) these soils are vulnerable for leaching of soil-applied chemicals and nutrients below the rootzone that may pose a potential threat to groundwater. Unpublished groundwater monitoring data collected in Highlands County (predominantly citrus growing area) in central Florida in 1990 by the Florida Department of Environmental Protection (FLDEP) revealed that 32% of home owner drinking water wells (mostly into the surficial aquifer with poor construction) contained NO_3-N in excess of the USEPA maximum contaminant level (MCL) of 10 mg NO_3-N L^{-1}. Thus, there was a need to investigate the fate of N applied with various N sources to bearing citrus trees under current production practices, specifically with careful irrigation management.

The objective of this study was to develop N best management practices (N-BMP) for bearing citrus trees on vulnerable soils to attain optimum fruit yield and quality with minimal leaching of NO_3-N below the rootzone that eventually minimize the NO_3-N in groundwater.

Materials and methods

A field experiment was conducted in a commercial grove with highly productive (with mean yield of 70-80 Mg ha^{-1}) using 25-yr-old 'Hamlin' orange trees on Cleopatra mandarin rootstock planted (7.62 x 6.57 m, 286 trees ha^{-1}) in a Tavares fine sand (hyperthemic, uncoated, Typic Quartzipsamments moderately well drained). The trees were irrigated with under-tree microsprinklers (Maxijet Inc., Haines City, FL). One emitter per tree was installed to irrigate with a delivery rate of 0.083 m^3 h^{-1} at 0.276 Mpa, with a wetted area of 18.7 m^2. Each plot consisted of four uniform trees in a row, with the middle two trees used for response evaluation.

Trees received nitrogen rates from 112 to 280 kg ha^{-1} yr^{-1} applied as either fertigation (FRT 15 appl yr^{-1}) water soluble granular (WSG 4 appl yr^{-1}), a combination of 50% WSG and 50% FRT (2 + 15 appl yr^{-1}) or controlled-release fertiliser (CRF 1 appl yr^{-1}) were evaluated for fruit yield, quality and spring flush N content. Tensiometers installed at 15 and 30 cm depths were used to schedule irrigation carefully and to minimize drainage below rootzone (Paramasivam *et al.*, 2000). Water and nitrogen leaching below the rootzone were evaluated using tensiometer data and analysis of nitrate concentration in suction lysimeter samples collected from the vadose zone below the rooting depth (Paramasivam *et al.*, 2001). Impact of various N sources and irrigation management on surficial water quality was evaluated by sampling groundwater using monitoring wells.

Results

Yield responses to N sources and rates were evaluated as the mean of last 4-yr data (Fig. 1) in which the N rates

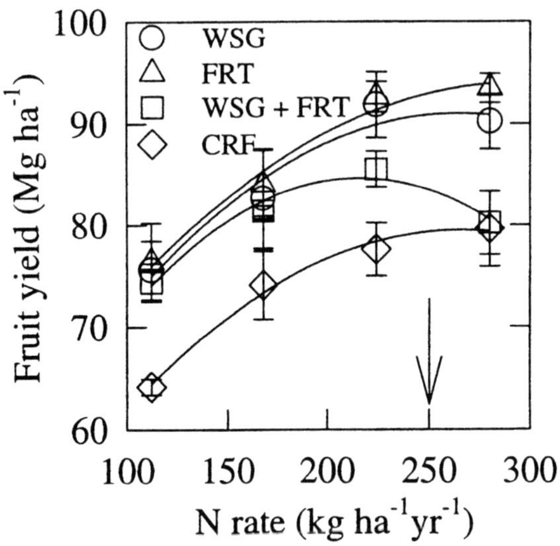

Figure 1. Effect of fertiliser N sources, rates and frequency of application on fruit yield.

were uniform across all N sources. Within an N rate, mean fruit yield over 4 years was greater with FRT or WSG forms of N than that with either WSG + FRT or CRF sources. Optimum fruit yield was obtained at about 250 kg N ha^{-1}, which supported the spring flush N concentrations within the established optimum range of 25 to 27 g kg^{-1}. Evaluation of drainage water throughout the study period indicated that water and nitrogen leaching below the rootzone occurred mostly during May through September, which coincides with heavy rainfall. Using careful scheduling of irrigation and fertiliser application, leaching of NO_3-N accounted for only 22 to 34 kg ha^{-1} yr^{-1} and the NO_3-N concentration in the leachate collected below the rootzone remained below 10 mg L^{-1}. Surficial groundwater monitoring over years also indicated the NO_3-N concentration in surficial groundwater (Fig. 2) remained below the maximum contaminant level (MCL, 10 mg L^{-1}) suggested by USEPA.

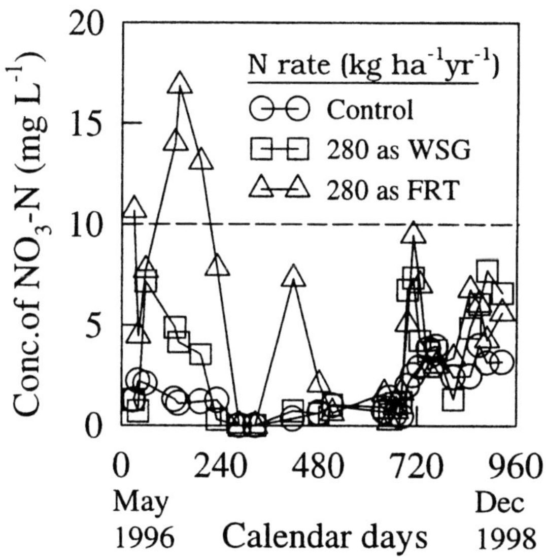

Figure 2. Effect of fertiliser sources, rates and frequency of application on surficial groundwater NO_3-N.

Discussion

Careful analysis of the results indicated that the N requirement per Mg of fruit varied from 1.34 to 2.80 (mean =2.07) kg N for the WSG, from 1.32 to 2.70 (mean=2.01) kg N for FRT, from 1.36 to 3.10 (mean=2.23) kg N for the 50:50 blend of WSG:FRT, and from 1.58 to 3.17 (mean=2.38) kg N for the CRF source. These values are substantially lower than the previously reported, i.e. 4.44 kg N Mg^{-1} fruit (Koo *et al.*, 1984). This study further suggests that the careful scheduling of irrigation and fertiliser management practices contributed to a reduction in N requirement for fruit production while maintaining spring flush N concentration within the optimum range. These improved management considerations used in this study are critical to minimize the N losses through leaching and subsequently reducing surficial groundwater contamination and to improve N uptake efficiency. Therefore, all these management considerations should be incorporated in the proposed N best management practices for citrus in sandy soils along the central Florida ridge area of USA.

References

Koo RCJ, Anderson CA, Stewart I, Tucker DPH, Calvert DV and Wutscher HK 1984 Recommended fertilisers and nutritional sprays for citrus. Fla. Agric. Exp. Stn. Bull. 536D. University of Florida, Gainesville, Florida, USA.

Paramasivam S, Alva AK and Fares A 2000 Soil Sci. 165, 343-353.

Paramasivam S, Alva AK, Fares A and Sajwan KS 2001 Soil Sci. Soc. Am. J. 65, (in Press).

W. J. Horst et al.(Eds.), Plant nutrition – Food security and sustainability of agro-ecosystems. 780~781, 2001

Effect of N fertilisation in a young 'Lane Late' orchard fertigated by drip irrigation

M. R. Menino[1)], J. C. Tomás[2)], C. Carranca[1)], M. L. Fernandes[1)] and A. de Varennes[3)]
[1)] *Estação Agronómica Nacional - Av. da República, Nova Oeiras, 2784-505 Oeiras, Portugal*
[2)] *Direcção Regional de Agricultura do Algarve - Apart. 282 - Patacão, 8000-904 Faro, Portugal*
[3)] *Instituto Superior de Agronomia - Tapada da Ajuda, 1300 Lisboa, Portugal*

Key words: fertigation, flower nutrient content, leaf nutrient content, N rates, orange tree, sandy soil, tree size

Abstract

The aim of this 2 year study was to investigate , the response of young 'Lane Late' trees grafted on 'Carrizo' citrange rootstock to N fertilisation in the South of Portugal, using fertigation, in order to establish proper N rates and optimal foliar nutrient levels according to the tree size/age. N rates significantly affected canopy width, volume and density, trunk perimeter, and flowers dry weight and N content. Foliar N content significantly increased with the N rates in the 1st year. In both years N concentration was in the high to excess ranges.

Introduction

To obtain well developed trees and to accelerate tree growth and production, citrus growers often overfertilise young trees with high N, P and K rates. However, adequate rates of these nutrients have not been established in Mediterranean countries. The optimal N rates for citrus trees are highly dependent on environmental conditions, tree age and size, tree N status, management practices and crop varieties. It is frequent to find large increases in trunk perimeter and canopy volume in response to N application during the first two years of the tree growth. Canopy density can vary among trees with similar canopy volume. Also, N deficient plants show smaller flowers production (Obreza and Rouse, 1993; Davies and Albrigo, 1994).

Citriculture is one of the most important agricultural activities in the south of Portugal, where Campina de Faro (Algarve) was classified as a Zone Vulnerable for Nitrate leaching (ZVN). This region represents 78% of the total acreage under orange production in Portugal (Guerra, 1994).

The aim of this study was to investigate the response of young 'Lane Late' trees grafted on 'Carrizo' citrange rootstock to N fertilisation, in Campina de Faro. Fertigation by drip irrigation was used in order to establish proper N rates according to the tree size/age and optimal foliar nutrient levels.

Material and methods

Month-old seedlings of *Citrus sinensis* L. Osbeck cv. Lane Late grafted on 'Carrizo' citrange rootstock were planted in April 1999 in a Gleyic Podzol, located at Campina de Faro. The trial was randomly designed, with five N rates and three replicates. Each replicate comprised four trees. The fertiliser experiment was conducted in 1999 and 2000. Nitrogen (Table 1) as ammonium nitrate, P as phosphoric acid and K as potassium sulphate were supplied by drip irrigation, between March and October (Table 2). Irrigation water calculated by class A pan evaporation was supplied through 2 emitters (3.75 L h^{-1}) $tree^{-1}$. Mean maximum and minimum monthly air temperatures were 24^{0}C and 12^{0}C, respectively Total precipitation was 248 in 1999 and 513 mm in 2000 during the two years.

Before and after fertilisation, the trees were measured for tree height, trunk perimeter, and canopy height and width to estimate the canopy volume, assumed as being a prolate spheroid. Leaf samples from spring flushes (7-8 month old) were collected each year to evaluate the nutritional status of the tree. At the end of the 2nd growth cycle, a visual ranking of the canopy density was conducted. In the 2nd year, flowers at flowering stage were taken from each tree to evaluate their response to applied N. Plant material was analysed for macro- and micronutrients. Data were analysed by covariance and canonical analysis and visual rankings were analysed by three tests (Wilcoxon/Kruskal-Wallis, Van der Waerden, and the equality of the variances) to evaluate the effect of N treatment.

Table 1. Nutrient rates (g tree^{-1}) for both years experiment

Orchard age	N_1	N_2	N_3	N_4	N_5	P	K
0-1	10	20	40	80	160	4	17
1-2	15	30	60	120	240	7	21

Table 2. Relative distribution of N, P and K application during the growth period

Nutrients	Mar.	Apr.	May	June	July	Aug.	Sept.	Oct.
N	5	10	10	15	20	20	15	5
P	10	10	15	15	15	15	10	10
K	5	5	10	10	15	20	20	15

Results and discussion

The soil showed a light texture (95.6% sand), pH 6, and low organic matter, macro- and micronutrients contents and low cation exchange capacity. Tree size and canopy height did not vary significantly with N treatment for both years. In contrast to results reported by Obreza and Rouse (1993), N rates significantly affected canopy width (increasing from 108 to 111 cm) and volume (ranging from 512 to 515 dm^3), and trunk perimeter (varying from 9.3 to 9.9 cm). Visual rankings in the 2nd year showed a significant effect of N treatment on canopy density. The data for leaf analysis were compared with the standards reported for mature trees by Davies and Albrigo (1994), although they should not be used to evaluate the nutritional status of vigorously growing young trees (Obreza and Rouse, 1993). Foliar N content significantly ($P \geq 0.01$) increased with the N rates in the 1st year (from 31.8 g N kg^{-1} to 36.8 g N kg^{-1}). In both years, N concentration was in the high to excess ranges (33.4 g N kg^{-1}). Uptake of P, Ca, Mn, Cu and B was independent of the treatment, whereas K significantly decreased and Zn significantly increased. Nevertheless, Mn, Zn and B were in the low to deficient ranges, probably due to the increase in soil pH from 6 to 8, caused by the high contents in bicarbonates (6.7 me L^{-1}), Ca (7 me L^{-1}) and Mg (2.5 me L^{-1}) in the irrigation water. Although K concentration in the leaves was in the optimum range in the 1st year (14.5 g K kg^{-1}), it decreased to (6.2 g K kg^{-1}) in the 2nd year. The data obtained in this study indicate that K, Mn, Zn and B should be corrected in the following season. For the largest N rate, flowers dry weight and their N content increased significantly, confirming the results from Obreza and Rouse (1993) and Davies and Albrigo (1994): from 5.8 g dry weight tree^{-1} in the lowest N level to 9.7 g dry weight tree^{-1} in the largest one, and from 28.8 g N kg^{-1} to 32.3 g N kg^{-1}, respectively. According to Tagliavini *et al.* (1997) optimum N concentration in the flowers should range from 22.2 to 44.5 g N kg^{-1}.

References

Davies F S and Albrigo L G 1994 Citrus, CAB International, Great Britain, p. 244.

Guerra JMA 1994 Estudos Técnicos (22), DRAALG, p. 107.

Obreza TA and Rouse RT 1993 Hort. Science 28, 111-114.

Tagliavini M, Quartieri M and Millard P 1997 Plant Soil 195, 137-142.

W. J. Horst et al.(Eds.). Plant nutrition – Food security and sustainability of agro-ecosystems. 782~783, 2001

Effect of fertigation timing and aeration on nitrogen dynamics in Chinese cabbage cultivation

H. T. Heuberger[1)], S. von Tucher[2)] and W. H. Schnitzler[1)]

[1)] *Chair of Vegetable Science, Crop Physiology and Quality Research, TU München-Weihenstephan, Dürnast 2, Freising, D 85350, Germany, heuberger@vegetable.de*

[2)] *Chair of Plant Nutrition, TU München-Weihenstephan, Am Hochanger 2, Freising, D 85350, Germany, tucher@pollux.weihenstephan.de*

Key words: aeration, Chinese cabbage, fertigation timing, fertiliser recovery, nitrate, plant sap, subsurface drip irrigation

Abstract

Different fertigation timing strategies and forced soil aeration in a subsurface drip irrigation system were evaluated for their effect on N availability and accumulation of N by Chinese cabbage (*Brassica rapa* L. ssp. *pekinensis* (Lour.) Hanelt.), as well as for their agronomic effects. In all strategies, 40% of the nutrients were applied broadcast prior to planting. After planting 60% of the nutrients were fertigated either in equal or increasing doses based on an uptake model. In one equal-dose treatment air was forced through the drip tapes when soil moisture was high.

Soil NO_3^- content was low after planting, increased to 23 and 14.5 mg NO_3^--N kg^{-1} soil at 0-30 and 30-60 cm soil depth (i.e. 110 and 69 kg N ha^{-1}, resp.) 4 weeks after planting (WAP) and dropped to 3.6 to 4.5 mg NO_3^--N kg^{-1} 6-8 WAP. The N status of the plants was adequate during initial growth and turned drastically deficient during the exponential growth phase in all treatments. 4 WAP plant N status was low despite high soil nitrate levels. This was probably caused by (1) inadequate access of the root system to the nutrients applied in fertigation due to soil compaction and emitters placed at 15 cm soil depth and (2) limited root functions due to soil moisture above field capacity for 2 weeks. Aeration had no effect on N uptake under the described conditions. 6 WAP the fertigated NO_3^-, which had not been taken up was lost after intensive rainfall. With fertigation according to the model, a slight increase of NO_3^-content in the cabbage head was observed before harvest. Under the conditions of this field experiment, the fertigation strategies with equal weekly rates resulted in more plant biomass and more marketable produce compared to strategies with increasing weekly rates following the N uptake model.

Introduction

Fertigation offers the possibility of 'spoonfeeding' plants according to their actual demand or the desired application. The key questions for a fertigation strategy are: 1) when to start, 2) how to split and 3) when to stop? The fertigation timing is particularly important for N because the soil nitrate concentration should be kept to a minimum in order to reduce potential leaching.

Forced soil aeration by means of injection of atmospheric air under pressure into the soil *via* a subsurface drip irrigation (SDI) system is thought to increase the O_2 concentration in wet soil. It has been postulated that reduced denitrification, enhancement of N mineralisation and uptake by the plants ensue.

Materials and methods

Different strategies of fertigation timing and the effect of aeration on N availability and uptake were studied in the field. Chinese cabbage 'Bilko' was transplanted (45 cm x 50 cm) on June 9 (2000) to a silty clay loam with a SDI system installed at 15 cm depth, 2 drip laterals per bed (1.5 m width) and 30 cm emitter spacing. P_2O_5 and K_2O were supplied at 80 and 340 kg ha-^{1}, respectively. N was supplied at 210 (based on N_{min} target value for Chinese cabbage, *i.e.* 250 kg N ha^{-1} minus 40 kg NO_3^-N ha^{-1} in the soil) or 160 kg N ha^{-1} (based on 80% of the N_{min} target, which is 200 kg N ha^{-1}). 40% of the nutrients were broadcast applied prior to planting (84 and 64 kg N ha^{-1}, resp.) with N applied as Ca cyanamide. The cabbage was transplanted 6 weeks later to avoid any damage from Ca cyanamide on the young seedlings. The remaining 60% of the nutrients were supplied by fertigation as follows:

(1) in equal portions weekly, starting at planting (0W linear),

(2) in equal portions weekly, starting 2 weeks after planting (WAP) (2W linear),

(3) as for (2) but with forced aeration at 0.6 bar for 90 minutes after each irrigation event or when water potential (by tensiometer) exceeded –9.0 kPa (2W linear air),

(4) rates adjusted weekly according to a N-uptake model starting 2 WAP (2W model), and

(5) as for (4) but with N according to 80% of the N_{min} target value (2W model 80%).

$Ca(NO_3)_2$ and a soluble fertiliser with 15 (11.3 NO_3-N + 3.7 NH_4^+-N), 5, and 30 % of N, P_2O_5, and K_2O, respectively were used for fertigation. The crop was drip-irrigated automatically with 5 l m^{-2} when soil water potential 15 cm away from the emitters fell below –12 kPa. Forced aeration was automatically started at soil water potential above –9 kPa, which is not extremely wet but safe in time to avoid oxygen deficiency.

Results

Nitrate in the soil

NO_3^--N contents of the soil sampled 10 cm apart from

the drip tape and between two plants are presented in Table 1. One WAP of Chinese cabbage 4.6 mg NO_3^--N kg^{-1} soil were available in the top 30 cm. This represents about 40 kg N ha^{-1} indicating that most of the NO_3^- applied and available before planting could not be retrieved. 4 WAP 23 and 14.5 mg NO_3^--N kg^{-1} soil were available at 0-30 and 30-60 cm soil depth, respectively. In the following 2 weeks rainfall amounted to 70 l m^{-2} and soil moisture was above field capacity. From 6 WAP until harvest, the soil in the top 60 cm contained only 3.1 to 4.5 mg NO_3^--N kg^{-1}, with no differences between treatment.

Table 1. Soil NO_3^--N content

NO_3^--N content (mg kg^{-1}) in 0-30 and 30-60 cm soil depth. Means across treatments at different weeks after planting (WAP).

Depth (cm)	1	3	4	6	8
			WAP		
0-30	4.6	3.5	23.0	3.8	4.5
30-60	-	3.7	14.5	3.4	3.1

Table 2. NO_3^- in the plant

NO_3^- concentration (mg kg^{-1} FW) in the petiole sap (3-6 WAP) and in the head (7 and 8 WAP) of Chinese cabbage as affected by fertigation timing strategy and aeration. The bottom line gives recommended values after Matthäus and Matthäus (1997)

Treatment	3	5	6	7	8
			WAP		
	Petiole sap			Head	
0W linear	4590 a	170a	170 a	200b	180b
2W linear	4090 a	100ab	130 a	140b	160b
2W model	-	30b	120 a	950a	520a
2W model 80%	-	40b	120 a	610ab	320ab
2W linear air	3830 a	110ab	70 a	340b	200b
Recomm.	5000	4000			

Values followed by the same letter in a column are not significantly different at 5% (Tukey test)

Plant N

Plant N status was monitored by NO_3^- analyses initially in the petiole sap of the youngest fully expanded leaf and after heading in the head (Tab. 2) as described by Matthäus and Matthäus (1997). The NO_3^- concentration of the petiole sap 3 WAP was 4000-4500 mg NO_3^- kg^{-1} FW in the linear treatments. At 5 and 6 WAP, NO_3^- concentration in the petiole sap was close to zero in all fertigation and aeration treatments. In the 7th and 8th WAP, NO_3^- concentration in the head was highest in the two 2W model treatments (320-950 mg kg^{-1} FW) where higher rates of NO_3^- were applied 5-6 WAP as the heads developed. However, the NO_3^- level in these heads, *i.e.* in the saleable product, was still beneath regulatory limits.

Approximately 52-83 kg N ha^{-1} had been accumulated in whole shoots at harvest 8 WAP (not shown). There were no treatment effects on shoot N accumulation. Limited N uptake and resulting N deficiency resulted in a low biomass of 48 t ha^{-1} FW and a marketable head yield of 29 t ha^{-1} (trial means, not shown). Within the limits of this trial, the linear fertigation strategies resulted in higher yields compared to the fertigation strategy based on the N uptake model ($P<0.05$, not shown).

Discussion

The low soil NO_3^- levels after planting as well as soil and plant NO_3^- levels from 5 WAP until harvest in contrast to the sufficiently high N input was striking and seeks for explanation.

The Ca cyanamide had been broadcasted 6 weeks before the Chinese cabbage was planted and within that time complete nitrification may have occurred. Part of the NO_3^- may have been leached below 60 cm soil depth where the soil was not sampled.

3 WAP N uptake was still within normal ranges since NO_3^- concentration in the petiole sap was 4000-4600 mg NO_3^- kg^{-1} FW, which is close to the recommended concentrations for Chinese cabbage at this stage (Tab. 2, Matthäus and Matthäus, 1997). Therefore, NO_3^- level in the root zone was possibly still adequate at this stage. Two weeks later (5 WAP) NO_3^- concentration in the petiole sap was less than 200 mg kg^{-1} indicating severe N deficiency. The low plant N status is contradictory to the high NO_3^- level in the top 60 cm of the soil measured 4 WAP. Roots may not have had adequate spacial access to soil nitrate pool or/and root functions might have been inhibited. Reduced root functions may have been caused by hypoxic conditions due to high rainfall intensity and soil moisture above field capacity from 4 to 6 WAP. Access to the fertigated nutrients may have been inadequate because N was supplied through the drippers in 15 cm depth in the compacted silty clay loam soil. Under these conditions, the plants' roots may not have penetrated to the depth that the N (and O_2) was supplied. This would also account for the lack of an effect of forced soil aeration.

6 WAP and later both soil and plant NO_3^- levels were again low. The high rainfall intensity and soil moisture above field capacity from 4 to 6 WAP are likely to have contributed significantly to N loss via leaching and probably to a minor degree via denitrification. Whether soil aeration affected the denitrification process is not yet analysed.

Under the conditions of this trial the fertigation strategies based on equal weekly rates of nutrient supply were agronomically superior to strategies with increasing weekly rates derived from a N uptake model. The model approach may even carry the risk of too high NO_3^- contents in the produce if sink size is low and N availability is not limited.

References

Matthäus K and Matthäus D 1997 Der Gemüsebau/Le Maraîcher (60) 6, 4-6.

W. J. Horst et al.(Eds.), Plant nutrition – Food security and sustainability of agro-ecosystems. 784~785, 2001

Optimization of *Anthurium andreanum* mineral nutrition in soilless culture under tropical conditions

L. Dufour
Institut National de la Recherche Agronomique, Centre Antilles Guyane, Station Agropédoclimatique de la zone Caraïbe, Domaine Duclos, 97170 Petit-Bourg, France, dufor@antilles.inra.fr

Key words : anthurium, cut flower, environmental damage, fertilisation, nutrient balance

Abstract

Soilless and sheltered cultivation of anthurium has been developed in tropical countries in order to improve yield and quality of cut flowers. Improved knowledge of plant nutrients requirements is necessary for optimizing yield and reducing environmental damages. Four nutrient solutions, differing in nitrogen and calcium concentrations, were tested on 'Cancan' hybrid plants. Plants growth, yield and the quality of the flower were determined. Nutrient, leaching and substrate solutions, substrate and different parts of the plants were analyzed during the second year after planting.
The best results regarding plants growth and yields were obtained with the solution having the lowest calcium (2.25 meq Ca l^{-1})and the highest ammonium (2.43 meq NH_4 l^{-1}) concentrations. The leaching rate was always very high: about ¾ of the provided nutriments were lost to the environment. Potassium is the major mineral component of the plants and plays an important role in the fertilisation balance.

Introduction

Anthurium cultivation is becoming increasingly important in cut flower production. In the tropical French islands, it is in more over important for agricultural diversification. In response to the market quality requests and because of sanitary problems, traditional cultivation systems have been replaced by sheltered and soilless systems during the past twenty years. These expensive installations have to provide high yields and quality. Since the nutrients requirements of the plant are not well known, the producers are inclined to apply great amounts of fertilisers. As there is no recycling system in Guadeloupe, so nutrients are lost, increasing the production costs and causing environmental damage. Therefore, the aim of this study was to improve the knowledge of the plant needs in order to optimize the yields and to limit fertiliser losses to the environment.

Materials and methods

Cultivation conditions

Tissue-cultured plants of *A. andreanum* 'Cancan' (size 8-10 cm) were planted in 2.5 litres pots filled with a subtrate consisting of a 1:2 ratio of composted and desinfected wood chips and 5-15 mm pouzzolane (volcanic gravel). The pots were settled in a 200 m² shade house. The covering materials stopped approximately 87 % of the Photosynthetic Active Radiation (PAR). One year after planting, the plants were transfered into 5 litres pots.

Using a fertigation system 4 nutrient solutions were applied (Tab. 1). There were 4 replications of each fertilisation treatment with 64 plants in each replication. The focus was set on calcium and nitrogen because of their effect respectively on the spathe coloration (Higaki *et al.*, 1980) and on the flower production (Boertje, 1978). Furthermore the influence of nitrogen form (nitrate vs. ammonia) was checked. P (14.1 mg/l) and Mg (37.8 mg/l) as well as micronutrients (B, Cu, Fe, Mn, Mo, Zn) were applied at the same rate for the 4 solutions.

Table 1. Concentrations of used the 4 nutrient solutions in the study (mg/litre)

	NO_3-N	NH_4-N	K	Ca
N_1Ca_1	100.8	24.9	125.6	90.0
$N_1Ca_{1.5}$	98.1	27.9	125.6	**135.0**
$N_1Ca_{0.5}$	91.3	**34.3**	125.6	**45.0**
$N_{0.5}Ca_{0.5}$	**41.0**	22.0	**62.8**	**45.6**

Measurements and analysis

Measurements of the growing leaves and of the flowers of ten plants of each replication were taken each week. The mature flowers of all the plants were harvested the same day and the yields were recorded.

Each month samples of nutrient solutions, of leaching solutions, and of induced percolates, that give an accurate assessment of the nutrient status of the substrate (Lemaire *et al.*, 1995) were taken and analyzed for NO_3^--N, NH_4^+-N, P, K, Ca and Mg. The electric conductivity (EC) of the solutions was also checked.

Four times between day 186 and 390 after planting, two plants from each replication were collected. Roots, stem, mature blades, mature petioles, young blade, young petiole, peduncle, spathe and spadix of each plant were weighted and dried at 80°C for 2 days. Then the dry matter was weighted and the samples were analyzed.

Results

Plants growth and yield

The vegetative period, during which no flower was produced, was longer for the plants that received the smallest amount of nitrogen (Fig. 1).

Harvested flowers were greater for the $N_1Ca_{0.5}$ treatment (Tab. 2) then for N_1 Ca_1. No color breakdown occurred in the spathes, even with the lowest calcium level (data not shown).

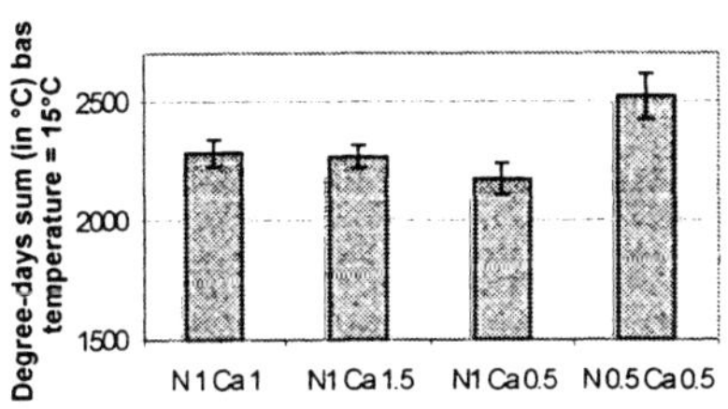

Figure 1. Duration of the vegetative growth stage

Table 2. *Number of flowers harvested between days 495 and 860 for 40 plants per treatment*

	preduncle length			
	5055 cm	55-60 cm	> 60 cm	big fl./ plant/yr
N_1 $Ca_{0.5}$	85	64	61	5.25
N_1 Ca_1	85	62	19	4.15

Analysis results

A nutrient balance (Tab. 3) was calculated for one year (Day 258 to 623) for the $Ca_{0.5}$ treatment. The average EC value of the solution used in this treatment was 1.10 $mS\times cm^{-1}$. The substrate solution EC average value was $0.95mS\times cm^{-1}$ while, the optimum recommended by Sonneveld and Voogt (1983) is 0.691 $mS\times cm^{-1}$.

Table 3. Nutrient balance for one year of cultivation

Nutrient ($g\times m^{-2}$)	N	P	K	Ca	Mg
Nutrient solution	16.9	7.3	18.0	8.8	5.9
Leachate	10.6	4.5	12.3	7.4	4.4
Substrate solution	0.1	0.0	0.2	0.0	0.0
Plants uptake	3.6	0.5	4.4	3.0	1.4

Substrate solution represents the composition variation of this solution during the year.

Boertje (1978) had shown the prevalence of potassium in leaves. Potassium was also the major element in almost all the organs (stem and spadix excepted; Tab. 4). The concentration of potassium in the whole plant was 3.1% during the vegetative growth phase and 2.5% during the flower production phase, while N concentration showed no differences in the 2 growth periods (1.8% vs 1.9% N)

Table 4. Nutrient concentrations in different plant organs from mature plants in treatment N_1 $Ca_{0.5}$

organs	N	P	K	Ca	Mg
roots	2.03	0.36	2.30	2.04	1.45
mature blades	1.93	0.16	2.46	1.62	0.58
mature petioles	0.97	0.26	3.49	1.39	0.45
young blade	2.79	0.33	3.07	1.55	0.53
young petiole	1.17	0.34	4.76	0.82	0.38
stem	2.50	0.28	1.68	2.38	0.75
peduncle	1.31	0.23	3.85	0.58	0.36
spathe	2.13	0.32	2.58	1.28	0.36
spadix	2.38	0.43	1.89	3.66	0.96

Discussion

The nitrogen fertilisation level influenced plants growth and production. A 63 mg N/l supply was inadequate and the best results were obtained with 1/3 NH_4-N and 2/3 NO_3-N. Higher NH_4 rates should be tested.

The levels of different macronutrients in young and mature leaves were higher (except for nitrogen) than those measured by Sonneveld and Voogt (1983). The N level in young and mature leaves in our study was also lower than the ones measured by Boertje (1978), Higaki *et al.* (1980) and Mills and Scoggins (1998). However, if the nitrogen supply would have been increased, the EC value might be too high compared to the levels reported by Sonneveld and Voogt (1983).

Taking into account the uptake of the mature plant (Tab. 3) the following composition of the nutrient solution can be calculated: 65 mg NO_3-N/l; 35 mg NH_4-N /l; 14 mg P/l; 120 mg K/l; 80 mg Ca/l; 40 mg Mg/l; calculated EC = 1.1 $mS\times cm^{-1}$.

The nutrient balance (Tab. 3) indicates that more than 100 kg of nitrogen and 45 kg of phosphorus per ha are lost during one year of cultivation in conditions as decribed in our study. In Guadeloupe the growers usually apply about 1/3 more nitrogen than tested in this study. It has to be hypothesized that it might be possible to reduce the leachate by decreasing the amount of solution supplied without any yield reduction. However, the most important way to limit the pollution by nitrate and phosphate is the recycling of the leachate solution.

References

Boertje GA 1978 Acta Hort. 82, 159-164.

Higaki T, Rasmussen HP and Carpenter WJ 1980 J. Am. Soc. Hort. Sci. 105, 441-444.

Lemaire F, Guerin V and Charpentier S 1995 Acta Hort. 401, 451-459.

Mills HA and Scoggins HL 1998 J. Plant Nutr. 1, 199-203.

Sonneveld C and Voogt W 1983 Acta Hort. 342, 61-67.

W. J. Horst et al.(Eds.), Plant nutrition – Food security and sustainability of agro-ecosystems. 786~787, 2001

Effect of fertigation timetable on growth and nitrogen content of an ornamental shrub (*Ligustrum ovalifolium* L.)

M. Salaün, S. Charpentier, V. Guérin and L. Huché-Thélier
National Institute of Agronomic Research, UMR SAGAH INRA/INH/Université d'Angers, Angers, 42 rue Georges Morel B.P. 57, Beaucouzé Cedex, F 49071, France, Morgane.Salaun@inh.fr

Key words: fertilisation, growth, *Ligustrum ovalifolium*, nitrogen, storage

Abstract

In order to limit nitrate leaching without affecting growth of soilless culture, a better adjustment between fertilisation timetable and nitrogen (N) seasonal demand of plants is required. One-year-old *Ligustrum ovalifolium* L. were grown outdoors in containers for two years. The first year plants received three different fertigation -water and nutrients- timings (continuous, during spring and autumn or only during spring). The second year, only water was provided, compelling plants to use their reserves for regrowth. Growth and total N were determined for nineteen destructive harvests along the two years. Stopping fertigation in summer had no immediate effect on growth. Restoring fertigation in autumn (did not) increase current growth but allowed N pool reconstitution. Plants receiving continuous fertigation and spring and autumn fertigation had an identical regrowth. Regrowth was linked to an identical N status of both plants in winter. Treatment combining spring and autumn fertigation allowed to limit nitrate supply and to ensure a sufficient N storage to support regrowth.

Introduction

Soilless ornamental cultures are an important source of pollutans. In order to limit nitrate leaching without affecting plant growth, a better adjustment between fertigation timetable and N seasonal demand of plants is required. Optimization of N fertilisation for perennial shrubs needs to take into account both of N uptake and N redistribution within the tree. Remobilisation of N reserves in the spring provides a supply of highly assimilable N for regrowth (Habib *et al.*, 1990). The amounts of N remobilised are affected by the N supply the previous year (Millard, 1996). The aim of the study was to determine fertilisation strategies that ensure the proper formation of storage pools and advantage regrowth of an ornamental shrub in containers.

Materials and methods

One-year-old *Ligustrum ovalifolium* L. were cultivated from March 1998 to July 1999. Plants were grown outdoors in pots (4L) containing a perlite-sphagnum peat mixture (50/50 v/v pH 6). Three identical plots of 108 plants (6 plants per m^2) were supplied with drip irrigation containing nutrients (mg L^{-1}: 158 NO_3^--N, 28 NH_4^+-N, 203 K, 118 Ca, 35 Mg, 40 P, and minor elements). Water and nutrients were supplied during (a) spring (13 May-10 Jul.), summer (10 July-8 Sept.) and autumn (8 Sept.-20 Nov.) (SSA) (b) spring and autumn (SA) and (c) only spring (S). Outside of these periods, plants received pure water. With this substrate, when fertiliser supply stops, available nitrogen disappears in substrate solution before 3 days (Charpentier *et al.*, 2000). Plants were harvested intermittently over the two-year period (19 harvests). At each date, five plants were selected at random from each plot. Fresh matter of the five plants was pooled, weighed and dried at 60°C for the determination of the total soluble nitrogen by the Kjeldhal method.

Results

Plants receiving continuous fertigation had a regular growth (Fig. 1). Stopping fertilisation in summer had no immediate effect on plant growth. Restoring fertilisation in autumn allowed neither growth recovery nor a better growth than plants receiving spring fertilisation only. During winter, leaves fell except for SA. In the second year, regrowth was identical for SSA and SA plants, and higher than for S plants. Plants received no fertilisation in spring so the regrowth relied on N stored.

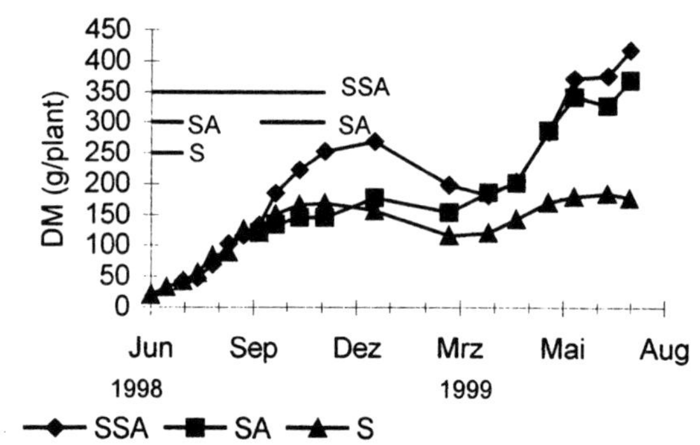

Figure 1. Dry matter (mg/plant) of whole plants of *Ligustrum ovalifolium* submitted to 3 fertigation timetables : SSA (13 May-20 Nov.), SA (13 May-10 July and 8 Sept.-20 Nov.) and S (13 May-10 July). Outside of these periods, plants received pure water. Values are means obtained on 5 plants.

Nitrogen absorption was regular when plants received continuous fertilisation (Fig. 2). Stopping fertilisation led to plant N use and dilution in growing tissues during summer. Restoring fertilisation in autumn allowed an immediate reconstitution of N pool. At the beginning of winter, the N concentration was the same in SSA and SA plants. For both, regrowth relied on an identical exhaustion of N pool. In treatment S, N concentration was low so it could explain the weak growth.

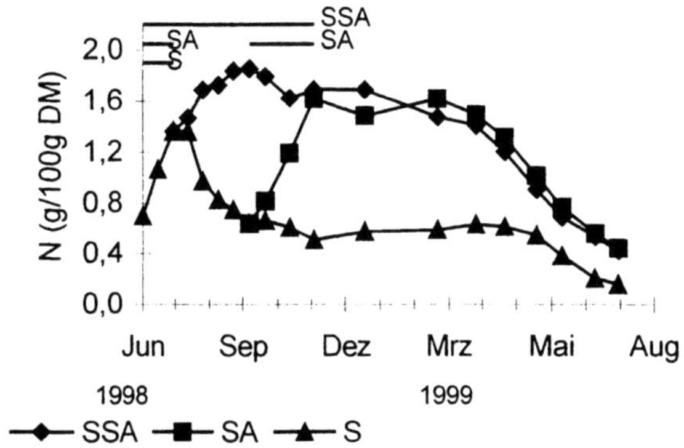

Figure 2. Total soluble N concentration (g/100g DM) of *Ligustrum ovalifolium* submitted to 3 fertigation timetables : SSA (13 May-20 Nov.), SA (13 May-10 July and 8 Sept.-20 Nov.) and S (13 May-10 July). Outside of these periods, plants received pure water. Values are means obtained on 5 plants.

Discussion

Autumn fertilisation for SA plants advantaged storage rather than current growth and allowed an identical regrowth to SSA plants. Uptake of N in the autumn has been shown to contribute more to storage and subsequent remobilisation of N the following spring than the N uptake during summer in several species of deciduous tree (Millard 1996, Tagliavini *et al.* 1999). In *Ligustrum*, autumn fertilisation provided a sufficient quantity of N that was remobilised for regrowth. So, treatment combining spring and autumn fertilisation seems to be adequate to limit nitrate leaching and to ensure a suitable N storage for regrowth.To optimize N use, fertigation must be restored after bud break and before the exhaustion of the N pool. Future experiments will determine the period where plant regrowth solely depends on N reserves.

References

Charpentier S, Guérin V and Huché-Thélier L 2000 Acta Hort., (in press).

Habib R, Loisel P and Wallach D 1990 Acta Hort. 276, 105-112.

Millard P 1996 Z. Pflanzenernähr. Bodenk. 159, 1-10.

Tagliavini M, Millard P, Quartieri M and Marangoni B 1999 Plant Soil 211, 149-153.

W. J. Horst et al.(Eds.), Plant nutrition – Food security and sustainability of agro-ecosystems. 788~789, 2001

Effect of foliar fertilisation with magnesium, sulfur, manganese and boron to sugar beet, oilseed rape, and cereals

K. Orlovius

Agricultural Division, Kali und Salz GmbH, PB 102029, D-34111 Kassel, Germany, kristian.orlovius@kalisalz.de

Key words: boron, cereals, foliar fertilisation, magnesium, manganese, oilseed rape, sugar beet, sulfur, yield response

Abstract

Sulfur foliar fertilisation in the form of epsom salt in pot trials on spring wheat resulted in considerable yield responses under S deficiency conditions. The yield increase depended on the available sulfur reserves of the soil. An early application at stem elongation was more effective than at heading.

In field trials (1998 to 2000) a foliar application of epsom salt with the addition of Mn and B resulted in an average yield increase of 3-7% for oilseed rape and 3-5% for sugar beets although soil analysis indicated a sufficient Mg, B, Mnsoil status. Yield increase at different sites and years could not be explained by differences in the plant nutrient content. As prognosis of fertiliser demand based on soil or leaf analysis seems to be inadequate for these nutrients, preventive foliar fertilisation for crops with a high B and Mn demand is (e.g. sugar beets) recommended.

Introduction

The use of foliar fertilisation for agricultural crops has increased substantially in the last 10 years in Europe. Particularly the nutrients magnesium (Mg) and sulfur (S) are applied as foliar fertiliser in the form of epsom salt. In addition foliar spray with boron (B) and manganese (Mn) became more important for crops such as oilseed rape and sugar beets to react on visible deficiency symptoms within the fields (Shorrocks 1997). However, with foliar spray only small amounts of nutrients are applied. In pot and field trials it was investigated whether a small application of the nutrients Mg, S, B, Mn in form of foliar sprays is able to improve the contents of these nutrients in the plants.

Methods

In pot trials the effect of a foliar fertilisation with S was examined on grain yield of spring wheat cropped to a sand (1.2% humus content) in Mitscherlich pots (6 kg pot^{-1}) at five different levels of S-supply in the soil (0, 5, 20, 50, 100 mg S pot^{-1}). To avoid deficiency in other nutrients than S, a base fertilisation with macro- and micronutrients was applied to each pot. The different sulfur contents in soil were established via K- and Mg-sulfate. Accordingly K- and Mg-chloride was applied in low S treatments to adjust K and Mg supply. The foliar fertilisation was carried out by using a 5 % solution of epsom salt (Mg: 10 %, S: 13 %) at the beginning of stem elongation or at heading (each 16 mg S pot^{-1}).

In field trials the effect of a new foliar fertiliser called 'Bittersalz microtop' (9 % Mg, 12% S, 1% B and 1% Mn in directly plant available forms) was investigated for oilseed rape and sugar beets from 1998 to 2000. The trials on oilseed rape were conducted on loamy sands and loam soils in Northern and North-Eastern Germany. The trials on sugar beet were carried out in typical sugar beet regions of Germany on silty soils. The foliar fertiliser was applied at two dates to oilseed rape (beginning of stem elongation and at inflorescence emergence), and to sugar beet (at row closure and at the end of July) with 25 kg ha^{-1} (3 kg S ha^{-1}, 2.26 kg Mg ha^{-1}, 250 g Mn ha^{-1}, 250 g B ha^{-1}) at each of the two dressings.

Results and Discussion

Pot trials

Without S and at low S supply rates in soil (5 and 20 mg S pot^{-1}) insufficient S-supply of the plants was indicated by visible deficiency symptoms such as yellow discoloration, disturbed growth, weak tillering and insufficient grain formation. Hence, S-foliar fertilisation increased seed yield in particular at low S supply in soil, when yield increases > 100% occurred (Fig. 1).

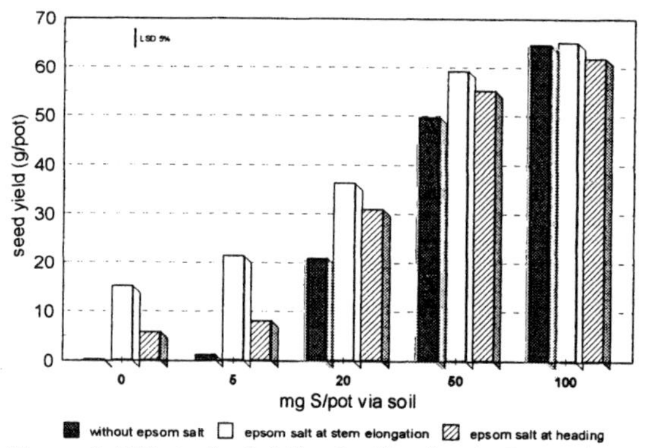

Figure 1. Effect of a S-foliar fertilisation in form of epsom salt on the yield of spring wheat in a pot trial with different S supply via the soil.

Even at S-soil-supply of 50 mg S pot^{-1} deficiency symptoms were not visible but yield still increased by 15% due to S-foliar spray. Only at the highest soil S-supply foliar fertilisation had no yield effect.

The S contents in the plants at stem elongation and in grains at maturity were increased by the treatments (data not shown). Crops achieved highest yields at a tissue S-content > 0.3% S or a N/S ratio of < 15 determined one week after the first spraying. Early foliar application at stem elongation seemed to be more effective than at heading. This confirms the recommendation of an "as-early-as-possible" S-application to cereals (Haneklaus *et al.* 1995). On the other hand the presented results show that severe S deficiency symptoms ocurring late at heading can be diminished by late S-application. However, optimum yields could not be achieved with this late S application.

Field trials

Despite soil analysis indicated sufficient Mg-, B-, Mn-supply for oilseed rape, tissue analysis revealed an insufficient B- and S-level at several trial sites. The application of the multi-nutrient foliar fertiliser caused an increase in B and Mn leaf contents of 10 mg kg^{-1}, each and in S content of 0.5 g kg^{-1}. The Mg concentration of the leaves did not change. The increase in the nutrient status could explain the average yield increases of 3-7 % for oil seed rape (Fig. 2). Matthey *et al.* (2000) found similar yield increases of 3% in oilseed rape due to foliar application of epsom salt with 2 x 20 kg ha^{-1}. Foliar fertilisation of 400 g B ha^{-1} led to yield increase of 8% in rape (Krauze *et al.* 1991).

In sugar beet increases in the nutrient content of leaves were not distinct because of the longer interval between foliar treatment and leaf sampling (6 weeks). The decrease in the nutrient content of the leaves during the period of most intensive plant growth after row closure could explain that no foliar treatment effect was detectable by tissue analysis 6 weeks after the treatment. Nevertheless, the foliar treatments at row closure and at the end of July resulted in yield increases between 3 and 5% (0.38 and 0.64 t ha^{-1} sugar) on average for 24 trials between 1998 and 2000 (Fig. 2).

However, considerable differences in yield responses between sites and years were observed which could not be explained by the differences in nutrient content either in the soil or in the plants.

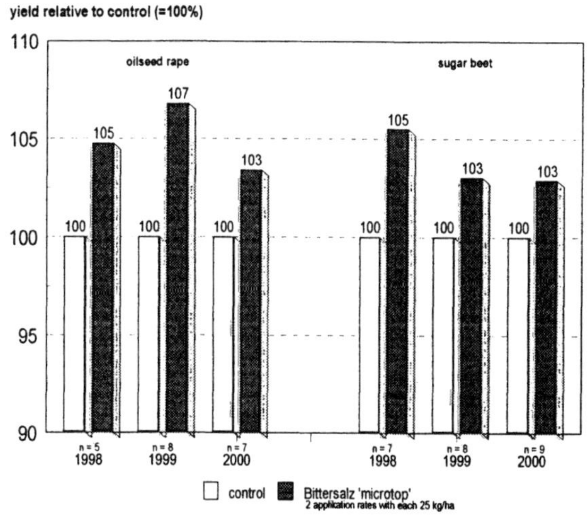

Figure 2. Effect of foliar fertilisation with Mg, S, Mn and B on the yield of oilseed rape and sugar beets in three-year filed trials

Conclusions

Even in situations of apparently adequate soil nutrient supply, multiple nutrient deficiencies are frequently found in crops under conditions of farming practice. The direct nutrient supplementation via foliar fertilisation may quickly alleviate such nutrient bottlenecks and reduce the risk of yield losses due to insufficient nutrient supply from the soil. The presented results show that preventive foliar fertilisation avoid yield losses of crops, like oilseed rape or sugar beets with a high B and Mn demand for site conditions where a low availability of B and Mn (high pH values, frequent drought periods) can be expected.

References

Haneklaus S, Murphy DPL, Nowak G, Schnug E 1995 Z. Pflanzenernähr. Bodenk. 158, 83-85.

Krauze A, Bowszys T, Bobrezecka D, Rotkiewicz D 1991 Proc. Eighth Int. Rapeseed Congress, Saskatoon, Sask., Canada p 547-553.

Matthey J, Sauermann W, Finck M 2000 *In* VDLUFA-Schriftenreihe 53, 72-79.

Shorrocks VM 1997 Plant Soil 193, 121-148.

W. J. Horst et al.(Eds.), Plant nutrition – Food security and sustainability of agro-ecosystems. 790–791, 2001

Effect of different methods of fertilisation on grape yield in Qazvin

M. Shahabian[1)], M. Mostashari[2)] and M. J. Malakouti[3)]

[1)] *Soil Fertility and Plant Nutrition Department, Soil and Water Research Institute, North Karegar, Jalal Ale Ahmad Highway, 14155/6185, Tehran, Iran, Mshahabian@yahoo.com*

[2)] *Soil and Water Research Center, Agricultural Research Center, Qazvin, Iran*

[3)] *Soil Sci. Dep. Agricultural Faculty, Tarbiat Modarres University, Tehran, Iran*

Key words: fertiliser application, fruit set, grape, spray, soil placement

Abstract

At present 2.3 million tons of grapes are produced annually from 280 thousand hectares of grapevines in Iran. Grape production is projected to increase to 5.6 million tons per year by the end of the third development plan. This increase will not be possible without appropriate management practices, particularly with respect to micronutrient supply in the context of a balanced fertiliser program. An investigation was carried out in 1998 at 2 places (Esmail Abad and Sultan Abad of Qazvin province, central Iran) to find the simplest way to increase yield of grapes (*Vitis vinifera;* variety Thompson seedless; syn Sultana). The experiment was carried out in a completely randomized block design with five treatments and three replications. The results showed that the grape yield in the control treatment was 4.55 t ha^{-1} in Esmail Abad and 4.94 t ha^{-1} in Sultan Abad, while it was 7.73 t ha^{-1} in the foliar application treatment in Esmail Abad and 19.20 t ha^{-1} in foliar plus soil application treatment in Sultan Abad. A comparison of the different methods showed that the foliar plus soil application treatment showed the best results. However, foliar application of K, Mg and micronutrients were the fastest and the most economical way to increase grape yield.

Introduction

The rapid increase in world population, especially in developing countries such as Iran, requires greater production of agricultural products. Appropriate use of fertilisers could be an important factor to enable this increased production. However, unbalanced mineral nutrition has to be regarded as a constraint on plant production in Iran (Malakouti, 2000).

Grapes are potentially a good source of foreign exchange if improvements in quality are achieved. About 282000 ha is planted to grape in Iran, and of that 92.5% is producing. Approximately 2315000 tonnes of grapes are produced in Iran, and the Qazvin province produces about 13% of the total.

Grape productivity in Iran stands on average at 8.9 t ha^{-1}. Relative to other grape producing countries, Iran is the 7th most productive grape producer (Anonyms, 1999). However, to compete effectively on the world markets yields per hectar must be increased.

Materials and methods

A trial was conducted using *Vitis vinifera* var. Thompson seedless (syn. Sultana) in a completely randomised block design with five treatments (Tab. 1) and three replications. The trial was carried out in two regions (Esmail Abad and Sultan Abad) located in the Qazvin province of central Iran. Some soil characteristics are given in Tab. 2.

At harvest cluster weight, fresh weight per vine and total grape yield per ha were measured. Differences between the treatments were tested using the Tukey HSD procedure at the 5 % significance level.

Table 1. Treatments applied to Thompson seedless grapes.

T1	Control (N+P)
T2	T1+foliar application of N, B and Zn for fruit set
T3	T1+foliar application of macro and micronutrients
T4	Soil application (deep placement) of NPK+Mg+Fe+Zn+Mn+Cu+B+ organic matter
T5	T3+T4

Table 2. Soil chacacteristics

	Esmail Abad	Sultan Abad
pH	7.8	7.5
OM [%]	0.5	0.4
P [ppm]	2.4	10
K [ppm]	225	220
B [ppm]	0.27	0.22
Cu [ppm]	1.5	1.7
Mn [ppm]	7.5	4.0
Fe [ppm]	10.5	3.8

Results

Esmail Abad

Cluster weights were not significantly affected by the treatments, but yield per vine and yield per hectare were increased by treatment T3 and T5 relative to the control treatment (Tab. 3).

Table 3. Effect of nutrient supply treatments on cluster weight and yields per vine in Esmail Abad

Treatments	cluster weight [g]	fresh weight [kg per vine]	yield [t ha^{-1}]
T1	160 a	1.39 b	4.55 b
T2	195 a	1.88 ab	6.21 ab
T3	206 a	2.34 a	7.73 ab
T4	178 a	1.79 ab	5.89 ab
T5	180 a	2.26 a	7.45 ab

Sultan Abad

Treatments T3 and T5 significantly increased cluster weights compared to T1. Furthermore T5 resulted in a significantly higher yield per vine compared to T1 and T2 (Tab. 4).

Table 4. Effect of nutrient supply treatments on cluster weight and yields per vine in Sultan Abad

Treatments	cluster weight [g]	fresh weight [kg per vine]	yield [t ha^{-1}]
T1	487 c	1.98 b	4.94 b
T2	650 bc	3.05 b	7.62 b
T3	878 ab	4.58 ab	11.46 ab
T4	725 abc	6.04 ab	15.10 ab
T5	1043 a	7.68 a	19.20 a

Discussion

In previous experiments conducted in Esmail Abad it was shown that soil application of macro and micro nutrients as Ca, Mg, N, P, K were effective in increasing grape yield (Shahbian, 1998). However, in that experiments a foliar treatment was not tested. The present study revealed that foliar application is also effective in increasing weight of grape clusters and vine yield.

In Esmail Abad chilling damages caused low grape yield and there was no significant difference between the fertiliser application strategies tested. However, foliar application (T3) and foliar application plus deep placement of nutrients and soil organic matter resulted in highest vine yields. Also for the Sultan Abad region treatments T3 and T5 resulted in higher yields and cluster weights compared to the control treatment. Hence the combination of foliar application of macro-, micronutrients and deep soil application of NPK + Mg + Fe + Zn + Mn + Cu + B seems to be most effective for yield production.

It can be concluded that unbalanced nutrient supply is one of the reasons for low vine productivity in Iran. Other factors might be the high amount of $CaCO_3$ in the soil profile, HCO_3^- concentration in the irrigation water, low soil organic matter and depletion of K and micronutrients. Previous excessive application of N and P in combination with no micronutrient inputs may also have contributed to mineral nutrient imbalances. Balanced supply of all nutrients and organic matter could lead to improved productivity and grape quality, and consequently to greater financial returns.

References

Anonymes 1999 Agricultural Statistical Report. Statistics and Information Center. Ministry of Agriculture, Tehran, Iran.

Faust M 1989 Physiology of temperate Zone fruit trees.

Malakouti MJ 2000 Sustainable agriculture and yield increases through balanced fertilisation. Tehran, Iran.

Shahbian M 1998 Effect of some macro and micro nutrients on the quality and yield of grape in Qazvin. Faculty of Agriculture, Tarbiat Modarres University, Tehran, Iran.

W. J. Horst et al.(Eds.), Plant nutrition – Food security and sustainability of agro-ecosystems. 792~793, 2001

Effects of foliar applied K on California cotton

B. L. Weir, B. A. Roberts and S. Stoddard
University of California, 2145 Wardrobe Ave., Merced, CA 95340, USA

Key words: cotton, foliar fertilisers, KNO_3, K_2SO_4, $K_2S_2O_3$, potassium deficiency

Abstract:

Potassium deficiency symptoms become prevalent on leaves of many cotton varieties during the boll filling stage of growth. Soil applications of even large amounts of K are often not adequate due to fixation by vermitulic soils. A multi-year study was conducted to evaluate the response of California cotton (*Gossypium hirsutum*) varieties to foliar K applications. Rates of 10 kg/ha of K_2O were used employing potassium nitrate (KNO_3), potassium sulfate (K_2SO_4), and potassium thiosulfate ($K_2S_2O_3$). Potassium applied to the foliage about two weeks after the first bloom stage consistently gave positive yield responses. Yield responses ranged from 50 to150 kg/ha over the untreated control, with no change in lint quality.

Introduction

Cotton plants require K at a rate of 2 to 3 kg/ha/day during boll fill. These levels of K can be difficult to maintain in the vermiculitic soils used for cotton production in California. Foliar applications of K to the subtending leaf of developing bolls can provide extra nutrients when they are required. Tests have shown that a positive response can be realised from foliar applications of these nutrients when rapidly developing bolls are present to provide strong sinks for potassium.

Methods and materials

Field tests have been conducted since 1992 in which K was applied in foliar sprays at various times during the growing season. Two to five tests were conducted each year using various carriers of K. Materials utilised were potassium nitrate, potassium sulphate, and potassium thiosulfate, and the cotton cultivar was Acala Maxxa. Test plots were treated using a tractor sprayer set up for research work and all tests were conducted employing a completely randomized block statistical design.

Results

Each of the three foliar materials used, (potassium nitrate, potassium sulphate, and potassium thiosulfate), resulted in increased lint yields, (figures 1, 2, & 3). Greatest yield advantages were measured from foliar applications made at 2 to 3 weeks after first bloom, and the effect was similar regardless of the K source. Yield responses were positive anytime after first bloom until the end of blooming, but foliar applications made later in the season gave lower yield responses. Multiple split applications were more effective than a single dose. When compared to an untreated control, lint increases ranged from 50 to 150 kg/ha, with no change in lint quality.

These results show that positive benefits can be obtained from foliar applications of K. These and other cultural practices have resulted in annual lint yield increases of 32 kg/ha/year. New, more determinant cotton varieties require larger amounts of some nutrients during the boll development stage of growth. Even though soil K levels are adequate by current guidelines, foliar applications at 2 to 3 weeks after first bloom can result in positive yield responses.

References

Oosterhuis DM, Albers D, Baker WH, Burmiester C, Cothren JT, Ebelhar MW, Guthrie DS, Hickey MG, Hodges SC, Howard DD, Janes LD, Mullins G, Roberts BA, Silvertooth J, Tracy P and Weir BL 1993 *In* Proc. Beltwide Cotton Prod Conf., New Orleans, LA. Vol 3:1351. Nat'l Cotton Council, Memphis, TN.

Roberts BA, Kerby TA and Weir BL 1993 PPI/FAR Tech. Bulletin, 1993-1, Soil Sci. Soc. Am., 64-76.

Weir BL, Roberts BA and Kerby TA 1992 *In* Proc. Beltwide Cotton Conf. Nat's Cotton Council, Memphis, TN.

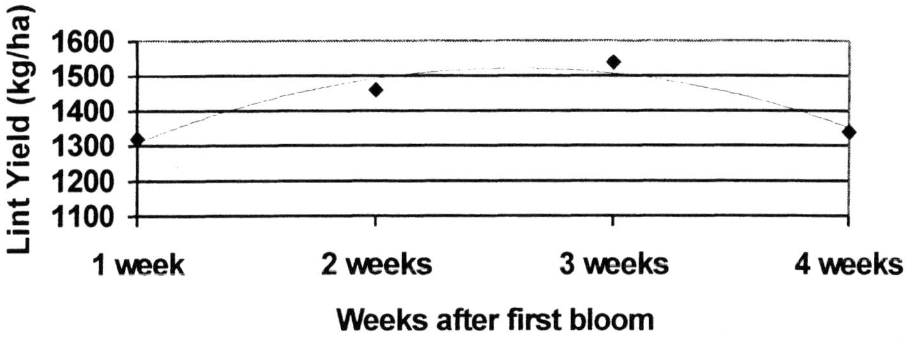

Figure 1. Effect of foliar KNO_3 on lint yields.

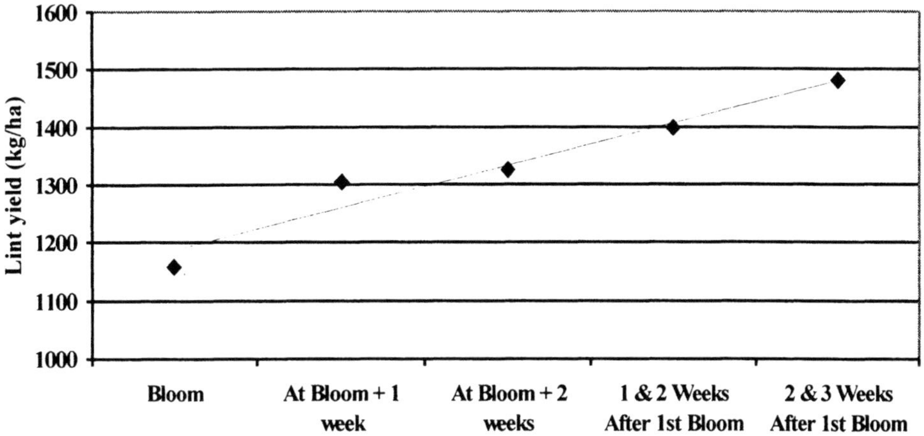

Figure 2. Effect of foliar potassium sulfate (K_2SO_4) on lint yields.

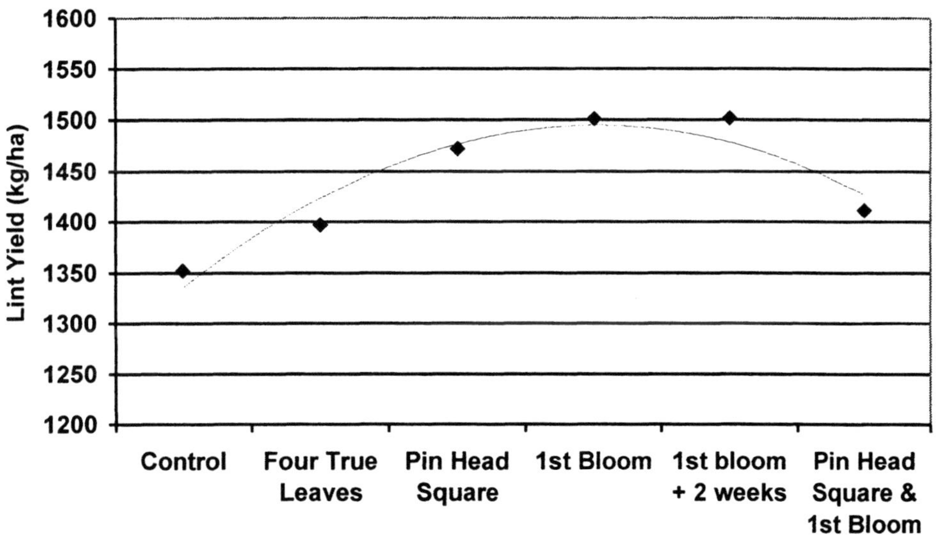

Figure 3. Effect of foliar potassium thiosulfate (KTS) on lint yields.

W. J. Horst et al.(Eds.), Plant nutrition – Food security and sustainability of agro-ecosystems. 794~795, 2001

Response of broad bean (*Vicia faba*) to method of phosphorus application and foliar application of micronutrients

A. T. Thalooth[1], G. M. Yakout[2] and A. O. M. Saad[1]
[1] *National Research Center, Field Crop Dept. , Dokki , Cairo, Egypt.*
[2] *Faculty of Agric. Suez Canal Univ. Agronomy Dept.,Esmeilia Egypt.*

Key words: broad bean, method of phosphorus application, micro elements, *Vicia faba*

Abstract

Broad bean plants sprayed with Fe, Mo, and Zn produced heavier pods, seeds, straw and whole plant when P was supplied via the foliage or the soil. Spraying the plants with the different micronutrients or their mixture with soil P application increased seed weight. The ratio of weight of pods/whole plant was increased due to foliar spraying with micronutrient elements. Shelling percentage was not affected neither spraying with different microelements nor method of P application. The highest seed yield was obtained by foliar spraying with Fe or Mo when P was either foliar or soil applied.

Introduction

The antagonistic effect between P and some micronutrient elements is well known and recorded. High P supply was found to be associated with incipient iron (Fe) chlorosis due to Fe precipitation in the soil (Mathan and Amberger, 1977). The antagonistic effect of P on Zn uptake was reported by Camak and Marchner (1986). Furthermore, Zn was rendered physiological immobile in the root tissue as revealed by substantial reduction in Zn concentration in the shoot as compared to its concentration in the root (Chahal and Ahluwalia, 1977). Thus, supplying nutrient elements to the plant as foliar spray is essential.

Productivity of leguminous crops is known to respond to P fertilisation combined with micronutrients. Walker *et al.* (1969) reported that when Zn concentration in the leaves was 60 ppm, soybean yield was increased with increasing level of P, but decreased without P addition. Furthermore, Saxena and Mehrotra (1984) recorded a positive interaction between P and Mo or B. Jain (1972) reported similar response with P and Fe. Total-N and protein-N contents were also found to be increased by the interaction between P and microelements such as Mo and Zn (Singh *et al*, 1984).

Materials and methods

In a greenhouse pot-experiment, broad bean seeds were sown in pots 40 cm in diameter, each containing 23-Kg clay soil. Three weeks after sowing, plants were thinned to four plants per pot and divided to two sets. 16 kg P_2O_5 / fad (1 fad = 4200 m^2) were added after 3 and 6 weeks from sowing to the first set via the soil (3 gm calcium super phosphate/pot), and to the other set via the leaves (13.6 ml 0.35% orthophosphoric acid/pot). Every set included seven treatments as follows; they were either sprayed with the micronutrients B, Zn, Fe, Mo, their mixture, distilled water (control), or P alone. Plant leaves of the five treatments were sprayed 3 and 6 weeks after sowing with 25 ml of one of the following microelement solutions: 0.1% H_3BO_3, 0.1% $ZnSO_4$, 0.1% $(NH_4)_2MoO_4$; 0.03% $FeSO_4$, or their mixture. At harvest time, plants were harvested and the weight of pods, seed yield, straw yield, weight of whole plant as well as ratio of weights of pods/whole plant were recorded. Total-N content in the seeds was determined according to A.O.A.C. (1965) procedure and then crude protein percentage was calculated by multiplying by 6.25. Analysis of variance of the complete randomized design was done according to the method described by Snedecor and Cochran (1967). Duncan's multiple range test has been used to indicate treatment differences (Duncan, 1955).

Results and discussion

Weight of pods significantly increased due to micronutrient elements (Zn, Fe, Mo, and B), irrespective of the mode of P application (Table 1). In this respect, Jain (1972) reported that application of Fe, Zn and Mo separately or in combination significantly increased pod yield of Moth *Phaseolus aconitifolius*. The yield of seeds as shown in Table 1 was not increased with foliar application of P. A significant increase in grain yield was found with the combined application of Zn, Fe, or Mo with P either via soil or foliage. These results agree with those obtained by David *et al* (1985), Kishnarajan *et al* (1977), and Singh *et al* (1972). It is also obvious from the same table that straw yield was significantly affected by soil P application as well as by Zn, Fe, and Mo. Though weight of the whole plant was not affected by mode of application, it was significantly affected by foliar application of Fe, Zn and Mo. As for the ratio of weight pods/whole plant, it was significantly affected by method of P and greatly increased with foliar application of P. However, the highest seed yield was obtained with P applied to the soil or foliage when combined with Fe or Mo. These results agree with those obtained by Reddy and Patil (1980) and Saxena and Mehrotra (1984).

Data presented in Table 2 show that total-N concentration in seeds was significantly increased with soil P application. Moreover it is of interest to note that the interaction of P and microelements was more pronounced and statistically significant with most of the tested micronutrient elements. These results confirmed the results obtained by Chatterjee (1977) and Singh *et al* (1984). The data presented in the same table also show that crude protein concentration followed the same trend of Total -N concentration where soil P significantly enhanced protein accumulation in the seeds In accordance with the data presented in the same table, it is clear that the interaction between soil P and these micronutrients significantly increased protein content. These results confirm the previous results of Chatterjee (1977) and Lixandru *et al.* (1979). From above mentioned results it could be concluded that the beneficial effect of foliar spraying of Fe, Zn and Mo in combination with soil P application on yield and protein content may be attributed to the role of certain micronutrients in N fixation and metabolism. In this concern Badwe and Badhe (1973) and Yakout *et al* (1982) suggested that Mo combined with P increased symbiotic N fixation. In addition Thalooth and Kabesh (1988) reported that treatment with Zn sulphate led to an increase in the amount of chlorophyll and the vegetative growth which in turn favorably affected the final yield.

References

AOAC 1965 "Official Methods of Analysis", 10th Ed Association of Official Agricultural Chemicals, Washigton D C.

Cakmak I and Marchner H 1986 Physiologia Plantarum 68(3), 483-490.

Chahal RS and Ahluwalia J 1977 Plant Soil 47, 54-59.

Chatterjee BN 1977 Indian J Agron. 21, 209.

David MS, Gasivonie NN and Kamath 1985 J. Nuclear Agric. Biol. 14, 52-56.

Duncan DB 1955 Multiple range and multiple F- Test Biometries 11, 1-42.

Jain GL 1972 Indian J. Agron. 17, 271.

Kadwe RS and Badhe NN 1973 Indian J Agric. Res. 7, 43.

Krishnarajan J, Chowdappan SR and Morachan YB 1977 Madras Agric. J. 60, 277.

Lixandru G, Tarnauceanu E and Ciurea G 1979 Iasi Agronomie 23, 63- 66.

Methan KK and Amberger A. 1977 Plant Soil 46, 413.

Reddy SCS and Patil SV 1980 Oleagenoux 35, 507-510.

Sexena HJ, and Mehrotra ON 1984 Int. J. Agric. Res. 28, 91-94.

Singh B, Laura RD, Gupta VK and Sigh I 1984 Int. J. Tropical Agric. 2, 159-165.

Sigh R, Sharma SP and Bhatle IS 1972 J. of Res. Punjab Agric. Univ. 9, 631.

Snedcor GW and Cochran WG 1976 Statistical Methods Iowa State Collage Press 6th Ed. USA, pp 369-375.

Thalooth AT and Kabesh MO 1988 J. Agric. Sci. Manssoura Univ. 13 , 2105-2112.

Walker WM, Peck TR Johnson PE and Boona LV 1969 Illinios Res. 11, 12-19.

Yakout GM, Saad AOM, Ashour NI aand Thalooth AT 1982 Bull. National Res. Centre Egypt 7, 462- 471.

Table 1. Effect of method of P application and foliar application of micronutrients on yield components of broad bean (*Vicia faba*).

Yield components/plant	Method of P application	Control	P	B	Micro Zn	Elements Fe	Mo	Mix	Mean
No. of									
	Foliar	8.0a	10.6a	11.8a	11.0a	13.0a	11.0a	10.4a	10.8a
Pods/plant	Soil	8.8a	8.7a	11.5a	11.8a	11.7a	11.6a	10.3a	10.6a
	Mean	8.4a	9.7b	11.7bc	11.4bc	12.4c	11.3bc	10.4b	
Wt. of pods,	Foliar	19.3a	26.5a	27.0a	27.2a	31.7a	30.4a	28.1a	27.2a
gm/plant	Soil	20.4a	22.3a	25.2a	28.3a	30.4a	30.8a	26.8a	26.3a
	Mean	19.8a	24.4b	26.1b	27.8bc	31.1c	30.6c	27.5bc	
Seed yield,	Foliar	15.7a	21.8a	22.3a	22.3a	25.8a	24.8a	23.1a	22.3a
gm/plant	Soil	16.6a	17.9a	20.7a	23.0a	24.6a	24.6a	21.7a	21.3a
	Mean	16.1a	19.4b	21.5c	22.7cd	25.2e	24.7cd	22.4c	
Straw yield,	Foliar	6.7a	8.3a	8.6a	10.2a	10.5a	9.5a	7.9a	8.8a
	Soil	8.8a	8.8a	9.4a	11.4a	11.0a	10.5a	10.6a	10.2b
	Mean	7.7a	9.1abc	9.0ab	10.8c	10.8c	10.0bc	9.4abc	
Wt. of whole	Foliar	26.3a	34.8a	35.6a	37.4a	42.2a	40.0a	36.0a	36.0a
Plant, gm	Soil	29.3a	32.1a	34.5a	39.5a	411.4a	41.3a	37.4a	36.5a
	Mean	27.8a	33.5b	35.1b	38.5cde	41.8e	40.7de	36.7bcd	
Wt. of pods/	Foliar	73.5de	76.2g	75.8fg	73.0cd	75.2fg	76.3g	78.1h	75.4b
Wt. of whole	Soil	70.6a	69.7a	72.9bcd	71.8bc	73.4de	74.6ef	71.5b	72.1a
Plant	Mean	72.0a	73.0a	74.4b	72.4a	74.3b	75.5c	74.8bc	

Table 2. Effect of method of P application and foliar application of micronutrients on yield components of broad bean (*Vicia faba*).

Nitrogen fraction	Method of Application	Control	P	B	Micro Zn	Elements Fe	Mo	Mix	Mean
Total- N	Foliar	38abc	40abcd	37ab	34a	54def	52bcde	57ef	45a
	Soil	39abcd	58ef	69f	58ef	57ef	55ef	53cdef	57ab
	Mean	38.5a	49ab	53b	46ab	55.5b	53.5b	55b	
Crude	Foliar	239a	249a	230a	210a	340b	325b	354b	278a
protein	Soil	242a	363b	430c	362b	357b	342b	334b	347b
	Mean	240.5a	306bc	330bc	286b	348.5c	333.5c	344c	

W. J. Horst et al., Plant nutrition – Food security and sustainability of agro-ecosystems. 796~797, 2001

Immobilization of fertilizer nitrogen and consequences for nitrogen fertilization of cereals

K. Blankenau, H.-W. Olfs and H. Kuhlmann
Hydro Agri, Centre for Plant Nutrition, Hanninghof 35, D-48249 Dülmen, klaus.blankenau@hydro.com

Key words: cereal, ammonium, nitrate, N immobilization, N fertiliser recommendation

Abstract

From field trials with winter cereals it is known that about half of the applied fertiliser N (N) is not recovered in crops nor in soil mineral N (N_{min}). In some experiments it has been shown that most of this non-recovered N was immobilized by soil microorganisms, which reduced fertiliser N availability for the crops. In field trials with winter wheat two strategies were tested to reduce fertiliser N immobilization and therefore increase N uptake and grain yields. Strategy 1: Between half and three quarters of the fertiliser N immobilization determined at harvest already takes place after application of the first N rate (i.e. between tillering and stem elongation, when crop N uptake is low). Therefore the first N rate at tillering was reduced and the second N rate at stem elongation increased compared to conventional practice. This resulted in higher grain yields and grain N contents compared to conventional N application. Strategy 2: Fertiliser N availability for crops can be increased by fertilisation of N in the NO_3^--form because of preferential immobilization of NH_4^+ versus NO_3^- by soil microbes. Fertilisation of N in the NO_3^--form at tillering resulted in a higher recovery of fertiliser N in crops and N_{min} at harvest and in higher N uptake and grain yields than use of NH_4NO_3. It can therefore be recommended to reduce the first N rate at tillering and increase the second N rate at stem elongation accordingly. An additional advantage can be expected by use of NO_3^--N fertilisers at tillering.

Introduction

From many field trials it is known that about half of the fertiliser nitrogen (N) applied to winter cereals is not recovered in crops or as soil mineral N (N_{min}) (e. g. Engels & Kuhlmann, 1993). It was also shown that most of this non-recovered N reflects nitrogen which was immobilized by soil microorganisms rather than N lost via volatilization or nitrate leaching.

Immobilization of (fertiliser) N is considerably higher from the first application at tillering in spring than from later N applications (Blankenau *et al.*, 2000). Therefore, it was hypothesized that fertiliser N availability for crops (and fertiliser N uptake and yields) could be increased if immobilization of fertiliser N is reduced between tillering and stem elongation.

One possible management strategy to achieve this, is to decrease N application rates at tillering, and increase them at stem elongation. As it is known that NH_4^+-N is preferentially assimilated by the microbial biomass compared to NO_3^--N (e. g. Schlegel, 1992), for the second strategy it was hypothesized that fertiliser N immobilization could be reduced if fertiliser N is applied as NO_3^- at tillering. To test both strategies field trials were conducted with winter wheat.

Materials and methods

Experiment 1: 16 winter wheat trials (8 sites in 1996 and 1997) were conducted in different areas of Germany to test the effect of varying N rates at tillering (Zadok's growth stage [GS] 25) and beginning of stem elongation (GS 31) on grain yield and grain N content. According to common farming practice in Germany, an N target value ("N-Sollwert") of 120 kg N ha^{-1} (= Σ N_{min} in 0 - 90 cm in spring + 1st N rate) was used, followed by applications of 40 kg N ha^{-1} at GS 31 and 60 kg N ha^{-1} at ear emergence (GS 49; = treatment 120-40-60). In treatment 90-70-60 the same total N supply of 220 kg N ha^{-1} was split into an N target value of 90 kg N ha^{-1} followed by 70 kg N ha^{-1} at GS 31 and 60 kg N ha^{-1} at GS 49. Each N treatment received calcium-ammonium-nitrate (CAN).

Experiment 2: To test for the effect of different fertiliser N forms on fertiliser N availability for crops, five field trials were conducted with winter wheat in 1998. The effect of different N forms was tested at a reduced N supply level of 170 kg N ha^{-1}. At GS 25, CN (calcium-nitrate with 14.4% NO_3^--N and 1.1% $NH4^+$-N) or CAN (13.5% NO_3^--N and 13.5% NH_4^+-N) was applied to achieve an N target value of 90 kg N ha^{-1}. In both treatments, 40 kg N/ha as CAN was applied at GS 31 and at GS 49. Treatment 120-40-60 (CAN) represented conventional farming practice. The N0 treatment was kept unfertilised.

In each of the 21 field trials the N treatments were replicated four times in a randomized trial layout. P, K, Mg and S and crop protection chemicals were applied according to common farm practice. Grain was harvested using a combine harvester (in exp. 2 straw was also collected). Plant samples were dried, milled and analyzed for total N after Kjeldahl digestion.

In spring, soil was sampled down to 90 cm depth from each field to determine N_{min} (= $N0_3^-$-N and NH_4^+-N) using standard lab methods. In exp. 2 N_{min} was also analyzed for each plot at harvest.

Non-recovered N indicates applied N that was not recovered in crop or N_{min} (= immobilized + lost N). Non-recovered N was calculated as: (N supply of fertilised treatment - N supply of N0) - ([Crop N + N_{min} of fertilised treatment] - [Crop N + N_{min} of N0]).

Results and discussion

Exp. 1: As expected, grain yields and grain N removal were higher in the 90-70-60 compared with the 120-40-60 treatment in both years on most of the 16 sites. Grain yield was increased on average by 2.4% (1996) and 3.5% (1997) and grain N uptake was 5.4% (1996) and 4.9% (1997) higher than in the 120-40-60 treatment (Figure 1).

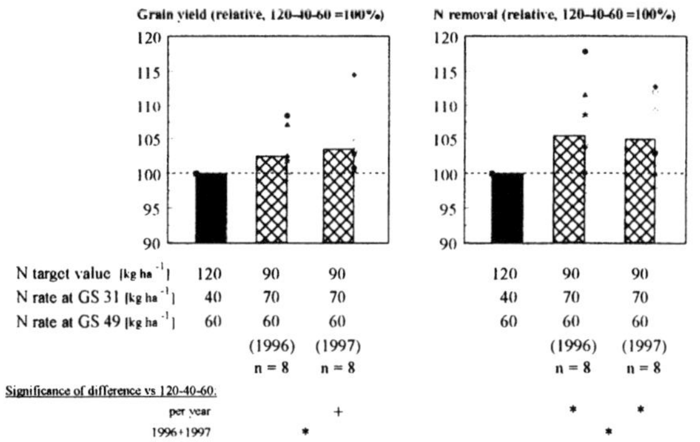

Figure 1. Effect of different N target values (GS 25) and N rates at GS 31 on grain yield and N removal with winter wheat grain in 1996 and 1997 (+= p<0,1; *= p<0.05). Symbols indicate different grain yield and N removal in the 90-70-60 vs the 120-40-60 treatments on single sites.

Exp. 2: On average, of the five trials at a reduced N supply level of 170 kg N ha^{-1} the sum of crop N + N_{min} was higher for the CN treatment than the CAN treatment (Figure 2). As a result, non-recovered fertiliser N was lower for CN compared to CAN. In addition, grain yield was significantly higher for CN than for CAN. Highest grain yields and N uptake by crops was found at an optimum N supply level of 220 kg N ha^{-1} (CAN).

Previous field trials with winter wheat have shown that during the growth period, highest losses but also highest immobilization of applied fertiliser N occur before stem elongation when crops are small (Limaux *et al.*, 1999; Blankenau *et al.*, 2000). Thereafter, N uptake is higher and N immobilization and N losses are lower. Indeed the combination of a reduced N application at tillering and higher N rates at stem elongation in the 90-70-60 treatment resulted in a higher fertiliser N availability for winter wheat compared to conventional practice (120-40-60, Figure 1). The higher N availability for crops in the 90-70-60 treatment was reflected in the higher N uptake (i.e. higher N removal) and higher grain yields than the conventional 120-40-60 treatment.

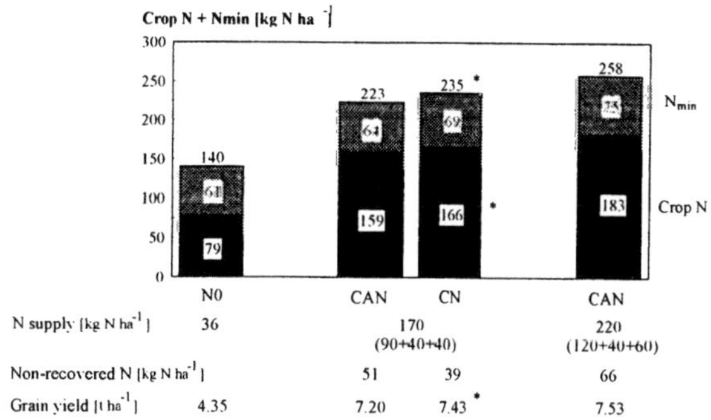

Figure 2. Effect of different forms of N at tillering on grain yield, N uptake, Nmin and non-recovered N on average of 5 sites with winter wheat (*= p< 0.05).

Net N immobilization in particular can be reduced if nitrate-based fertilisers are applied at tillering because NO_3^- is immobilized to a lower extent than NH_4^+ (Recous *et al.*, 1988). NH_4^+-fixation and NH_3 volatilization can be assumed to have been negligible on the sites due to low clay contents and soil pH values below 7. As the amounts of non-recovered N were lower in CN than in the CAN treatments until harvest (Figure 2), even though both treatments received CAN at 2nd and 3rd N applications, the use of CN with the 1st N dressing at tillering increased fertiliser N availability during the entire growth period. This resulted in higher N uptake and consequently also higher grain yields for CN than for the CAN treatment.

In conclusion, N application systems could be modified towards strategies with lower N applications at tillering compensated by higher N dressings applied later in the growth period. An additional advantage can be expected by use of nitrate-N fertilisers at tillering, at least at reduced N supply (e.g. in nitrate sensitive areas).

References

Blankenau K, Kuhlmann H and Olfs H-W 2000 Biol. Fertil. Soils 32, 157-165.

Engels T and Kuhlmann H 1993 Z. Pflanzenernähr. Bodenkd. 156, 149-154.

Limaux F, Recous S, Meynard J-M and Guckert A 1999 Plant Soil 214, 49-59.

Recous S, Fresneau C, Faurie G and Mary B 1988 Plant Soil 112, 205-214.

Schlegel H 1992 Allgemeine Mikrobiologie. Thieme-Verlag. Stuttgart, New York.

W. J. Horst et al., Plant nutrition – Food security and sustainability of agro-ecosystems. 798~799, 2001

Yield and quality of winter oilseed rape related to nitrogen supply

G.-W. Rathke[1)] and C. Schuster[2)]

[1)] *Institute of Agronomy and Crop Science, Martin-Luther-University Halle-Wittenberg, Lúdwig-Wucherer-Str. 2, D-06108 Halle (Saale), Germany, rathke@landw.uni-halle.de*

[2)] *SKW Stickstoffwerke Piesteritz GmbH, Landwirschaftliche Anwendungsforschung, Am Wieseneck 7, D-04451 Cunnersdorf, Germany*

Key words: nitrogen nutrition, oil content, pre-crop, winter oilseed rape, yield

Abstract

In a field experiment the influence of nitrogen supply on seed yield and oil content of winter oilseed rape was examined. The N treatments were achieved by different pre-crops (winter barley, peas) and different forms and rates of N fertiliser (mineral and organic fertiliser; 0 to 240 kg N/ha). The effect of mineral fertilisation on yield and oil content could not be substituted adequately by organic N sources under the given site conditions.

Introduction

Among the oil crops grown in Germany rape occupies a remarkable position. Rape is not only grown for food, but increasingly also as a renewable resource. The major criteria for cultivation are yield and oil content. An important precondition of rape growing (from both the ecological and economic point of view) is the optimum nutrition of the plants. Here, nitrogen supply takes the priority. The efficiency of N fertilisation may be influenced by various agronomic factors, but also by the soil type and weather conditions. The paper offers a survey of the effects of differentiated N application on yield and quality of rape under the conditions of Central Germany.

Materials and methods

The results were obtained in field experiments carried out in the Hercynian dry region of Central Germany since 1994. The soil type is sandy loam (FAO-classification: Stagno-Luvic Gleysol). The long-term mean annual rainfall is 552 mm, the mean air temperature 9.1 °C (1961-1990).

The trials were arranged in a three factorial split plot design with four replications. As pre-crop (treatment factor A) of winter oilseed rape, peas and winter barley were chosen. The fertiliser (treatment factor B) was applied either in mineral form as calcium ammonium nitrate (27 % N), or in organic form as slurry (~ 0.40 % N_t). The amounts of nitrogen (N levels; treatment factor C) were 0 (N0), 80 (N1), 160 (N2) and 240 (N3) kg N/ha. Further information is given by *Rathke* (1999) and *Rathke* and *Schuster* (2001).

The statistical analysis was made with the MIXED-Procedure of the software package SAS. In the analysis of the whole trial series, the years were accidental, and the location range was fixed. Based on variance analysis, tests of means (Newman-Keuls-test) were carried out with consideration of interactions (p = 5 %).

Results

The yield (Fig. 1) was not influenced by the pre-crop but was influenced both by fertiliser form and N rate. Since no significant interactions between the treatment factors were found, only main effects are described. Winter rape after peas (4.4 t/ha) yielded only slightly higher than after winter barley (4.2 t/ha). The comparison of fertiliser forms revealed that mineral fertiliser (4.5 t/ha) led to higher yields than organic fertiliser (4.1 t/ha). An increased N rate raised the yield as expected, whereas yields rose significantly up to the level N2 (4.7 t/ha). N doses beyond this produced hardly any noticeable yield effects (4.9 t/ha at N3).

The oil content of rape was not significantly influenced by the pre-crop. After peas, the oil content was 46.5 %, and after winter barley 46.4 %. In contrast to this, fertiliser form and N rates as well as interactions between them affected the oil content. As a result of interactions, the fertiliser forms (Fig. 2) were compared for the same N level, and the N levels (Fig. 3) for the same fertiliser form.

Discussion

The results confirm the central importance of nitrogen supply in winter rape cropping. In the field, yield and quality of winter oilseed rape were primarily influenced by different N rates and secondly by the type of fertiliser. There was no influence of the pre-crop. Overall, the highest yield was produced after peas at medium to high (160-240 kg N/ha) mineral N fertilisation. In contrast, the highest oil content was achieved independently of the pre-crop and the fertiliser form in the non-fertilised (0 kg N/ha) plot. Which form and rate of N supply is given to winter oilseed rape depends on the production objective. Whereas the economically relevant parameter in winter rape cropping is the yield, the oil content (or oil yield) is the main criteria for the production of renewable resources (e.g. "biodiesel", rape methyl ester).

References

Rathke G-W 1999 Auswirkung von Vorfrucht und Stickstoff (N)-Düngung auf die Energiebilanz des Winterrapsanbaus. Cuvillier Verlag Göttingen.

Rathke G-W and Schuster C 2001 *In* Gestaltung der Anbauverfahren landwirtschaftlicher Kulturpflanzen.

Ed. W Diepenbrock. pp. 1-13. Shaker Verlag, Aachen.

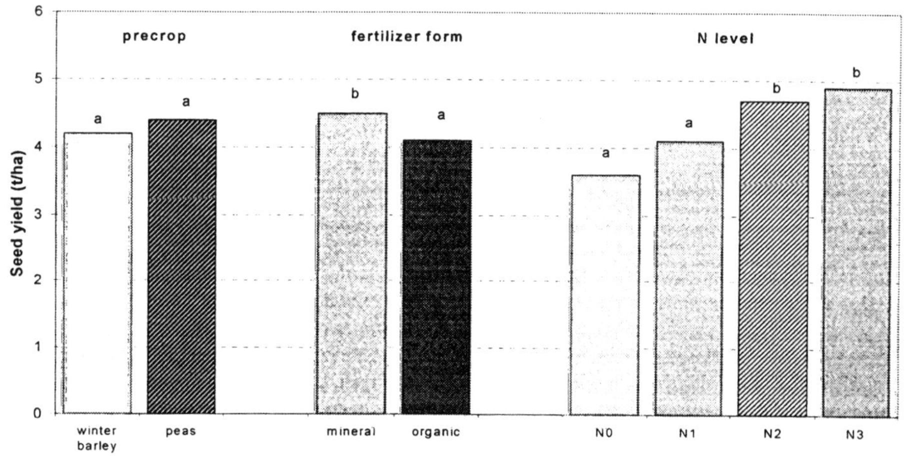

Figure 1. Seed yield of winter oilseed rape (t/ha; 91 % DM) related to N supply (columns with same letter at same examining factor are not different significantly, p = 5%)

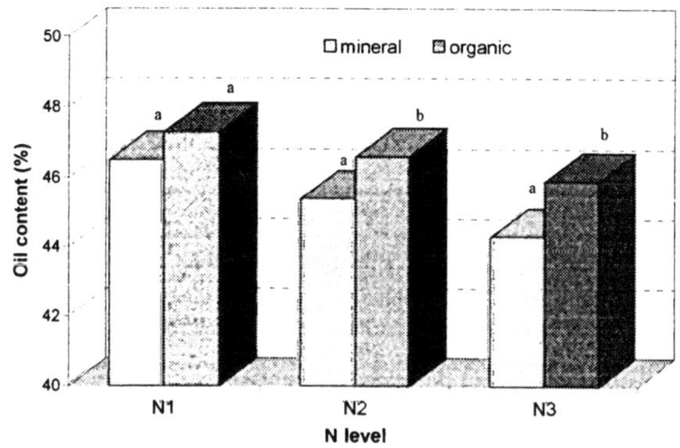

Figure 2. Oil content of winter oilseed rape (%; 100 % DM) related to the fertiliser form (columns with same letter at same N level are not different significantly, p = 5%)

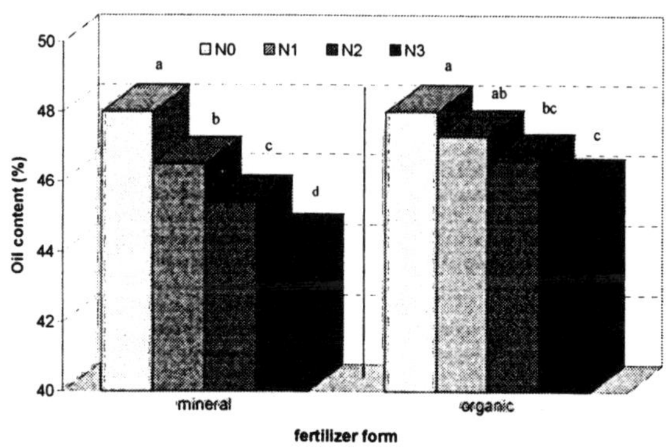

Figure 3. Oil content of winter oilseed rape (%; 100 % DM) related to the N levels (columns with same letter at same fertiliser form are not different significantly, p = 5%)

W. J. Horst et al., Plant nutrition – Food security and sustainability of agro-ecosystems. 800–801, 2001

Effect of rate, timing and form of nitrogen application on yield formation and nitrogen balance in oilseed rape production

T. Behrens, W.J. Horst and F. Wiesler

Institute of Plant Nutrition, Department of Horticulture, University of Hannover, Herrenhaeuser Str. 2, D 30419 Hannover, Germany. Torsten.Behrens@mbox.pflern.uni-hannover.de

Key words: *Brassica napus*, N balance, N form, N uptake, oilseed rape, timing of N application, yield, yield components

Abstract

In 4-years field experiments the effects of rate (0, 96, 120, 216 kg N ha^{-1}), timing (beginning of the growing season / beginning of shooting / beginning of flowering) and form (CAN: ammoniumnitrate + $CaCO_3$; BA: 73 % ammonium N, 19 % nitrate N, 8 % DCD N) of nitrogen (N) application on yield formation and N balance in oilseed rape (*Brassica napus* L.) production were studied. Seed yield increased up to the highest N rate. Under conditions of limited N supply delayed N application resulted in significantly higher yields compared with early application, despite impaired shoot growth until the beginning of flowering. N balance studies confirmed that high N application to achieve maximum seed yields may result in a considerable N surplus. The N form affected neither seed yield nor N uptake and N balance.

Introduction

Winter oilseed rape is characterized by high N uptake (i.e. high N fertiliser demand) but low N export with harvested seeds resulting in large N balance surpluses which increase the risk of N losses to the environment (Aufhammer *et al.*, 1994). Numerous studies showed that timing and form of N application may significantly influence grain yield, yield components and N recovery of cereals (Wiesler, 1998). However, contradictory results were obtained with rape in this respect (e.g. Boelcke *et al.*, 1991; Hocking *et al.*, 1997). In our field experiments the effects of rate, timing and form of N application on growth, yield and N uptake of winter oilseed rape were studied. Results should contribute to a better knowledge of the N demand of rape in various growth stages under conditions of reduced N fertilisation.

Materials and methods

Field experiments were conducted on 2 Luvisols near Göttingen, northern Germany, in co-operation with the Institute of Agronomy and Plant Breeding, University of Göttingen (H. Becker). In 7 different treatments (s. Tab. 1) effects of rate (0, 96, 120, 216 kg N ha^{-1}), timing (beginning of the growing season / beginning of shooting / beginning of flowering) and form (CAN: ammoniumnitrate + $CaCO_3$; BA: 73 % ammonium N, 19 % nitrate N, 8 % DCD N) of N application on growth, yield and N uptake of winter oilseed rape were studied. N rate at the beginning of the growing season considered the initial N_{min} content in 0-60 cm soil depth (averaged over sites: 24 kg N ha^{-1}). Plant samples were taken at the beginning of flowering (1 m^2) and at maturity (5 m^2). N concentration in the plant dry matter was measured with a CNS analyser and shoot N uptake as well as N balances were calculated. At maturity yield components were determined at one site. As an example for the 4 growing seasons results of the growth period 1997/98 are presented (means of the 2 different field sites).

Results and discussion

Shoot dry weight at the beginning of flowering and seed yield at maturity increased with increasing N supply (Fig. 1). Under conditions of reduced N supply early N application (96/0/0) improved shoot growth until flowering but reduced seed yield at maturity compared with delayed or splitted N application (0/120/0, 16/40/40). The N form had hardly an effect on seed yield. As shown in other field experiments (e.g. Hocking *et al.*, 1997), increasing seed yields with increasing N supply were mainly achieved by increasing pod numbers whereas other yield components were negatively (mean seed number per pod) or not consistently (1000 seed weight) affected by N supply (Tab. 1). Our studies further show that increased pod number mainly results from an increased number of pod bearing branches and not from an increased number of pods per branch. Increased seed yields with delayed (0/120/0) or threefold splitted (16/40/40) N application were due to increased seed numbers per pod and 1000 seed weight whereas pod number was higher in the treatment with early N application (96/0/0). A comparison of the two N forms shows that BA decreased pod number and increased seed number per pod slightly.

Shoot N uptake in the various treatments showed a similar pattern than shoot dry weight, i.e. straw and seed N increased with increasing N supply (Fig. 2). Delayed or threefold splitted N application increased N uptake compared with early application whereas the N form had little effect on N uptake. Interestingly, timing of N application had considerable effects on N uptake during the reproductive growth phase. In the treatments with conventional twofold fertiliser splitting and the unfertilised control N uptake in this growth period was low, being in the range of 9 – 17 kg ha^{-1}. In clear contrast, there was a net loss of 32 kg N ha^{-1} in the 96/0/0 treatment whereas net N uptake during the reproductive growth phase significantly contributed to total N uptake in the 0/120/0 (58 kg ha^{-1}) and 16/40/40 (38 kg ha^{-1}) treatments.

N balance was clearly influenced by N fertiliser rate whereas timing and form of N application had little effect (Fig 3). In the 96/120/0 treatment more than 50 % (117 kg ha^{-1}) of the fertiliser N input were not exported with harvested seeds resulting in both relatively high contents of residual soil mineral N (67 kg ha^{-1}) and N in crop residues (8 kg ha^{-1} in dropping leaves and 40 kg ha^{-1} in straw). With limiting N supply fertiliser N input exceeded N export with harvested seeds only by 16 – 31 kg ha^{-1} resulting in comparatively low soil N_{min} residues (25 - 32 kg ha^{-1}) and N in crop residues (4 kg ha^{-1} in dropping leaves and 24 - 28 kg ha^{-1} in straw).

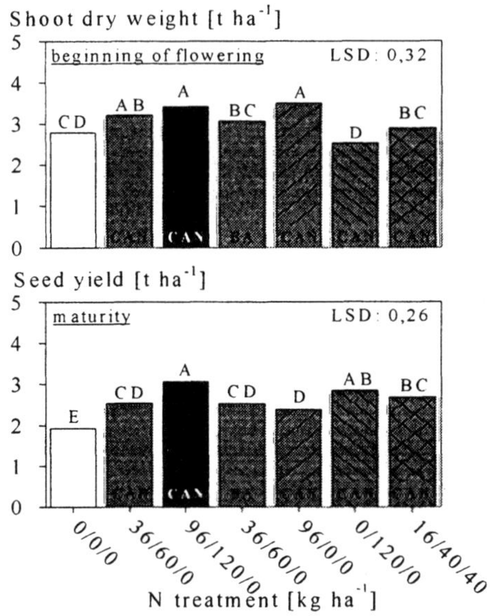

Figure 1. Shoot dry weight at the beginning of flowering and seed dry weight at maturity as influenced by rate, timing (beginning of the growing season / beginning of shooting / beginning of flowering) and form (CAN, BA) of N application (means of 2 sites).

Table 1. Yield structure as influenced by rate, timing and form of N application (one site).

N treatment		Pods per m^2 [n]	Pod bearing branches per m^2 [n]	Pods per branch [n]	Seeds per pod [n]	1000 seed weight [g]
0/0/0		2831	230	12.3	17.1	4.6
36/60/0	CAN	5071	341	14.9	11.3	4.4
96/120/0	CAN	5812	457	12.7	11.6	4.8
36/60/0	BA	4632	319	14.5	12.7	4.4
96/0/0	CAN	5519	368	15.0	10.4	4.4
0/120/0	CAN	4756	423	11.3	13.1	4.9
16/40/40	CAN	4674	321	14.6	13.2	4.6
LSD		*1665*	*111*	*2.9*	*4.1*	*0.2*

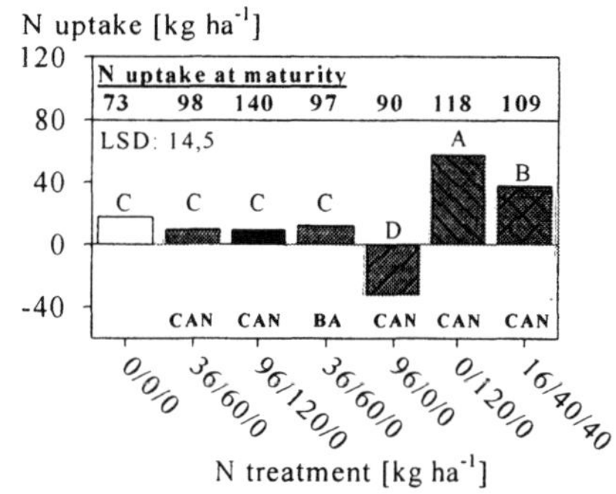

Figure 2. Total N uptake at maturity and N uptake during the reproductive growth phase (columns) as influenced by rate, timing and form of N application (means of 2 sites).

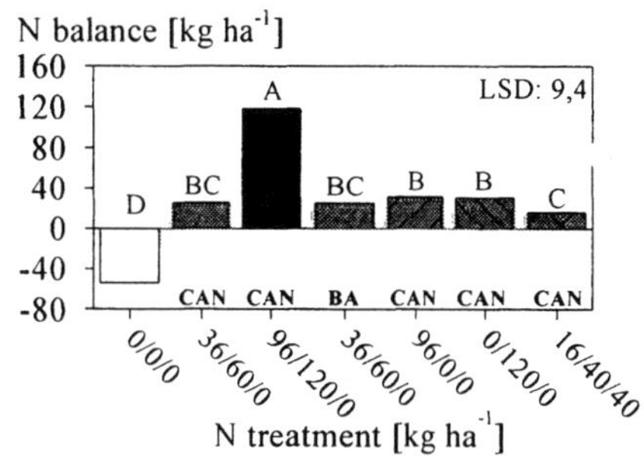

Figure 3. N balances (fertiliser N – N export with seeds) as influenced by rate, timing and form of N application (means of 2 sites).

Conclusions

More efficient use of fertiliser N in rape production is essential for reducing the risk of environmental pollution through application of excess N. As expected, N balance surpluses can be significantly reduced by a reduction of N fertilisation rates. Under such conditions delayed N application is an important measure to meet the N demand of the crop during the reproductive growth phase. N demand in this growth phase should also be considered when breeding strategies are developed to improve overall N efficiency of rape (see Wiesler *et al.*, this issue).

References

Aufhammer W, Kübler E and Bury M 1994 J. Agron. Crop Sci. 172, 255-264.

Boelcke B, Léon J, Schulz RR, Schröder G and Diepenbrock W 1991 J. Agron. Crop Sci. 167, 241-248.

Hocking PJ, Randall PJ and DeMarco D 1997 Field Crop Res. 54, 201-220.

Wiesler F 1998 J. Crop Prod. 1, 81-114.

W. J. Horst et al., Plant nutrition – Food security and sustainability of agro-ecosystems. 802~803, 2001

Tânzania grass: Nitrogen distribution in shoots and utilization

F. A. Monteiro, I. Benetti and Q. A. C. Carmello
Escola Superior de Agricultura "Luiz de Queiroz", Universidade de São Paulo, Av. Padua Dias,11, 13418-900, Piracicaba, SP, Brazil, famontei@esalq.usp.br

Key words: forage, mineral nutrition, nitrogen supply, *Panicum maximum*, plant parts

Abstract

Brazilian pastures are mainly based on forage grasses either in natural systems or through introduced species. *Panicum maximum* cv. Tanzania is one recently selected grass cultivar which is expanding the cultivated area. A hydroponic experiment with ground quartz as the substrate was carried out in a greenhouse to study the nitrogen distribution in plant shoots and the efficiency of nitrogen utilization by Tanzania grass. Eight nitrogen rates (0, 14, 42, 126, 210, 294, 378 and 462 mg L^{-1}) were studied in a completely randomized block design with four replications. Plants had three growth periods and were harvested 35 days after transplanting (DAT) at 63 DAT and at 91 DAT. Plant shoots were separated into expanding leaves, newly expanded laminae, mature laminae and culms plus sheaths. Nitrogen mainly accumulated in the laminae of the two newly expanded leaves. Plants absorbed between 46 and 54% of the N supplied in the nutrient solution during the experiment.

Introduction

Forage grasses are widespread in Brazil and *Panicum* is one of the most common palnt species in pastures. *Panicum maximum* cv. Tanzania was selected by Embrapa (Brazilian Research Institution) and released to the farmer in 1990 (Jank, 1995). Tanzania grass has been sown in many areas and only a few studies on its mineral nutrition are available

Nitrogen is a major nutrient for forage production throughout the world. Grasses rely mainly on mineral nitrogen supply to have adequate growth and productivity. Yield increases were shown in Coloniao, Vencedor and Tanzania cultivars of *Panicum maximum* (Gomide,1984, Santos *et al.*, 1995, Correa, 1996, Abreu, 1999).

Materials and methods

A greenhouse experiment was carried out with 3.6 L plastic pots filled with ground quartz. Seedlings of *Panicum maximum* cv. Tanzania were transplanted to the pots and five plants were grown in each. One litre of nutrient solution was supplied to each pot, and nutrient concentrations (except nitrogen) followed that reccomended by Hoagland and Arnon. Eight nitrogen concentrations (0, 14, 42, 126, 210, 294, 378 and 462 mg L^{-1}) were tested in a completely randomized block design with four replications.

Just after transplanting the seedlings to the pots, nutrient solutions with one-fourth strength of that corresponding to each treatment was supplied for three days. After that, the nutrient solutions were set to the treatments and replaced every 14 days and also after each harvest. Throughout the experiment full-strength solutions were supplied seven times.

Plants were harvested three times. The first harvest was at 35 days after transplanting (DAT), the second at 63 DAT, and the third at 91 DAT. At each harvest, plant shoots were separated into expanding leaves (EL), newly expanded laminae (NL = the two youngest expanded laminae), mature laminae (ML = all the expanded laminae except NL), and culms plus sheaths (CS). Roots were collected after the third harvest.

All plant material was oven-dried at 65°C, weighed, ground and analysed for nitrogen by a semi-micro-Kjeldahl procedure. Analysis of variance and regression analysis were performed according to the procedures given by Freund and Littell (1981).

Results and discussion

Plants in the no-nitrogen treatment did not grow and died during the experiment.

More than two-thirds of the nitrogen in plant shoots was found in the laminae of Tanzania grass. Among the three types of leaf tissue collected at each harvest, the two newly expanded laminae (NL) accumulated most nitrogen (35 to 45% of the total nitrogen in plant shoots). Culms plus sheaths contained at most 32.5% of the total amount of nitrogen (Table 1).

For plants growing at N rates from 14 to 462 mg L^{-1} had from 101.5 to 3349.5 mg N were available during the entire experiment. From 45.8 up to 1732.0 mg nitrogen was found into the Tanzania plants during the three growth periods. The percentage of nitrogen utilization (from that applied) significantly changed with N rates and ranged from 45.1 to 54.4% (Table 2).

Table 1. Percentage nitrogen distribution in four shoot parts of *Panicum maximum* cv. Tanzania as related to nitrogen concentration in the solution, at each plant harvest.

N	First harvest				Second harvest				Third harvest			
[mg/L]	EL	NL	ML	CS	EL	NL	ML	CS	EL	NL	ML	CS
14	16.8	37.6	19.6	26.0	11.2	45.4	15.9	27.5	16.4	39.0	19.0	25.6
42	16.4	38.8	19.8	25.0	13.9	41.2	21.5	23.4	14.1	41.5	23.2	21.2
126	13.5	38.8	19.9	27.8	13.2	40.7	20.6	25.5	15.4	39.6	22.0	23.0
210	13.8	39.2	20.6	26.4	13.1	40.0	24.5	22.4	15.6	40.0	21.2	23.2
294	14.7	36.8	21.5	27.0	12.1	39.3	26.6	22.0	16.4	37.6	23.1	22.9
378	17.1	38.2	16.0	28.7	12.8	36.4	24.1	26.7	13.7	41.1	23.1	22.1
462	15.6	36.2	15.7	32.5	12.8	35.2	21.1	30.9	13.3	33.7	24.0	29.0

EL = expanding leaves; NL = newly expanded laminae; ML = mature laminae and CS = culms + sheaths

Table 2. Nitrogen supplied and taken up, and percentage of nitrogen utilization by *Panicum maximum* cv. Tanzania as related to nitrogen concentration.

N	N supply	N uptake	N utiliz.
[mg/L]	mg	mg	%
14	101.5	45.8	45.1
42	304.5	165.1	54.2
126	913.5	497.0	54.4
210	1522.5	737.0	48.4
294	2131.5	985.2	46.2
378	2740.5	1352.4	49.3
462	3349.5	1732.0	51.7

References

Abreu JBR 1999 Produção e nutrição dos capins Tanzania-1 e Marandu em função de estádios de crescimento e adubação nitrogenada. Doctoral Dissertation. ESALQ-USP.

Corrêa BD 1996 Doses de nitrogênio e de magnésio afetando aspectos produtivos e bioquímicos dos capins colonião, Tanzania-1 e Vencedor. Master Thesis. ESALQ-USP.

Freund RJ and Littell RC 1981 SAS for linear models : a guide to the Anova and GLM procedures. Cary : NC, SAS Institute.

Gomide JA 1984 *In* Simpósio sobre Manejo da Pastagem, 7. Eds. AM Peixoto, JC Moura and VP Faria. pp. 33-60. FEALQ, Piracicaba, SP.

Jank L 1995 *In* Simpósio sobre Manejo da Pastagem, 12. Eds. AM Peixoto, JC Moura and VP Faria. pp. 21-58. FEALQ, Piracicaba, SP.

Santos AR, Corrêa BD and Monteiro FA 1995 Congresso Brasileiro de Ciência do Solo 25, 741-742.

W. J. Horst et al., Plant nutrition – Food security and sustainability of agro-ecosystems. 804~805, 2001

Agro-physiological responses of tropical maize cultivars to nitrogen fertilization in the moist savanna of West Africa

S. O. Oikeh[1)] and W. J. Horst[2)]

[1)] *US. Plant Soil and Nutrition Lab, Cornell University, Tower Road, Ithaca,, NY 14853, United States, sooikeh@yahoo.com*

[2)] *Institute of Plant Nutrition, University of Hannover, Herrenhaeuser Strasse 2, Hannover, D 30419, Germany, horst@mbox.pflern.uni-hannover.de*

Key words: agro-physiological traits, maize cultivars, moist savanna, nitrogen efficiency, nitrogen fertilisation, *Zea mays*

Abstract

The moist savanna of West Africa has a great potential for maize production, but this is not often realised because the soils are particularly poor in available nitrogen. In two-year field experiments conducted on a Plinthic Luvisol in Zaria, northern Nigeria, agro-physiological traits of five tropical maize cultivars were assessed under four N levels (0, 0.56, 1.12 and 2.26 g plant^{-1}). Application of 1.12 g plant^{-1} was optimum for maize production under the prevailing conditions. The most widely grown open-pollinated cultivar (TZB-SR) was the least efficient in utilizing available nitrogen for grain production especially under nitrogen stress because of its limited sink capacity (small harvest index and kernel size) and a slow kernel growth rate. But it ranked among the highest in whole shoot biomass accumulation and in number of kernels per plant. Under nitrogen stress, silking was delayed in cultivars 87 TZPB-SR and TZB-SR by 5-7 days compared to the other cultivars. Cultivars SPL and EV8728-SR showed better synchronization of anthesis and silking (ASI $\leq$ 4.5 days) which indicates a better ability to adapt to nitrogen and moisture stresses than the other cultivars. Nitrogen-utilisation efficiency of the cultivars was the most important component of N efficiency and could be predicted from harvest index.

Introduction

The moist savanna of West Africa has a great potential for maize production. But this is not often realized because the soils are generally poor in fertility, particularly in available N. Nitrogen deficiency is common in the savanna because N is applied below-optimal levels due to the high cost of fertilisers. One strategy to reduce the impact of N deficiency on maize production under sub-optimal N fertility is to identify cultivars with improved N efficiency, defined as grain yield per unit of available soil N (Moll *et al.*, 1982). The identification of secondary agro-physiological traits in tropical maize as influenced by N is useful for crop improvement programs aimed at developing N-efficient maize cultivars.

Materials and methods

In a field study conducted on a Plinthic Luvisol (Oikeh *et al.*, 1999) in Zaria, northern Nigeria, four improved, late maturing (120 days) maize cultivars and a widely grown open-pollinated, late maturing cultivar developed at IITA, Nigeria were grown with N application levels of 0, 0.56, 1.12, and 2.26 g plant^{-1}. The treatments were arranged as a split-plot in randomized complete block design with four/five replications in the first/second year. Nitrogen levels were the main plots and cultivars were the subplots. Each subplot consisted of seven maize rows (ridges) 6 m in length and 0.75 m apart. Sowing was done early in the season to take advantage of the soil mineral N released at the commencement of the wet season (Wild, 1972; Weber *et al.*, 1995). Plant population was 53 333 plants ha^{-1}. 0.49 g P plant^{-1}, 0.94 g K plant-1, and 18.8 mg Zn plant^{-1} were applied based on standard soil-test values. Nitrogen was applied at 2 weeks after planting (WAP) as calcium ammonium nitrate. The highest rate of 2.26 g plant^{-1} was applied in two equal splits at 2 and 5 WAP. Weed was controlled by hand.

Data on agronomic and physiological traits were collected. General Linear Model procedures of SAS statistical program were used to analyze the data collected (SAS Institute, 1992). Data were analyzed by N levels and across N levels and years. Mineral N in 0-90 cm soil depth was used as a covariate.

Results and discussion

Yield and yield components

Aboveground dry-matter yield and grain-yield components were significantly lower in 1994 than in 1993 (Table 1), probably due to the midseason drought that occurred during grain filling period in 1994 (Oikeh *et al.*, 1999).

Averaged across years and N, the most widely grown open-pollinated cultivar (TZB-SR) in the savanna of Nigeria had the lowest grain yield, kernel size, and kernel growth-rate even though it ranked among the greatest in total aboveground dry matter production (Table 1). Grain yield was similar among the improved cultivars except TZPB-SR. Only the semiprolific cultivar (SPL) attained an average of more than one ear per plant, and had the greatest sink potential as indicated by the production of the highest number of kernels per plant. At low N supply, for all cultivars grain yield, total aboveground dry-matter yield, and grain-yield components were significantly depressed (Table 1).

Nitrogen efficiency

The N-utilization efficiency was highest in 1994 and under low N supply (Table 1). Differences among cultivars were observed for efficiency traits and harvest indices for dry matter (Table 1). These traits were highest for hybrid 8644-27. A strong positive relationship has been reported between harvest index for dry matter with N-utilization efficiency (Moll *et al.*, 1982; Heuberger, 1998). Harvest index for dry matter could be a good indicator of N-utilization efficiency in maize under the moist savanna. There were no significant differences among cultivars in N uptake efficiency (data not shown).

Days to phenological stages

Nitrogen stress delayed silking, increased the interval between anthesis and silking (ASI), and decreased the duration of grain filling by 3 to 4 days (Table 1). These are major symptoms of N deficiency in tropical maize that reduce sink capacity and grain yield (Heuberger, 1998). Cultivars SPL and EV 8728-SR showed better synchronization of anthesis and silking, which indicate a greater ability to adapt to N and moisture stresses (Lafitte and Edmeades 1994). EV 8728-SR and hybrid 8644-27 had a longer duration of grain filling than the other cultivars (Table 1). In conclusion, cultivar differences in nitrogen efficiency existed at yield-limiting N supply. Nitrogen-utilization efficiency was the most important component of N efficiency and could be predicted from harvest index, increased ASI and decreased length of grain filling.

Acknowledgment

The authors thank Dr. Georg K. Weber for valuable discussions and BMZ for financial support to WJH.

References

Carlson RE and Shaw RH 1982 Agron. J. 74, 275–278.

Heuberger H 1998 Nitrogen efficiency in tropical maize: Indirect selection criteria with special emphasis on morphological root characteristics. Verlag Ulrich E. Grauer, Stuttgart, Germany.

Lafitte HR and Edmeades GO 1994 Field Crop Res. 39, 1–14.

Moll RH, Kamprath EJ and Jackson WA 1982 Agron. J. 74, 502-564.

Ogunlela VB Amoruwa GM and Ologunde OO 1988 Fert. Res. 17, 189-196.

Oikeh SO, Kling GJ, Horst WJ, Chude VO and Carsky RJ 1999 Field Crops Res. 62, 1-13.

Wild A 1972 Exp. Agric. 8, 91-97.

Weber G Chude V Pleysier J and Oikeh S 1995 Exp. Agric. 31, 333-344.

Table 1. Influence of N on phenology, aboveground dry-matter yield, and grain yield components and nitrogen efficiency of five maize cultivars in 1993 and 1994.

Year	Cultivar	N rate (g plt^{-1})	ASI (d)	LGF (d)	Grain yield (g plt^{-1})	Total DM (g plt^{-1})	HI (%)	Kernel (no plt^{-1})	KGR (mg k^{-1}d^{-1})	N-util. eff. (%)	N eff. kg grain/ kg avail-N
1993			4.3	45.5	98.6	278.6	35	482	5.1	42	18.9
1994			4.1	47.1	86.8	201.9	42	404	4.7	55	16.6
SE			0.3^{ns}	0.9^{ns}	5.3^{+}	12.3**	1.1**	13**	0.2*	0.8**	1.5^{ns}
		0	5.8	43.4	53.9	165.6	33	325	4.1	50	11.3
		0.56	3.8	46.7	83.7	221.5	38	425	4.6	52	15.9
		1.12	3.9	47.0	108.7	121.1	42	490	5.3	49	21.9
		2.26	3.5	47.8	121.8	295.9	42	515	5.6	45	21.1
		SE	0.3**	0.7**	3.9**	10.3**	1**	15**	0.1**	1**	0.5*
	EV8728-SR		3.4	48.0	92.9	232.1	40	428	5.1	50	18.5
	87TZPB-SR		5.3	45.6	89.9	247.3	36	405	5.2	46	16.6
	SPL		3.1	43.4	92.8	228.7	41	482	4.8	50	17.4
	8644-27		5.2	47.9	96.9	225.5	43	412	5.2	54	18.5
	TZB-SR		4.2	46.1	87.7	246.5	35	467	4.2	45	16.6
	SE		0.2**	0.5**	2.4^{+}	5.6**	0.5**	9**	0.1**	0.7**	0.5*
	CV (%)		3	3	16	15	9	14	14	9	20
ANOVA			**F-test**								
Year × Nitrogen			ns	ns	ns	ns	ns	ns	ns	ns	ns
Year × Cultvar			ns	**	ns	ns	*	+	**	+	ns
Cultivar × Nitrogen			ns	ns	ns	ns	+	ns	**	**	ns
Year × Cultivar × Nitrogen			ns	ns	ns	ns	+	ns	+	*	ns

^{ns}, $^{+}$, * ,** indicate *F*-test for main effect not significant, significant at $P<0.1$, 0.05, and 0.01 level

Abbreviations: ASI = anthesis-silking interval; LGF = length of grain filling; HI = harvest index for dry matter (DM); KGR = kernel growth rate; N-util eff. = N-utilization efficiency; N eff. = N efficiency.

W. J. Horst et al., Plant nutrition – Food security and sustainability of agro-ecosystems. 806–807, 2001

Nitrogen and chlormequatchloride on cotton cultivar

P. K. Karthikeyan[1)] and R. Jayakumar[2)]

[1)] *TamilNadu AgriculturalUniversity, Ph.D Student, Department of SoilScience &AgriculturalChemistry,TamilNadu State., Coimbatore, 641003, India. pk.kl@usa.net*

[2)] *TamilNadu AgriculturalUniversity, Professor, PesticideResidue laboratory, TamilNadu State., Coimbatore, 641003, India.*

Key words: active ingredient, aqueous solution, chlormequatchloride, cotton, *Gossipium hirusutum*, Madras Combodia Uganda

Abstract

A Field experiment was conducted at Agricultural Research Station, Bhavanisagar, Tamil Nadu Agricultural University, India on Typic Ustrocrept with cotton cultivar MCU.5 during 1998-99 to find out the influence of fertiliser N rates and chlormequatchloride on growth, yield and quality of cotton, The treatments consisted of three N rates (60, 80 and 100 kg ha-1) and four levels of chlormequatchloride formulations (50% AS, 30 ml ha^{-1}, 50% AS, 60 ml ha^{-1}, 10% AS, 150 ml ha^{-1} and 10% AS, 300 ml ha^{-1}). The field experiment was laid out in split plot design with three replications. The spraying of the chlormequatchloride was done during flowering phase of cotton viz., 60, 70, 80 and 90 DAS. Adopting all the recommended package of practices the crop was grown to maturity and harvested. The results showed that the height of the cotton plant was markedly increased with graded doses of N and reduced to the tune of 17.6 to 33.6 cm in chlormequatchloride treatments. There was profuse vegetative growth of the test crop of MCU- 5 cotton at 100 kg of N ha^{-1}. When chlormequat chloride either at lower concentrations or at higher concentrations were sprayed along with N at 100 kg ha^{-1} there was a profound height reduction leading to the development of short and sturdy plant. With reference to the yield attributes, chlormequatchloride sprayed in two formulations (chlormequatchloride 50 % AS, 30 ml ha^{-1} x 4 applications at 60,70,80,90 days interval and 10 % AS, 150 ml ha^{-1} x 4 applications at 60,70,80,90 days interval) and with increased N rates had produced more number of monopodia, sympodia and bolls as compared to control that did not receive chlormequatchloride. The boll weight was also profoundly increased. The favourable influence of chlormequatchloride alone at lower concentration and with N on yield contributing factors resulted in marked increase in the yield of seed cotton. However the quality parameters remained unaffected due to the spraying of chlormequatchloride.

Introduction

From time immemorial cotton '"The white gold" is an integral part of the Socio-economic activities in the Indian sub continent. Unfavourable agroclimatic conditions, non-judicious use of fertilisers and chemicals coupled with various other agro economical factors have been attributed to the productivity decline (Bardhan 1998). Nitrogen, phosphorous and potassium are required for cotton to produce its normal growth and yield. Investigations on fertiliser use have conclusively proved the efficacy of nitrogen in augmenting the yield of cotton (Chand et al 1997). A short duration compact plant type with synchronous flowering, boll bushing and possessing resistance/tolerance to pests and diseases are found to be ideal (Kairon *et al.*,1999). Application of growth retardant chlormequatchloride, counteracts the tendency for higher vegetative growth and may induce a redistribution of phytosynthates in favour of reproductive organs (Scraibrick et al 1982). Extensive work has been done in TamilNadu and other states of India and to optimize the right dose and time of application of chlormequatchloride for cotton to increase its production potential. The present investigation is also one among them to assess the effect of nitrogen and chlormequatchloride on quality of yield parameter of cotton on MCU 5.

Materials and methods.

A field experiment was conducted during 1998-99 at Agricultural Research Station, Bhavanisagar to find out the influence of N and chlormequatchloride on the quality parameters of cotton cultivar MCU 5. The treatments were 3 nitrogen levels N_1, 60 kg ha^{-1}, 80 kg ha^{-1} N_2, and 100 kg ha^{-1} N_3 in the Main plots and two chlormequatchloride formulations at 2 levels (50% AS 30 ml ha^{-1} (T_1), 50% AS 60% ml ha^{-1} (T_2); 10% AS 150 ml ha^{-1} (T_3), 10% AS 300 ml ha^{-1} (T_4), water spray (T_5) in the sub plots. The spraying of chlormequatchloride was done during 60, 70, 80 and 90 days after sowing. The field experiment was laid out in split plot design with three replications. All the recommended package of practices the crops were adopted. The crop was grown to maturity and yield of seed cotton was recorded. Quality of cotton has been defined as the physical attributes that affect its usefulness. The quality attributes viz. ginning %, lint index, fibre length, bundle strength, seed index, and uniformity ratio were assessed for the samples collected from the various treatments. The fibre properties of cotton lint were analyzed according to the procedures of Sundaram and Iyengar (1968).

Results

The yield of seed cotton was increased with increasing rates of N, spraying of chlormequatchloride at lower concentrations and their interactions with N rates. The increase in seed cotton yield was observed up to 80 kg ha^{-1} level and beyond that there was diminishing response in terms of yield. Upto 80 kg ha^{-1} of N the yield increased and further increase in N upto 120 kg ha^{-1} reduced the yield. However the quality parameters remained unaffected after the application of chlormequatchloride.

Discussion

Similar to results of this study trend of diminishing of the yield increase with increased N rates was reported by Chand *et al.* (1996; Baskar *et al.* (1993).The yield of seed cotton was increased due to spraying of chlormequatchloride at lower concentrations sprayed, during flowering stage of cotton as evidently shown by Zur *et al.* (1972). The application of chlormequatchloride @ 50 g a.i. ha^{-1} produced higher yield than that by 100 and 150 g. a.i. ha $^{-1}$. They had also shown that early spraying of chlormequatchloride produced insignificant effect on yield due to low bolls per plant but intermediate spraying (beginning of flowering) was found to yield better. There again they insisted on low concentrations of chlormequatchloride (50 g a.i. ha^{-1}). Seshadri (1989) reported yield increase in hybrid cotton for the application of chlormequatchloride upto 80 ppm after 60 DAS (flower initiation stage). The increased boll number and boll weight on account of more number of sympodia the boll bearing branches of cotton plant and the reduced boll shedding per cent in chlormequatchloride sprayed at lower concentrations were finally reflected in the productivity of seed cotton. The maximum seed cotton yield obtained for the interaction of lower concentration of chlormequatchloride applied during flowering with N_3 suggested that it could be possible to accentuate the productivity of cotton. The higher N dose, might result in "luxury" N uptake and may cause excessive vegetative growth. Which might not have any bearing on economic yield. Under such circumstances, the integration of chlormequatchloride spray would help in arresting the vegetative growth and assisting the reproductive parts. Among the various quality parameters ginning % alone was affected markedly by the nitrogen rates. The control was as good as those treatments that received chlormequatchloride. The ginning % was significantly higher in N_1, than in N_2 and N_3. Similar results were reported by Rathanparke et al (1984) who observed that high dose of nitrogen significantly lowered the ginning %. The bundle strength or the fibre length was favourably affected by the nitrogen rates, thereby inducting the influence of N on producing strong fibres. The other parameters fibre length, and uniformity ratio were unaffected by rates of N and chlormequatchloride sprays. Kubde and Lakhdive (1993) and Chand et al (1997) reported that fibre length was not affected by rates of N. The fibre length was found to be increased at lower concentrations of chlormequatchloride. Bhatt and Ramanujam (1972) reported increased fibre length of MCU varieties by chlormequatchloride spray (20 ppm at the flowering stage.) However, the bundle strength remained unaffected by the foliar application of chlormequatchloride sprays. The application of chlormequatchloride at lower concentrations with high rates of nitrogen (100 kg ha^{-1}) contributes towards the augmenting the seed cotton yield.

Acknowledgment

Thanks are due to **BASF** -INDIA limited and Tamil Nadu Agricultural University for funding the research programme under which this work was carried out.

References

Bardhan S 1998 Pesticide World 2 ,34-38.

Baskar KS, Gaikwad ST and Ananthakumari P 1993 Indian J. Agron. 38, 89-92.

Bhatt JG and Ramanujam T 1972 Cotton Grow. Rev. 49, 354-358.

Chand M, Prasad M and Singh H 1997 Indian J. Agron. 42, 528-530.

Kairon MS, Ramasundaram P and Venugopalan MV 1999 Survey of Indian Agriculture 1, 113-115.

Kubde KJ and Lakhdive BA 1993 P K V. Res. J. 17, 231-232.

Rathanaparke BR, Sinha SB and Bhargava BS 1984 Jawaharlal Nehru Krishi Vishwa Vidyalaya Research Journal 18, 85-90.

Scarisbrick DH, Daniels RW and Rawi ABN 1982 J. Agri. Sci. 99, 453-455.

Seshadri V 1989 Indian J. Agri. Sci. 59, 107-108.

Sundaram V and RLN Iyenger 1968 Hand book of methods of tests for cotton fibre, yarn and fabrics. Pub; Cotton Technological Research Laboratory, ICAR, New Delhi.

Zur M, Marani A and Kara David B 1972 Cotton Grow Rev. 49, 250-257.

W. J. Horst et al., Plant nutrition – Food security and sustainability of agro-ecosystems. 808~809, 2001

Nitrogen dynamics in irrigated grapevines (*Vitis vinifera*)

M. T. Treeby[1)] and D. M. Wheatley[1),2)]

[1)] *CSIRO Plant Industry – Horticulture Unit, River Avenue, Merbein VIC 3505, Australia. michael.treeby@pi.csiro.au*

[2)] *Current address: Paschendale Producers, Middleton Road, Merbein VIC 3505, Australia*

Key words: annual, grapevines, nitrogen, perennial, Sultana, *Vitis vinifera*

Abstract

The size of annual and perennial N pools of irrigated grapevines (*Vitis vinifera* var. Sultana) in response to N supply (0 and 100 kg ha^{-1} y^{-1}) were quantified for 2 consecutive seasons. Nitrogen present in the leaves was the largest pool of N amongst the annual structures for most of the growing season. Leaf N pool maxima were 90 and 190 kg N ha^{-1} for the 1st and 2nd season, respectively. Approximately 80 extra kg N ha^{-1} were present in the leaves in the 2nd season in response to N supply at 100 kg ha^{-1} y^{-1}. The stem N pool increased throughout each season. This pool was only responsive to N supply in the 2nd season, and this was due to more shoots per vine, implying that the capacity of this pool to accumulate N is limited by sink size. Stem N in vines not receiving N appeared to be strongly mobilised just prior to bud burst.

Approximately 10-70 kg N ha^{-1} were removed in the clusters at harvest, depending on N supply and season and crop size. More N was accumulated by the clusters in response to N supply. In the 2nd season a larger sink size accentuated the effect of N supply on N accumulation by the clusters. In response to N supply crop size increased by 114% in the 2nd season, but N accumulation increased by 173%. Thus sink size is less of a limitation to N accumulation by clusters.

The perennial N pool varied from 70 kg ha^{-1} at the start of the trial to 120 kg ha^{-1} at the end. Following veraison, the perennial N pool increased, particularly if N was supplied. This extra N could be accounted for by remobilised leaf N.

Introduction

N is frequently the most limiting nutrient for grape production in the irrigated sandy soils of southeast Australia. To overcome this, high N fertiliser inputs are often made, but are not based on an understanding of the dynamics of N in the vine throughout each production cycle. Scaling up from studies on single organs (Alexander 1957) and potted vines (Conradie 1980) to quantify the annual and perennial N pools in field grown grapevines was needed to devise N supply strategies to minimize potential leaching losses, control grapevine vigour and manipulate yields and composition.

Materials and methods

Fifty-60 year-old furrow irrigated *Vitis vinifera* cv Sultana (syn. Thompson seedless) vines on own roots were supplied with NH_4NO_3 at 2 rates (0 and 100 kg N ha^{-1} y^{-1}) in 5 equal split surface applications in the 1st week of each of the 1st 5 months of 2 consecutive growing seasons. There was a light irrigation to wash the fertiliser into the rootzone after each application. Biomass and N content of the annual structures were estimated by destructively sampling whole shoots (leaves and stems) and clusters at *ca.* 14 day intervals over 2 growing seasons. N levels in the perennial structures (roots, trunk and canes) were estimated from samples collected over 24 months. The biomass of the perennial structures was estimated by destructive sampling of whole vines at the completion of the study. The amount of N in each pool was compiled for the 2 seasons and used to estimate the total amounts of N in the annual and perennial structures. Data are presented relative to days from bud burst in each season. The key phenological stages, namely flowering, veraison (berry softening), harvest and leaf fall occurred 61-63, 115, 162-170 and 235-249 days after bud burst (which occurred *ca.* Sept. 8 each season).

Results and discussion

The annual structures accumulated *ca.* 100-150 kg N ha^{-1} in the absence of N fertiliser and *ca.* 120-250 kg N ha^{-1} when N was supplied (Fig. 1). The total amount of N in the annual structures was mainly influenced by the size of the leaf N pool up to *ca.* 180 days after bud burst, and thereafter increasingly became a function of the amount of N in the stem pool.

The amount of N in the leaves peaked 4 months after bud burst in both seasons, but substantial differences in response to N supply were only apparent during the 2nd growth season (Figs 1a and 1b). This pool's maximum size in the 1st season was approximately 90 kg ha^{-1}. Vines not supplied N had a maximum pool size of 110 kg ha^{-1} in the 2nd season compared to 190 kg ha^{-1} for those vines supplied N. The difference in size of the leaf N pool between seasons probably reflects seasonal differences in soil N *mineralis*ation and growth conditions. During fruit maturation (*ca.* 120-170 days after bud burst) the leaf N pool remained relatively static during the 1st season for both N supply treatments, and for the control treatment (0 kg ha^{-1} y^{-1}) during the 2nd season. The size of the leaf N pool of vines supplied N decreased by *ca.* 46 kg ha^{-1} during the 2nd season's fruit maturation stage.

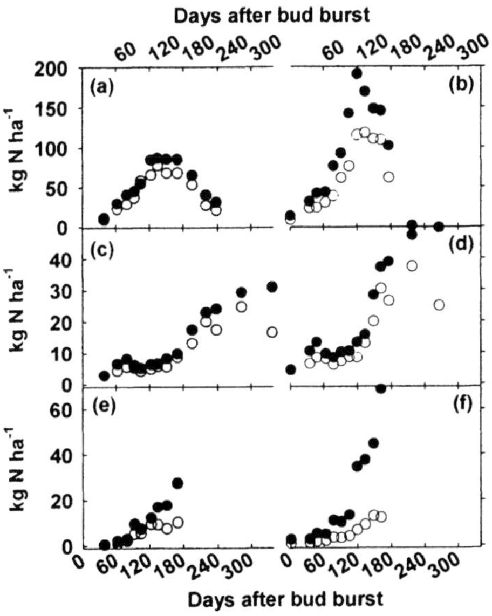

Figure 1. Effect of N supply on N in annual structures of irrigated Sultana grapevines over two consecutive seasons. Key to symbols: μ, 0 kg N ha^{-1} y^{-1}; λ, 100 kg N ha^{-1} y^{-1}. Left panel, Season 1; right panel, Season 2. Leaf, (a) and (b); stem, (c) and (d); cluster, (e) and (f).

The size of the N pool present in the stems (Figs. 1c and 1d) increased over both growing seasons. There was some indication of two phases in the accumulation of N by this pool: a peak at *ca.* 50-80 days after bud burst followed by a sigmoid accumulation pattern to leaf fall. By leaf fall in the 2nd season N application had doubled the size of this pool due to more shoots per vine, implying that sink size is a limitation to the amount of N that can be accumulated by this pool. Prior to bud burst in the 2nd season stem N of vines not supplied N decreased, indicating substantial mobilisation of N.

In both seasons N supply increased the cluster N pool size (Figs 1e and 1f) due to greater N transport in the 1st season and in the 2nd season a larger sink size (more clusters) as well as more N transport. Substantial increases in the cluster N pool began at around veraison in both seasons and continued to fruit harvest. Gains in the cluster N pool in the 1st season during this period were not at the expense of the leaf N pool, which remained relatively constant. In the 2nd season a similar pattern was observed for clusters on vines not supplied N. An extra 57 kg N ha^{-1} was found in the clusters at harvest in vines supplied N in comparison to vines not supplied N. The magnitude of this increase (173%) was greater than the magnitude of the increase in the size of the sink (114%). This suggests that sink size is less of a limitation to N accumulation by the clusters.

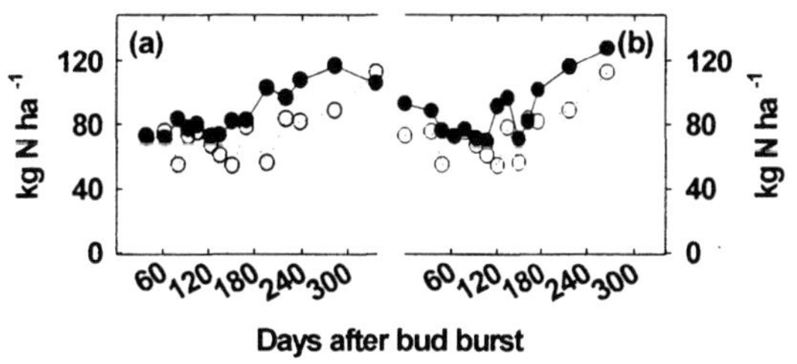

Figure 2. Effect of N supply on N in perennial structures of irrigated Sultana grapevines over two consecutive seasons. Key to symbols: μ, 0 kg N ha^{-1} y^{-1}; λ, 100 kg N ha^{-1} y^{-1}. (a) Season 1 and (b) Season 2.

Estimates of the size of the perennial N pool over time are less robust due to the difficulty in estimating the biomass of the perennial structures over time. Measurements collected at the end of the second season and tissue %N over time allow approximations of the size of the total perennial N pool to be made (Fig 2). The perennial N pool amounted to 50-130 kg ha^{-1}, depending on N supply and vine phenological stage. Up to *ca.* 100-120 days after bud burst in both seasons N supply generally did not affect the size of the perennial N pool. The perennial N pool generally increased from about that time through to leaf fall. During this period, those vines supplied 100 kg N ha^{-1} y^{-1} generally had a larger perennial N pool. Winter perennial N of vines not supplied N continued to increase coinciding with the loss of N from the stems in both seasons. In the 2nd season the perennial N pool lost 35 kg N ha^{-1} by *ca.* 100 days after bud burst. This amount only partially accounted for the additional 80-90 kg N ha^{-1} in the annual structures above that in vines not supplied N. Clearly mobilisation from perennial reserves only met a portion of those vines' N requirement during the period up to *ca.* 100 days after bud burst.

References

Alexander DMcE 1957 Aust. J. Agric. Res. 8, 162-178.
Conradie WJ 1980 Sth. Afr. J. Enol. Vitic. 1, 59-65.

W. J. Horst et al., Plant nutrition – Food security and sustainability of agro-ecosystems. 810~811, 2001

Effect of plant population density and nitrogen fertilization on yield and yield components of some white and yellow maize hybrids under drip irrigation system in sandy soil

M. H. Griesh[1)] and G. M. Yakout[2)]
[1)] Agricultural Research Station, Agricultural Research Center at Ismailia, Egypt
[2)] Agron. Dept. Fac. Agric., Suez Canal University Ismailia, Egypt

Key words : maize hybrids, nitrogen fertilisation

Abstract

Two field experiments at Private Farm in Middle Sinai, Eastern Bitter Lakes, Ismailia Governorate to study the effect of three plant population densities, i.e. 175000,21000 and 28000 plants per feddan and three nitrogen levels, i.e. 60,90,120 Kg/fed. On the yield and yield components of for maize hybrids (single cross 124,155, three way cross 320 and 352 productivity under drip irrigation system in sandy soil.
The main results can be summarized as follows:
The hybrid single cross 124 was superior to the other three hybrids (single cross 155, three way crosses 320 and 352) in grain yield and yield components and followed by single cross 155, three way crosses 320 and 352.
Increasing N levels from 60 to 120 Kg/fed significantly increased plant height, ear height ear length and diameter, number of rows, grain per row, ear per plant, 100 kernels weight, yield per plant and per feddan in both seasons of the study.
Increased plant densities significantly decreased all traits of yield and yield components except ear position.
A significant interaction effect between densities and hybrids was detected for yield per plant and yield per feddan in both seasons and ear diameter, grain per row and 100-kernel weight in the first season only. Meanwhile, a significant interaction effect between hybrids and nitrogen levels was detected for plant height, ear length ear position and ear diameter grain per row, 100-kernel weight, grain per plant and per feddan in both seasons.

Introduction

Maize is one of the most strategically crop in Egypt. To overcome the increasing requirement from maize grain for bread industry (20% mixed with wheat flower to reduce the imported quantity of wheat), animal and poultry feeding as well as many industrial purposes.

Several researches reported that increasing plant population density within certain limits delayed number of days to tasseling and silking (Bonaparte and Brown, 1976, Mourad et al, 1986), and decreased plant height, number of leaves and leaf area (LA) per plant) Badr et al; 1993.

On the other hand, El-Agamy et al (1986), Mourad et al (1986), Faisal et al (1996), demonstrated that leaf area index (LAI) and grain yield per unit area significantly increased by increasing plant population densities.

Increasing nitrogen fertiliser rates up to certain levels delayed silking dates and increased plant height, number of leaves per plant and grain yield per area (Nawar et al; 1992, Abdrabou 1996).

The purpose of the present investigation was to study the effect of three population densities and three nitrogen levels on the yield and yield components of four maize hybrids.

Material and methods

Two field experiments were conducted during 1999 and 2000 seasons at Private Farm in Middle Sinai, Eastern Bitter Lakes, Ismailia Governorate to study the effect of three plant population densities, i.e. 17500,21000 and 28000 plants per feddan and three nitrogen levels, i.e. 60,90,120 Kg/fed. On the yield and yield components of maize hybrids (single cross 124,155, three way cross 320 and 352 productivity under drip irrigation system in sandy soil.

At harvest time four months from sowing ten plants were taken at random from each sub-sub plot from measuring plant height, ear height, length, diameter and weight, 1000-grain weight and grain yield/ plant. Mean while, the plants of each sub-sub plot were harvested and grain yield (ard/fad) was recorded.

Results and discussion

The main results can be summarized as follows:

- The hybrid single cross 124 was superior to the other three hybrids in grain yield and yield components and followed by single cross 155, three way crosses 320 and 352.
- Increasing N levels from 60 to 120 Kg/fed significantly increased 100 kernels weight, yield per plant and per feddan in both seasons of the study.
- Increased plant densities significantly decreased all traits of yield and yield components. A significant interaction effect between densities and hybrids was detected for yield per plant and yield per feddan in both seasons. Meanwhile, a significant interaction effect between hybrids and nitrogen levels was detected for 100-kernel weight, grain per plant and per feddan in both seasons.

Amer et al (1995), Abdrabou (1996), Faisal et al (1996) reported similar results.

References

Abdrabou RT 1996 Annals Agric. Sci. Fac. Agric. Ain Shams Univ. Egypt 41, 683-695.

Amer SA, Mahgoub GMA and Khedr SAF 1995 Zagazig J. Agric. Res. 22, 387-398.

Mourad SB, El-Lakany MA, El-Agamy AI and Soliman FH 1986 Al-Azhar J. Agric. Res. 6, 401-411.

Nawar AA, Ibrahim ME and Attia MB 1992 Grain yield, yield components and infestation rate of corn borers and aphid of maize genotypes as influenced by nitrogen.

Table 1. Effect of maize hybrids, nitrogen fertilisation, plant population density and their interaction on yield and yield components during 1999 and 2000 seasons.

Treat	Y/P/gm		100 KW		Y/A/fad.	
	1999	2000	1999	2000	1999	2000
Var:						
TWC 320 1	156.9	159.9	34	34	23.1	23.5
SC 124 2	182.7	181.9	34	34	26.8	27.0
TWC 352 3	155.7	158.1	31	32	23.0	21.1
SC 352 4	159.8	143.7	29	30	23.5	23.4
F	**	**	**	**	**	**
L.O.D	4.08	11.99	0.32	0.47	0.51	1.58
Nit.						
60 kg N 1	150.8	151.3	31	31	22.1	22.3
90 kg N 2	162.7	161.9	32	33	24.0	24.0
120kg N 3	177.8	169.5	33	34	26.2	25.0
F	**	**	**	**	**	**
L.S.D.	2.4	1.9	0.13	0.25	0.33	0.27
Dens						
19600 plant 1	178.1	176.5	33	33	23.3	23.0
22240 plant 2	162.4	158.3	32	33	24.1	23.4
25200 plant 3	150.7	147.7	31	32	25.0	25.0
F	**	**	**	**	**	**
L.S.D	2.4	2.3	0.14	0.19	0.36	0.35
I						
nter						
VXN	**	**	**	**	**	**
VXD	**	**	**	**	**	**
NXD	*	N.S	**	N.S	N.S	NS
VXNXD	**	*	**	**	N.S	N.S
L.S.D.	3.17	3.04	0.93	1.12	3.18	3.17

W. J. Horst et al., Plant nutrition – Food security and sustainability of agro-ecosystems. 812~813, 2001

Effectiveness of mineral fertilizer use on calcareous chernozems

N. E. Kravtsova[1)], O. A. Birukova[1)] and I. A. Nagabedian[2)]

[1)] *Department of Soil Science and Agrochemistry, Biological Faculty, Rostov State University, 105, Bolshaya Sadovaya Street, Rostov-on-Don, 344006, Russia, kravtsova@mis.rsu.ru*

[2)] *Federal State Establishment State Center of Agro-Chemical Service "Rostovsky", 105, Bolshaya Sadovaya Street, Rostov-on-Don, 344006, Russia, agrosoil@mis. rsu.ru*

Key words: clacareous chernozem, effectiveness, mineral fertilisers, sunflower, winter wheat, yield

Abstract

In order to ensure an efficient use of mineral fertilisers and to increase yield and quality of crops, a thorough evaluation of various combinations of mineral fertilisers under field condition is necessary. A three year field experiment was conducted on calcareous chernozem soils with wheat and sunflower in Rustov state of Russia (i) to evaluate the effect of various combinations of mineral fertilisers on yield and quality and (ii) to determine optimum mineral fertiliser combination, which will finally result in a bio-energy efficient and economically feasible increase in yield and quality. Results of this study indicate that maximum yield and optimum quality of wheat could be obtained at $N_{90}P_{60}K_{60}$. For sunflower the treatments $N_{60}P_{60}K_{60}$ and $N_{90}P_{60}K_{60}$ resulted in the highest yields. At medium input of NPK fertiliser ($N_{60}P_{60}K_{60}$) the bio-energetic coefficient for wheat and sunflower ranged from 3.2 to 3.5 in this experiment. These calculations also indicated that the bio-energetic coefficient tended to decrease with increased levels of mineral fertiliser application.

Introduction

Increased fertiliser cost and extreme fluctuation in prices of agricultural commodities resulted in a drastic decline in mineral fertiliser application for agricultural production during the last decade. This resulted in a decline in production of all agricultural crops in Rostov state of Russia. Yields of cereal and legume crops recorded an average decrease of 38 % and 35%, respectively. Therefore, the objectives of this study were to (i) evaluate yield and quality of wheat and sunflower under various combinations of mineral fertilisers and (ii) to determine the optimum mineral fertiliser combination, which will result in an economically optimal yield and quality.

Materials and methods

A three-year field experiment was conducted with wheat and sunflower in Rostov state of Russia. Soils of this region are calcareous chernozems with ca. 3.6 % humus, 25 mg N kg-1, 28 mg phosphorus kg-1, and 362 mg exchangeable potassium kg-1 in the top 0-20 cm soil layer. In this evaluation study winter wheat ("Donskaya besostaya") and sunflower ("Donskoy - 60") were used. Prior to the seeding of these crops, this field was used to cultivate corn for silage. Fertiliser (ammonium nitrate, single super phosphate, and potassium chloride) were tested in the following combinations: control, N60P60, N60K60, P60K60, N60P60K60, N90P60K60, N120P60K60, N60P30K60, N60P30K60, N60P90K60, N60P120K60, N60P60K30, N60P60K90, N60P30K120 (see Table 1). Effect of fertilisers on content of N, P, K, protein, gluten, 1000 grain mass and glassy factor in winter wheat grain was tested. For sunflower only the yield at harvest was monitored. A bio-energetic coefficient was calculated from the following equations (Mineev, 1990):

$$\eta = \frac{vf}{A},$$

where η is the bio-energetic coefficient,

vf is the quantity of energy obtained as crop yield and

A is the energy input via the use of the fertilisers.

Results and discussion

Yield and grain quality for winter wheat

Results of this study indicated that yield of winter wheat at our experimental field with medium level of available soil P was determined by application of nitrogen fertilisers. Maximal yields of 4.20 and 4.31 t ha^{-1} were found at $N_{90}P_{60}K_{60}$ and $N_{120}P_{60}K_{60}$ treatments, respectively.

Application of mineral fertilisers in this study resulted in a to substantial improvement in quality parameters of winter wheat. As expected application of higher N rates led to higher protein content in wheat grain. However, protein content has to be seen in context with the yield. Although treatment $P_{60}K_{60}$ resulted in a protein content of 12.5 % (which is even higher than the protein content in $N_{120}P_{60}K_{60}$), yield for this treatment is statistically lower and the overall protein production per ha is quite low. Various mineral fertiliser combinations evaluated in this study indicated a tendency for improvement in gluten content (25.1 to 25.9%), 1000 grains mass (from 36.2 to 39.1 g), and glassay factor. For all these quality parameters an adequate N supply seems to be the most important aspect. In summary it can be concluded that mineral fertilisers increase yield and improve quality parameters of winter wheat.

Sunflower yield

Maximum sunflower yield of 3.08 t ha^{-1} was obtained at $N_{60}P_{60}K_{60}$ and $N_{90}P_{60}K_{60}$ fertiliser levels. When the N rate was increased to 120 kg N ha^{-1} (treatment $N_{120}P_{60}K_{60}$) yield of sunflower was significantly reduced and even lower than the yield that was obtained with no fertiliser (i.e. control treatment).

Bio-energetic coefficient for crop production

Overall the bio-energetic coefficient for wheat production under the conditions of our experiment ranged from 2.29 ($N_{60}P_{120}K_{60}$) to 4.99 ($N_0P_{60}K_{60}$). However, this treatment did not increase the grain yield compared to the unfertilised control (Tab. 1). For the high yielding treatment $N_{90}P_{60}K_{60}$ a bio-energetic coefficient of 3.40 was obtained. Application of a higher dose of nitrogen (120 kg N ha^{-1}) creates additional energy cost, and as a consequence decreases its bio-energetic coefficient. For sunflower the highest bio-energetic coefficients were obtained at $N_{30}P_{60}K_{60}$ and $N_{60}P_{60}K_{60}$ (Tab. 2). Even in a year with drought, application of fertilisers at the rates of $N_{30}P_{60}K_{60}$ and $N_{60}P_{60}K_{60}$ was energy efficient. According to Konstantinov (1978) and Johnston and Fowter (1994) high correlations (r^2 = 0.5 to 0.86) might be expected between effectiveness of medium levels of mineral fertilisers and weather conditions depending on the zone of cultivation.

Calculated bio-energetic coefficient from our 3-year study ranged from 0.9 to 4.99 (Tab. 1 and 2). The calculation also indicated that the bio-energetic coefficient tended to decrease with increased levels of mineral fertiliser application. This was due to an increase in energy use for fertiliser production, crop harvest and processing of surplus crop harvested as compared to that of energy accumulated in the harvested product. Fertiliser use should be considered as an important factor (along with weather conditions) to maximize the bio-energetic coefficient of the harvested crop.

Table 2. Sunflower yield and the bio-energetic coefficients for different fertiliser regimes

treatments N	P	K	yield	η
kg ha^{-1}			(t ha^{-1})	
control			1.92	
60	60		2.48	1.7
60		60	2.50	1.8
	60	60	2.03	1.5
60	60	60	3.08	3.2
30	60	60	2.62	3.3
90	60	60	3.08	2.3
120	60	60	1.81	1.4
60	30	60	2.51	1.7
60	90	60	2.78	2.3
60	120	60	2.60	1.7
60	60	30	2.34	1.2
60	60	90	2.43	0.9
60	60	120	2.55	1.6
GD 5 %			0.13	

References

Johnston AM and Fowter DB 1994 Crop. Dev. Center Res. Rep. 1992-1993. 1-23. Saskatoon.

Konstantinov AP 1978 Weather, soil and winter wheat yield. 1-264. Gidrometeoizdat, Leningrad.

Mineev VG 1990 Agrochemistry. 480. Moscow State University, Moscow.

Table 1. Winter wheat yield and grain quality parameters and bio-energetic coefficients for different fertiliser regimes

treatments N	P	K	yield	N	P	K	Protein	Gluten	1000 grain	Glassy	η
kg ha^{-1}			(t ha^{-1})	(%)	(%)	(%)	(%)	(%)	(g)	(%)	
control			2.53	2.14	0.96	0.68	12.2	25.2	37.0	33.7	
60	60		3.47	2.21	0.98	0.64	12.2	25.7	36.9	30.5	3.16
60		60	3.16	2.05	0.96	0.65	12.1	25.4	37.8	32.0	2.33
	60	60	2.71	2.15	0.94	0.65	12.5	26.0	38.4	31.0	4.99
60	60	60	3.69	2.09	0.95	0.65	12.3	25.5	37.5	30.5	3.48
30	60	60	3.26	2.06	0.94	0.69	12.1	25.8	37.4	29.0	3.97
90	60	60	4.20	2.16	0.96	0.66	12.1	25.2	38.7	32.0	3.40
120	60	60	4.31	2.13	0.98	0.68	12.4	25.8	39.1	33.0	2.79
60	30	60	3.51	2.18	0.97	0.64	12.5	25.9	37.6	33.0	3.21
60	90	60	3.69	2.13	0.96	0.67	12.1	25.8	36.2	33.5	3.27
60	120	60	3.33	2.11	0.96	0.64	12.1	25.7	38.1	32.5	2.29
60	60	30	3.54	2.17	0.92	0.63	12.4	25.9	37.5	31.5	3.22
60	60	90	3.46	2.09	0.98	0.65	12.2	25.6	37.2	32.0	2.77
60	60	120	3.34	2.17	0.94	0.64	12.1	25.2	37.7	32.0	2.40
GD 5 %			0.21	0.22	0.17	0.08	0.4	1.2	3.4	5.1	

W. J. Horst et al., Plant nutrition – Food security and sustainability of agro-ecosystems. 814~815, 2001

Influence of weed competition and fertiliser application on soil nutrients and performance of weeds and maize

G. O. Iremiren, D. O. Ikhu-Omoregbee, E. Oserogho and E. O. Odile
Department of Crop Science, University of Benin, PMB 1154, Benin City, Nigeria. cropsci@uniben.edu

Key words: biomass yield, chlorophyll a and b, ear leaf, grain yield, K, N, P, weed

Abstract

Weeds constitute a major problem in crop production in the humid tropics. Small-scale farmers use inadequate amount of fertilisers because of their expense. The performance of maize in an acid sand soil in the humid forest zone of Nigeria was evaluated in two field experiments. Treatments consisted of either weeding or no weeding and a combination of 90kg N ha-1, 45kg P205 ha-1 and 45kg K20 ha-1 or no fertiliser application. Fertiliser application enhanced weed biomass yield and decreased both vegetative growth and maize grain yield by as much as 67.5% in the unweeded maize plots. Nutrient content of weeds was greater in the unweeded than the weeded maize plots. Both chlorophyll a and b in the ear leaf increased when maize was both weeded and fertilised. Residual soil nutrients in the topsoil at maize harvest were similar in all treatments. Fertiliser application enhanced the ability of maize to compete with weeds and thereby improved maize performance.

Introduction

Small-scale farmers in tropical Africa often apply little or no fertiliser to maize (*Zea mays* L.) (Onwueme and Sinha, 1991) although maize responds well to fertiliser application, particularly nitrogen (Iremiren *et al.*, 1997; Oikeh *et al.*, 1998). Weeds constitute a serious problem in agriculture in the humid tropics, and insufficient weed control greatly reduces grain maize yield in Africa (Onwueme and Sinha, 1991). This study reports on the impact of fertiliser application and weed competition on the performance of maize in the humid forest of Nigeria.

Materials and methods

Two field experiments were conducted on a sandy loam soil (Rhodic Paleudult) in Benin City (6° 19'N, 50° 39'E), Nigeria in 1999 using early maturing and downy mildew resistant maize cv. DMR White. Maize was sown on 15 May (early season) and 2 August (late season) at a spacing of 75cm x 25cm in 18m^2 plots.

The four treatments replicated four times were: no weeding and no fertiliser application (W_0F_0), no weeding but application of fertiliser (W_0F_1), weeding without application of fertiliser (W_1F_0), and weeding with application of fertiliser (W_1F_1). All plots were weed-free at sowing. (W_1F_0 and W_1F_1) plots were weeded at 3 and 7 weeks after sowing. Weeds were also harvested from all plot at maize harvest. At each weeding, the weeds from a plot were weighed and sub-sampled for oven-drying and nutrient analysis. The fertilised plots each received a total of 90kg N ha^{-1}, 45kg P_20_5 ha^{-1} and 45kg K_20 ha^{-1} in equal split applications at 3 and 7 weeks after sowing. Observations for the growth and yield of maize at harvest, including chlorophyll content of ear leaf, were as reported by Iremiren *et al.* (1997).

Results and discussion

The primary weed species in the maize plots were *Mimosa pudica, Eleusine indica, Cynodon dactylon, Mariscus alternaria, Commelina benghalensis* and *Eragrostis tenella.* In both experiments, weed biomass yield was greatest in the unweeded plots and the application of fertiliser to maize further enhanced weed infestation by 9 to 15%. Weed infestation was, however, decreased in weeded maize plots with or without fertiliser application by about 58% in each experiment. Even though no differences occurred between treatments in concentrations of N, P, K Mg and Ca in weeds, weed biomass yield exerted a significant ($P< 0.01$) influence on the amount of these nutrients in weeds (Table 1).

Both weeding and fertiliser application (W_1F_1) increased plant dry weight, ear growth and grain yield of maize ($p< 0.05$). Mason and Madin (1996) reported similar results for wheat. The greatest reduction for these traits occurred in the weeded but unfertilised maize (W_1F_o) in Experiment 1 and the unweeded and unfertilised maize (W_0F_0) in Experiment 2 (Table 1). Compared to the treatment both weeded and fertilised (W_1F_1), maize grain yield was reduced by 40, 8 and 52% in W_0F_0, W_0F_1 and W_1F_0 treatments respectively, in Experiment 1 corresponding reductions for these treatments in Experiment 2 being 76, 47, and 29% respectively. Fertiliser application to the unweeded plots (W_0F_1) reduced the impact of weeds on both biomass and grain yield loss of maize compared to the unweeded and unfertilised plots (W_0F_0 $p< 0.05$). This indicates that fertiliser application increases the competitiveness of maize with weeds. King and Purcell (1997) found that fertiliser N increased competitiveness by soybean grown in association with hemp sesbania although this was reflected only in biomass production and not soybean seed yield. The performance of maize was considerably reduced when only weeding or fertiliser application occurred, indicating the need for one of these management practices to complement the other for a successful maize crop. However, the absence of both weeding and fertiliser application led to a poor maize crop as highlighted particularly in Experiment 2.

Both chlorophyll a and b in the ear leaf at maize harvest were greater ($P< 0.01$) in the fertilised plots, particularly when this was accompanied with weeding (Table 1). The improved greenness of maize leaves (stay-green phenomenon) which translated into greater vegetative growth and grain yield in W_1F_1 plot arose from a probable enhanced N nutrition of maize in this treatment. Esechie *et al.* (1995) found that increased ear leaf N in maize was associated with increased in both leave canopy size and biomass yield per plant.

Nutrients in the topsoil after maize harvest were similar in all treatments and the values averaged across treatments for both experiments were: N, 0.1 ± 0.1 (g kg^{-1}), P, 12.33 ± 5.62 (mg kg^{-1}), K, 0.21 ± 0.08 (mol_c kg^{-1}), Mg, 1.20 ± 0.04 (mol_c kg^{-1}) and Ca, 2.91 ± 0.06 (mol_c kg^{-1}), an indication that weeds were as efficient as maize in the removal of nutrients from soil independent of fertiliser application.

References

Esechie HA, Elias S and Magpantay J 1995 J. Agron. Crop Sci. 175, 271-278.

Iremiren GO, Adewumi PO, Aduloju SO and Ibitoye AA 1997 J. Agric. Sci. Camb. 128, 425-430.

King CA and Purcell LC 1997 Weed Sci. 45, 91-97.

Mason MG and Madin RW 1996 Aust. J. Expt. Agric. 36, 443-450.

Oikeh SO, Kling JG and Okoruwa AE 1998 Crop Sci. 38, 1056-1061.

Onwueme IC and Sinha TD 1991 Field Crop Production in tropical Africa Technical Centre for Agricultural and Rural Co-operation (CTA), Ede. The Netherlands.

Table 1. Effects of weeds and fertiliser application on the growth of weeds and maize

	Experiment 1					Experiment 2				
	[a]W_0F_0	W_0F_1	W_1F_0	W_1F_1	SE	W_0F_0	W_0F_1	W_1F_0	W_1F_1	SE
Weed weight (g m^{-2})	360	410	110	130	35.1**	213	228	128	143	37.7^{ns}
Weed N (g m^{-2})	2.5	1.8	0.7	0.8	0.39**	3.9	4.3	2.5	2.8	0.29^{ns}
Weed P (g m^{-2})	0.21	0.37	0.05	0.06	0.076**	0.27	0.13	0.07	0.05	0.081^{ns}
Weed K (g m^{-2})	4.8	5.4	1.4	1.8	0.45**	3.0	3.0	2.0	2.1	1.63^{ns}
Weed Mg (g m^{-2})	1.4	1.7	0.6	0.5	0.24**	0.7	0.7	0.4	0.4	0.15^{ns}
Weed Ca (g m^{-2})	0.8	0.8	0.2	0.2	0.08**	1.2	1.2	0.7	0.8	0.31^{ns}
Plant height (m)	1.7	1.8	1.5	1.9	0.10*	1.2	1.5	1.7	1.8	0.31^{ns}
Mean internode length (m)	0.12	0.13	0.11	0.13	0.008*	0.12	0.14	0.13	0.14	0.004**
Stem girth (cm)	5.1	5.4	4.5	5.4	0.40^{ns}	4.2	5.2	5.4	6.3	0.14**
Plant dry weight (g m^{-2})	512	703	431	756	90.2^{ns}	217	431	534	744	81.2**
Ear length (cm)	22	29	24	31	2.2**	19	23	24	27	1.2**
Ear girth (cm)	14.8	17.0	14.7	16.6	0.69**	10.8	13.5	14.5	16.0	1.06**
Mean ear weight (g)	42	58	34	61	6.5**	17	35	39	53	4.9**
Grain yield (Mg ha^{-1})	1.5	2.3	1.2	2.5	0.26**	0.40	0.90	1.2	1.7	0.21**
Chlorophyll a (mg g^{-1})	0.86	0.99	0.67	1.03	0.084**	0.42	0.72	0.50	0.93	0.016**
Chlorophyll b (mg g^{-1})	1.03	1.20	0.79	1.32	0.1**	0.54	0.85	0.81	1.07	0.042**

[a]Key to treatments: W_0F_0, no weeding and no fertiliser application
W_0F_1, no weeding but fertiliser application
W_1F_0, weeding with no fertiliser application
W_1F_1, weeding with fertiliser application. SE, standard error of single comparison of treatment means
ns, *, and ** indicate non-significant and significant F values at *P*< 0.05, and 0.01, respectively.

W. J. Horst et al., Plant nutrition – Food security and sustainability of agro-ecosystems. 816~817, 2001

Effect of S and N nutrition on N-accumulation and N-harvest in rapeseed-mustard (*Brassica juncea* L. and *Brassica campestris* L.)

M. Z. Abdin[1], A. Ahmad[2], I. Khan[1], M. I. Qureshi[1] and Y. P. Abrol[2]
[1] *Centre for Biotechnology, Faculty of Science, Hamdard University, New Delhi-110 062, India*
[2] *Division of Plant Physiology, Indian Agricultural Research Institute, New Delhi-110 012, India*

Key words: nitrogen accumulation, nitrogen harvest, nitrogen, rapeseed, sulphur

Abstract

The effect of S and N application on N-accumulation and N-harvest in rapeseed mustard (Brassica juncea L., V1; and Brassica campestris L., V2) was studied under field conditions with a randomized block design. Three treatments viz. 0S+100N (T1), 40S+100N (T2) and 60S+150N (T3) were applied. The results showed that application of S along with N increased N-accumulation in both genotypes at all growth stages compared with N applied alone. This increase in N-accumulation was due to an increase in nitrate assimilation from reduced nitrogen forms as evidenced by the higher nitrate reductase (NR) activity in the leaves of plants grown with both S and N. This showed that application of S along with N considerably reduced the nitrate content in the leaves due to higher NR activity. The decline in nitrate assimilation was followed by an overall increase in N-accumulation in plants. Consequently, N content in the plant was increased from 22.1- to 146.2% and from 25.5 to -249.5% in V1 and V2 respectively, with the application of S along with N, compared with the application of N alone. N-harvest index was also higher in plants grown with both S and N compared with N alone. It is concluded that application of S along with N increases the accumulation of N. The treatment, T2 resulted in maximum improvement in these parameters.

Introduction

Nitrogen (N) is an essential nutrient for plant growth. Crop productivity is largely determined by N and hence, soil fertility and soil nitrogen have almost become synonyms. The efficiency of nitrogenous fertilisers applied to the farmer's field is generally low. One of the causes of low N-use efficiency is the increase in S-deficiency in the soil. Schung *et al.* (1993) have reported that 4000 to 6000 tonnes of nitrogen are added annually to the external environment with rapeseed-mustard cropping systems in south Germany due to insufficient S supply. The need for S fertilisation could be particularly related to the amount of N applied, since both nutrients are required for protein formation and their assimilatory pathways are interdependent (Friedrich *et al.*, 1977).

In this communication, results of the interactive effect of nitrogen and sulphur on N-accumulation and nitrogen harvest in rapeseed-mustard are discussed in relation to improved N-utilization efficiency of rapeseed-mustard.

Materials and methods

Field experiments, using a randomized block design, were conducted with rapeseed-mustard (Brassica juncea L. Czern. and Coss cv. Pusa Jai Kisan, V1 and Brassica campestris L. cv. Pusa Gold, V2) at the experimental site (16 m2) of Hamdard University, New Delhi, India during winter with three treatments: viz. 0S+100N (T1), 40S+100N (T2) and 60S+150N kg ha-1 (T3) replicated thrice. Phosphorus and potassium were applied at the rate of 40 kg ha-1 each to all plots as the basal dressing. The sources of nitrogen, sulphur, phosphorus and potassium were urea, gypsum, single super phosphate and muriate of potash, respectively. The samples were taken at 30, 60, and 90 days after sowing (DAS), and at harvest. All other cultural practices were followed as described earlier (Ahmad *et al.* 1998).

For analysis of nitrogen content, 100 mg of dried tissue was digested with sulphuric acid and o-phosphoric acid in a Kjeldahl digestion unit, estimated by the method of Linder (1994). Nitrate from the leaves was extracted by the procedure of Grover *et al.* (1978), reduced by the method of Kamphake *et al.* (1967) and estimated by using the protocol of Evans and Nason (1953). In vivo nitrate reductase activity in the leaves was performed according to the procedure of Hageman and Hucklesby (1971). Nitrogen harvest index (N-HI) was calculated by the equation: N-HI = (seed nitrogen/total nitrogen) x 100.

Results and discussion

Nitrate reductase activity (NRA) in the leaves and total nitrogen accumulation was higher in plants grown with the combination of S and N compared with those grown with N alone (Table 2, 3). Contrary to this, nitrate content in the leaves of plants grown with N alone was higher (Table 4). The maximum NRA and N-accumulation was observed in plants grown with T2. These plants had the lowest nitrate content in the leaves compared with those grown with T1 or T3. The increase in NRA with T2 is an indicator of increased metabolic activity related to N-assimilation as a result of a balanced and adequate supply of S and N. A positive role of S in regulating NRA has been reported earlier in various plants (Reuveny *et al.*, 1980; Clarkson *et al.*, 1989; Bell *et al.*, 1995). Nitrogen harvest index (N-HI) reflects the N-utilization efficiency of the crop during the reproductive phase of growth and the efficient partitioning of reduced nitrogen to the grain. Application of S along with N also improved the N-HI, which is maximised with the treatment T2 (Table 1). This suggest that balanced doses of S and N not only increased the accumulation of N, but also improved the partitioning of nitrogen towards the economic sink.

In conclusion, application of S along with N in balanced and split doses increased the uptake and assimilation of nitrate and thereby, enhanced the total N-content in the plant and better partitioning to the grain. Thus, in order to maximize and reconcile the economic and environmental demands in agricultural practices, the vital role of S for agro-ecosystem can not be overlooked.

Table 1. Interactive effect of S and N on Nitrogen harvest index of rapeseed-mustard

Treatments	Cultivars	
	V_1	V_2
	Mean ± S.E	Mean ± S.E
T_1	52 ± 2.6	51 ± 3.0
T_2	73 ± 5.5	75 ± 3.1
T_3	65 ± 2.0	61 ± 2.4

DAS = Days After Sowing
T_1 = 0S + 100N kg ha^{-1}
T_2 = 40S + 100N kg ha^{-1}
T_3 = 60S + 150N kg ha^{-1}

References

Ahmad A, Abraham G, Gandotra N, Abrol YP and Abdin MZ 1998 J. Agron. Crop Sci.181, 193-199.

Bell CI, Clarkson DT and Cram WJ 1995 J. Exp. Bot. 46, 73-81.

Clarkson DT, Sarker LR and Purves JV 1989 J. Exp. Bot. 40, 953-963.

Evans HJ and Nason A 1953 Plant Physiol. 28, 233-254.

Friedrich JW, Smith D and Schrader LE 1977 Can. J. Plant Sci. 57, 1151-1157.

Grover HL, Nair TVR and Abrol YP 1978 Physiol Plant. 42, 287-292.

Hageman RH and Hucklesby DP 1971 *In* Methods in Enzymology. Ed. A San Pietro. pp. 491-503. Academic Press, New York.

Kamphake LJ, Hannah SA and Cohen JM 1967 Water Res. 1, 205-216.

Linder RC 1944 Plant Physiol. 19, 76-84.

Reuveny Z, Dougall DK and Trinity PM 1980 Proc. Natl. Acad. Sci. USA 77, 6670-6672.

Schung E, Haneklaus S and Murphy D 1993 *In* Agriculture, The Sulphur Institute, Washigton DC.

Table 2. Interactive effect of S and N on NRA in the leaves of rapeseed-mustard

Treatments	*Brassica juncea* cv. Pusa Jai Kisan (V_1)			*Brassica campestris* cv. Pusa Gold (V_2)			LSD at 5% P	
	Nitrate reductase activity (μmol NO_2 g^{-1} FW h^{-1})							
	30 DAS	60 DAS	90 DAS	30 DAS	60 DAS	90 DAS	T	V
T_1	5.13	4.23	1.23	4.86	3.94	1.35	1.024	0.271
T_2	7.08	7.01	3.34	7.34	7.24	4.08	1.093	0.315
T_3	6.67	5.11	2.89	6.76	6.67	3.64	0.569	0.106

Table 3. Interactive effect of S and N on total nitrogen accumulation in rapeseed-mustard

Treatments	*Brassica juncea* cv. Pusa Jai Kisan (V_1)			*Brassica campestris* cv. Pusa Gold (V_2)			LSD at 5% P	
	Total Nitrogen accumulation (kg ha^{-1})							
	30 DAS	60 DAS	90 DAS	30 DAS	60 DAS	90 DAS	T	V
T_1	12.3	41.6	138.3	10.7	28.9	128.4	1.61	1.25
T_2	15.1	102.4	237.6	13.4	101.4	228.1	14.36	5.64
T_3	13.6	89.6	211.1	11.9	83.2	204.9	20.36	14.64

Table 4. Interactive effect of S and N on nitrate content in the leaves of rapeseed-mustard

Treatments	*Brassica juncea* cv. Pusa Jai Kisan (V_1)			*Brassica campestris* cv. Pusa Gold (V_2)			LSD at 5% P	
	Nitrate content (μmol g^{-1} DW)							
	30 DAS	60 DAS	90 DAS	30 DAS	60 DAS	90 DAS	T	V
T_1	321	425	327	342	401	267	43.2	6.8
T_2	179	181	107	163	169	87	52.4	15.6
T_3	224	238	204	205	229	188	78.3	22.5

W. J. Horst et al., Plant nutrition – Food security and sustainability of agro-ecosystems. 818~819, 2001

Effect of increasing N and P application rates on grain yield of winter wheat in long-term fertilization experiments

K. Debreczeni[1)] and T. Kismányoky[2)]

[1)] *Institute of Agrochemistry and Soil,Science, Department of Agrochemistry, Georgikon Faculty of Agriculture, University of Veszprém, Deák Ferenc Str. 16, Keszthely, H 8360, Hungary,dbk@georgikon.hu*

[2)] *Institute of Agronomy, Department of Soil Management and Land Use, University of Veszprém, Deák Ferenc Str. 16, Keszthely, H 8360, Hungary*

Key words: agro-ecological conditions, chemical fertilisation, correlation analysis, long-term field experiments, 3-D surface charts, winter wheat

Abstract

In 1967, a network of long-term fertilisation trials (National Long-term Field Fertilisation Trials) was set up on 26 of the most characteristic land types in Hungary. These experiments have been maintained at nine experimental sites representing different agro-ecological regions of the country. The trials were established according to a uniform experimental design with a four–year crop rotation using the same nutrient treatments, plant types and varieties at every location. The fertiliser treatments included different combinations of increasing N-, P-, and K-rates. One unfertilised control plot was cultivated at each experimental sites. Different treatments and control plots provided opportunity for fertiliser response analyses of winter wheat. The long-term effect of 28-year N, P fertilisation was investigated on wheat grain yields at 9 sites of the network. Our results could be very useful for forecasting the fertiliser effects to be expected under the given growing conditions and could contribute to site-specific, environmentally friendly but economic nutrient management.

Introduction

Half of Hungary's total area is cultivated as cropland. In the past decades winter wheat areas have stabilized at around 1 200 000 hectares-, representing about 26 % of the total cultivated land. National average yields in 1988 were 5.4 t/ha, demonstrating a significant increase from the low 2.1 t/ha of the early 70's. Hungary first reached then exceeded the mean European values (Bedő and Láng, 1997).

These yields improvements were the result of appropriate crop variety selection, regular mineral fertilisation based on extension systems, up-todate plant protection and modern mechanized agrotechnology.

Yield potential of high-yielding winter wheat varieties could result in even higher average yields. However, agro-ecological conditions for plant growth are quite variable. Harnos (1991) studied the relationships between the yield-potential and meteorological year type for the period 1981-1990. He concluded, that cumulative heat values and water deficiency (dry/cold climate type) influenced wheat production to a great extent. The frequency of this dry/cold year type increased in the second half of the 20th century.

Maintaining sufficient nutrient levels in cultivated soils is a key element of agro-technology. Regular mineral fertilisation of cultivated lands in Hungary became common from the mid-20th century. In the agricultural land area statistical data show (KSH, 1999), that by the years from 1971 to 1975, the average amount of applied NPK mineral fertiliser (combined) reached the level of 218 kg/ha. Between 1981 and 1985, the amount of NPK nutrient applied reached the 282 kg/ha the highest so far.

The effect of fertilisation on yields depends on agro-ecological land characteristics to a great extent. Results of experiments conducted under different climates and distant locations from Hungary provide important scientific information, but they cannot be directly adapted to Hungarian cropping situations. Recognising this fact, G. Láng organized the network of uniform field trials in 1967. Debreczeni and Debreczeni (1994) published the detailed overview and evaluation of the long-term field experiments.

Materials and methods

The Hungarian Long-term Field Experimental Network operates on 9 research locations. The results of winter wheat experiments discussed below reflect the long-term nutrient effect of 28-years of continuous treatments. The discussion includes the analysis of the fourth year's results in the seventh rotation of a winter wheat-maize-maize-winter wheat sequence. The parallel experiments have identical crop types and varieties, rotation patterns and nutrient treatments. The fertiliser treatments include different combinations of increasing N-, P-, and K-rates (0-50-100-150-200 kg N/ha/year, 0-50-100-150 P_2O_5/ha/year, 0-100 kg K_2O/ha/year) and apply standard N,P,K mineral fertilisers. One unfertilised control plot is cultivated at each experimental site. Different treatments and control plots provide opportunity for fertiliser response analysis of winter wheat.

Soil types - even soil types – grouped under the same classification often - differ in several characteristics (e.g. pH, texture, nutrient and water regime), therefore they are affected differently by the long-term effects of the nutrient treatments.

Sites and the main soil characteristics of the control plots of the long-term fertilisation trials:

1. Karcag (KA): Luvic chernozem, clayey loam, acidic in the upper layer, pH_{KCl} = 4.6, below 50 cm $CaCO_3$ >5%
2. Iregszemcse (IR): Calcaric phaeozem, loam, pH_{KCl} = 7.3
3. Nagyhörcsök (NH): Calcaric phaeozem, loam, pH_{KCl} = 7.3
4. Bicsérd (BI): Luvic phaeozem, clayey loam, pH_{KCl} = 5.8
5. Kompolt (KO): Haplic phaeozem, clayey loam, pH_{KCl} = 4.3
6. Keszthely (KE) Eutric cambisol, loam, pH_{KCl} = 6.4, $CaCO_3$ under 30 cm
7. Putnok (PU): Ochric phaeozem, clay, pH_{KCl} = 4.6
8. Mosonmagyaróvár (MO): Calcaric fluvisol, loamy sand, carbonated soil
9. Hajdúböszörmény (HB): Luvic phaeozem, clayey loam, pH_{KCl} = 6.0

Results

Taking the long-term effects into account, the following NP-interactions can be identified in the 28th year:

As an effect of fertilisation, winter wheat yields on clayey loam Luvic chernozem soil in KA increased from 3.5 t/ha to 5.2 t/ha. N rates higher than 150 kg combined with P fertilisation resulted in highest wheat yields regardless of the amount of P-input.

Winter wheat grown on Eutric cambisol in KE yielded between 2.8 t/ha and 4.1 t/ha. There was a significant effect of applied N. Winter wheat yields exceeded 4 t/ha only in the case of high application rates of combined NP fertiliser (N>150 kg/ha and P_2O_5>100 kg/ha) in the given year of the experiment.

Haplic phaeozem of KO produced wheat yields between 3.9 and 5.3 t/ha. When 50 kg N and 50 kg P (P_2O_5) fertiliser were applied, yields varied between 4-5 t/ha, higher combined applications of N and P resulted 5-6 t/ha wheat yields.

Wheat yields on a Calcaric fluvisol in MO were between 3.7 and 6.5 t/ha. Winter wheat yields reached their highest value when 100 kg N was applied. Further increases in combined NP application rates did not result in significant increases in yields.

Calcaric phaeozem in NH had yields between 1.0-3.6 t/ha. The low yields were due to the limiting effect of unfavorable weather conditions in the investigated dry year.

In BI 2.5-6.3 t/ha yields were harvested on Luvic phaeozem. Different N treatments, up to 150 kg had positive effects on yields. Each unit of increase in P fertilisation resulted in 1.0 t/ha increase in wheat yields. On plots fertilised with 200-250 kg N/ha/year 6.0-7.0 t/ha yields were harvested.

Luvic phaeozem in HB produced the highest average wheat yields (6.9-7.9 t/ha). Due to inherent high soil fertility and long-term nutrient input effects, similar yields were produced on the plots with different levels of NP combinations.

Due to low rainfall (similar to those in NH) Calcaric phaeozem of IR had low wheat yields (1.5-2.5 t/ha).

The most significant fertiliser response occurred on an Ochric phaeozem in PU. Winter wheat yield varied between 3.2 and 8.8 t/ha. Each 50-100 kg/ha/year increase in the input of combined NP fertiliser was followed by 1 t/ha increase in yields. In treatments with 150-200-250 kg N/ha/year, wheat yields reached 8-9 t/ha.

Discussion

As mentioned earlier, winter wheat production in Hungary is highly influenced by annual climate variation. Unfortunately, dry growing seasons and high temperatures in the 28th year of the experiments resulted in low yields on loamy Calcaric phaeozems in Central Hungary. Even the effect of long-term nutrient input was not able to compensate for the harmful effect of drought. Because of the low yields, fertiliser response of winter wheat could not be evaluated for these locations, in the given year.

In the case of the other experiments (where the amount of precipitation was favorable), positive NP interaction was detected. This general tendency was valid even for highly acid soils.

A mean of the precipitation for each location would be appreciated.

References

Bedő Z and Láng L 1997 *In* Az agrárgazdaság jövőképe. Ed. L Csete. pp. 8-28. "AGRO-21" füzetek, 14. Kutatási Programiroda, Budapest.

Debreczeni B and Debreczeni K 1994 Trágyázási kutatások 1960-1990. Akadémiai Kiadó, Budapest.

Harnos Zs 1991 Az alkalmazkodó mezőgazdaság rendszere. Kertészeti és Élelmiszeripari Egyetem, Budapest.

Központi Statisztikai Hivatal (KSH) Kiadványa 1999 Budapest.

W. J. Horst et al., Plant nutrition – Food security and sustainability of agro-ecosystems. 820–821, 2001

Response of selected crops to K fertilization on major soil types in South Vietnam

P. Cong[1)], C. Sat[1)] and R. Härdter[2)]

[1)] *Institute of Agricultural Science of South Vietnam, 121 Nguyen Binh Khiem, District 1, Ho Chi Minh City, Vietnam, congphanthi@hcm.vnn.vn*

[2)] *International Potash Institute, c/o Kali und Salz GmbH, P.O.Box 10 20 29, 34111 Kassel, Germany, rolf.haerdter@kalisalz.de*

Key words: Acrisols, cabbage, coffee, ferralsols, fluvisols, limiting nutrient factor, *Nephelium lappaceum*, nitrogen, potassium, rice

Abstract

Potassium has recently been recognised as an important nutrient limiting factor to crop yield and quality due to nutrient mining and soil degradation in Vietnam. Experiments on the response to K fertilisation of different crops on major soil types in South Vietnam were carried out. K application improved yield of rice grown on an alluvial soil in the Mekong Delta, coffee, cabbage, and important tropical fruit trees on typical tropical upland soils of South Vietnam. Overall quality of crops was improved with K application though there is no generally accepted indicator according to which fertiliser effects on quality can be measured. Therefore, there is a need to refine parameter selection for monitoring effects of K on quality, especially those that are paid by customers and hence improve growers' competitiveness in the markets.

Introduction

In South Vietnam, Acrisols and Ferralsols are the dominant soil groups covering 31.8% and 45.6% of the uplands, respectively. Prolonged weathering has resulted in kaolinite-rich, coarse textured soil with low organic matter content. On the other hand, in the lowlands of the Mekong Delta, rice is extensively grown on Fluvisols with up to three crops per year. The rice cultivation essentially contributes to the population's food supply and the nation's economic growth, due to the large amounts of grains exported. Owing to intensive pressure on land due to a steadily growing population, all the suitable land for rice production is under cultivation. Therefore, larger emphasis is laid on the use of uplands for food production during recent years. On these uplands, especially in the remote areas, however, few farmers had any knowledge about potassium and its effect on crop growth and soil fertility. In most cases K was unintentionally applied to soils as a component of multinutrient fertilisers or in form of crop residues until the late eighties. Owing to further increased population pressure, especially in the uplands, nutrient mining and improper cultivation practices led to a widespread degradation of the country's major ecosystems. K is found to be the most limiting nutrient for crop production. In early nineties, the use of K fertilisers gradually increased. Momentum to the use was brought by researchers and farmers in recent years when the importance of K fertilisation on crop production and product quality was realised.

The objective of the research was to identify main yield and quality limiting nutritional factors to crop production in major cropping systems on representative soils. The response of K, either alone or in combination with other factors (nutrients and organic manure) on yields and quality of different crops was studied.

Materials and methods

Rice, coffee, cabbage and rambutan (*Nephelium lappaceum* L.) were grown on major soil types of South Vietnam, i.e. Fluvisols, Acrisols and Ferralsols. Average exchangeable K contents of these soils are 0.20, 0.04 and 0.02 $cmol_c\ kg^{-1}$, respectively. Potassium saturation is less than 1% except in some Fluvisols. Potassium fertilisers used were either in KCl or K_2SO_4 forms. Crop growth, yield, yield components and quality indices were determined.

Results

In the Mekong Delta, Fluvisols generally bear large amounts of K due to annual alluvial depositions. Despite relatively high soil K (Kexch.=120 mg kg^{-1}) two crops of the 1999 seasons showed positive response to K application (Figure 1). Analysis of panicles (not shown) indicated that at these rates the proportion of filled-grains also increased, whereas there was no K effect on the broken-grain ratio.

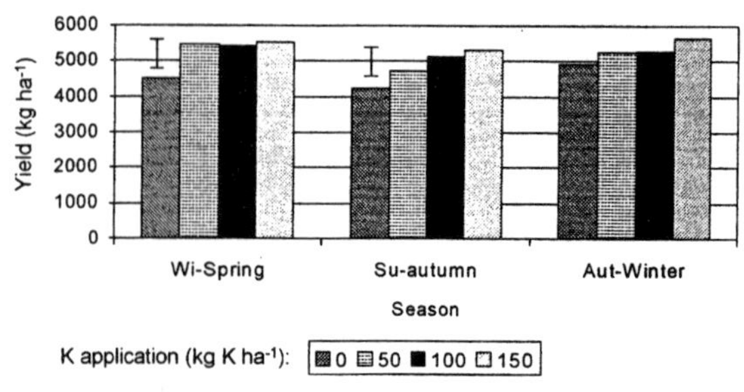

Figure 1. Effect of K application on yield of rice grown on an alluvial soil in Tien Giang Province.

Coffee was grown on a typical Ferralsol ($K_{exch.}$ = 85 mg kg^{-1}). N application (150, 250 and 500 kg ha^{-1}) tended to increase coffee yields up to N2. K application at rates of at 125, 250 kg K ha^{-1} increased yield only in the combination N3K2. K, however significantly improved bean quality. Independent from N supply best quality was obtained at the largest K amount applied (Figure 2).

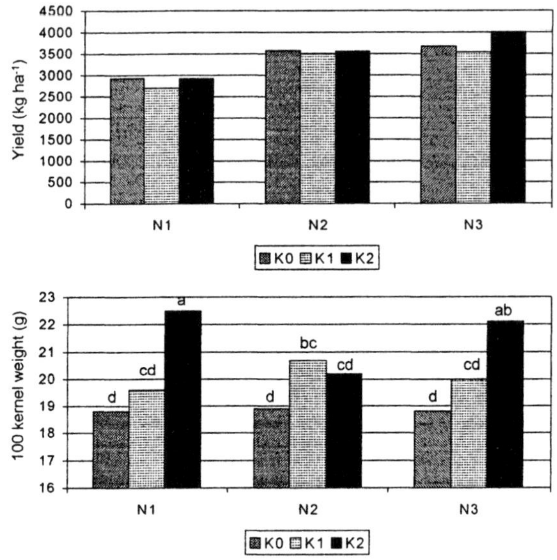

Different letters denote significant differences at P = 0.05

Figure 2. Effect of N and K application on yield and quality (100 grain weight) of robusta coffee.

Cabbage was grown on a grey degraded soil ($K_{exch.}$ = 78 mg kg^{-1}) and fertilised with three different rates of N and K respectively. Yields and the amount of rotten heads during harvest were analyzed (Figure 3).

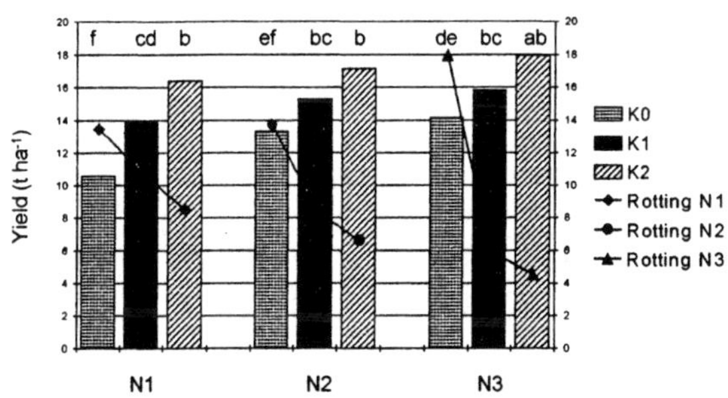

Different letters denote significant differences at P = 0.05

Figure 3. Effect of N and K application on yield and quality (% of rotten heads at harvest) of cabbage.

Both N and K application led to yield increases. At the same level of K application differences between the N treatments where not significant. At all N levels K increased cabbage yields up to the highest rate of K application (at N2 and N3 only tentatively). K application drastically reduced the percentage of rotten heads at harvest, hence improved quality, especially at N3.

Rambutan (*Nephelium lappaceum* L.) a tropical fruit tree which is widespread in the upland areas was grown on a typical Acrisol ($K_{exch.}$ 70 mg kg^{-1}). Three different rates namely 0, 0.6 and 1.2 kg K $tree^{-1}$ were applied either as KCl or K_2SO_4. Results show that fruit yields increased with increasing K application rates. Differences between the two K sources were not observed (Figure 4).

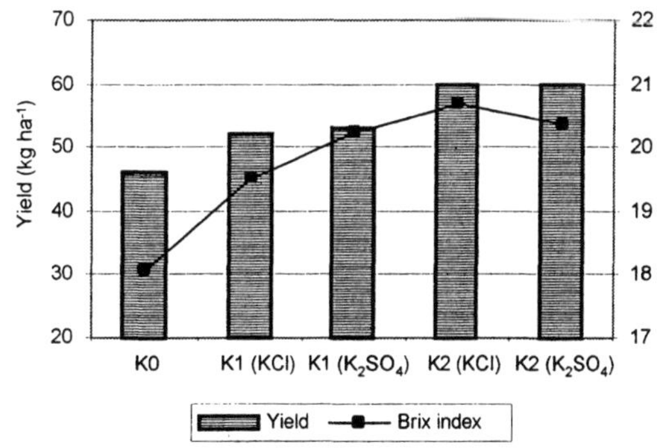

Figure 4. Yield and quality (sugar content, expressed as brix index) of rambutan as affected by K application.

Discussion

Yield response to K application in Vietnam became increasingly evident due to the high intensity of crop production on limited land. In the past hardly any yield response was reported in rice grown on the alluvial soils of the Mekong Delta (Cong, 1998). This trend is changed with the introduction of high yielding varieties and new irrigation technologies, allowing three crops per year. Continuous K removal without replenishment led to a depletion of this nutrient so that K responses even under favourable alluvial conditions are widespread. The uplands. which are important regions for the production of cash crops, e.g. fruits, vegetables and coffee, are highly weathered and heavily degraded (Vu and Pham, 1997). In these soils, K has become the main nutritional factor limiting production. This is evident from the distinct yield increases observed in coffee cabbage and rambutan. The dramatic quality improvements through K, need further study, especially with regard to their importance in improving growers' competitive edge on an increasingly dynamic market.

References

Cong DS 1998 Potassium dynamics and requirements of the major soil groups of South Vietnam. Annual report.

Cong DS 1999 Potassium dynamics and requirements of the major soil groups of South Vietnam. Annual report.

Vu CT and Pham QK 1997 Workshop on Integrated Management of Water and Nutrients. Hochiminh City, Vietnam, Jan.15-16, 1997.

W. J. Horst et al., Plant nutrition – Food security and sustainability of agro-ecosystems. 822~823, 2001

Potassium distribution and use efficiency in *Brachiaria decumbens*

W. T. Mattos, F. A. Monteiro and A. R. Dechen
Escola Superior de Agricultura "Luiz de Queiroz", Universidade de São Paulo, Av. Padua Dias,11, 13418-900, Piracicaba, SP, Brazil, wtmattos@carpa.ciagri.usp.br

Key words: forage, mineral nutrition, potassium supply, *Brachiaria decumbens*, plant parts

Abstract

A greenhouse hydroponic experiment was carried out with the forage *Brachiaria decumbens* Stapf. to study the effects of potassium supply on potassium distribution in the plant shoots and the potassium use efficiency. A nutrient solution experiment with eight potassium rates (0, 9.75, 39, 78, 156, 234, 312 and 468 mg L^{-1}) was conducted in a complete randomized block design with four replications. Plants were first harvested 38 and 71 days after transplanting the seedlings to the pots and the second harvest 33 days after the first harvest. At each plant harvest plant shoots were split into, expanding leaves, newly expanded laminae, mature laminae and culms plus sheaths. Potassium was mainly accumulated in the laminae of the two newly expanded leaves. Plants absorbed potassium between 72 and 91% of the K supplied in the nutrient solution during the experiment.

Introduction

In Brazil it is worthwhile exploring economically the productive capacity of tropical forages, among which stands out *Brachiaria decumbens* stands out.

Potassium supply has been a limiting factor for growth in some pasture areas. The roles of potassium in plant metabolism is well known: it is important on the process of photosynthesis and in the translocation of carbohydrates, and in enzymatic reactions (Epstein, 1975).

Tropical forage (other than Brachiaria) responses to the potassium supply can be found in the literature (Vicente-Chandler *et al.*, 1962; Werner and Haag, 1972; Monteiro *et al.*, 1980 and Toledo, 1984). More recently, other studies added subsidies were conducted on the K nutrition of *Brachiaria decumbens* regarding to potassium (Carriel *et al.*, 1989; Faquin *et al.*, 1995).

Materials and methods

An experiment was carried out in a greenhouse with *Brachiaria decumbens*. Seeds of the grass were germinated in washed sand and 15 seedlings were transferred to plastic pots (3.6 L each) filled with ground quartz. Five plants were grown in each pot. One litre of nutrient solution was added to each pot. In the first week, nutrient solutions (except potassium) were diluted to one-third of the normal concentration (Hoagland and Arnon, 1950). Eight potassium rates (0, 9.75, 39, 78, 156, 234, 312 and 468 mg L^{-1}) were tested in a complete randomized block design with four replications. Each solution was circulated through the substrate three times a day and the one litre volume was re-established everyday by adding deionized water.

After that the nutrient concentrations were set to the treatments. They were replaced every 14 days and also after each plant harvest. Throughout the experiment full-strength solutions were supplied six times.

Plants were harvested 38 and 71 days after transplanting the seedlings to the pots. At each harvest plant shoots were separated into expanding leaves (EL), newly expanded laminae (NL - the two youngest expanded laminae), mature laminae (ML - all the expanded laminae except NL), and culms plus sheaths (CS). At the second harvest roots were also collected.

All plant material was oven-dried at 65°C, weighed, ground and analysed for potassium by flame photometry. Analysis of variance and regression analysis were performed according to the procedures given by Freund and Littell (1981).

Results and discussion

Potassium amounts in the expanding leaves (EL) decreased with increasing K supply from 17% to 10% and from 25% to 6% in the first and second harvests, respectively (Table 2). In the newly expanded laminae (NL) and mature laminae (ML) the changes in the potassium amount were smaller than in EL. Potassium accumulation in the culms plus sheaths fractionwas much greater than in any other shoot part.

Plants growing at K rates from 9.75 to 468 mg L^{-1} had available from 58.5 to 2808.0 mg potassium during the whole experiment. From 51.2 up to 2009.2 mg potassium was found in the plants in total for the two harvests. The percentage of potassium utilization (from that supplied) significantly changed with K rates and ranged between 71.5 and 91.0% (Table 1).

Table 1. Potassium supplied and taken up, and percentage of potassium utilization by *Brachiaria decumbens* Stapf. as related to potassium rates in the solution.

K rate mg L^{-1}	K supply mg	K uptake mg	K utiliz. %
9.75	58.5	51.2	87.5
39	234.0	209.5	89.5
78	468.0	425.8	91.0
156	756.0	671.5	88.8
234	1404.0	1144.1	81.5
312	1872.0	1428.3	76.3
468	2808.0	2009.2	71.5

References

Carriel JM, Werner JC, Abramides PLG, Monteiro FA and Meirelles NMF 1989 Boletim da Indústria Animal. 46, 61-73.

Epstein E 1975 Nutrição mineral de plantas: princípios e perspectivas. São Paulo: Editora da Universidade de São Paulo. 341p.

Faquin V, Hoffman CR, Evangelista AR and Guedes GAA 1995 Revista Brasileira de Ciência do Solo. 19, 87-94.

Freund RJ and Littell RC 1981 SAS for linear models : a guide to the Anova and GLM procedures. Cary : NC, SAS Institute.

Hoagland DR and Arnon DI 1950 The water-culture method for growing plants without soil. Berkeley: California Agriculture Experimental Station (Circular nº 347).

Monteiro FA, Lima SAA, Werner JC and Mattos HB 1980 Boletim da Indústria Animal. 37, 127-147.

Toledo JM 1984 Simpósio do Trópico Úmido 1, 315-316.

Vicente-Chandler J, Pearson RW, Abruña F and Silva S 1962 Agr. J. 54, 450-453.

Werner JC and Haag HP 1972 Boletim da Indústria Animal. 29, 191-245.

Table 2. Percentage potassium distribution in four shoot parts of *Brachiaria decumbens* Stapf. as related to potassium rates in the solution, at two plant harvests.

K rate $mg\ L^{-1}$	First harvest				Second harvest			
	EL	NL	ML	CS	EL	NL	ML	CS
9.75	16.9b	30.7a	13.9b	38.5a	15.7c	29.6ab	17.3bc	37.4a
39	16.8c	32.9b	6.5d	43.8a	25.0b	29.4b	5.4c	40.2a
78	11.1c	28.3b	11.6c	49.0a	15.2c	27.1b	7.5c	50.2a
156	8.3b	21.2b	13.8b	56.7a	9.7c	24.6b	8.2c	57.5a
234	9.5b	26.7b	19.6b	44.2a	7.0c	19.4b	9.0c	64.6a
312	11.3c	24.7b	14.4c	49.6a	6.1c	18.2b	10.6bc	65.0a
468	10.0c	27.0b	16.3c	46.7a	6.1c	19.5b	12.3c	62.0a

a) EL = expanding leaves; NL = newly expanded laminae; ML = mature laminae and CS = culms + sheaths

b) Means within rows for each harvest followed by the same letters are not significantly different at the 5% level

W. J. Horst et al., Plant nutrition – Food security and sustainability of agro-ecosystems. 824~825, 2001

Corn yield and potassium accumulation related to potassium fertilizer rates

Z. Khademi[1)], M. R. Balali[1)] and M. J. Malakouti[2)]
[1)] *Soil and Water Research Institute, Tehran, I. R. Iran.*
[2)] *Soil Science Department. Tarbiat Modarres university, Tehran, I. R. Iran.*

Key words: *Corn Yield*, K-accumulation, K-fertiliser

Abstract

A 3-years Field experiment was conducted in Khuzestan to determine the effect of Potassium (K) on corn yield and K-accumulation. Above ground whole plant parts were sampled at five different times during the growing season from irrigated field experiment conducted on silty clay soil at Shavoor research station. The potassium applied as K_2SO_4 to supply 0, 50, 100, 150, 200, 250, 300, 350, 400, and 700 kg K_2O ha^{-1}. The treatments were arranged in completely randomized block design with four replicates. Application of K increased grain yield over that of the control. The highest and significant yield increase in grain and dry matter was obtained at 700 kg K_2O ha^{-1}. The highest plant K accumulation was highest at 327.3 and the lowest was at 162.3 kg K_2O ha^{-1}. The peak k-accumulation rate was 11.35 kg K_2O ha^{-1} d^{-1}. N uptake by corn was highly affected by K application rate, but P uptake was unaffected.

Introduction

The adoption of more intensive crop production practices has increased the need to re-examine K fertiliser practices for corn on silty clay soil of Khuzestan in south of Iran. The practices include irrigation, plant population and N applications, have showed that yields more than $10 Mgha^{-1}$ are possible (karlen *et al.*, 1987; stanley and Rhoads, 1975). Although a lack of corn response to K fertilisation on soils of this e region has been noted, more intensive production practices and higher yield goal may increase requirement for reponse to K fertiliser. Intensive management practices may increase the rate of K accumulation and the total K accumulation by corn at maturity. Potassium fertilisation should be related to the soil's capacity to supply K with sufficient rapidity to match the maximum rate of K accumulation by the crop and to supply a sufficient amount of K to meet total K requirement of the mature crop (Heckman *et al.*, 1992). The K-accumulation-rate is the level of K accumulation per unit of time. The K-accumulation rate requirement changes during the growing season and is at its maximum during pretasseling period of most rapid growth at approximately 50 days after planting (Hanway, 1962; Karlen *et al.*, 1987). The objectives of this study were to determine the effect of various K rates on dry matter, K accumulation and yield of corn under intensieve crop production.

Materials and methods

Field experiments were conducted with corn for 3 years on silty clay soil at Shavoor research station in Khuzestan. Treatments were 0, 50, 100, 150, 200, 250, 300, 350, 400 and 700 Kg K_2O ha^{-1}. The K source was K_2SO_4. Treatments were arranged in a completely randomized block design with 4 replications. Four whole above ground plants were collected from the inner four rows in five different times during the growing season to determine seasonal dry matter, K-accumulation for corn. Potassium accumulation was calculated from drymatter accumulation and K-concentration data. At maturity, corn ears were hand harvested from 4-m-long sections of each of the four center rows of each plot and weighed.

Results and discussion

K application increased grain yield compared to the control and the highest and significant grain yield and dry matter were obtained with 700 kg K_2O ha^{-1} (Table 1). Our results were in complete agreement with the results of kemmler and hobot (1985). The highest K accumulation of 327 kg K_2O ha^{-1} was observed at the highest rate of K application (Table 2). At early growth stages, the growth of corn plant was not highly affected by K fertiliser rates except for the last treatment. The peake K accumulation rate was 11.3 kg K_2O $ha^{-1}d^{-1}$ and that coincided with rapid growth stage of corn. This is in complete agreement with other researchers (Welch and Flanery, 1985). Kovacevic and Grgic, (1995) obtained a significant yield increase with 1000 kg K_2O ha^{-1} applied to soybean. Total N uptake was highly affected by K- rates but there was no effect on P uptake by K- rates.

References

Hanway JJ 1962 Agron. J. 54, 217-222.

Heckman JR and Kamprath EJ 1992 Soil Sci. Soc. Am. J. 56, 141-148.

Karlen KL, Sadler EJ and Camp CR 1987 Agron. J. 79, 649-656.

Kemmler G and Hobot H 1985 Kali und Salz GmbH, Kassel, Germany.

Kovacevic V and Grgic D 1995 Rostlimma vyroba. Vol. 41, No. 5, 246-248.

Welch LF and Flannery RL 1985 *In* Potassium in agriculture. Ed. RD Munson. pp. 647-667. ASA, CSSA, and SSSA, Madison, WI.

Table 1. Grain yield as affected by K fertiliser rates

Treatment K_2O (kg ha^{-1})	*1995* t ha^{-1}	*1996* t ha^{-1}	*1997* t ha^{-1}	*Average* t ha^{-1}
0	10.26	10.38	8.66	9.77
50	11.38	11.2	9.42	10.67
100	10/81	11.42	9.98	10.74
150	11.18	11.48	10.19	10.95
200	11.19	11.67	10.48	11.11
250	12.13	11.79	10.59	11.50
300	10.59	10.91	9.48	10.45
350	11.24	11.44	10.91	11.20
400	11.55	11.70	10.90	11.38
700	12.37	12.35	11.35	12.02
LSD(0.01):	*.7514*	*.6794*	*.5899*	*.6613*

Table 2. Average K-uptake by corn plants as affected by K-fertiliser rates at different days after planting (kg K_2O ha^{-1})

Teatment K_2O (kg ha^{-1})	*11 Day*	*21 Day*	*35 Day*	*49 Day*	*77 Day*
0	0.54	2.97	16.54	74.43	162.30
50	0.58	2.89	17.42	93.97	193.60
100	0.57	3.24	17.46	119.20	221.90
150	0.66	3.99	20.37	119.20	189.30
200	0.78	3.61	21.18	117.30	231.40
250	0.69	4.03	23.24	111.30	229.20
300	0.69	3.58	23.98	133.70	228.10
350	0.68	4.09	21.70	125.50	234.60
400	0.72	4.15	23.68	131.90	266.80
700	0.82	5.01	29.08	204.70	327.30
LSD (0.05)	*.2099*	*1.451*	*8.143*	*37.19*	*88.66*

W. J. Horst et al., Plant nutrition – Food security and sustainability of agro-ecosystems. 826~827, 2001

Direct, cumulative and residual effect of P-fertilizer applied at different times to wheat and barley under the dryland conditions of Turkey

B. Firat[1)], S. Gümüs[2)] (†)

[1)] Dept of Soil Science, Faculty of Agriculture, University of Selcuk Konya, Turkey,

[2)] Office of Forest, Adana, Turkey

Key words: diammonium phosphate, dryland condition, time, triple super phosphate

Abstract

In order to investigate the fertiliser efficiency in dryland farming agriculture a two years experiment with increasing P-doses at different times was carried out with wheat in first year and barley in second year in a alluvial soils from the area of Konya/Türkiye (200-250 mm precipitation). P-fertiliser as triple super phosphate (TSP) and diammonium phosphate (DAP) were applied in doses of respectively 0, 4, 8, 12, and 16 kg P_2O_5 and at different times (5 and 2,5 months before planting and at planting) in the first year. In the second year the plots were divided in three equal parts, in two of them 50% and 100% of fertiliser treatment were applied to obtain the cumulative effect whereas on the third part no fertilisers were applied in view of assessing the residual effects. The results showed that grain yield wasn't significantly increased by different times of application of P-fertiliser and that the residual effects of DAP affected slightly the grain yields but not significantly. Continuous fertilisation generally reduced the yield of grain as well as the thousand of grain weight and the protein content of the grain. In all case, the presence of P-fertiliser applications increased the percent of protein in grain (not clear). The residual effect and the continuous application of P-fertiliser increase the amount of available P in soil and the P concentrations in plants.

Introduction

In dry land climate regions crop production is mainly determined by water and fertilisation practices. Great care must be taken about the balance between fertilisation and production , where irrigation isn't possible. Because of this, a two years field experiment with different doses of diammonium phosphate (DAP) and triple super phosphate (TSP) was carried out on a alluvial soil in the area of Konya-Türkiye.

Materials and methods

In order to investigate the P-fertiliser response in dryfarming agriculture a two years experiment with increasing P-doses at different times was carried out on a alluvial soil in the area of Konya/Turkiye. Konya is in Central Anatolia with 2.366.413 km^2 ^{of} agricultural area between 38^0 east latitude and 32^0 north longitude, 1016 over NN. According to previous data recorded for 43 years, the average rainfall in this area is 324 mm/year, average temperature is 11,4^{0}C and low relative humidity is 59,3 %. Because of the arid climate most of the rainfall falls in winter. During the experiment recorded rainfalls were 206,7 and 251,8 mm in 19XX and 19XX respectively Characteristics of the experimental soil are: pH 7.7, clay content 32,2%, sand 29,1%, $CaCO_3$ 35,2 % organic substance 1,35%, available P 2,57 ppm, exchangeable K 18,72 ppm and CEC 17,99 me/100g. In the experiment DAP and TSP were used in doses of respectively 0, 4, 12, and 16kg P_2O_5/ha. In addition 6 kg N/ha and 6 kg K_2O/ha were used as base fertiliser.The test plant was wheat in the first year and barley in second. In the second year the plots were divided in three equal parts 50% and 100% of P-fertiliser were applied according to the experiment design to the first and second part to obtain the cumulative effect.In the third part no P-fertiliser was applied to recordthe residual effect. Soil samples were taken at different depth between 0 and 20 m durins the two years of the experiment and analysed in the laboratory.

Results

Grain Yield (Effect of Time):

In dryfarming agriculture, applying P-fertilisers at different times increased the barley and wheat yield significantly at 5% level, except in the case of the residual effect design. The application of 50% or 100% of DAP 2,5 months before planting caused the maximum grain yields of wheat and barley during this experiment.

Effect of Fertilizer Doses and Time:

The increase in the rate of P-fertilisers at different times, caused the grain yield to form significant groups (Not clear, need precision). The first group was composed by the addition of 12 kg P2O5/ha 2,5 months before planting and reached the top wheat yield of 120,70 kg/ha . The same rate of P-fertiliser gives the maximum yield of 96,96 kg/ha of barley in the residual effect design. In dry-farming agriculture, the negative effect of fertilisation on efficiency without taking attention to plant's water needs, is seen clearly in continuous fertilisation practice. As a matter of fact, while 123,20 kg/ha and 103,80 kg/ha grain yield formed the first group (a) the maximum dose of P-fertiliser took place at the last group . The results underlined the importance of water as much as plant nutriment substance for crop production. The same type of results were obtained by by Mergen (1984) and Christiansen-Weniger (1970).

Thousand Grain Weight (TGW)

In the first year, the treatment in which P-fertiliser was applied 5 months before planting, gave33,08 gramme of TGW and 40,40 gramme were obtained when P-fertilisers were applied 2.5 month before planting In general, the application of P fertiliser at planting time induced a decreased in TGW . These results agreed with those from Frat 1992,1998, Katkat et al 1984). Our results suggest that DAP increased more the TGW than others products

Protein Content (%)

The protein content in the controls treatement were about 1% lower than in the P-fertiliser treatements . In the first year, no differences were found between the applications of P-fertilisers at different times, and between the fertilisers themselves. In the second year, CPA of barley grain decreased in the residual effect treatment, when the whole P-fertiliser was applied 2,5 months before planting. Asalready shownfor continious fertilisation, the increase in the amount of P conduced generally to an increase in the CPA . DAP had increased the CPA on grain by about 1% more than TSP. (Not clear). In presence of DAP, an increase in the rate of application of P leaded to a decresed in grain yield and conversely an increase in CPA Phosphorus has theknown property of increasing the protein amount of the grain. However the insufficiency of the water reserves in the soil and impossibility of irrigation caused the grain to shrivel up, as a result, the CPA increased (Not clear). These results agreed with those found by Porschew at. al. 1968 and Saraçolu 1984).

Amount of P taken up by the plants

Fertilisers increased the amount of P that is taken up by the grain and by the straw of wheat and barley as expected. These uptake were also related to the yield, for both fertiliser types. (not clear). This situation was also observed with the residual effect of TSP. Even if the data obtained from the continue application of TSP where variables, in general, uptake of Pwere lower in this treatment than in residual effect treatment. This situation again suggest that plant couldn't improve their P statut during their growth phase and this probably because of lack of water . Values obtained from the DAP treatment showed that a continuefertilisation increased the amount of P taken up by plants, but this effect remains limited.

Available P in Soil

Application of P-fertilisers affected the amount of available P in soil differently: On the controls, the amount of P after reaping had increased and this can probably be explain by the biologic activity. In the first year, DAP and TSP applied 2.,5 months before planting increased the amount of available P. With the residual effect treatmentthe amount of P decrease and on average, falled to 3,83 ppm. The continue fertilisation treatment increased the available P

References

Barekzai AM 1987 DLG-Verlag Frankfurt/M.

Firat B 1992 Deutsch-Türkisches Symposium in Hohenheim, Türkische Agrarforschungen, pp. 237-244.

Firat B 1998 Deutsch-Türkisches Agrarforschung Symposium Antalya 1997, Verlag Ulrich E. Grauer Stutgart, pp. 223-228.

Katkat V, Yürür N and Kaplan M 1984 Zýr. Fak. Dergisi 3, 55-62.

Li ZY 1998 Dep. Agrochem, Nortwest Agric. Uni. Wupong shaorxi (Turang) 20, 255-258, China.

Mergen C 1984 TZDK mesleki yayýnlar No 34, Ankara.

Porschew LA, Lesogorova AL and Muzychicin ET 1968 Agrochemiya No 7, 11-16.

Saraçoðlu H 1984 TZDK mesleki yayýnlar No 32 Ankara.

W. J. Horst et al., Plant nutrition – Food security and sustainability of agro-ecosystems. 828~829, 2001

Physiological and biochemical responses of two precocious varieties of wheat to phosphate rock and TSP fertilisation in semi-arid land

C. Rahmoune[1)], S. Maâlem[1)], F. Redjel[2)], S. Hioun[2)] and M. Bennaceur[3)]

[1)] *Laboratoire d'Ecotoxicologie, Biologie, Faculté des Sciences, Univérsité de Constantine, Algeria.*
[2)] *ISN, Dpt de Biologie, Centre Univérsitaire de Tebessa, Algeria.*
[3)] *Physiologie Végétale, INRAT, 2049 Tunis, Tunisia*

Key words: Algeria, chlorophyll, growth, phosphate rock, phosphorus, semi-arid area, *Triticum durum*, TSP, wheat, yield

Abstract

The study was conducted in the semi arid area of Northeast of Algeria to characterize the physiological and biochemical effects of phosphate fertilisers on two precocious varieties of wheat (*Triticum durum* Desf.).
The experiments were conducted on the experimental plots with the addition of different levels of the commercial fertiliser TSP at 15, 20 and 25 g m^{-2} or natural phosphate rock (DO_{20}) at 35, 50 and 65 g m^{-2} containing 46% and 20% of P_2O_5, respectively, whereas potassium (K_2SO_4) and nitrogen $(NH_4)_2SO_4$) were applied at constant split rates to all plots at 17.83 g m^{-2} for the first and 47.5 g m^{-2} for the second rate.
The crop was irrigated weekly during the wet period and 2 times during the dry period.
Ten samples from each treatment were analyzed for leaf chlorophyll concentration, dry weight accumulation (shoots and roots), grain yield, and phosphor accumulation.
The results showed that addition of phosphate fertiliser were equally effective with regard to all parameters irrespective of the form (TSP and DO_{20}).
Therefore, we conclude that the commercial water soluble TSP can be substituted by the sparingly soluble natural phosphate fertiliser (DO_{20}) without significant negative effects on growth and yield of the two wheat varieties. , whereas varietal difference was observed towards these two fertilisers.

Introduction

The use of organic and the inorganic fertilisers in the modern intensive agriculture is primordial if we consider their fundamental contribution to the yield increase of crops, worldwide (Emberger, 1995). This development of intensive agricultural production has markedly enhanced the nutrient losses from soils through increased runoff, erosion and leaching, which in turn can have adverse effects on environmental quality (Frossard *et al.*, 2000).

Besides nitrogen and potassium, increased application of P fertilisers have greatly contributed to meet the increasing food needs of humans and lifestock, worldwide. (Bennaceur, 1991; Maene, 1999).

The main commercially available P fertilisers applied to agricultural soils are: super phosphate (TSP), di-ammonium phosphate (DAP), concentrated super phosphate (CSP). Sometimes P is applied also as natural phosphate rock (PR) and as urban and agricultural wastes and sludge (Mortvedt and Osborn, 1982; Pommel ,1982; Ayuso *et al.*, 1996; Frossard et Dufour, 1996).

Though the application of P fertilisers in many cases improves yields of most crops, this also may lead to environmental contamination. In this context, the input of heavy metals with the P fertilisers have attracted special attention (Emberger, 1995; Paul *et al.*, 1992; Bahtimi, 1995) particularly under semi-arid climatic conditions (Paul *et al.*, 1992; Mhiri, 1995). We have addresses this problem by conducting a field experiment with wheat, a principal cereal cultivated in this area.

Materials and methods

Seeds of two wheat varieties, Waha and Hoggar, were sown directly into the experimental field (split plot, three repetitions) with 119 and 145 Kg ha^{-1} of seeds, respectively. Phosphate was applied in two forms, TSP containing 46% of P_2O_5 and phosphate rock (DO_{20}) containing 18% at three rates (D_1, D_2, D_3) of 15, 20 and 25 g m^{-2} and 35, 50 and 65 g m^{-2}, respectively. During the culture cycle, the plots were irrigated once weekly during the wet and twice during the dry season.

Ten samples were taken for the measurement of leaf chlorophyll concentration, dry matter accumulation of shoots and roots, grain yield (1000 seeds weight) and total phosphorous accumulation in plants.

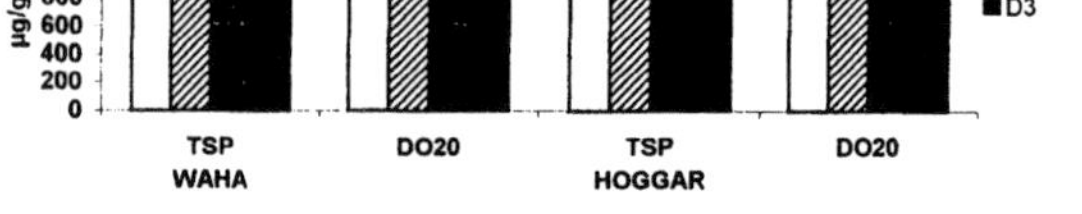

Figure 1. Leaf clorophyll concentrations in the fresh matter (FM) of two wheat cultivars as affected by form and rate of P fertilisers.

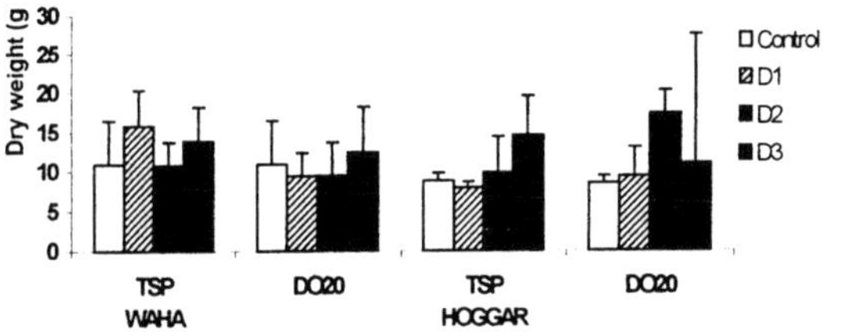

Figure 2. Dry weight production of two wheat cultivars as affected by form and rate of P fertilisers.

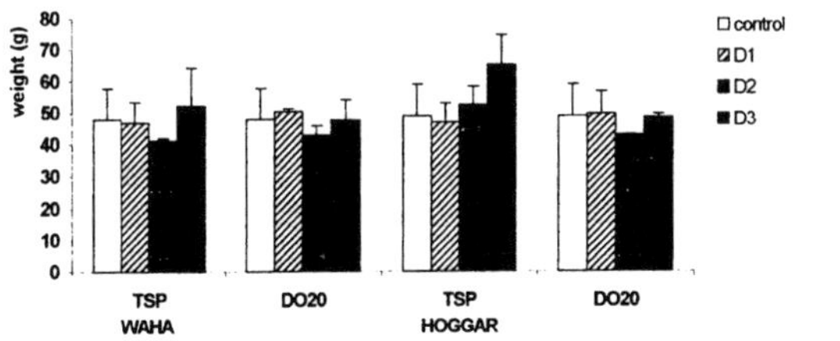

Figure 3. One thousand seeds weights of two wheat cultivars as affected by form and rate of P fertilisers.

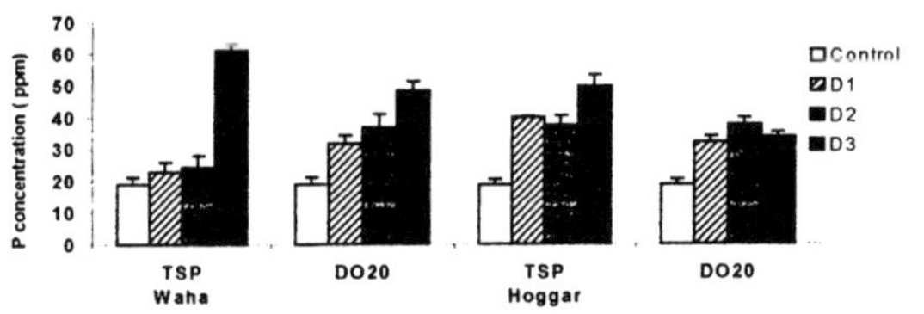

Figure 4. Phosphorus concentrations in the dry matter of two wheat cultivars as affected by form and rate of P fertilisers.

Results

Figure 1 shows that P fertiliser application did not significantly affect (Fisher test) chlorophyll (a+b) content of the two wheat cultivars. However, it appeared that cv Hogga responded more sensitively to low P supply then cv Waha. Also dry matter production (Fig. 2) was not significantly different between cultivars and treatments. But again, cv Hoggar tended to be more reduced in dry weight at limiting P supply. One thousand seed weight (Fig. 3) commonly used to describe P effects on seeds yield (Delbaugh et al.,1988; Bennaceur, 1991) did not show a significant variation (Fisher test) in the response of the two cultivars of wheat used in our experiment. The plant P concentrations (Fig. 4) clearly responded to the rate of P fertiliser application, however, there was no statistically significant difference between the forms of P fertiliser (Fisher test).

Discussion

From the results presented we conclude that natural phosphate fertiliser (DO_{20}) can substitute the commercial TSP when applied to wheat without significant effects on the parameters measured. There appear to be differences in P efficiency between the two wheat cultivars included in the experiment. The substitution of TSP by natural fertiliser P has adventages with regard to the reduction of cost of for P application and to the protection of the environment.

References

Amar B 1999 Diagnosis and prescription of phosphate fertilisation practices within the Mediterranean Region. The IFA agricultural Conference on Managing Plant Nutrition, 29 June-2 July, 1999, Barcelona, Spain.

Amberger A 1995 *In* Proc. Seminar: Production and use of the Chemical Fertilisers and Environment, 17-21 Dec. 1994. Eds. MM El Fouly and FE Abdalla. pp. 93-99. Cairo.

Ayuso M, Pascual JA, Garcia C and Hernandez T 1996 Soil Sci. Plant Nutr. 42, 105-113.

Bennaceur M 1991 Tropicultura 9, 77-80.

Cisse L and Amar B 1999 The Management of P Fertilisers for Sustained crop yield and Quality in Western European Countries. The IFA agricultural Conference on Managing Plant Nutrition, 29 June-2 July1999, Barcelona, Spain.

Delbaugh M, Chaudry GA, Mackhdum MI and Shababuddin M 1988 J. Sci. Ind. Res. 31, 206-211.

El Bahtimy M 1995 *In* Proc. Seminar: Production and use of the Chemical Fertilisers and Environment, 17-21 Dec. 1994. Eds. MM El Fouly and FE Abdalla. pp. 273-283. Cairo.

Frossard E, Condron LM, Oberson A, Sinaj S and Fardeau JC 2000 J. Environ. Qual. 29, 15-23.

Frossard E, Sinaj S, Zhang L-M and Morel JL 1996 Soil Sci. Sic. Am. J. 60, 1248-1253.

Maene LM 1999 Phosphate Fertiliser Production, Consumption and Trade: The Present situation and Outlook to 2010. The Sulphur Institute: 17[th]Sulphur Phosphate Symposium, Jan. 17-19, 1999, Boca Raton, Florida.

Mhiri A 1995 *In* Proc. Seminar:Production and use of the Chemical Fertilisers and Environment, 17-21 Dec. 1994. Eds. MM El Fouly and FE Abdalla. pp.65-84. Cairo.

Mortvedt JJ and Osborn G 1982 Soil Sci. 184, 185-192.

Paul R, Rahmoune C and Délcare E 1992 Annales de Gembloux 97, 133-147.

Pommel B 1982 Agronomie 2, 851-857.

W. J. Horst et al., Plant nutrition – Food security and sustainability of agro-ecosystems. 830~831, 2001

Role of phosphorus on mineral nutrient concentrations and their ratios in rice

M. A. Saleque[1)], K. M. Akther[2)] and G. M. Panaullah[2)]

[1)] *Bangladesh Rice Research Institute, BRRI, Gazipur, 1701, Bangladesh, enrecabd@bdonline.com*

[2)] *Soil Science Division, Bangladesh Rice Research Institute, BRRI, Gazipur, 1701, Bangladesh, enrecabd@bdonline.com*

Key words: dry matter, Fe/Mn ratio, root dry weight, varietal differences

Abstract

A greenhouse experiment was conducted to determine the significance of soil P on mineral nutrition of lowland rice. Phosphorus deficient soil was collected from a long-term P management experimental field. A portion of 3.5 kg dry soil was packed into PVC pot leaving a rim of 7.5 cm for irrigation. Six modern rice varieties (MV) were tested at two P levels - 0 and 20 mg P/kg soil. Pre-germinated rice seeds of each variety were planted to the respective pots were grown for 60 days. The application of P fertiliser increased the dry matter, tissue P concentration, and P uptake by rice. The increase in dry matter due to the P application varied from 80% in BR28 to 282% in BR9. The N and Fe concentrations were decreased, and those of Mg and Mn in shoot were increased considerably due to the P application. The application of P decreased K concentration in BR8 and BR9 while increased it in other varieties. The application of P decreased N/P, N/K, K/P, K/Mg, Ca/Mg, Fe/P and Fe/Mn ratios in all the six test varieties. The Fe/ Mn ratio in P-control pots varied between 1.32 (in BR28) to 2.66 (in BR29) while in P treated pots, it ranged from 0.96 (in BR29) to 1.62 (in BR8). There was a significant negative correlation between Fe/ Mn ratio and root dry weight of rice, suggesting that the increase in the Fe/Mn ratio decreased the root dry weight of rice. It seems from the simple experiment that P deficiency affects not only the P nutrition but also the N, K, Mg, Fe and Mn nutrition of rice.

Introduction

Rice is the staple food of the vast majority of Asians, including the people of Bangladesh. Intensive and continuous cultivation of rice without use of appropriate fertilisers has resulted in nutrient deficiencies like N, P, K, S deficiencies in many rice soils of Bangladesh (Ali *et al.*, 1997). The question of P efficiency in rice is becoming especially important in Bangladesh as deficiency of rice soils of Bangladesh, which was quite unknown only two decades ago is, of late, being observed to limit rice yields in many farmers' fields. Nutrient synergism or antagonism in crop plants is well known. An optimum P nutrition was found to increase the uptake and leaf concentrations of both Mg and Ca in wheat seedlings (Reinbott and Blevins, 1991). The objective of the present investigation was to understand the effect of P deficiency on the nutrition of other minerals in rice.

Materials and methods

Phosphorus deficient soil was collected from a long-term P management experimental field. The soil was air-dried, crushed and potted. The experimental soil has a clay loam texture, pH 7.1, organic carbon concentration 0.66%, 4 mg/kg available P (0.5 M $NaHCO_3$ extractable), exchangeable Ca 6.02 meq/100 g soil, Mg and K 1.80 and 0.19 meq/100 g soil, respectively. A portion of 3.5 kg dry soil was packed into PVC pots leaving a rim of 7.5 cm for irrigation. The potted soil was puddled and left for one week under saturated conditions.

Six rice varieties – BR8, BR9, BR15, BR16, BR28 and BR29 were tested at two P – levels: 0 and 20 mg P/kg soil in CRD design with three replications. Each pot received a flat dose of N, K and S at the rate of 150, 50 and 10 mg/kg, respectively. The average maximum/minimum temperatures in the glasshouse over the period of plant growth were 30 °C/20 °C.

Ten pre-germinated rice seeds of each variety were planted in the respective pots. After 15 days the plants were thinned to four and were grown for 60 days. At harvest, the plants were uprooted carefully and the roots were washed free of soil with distilled water.

The shoot and root samples were dried at 68 °C for 72 hours and weighed. Shoot sample (200 mg each) was digested in 5:2 conc. HNO_3 and $HClO_4$ mixture (modified from Yoshida *et al.*, 1976) and analyzed for K, Ca, Mg, Cu, Fe, Mn and Zn by atomic absorption spectrophotometer and P by colorimeter (Yoshida *et al.*, 1976). Further 200 mg samples were digested in conc. H_2SO_4 and analyzed for N by micro-Kjeldahl distillation.

Results

The application of phosphorus dramatically increased dry matter of rice by 2-3 fold (Table 1). The increase in dry matter due to the P application varied between 80% in BR28 to 282% in BR9. At P_0, BR28 produced the highest dry matter yield (4.25 g/pot) followed by BR8 (3.17 g/pot) and the lowest yield of 2.03 g/pot was obtained from BR9. In P fertilised pots, the highest dry matter of 9.95 g/pot was obtained from BR29 followed by BR15 which yielded 8.73 g/pot and BR28 produced the least (7.67 g/pot). Genotypic differences in dry matter production was more pronounced at P-control conditions than in the pots which received P treatments. In P_0 pots, there was about 109%

variation in dry matter production among the varieties but in P_{20} pots, these varietal differences converged to only 29%.

Nitrogen concentration in shoots was higher in the plants grown on P_0 pots than that of P_{20} pots (Table 1). In P_0 pots, N concentration varied between 2.62% (in BR28) to 3.36% (in BR29). The application of P increased shoot P concentration significantly (Table 1). There was about a 2-3-fold increase in P concentration in different varieties due to P application. The application of P decreased K concentration in BR8 and BR9 whilst increasing it in other varieties. The application of P increased Ca concentration in five tested varieties, but in BR8, it was reduced slightly. The application of P increased Ca concentration significantly in BR15, BR16 and BR29.

The application of P increased Mg concentration remarkably in shoots in all the tested rice varieties (Table 1). In P_0 pots, Mg concentration varied from 0.33% in BR9 to 0.64% in BR16. There was about 14% reduction in Mg concentration in BR8 due to P deficiency. In BR9 the reduction in Mg was 51%, in BR29 it was 21% but in BR16 it was only 3%. In BR8, BR15 and BR29, Fe concentration in shoots decreased due to P application, however, in BR28, the P application tended to increase Fe concentration.

There was a significant increase in Mn concentration due to P application in the tested rice varieties, however, the magnitude of the increase varied greatly among the genotypes. The decrease in Fe concentration and increase in Mn concentration in rice plants due to the P application indicated that P deficiency influenced the Fe and Mn nutrition of some rice varieties. Phosphorus deficiency increased N/P, N/K, K/P, Ca/Mg, Fe/P and Fe/Mn ratios in rice. There was a good correlation between Fe/ Mn ratio and root dry weight suggesting that the increase in the Fe/Mn ratio decreased the root dry weight of rice.

Discussion

The increase in N concentration in tissues is an important adaptive mechanism of rice plants to P deficient conditions. In lowland conditions, rice plants absorb N in the form of NH_4^+ and to maintain electro-neutrality they efflux H^+. The exuded acid diffuses out into the soil where it reacts, raising the concentration of P in the soil solution above its initial value (Saleque and Kirk, 1995). The export of H^+ to the soil increased, largely in association with increased N uptake per unit dry weight. This result indicated that there was some association between P and Mg nutrition of some rice varieties. The relationship between P and Mg has been noted in rice grain by Tanaka *et al.* (1974).

References

Ali MM, Shahid SM, Kubota D, Masunaga T and Wakatsuki T 1997 Soil Sci. Plant Nutr. 43, 879-890.

Reinbott TM and Blevins DG 1991 Agron. J. 83, 1043-1046.

Saleque MA and Kirk GJD 1995 New Phytol. 129, 325-336.

Tanaka K, Yoshida T and Kasai Z 1974 Soil Sci. Plant Nutr. 20,87-91.

Yoshida S, Forno DA, Cock JH and Gomez KA 1976 Laboratory manual for physiological studies of rice 3rd edn., International Rice Research Institute, Manila, Philippines.

Table 1. Effect of P fertiliser on N, P, K, Ca and Mg contents in shoots at 60 days after sowing in six rice varieties under green house conditions.

Variety	Dry weight (g/pot)		N	P	K	Ca	Mg	Fe	Mn	Cu	Zn
	Shoot	Root	(%)					(mg/kg)			
					0 mg P/kg soil						
BR8	3.17	0.51	2.64	0.17	5.03	0.75	0.36	854	283	37	46
BR9	2.03	0.39	3.03	0.26	5.30	0.66	0.33	547	219	42	28
BR15	2.38	0.56	3.00	0.18	4.29	0.78	0.43	887	344	37	44
BR16	2.98	0.55	3.04	0.13	4.02	0.74	0.64	447	234	47	40
BR28	4.25	1.16	2.62	0.13	4.23	0.77	0.63	372	292	42	52
BR29	2.83	0.73	3.36	0.20	4.36	0.75	0.55	655	246	42	40
					20 mg P/kg soil						
BR8	8.34	1.06	2.18	0.30	4.76	0.66	0.42	522	378	32	59
BR9	7.76	1.00	2.01	0.36	4.49	0.78	0.67	505	433	37	32
BR15	8.73	1.51	2.28	0.38	5.03	0.88	0.76	563	393	32	33
BR16	8.56	1.34	2.63	0.38	5.57	0.87	0.66	447	464	32	28
BR28	7.67	1.57	2.08	0.31	5.37	0.79	0.76	414	433	32	40
BR29	9.95	1.28	2.52	0.48	5.50	0.87	0.70	514	439	37	47
$LSD_{0.05}$ for P	1.39	0.26	0.20	0.06	0.47	0.06	0.12	92	50	NS	4
$LSD_{0.05}$ for V	NS	NS	0.35	NS	NS	NS	0.22	159	NS	8.1	7
$LSD_{0.05}$ for PxV	NS	NS	NS	NS	1.15	NS	NS	NS	NS	NS	10

W. J. Horst et al., Plant nutrition – Food security and sustainability of agro-ecosystems. 832~833, 2001

Fluid fertilisers - An efficient source of P for calcareous soils

R. E. Holloway[1)], I. Bertrand[2)], A. J. Frischke[1)], D. M. Brace[1)], M. J. McLaughlin[2)], W. Shepperd[1)] and R. D. Graham[3)]

[1)] *South Australian Research and Development Institute (SARDI), Minnipa Agricultural Centre, PO Box 31, Minnipa SA 5654, Australia*

[2)] *CSIRO Land and Water, PMB 2, Glen Osmond, SA 5064, Australia*

[3)] *University of Adelaide, SA 5064, Australia*

Key words: calcareous, fluid, granular, phosphorus, wheat

Abstract

A range of fluid products together with granular Mono-ammonium phosphate (10N:22P:0K:) + Ammonium nitrate (34N:0P:0K) were used in field experiments to provide 12 kg P ha^{-1} and 15 kg N ha^{-1} in 1998 on a highly calcareous grey sandy loam and a calcareous red brown sandy loam. Fluid sources of P had large impacts on early growth, P uptake and grain yield at both sites. The best fluid fertiliser on the highly calcareous grey soil produced 2.5 times more dry matter than the granular MAP + AN plots. The least biomass produced by fluid sources of P was still 50% greater than that produced by granular MAP + AN. P concentrations in tissues were significantly higher with all fluid treatments than granular. Grain yield was increased by 23% by applying fluid P compared with granular P on the grey calcareous soil and by 19% on the calcareous red brown sandy loam. Fluid also significantly increased grain yields in P response trials on highly calcareous soils in 2000. Fluid fertilisers have been shown to be much more efficient sources of P than granular products on the calcareous soils of Eyre Peninsula and offer a large potential for improving the sustainability and productivity of agriculture in this region.

Introduction

Low rainfall Eyre Peninsula in South Australia is characterised by large areas of highly calcareous soils (8-90% carbonate), which occupy more than a million arable hectares. Early cereal growth is slow, tillering is poor and grain yields low - often less than 1.0 t ha^{-1}. Phosphorus deficiency in wheat is endemic on these soils, despite a long history of fertiliser application and apparently "adequate" concentrations of extractable P in soil tests. Field experiments indicate that responses to granular P are best represented by relatively "flat" response curves, with poor responses continuing up to very high levels of applied P, *i.e.* 100 kg P ha^{-1}. These highly calcareous soils are able to rapidly immobilise P and alternative fertilisation strategies are needed.

Materials and methods

A range of fluid products were compared with granular products in field experiments with spring wheat on highly calcareous grey sandy loams (>50% carbonate) and a calcareous red brown sandy loam (5-10% carbonate). Sites were located on Upper Eyre Peninsula in South Australia.

In 1999, treatments were ammonium polyphosphate (16:23), orthophosphoric acid with either UAN (32N:0P:0K) or ammonium sulphate added in solution, "Agstar®", a commercial granular fertiliser (15:13) which was dissolved in water with added AN, and "Agras®" (18:8:0), another commercial granular product. Each plot received 12 kg P ha^{-1} and 15 kg N ha^{-1} except the nil fertiliser plot and the "Agras" plot . Because "Agras" is relatively soluble in water and has a low P:N ratio, two forms were applied. "Agras" was dissolved in water and phosphoric acid to give 12 kg P and 15 kg N ha^{-1}. The product was also applied in granular form at 150 kg ha^{-1} to give 12 kg P and 27 kg N ha^{-1}. Micronutrients (Zn, Mn and Cu) were sprayed onto the soil surface before sowing and all fertiliser treatments were placed in a single band 3 cm below the seed at sowing. All plots (including granular treatments) received equal amounts of liquid at sowing (400 L ha^{-1}).

In 2000, fluid and granular fertilisers providing 10 kg P ha^{-1}, 0, 10, or 20 kg N ha^{-1} and 0 or 2.5 kg Zn ha^{-1} were compared in terms of shoot growth of Frame wheat, nutrient uptake at the Haun 5 growth stage, grain and other yield parameters on grey and red brown soils.

Results

Dry weights of shoots and concentrations of P in whole shoots at early tillering and grain yield data for 1999 are show in Table 1. Fluid sources of P had large impacts on early growth, P uptake and grain yield at both sites. The best fluid fertiliser on the highly calcareous grey soil produced 2.5 times more dry matter than the granular MAP + AN plots.

The least biomass produced by fluid sources of P was still 50% greater than that produced by granular MAP + AN. P concentrations in tissues were significantly higher with all fluid treatments than granular. Grain yield was increased by 23% by applying fluid compared with granular fertiliser on the grey calcareous soil and by 19% on the calcareous red brown sandy loam.

In 2000, the fluid fertiliser on the grey highly calcareous soil increased shoot dry weight by 38%, P uptake in shoots by 37%, N uptake by 35%, Zn uptake by 31% and doubled the Mn uptake compared with granular fertiliser providing the same nutrients. Grain yield was 14% higher with fluid fertiliser. On the red brown soil, the application of 10 kg N ha^{-1} in fluid increased shoot weight (Haun 5) by 27%, P uptake in shoots by 29%, Mn uptake by 31% and N uptake by 30% compared to granular products. Adding N or Zn had no effect on grain yield with granular fertilisers but with fluids, significant positive yield responses occurred to both 10 and 20 kg N ha^{-1}. Grain yield at the 20 kg N ha^{-1} application rate was 11% higher with the fluid fertiliser than the granular. Similarly, the addition of 2.5 kg Zn ha^{-1} in fluid increased grain yield by 11%, but not with the granular treatments.

Table 1. Dry weight of, and concentrations of P in, whole shoots at early tillering and grain yields of Frame wheat on a grey highly calcareous soil and a red-brown calcareous soil on Eyre Peninsula, South Australia, 1999. Values in the same column appended by the same letter are not significantly different ($P < 0.05$).

	Grey highly calcareous soil			Red brown calcareous soil		
	Shoot wt (g $plant^{-1}$)	[P] shoots (μg g^{-1})	Grain yield (kg ha^{-1})	Shoot wt (g 20 $plants^{-1}$)	[P] shoots (μg g^{-1})	Grain yield (kg ha^{-1})
Fluid fertilisers						
Agstar	0.207a	2607a	907a	0.172a	4050a	679a
Agras	0.178ab	2308bcd	842b	0.168ab	3828a	639abc
16:23+UAN	0.170b	2468ab	856ab	0.159abc	4097a	663ab
Phosphoric acid+UAN	0.157bc	2110e	817bc	0.149bcd	3505b	631abcd
Phosphoric acid+AS*	0.174b	2387bc	814bc	0.136d	3545b	577bcd
TGMAP	0.169b	2267cde	841b	0.168ab	3902a	567cd
Granular fertilisers						
Agras granular	0.086d	1767f	737d	0.098e	3367b	545d
MAP	0.084d	1740f	657e	0.082ef	3022c	549d
Nil	0.077d	1638f	442f	0.062f	2750c	401e

W. J. Horst et al., Plant nutrition – Food security and sustainability of agro-ecosystems. 834~835, 2001

Manganese nutrition of wheat as affected by phosphorus and manganese application to a calcareous soil

N. Karimian and S. M. Hashemi

Department of Soil Science, College of Agriculture, Shiraz University, Shiraz, Iran. nkarimian@excite.com

Key words: calcareous soils, fars province, Iran, manganese, micronutrients, phosphorus, wheat

Abstract

A buildup of phosphorus (P) is common in the plow layer of soils of Fars Province of Irán due to high rates of application of P fertilisers. The interaction between P and zinc (Zn) and iron (Fe) in plant nutrition is well doccumented in the literature. This pot experiment was conducted to investigate the effect of a high rate of P application on nutrition of wheat (*Triticum aestivum* L.) grown on calcareous soils. A 6x3 factorial design in randomized blocks with 3 replications was emoloyed. Treatments consisted of 6 levels of P (0, 10, 20, 40, 80, and 160 mg P kg^{-1} soil) applied as $Ca(H_2PO_4)_2.H_2O$ and 3 levels of Mn (0, 10, and 20 mg Mn kg^{-1} soil) applied as $MnSO_4.4H_2O$. Nitrogen, Fe, and Zn were applied uniformly as $CO(NH_2)_2$, Fe-EDDHA, and Zn-EDTA, respectively. A soil from Fars province classified as Calcic Haploxeralf was used. The main chemical characteristics of this soil were pH 7.7, 35% clay, 34% $CaCO_3$-equivalent, 9 mg kg^{-1} $NaHCO_3$-extractable P (Olsen), and 14 mg kg^{-1} DTPA-extractable Mn. Wheat (cultivar Falat) was grown in a greenhouse in pots containing 2500 g of soil. Grains were harvested at maturity and their weight per pot, after drying at 65 °C, were recorded as yield. Grains were dry-ashed, dissolved in HCl and used for the determination of P, Mn, Fe, and Zn. Analysis of variance, F test, Duncan's multiple range test, and regression equations were used to determine the effect of treatments on nutrition of Mn. Results showed that the interaction effect of P and Mn on yield was statistically significant. The highest yield was obtained when 80 mg P kg^{-1} was combined with 20 mg Mn kg^{-1}. Increasing the application rate of P resulted in an increase in P concentration in grain while decreasing those of Zn and Fe. Application of Mn increased P but decreased Zn and Fe concentrations in grain. It is concluded that a buildup of P in the plow layer of soils may interfere with nutrition of wheat, increasing the need for supplemental Mn, Fe, and Zn.

Introduction

Fars province is the leading wheat producer in Iran. Soils of this region are calcareous with pH > 7.0 and $CaCO_3$-equivalent mostly > 20%. A buildup of phosphorus (P) is common in the plow layer of soils of the province, due to high rates of application of P fertilisers in the past (Karimian, 1999). However, some farmers are still applying P fertilisers to these soils. Interference of high amounts of P with zinc (Zn) and iron (Fe) nutrition of plants is well documented in the literature (Moraghan and Mascagni, 1991). Karimian (1995) reported the depressing effect of high rate of P on Zn nutrition of corn (*Zea mays* L.) grown on a calcareous soil in the region. Gholamalizadeh Ahangar *et al.* (1995), using 23 calcareous soils of the region observed the beneficial effect of Mn application in almost all of the soils. This study was conducted to investigate the effect of high rates of application of P on manganese (Mn) nutrition of wheat grown on a calcareous soil.

Materials and methods

The experiment was a 6x3 factorial design in randomized complete blocks with 3 replications.

Treatments consisted of 6 levels of P (0, 10, 20, 40, 80, and 160 mg P kg^{-1} soil) applied as $Ca(H_2PO_4)_2.H_2O$ and 3 levels of Mn (0, 10, and 20 mg Mn kg^{-1} soil) applied as $MnSO_4.4H_2O$. Nitrogen, Fe, and Zn were applied uniformly as $CO(NH_2)_2$, Fe-EDDHA, and Zn-EDTA, respectively. The soil used was calcareous (fine, mixed, thermic, Calcic Haploxeralf) from Fars province with a pH of 7.7, 35% clay, 34% $CaCO_3$-equivalent, 9 mg kg^{-1} $NaHCO_3$-extractable P (Olsen), and 14 mg kg^{-1} DTPA-extractable Mn. Wheat (cultivar Falat) was grown in pots containing 2500 g soil in greenhouse. Grains were harvested at maturity and their weight per pot, after drying at 65 °C, were recorded as yield. Grains were dry-ashed, dissolved in HCl and used for determination of P, Mn, Fe, and Zn. Analysis of variance, F test, Duncan's multiple range test, and regression equations were used to determine the effect of treatments on Mn nutrition.

Results and discussion

Results are shown in Tables 1 and 2 and Equations [1] through [3].

$$GP = 2675 + 17.4\,P + 13.9\,Mn \qquad R^2 = 0.743 \quad p < 0.001 \qquad [1]$$

$$GZN = 29.1 - 0.0358\,P + 0.0278\,M \qquad R^2 = 0.435 \quad p < 0.001 \qquad [2]$$

$$GFE = 84.5 - 0.0935\,P - 0.039\,Mn \qquad R^2 = 0.404 \quad p < 0.001 \qquad [3]$$

where GP, GZN, and GFE are the grain concentrations (mg kg^{-1}) of P, Zn, and Fe, respectively; and P and Mn are the applied rates of P and Mn in mg kg^{-1}, respectively.

The interaction effect of P and Mn on yield was statistically significant (p=0.05). The highest yield (7.98 g pot^{-1}) was obtained when 80 mg P kg^{-1} was supplemented with 20 mg Mn kg^{-1} (Table 1). Higher rates of P depressed the yield, even with supplemental Mn. The Mn concentration in grain was also affected by application of P and Mn (Table 2), increasing as a result of Mn application while generally decreasing due to P application. The lowest Mn concentration in grain was obtained with a rate of P of 80 mg P kg^{-1} (with no applied Mn), while the highest Mn concentration of grain occurred when 20 mg P kg^{-1} was supplemented with 20 mg Mn kg^{-1}. Apparently, the effect of excess P on plant Mn concentration depends on soil and plant conditions. Moraghan and Mascagni (1991) point to some conflicting reports in this regard, showing that in some cases excess P caused a reduction in plant Mn concentration whereas in the others it was increased, leading to intensified Mn toxicity.

Application of P increased P concentration in grain (Equation [1]) but decreased those of Zn (Equation [2]) and Fe (Equation [3]). The soils of the region generally suffer from the low availability of micronutrients. They also have a high capacity for retaining Mn (Karimian and Gholamalizadeh Ahangar, 1998), Zn (Karimian and Moafpouryan, 1999), and some other micronutrients. High rates of fertilisation with P, therefore, will aggravate the situation, causing an imbalanced plant nutrition. Application of Mn increased P but decreased Zn and Fe concentrations in grain. Roomizadeh and Karimian (1996) observed that application of Fe to calcareous soils of Fars province drastically reduced the Mn concentration of plant tops.

It is concluded that buildup of P in the plow layer of calcareous soils in Iran may interfere with the Mn, Fe, and Zn nutrition of wheat, increasing the need for supplemental Mn, Fe, and Zn. It is suggested that P fertilisers be applied only when soil testing indicates the need for such a practice and that application of Mn, Fe, and Zn are encouraged where there is a buildup of P and/or where soil testing shows that there is a deficiency of these micronutrients. Further research on the interaction of different nutrients in calcareous soils of Iran (i.e, soils with very large amounts of $CaCO_3$-equivalent) is recommended.

Acknowledgement

The authors would like to thank the National Research Council of the Islamic Republic of Iran for financial support given to this study through Grant No. NRCI-540.

References

Gholamalizadeh Ahangar A, Karimian N, Abtahi A, Assad, MT and Emam, Y 1995 Commun. Soil Sci. Plant Anal. 26, 1441-1454.

Karimian N 1995 J. Plant Nutr. 18, 2261-2271.

Karimian N 1999 J. Soil Water Sci. 12 (4), 1-14 (in Farsi).

Karimian N and Gholamalizadeh Ahangar A 1998 Commun. Soil Sci Plant Anal, 29, 1061-1070.

Karimian N and Moafpouryan GR 1999 Commun. Soil Sci. Plant Anal. 30, 1721-1731.

Moraghan JT and Mascagni HJ Jr 1991 *In* Micronutrients in Agriculture. Eds. JJ Mortvedt, FR Cox, LM Shuman and RM Welch. pp 371-425. Soil Science Society of America, Inc., Madison, WI.

Roomizadeh S and Karimian N 1996 J. Plant Nutr. 19, 397-406.

Table 1. Wheat grain yield (g pot^{-1}) as affected by P and Mn application*.

Applied P, mg kg^{-1}	Applied Mn, mg kg^{-1}			
	0	10	20	Average
0	2.70i	4.37h	3.99h	3.69D
10	6.82bcde	6.00efg	5.45g	6.09C
20	6.23defg	6.46cdef	6.06defg	6.25C
40	5.89fg	5.93fg	7.09bc	6.30C
80	7.51ab	7.64ab	7.98a	7.71A
160	7.13bc	7.19abc	6.87bcd	7.06B
Average	6.05A	6.26A	6.24A	

*Means followed by the same letters (capital letters for the main effects, small letters for the interaction effects) are not significantly different using Duncan's multiple range test (p=0.05).

Table 2. Concentration of Mn in wheat grain (mg kg^{-1}) as affected by P and Mn application*.

Applied P, mg kg^{-1}	Applied Mn, mg kg^{-1}			
	0	10	20	Average
0	25ef	29bcde	31abc	29AB
10	24fgh	30bcde	34ab	29AB
20	24fgh	32abc	35a	30A
40	20gh	24fg	28cdef	24C
80	19h	26def	30bcde	25BC
160	20gh	26def	28cdef	25BC
Average	22C	28B	30A	

*description see table 1

W. J. Horst et al., Plant nutrition – Food security and sustainability of agro-ecosystems. 836–837, 2001

Sulphur supply and the optimisation of the yield of wheat

R. V. Palmer, F. J. Zhao, S. P. McGrath and M. J. Hawkesford
Agriculture and Environment Division, IACR-Rothamsted,, West Common, Harpenden, Hertfordshire, AL5 2BG, United Kingdom, rachel.palmer@bbsrc.ac.uk

Key words: sulphate, sulphur, *Triticum aestivum*, yield

Abstract

With an increased incidence of S deficiency occurring in many crops due to reductions in S deposition from atmospheric sources there is a need to address the questions of optimum quantity and timing for application of S fertilisers. Recent reports have demonstrated that S supply is important for optimum yield and quality of *Triticum aestivum*. Field trials on a known deficient site were completed with S applied at times between March and June. Sulphur uptake and the fate of this sulphate was monitored with final yield also being measured. In separate greenhouse experiments using controlled environment conditions and measured nutrient solutions to provide regulated amounts of sulphate within a wide range of applied-S concentrations the measured internal pools were remarkably similar. Field trial results showed that the timing of application of S to a crop could influence the yield of the crop. Pot experiments showed that excess S could effect the basic sulphate storage capacity of wheat plants and that under both S deficient and S excess circumstances there was a reduced yield.

Introduction

Cereals have an S requirement of 10-30 kg/ha (Zhao *et al.*, 1999). Sulphur nutrition in arable crops has, until fairly recently, been ignored, as the inputs of S from atmospheric sources and other fertilisers were sufficient to facilitate growth with acceptable quality and yield.

It is probable that S is required throughout the growth period of cereals if optimum grain quality and yield is to be achieved. The role of S use, in plants, is a continuous and changing one that is influenced by external factors such as sulphate availability (Adiputra *et al.*, 1995; Sunarpi *et al.*, 1995; Fitzgerald *et al.*, 1999).

Sulphur is necessary for optimal grain production, particularly if the grain is to be used for the manufacture of bread grade flour (Zhao *et al.*, 1999). Much of the S available to wheat plants is assimilated into amino acids and proteins: cysteine, methionine and glutathione. These amino acids contain about 90% of the total S found in plant storage tissues (Allaway *et al.*, 1966) although the levels are much lower in the vegetative tissues. The majority of the cysteine and methionine is bound up in the proteins of the plant, and as such provides an important dietary source of these amino acids.

When the level of sulphate supply to the plant exceeds requirements, then sulphate is stored in the vegetative tissues, particularly in the leaves (Brunold, 1993; Blake-Kalff *et al.*, 1998). These reserves may be mobilised during grain development if S supply is limited. In contrast to the synthesis of leaf proteins, both genotype and nutritional conditions influence the synthesis of storage proteins in grains. When S supply is limiting to growth there is a shift in the balance of synthesis from S-rich to S-poor proteins (Randall *et al.*, 1986) and the ratio of N:S in the grain protein increases. The aims of this study were to determine the impact of S timing on the yield of field grown crops and to establish reasonable limits for the amount of S that can be applied to wheat.

Materials and methods

In greenhouse pot-experiments, *Triticum aestivum* L. Axona plants were grown to maturity on a 50:50 (v/v) sand and perlite mixture. Plants were supplied with sulphate S at 0.1, 1.0, 5.0, 20.0 or 50.0mM. Plants were grown under glass house conditions of 12-16°C with a photoperiod of 16 hours and were watered from above with either demineralised water or nutrient solution consisting 7mM KNO_3, 1mM KH_2PO_4, 1.71mM $Mg(NO_3)_2$, 0.01mM NaCl, 0.5mM EDTA FeNa, 5mM $Ca(NO_3)_2$, 2mM $CaCl_2$, 1.9μM Zn acetate, 49.0μM H_3BO_3, 2.2μM $Cu(NO_3)_2$, 1.6μM ammonium molybdate, 100.3μM FeNa-EDTA, 19.7μM $Mn(NO_3)_2$ with the sulphate concentrations being achieved via $MgSO_4$. Plants were grown to maturity with samples taken at GS 21, GS 31, GS 57, anthesis and maturity. Tissues were weighed, dried and analysed for sulphate and total S.

The field trial consisted of twenty 10m by 4m plots, in four blocks of 5 plots, with each block containing one of each of the treatments. Plots were sown at Woburn Experimental Farm, a known S deficient site, with Triticum aestivum L. Hereward, a common bread making winter wheat variety. Each of the plots received either a measured dosage of sulphur, by hand as gypsum (calcium sulphate) at 30 kg S ha^{-1} at one of four different times during growth: March, April, May or June, or no treatment. No S-containing agrochemicals were used on this site at any time during the study.

Samples were taken from all the plots approximately 1 month after treatment, with yields and a final grain sample being taken at final harvest on the 18th August 2000. Samples were dried and measurements of total S and sulphate content were completed on the vegetative tissues.

Results

Pot trials with wheat showed that, under different S treatments there were differences in biomass, total S, and sulphate content. An increase in final biomass was observed in plants grown on up to 5mM S: 70.7 ± 6.5, 78.3 ± 3.4, 85.1 ± 8.5 g (DW) grown at 0.1, 1 and 5mM respectively.

Increasing S supply beyond 5mM resulted in a reduction in biomass: 53.3 ± 17.2 and 59.81 ± 9.5 g (DW) when grown at 20 and 50mM sulphate respectively. Plants receiving in excess of 5mM sulphate were chlorotic and stunted with a reduced number of small, non-seed bearing tillers. Plants grown on 0.1mM sulphate had reduced growth rates, were stunted and had reduced yields (data not shown).

With increased S supply, from 0.1mM to 20mM, the total S remained relatively constant, however the proportion of sulphate S to non-sulphate S increased. Plants grown on 0.1mM sulphate contained less sulphate S proportionally than non-sulphate S when compared to plants grown on 20 and 50mM sulphate (Fig.1).

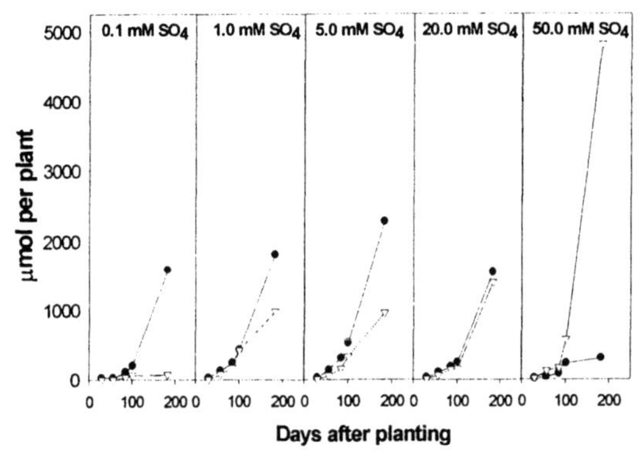

Figure 1. Sulphur pools in wheat plants grown in pots on nutrient solutions containing 0.1 to 50mM sulphate. Open triangles: sulphate-S, closed circles: non sulphate-S.

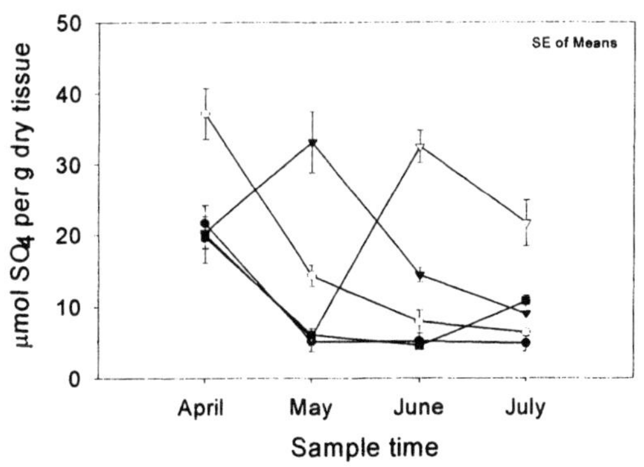

Figure 2. Sulphate pools in wheat plants as affected by timing of S-fertilisation. Plots receiving no treatment, closed circles, March treatment, open circles, April treatment, closed triangles, May treatment, open triangles, June treatment, closed squares.

In field grown plants the timing of S application had an effect on grain yield: 5.76 ± 0.48, 7.56 ± 0.56, 6.89 ± 0.29, 7.41 ± 0.40 and 5.49 ± 0.41 tonnes ha^{-1} with no treatment or treatments in March, April, May and June respectively. There was a transient effect on the sulphate pool of the different plots after S application at the different times during the study (Fig. 2). Irrespective of treatment all plots showed an increase in the tissue sulphate concentrations one month after application, the level of sulphate in the tissues however subsequently decreased with time. Plants given S in May had more sulphate available during grain filling (GS 70 onwards, occurring in July) than the other treated plants. S applied in June had little effect on the sulphate content of the tissues with levels remaining similar to those of plots receiving early S treatments (March and April) and grain yields remaining close to that of the control plots (no applied S). Yields were similar in the March, April and May applied plots with the June plots showing no recovery of yield over the control treatment.

Discussion

The data from the pot experiment suggest that there is a regulation of sulphate uptake and storage that fails to operate at high levels of supplied S. These data indicate that wheat plants are capable of storing large quantities of S as sulphate in their tissues although there is a detrimental effect on the plants at the highest exposure.

In the field trial, yields responded to applied S at all application times except the June application. Fitzgerald *et al.*, (1999) reported that S was redistributed from the older vegetative tissues to developing tissues in young plants of wheat grown under deficient conditions. These data suggested that the capacity for storage is regulated and may limit sulphate availability during grain filling. The stored pools fluctuate widely depending on availability. These data also show that there was a large window, during the growing season, for S deficiency correction by S application (early March until May).

Acknowledgement

This project is supported by the HGCA. IACR-Rothamsted receives grant-aided support from the Biotechnology and Biological Science Research Council (BBSRC) of the UK.

References

Adiputra IGK and Anderson JW 1995 Physiol. Plant. 95, 643-650.

Blake-Kalff MMA, Harrison KR, Hawkesford MJ, Zhao FJ and McGrath SP 1998 Plant Physiol. 118, 1337-1344.

Brunold C 1993 *In* Sulfur Nutrition and Assimilation in Higher Plants. Eds LJ De Kok, I Stulen, H Rennenburg, C Brunold and WE Rauser. pp 61-75. SPB Academic, The Hague.

Fitzgerald MA, Ugalde TD and Anderson JW 1999 J. Exp. Bot. 50, 1587-1592.

Randall PJ and Wrigley CW 1986 Advances in Cereal Science and Technology, 171-206.

Sunarpi and Anderson JW 1995 Physiol. Plant. 94, 143-150.

Zhao FJ, Hawkesford MJ, McGrath SP 1999 J. Cereal Sci. 30, 1-17.

W. J. Horst et al., Plant nutrition – Food security and sustainability of agro-ecosystems. 838~839, 2001

Yield losses in groundnut due to micronutrient deficiencies in calcareous soils of India

A. L. Singh

National Research Centre for Groundnut, P. B. 5, Junagadh 362001, India, alsingh@nrcg.guj.nic.in

Key words: calcareous soil, chlorosis, groundnut, micronutrient deficiencies, yield losses

Abstract

Field experiments were conducted to determine the extent of yield losses in groundnut (*Arachis hypogaea* L.) due to the deficiencies of various micronutrients, appearing as chlorosis of leaves, in calcareous soils of India. The crop was raised by providing all the macro- and the micro-nutrients and also by withholding individual micronutrient (Fe, Mn, Zn, Cu, B or Mo), in calcareous soils, and compared for determining the effect of individual micronutrient.
Severe chlorosis was observed in the unfertilised plots while it was relatively less severe in the plots that received only macronutrients and chlorosis was not seen at all in the plots that received all the macro- and the micro-nutrients. Application of micronutrients successfully prevented occurrence of chlorosis and increased chlorophyll content, pod yield and micronutrient uptake. The three years of experimentation showed that application of macro- as well as micro-nutrients caused increase in pod yield by 60% of which 17% could be ascribed to the micronutrients. Further, considerable decrease in pod yield occurs even if a single micronutrient is omitted. On an average, 15, 12, 16, 13, 19, and 17% yield losses could be attributed to the deficiencies of Fe, Mn, Zn, Cu, B, and Mo, respectively. Hence for maximum realization of yield potential of the crop, in calcareous soils, application of micronutrients may be incorporated in the package of practices for cultivation of groundnut.

Introduction

Because of high calcium requirement the farmers in India prefer to grow groundnut in calcareous soil where chlorosis, due mainly to the lime-induced deficiencies of micronutrients, is the main problem causing reduction in pod yield (Singh 1999). The deficiency symptoms show mainly on young and developing leaves as an interveinal-to complete leaf-blade- chlorosis, which when further aggravated turns pale yellow or white causing death of leaflets and consequently of plants (Papastylianou, 1990; Singh *et al.*, 1990, 1993). These multi-micronutrient deficiencies are so intermingled that it becomes difficult to diagnose and single them out in a field-grown crop, however, these could be detected through their correction by applying appropriate micronutrients (Singh and Chaudhari, 1997; Singh *et al.*, 1993). Several efforts have been made to develop corrective measures for these deficiencies, however, the extent to which this causes loss of yield is not known with any degree of certainty. In this study, an effort was made to determine the extent of yield losses in groundnut that occur due to deficiencies of various micronutrients in calcareous soil.

Materials and methods

Field experiments were conducted during three consecutive (wet 1993, dry 1994 and dry 1995) years at the National Research Centre for Groundnut, Junagadh, India in a medium black, calcareous (21 % $CaCO_3$) clayey soil containing, 7.8 pH, 0.81 % organic carbon, 710 mg kg^{-1} total N, 6.2 mg kg^{-1} available P, 11 mg kg^{-1} heat soluble S, 3.02, 4.9, 0.61 and 0.60 mg kg^{-1} DTPA extractable Fe, Mn, Zn, and Cu, and 0.36 and 0.26 mg kg^{-1} available B and Mo respectively. The experiment was laid out in a randomized block design with nine treatments as detailed in Table 1 and three replications. The field was prepared, and divided into 27 plots of 25 m^2 (5 m x 5 m).

The macronutrients N, P, K, Ca, S and Mg at 40, 40, 60, 100, 30 and 10 kg ha^{-1}, respectively, were applied, 50% at the time of sowing in the furrows and the remaining 50% at 30 days after emergence (DAE) in all the treatments except control (T_1). The micronutrients Fe, Mn, Zn, Cu, B and Mo at 10, 10, 5, 2, 2 and 1 kg ha^{-1}, respectively were applied, 50% as basal in the furrows at the time of sowing along with macronutrients and the remaining 50% as two foliar sprays at 40 and 60 DAE, respectively. The treatment T_1 was unfertilised control, in T_2 all the macronutrients were provided, while in T_3 all the macro- as well as micro-nutrients were provided. In the remaining T_4 to T_9 treatments (for determining the effect of the individual micronutrient), one particular micronutrient (Fe, Mn, Zn, Cu, B or Mo) was excluded while the rest of macro- and micro-nutrients were applied. Following standard package of practices grew the groundnut genotype FeESG 10-1.

The chlorophyll content, in the fully emerged top five leaves, was estimated at 60 and 80 DAE. The crop was harvested at maturity, dried and observations on yield and related attributes were recorded. Yield losses due to exclusion of individual micronutrients in T_4 to T_9 treatments were calculated over the yield recorded for T_3. The plant samples were analyzed for micronutrients (Fe, Mn, Zn and Cu) by atomic absorption spectrophotometry while B and Mo by colorimetry using curcumine and thiocynate methods, respectively.

Results and discussion

The crop, during 40-80 DAE, showed severe

deficiency symptoms (chlorosis of leaf), in unfertilised plots, and relatively less severe in the plots that received only macronutrients (T2), whereas chlorosis was absent in the plots that received macro- as well as micro-nutrients (T3). Application of micronutrients, along with macronutrients, thus prevented the occurrence of chlorosis by maintaining or enhancing chlorophyll content in leaves, besides enhancing nutrient uptake and improving yields. Exclusion of any one of the micronutrients lowered the chlorophyll content of leaves at 60 and 80 DAE, maximum reduction, however, was caused by exclusion of Fe (Table 1).

On an average, application of all the macronutrients (T2) increased pod yield by 36 % (range 27-47 %) over control, however, application of all macro- and all micro-nutrients (T3) produced 60 % (Range 40-94 %) more pod yield than the control (Table 1). The haulm yield was also increased by 14 % due to macronutrients, but addition of micronutrients did not increase it further indicating thereby that micronutrients exert their beneficial effect more by improving the partition of photosynthates. Though the micronutrients (T3) collectively increased 17 % (range 9-40 %) pod yield over the macronutrients (T2), their contribution appeared to be of concerted type as there was considerable decrease in pod yield even if any one of the micronutrients was withheld (Table 1).

The yield data of treatments lacking in any one of the micronutrient (T4-T9) showed that there was a significant loss of pod yield in these treatments than macro- plus micro-nutrients (T3). This indicated that micronutrient deficiencies resulted in loss of yield and this loss could be recovered if the relevant micronutrient was applied. The average loss of pod yield due to individual deficiency of Fe, Mn, Zn, Cu, B, and Mo was 15, 12, 16, 13, 19, and 17% and the corresponding range were 10-22, 8-17, 13-20, 12-14, 14-26, and 14-19%, respectively. The critical levels of these micronutrients for groundnut crop, in calcareous soil, have been reported to vary in the range of from 3-5 ppm for Fe, 3-5 ppm for Mn, 0.5-0.8 ppm for Zn, 0.2-0.5 ppm for Cu, 0.2-0.4 ppm for B and 0.4-0.2 ppm for Mo (Singh, 1999). Accordingly, most of the micronutrients, except Cu and Mo, were either deficient or at near critical level in this soil and clear response of micronutrients was observed. Moreover, in calcareous soils, there is high uptake of Ca by plants that restricts uptake of other micronutrients (Singh *et al.*, 1990) and thus constrains the plant to suffer from micronutrient deficiencies.

In calcareous soils, occurrence of Fe, Mn and Zn deficiencies is well documented (Singh *et al.*, 1990, 1993). But in this study, the groundnut crop also showed loss of yield due to deficiencies of Mo and B, which can be explained on the basis of its being a leguminous crop, which are know to require Mo for nitrogen-fixation and B for partitioning. Micronutrients application increased the uptake of most micronutrients, except Fe, to the extent of doubling and withholding any one of the micronutrients (T4-T9) caused severe reduction in the uptake of not only that particular micronutrient but also mutual interaction. Of all the micronutrients, omission of Fe and Mn caused maximum reduction in uptake of all the micronutrients, whereas withholding Mo caused reduction in uptake of only Fe and Mo.

It was concluded that the micronutrient deficiencies (in form of chlorosis), which are of regular occurrence in groundnut crop grown in calcareous soil, unless corrected by external application of micronutrients would result in significant loss of crop yield.

References

Papastylianou I 1990 J. Plant Nutr. 13, 555-566.

Singh AL 1999 *In* Advances in Plant Physiology. Ed. A Hemantaranjan. pp. 161-200. Scientific Publishers (India), Jodhpur, India.

Singh AL and Chaudhari V. 1997 J. Agr. Crop Sci. 179, 107-114.

Singh AL Chaudhari V. and Koradia VG 1993 Indian J. Plant Physiol. 34, 35-39.

Singh AL, Joshi YC, Chaudhari V. and Zala PV 1990 Fert. Res. 24, 85-96.

Table 1. Influence of micronutrient on the chlorophyll content of leaves, pod and haulm yields and micronutrient uptake by groundnut grown in calcareous soil (values are average of three seasons).

Treatments	Chlorophyll at DAE ($mg\ g^{-1}$ dw)		Yield ($kg\ ha^{-1}$)		Pod yield loss over T3 (%)		Micronutrient uptake ($g\ ha^{-1}$)					
	60	80	Pod	Haulm	Mean	Range	Fe	Mn	Zn	Cu	B	Mo
T_1: Control	2.14	4.52	1425	3125			3048	281	108	43	52	34
T_2: Macronutrients	3.54	6.16	1941	3577			3387	301	122	58	62	42
T_3: T_2 + micronutrients	6.54	8.98	2276	3588			4545	804	493	184	105	71
T_4: T_3 minus Fe	3.58	6.78	1926	3266	15.4	10-22	2487	436	314	119	80	56
T_5: T_3 minus Mn	4.55	7.69	2012	3164	11.6	8-17	2002	374	310	92	86	47
T_6: T_3 minus Zn	4.38	8.7	1905	2979	16.3	13-20	2233	562	161	125	106	60
T_7: T_3 minus Cu	4.49	7.37	1976	3073	13.2	12-14	2969	470	319	82	92	55
T_8: T_3 minus B	4.41	7.33	1843	3004	19.0	14-26	3960	679	410	120	64	57
T_9: T_3 minus Mo	5.02	7.41	1888	3329	17.0	14-19	3361	819	479	163	102	42
LSD (0.05)	0.52	0.61	210	320			330	121	70	36	15	7

W. J. Horst et al., Plant nutrition – Food security and sustainability of agro-ecosystems. 840–841, 2001

Effects of Fe, Mn, Zn and Cu fertilization on the yield and grain quality of wheat in the calcareous soils of Iran

A. H. Ziaeian[1)] and M. J. Malakouti[2)]

[1)] *Soil Fertility and Plant Nutrition Dept., Soil and Water Research Institute, P.O. Box 14155/6185, Tehran, Iran, Ziaeian@areeo.or. ir.*

[2)] *Soil Sci. Dept. Agricultural Faculty, Tarbiat Modarres University, Tehran, Iran.*

Key words: calcareous soil, critical level, fertilisation, micronutrient, *Triticum aestivum*, wheat

Abstract

In order to study the effects of micronutrients on wheat production in calcareous soils, 25 field experiments with wheat were conducted in the Dorodzan region, about 50 km west of Shiraz in south of Iran in 1998-99. Five fertiliser treatments including control (nitrogen, phosphorous and potassium based on soil tests), control + Fe, control + Mn, control + Zn and control + Cu were tested in a completely randomized block design with three replications. The soils of the selected fields contained high amounts of calcium carbonate equivalent (31-40%). The results showed that Fe, Mn, Zn and Cu fertilisation caused significant increase in grain yield, straw yield, 1000 grain weight, and the number of seeds per spiklet. With the application of these nutrients, their concentration and total uptake in grain and flag leaves and the grain protein content increased significantly. Based on the DTPA method, the critical soil levels were found to be 4.3 mg kg^{-1} for Fe and Mn and 0.9 mg kg^{-1} for Zn and Cu.

Introduction

Micronutrient-deficient soils are widespread: millions hectares of arable land world-wide are deficient in one or more micronutrient (Welch *et al.*, 1991). Usually micronutrient-deficiency problems are found in calcareous soils of arid and semi-arid regions. Most soils of Iran are highly calcareous and therefore are deficient in micronutrients, notably in Zn. Wheat (*Triticum aestivum L.*) grown under such deficient conditions has low yield and produces seeds with low micronutrient content. Wheat is the most important crop in Iran. It is the people's main staple and supplies more than 45% of protein and 55% of the calorie requirements of the people. Despite a domestic production of 11 million tons a year, Iran is still an importer of wheat.

Trace elements play a major role in wheat production (Kashirad, 1970; Hemantaranjan & Garg, 1988; Agrawal, 1992; Cakmak *et al.*, 1997; Yilmaz *et al.*, 1997). Positive effects of micronutrient application on growth of corn in calcareous soils of southern and central Iran has been reported by Malakouti *et al.* (1988), Maftoun and Karimian (1989), and Karimian (1995). But there is little information about the effect of micronutrient nutrition on wheat available for Iranian soils. The purposes of this investigation were (a) to determine the effects of micronutrient application on the yield and grain quality of wheat, and (b) to determine the critical levels of these elements in highly calcareous soils.

Materials and methods

Twenty-five fields were used in this experiment. The soils contained high amounts of calcium carbonate equivalent (31 - 40 %). The DTPA extractable micronutrient contents and other selected physicochemical properties of the soils are presented in Table 1. The wheat variety *"Falat"* was sown at a rate of 180 kg ha^{-1}. In each field five fertiliser treatments arranged in a completely randomized block design with three replications were compared. The fertiliser treatments included: Control (N, P, K based on soil test); control + 20 kg ha^{-1} Fe-EDDHA; control + 40 kg ha^{-1} $ZnSO_4*7H_2O$; control + 40 kg ha^{-1} $MnSO_4*H_2O$; control + 20 kg ha^{-1} $CuSO_4*5H_2O$. The concentration of N, P, K, Cu, Fe, Mn, and Zn in grains and in flag leaves, yield components, straw and grain yields were measured for each site. Final harvesting was done on 22 fields and therefor statistical analysis is based on these data. The critical limits obtain are based on Cate and Nelson (1965) procedure. F tests, Duncan multiple range tests, and regression equation using MSTATC computer programs were used to statistically analyse the data obtained.

Table 1. Selected soil properties

Soil Characteristic	Mean	Range
P avail. (mg kg^{-1})	16.2	5.0 – 41.0
K avail. (mg kg^{-1})	308	120 – 536
Fe avail. (mg kg^{-1})	6.2	2.9 – 10.5
Mn avail. (mg kg^{-1})	6.4	2.8 – 13.2
Zn avail. (mg kg^{-1})	1.04	0.44- 1.65
Cu avail. (mg kg^{-1})	1.50	0.36 – 2.30
O.C. (%)	0.8	0.3 – 1.7
Clay (%)	34	24 – 46
CCE (%)	38	31 – 49
CEC (cm kg^{-1})	16.6	12.0 – 21.5
pH	7.8	7.1 – 8.1

Results

The results showed that application of Fe, Mn, Zn and Cu significantly increased grain yield compared to the control (Tab. 2). The application of these micronutrients also caused significant increases in straw yield and grain protein content. Furthermore, higher concentrations and increased total uptake in grains and flag leaves were found. Concerning the yield contributing components a significant increase in 1000 seed weight and the number of seeds per spiklet was found while only a slight effect on the number of spiklet per m^2 occured. Regression analysis showed that soil characteristics, especially the calcium carbonate equivalent, had significant effects on the plant responses. Based on the DTPA method the critical values were 4.3 mg kg^{-1} for Fe and Mn, and 0.9 mg kg^{-1} for Zn and Cu (data not shown).

Discussion

Grain yield is a function of the number of spiklet per m^2, number of seeds per spiklet and the 1000 grains weight. Factors, which can effect these parameters, will change the total grain yield. The results show that micronutrient application causes a significant increase in the number of seeds per ear and the 1000 grains weight. Hemantranjan and Grag (1988) also reported that Fe and Zn application caused a significant increase in the number of ear heads per tiller, number of seeds per ear, grain yield per plant and 1000 grains weight. They reported that this was coupled with the increase in total carbohydrate, starch and crude protein content of wheat grain. The results also show that with the application of these nutrients, their concentration and total uptake in shoots and grain increased significantly. These results are in agreement with Agrawal (1992), Karimian (1995) and Yilmaz *et al.* (1997). In addition micronutrient application causes a significant increase in protein content. Marschner (1995) reported that zinc deficient plants drastically reduce the rate of protein synthesis and the protein content, but increase the accumulation of amino acids due to higher rates of RNase activity.

Based on the DTPA method, the critical soil levels were 4.3 mg kg^{-1} for Fe and Mn, and 0.9 mg kg^{-1} for Zn and Cu. According to Agrawal (1992) critical limits of soil DTPA Cu, Zn, Fe and Mn were 0.78, 0.80, 5.0 and 5.5 mg kg^{-1}, respectively. Regression analysis showed that soil characteristics, especially the calcium carbonate equivalent, had significant effects on the plant response. In general, the combination of high pH, $CaCO_3$ and clay in soils together with low annual precipitation can be considered to be the major factors causing micronutrients deficiencies, especially Zn, in plants grown in these soils.

References

Agrawal HP 1992 Commun. Soil Sci. Plant Anal. 23, 2555-2568.

Cakmak I, Yilmaz A, Kalayci M, Ekiz E, Torun B, Erenoglu B and Braun HJ 1996 Plant Soil 180, 165-172.

Cate RB Jr and Nelson LA 1965 North Carolina State Univ. Int. Soil Testing Series Tech. Bull. 1.

Hemantaranjan A and Grag OK 1988 J. Plant Nutr. 11, 1439-1450.

Kariminan N 1995 J. Plant Nutr. 18, 226-221.

Kashirad A 1970 J. Agric. Res. 20, 179-182.

Mafton M and N Karimian 1980 Agronomie 9, 771-775.

Malakouti MJ 1992 Commun. Soil Sci. Plant Anal. 23, 2687-2695.

Marschner H 1995 2nd ed. Academic Press, San Diego, CA.

Yilmaz A, Ekiz H, Torum B, Gullekin I, Karanlik S, Bagei SA and Cakmak I 1997 J. Plant Nutr. 20, 461-471.

Table 2 Effect of micronutrient application on yield, yield contributing components, nutrient concentrations in grain and leaves

	Treatments					
	Ctrl.	+ Fe	+ Mn	+ Zn	+ Cu	LSD
straw yield (kg ha^{-1})	8765	9772	9673	9932	9750	420 **
Grain yield (kg ha^{-1})	7064	7865	7930	8132	7790	310 **
1000 grain weight (g)	38.5	46.1	45.1	44.7	44.5	3.2 **
No. of seeds per spiklet	36.3	40.8	40.6	41.3	42.4	5.2 *
No. of spiklet per m^2	436	444	451	462	444	ns
Grain Protein (%)	11.3	12.8	12.6	13.1	12.6	0.5 **
Grain P (%)	0.47	0.44	0.44	0.44	0.45	ns
Grain K (%)	2.27	2.32	2.35	2.29	2.29	ns
Grain Fe (µg g^{-1})	44.1	54.9	48.1	40.6	41.0	11.5 **
Grain Mn (µg g^{-1})	42.5	42.7	54.6	38.7	39.9	15.9 **
Grain Zn (µg g^{-1})	25.2	22.0	22.4	47.1	22.6	17.0 **
Grain Cu (µg g^{-1})	6.3	5.8	5.9	6.3	8.8	0.3 **
Leaves N (%)	2.74	2.82	2.69	2.95	2.82	0.26 *
Leaves P (%)	0.27	0.24	0.24	0.24	0.23	0.01 **
Leaves K (%)	3.01	3.10	3.18	3.16	3.15	ns
Leaves Fe (µg g^{-1})	85.5	101.4	89.9	82.5	82.0	4.9 **
Leaves Mn (µg g^{-1})	53.3	59.0	74.0	52.9	52.6	4.1 **
Leaves Zn (µg g^{-1})	18.7	16.5	17.2	35.3	17.3	2.7 **
Leaves Cu (µg g^{-1})	5.6	5.2	5.4	5.0	7.7	2.6 **

ns: not significant, *: Significant at 5% level, **: Significant at 1 % level.

W. J. Horst et al., Plant nutrition – Food security and sustainability of agro-ecosystems. 842~843, 2001

Influence of N, P, K, Ca and Mg rates on leaf micronutrient concentration of 'Navaho' blackberry

J. M. Spiers[1)] and J. H. Braswell[2)]
[1)] *USDA ARS Small Fruit Research Station, Poplarville, Mississippi 39470, United States, jspiers@ars.usda.gov*
[2)] *Mississippi Cooperative Extension Service, Poplarville, MS 39470, United States*

Key words: fertilisation, pant nutrition, *Rubus*

Abstract

To determine the effect of wide ranges of macronutrient fertilisation on leaf micronutrient concentrations and plant growth characteristics 'Navaho' blackberry, plants were grown in sand culture for two years and fertilised with solutions consisting of five macro nutrients (N, P, K, Ca, and Mg) applied at three different rates (0 mg/l, control and 10X control). Foliar samples were collected at 4, 8 and 16 weeks after treatment initiation during the second growing season. Leaf Fe was highest at recommended rates of all macronutrient fertilisations. Mn leaf concentrations were increased with increasing rates of Mg, but decreased with 10x rates of N and Ca. High concentrations of Zn were found in plants receiving 10x rates of P fertilisation. Leaf Cu was increased by increasing N or K, but was decreased by P, Ca or Mg. Toxicity and deficiency symptoms were more evident with the N, K, and Mg fertiliser treatments.

Introduction

Information on nutritional requirements for blackberry production in the Gulf Coast region of the United States is very limited. Nitrogen is the nutrient used in largest amounts and is often used to regulate cane size and number in brambles (*Rubus* species) and has been shown to decrease sugar content while increasing leaf N concentrations (Galletta & Himelrick, 1990; Papp *et al.*, 1984). Potassium fertilisation has induced Ca and Mg deficiencies in several brambles (Ljones, 1966). Potassium fertilisation increased, K but decreased Mg and Zn in the leaves of 'Shawnee' blackberry (Spiers, 1993). Increased N + K fertilisation decreased leaf content of Ca in 'Thornless Evergreen' blackberry (Nelson & Martin, 1986). Kowalenko (1981) working with 'Willamette' raspberry found a positive correlation between leaf Mg and soil Mg. In 'Cheyenne' blackberries, Mg and Ca fertilisation increased plant growth after two seasons (Spiers, 1987). This study was initiated to determine the influence of wide ranges N, P, K, Ca, and Mg fertilisation on leaf micronutrient concentrations and plant growth characteristics of 'Navaho' blackberry on erect thornless cultivar released from Arkansas (Clark, 1992).

Materials and methods

The study was initiated June 1, 1993 at the USDA Small Fruit Research Laboratory in Poplarville, Mississippi. The experiment consisted of 5 macro nutrient elements (N, P, K, Ca and Mg) applied at 3 different rates (0 mg/l, control and 10X control; Table 1). The control treatment consisted of a standard nutrient solution consistent with recommended rates (Spiers, 1978). Experimental design was a split plot with 4 replications of the low and control rates and 3 replications of the high rate. In 1992, healthy two-year-old 'Navaho' blackberry plants were potted in 8 liter containers with a sand media and fertilised with a complete liquid fertiliser solution throughout the summer of 1992 and during May of 1993. All plants were cut back to 20 cm on May 28, 1993. Plants were grown on benches in a shade house (20% shade) with natural daylength and temperature. The shade house had a transparent fiberglass roof to eliminate influence of rainfall. Nutrient solutions were applied daily, Monday through Friday, and all pots were flushed with water Saturday and Sunday. Treatments 1 to 6, began June 1, 1993 and treatments 7 to 11, began July 1, 1993. Each pot was filled to the rim with the appropriate fertiliser treatment, completely drenching the media. Leaf samples were taken 4, 8 and 16 weeks after treatment initiation during the second growing season. At the end of the second year, plants were cut to ground level and dry weight determined. Analysis of variance and regression were performed on treatment effects. Correlations (Pearson, 1990) between elemental leaf concentrations were determined.

Results

Increased N fertilisation resulted in decrease leaf Mn and increase leaf Cu. Leaf Fe was highest when plants received the recommend N rates and was reduced when plants were stressed by deficient or excessive N rates (Table 2). Plants receiving 0 mg N/l exhibited stunted growth with small yellow leaves. All plants fertilised at the 10X N rate were dead by the end of this study.

Increased P fertilisation resulted in a linear increase in leaf Zn and a linear decrease in leaf Cu. P fertilisation levels did not affect leaf Fe or Mn. No P deficiency symptoms were evident. Leaves from plants fertilised at the 10X rate of P exhibited interveinular chlorosis and pale leaf tips.

Increased K fertilisation caused an increase in leaf Cu but did not significantly influence any other micronutrients. Plants receiving 0 mg K/l exhibited small leaves and stunted growth and two of the four plants were dead at the end of the study. Calcium fertilisation linearly reduced leaf Cu. At the 10X rate of applied Ca, leaf Mn was significantly lower than at the other Ca fertilisation rates. No visual deficiency symptoms were present on plants fertilised with the low rate of Ca while leaves from plants receiving the 10X Ca rate were small with interveinular chlorosis and had white tips.

Increasing Mg fertilisation rate linearly reduced leaf Cu as did Ca fertilisation. Leaf Mn was lowest at the 0 mg Mg/l rate. Iron leaf content was higher in plants receiving the recommended rates of Mg. Plants fertilised with the low rate of Mg remained vigorous in growth, but leaves showed interveinular chlorosis.

Discussion

Leaf Fe was highest at recommended rates of all macronutrient fertilisation. Mn leaf concentrations were increased with increasing rates of Mg, but decreased with higher rates of N and Ca. High concentrations of Zn were found in plants receiving 10X rates of P fertilisation. No other macronutrient fertilisation effected Zn. Leaf Cu was increased by increasing N or K, but was decreased by P, Ca or Mg. Compared to foliar tissue analysis standards of red raspberry (Crandall, 1995), 'Navaho' blackberry leaf concentrations of Zn and Fe were in the normal range and leaf Cu was below normal regardless of fertilisation treatments. Leaf Mn concentrations were in the above normal range regardless of fertilisation treatments. High N fertilisation resulted in leaf concentrations of Mn in the toxic range and eventually dead plants.

References

Clark JR 1992 Fruit Var. J. 46, 217-222.

Crandall PC 1995 Bramble Production: Haworth Press, Inc., New York, NY, p. 213.

Galletta CJ and Hemilrick DC 1990 Small Fruit Crop Management, 157-213, Prentice Hall Inc., Englewood Cliffs, NJ.

Kowalenko CG 1981 Soil Sci. Plant Anal. 12, 795-809.

Ljones B 1966 *In* Nutrition of Fruit Crops.Ed. Nf Childers. pp. 130-142. Somerset Press, Somerville, NJ.

Nelson E and Martin LW 1986 Hort. Sci. 21, 1153-1154.

Pearson 1990 SAS Procedures Guide. Version 6, Third Edition. SAS Institute Inc.,Cary, NC.

Papp J, Kobzos-Papai I and Nagy J 1984 Acta Agron. Acad. Sci. Hungaricae 33, 337-343.

Spiers JM 1987 Hort Sci. 22, 576-577.

Spiers JM 1993 J. Plant Nutr. 16, 297-303.

Table 1. Chemical content (mg/l) of treatment solutions

		Treatment										
		1	2	3	4	5	6	7	8	9	10	11
Element	Chemical	Control	-N	-P	-K	-Ca	-Mg	10 N	10 P	10 K	10 Ca	10 Mg
N	NH_4NO_3	100		100	100	100	100	1000		100	100	100
N	$(NH_4)_2PO_4$								100			
P	$(NH_4)_2PO_4$								111			
P	$NaH_2PO_4\ H_2O$	20	20		20	20	20	20	89	20	20	20
K	KCl	50	50	50		50	50	50	50	250	50	50
K	K_2SO_4									250		
Ca	$CaCl_2\ 2H_2O$	80	80	80	80		40	80	80	80	400	80
Ca	$CaSO_4\ 2H_2O$						40				400	
Mg	$MgSO_4\ 7H_2O$	24	24	24	24			24	24	24	24	121
Mg	$MgSO_4\ 7H_2O$											121
	STEM	67	67	67	67	67	67	67	67	67	67	67

Control = recommended fertiliser solution complete with all elements.

STEM = S, 14.00%; B, 1.35%; Cu, 3.20%; Fe, 7.50%; Mn, 8.00%; Mo, 0.04%; Zn, 4.50%.

Table 2. The influence of macronutrient fertiliser rates on leaf elemental concentrations of 'Navaho' blackberries (1994).

Treatment	Macronutrient mg/l	Fe ppm	Mn ppm	Zn ppm	Cu ppm
N	0	53b^z^	1734a	80a	7.14b
	100	101a	709b	66a	8.75ab
	1000	49b	464c	64a	10.33a
	Linear	ns^y^	*	ns	*
P	0	73a	776a	59b	9.50a
	20	101a	709a	66b	8.75a
	200	60a	732a	281a	3.33b
	Linear	ns	ns	*	***
K	0	69a	563a	56a	7.67a
	50	97a	702a	67a	8.33a
	500	61a	496a	69a	11.75a
	Linear	ns	ns	ns	*
Ca	0	77a	746a	58a	11.00a
	80	101a	709a	66a	8.75ab
	800	95a	483b	64a	6.50b
	Linear	ns	ns	ns	*
Mg	0	64b	464b	53a	13.00a
	24	101a	709a	66a	8.75b
	240	77ab	665a	63a	3.83c
	Linear	ns	ns	ns	***

z^Means separation by LSD, 0.05 level. ^y^ns, *, **, *** Non-significant, and significant at 0.05, 0.01, and 0.001 levels, respectively.

W. J. Horst et al., Plant nutrition – Food security and sustainability of agro-ecosystems. 844–845, 2001

Plant availability of sulfur from organic fertilisers

R. Gutser and S. von Tucher

Department of Plant Sciences, Chair of Plant Nutrition, Technical University of Munich, Am Hochanger 2, 85350 Freising, Germany, gutser@weihenstephan.de.

Key words: C/S ratio, organic fertiliser, S availability, S immobilisation, S mineralisation

Abstract

Short term S availability markedly depends on C/S ratio (r2=0.95), and to a less degree on SO42- content (r2=0.50) of organic fertilisers. In pot experiments highest mineral fertiliser equivalents of 77 and up to 59% were obtained with urine (C/S 11) and sewage sludge (C/S 11-28) followed by slurry (C/S 81), stable manure (C/S 84) and biocompost (C/S 70) with equivalents of 17, 13, and 8%. Long term S effect is related to an increased S release potential in soil and in consequence a higher net S mineralisation.

Introduction

During the last 15 years S immissions were reduced from 30 to 6 kg ha-1 (average in Bavaria, 1999). Therefore sufficient S supply of plants (20 –70 kg ha-1) requires additional S fertilisation. For optimising the use of organic fertilisers, the short and long term S availability has to be considered. The S release is thought to correlate with SO42- content and C/S ratio (Eriksen, 1997). Critical C/S ratios for a S net mineralisation from organic materials vary from 200 (Barrow, 1960) to 60–80 (Janzen and Ellert, 1998). Sulfur from catch crops (C/S ratio 167-234) proved to be unavailable for the following crop and induced S immobilisation (Gutser and v. Tucher, 2000).

Subject of this study was to investigate the relationship between C/S ratio and short term S availability of organic fertilisers (pot experiments). In addition, long term S release was deduced from net mineralisation in field experiments.

Materials and methods

Short term S availability of organic fertilisers (characterised in Table 1) is determined by S uptake of optimally fertilised (N, P, K, Mg) wheat (growth period 110 d) followed by mustard (30 d) in pot experiments (5.5 kg loamy sand, pH 6.0, 0.1 mg SO4-S 100 g-1 pot-1). Organic fertilisers are compared with mineral fertilisation. S dose: 100 mg St and for K2SO4 in addition 45 mg S pot-1. For N fertilisation NH4+-N content of organic fertilisers is taken into account.

Table 1. Characterisation of organic fertilisers

Organic Fertiliser	*Dry M.* %	S_t $mg\ g\ TS^{-1}$	SO_4-S $\%\ S_t$	*C/S*
urine	*2.4*	*2.4*	*64*	*11*
sewage sludge 1	*25.5*	*8.1*	*2*	*11*
sewage sludge 2	*3.1*	*10.0*	*30*	*28*
slurry (cattle)	*8.3*	*5.2*	*16*	*81*
stable manure	*18.9*	*5.1*	*3*	*84*
biocompost	*82.2*	*3.3*	*4*	*70*
*catch crops**	*9.5-16.1*	*1.6-2.5*	-	*167-233*

** Gutser and v. Tucher (2000)*

S effect is calculated as mineral fertiliser equivalent:

S uptake from organic fertiliser in % of S uptake from mineral fertiliser (control subtracted).

Net S mineralisation is determined in incubation studies without plants over a period of 255 days. The additive S effect of long term fertilisation is calculated in field (March–July 1999) by S net mineralisation on foil covered fallow plots (silty loam, 0-25 cm). Treatments: control and slurry (Æ 83 kg NH4+-N equal to 1200 Corg ha-1 a-1) since 1980, biogenic waste compost (biocompost) (Æ 100 and 200 kg Nt equal to 1450 and 2900 kg Corg ha-1 a-1) since 1992.

S is analysed in plants by ICP after microwave digestion, in soil by elemental analysis (St), and by anionic chromatography (SO42-) after extraction with CaCl2.

Results

Short term S effect in incubation experiments

Comparable to mineral fertilisation, sewage sludge 2 and in particular urine markedly increase SO42- content of soil (Figure 1). This rise in SO42- starts within the first 30 days and continues during the whole period (255 d). Slurry, biocompost, and manure release significantly less S. Biocompost even results in a weak S immobilisation on soils with a higher SO42- level (data not shown). Mineral fertiliser equivalents after 120 days are 86 and 50 % for urine and sewage sludge and 13, 12, and 2 % for slurry, manure and biocompost, respectively.

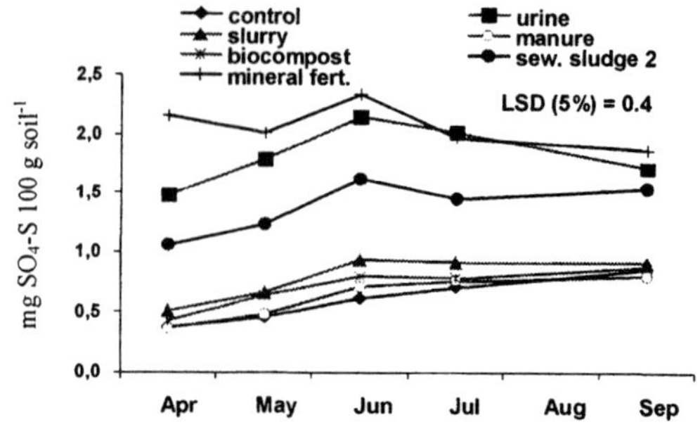

Figure 1. Sulfate content of soil following S fertilisation.

Short term S effect in pot experiments with plants

Wheat accounts for 85-90% of the total S uptake (Figure 2). The uptake significantly correlates with the level of S fertilisation (K2SO4). Within the organic fertilisers, urine application (mineral fertiliser equivalent: 77%) results in a very high S utilisation, followed by sewage sludges (59 and 50 %). Slurry (17%), manure (13%), and biocompost (8%) show only a weak effect. Fertiliser SO42- content only explains 51 % of the total S uptake. Sludge 1 and 2 differ in SO42- content (2 and 30% of total S), although S uptake is similar. Compared to SO42- content, C/S ratio of organic fertilisers (Figure 3) is much more correlated to the differences in S uptake (r^2=0.95). This regression also integrates the short term S immobilising effect of catch crops (C/S ratio 167-233, Gutser and v. Tucher, 2000). The critical C/S ratio of organic fertilisers is calculated to be about 80.

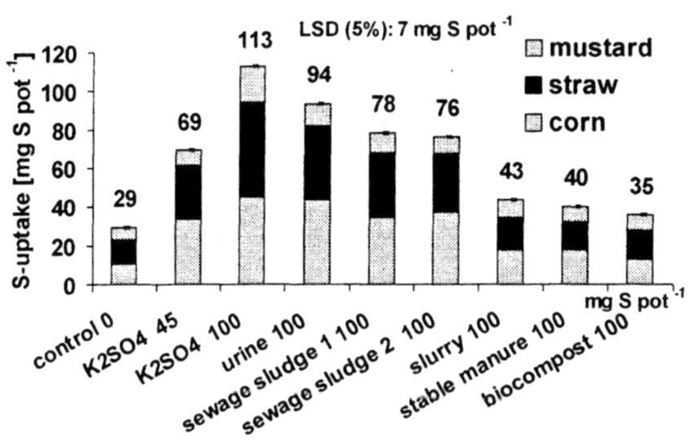

Figure 2. Utilisation of sulfur from organic fertilisers by wheat (grain and straw) and mustard.

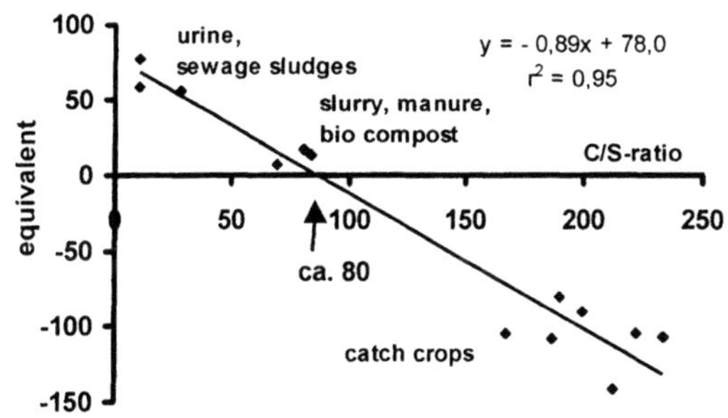

Figure 3. Regression between C/S ratio and short term S availability (mineral fertiliser equivalent) of organic fertilisers.

Long term S effect in field experiments

Long term application of organic fertilisers increases Corg content of surface soils (slurry 1.20 %, biocompost 1.51 and 1.64 %, control 1.15%) and consequently the S release potential. Enhanced S net mineralisation from manured soils reached its maximum in the end of June with 2-4 kg for slurry and biocompost (1x) and up to 8 kg S ha-1 for biocompost (2x) (Figure 4). In July S turnover is influenced by cultivation practise (irrigation of dry soil, harvest, straw manuring).

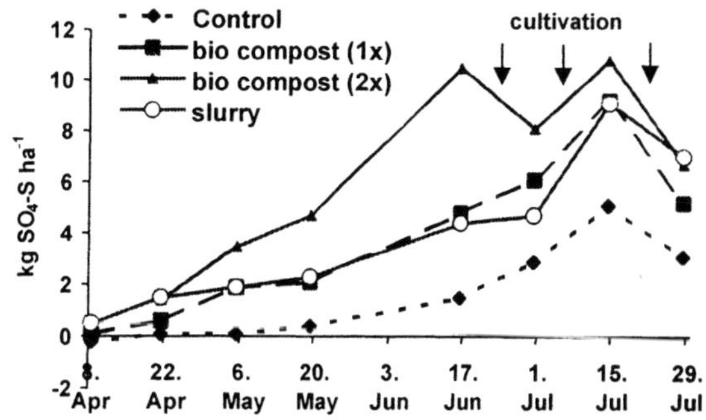

Figure 4. Net S mineralisation in surface soils after long term application (since 1980) of slurry and biocompost. Field measurements March - July 1999, fallow covered with foil.

Discussion

Short term S availability of organic fertilisers in incubation and pot experiments confirm the results of Eriksen *et al.* (1995) for slurry and Tabatabai and Chae (1991) for sewage sludge (pot experiments) and Janzen and Ellert (1998) in field. The outstanding effect of urine is based on its high SO42- content. S utilisation from sewage sludge is thought to be due to the high total S content, resulting from the treatment of sewage sludge with SO42-containing precipitants. There is a significant negative relationship between the short term S fertilisation effect and the C/S ratio of organic fertilisers. The critical C/S ratio for net S mineralisation of about 80, which can be deduced from the regression in Figure 3, is in agreement with results of Janzen and Ellert (1998).

The long term S availability is defined as net S mineralisation within one year and is controlled by the soil organic substance enriched S pool. S release from this source occurs in a later stage of the vegetation period (June/July) when soil temperature increases. This S release is to be considered in the S balance of agricultural crop management.

References

Barrow NJ 1960 Austr. J. Agric. Res. 11, 960-969.

Janzen HH and Ellert BH 1998 *In* Sulfur in the Environment. Ed. DG Maynard. pp. 11-43. Marcel Dekker, Inc. New York.

Eriksen J, Mortensen JV, Kjellerup VK and Kristjansen O 1995 Z. Pflanzenern. Bodenkde 158, 113-116.

Eriksen J 1997 Soil Biol. Biochem. 29, 1379-1385.

Gutser R and v. Tucher S 2000 *In* Proceedings 13th Int. IFOAM Scient. Conf., Basel. Eds. Alföldi T, Lockeretz W and Niggli U. p 30. VDF Hochschulverlag, Zürich.

Gutser R and v. Tucher S 2000 VDLUFA Schriftenreihe 53, 48-63.

Tabatabai MA and Chae YM 1991 J. Environ. Qual. 20, 684-690.

Phosphorus availability to maize plants from sewage sludge treated with Fe compounds

I. F. Samie and W. Römer

Institute of Agricultural Chemistry, Faculty of Agriculture, University of Goettingen, Von - Siebold Str. 6, D-37075 Goettingen, Germany, isamie@gwdg.de

Key words: iron phosphate, maize, oxalate soluble iron, phosphate sorption, phosphate uptake, sewage sludge, soil solution

Abstract

Five loamy and four sandy soils were incubated with either $CaHPO_4$ or two sewage sludges with the P:Fe ratios 1:1.5 (sludge Sh) and 1:5.4 (sludge Gö) for 1 and 6 months. P sorption and $_{inorganic}$ P (P_i) concentration of the soil solution were measured. In a pot experiment P uptake of maize was examined. The application of sludge Sh did not modify P sorption compared to soils without sludge. However, after application of sludge Gö the P sorption increased to 116 % in loamy and 112 % in sandy soils. After application of sludge Sh the mean P_i concentration increased in loamy soils to 124 % and in sandy soils to 113 %. On the other hand the P_i concentration decreased after applying sludge Gö to 82 % (loamy soils) and 62 % (sandy soils) as compared to the controls of the respective soils. One month after the application of increasing amounts of sludge Gö (5, 10, 15 t DM ha^{-1}) both the concentration of oxalate-soluble Fe in the soil and the P sorption were increased. The relationships between these two parameters were highly significant (r^2 = 0.87 – 0.97). Plant uptake of P was lower after application of sludge Gö than after application of sludge Sh and much less than P uptake from $CaHPO_4$. Sewage sludges with a P : Fe ratio of 1 : 5 should not be recommended for agricultural use, as the P availability is significantly reduced. Iron salts should not be used for conditioning of sludges.

Introduction

In Germany 500 000 t of iron salts are used every year in waste water purification plants to precipitate P. 38% of the $2.4*10^6$ t DM of sewage sludge are used in agriculture. The aim of this investigation was to study the P availability of soils treated with sludges containing different iron contents. Five loamy and four sandy soils were incubated with two sludges (separately) for periods of 1 and 6 months. Following these treatments, P sorption, the soil content of oxalate soluble Fe and the P_i concentration of soil solution were measured. The P uptake of maize plants from one soil treated with two sludges was determined in a separate experiment.

Materials and methods

Table 1. Characteristics of the sludges

sludge	DM cont. g kg^{-1}	pH	Ca	P kg t^{-1}DM	Fe	P : Fe
S Gö	320	11,5	197*	12	65*	1:5.4
S Sh	40	7,0	21	25	39**	1:1.5

* Application of $FeSO_4$ at aerobic basin and $FeCl_3$ and $Ca(OH)_2$ for conditioning process.

** Application of $FeCl_3$ at aerobic basin.

Incubation experiment 1

Five loamy soils and four sandy soils were each mixed with S Gö , S Sh or $CaHPO_4$ (30 mg P kg^{-1} soil = 60 kg P ha^{-1}) and incubated at 20°C and 50% of maximum water holding capacity in pots (6.5 kg) for 1 and 6 months. Soil solution for P_i determination were collected (Adams, 1974). P sorption ability was determined according to Fritsch and Werner (1988): extraction of 10 g soil with 100 ml of a 10 mg P L^{-1} solution for 2 hours.

Incubation experiment 2

Experiment 1 was repeated but sludge equivalent to 30, 60, 90 mg P kg^{-1} soil was added. P sorption was determined as in experiment 1. The amorphous Fe of soil was extracted with NH_4 oxalate (Fe_{ox}) and determined according to Schwertmann (1964).

Pot experiment with maize

Four maize plants were grown in pots filled with 6.5 kg of a loamy soil (sand 5%, silt 75%, clay 20%, C_{org} 13 g kg^{-1} , pH 7.2 ($CaCl_2$) , two P levels: lactate soluble P: 46 or 110 mg kg^{-1}) for 90 days. 30 mg P kg^{-1} soil were applied as $CaHPO_4$ or sludge. P content of above ground biomass was measured.

More details are given by Römer and Samie (2001).

Results

One month after application of S Gö the P_i concentration was reduced to 82% in the 5 loamy and 62% in the 4 sandy soils compared with the controls without P (Tab.2), whereas the P sorption was increased to 116% and 112%, respectively (Tab.3). For S Sh P_i concentration increased to 113 and 124 %, respectively, but the P sorption ability remained similar in all soils (95 to 103 %). 6 months later results were quite similar date not shown.

Table 2. Mean values of P_i concentration of the soil solutions 1 month after P application (µg P L^{-1})

Soils	Without P	$CaHPO_4$	S Gö	S Sh
5 loamy	222 (100)	438 (198)	182 (82)	275 (124)
4 sandy	118 (100)	275 (232)	73 (62)	133 (113)

Table 3. Mean values of P sorption 1 month after P application (mg P kg^{-1})

Soils	Without P	$CaHPO_4$	S-Gö	S-Sh
5 loamy	43 (100)	35 (81)	50 (116)	41 (95)
4 sandy	58 (100)	53 (91)	65 (112)	60 (103)

With increasing rates of sludge applications a linear relationship was found between Fe_{ox} and P- sorption for one loamy soil well as for sandy soils (Fig. 1a, b).

At low soil P level plants apparently took up P from all three P sources, but the lowest amount from S Gö. At high soil P level plants variable to utilize P from $CaHPO_4$ and S Sh only but not from S Gö (Tab. 4).

Table 4. P uptake of maize plants grown on a loamy soil at two P levels

Treatment		Low P[1] 46 mg kg^{-1}	High P[1] 110 mg kg^{-1}
Without P	mg P pot^{-1}	160	231
Without P	rel.	100 c	100 c
$CaHPO_4$	rel.	160 a	131 a
S Gö	rel.	131 b	98 c
S Sh	rel.	167 a	115 b

[1] Ca acetate lactate soluble P
Different letters in one columne show significant differences at p =5% (Tukey test)

Discussion

The P uptake by the plants was significantly lower after the application of S Gö as compared to P uptake after the application of $CaHPO_4$ or S Sh (Table 4). This lower P uptake is related to the low P_i concentration in the soil solution after the application of S Gö (Table 2). The low P_i concentration is a result of the significantly higher P sorption in the soil (Table 3). Obviously application of S Gö results in an increase of sorption sites for phosphate ions, due to its wide P:Fe ratio (1 : 5). Figure 1 shows that the P sorption capacity in soil increases with the increasing rate of applied Fe. This process has a negative impact on plant growth since it decreased the P availability. The sludge Sh has a P:Fe ratio of 1 : 1.5.and did not cause a significant increase in P sorption. As a result plants were able to take up P from the soil treated with sludge Sh nearly as good as from the soil where $CaHPO_4$ had been applied (Table 4).

Conclusions

Maize plants took up the lowest P amounts from the soil treated with S Gö with a P:Fe ratio (1 : 5.4). It has to be assumed that this is due to the increased P sorption ability of the soil and the decreased P_i concentration of soil solution. Such sewage sludges are not suitable for use as fertilisers on arable land.

Acknowledgement

The project was partially funded by the Ministry of Agriculture and the Environmental Foundation of Lower Saxony.

References

Adams F 1974 Soil solution. In: The plant root and its environment, Carson EW (ed.), University press of Virginia, Charlottesville, 441 – 482.

Fritsch F and Werner W 1988 Landwirtschaftliche Forschung 41, 305–311.

Römer W and Samie IF 2001 Influence of iron content in sewage sludges on parameters of phosphate availability in arable soils. J. Plant Nutr. Soil Sci., (in press).

Schwertmann U 1964 Differenzierung der Eisenoxide des Bodens durch Extraktion mit Ammoniumoxalat-Lösung. Z. Pflanzenernähr. Bodenk. 105, 194-202.

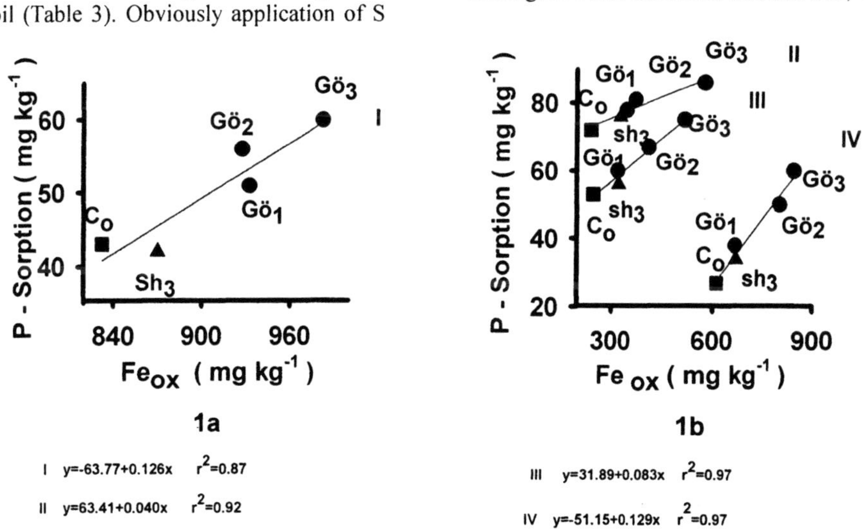

Figure 1. Relationship between Fe_{OX} content of soil and P sorption (1a: loamy soil " Börry"; 1b: 3 sandy soils) C_0 = control (without P), $Gö_1$ = 157, $Gö_2$ = 214, $Gö_3$ = 471, Sh_3 = 132 mg Fe kg^{-1} as sludge.

Degradation of chlorinated aromatic acids from sludge compost in soils

T. Nagaoka, J. Tanaka, K. Kouno and T. Ando
Faculty of Applied Biological Science, Hiroshima University, Kagamiyama 1-4-4, Higashi-Hiroshima, 739-8528, Japan, tnagaok@hiroshima-u.ac.jp

Key words: 3,4-dichlorobenzoic acid, 3,4-dichloro-phenylacetic acid, degradation, 3,4,5-trichlorobenzoic acid, 3,4,5-trichlorophenylacetic acid, sludge compost

Abstract

Degradation of four chlorinated aromatic acids, i.e. 3,4-dichlorophenlyacetic acid (3,4-DCP), 3,4-dichlorobenzoic acid (3,4-DCB), 3,4,5-trichlorophenylacetic acid (3,4,5-TCP) and 3,4,5-trichlorobenzoic acid (3,4,5-TCB) in regosol and andosol amended with sludge compost was investigated during 90 days in incubation experiments. Degradation rates of these compounds including both free acids and conjugates were dependent on types of compounds, and faster in regosol than in andosol. In regosol, 80 % of 3,4-DCB and 3,4,5-TCB were gradually degraded in 60 days, while about half of 3,4-DCP and 3,4,5-TCP remained in 90 days. Amounts of chlorinated aromatic acids eluted from soil columns were significantly larger in regosol than andosol, suggesting stronger sorption in andosol. Lower degradability of chlorinated aromatic acids in andosol may result from stronger sorption in soil.

Introduction

Conversion of organic wastes into composts and their restoration to agricultural ecosystems are efficient to protect environments and utilize efficiently natural resources. In our previous studies (Nagaoka *et al.* 1996), however, some sludge compost was found to selectively inhibit seed germination of some legumes, especially white clover. Four germination inhibitors were isolated from sludge compost and identified as 3,4-dichlorobenzoic acid (3,4-DCB), 3,4-dichlorophenylacetic acid (3,4-DCP), 3,4,5-trichlorobenzoic acid (3,4,5-TCB) and 3,4,5-trichlorophenylacetic acid (3,4,5-TCP). These chlorinated aromatic acids were suggested to be generated from some alkali-hydrolyzable conjugates existed in sludge during composting process (Nagaoka *et al.* 1997). For practical use of this compost, degradation and mobility of these chlorinated aromatic acids in soils were investigated.

Materials and methods

Granitic regosol and andosol were used for the experiments. Each soil (100 g) amended with sludge compost (2 g) containing chlorinated aromatic acids was put into a glass bottle for each treatment and incubation period with three replicates. Amounts of chlorinated aromatic acids extracted with some extractants from sludge compost were shown in Table 1. Moisture content of soil amended with sludge compost was adjusted to 40 % (regosol) or 50 % (andosol) of water holding capacity, and incubated at 25 °C for 30, 60 and 90 days. Sludge compost was also incubated at 50 % of moisture content in a similar way. Samples after incubation were extracted with 2*N*NaOH-MeOH (1:1), and evaporated under reduced pressure after filtration and neutralization. Four chlorinated aromatic acids in soil extracts were determined with GC-MS after ether extraction and methylation of extracts.

Mobility of 3,4-DCB and 3,4-DCP were examined with glass columns (2.5 cm internal diameter) that were filled with regosol and andosol to 1, 2, 3, 4 and 5 cm height. Aqueous solutions (100 μg/ml) of 3,4-DCB and 3,4-DCP were used for column experiment. Each solution (1 ml) was subjected to soil columns, and eluted with 25 ml of distilled water. 3,4-DCB and 3,4-DCP in eluted solution were determined with HPLC after filtration. All treatments were duplicated.

Results

Relative amounts of chlorinated aromatic acids extracted with 2*N* NaOH-MeOH (1:1) from regosol and andosol amended with sludge compost after incubation were shown in Figure 1. Amounts of 3,4-DCB and 3,4,5-TCB extracted from regosol amended with compost decreased gradually with time to about 20 % in 60 days, and were constant after that. 3,4-DCP and 3,4,5-TCP reduced to 54% and 65 % levels in 30 days, respectively, and remained at constant levels. In andosol, amounts of chlorinated aromatic acids except 3,4,5-TCP decreased gradually to 40 - 50 %, while 3,4,5-TCP reached 77 % level in 90 days. Degradation rates of 3,4-DCB, 3,4-DCP and 3,4,5-TCB in compost were similar to those in regosol, but 3,4,5-TCP did not disappear in 90 days.

Table 1. Chlorinated aromatic acids extracted from sludge compost with various extractants

Extractant	Chlorinated aromatic acids ($mg\ kg^{-1}$ dw)			
	3,4-DCB	3,4-DCP	3,4,5-TCB	3,4,5-TCP
H_2O	9.0	12.5	7.9	18.4
MeOH-H_2O (1:1)	11.7	12.7	10.2	15.3
2*N*NaOH-MeOH (1:1)	14.6	26.8	11.7	35.5

The percentages of 3,4-DCB and 3,4-DCP eluted through soil columns were shown in Table 2. The percentages of 3,4-DCB and 3,4-DCP eluted through regosol columns were about 85 - 95 %, not related to column length, and significantly larger than those through andosol columns. Amounts of 3,4-DCB and 3,4-DCP eluted through andosol columns reduced as column was longer, and were 1.7% and 0 % at 5 cm length, respectively.

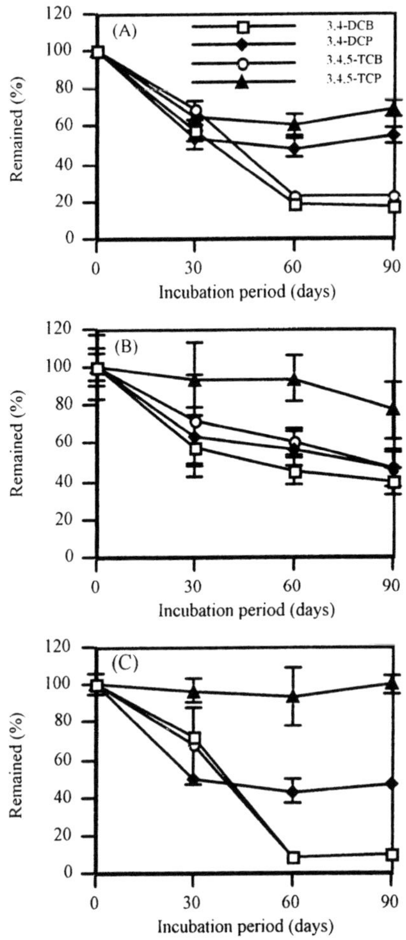

Figure 1. Relative amounts of chlorinated aromatic acids extracted with alkali from soils (A, granitic regosol B, andosol) amended with sludge compost and compost (C) after incubation. Vertical bars indicate standard deviations.

Table 2. The percentages of 3,4-DCB and 3,4-DCP eluted through soil columns

Column length	3,4-DCB (%)		3,4-DCP (%)	
(cm)	Regosol	Andosol	Regosol	Andosol
1	88.6	31.9	93.6	17.1
2	85.0	16.3	89.2	6.8
3	91.8	7.0	88.3	0.0
4	86.9	2.9	88.5	0.0
5	93.8	1.7	88.2	0.0

Discussion

Degradation rates of chlorinated aromatic acids were different among types of compounds, and larger in regosol than in andosol. In regosol, degradation rates of chlorinated aromatic acids extracted with alkali were almost same till 30 days, and about 80 % of 3,4-DCB and 3,4,5-TCB were linearly degraded with time till 60 days, while 3,4-DCP and 3,4,5-TCP remained at constant levels after 30 days. Free forms of 3,4-DCB, 3,4-DCP, 3,4,5-TCB and 3,4,5-TCP extracted from sludge compost with MeOH-H_2O (1:1) were 80%, 47%, 87% and 43 % of both free and conjugated forms extracted with 2*N*NaOH-MeOH (1:1), respectively (Table 1). These data suggested that free forms might be decomposed more rapidly than conjugated forms. Degradation rates of chlorinated aromatic acids except for 3,4,5-TCP in sludge compost were similar to those in regosol. 3,4,5-TCP was more stable in compost and andosol than in regosol.

In column experiment, little of free 3,4-DCB and 3,4-DCP were sorbed to regosol even at 5 cm length, but most of them were sorbed to andosol at more than 3 cm length (Table 2), although 3,4,5-TCB and 3,4,5-TCP were not tested. Lower degradability of chlorinated aromatic acids in andosol may be caused by sorption of free forms in soil. Degradation processes of chlorinated aromatic acids should be further investigated.

References

Nagaoka T, Umezu K, Kouno K, Yoshida S, Ishiguro Y and Ando T 1996 Plant Growth Regulation 20, 295-302.

Nagaoka T, Yoshida S, Tanaka H, Kouno K and Ando T 1997 *In* Plant Nutrition for Sustainable Food Production and Environment. Eds. Ando T *et al.*. pp. 605-606. Kluwer Academic Publisher, Dordrecht.

W. J. Horst et al., Plant nutrition – Food security and sustainability of agro-ecosystems. 850~851, 2001

Responce of faba bean crop to phosphatic, foliar and bio-fertilization under new reclaimed sandy soil conditions

G. M. Yakout[1] and M.H.Greish [2]

[1] *Agron. Dept., Faculty of Agriculture, Suez Canal University, Ismailia, Egypt, gmyakout@yahoo.com*

[2] *Research station, Ismailia , Egypt, Agron. Dept., Faculty of Agriculture, Suez Canal University, Ismailia, Egypt, Shendawy@yahoo.com*

Key words: cover crops, faba bean, foliar spray, *Lupinus albus*, phosphorus efficiency

Abstract

To meet the growing demand for protein food, increase of arable land and improving utilization efficiency of applied fertilisers are necessary. A two years field study conducted on sandy soils in Egypt showed that soil application of phosphorus with or without biofertiliser along with foliar fertilisation significantly increased yield, yield components and quality of faba bean. Results also indicated that incorporating biofertilisers with soil application of fertiliser improved utilization efficiency thereby resulted in saving of significant amount of P fertilzer use. This effect was more pronounced by including foliar nutrition.

Introduction

Faba bean is the most important legume crop in Egypt. It is used for human consumption as a source of cheap protein. Therefore, its cultivation should be extended in more areas particulary in the newly reclaimed sandy soils. Phosphatic, foliar and bio fertilisation can be considered among the most important cultural treatments for increasing faba bean yield, its attributes and quality. Many investigators reported that these cultural practices increased faba bean productivity and quality (Abdel-Aal and Ebaid, 1989; El-Fieshawy and Fayed, 1990; Ali, 1997 and Hanna, 1999). Therefore, the aim of this work was to study the response of faba bean crop to phosphatic, foliar and bio fertilisation under newly reclaimed sandy soil conditions.

Materials and methods

Two field experiments were conducted on sandy soils in Ismailia, Egypt, to study the effect of bio, phosphatic and foliar fertilisers on (i) yield (grain yield/ha), (ii) yield components (number of pods/plant, grain yield/plant and grain index) and (iii) yield quality (grain protein content %) during 1998/1999 and 1999/2000 seasons. Each experiment consisted of 16 treatments which were the combination of two treatments of spraying with tap water (control) and foliar fertilisation (stimphol) and four levels of P fertiliser (0, 36.9, 73.8 and 110.7 kg P_2O_5/ha) with or without biofertiliser (microbein). Microbein functions as biodenitrogen fixer, nutrient mobilizer and growth promoter and it is composed of a selected group of micro-organisms having definite beneficial roles in supporting plant growth. A factorial experiment in split plot design with four replications was used in which foliar spray with stimphol were allocated in the main plots and P levels with or without microbein in the sub plots. Combined analysis of variance was performed for across the seasons and Duncan multiple range test was used to find treatment differences. The interaction among the factors were compared by LSD at 5%.

Results and discussion

Results of this study indicated that the increase in faba bean grain yield /ha was significant with increasing P levels from 0 to 110.7 kg P_2O_5/ha. However, other measured yield components and bean quality increased significantly with increasing P levels only up to 73.8 kg P_2O_5/ha, and there after addition P had only insignificant effects (Table 1). Also, foliar nutrition with stimphol increased yield and its attributes and quality. The favourable effect of increasing P levels on faba bean yield and quality was more pronounced when microbein and stimophol were applied together (Table 2). Fertilisation with microbein saved significant amounts of phosphatic fertiliser as yield in plots with microbein at 0 and 36.9 kg P_2O_5/ha resulted in the same yield level as the application of 110.7 kg P_2O_5/ha without microbein (Table 2).

These results are in agreement with those obtained by others (Abdel-Aal and Ebaid, 1989; El-Fieshawy and Fayed, 1990; Ali, 1997; Hanna, 1999 and Yakout *et al.* 2000).

The results of this study support the successful efficient use of foliar and biofertilisers in obtaining increased yield and quality of faba bean. This will help to reduce the use of soil application of mineral fertilisers, which in turn help to protect the environment from chemical pollution and its harmful effects on human and animal health.

References

Abdel-Aal SM and Ebaid MA 1989 Egypt. J. Agron. 14, 229-239.

Ali EA 1997 Ann. Agric. Sci., Moshtohor 35, 1099-1108.

El-Fieshawy MA and Fayed EH 1990 Zagazig J. Agric. Res.17, 227-233.

Hanna AM 1999 Zagazig. J. Agric. Res. 26, 229-237.

Yakout GM, Greish MH and Ata-Alla RA 2000 Mansoura Univ. J. Agric. Sci. 25, 4299-4306.

Table 1. Effect of bio, phosphorus and foliar fertilisers on yield, yield components and yield quality

Treatments Characters	Foliar nutrition		P_2O_5 levels (kg/ha) without microbien				P_2O_5 levels (kg/ha) with microbien			
	0	Stimo phol	0	36.9	73.8	110.7	0	36.9	73.8	110.7
No. of pods/plant	10.31A	11.36B	7.53a	8.93b	11.08c	11.50cd	10.63c	11.01c	12.77de	13.09e
Grain yield (g/plant)	27.87A	31.79B	21.36a	24.93b	26.31b	38.01bc	26.49b	30.84c	39.69d	41.03d
Grain index (g)	67.29A	79.67B	66.93a	71.48b	73.14c	73.46c	72.82c	75.21d	77.28e	77.97c
Grain yield (t/ha)	2.68A	3.18B	2.14a	2.45b	2.87c	2.88c	2.76c	3.16d	3.59e	3.60e
Grain protein (%)	21.93A	24.23B	20.17a	22.12b	22.61bc	22.86bc	22.98bc	24.12cd	24.72d	25.70d

Values in one row followed by the same (lower or upper) letter are not statistically different at the 5% level

Table 2. Effect of Stimophol at various P levels with and without microbein on yield and its attributes

Foliar Nutrition	P_2O_5 levels (kg/ha) without microbien				P_2O_5 levels (kg/ha) with microbien				LSD at 5%
	0	36.9	73.8	110.7	0	36.9	73.8	110.7	
	Number of pods/plant								
0	7.26	9.11	10.03	10.33	10.87	10.72	12.47	11.68	2.05
Stimo phol	7.79	8.74	12.37	12.64	10.40	11.29	13.07	14.52	
	Grain yield (g/plant)								
0	20.33	23.06	24.97	26.42	25.76	27.94	36.94	37.37	5.43
Stimo phol	22.40	26.82	27.68	29.59	27.19	33.77	42.46	44.33	
	Grain yield (t/ha)								
0	2.117	2.201	2.649	2.608	2.703	2.902	3.183	3.156	0.23
Stimo phol	2.263	2.686	3.083	3.148	2.808	3.421	3.999	4.031	

W. J. Horst et al., Plant nutrition – Food security and sustainability of agro-ecosystems. 852~853, 2001

Boron uptake by plants as affected by organic matter

U. Yermiyahu[1], R Keren[2] and Y. Chen[3]

[1] *Gilat Experimental Station, Agricultural Research Organization, Mobile Post Negev, 85280, Israel, uri4@netvision.net.il*

[2] *Institute of Soil, Water and Environmental Sciences, Agricultural Research Organization, The Volcani Center, Bet Dagan, P.O.B. 6, Israel*

[3] *The Seagram Center for Soil and Water Sciences, Faculty of Agricultural, Food and Environmental Quality Sciences,The Hebrew University of Jerusalem, P.O. Box 12, Rehovot, 76100, Israel*

Key words: boron uptake, organic matter, bell pepper

Abstract

Boron (B) uptake by plant is controlled by B level in soil solution rather than total B content in soil. The affinity of organic matter for B can affect B uptake by plants due to changing B concentration in soil solution. The role of soil organic matter content on B soil solution concentration and uptake by bell pepper were studied. The organic matter used was mature compost (COM), produced from the solid fraction of separated straw-containing cattle manure. Plants were grown in five soil-sand-COM mixtures containing 0, 1, 3, 6 or 10% COM by weight. Four levels of B were applied. The soil was analyzed for B content at the beginning of the experiment and at harvest. Boron concentration in the leaves was determined 45 d from planting. Boron concentration in the soil solution, at the beginning of the experiment, decreased with increasing levels of COM. This decrease was most prominent at high levels of B application. The effect of COM level on leaf B concentration was also prominent at high B application rates, increasing levels of COM resulting in less B in the leaf tissues. Boron concentration in the leaves was highly significantly correlated ($r^2 = 0.88$) with the B concentration adjusted in the soil solution. The results presented herein indicate that organic matter plays an important role in controlling B concentration in the soil solution and that it has a prominent effect on reducing B uptake by plants.

Introduction

Boron is one of the seven essential micronutrients required for the normal growth of most plants. The amount of B adsorbed by soil varies greatly with the contents of various soil constituents (Keren and Bingham, 1985). Boron adsorption and desorption from soil adsorption sites regulate the B concentration in the soil solution. This regulation depends on the changes in solution B concentration and on the affinity of the soil constituents for B (Mezuman and Keren, 1981).

The role of soil organic matter content on B soil solution concentration and B uptake by plants is not fully understood. Boron uptake by plants is, generally, controlled by the B level in the soil solution rather than total B content in the soil (Keren *et al.*, 1985). Because the affinity of organic matter for B is higher than that of clay minerals, at a given total B content (Yermiyahu *et al.*, 1988), B uptake by plants is expected to decrease as the organic matter content in the soil increases. This hypothesis is tested in the current study, the objective of which to determine the effects of organic matter in soil on B availability to the test plant bell pepper.

Materials and methods

Soil (Loess, calcic Haploxeralf) was collected from an uncultivated field in the southern coastal plain of Israel. Soil characteristics obtained by routine procedures were: contents of clay, silt and sand: 16, 43 and 41 %, respectively; $CaCO_3$, 45 g kg^{-1}; organic matter, 11 g kg^{-1}; cation exchange capacity, 168 $mmol_c$ kg^{-1}; pH of saturated paste, 7.8; and B concentration in saturated soil extract, 0.028 mmol L^{-1}. Mature compost produced from the solid fraction of separated straw-containing cattle manure was used in this study. The COM was leached in a column with deionized water at a solid-to-liquid ratio of 1:60 and then dried in an oven at 30°C.

The soil was sieved through a 2-mm sieve and mixed thoroughly with washed COM and pure sand added as an inert substance with respect to B adsorption. The total amount of added COM in the mixture (soil-sand-COM on a dry-weight basis) was 0, 10, 30, 60, or 100 g kg^{-1} of mixture and the added sand content was 100, 90, 70, 40, or 0 g kg^{-1} of mixture, respectively. The loess soil content was constant in all mixtures. Each soil-sand-COM mixture was tested at four levels of added B (0, 0.1, 0.5, or 1.0 mmol kg^{-1}). Boron was added as boric acid ($B(OH)_3$) in solution. Each treatment consisted of five pots, filled with 2 kg of soil-sand-COM mixture, and arranged in the randomized complete block experiment design. Two days after B addition, 2-wk-old pepper seedlings (*Capsicum frutescens*), were planted in each pot. The water contents of the soil mixtures in the pots were maintained between 25 to 30 % (w/w) by adding distilled water at least daily. Boron-free half-Hoagland solution of 100 mL per pot was added weekly. The plants were harvested 45 d after planting. After plants harvest, the moist soil from each pot was well mixed and a saturated paste was prepared and allowed to reach an equilibrium. Boron in the leaves and saturated paste solution was determined by the azomethine-H procedure.

Results

Boron concentration in the soil-sand-COM mixture solution was affected by the total amount of B added and the COM content (Fig. 1). The higher the amount of B added, the higher was its concentration in the soil-sand-COM mixture at any level of COM, whereas the effect of COM varied with B level. When B was added at 0.1 mmol kg^{-1} of mixture, B concentration in the soil solution increased only slightly above the values obtained for the mixtures without added B. The B concentration in the soil-sand-COM mixture solution decreased with increasing COM content at higher levels of applied B. At these B levels, the amount of B contributed by the COM compared to the amount of added B is negligible. The decrease of the B concentration in soil solution was more pronounced at higher B levels and at low levels of COM (Fig. 1). This reduction was due to B adsorption by COM, which has a higher adsorption capacity than soil (Yermiyahu *et al.*, 1995).

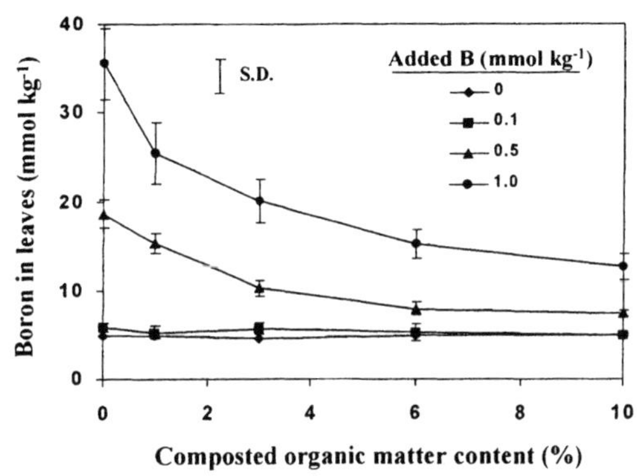

Figure 1. Adjusted B concentration in soil solution at a water content of 30% (w/w) after 2 d of equilibration vs. the COM content of the soil-sand-COM mixture at four levels of B application (B concentration was measured in the saturated paste extract and adjusted to a water content of 30% using sorption model [Yermiyahu *et al.*, 1988]).

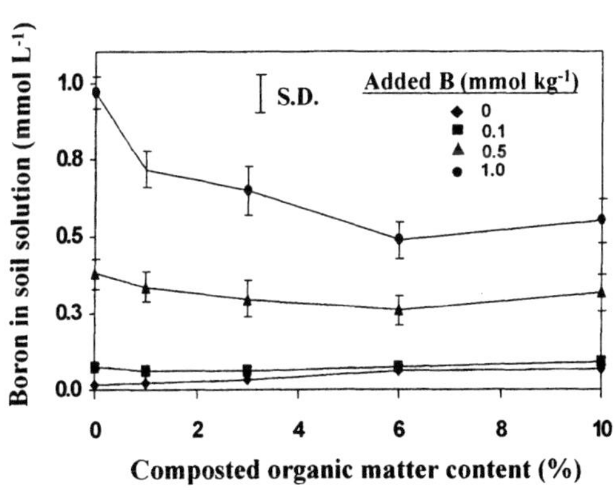

Figure 2. Boron concentration in pepper leaves vs. COM content in the soil-sand-COM mixture at four levels of B application.

B concentrations in the soil solution after harvesting the plants was equal or somewhat higher than that observed at the beginning of the experiment at the zero and lowest added B levels (data not shown). This small increase was probably due to release of native B existing in sparingly soluble minerals in the soil and COM during the growing season. At the highest B levels, B concentration in the soil solution after harvesting was substantially lower than that observed at the beginning of the experiment. This decrease was due to B uptake by plants (data not shown).

Boron uptake by the plant was affected by the amount of B applied and COM content (Fig. 2): the higher the level of added B, the higher the B concentration in the leaves. Boron concentration at the leaves exceeded the potentially toxic level of 30.5 mmol kg^{-1} dry matter only at the highest B application rate for the system without COM. The COM only had a significant effect on B uptake by the plant at the two highest added B levels. Boron concentration in the leaves was highly significantly correlated ($r^2 = 0.88$) with the B concentration adjusted in the soil solution. The linearity of the relationship suggests that plant uptake of B was controlled by soil solution B concentration.

Discussion

The experimental results indicate that organic matter plays an important role in B distribution between the solid and liquid phases in soils and that it can have a prominent effect on reducing B uptake by plants. The ability of organic matter to adsorb B is much higher than that of clay mineral components in soils (Yermiyahu *et al.*, 1995). Therefore, even low organic matter content in the soils can affect the concentration of B in the soil solution and, hence, B availability to plants. When a given soil contains high total B levels, a reduction of soluble B level can be achieved by adding organic matter to the soil. Thus, organic matter can be used as a remedy to control excessive B concentrations in the soil solution, thereby reducing B toxicity to plants. In case of B deficiency, addition of OM may help to retain B in topsoil and reduced its leaching, particularly on the sandier soil. A consequence is improved B nutrition of crop plants.

References

Keren R and Bingham FT 1985 *In* Advances in Soil Science. Ed. BA Stewart. pp. 229-276. Springer-Verlag, New York.

Mezuman U and Keren R 1981 Soil Sci. Soc. Am. J. 45, 722-726.

Yermiyahu U, Keren R and Chen Y 1988 Soil Sci. Soc. Am. J. 52, 1309-1313.

Yermiyahu U, Keren R and Chen Y 1995 Soil Sci. Soc. Am. J. 59, 405-409.

W. J. Horst et al., Plant nutrition – Food security and sustainability of agro-ecosystems. 854~855, 2001

Influence of different fertilization systems on the sulphur content of podzoluvisols and wheat grains

A. Kanal

Institute of Soil Science and Agrochemistry, Faculty of Agronomy, Estonian Agricultural University, Eerika, Faculty of Agronomy, Tartu, 5041, EE: Estonia, kanal@eau.ee

Key words: fertilisation, soil, straw, sulphur, wheat grain

Abstract

One of the International Organic Nitrogen Long-term Fertilisation Experiments has been established in Tartu (Estonia) in 1989 with the crop rotation potato - spring wheat - spring barley. Three treatments are compared: without organic manure (0); straw manure + mineral nitrogen (S+N); cattle manure (CM). Two mineral-N fertilisation levels were selected for this study: NO (non-fertilised) and N8 (i.e. 8 g NH_4NO_3-N m^{-2} y^{-1}). The highest amount of water soluble sulphur (S_{ws}) was determined in the soil amended with straw (10.2 mg kg^{-1}, i.e. 6.2% of total S). The highest S content (1.72 g kg^{-1}) in dry matter of spring wheat (*Triticum aestivum* L. var. Munk) grains was obtained in the treatment S+N8. Sulphur removal was 0.41-0.55 g m^{-2} on soils not fertilised with mineral N and 0.72-0.78 g m^{-2} on soils fertilised with mineral N. The N:S ratio of grains correlated better with total N than with total S content, but remained in the acceptable range (12.2-14.8.). The lowest ratio was found for of non-fertilised and unmanured soils and the highest for treatment N8 with straw application.

Introduction

The developments in Estonian agriculture over the past decade, such as straight N-application without any other mineral nutrients, as well as reduced manure and lime application have decreased soil fertility and yield. The recent reintensification of agricultural production will create a need for S nutrition. Sulphur is an essential plant nutrient. The amount of S needed for an adequate S status in the plant is comparable with phosphorus. Its physiological functions are analogous to those of nitrogen.

Especially in South Estonia, where soils have developed on slightly acid coarse textured parent material with a low organic matter content application of S containing fertilisers seem to be important in the future.

All soil management factors that affect soil organic matter also influence the S supply in soil. Especially grain quality of spring cereals could suffer under the conditions of exclusive N-fertilising due to the short growing period resulting in a reduced availability of S from soil reserves. S deficiency is known to decrease the fraction of S containing amino acids in the grain proteins (Eppendorfer, 1968). Therefor the aims of this investigation were: (a) to detect alterations in the grain N:S ratio depending on mineral-N supply and (b) to study the co-effects of different combinations of organic manure and mineral N.

Materials and methods

A long-term field experiment is carried out on a Stagnic Podzoluvisol (FAO-UNESCO) with a sandy loam topsoil. This so-called "International Organic Nitrogen Long-term Fertilisation Experiment" is designed as a two-factorial experiment: main factor is the organic manure application with 3 levels ([1] without, [2] straw + mineral N and [3] cattle manure) while different application rates for mineral N is the second factor. Three replicated plots sized 50 m^2 are used. Two mineral-N fertilisation levels were selected for this study: N0 (non- fertilised) and N8(= 8 g NH_4NO_3-N m^{-2} y^{-1}). The latter has been identified as the optimal N fertilising level for spring cereals during previous years. Straw was applied on 4 Mg ha^{-1} with additional mineral fertilisers (4 g N m^{-2}), while the application rate for manure was 60 Mg ha^{-1} every three years. Soil samples from the top soil layers (25 cm depth) were taken in early spring 1996. The total S content in soil was measured with a LECO SC 132 by dry combustion and water soluble (S_{ws}) using an extraction ratio of 1:5 (w:v) by continuous flow technique with a Tecator FIAstar analyzer. Grains were milled, digested with HNO_3 and analysed via ICP-AES. Total-C in soil was determined by Tjurin method. The total nitrogen content in grains and in soils was analysed following the Kjeldahl procedure.

Results

The average total-S content of the soil was low (153 mg S kg^{-1}) with an average C:N:S ratio 65:7.6:1. Only a weak correlation (r =0.50) was found between total C and total S in the topsoil. No relationship was found between total S and total N. After 7 years without addition of organic material (treatment "O") significant lower C_t values were found for both N levels. As sulphur occurs in soils mainly in organic forms, it is important to secure a continuous input of sufficient organic material for marginally fertile soils. Manure as a partly decomposed material increased soil S-supply potential mainly via increased organic matter content, but straw and mineral-N combination increased also plant available S in soil. The highest S_{ws} value was determined in the soil amended with straw + mineral N (10.2 mg kg^{-1}, i.e. 6.2% of total S).

No relationship was found between the S-content of soils and grains. Grain yield and S-content of grainscorrelated weekly (r = 0.51). A close correlation (r =0.92) was found between N-content and S-content of grains. An higher nitrogen uptake increased the sulphur uptake although no extra sulphur has been applied. The significant differences in S removal can be explained by differences in grain yields as well as in S content.

Discussion

In cropping systems where organic matter is the sole S source, it is obvious that the relative amounts of S released from soil organic matter during mineralisation depend on soil conditions. Straw application on N0 plots led to a high organic matter content and a low C: N: S ratio, which are seen as prerequisites for enhanced S mineralisation (Harvard *et al.*, 1962). Consequently a higher S_{ws} content was found in soil. In our experiment we did not detect a decrease in the availability of S caused by the incorporation of straw into soils with low S content, which is in contrast to results from Chapman (1997). We calculated that the lowest value of ca 7 mg S_{ws} kg^{-1} soil might result in ca 2.8 g m^{-2} in the plough layer in spring, which should meet the S demand of cereals at the current level of production in Estonia. It is especially important that a sufficient S supply is guaranteed in the first part of the growing season, as wheat accumulates half of the grain S already before anthesis (Monaghan *et al.*, 1999). Furthermore it might be considered that for crops grown under conditions of low S supply nearly half of the S uptake might be from the atmosphere (Siman and Jansson, 1976). According to the literature the critical values for S content in grain are 0.12% and for the N:S ratio over 17:1 (Randall *et al.*, 1981; Haneklaus *et al.*, 1992). In our experiment mature grain had S concentrations above these threshold values (independent of fertilisation level). Obviously no imbalance appeared in the N and S supply. It is more probable that a certain N shortage may be found in treatment N0 than a shortage of S in treatment N8. Our results showed that S-deficiency is not an acute problem in spring wheat at todays production level in Estonia. In intensively managed soils, however, more attention should be paid to the application of nutrients bound in soil organic matter through organic amendments.

References

Chapman SJ 1997 Soil Biol. Biochem. 29, 109-114.

Eppendorfer W 1968 Plant Soil 29, 424-438.

Haneklaus S, Evans E and Schnug E 1992 Sulphur in Agriculture 16, 31-34.

Harvard ME, Chao TT and Fang SC 1962 Agron. J. 54, 101-106.

Monaghan JM, Scrimgeour C, Zhao FJ and Evans EJ 1999 Plant Cell Environ. 22, 831-840.

Randall PJ, Spencer K and Freney JR 1981 Austr. J. Agric. Res. 67, 224-228.

Siman G and Jansson SL 1976 Swed. J. Agric. Res. 6, 135-144.

Table 1. Total C, total S and water soluble S (S_{ws}) in soils from selected plots of the International Organic Nitrogen Long-Term Fertilisation Experiment

	Total C [g kg^{-1}]		Total S [mg kg^{-1}]		S_{ws} [mg kg^{-1}]		C:S ratio		N:S ratio	
	N0	N8	N0	N8	N0	N8	N0	N8	N0	N8
0*	9,1 [a]	9,5 [a]	136 [a]	147 [a]	7,5 [a]	6,9 [b]	66,2 [b]	64,9 [b]	8,1 [b]	7,3 [a]
S+N	10,3 [b]	10,3 [b]	167 [b]	146 [a]	10,4 [b]	9,2 [ab]	61,7 [a]	70,9 [c]	7,2 [a]	8,3 [b]
CM	10,0 [b]	10,3 [b]	155 [b]	168 [b]	9,2 [b]	8,6 [ab]	64,1 [b]	61,6 [a]	7,6 [ab]	7,2 [a]

* **0:** non-manured
S+N: straw + mineral N (4 g N m^{-2} y^{-1})
CM: cattle manure with straw
N0: no mineral N
N8: mineral-N (8 g N m^{-2} y^{-1})
a: values followed by the same letter are not significantly different ($p < 0.05$)

Table 2. Grain yield, N and S contents in dry matter, S removed and N:S ratio (for explanation see Tab. 1)

	Grain [g m^{-2}]		Total N [g kg^{-1}]		Total S [g kg^{-1}]		Removal of S [g m^{-2}]		N:S ratio	
	N0	N8	N0	N8	N0	N8	N0	N8	N0	N8
0*	318 [a]	460 [c]	16,3 [ab]	22,5 [c]	1,30 [a]	1,63 [b]	0,41 [a]	0,75 [c]	12,6 [a]	13,8 [b]
S+N	430 [c]	456 [c]	17,2 [b]	25,4 [d]	1,28 [a]	1,72 [c]	0,55 [b]	0,78 [c]	13,5 [b]	14,8 [c]
CM	343 [b]	448 [c]	16,1 [a]	22,7 [c]	1,32 [a]	1,60 [b]	0,45 [a]	0,72 [c]	12,2 [a]	14,2 [b]

W. J. Horst et al., Plant nutrition – Food security and sustainability of agro-ecosystems. 856~857, 2001

Prominent nitrogen uptake response to cattle feces by specific plant species

M. Yamagata[1)], S. Matsumoto[2)], N. Koga[3)] and N. Ae[4)]

[1)] *Hokkaido National Agricultural Experiment Station, Shinsei Memuro-cho, Kasai-gun, 082-0071, Japan, yamagata@mmr.affrc.go.jp*

[2)] *Shimane Agricultural Experiment Station, St2440, Ashiwata-cho, Izumo-shi,, Simane-ken, 693-0035, Japan, smatsu@mx.miracle.ne.jp*

[3)] *Hokkaido National Agricultural Experiment Station, Shinsei Memuro-cho, Kasai-gun, 082-0071, Japan, nkoga@mmr.affrc.go.jp*

[4)] *National Institute of Agro-Environmental Sciences, 3-1-1, Kannondai, Tsukuba-shi, Ibaraki-ken 305-8604, Japan, noriae@niaes.affrc.go.jp*

Key words: cattle feces, nitrogen mineralisation, nitrogen uptake, organic nitrogen

Abstract

We examined differences in nitrogen uptake by various plant species from soil to which fresh organic matter had been applied, and tried to find plants with a high response of nitrogen uptake to organic matter.
Seventeen crop species were cultivated in a Brown Andosol upland field with 50 t FW ha^{-1} (149 kg N ha^{-1}) of fresh cattle feces (14.8 g N kg^{-1} DW, C/N 28). Crops were: turnip, head lettuce, onion, spinach, soybean, carrot, adzuki bean, chingensai, maize, potato, sugar beet, pimento, cabbage, broad bean, squash, cucumber, and spring wheat. Plant Nitrogen uptake and nitrate and ammonium contents in soils were determined.
Inorganic nitrogen was leached down the soil profile with time. The peak nitrate content was found in the soil at 0~15cm depth at 13 days after sowing (DAS), 15~45 cm at 79 DAS, and 45~60 cm at 108 DAS. The order of nitrogen uptake by squash, maize, and sugar beet was organic matter > chemical fertiliser > no-nitrogen at 108 DAS. On the other hand, the order of inorganic nitrogen contents in the soils were chemical fertiliser > no-nitrogen > organic matter at 79 DAS. There was a contradiction in the order between plant nitrogen uptake and inorganic nitrogen content in the soil. These facts indicate that some crops such as squash markedly took up nitrogen under soils with low inorganic nitrogen levels when organic matter was present. This study suggests that in some cases the nitrogen uptake does not reflect the amount of inorganic nitrogen in soils. We examined the possibility of nitrogen uptake in organic form and enhancement of nitrogen mineralisationbyplants.

Introduction

Livestock production waste such as cattle feces has increased recently in the Hokkaido area with increasing dairy food production in response to the change of food tastes in Japan. It is becoming hard to throw away livestock waste without processing due to problems with environmental conservation. There are many ways of processing livestock wastes such as compost production, making slurry and methane fermentation. Compost production by piling up the livestock waste with some other materials is common because of the simplicity of the equipment. But in Hokkaido, compost production of high quality needs a building with a roof since the temperature in the material must rise above 60 °C to kill bacteria and weed seeds during the process. Such buildings are inevitably costly, so livestock feces are more often spread on forage crop fields. However, the amount of feces are exceeding the capacity of the forage crop fields. The objective of this study was to find plants of high nitrogen uptake ability when organic matter such as cattle feces was applied, so that the waste is returned to the material cycle.

Materials and methods

Seventeen crop species were grown in a Brown Andosol upland field at the Upland Crop Research Center located in Memuro-cho, Hokkaido. A mixture of wheat straw and cattle feces produced within a few days was applied to the field at a rate of 50 t FW ha^{-1} (149 kg N ha^{-1}) 13 days before sowing to avoid emergence difficulty. This mixture contained 14.8 g N kg^{-1} DW and the C/N ratio was 28. The organic matter was applied to a depth of 12 cm by mixing surface soil with a rotary implement. Ammonium sulfate was applied to the control plot at a rate of 60 kg N ha^{-1}. Phosphorus (P_2O_5) and potassium (K_2O) were added at rates of 200 and 150 kg ha^{-1} respectively. Crops used were as follows : turnip (*Brassica campestris* L.), head lettuce (*Lactuca sativa* L.), onion (*Allium cepa* L.), spinach (*Spinacia oleracea* L.), soybean (*Glycine max* Merr.), carrot (*Daucus carota* L.), adzuki bean (*Vigna angularis*), chingensai (*Brassica campestris* L.), maize (*Zea mays* L.), potato (*Solanum tuberosum* L.), sugar beet (*Beta vulgaris* L.), pimento (*Capsicum annuum* L.), cabbage (*Brassica oleracea* L.), broad bean (*Vicia faba* L.), squash (*Cucurbita moschata* Duch.), cucumber (*Cucumis sativus* L.), and spring wheat (*Triticum aestivum* L.). The plant densities were 5.1 m^{-2} for pimento, cabbage, and broad bean, 2.6 m^{-2} for squash and cucumber, and 6.2 m^{-2} for all other crops except for spring wheat at 14 g m^{-2} seeds. The row spacing was 32.5 cm in spring wheat, and 65 cm in all other crops. Crops were sowed on May 26 in 1998, and sampled for nitrogen uptake on August 11 and September 11. The plant samples were dried at 80 °C, weighed, ground to pass 1 mm sieve, and analyzed for nitrogen content by a combustion method using a NC-analyzer. Soils in the fallow plot were taken to 60 cm depth in 15 cm increments on June 8, August 4 and 13, and September 11. The samples were sieved to pass 2 mm, nitrate was extracted with distilled water, and determined by ion chromatograph. Ammonium was extracted with 2 N KCl solution, and measured by flow injection method.

Results

During plant growth, a large amount of inorganic nitrogen leached down the soil profile from 0~15cm depth at 13 DAS to 15~45 cm at 79 DAS (Fig.3), and to 45~60 cm at 108 DAS (Fig.4). Nitrogen uptake by chingensai, radish, turnip, spinach, cabbage, spring wheat, squash, and sugar beet was generally higher in each nitrogen treatment at 77 DAS compared to other crops. The order of nitrogen uptake by these crops in terms of the treatments was chemical fertiliser > organic matter > no-nitrogen. But at 108 DAS, the order for squash, maize, and sugar beet was organic matter > chemical fertiliser > no-nitrogen. Other crops than 4 crops including these three crops were already harvested before 108 days. In contrast to the nitrogen uptake by plants, the amount of inorganic nitrogen in the soils was in the order chemical fertiliser > no-nitrogen ≥ organic matter.

Discussion

The levels of inorganic nitrogen in the organic matter treated soil were lower than in the chemical fertiliser treatment, and almost the same or less than in the no-nitrogen plot throughout the growth period. This means the nitrogen status of the soil with organic matter was as poor as the no-nitrogen soil. There was an obvious contradiction in the order of treatments between plant nitrogen uptake and inorganic nitrogen content in the soil. It is emphasized that high nitrogen uptake occurred in some plants under the low soil inorganic nitrogen status with the organic matter. Such phenomena was previously found when organic matter with a high C/N ratio was applied (Yamagata and Ae, 1997). Especially in vegetable cultivation, the difference in growth response to organic matter was dramatic (Matsumoto *et al.*,1998).

In conclusion, this study suggests that soil nitrogen fertility is not precisely assessed by the amount of inorganic nitrogen in the soil when organic matter with a high C/N ratio, such as fresh cattle feces, is applied. Also in view of environmental conservation, the application of organic matter with high C/N ratio including a mixture of cattle feces and wheat straw is less likely to pollute ground water compared to conventional fertilisation. In addition, some crops such as chingensai and squash can remove a great amount of nitrogen from soil treated with organic matter.

References

Yamagata M and Ae N 1997 *In* Plant nutrition - for sustainable food production and environment. pp. 509~510, Kluwer Academic.

Matsumoto S, Ae N and Yamagata M 1998 Jpn. J. Soil Sci. Plant Nutr. 70, 31-38.

W. J. Horst et al., Plant nutrition – Food security and sustainability of agro-ecosystems. 858~859, 2001

Mineralization patterns of nitrogen in waste-compost and responses of rice crop

S. Matsumura, M. Kato, T. Motobayashi and H. Isshiki
Field Science Center, Faculty of Agriculture, Tokyo University of Agriculture and Technology, Fuchu, Tokyo 183-8509, Japan, matsum@cc.tuat.ac.jp

Key words: cattle manure, leftover compost, mineralisation pattern, *Oryza sativa,* rice, seafood waste

Abstract

Mineralisation patterns of nitrogen for three kinds of waste-compost were investigated under anaerobic conditions, aiming to clarify the relationship between the original materials and the nitrogen mineralisation of the waste-compost. Results of an incubation experiment showed significantly different mineralisation patterns among the three composts. The patterns were successfully modelled by a simulation formula using some characteristic values. The yields of rice plants grown in the field under the application of the composts confirmed the mineralisation patterns.

Introduction

Efficient use of waste-compost in crop production is quite important in maintaining a reasonable circulation of materials and in mitigating environmental problems. Difficulties in achieving efficient use include variability of compost materials and processing procedures. Therefore, it is necessary to comprehend the characteristics of composts, particularly mineralisation patterns of nitrogen, to develop a method for controlling mineralisation patterns, and finally to improve application methods. Janssen (1996) reported that nitrogen mineralisation of organic materials was closely related to C:N ratio and decomposability. Sugihara *et al.* (1986) developed a kinetic method for estimating the nitrogen mineralisation from soil organic matter. Since we had already examined seven composts under aerobic conditions for a similar purpose (Matsumura *et al.* 2000), we examined three of them under anaerobic conditions.

Materials and methods

Three kinds of waste-compost, cattle manure mixed with wood chips (CMW, moisture 470 g kg^{-1}, C 145 g kg^{-1}, N 7.6 g kg^{-1}, C/N 19.7), seafood waste compost (SWC, moisture 240 g kg^{-1}, C 285 g kg^{-1}, N 35.7 g kg^{-1}, C/N 8.0) and leftovers compost (LC, moisture 42 g kg^{-1}, C 337 g kg^{-1}, N 32.5 g kg^{-1}, C/N 10.5) and ammonium sulfate as a control, were put into 250 mL plastic bottles at a rate of 50mgN. Then 100 g of paddy-field-soil (moisture 311 g kg^{-1}, C 206 g kg^{-1}, N 2.6 g kg^{-1}, C/N 7.9) was added to each bottle and the contents mixed. The bottles were divided into three groups and incubated at 20, 25 and 30 °C for 150 days. Mineralised nitrogen was analyzed regularly (Arima 1978) and mineralisation characteristic values of the waste-compost were calculated using the Sugihara's simulation program (Sugihara *et al.* 1986).

These waste-composts were also applied to a paddy rice field at the rate of 80 kgN ha^{-1} in duplicate at the university farm in Tokyo and a rice crop "Tsuki-no-hikari" was grown in 1999. In addition to these three treatments, no-nitrogen-fertiliser plots (NN) and control plots (CTRL, conventional method) were designed. Every treatment was arranged in duplicate. For CTRL, 40 kgN ha^{-1} of fertiliser was used as a basal application and 20 kgN ha^{-1} was applied twice, Jul. 26 and Aug. 17, as a topdressing. After transplanting on May 11, several plants were sampled regularly and the dry matter and the nitrogen content were measured. Soil temperatures at 10 cm depth were monitored throughout the growing season. All plants were harvested on Oct. 7.

Results

The mineralisation patterns of nitrogen and the amounts of nitrogen mineralised during the 150 day incubation varied significantly among these three waste-composts (Figure 1and Table 1). The mineralisation curve for SWC was similar to that for LC, but the amount of mineralised nitrogen was about half that of LC. Apparently, mineralisation for CMW did not occur during the study period.

The yield of hulled rice for LC was as high as for CTRL, while CMW was as low as NN (Table 2). This trend among the three waste-compost coincided with the mineralisation patterns of nitrogen in the incubation experiment.

Discussion

Mineralisation patterns of nitrogen for the three composts under anaerobic conditions were similar to those under aerobic conditions. The patterns and the amounts of mineralised nitrogen were quite different, and these differences may have been due to the different materials. $N_0 1$, $N_0 2$, $N_0 3$ in the equation in Figure 1 mean the amount of available organic nitrogen. $N_0 1$ means the most decomposable fraction, while $N_0 3$ means the least decomposable fraction. It is considered that values of ($N_0 1$+ $N_0 2$+ $N_0 3$) are closely related to the amounts of mineralised nitrogen. Because the $N_0 3$ fraction still contains more complicated constituents that have different decomposability more elaborate fractionation will give more reasonable values for k, the velocity constant.

It is concluded that characteristic values calculated from the incubation data are useful in estimating the mineralisation patterns of various kinds of waste-compost and in determining more appropriate compost applications in practical crop production. For this purpose, more simple methods for measuring these values should be developed.

References

Arima Y 1978 Jpn. J. Soil Sci. Plant Nutr. 49 304-308.

Janssen BH 1996 Plant Soil 181, 39-45.

Matsumura S, Fujita Y and Koizumi R 2000 Proceedings of international symposium between TUAT-Japan and ESKU-Korea, 25-35.

Sugihara S, Konno T and Ishii K 1986 Bulll. National Institute Agro-Environment. Sci. Japan, 127-166.

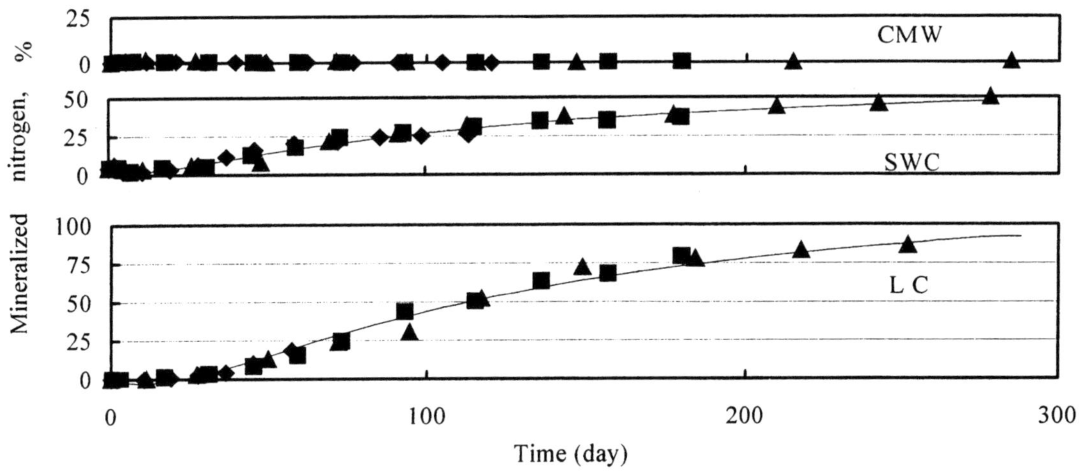

Figure 1. Mineralisation patterns of organic nitrogen in three kinds of waste-composts at 25 °C Data of mineralised nitrogen at different temperatures, 20, 25 and 30 °C, were transformed to the same base at 25 °C.

Table 1. Characteristic values for a model simulated from the data in Figure 1.

	N_01	N_02	N_03	k1	k2	k3	Ea1	Ea2	Ea3	C
CMW	-2.1	1.6	29.2	0.027	0.233	0.000	4,040	26,840	15,000	0.30
SWC	-11.2	58.9	-	0.108	0.008	-	15,650	16,490	-	5.27
L C	-34.6	50.0	90.0	0.005	0.008	0.008	17,490	15,210	18,030	0.21

Simulation model; $N = N_01(1-\exp(-k1 \cdot t))+N_02(1-\exp(-k2 \cdot t))+N_03(1-\exp(-k3 \cdot t))+C$

N_01, N_02, N_03; available organic nitrogen (mgN $100g^{-1}$): k1,k2,k3; velocity constant (day^{-1}):

Ea1,Ea2,Ea3; activation energy (cal mol^{-1}): t; time (day): C; constant

Table 2. Yield and yield components of rice

Treatment	No. of ears m^{-2}	No. of grains m^{-2}	Grains per head	Percentage of Ripened grains	Grain weight g (10^3 hulled rice)$^{-1}$	Yield of hulled rice, kg ha^{-1} Calculated yield	Actual yield
NN	256a	27,947ab	110.5a	91.4a	22.3b	5,704ab	5,567ab
CTRL	330b	34,364cd	104.1ab	89.0ab	21.8ab	6,653c	6,367bc
CMW	248a	26,006a	104.8ab	89.6ab	22.0ab	5,122a	5,049a
SWC	324b	31,642bc	97.7b	87.9ab	21.5a	5,956b	5,877b
LC	342b	35,978d	106.3ab	84.1b	21.6ab	6,521c	6,601c

a, b ; Same symbols within each column mean no significant difference at $P < 0.05$ level by Tukey's method.

W. J. Horst et al., Plant nutrition – Food security and sustainability of agro-ecosystems. 860~861, 2001

Root partial contact with localized organic matter increased ^{59}Fe uptake and alleviated lime-induced chlorosis of young apple trees

S. M. Samar[1)], M. J. Malakouti[2)], H. Siadat[3)], A. Sadjadi[4)] and H. Ghafourian[5)]

[1)] *Soil Fertility and Plant Nutrition Dept., Soil and Water Research Institute, P.O.Box: 14155/6185, Tehran, Iran. m-samar@areeo.or.ir.*

[2)] *Soil Science Dept., Agricultural Faculty, Tarbiat Modarres University, Tehran, Iran.*

[3)] *Soil Reclamation Dept., Soil and Water Research Ins., Tehran, Iran.*

[4)] *Agricultural Research, Education, and Extension Organization, Tehran, Iran.*

[5)] *Radio Isotopes Division, Nuclear Research Center, Tehran, Iran.*

Key Words: apple tree, ^{59}Fe, iron chlorosis alleviation, lime-induced chlorosis, localized placement of amendments

Abstract

Correction of lime-induced chlorosis (LIC) was examined in potted apple (Malus pamila Mill) trees. Manure, as a lime-free organic matter, was added to a calcareous soil by two methods. In the first method, about 1.4 percent of soil was substituted with manure in an auger hole near the tree (localized-placement treatment). Also radioactive iron (^{59}Fe) was added to the hole as a tracer. In the second method, the same amount of manure and ^{59}Fe, were added to the surface soil (surface mixed treatment). The effects of these two methods on some plant parameters were compared with the control treatment which received only ^{59}Fe on the soil surface. Localized-placement, significantly increased root surface area and dry weight near the holes, leaf length, leaf chlorophyll content, leaf iron specific activity (^{59}Fe), and maximum potential efficiency of photosystem II (PS II) of dark adapted leaves, but decreased leaf Fe content. Therefore, results suggests that apple LIC could be corrected by improving localized root growth and development, without using iron fertiliser.

Introduction

A number of different practices have been recommended for LIC correction (Wallace, 1995). One of these methods is to put a portion of plant roots in contact with a small volume of lime free soil. Kalbasi *et al.* (1986) succeeded in alleviating quince tree chlorosis by neutralizing lime with concentrated H_2SO_4, in two auger holes near each tree. Horesh *et al.* (1991) substituted a small volume of soil with peat (alone or enriched with Fe), instead of neutralizing lime with H_2SO_4. They showed that Fe enriched peat, could correct iron chlorosis of citrus trees. Alhendawi *et al.* (1997) reported that increasing bicarbonate concentration in the nutrient culture, depressed root length density in barley (resistant to LIC) to a lesser extent than that in sorghum and maize (sensitive crops). They concluded that lowering of Fe acquisition might be expected as a consequence of the root growth inhibition.

In the present study, we investigated the effects of localized-placement of manure on root development and iron absorption on young apple trees grown in a calcareous soil.

Materials and methods

In 1997, one year old apple trees (red delicious scion on Maling Merton 106), were planted in plastic pots containing 40 kg of a fine loamy carbonatic mesic Calcixerollic Xerochrept soil (pH 7.8, 42.5% total $CaCO_3$). Treatments were initiated after one year as follows:

1- Control: without any manure application.
2- Surface mixed: 100 gr. air-dried manure, mixed with surface soil.
3- Localized-placement: a hole was dug 5 cm away from the trees (3 cm in diameter and 14 cm deep), and filled with 100 gr. manure. In September 1998, 7 micro Curies of radioactive iron (^{59}Fe as $FeSO_4$) were added to each pot. In the control and surface mixed treatments, ^{59}Fe was mixed with the surface soil. In the localized-placement treatment, ^{59}Fe was injected into the hole. Iron specific activity of leaves was determined by a β-counter instrument at the end of the experiment (Zapata, 1990). Dark-adapted leaves fluorescence was determined by fluorescence kinetics apparatus (Hanstech Co.), as a measure of maximum potential efficiency of PS II. Leaf chlorophyll content was determined by a chlorophyll meter (SPAD 502, Minolta Co). At the end of the experiment, leaf samples were taken for nutrient analysis. Also, with the aid of a stainless steel cylinder (10 cm in diameter and 25 cm length), soil samples (with root) were taken in all treatments to study root development near the hole filled manure and else where.

Results and discussion

Effects of treatments on plant parameters are shown in Table 1. Localized placing of manure increased root surface area and root air-dried weight, inside and near the zone of application. Root development was not affected in treatments without localized application and further away from this zone (data not shown). Alhendawi *et al.* (1997) suggested that the high concentration of bicarbonate in nutrient culture is the reason for the depressed root growth. It seems that in our experiment, due to better aeration and lower concentration of bicarbonate in the manure application zone, root growth improved.

Both methods of manure application were effective in increasing of chlorophyll of leaves, but localized application was more effective and indicated the absence of chlorosis in the leaves. This treatment also significantly increased leaf length, which could be considered as another criterion for LIC correction.

Maximum potential efficiency of PS II was improved by manure application. Localized placing was more effective in enhancing this value. This method might improve the proportion of PS II reaction centers and reducing photoinhibition (Abadia, 2000).

Localized-placement method decreased leaf total iron, but on the other hand, ^{59}Fe concentration (data not shown) and specific activity of iron increased. Reduction of leaf iron content, after correction of LIC, has been observed by other researchers too (Morales *et al.*, 1998), and this is described as a consequence of more Fe mobility in the plant tissues following LIC correction. As leaves expand more, dilution effect must be also considered. But this decreasing trend (for total iron), could not affect ^{59}Fe increasing concentration in the leaves. It seems that root ability for iron absorption improved enormously in the localized-placement method, resulting in an inverse of iron specific activity. This was not observed for the original soil iron absorbed by the plant. The lower rate of ^{59}Fe precipitation in organic matter, or the higher dissolution rate of freshly precipitated ^{59}Fe inside the organic matter (Lindsay, 1991) together with better root development, can explain the increase in specific activity of ^{59}Fe in leaves, while total Fe concentration is reduced.

References

Abadia J, Morales F and Abadia A 2000 Plant Soil 215, 183-192.

Alhendawi RA, Romheld V, Kirkby EA and Marschner H 1997 J. Plant Nutr. 20 (12), 1731-1754.

Horesh I, Levy Y and Goldschmit EE 1991 Iron Nutrition and Interactions in Plants. Eds. Y Chen and Y Haddar. Kluwer Academic Publishers.

Kalbasi M, Manuchehri N and Filsoof F 1986 J. Plant Nutr. 9, 1001-1007.

Lindsay WL 1991 *In* Iron Nutrition and Interactions in Plants. Eds. Y Chen and Y Haddar. Kluwer Academic Press.

Morales F, Grasa R, Abadia A and Abadia J 1998 J. Plant Nutr. 21 (4), 815-825.

Wallace A 1995 *In* Advancements in Iron Nutrition Research. Ed. A Hemantaranjan. pp. 1-14. Scientific Publishers, Jodhpur, India.

Zapata F 1990 *In* Use of Nuclear Techniques in Studies of Soil-Plant Relationships. Ed. G Hadarson. pp.61-128. International Atomic Energy Agency.

Table 1. Effects of treatments on plant factors

Treatments	Root Surface Area Index (cm^2L^{-1}) **[i]	Root Weight Index (grL^{-1}) *	Leaf Chlorophyll Concent. (SPAD Reading) **	Leaf Length (mm) *	Maximum Potential Efficiency of PSII (Fv/Fm) **	Leaf Total Fe Concent. (mg Kg^{-1}) **	Leaf Iron Specific * Activity (DPS)[ii]
Control	23.6 B[iii]	0.77 B	11.5 C	44.2 B	0.545 C	172 A	950 B
Surface Mixed	30.2 B	0.90 AB	20.4 B	49.0 B	0.621 B	121 B	857 B
Localized-placement	65.4 A	1.68 A	42.4 A	66.8 A	0.688 A	93 B	1444 A
L.S.D. (a=0.05)	20	0.79	6.6	16.5	0.062	38	153

i * and **, significant at 0.05 and 0.01 levels respectively
ii disintegration per second
iii capital letters show statistical groupings of the results

W. J. Horst et al., Plant nutrition – Food security and sustainability of agro-ecosystems. 862~863, 2001

Non-root nutrition with vermicompost extracts as the way of ecological optimisation

A. V. Gamaley, M. A. Nadporozhskaya, A. I. Popov, O. G. Chertov, N. V. Kovsh and O. A. Gromova

Biological Research Institute, St. Petersburg State University, 2 Oranienbaum Rd., 198904 St. Petersburg-Peterhof, Russia. Email: adm@paloma.spbu.ru

Key words: earthworm compost, humic substances, non-root nutrition, vermicompost

Abstracts

Based on a theoretical approach for the optimisation of plant production a so-called "method of biological correction" was developed. It was hypothesized that foliar applied compost extracts have a promoting effect on plant physiology and thereby on yield and quality. For further evaluation field experiments were carried out on agricultural farms near Leningrad (Northwest Russia) and in the Tambov regions (Central Russia) during the last 6 years. The results of these experiments support that vermicompost extracts (i.e. extracts from compost treated with earthworms) was very successful for increasing yield and improving plant quality.

Introduction

The alarming ecological situation in many regions around the world requires to optimise mineral and organic fertilisation. In Russian agriculture during the last ten years the situation was even more complicated due to shortage of organic and mineral fertilisers, which led to a substantial decrease in crop production. Against this background there is an increasing need to apply the available fertilisers in an efficient way. One of such ways is "non-root" plant nutrition, which means plant nutrition via foliar application of water soluble nutrients (e.g. N, P, Fe, etc). The amounts of nutrients needed for non-root nutrition is less than for standard soil application. Non-root treatment allows the most rapid improvement of nutrient imbalances. Solutions containing organic matter are even more effective than those containing only mineral nutrients. Natural humic substances provide "physiological activity" for a wider range of concentrations than artificial compounds (Chukov *et al.*, 1996). The purpose of this work was the experimental verification of some theoretical claims for the use of organic substances as yield enhancing chemicals.

Materials and methods

Effects of vermicompost extracts have been studied on agricultural farms near Leningrad (Northwest Russia) and in the Tambov regions (Central Russia) during the last 6 years. Vermicompost is derived when cow manure is treated by earthworms (Eisenia fetida) during 1.5-2 months. Large-scale fields (10-30 hectares each) were treated with this extract (about 5-10 l/ha). Working concentration of organic carbon in alkali extract solution was about 0.001-0.002 g/l. The application of the vermicompost extract was done in combination with other sprayings (mineral fertilisers, herbicides, etc.). An untreated part of the field was used as a control plot. Crops used in this study were: cereals (oats, barley, wheat, corn), beets, carrots, potatoes, vegetables (cucumber, tomato), and perennial grasses (for hay or for silage). Total crop yield was measured and nutrient composition of plants was analysed. In addition selected plant growth parameters and soil agrochemical characteristics were measured.

Results

Positive effects of vermicompost extracts on yield have been found for all studied agricultural plants (Fig. 1).

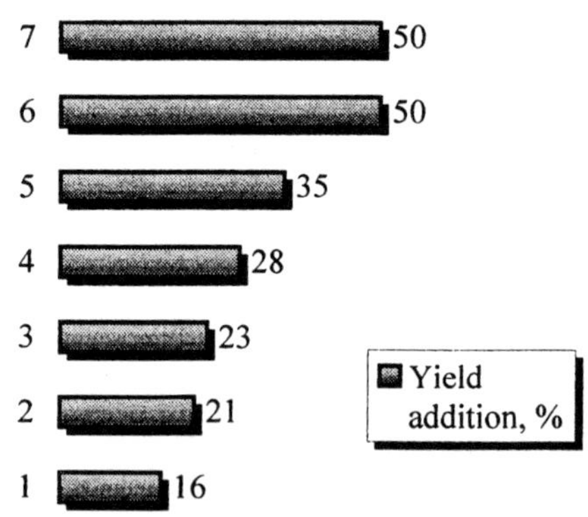

Figure 1. Percent increase in yields of various crops with vermicompost extracts. Plants: 1 – oats, 2 – tomato, 3 – carrot, 4 – beat, 5 – cabbage, 6 - corn, 7 – sunflower

Under adequate weather conditions yield increased by 10-50% compared to the control. Under water stress and/or lack of some soil nutrients the positive effect reached up to 100% in comparison to the control. Vermicompost extracts increased sunflower yield, which is in contrast to results obtained in earlier studies (Khristeva, 1951), where peat extracts have been used as a foliar spray.

An increase of root mass is the most important factor for soil carbon content. We found 15-29% higher values for rootmass in oat crops using vermicompost extracts (data not shown).

Vermicompost extracts were effective not only in increasing yield, but also in improving plant quality. The feed nutritional value of beats increased by 20-40% with the application of the compost extract as compared to the control treatment (Fig. 2).

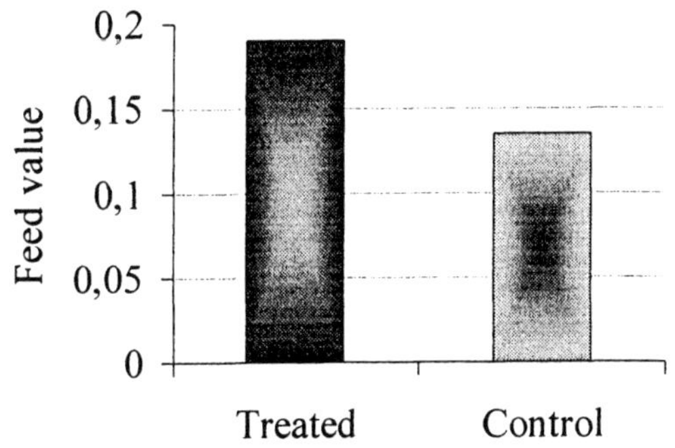

Figure 2. The influence of vermicompost extracts on the feed nutritional value of beat

The application of vermicompost extract did not influence the quality of oat seeds, but influenced the shoot composition (Tab. 1). Protein content in the leaves as well as lignin contents in leaves and straw were higher in treated plants compared to those in the untreated plants.

Table 1. Leaf and straw composition of oat plants, which received vermicompost extract (treated) and those that were untreated

Sample	Hemi-cellulose	Cellulose	Lignin	Protein
Leaves, untreated	22.3	18.1	21.8	19.7
Leaves, treated	18.6	15.6	25.2	22.8
Straw, untreated	23.4	23.1	27.4	12.2
Straw, treated	20.4	20.9	33.9	12.7
LSD_{05}	4.3	4.8	1.3	1.6

In the 1999 growing season two foliar applications with vermicompost extract to carrots resulted in a better product and higher yields even in comparison to a control, which received a higher amount of mineral fertilisers and a foliar application with "Kemira-combi". Furthermore the vermicompost extract treatment improved plant resistance against insects and drought (data not shown).

Discussion

The results of these experiments support the "method of biological correction" concept (i.e. using compost extracts as foliar sprays) in agricultural production systems (Gamaley *et al.*, 1996; Popov, 2000; Popov and Chertov 1994, Popov *et al.*, 1993).

The main targets of this concept are the creation of optimal conditions for 1) photosynthesis, 2) organic-mineral soil status, as well as 3) uptake and internal translocation of nutrients in plant.

Agricultural management may improve the first two. The third may be effectively increased by foliar application with physiologically active compounds with natural organic matter being more valuable for foliar nutrition. Vermicompost extracts suppose to have the highest physiological activity due to their richness in humus substances produced by earthworms, and the presence of vitamins, ferments and natural growth stimulators. According to Popov *et al.* (1993) the labile part of soil organic matter might have been lost in degraded arable soils. The treatment with natural organic compounds can provide an important input in the carbon cycle in these soils (Popov and Chertov, 1997).

References

Chukov SN, Talashkina VD and Nadporozhskaya MA 1996 Eurasian Soil Sci. 28, 30-39 (in Russian with English summary).

Gamaley Yu V, Popov AI and Gamaley AV 1996 *In* Proc. of All Russia Conf. "Management of productive process of plants in control conditions" 21-23 (in Russian).

Khristeva LA 1951 *In* Works of Soil Science Inst. V.38 Moscow, 10-21 (in Russian).

Popov AI 2000 *In* Eds. Efimov V N *et al.* Humus and soil formation. St. Petersburg, pp 31-47, (in Russian).

Popov AI and Chertov OG 1994 *In* Degteva SV, Kuznetsova EG, Archegova IB, Eds. North Development and Problems of Recultivation. Syktyvkar: 177-183 (in Russian); 467-468 (in English).

Popov AI and Chertov OG 1997 *In* Eds Drozd J *et al.* The role of humic substances in the ecosystems and in environmental protection Wroclaw, PTHS Polish Society of Humic Substances Proc. of the 8th Meeting of the IHSS, 993-998 (in English).

Popov AI, Chertov OG, Kerzum PP and Nadporozhskaya MA. 1993 *In* Proc of Int. Symp. "Soil Cover Structure". Moscow. Vol. 2, 204-207 (in English).

W. J. Horst et al. (Eds.), Plant nutrition – Food security and sustainability of agro-ecosystems. 864–865, 2001

Mineral fertiliser consumption and crop yields in Slovakia in the period 1989-1999

S. Torma

Soil Science and Conservation Research Institute Bratislava, regional work place Reimannova 1, SK 080 01 Presov, Slovak Republic. torma@vupop.sk

Key words: crop yields, mineral fertilisation, nutrient soil supply

Abstract

Mineral fertiliser consumption in Slovakia is approximately 42-57 kg NPK per hectare of farmland, while 10 years ago in 1989 it was 240 kg. Nitrogen fertilisers account for 66-71 % of this amount and phosphorus and potassium the remaining 29-34 %. The yields of the main crops (cereals, potatoes, sugar beet, oil plants, annual and perennial fodder crops) did not decrease with a similar tendency. Cereals, oil crops and fodder crops yields have decreased only by 10-30 %, while potato yields have been increased by 10 %, sugar beet yields by 10-15 % and corn maize by 25 %. This can be due to the relatively high rations of nitrogen fertilisers even to these plants and by sufficient supply of available forms of phosphorus and potassium in the soils.

Introduction

In present social conditions, when inputs into the agricultural production increase faster than the growth of prices of products, the optimisation of plant nutrition introduces a substantial economical and ecological factor. Crop yields have decreased in recent years with decreasing application of organic and mineral fertilisers. The natural nutrient supply in the soil is limited and the nutrient balance becomes negative due to inadequate return of withdrawn nutrients. It is necessary to study fertilisation, as there will continue to be a decrease in available nutrient supply from the soil.

The aim of this paper is to evaluate mineral fertiliser consumption and the major crop yields obtained in the recent decade in Slovakia. The long-term future nutrient status of Slovakian soils and crop yields are discussed.

Material and methods

The data on yields of crops and mineral fertilisers consumption were taken from Statistical yearbooks of the Slovak Republic.

Results and discussion

The period between 1989-1999 is characterised by a decrease in mineral fertiliser consumption all over the world. For example, in Western Europe NPK consumption has been decreased from 175 kg ha^{-1} in 1989 to 118 kg ha^{-1} of farmland in 1996 (Uebel, 1998). While in these countries ecological aspects came into the foreground, the NPK consumption decrease in countries such as Slovakia and other Central and Eastern European countries has been driven by an unfavourable economical situation.

Mineral fertiliser consumption (given as pure nutrients in kg ha^{-1} of farmland) in the period 1985-1999 is presented in Figure 1.

Fertiliser application is closely linked to soil nutrient supply. Chemical fertilisation is one of the most effective measures to increase soil nutrient content, but at the same time it is one of the most expensive measures. In the present economic situation the majority of farmers must carefully consider all inputs into the production process and into the soil. A saving in fertiliser input into the soil in the long-term may be unfavourable. The latest results of the Central Control and Testing Institute of Agriculture document the definite decrease in potassium and phosphorus content in the soils. For example the share of arable land with high phosphorus content (more than 95 mg P per kg of soil according to Mehlich II) has decreased from 36 % to 16 % in a period of 5 years. The changes in potassium content are less drastic due to the natural potassium reserves in the soils of Slovakia. The share of arable land with high potassium content (more than 300 mg K per kg of soil) has decreased from 29 % to 19 % during the last 5 years (Table 2).

Crop yields mainly depend on fertiliser application and on nutrient supply in the soil. Reductions in yields especially of cereals, but also of oil crops and fodder crops have been seen in the last 10 years in Slovakia (Table 1). Fertilisation is not the only reason for this decrease. Other factors play also a role, e.g. available nutrients in the soil, insufficient use of pesticides, increasing soil acidity, insufficient application of farmyard manure (negative carbon balance) and other factors that are in conflict with good agronomic practice.

Farming with a high negative nutrient balance leads to soil fertility loss. According to preliminary calculations (Torma, 2000) the financial cost of the nutrients that are lost each year from the soil reaches about 2500 million Slovak crowns annually (e.g. 55 million US $). However, this figure is not final because the financial losses caused by lower crops yields are not included. Additionally the nutrients losses caused by water erosion are not included. With 55 per cent of agricultural soils in Slovakia (e.g. more than 1.3 million ha) endangered by erosion this will also have a large impact on the financial cost of nutrient loss. The mentioned huge losses can be reduced by increased consumption of fertilisers (both mineral and organic), by increased sown area of leguminous plants and by accurate measures against water erosion.

Table 1. Crop yields in Slovakia, 1989-1999.

	Yield (t ha^{-1})		
Year	1989	1994	1999
Winter wheat	5.53	4.85	4.16
Spring barley	4.70	3.67	3.04
Corn maize	5.55	4.14	6.00
Winter rape	2.45	2.11	2.09
Sunflower	2.31	1.64	1.31
Annual fodder	31.82	18.69	21.96
Perennial fodder	8.98	6.73	5.75
Potatoes	13.56	9.67	14.33
Sugar beet	34.34	34.53	40.77

References

Statistical yearbook of Slovak Republic. Statistical office of Slovak Republic Bratislava, various years.

Torma, S. *In* Potassium and Phosphorus: Fertilisation effect on soil and crops. Proceedings of the Regional IPI Workshop Lithuania. IPI - LIA Dotnuva-Akademija, 2000, p. 37-42.

Uebel E. 1998 *In* Proceedings of the 11th CIEC Symposiums on Codes of Good Fertiliser Practice and Balanced Fertilisation. IUNG Pulawy, 1998, p. 409-415.

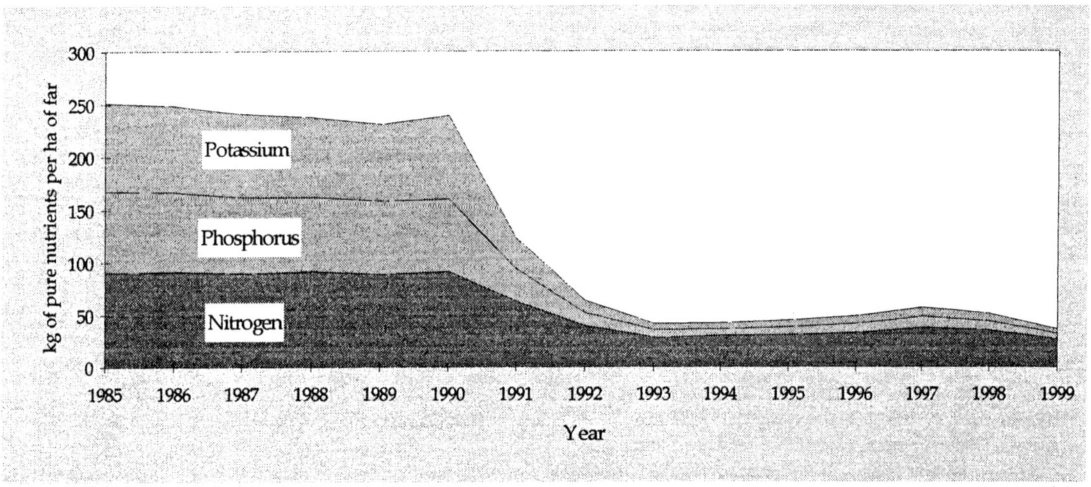

Figure 1. Mineral fertiliser consumption in Slovakia between 1985-1999

Table 2. Available phosphorus and potassium supply in the arable land of Slovakia

	Percentage of arable land in Slovakia (%)									
	Very little content		Little content		Medium content		Good content		High content	
	1990-94	1995-99	1990-94	1995-99	1990-94	1995-99	1990-94	1995-99	1990-94	1995-99
Available phosphorus	2.3	3.1	9.5	16.5	38.0	42.1	14.3	22.5	35.9	15.8
Available potassium	1.0	1.3	7.3	9.6	25.3	31.3	37.0	38.9	29.4	18.9

W. J. Horst et al. (Eds.), Plant nutrition – Food security and sustainability of agro-ecosystems. 866~867, 2001

Nutrient cycling and management in different agro-eco regions of China

J. Zhou[1)], J. Xie[1)], X. Chen[1)] and R. Härdter[2)]
[1)] *Institute of Soil Science, Academia Sinica, 71, Beijing Str. East, Nanjing, 210008, China, jmzhou@issas.ac.cn*
[2)] *International Potash Institute, P.O.Box 10 20 29, Kassel, 34111, Germany, rolf.haerdter@kalisalz.de*

Key words: cropping systems, nitrogen, nutrient balance, deficit, input, losses, on farm trial phosphorus, potassium, surplus, transfer

Abstract

A survey was carried out on different levels (field, farm, county, province level) and compared with on farm trials to describe nutrient inputs, transfers, losses and balances with a higher precision. On average nitrogen inputs of 320 kg ha^{-1} N against outputs of 250 kg ha^{-1} N created a surplus of 70 kg ha^{-1} N. The range was wide and varied between a deficit of -88 kg ha^{-1} N in Guangxi and a surplus +330 kg ha^{-1} N in the adjacent Hunan province. Phosphorus balance with 12 to 66 kg P ha^{-1} was positive throughout the studied areas. In all agro-eco regions potassium inputs did not match the outputs. The annual K deficit with -158 kg ha^{-1} K was most pronounced in Guangxi. Surplus application of nitrogen led to partly large N losses. On the other hand, positive balances of P especially on the highly weathered soils in southern China is attributed to the larger requirement to compensate for the widely observed P fixation in soils. In the central and northern regions without P fixing soils, a continued P surplus, however, also bears potential environmental risks. Insufficient K application together with pronounced leaching losses, e.g. in light soils, caused a drastic soil K depletion. Imbalances between N and K inputs caused poor N use efficiency.

Introduction

It is generally accepted that with an increase of mineral nutrient inputs the incremental yield increase per unit of nutrient is continuously declining. Reasons for this are related to crops' yield potentials as well as environmental factors. Besides these, especially in China, inappropriate nutrient input ratios have greatly contributed to poor nutrient efficiency, reduction in soil productivity and hence environmental risks, e.g. pollution of water reservoirs. Better nutrient efficiency by improving and sustaining soil productivity is therefore a crucial aspect of China's agriculture. In this context the reported study tries to evaluate the nutrient inputs, outputs, transfers and losses depending on agro-ecological regions and cropping systems as a guideline for fertiliser recommendation and the fertiliser policy of the country.

Materials and methods

With the aid of specifically designed questionnaires and an evaluation of existing statistics on crop production and nutrient inputs, nutrient budgets for whole counties and provinces were set up. The selected eco-regions covered temperate climates in the north with one crop per year across the subtropical to the tropical climate in the south with up to more than two crops per year. In the representative areas soil and plant analyses were carried out in farmers field and specifically designed field trials.

Results

Nutrient pathways in typical farmland ecosystems

Major nutrient inputs of a typical farmland ecosystem are through inorganic fertiliser, organic manure (recycling of agricultural by-products, e.g. crops residues and/or farm yard manure), biological nitrogen fixation, atmospheric wet and dry deposition, as well as by the irrigation water. Nutrient losses comprise mainly removal by the crop, runoff, leaching, volatilization, denitrification (Figure 1).

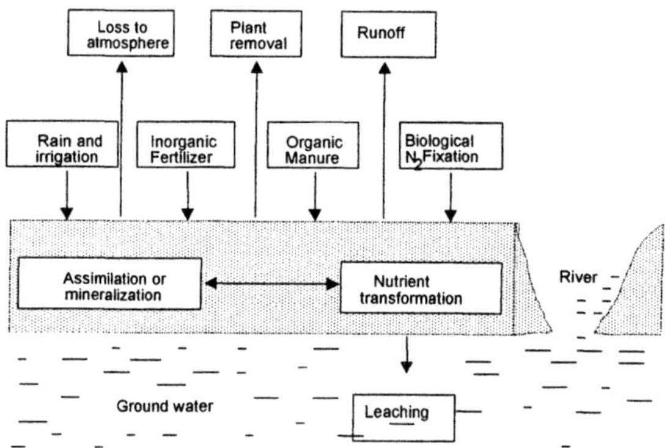

Figure 1. Main nutrient pathways in a farmland ecosystem

Nutrient inputs into farmland during the past decades.

Before the 1970s nutrient inputs into Chinese farmlands were mainly organic manures. With the development of the mineral fertiliser industry there was a continuous increase in the use of nutrients from inorganic sources. Since 1975 the use of inorganic N, and since 1980 the use of inorganic P exceeded that from organic origin. In contrast, the K supply until today remained relying primarily on organic manure (Figure 2). The total input from organic sources has increased despite the use of inorganic fertilisers for all nutrients. This is explained by a larger biomass production through mineral fertilisation and the consequently greater nutrient recycling. With the exception of K, the relative contribution of organic manure to the total nutrient supply, however, has drastically declined.

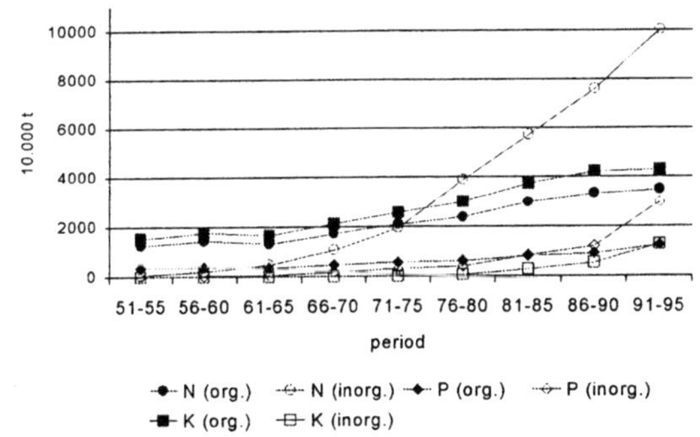

Figure 2. Changes in the sources of nutrient input to Chinese farmlands during the past decades

Nutrient balances and budgets in agricultural fields

Until the 1970s the budget between nutrient input and output was almost balanced for phosphorus, whereas for K and particularly N the balance was negative. This changed dramatically with the introduction of mineral fertilisers during the period 1971-75. During this period and the following time until today, inputs of these two nutrients by far exceeded the outputs. Throughout China soil P increased by more than 160 kg P ha^{-1} or 4-8 mg kg^{-1} soil. In contrast, for K the negative balance remained and even further widened, indicating a soil K depletion during the same period (Figure 3).

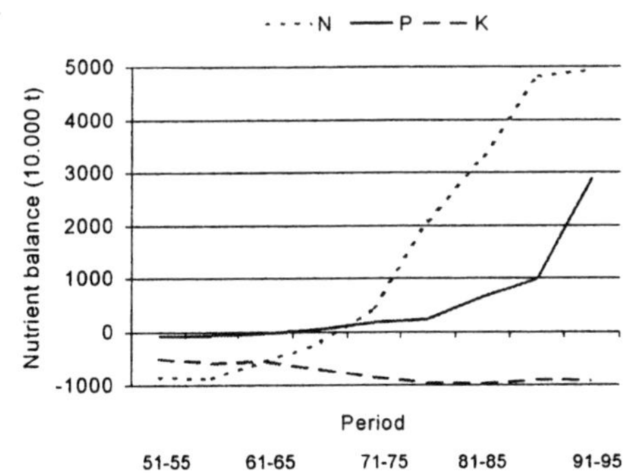

Figure 3. Changes in the nutrient balances of Chinese farmlands in the past decades

Between provinces, however, there are distinct differences in the nutrient balances (Table 1). In the north Helongjiang and Shaanxi, the nutrient budgets are almost balanced. This changes in the southern eco-regions with more rainfall and higher cropping intensity. The field trial in Table 2 shows that substantial nutrient losses occur by leaching and runoff, especially in the case of K, aggravating the imbalance between input and output. The small N losses may be due to the fact that in paddy soils main losses are caused by denitrification which, however, were not recorded in this experiment.

Table 2. Nutrient losses in a paddy soils of South China

Region	Type of loss	Water balance (t ha^{-1})	N	P (kg ha^{-1})	K
Mountain	Leaching	1650	5.1	0.15	17.0
	Runoff	1050	3.9	0.45	10.2
Alluvial	Leaching	1800	6.9	0.15	12.2
Soil	Runoff	675	2.0	0.30	3.2
Red soil	Leaching	1650	6.8	0.06	9.6
	Runoff	450	1.5	0.015	2.0

Discussion

Changes in fertilisation practices during the past decades resulted in an oversupply of N and P and an undersupply of K. The observations confirm those of Li *et al.* (1990) regarding N and Lu and Jiang (1990) for P, reporting very poor nutrient efficiencies. The inadequate replacement of removed K from the soil by harvest and runoff as well as leaching increases K losses. Xie *et al.* (1998) hence referred to drastic K depletion of soils in different eco-regions. Improving the nutrient balance is needed for better nutrient efficiency and soil productivity.

References

Li QK, Zhu ZL and Yu TR 1990 Fertilisation problems in sustainable development of Chinese Agriculture. Jiangxi Science and Technology Press, Nanchang.

Lu RK and Jiang BF 1990 *In* Soils of China. Eds. Y Hseung *et al.* pp. 486-501. Science Press, Beijing.

Xie JC, Xing WY and Zhou J 1998 *In* Nutrient management for sustainable Crop Production in Asia, Eds. AE Johnston and JK Syers. pp. 267-277. University Press, Cambridge UK.

Table 1. Nutrient budget of farmlands in seven typical agro-eco regions in China (kg ha^{-1}) in 1999

Province	N Input	N Output	N Balance	P Input	P Output	P Balance	K Input	K Output	K Balance
Heilongjiang	108	102	6	47	11	36	27	47	-20
Shaanxi	194	123	71	23	11	12	33	44	-11
Henan	136	123	13	75	50	25	70	118	-48
Jiangsu	481	394	87	155	91	64	163	196	-33
Sichuan	322	249	73	121	67	54	75	196	-121
Guangxi	416	504	-88	151	84	67	285	443	-158
Hunan	583	253	330	188	156	32	318	361	-43

W. J. Horst et al. (Eds.), Plant nutrition – Food security and sustainability of agro-ecosystems. 868~869, 2001

Nutrient balances and present state of available nutrients in the soil

P. Cermak

Central Institute for Supervising and Testing in Agriculture in Brno,
Department of Plant Nutrition, Hroznova street 2, Brno, 656 06, Cz : Czech Republic, pavel.cermak@oapvr.zeus.cz

Key words: nutrient outputs, inputs, balance, nutrient contents, soil monitoring, soil testing

Abstract

A widely accepted approach to evaluate the sustainability of agricultural production concerning the nutrient status is to calculate nutrient balances (= input – output of nutrients). For this it is necessary to register all nutrient outputs (uptake by main product and by-product, leaching) and nutrient inputs (use of fertilisers, atmospheric deposition, use of sewage sludge, nitrogen fixation, use of seeds and planting material). Based on statistic data and results from field experiments decreasing nutrient balances of N, P and K were calculated for the Czech Republic for the period 1985 to 1998. This was mainly due to decreasing fertiliser input. Since 1992 negative balances are obtained for P and K, positive balances for nitrogen of more then 25 kg N ha^{-1} year^{-1} remained until 1998. Negative balances for P and K tend to correspond with decreasing P and K contents in soil.

Introduction

One of the indicators that has been proposed and applied to evaluate the sustainability of production systems is the nutrient balance (input – output of nutrients). A nutrient balance is one of the priority issues in developing a set of agri-environmental indicators, as part of the contribution to the analysis of the interactions between agriculture and the evironment and impact of changes in agricultural policy on the environment.

Materials and methods

Nutrient balances can be based either on national statistics or on more-less precise data obtained in the field. In the Czech Republic both methods were used and compared.

For nutrient balance determination at national level, we used statistics on the acreage of individual crops, on mineral fertilisers consumption and on manure production based on number of animals. Estimated values for nitrogen fixation by leguminous plants and for atmospheric deposition were used. The balances were calculated for 1985, 1986, 1992, 1995 and 1998 respectively.

To compare these balances with nutrient contents in soils, results of agrochemical soil testing for the following periods were used: 1984-1986, 1987-1989, 1990-1992 and 1993-1998.

For calculation of nutrient balance "based on field data" we used 140 soil monitoring plots (Table 1 and Table 2). These data sets were selected from the national network of 217 plots in order to represent the main soil types, soil texture classes, production areas and climatic conditions. The whole monitoring network is in operation since 1992. Each plot has an area of 1000 m^2 and all inputs and outputs are monitored on a yearly basis. The atmospheric deposition was measured individually on almost all monitoring plots. Furthermore leeding studies was carried out on 15 plots.

Results and discussion

- Nutrient balances determined for the area of agricultural land from national statistics shows similar trends as balances derived from the data of the soil monitoring plots.
- Decreasing of the final balance in all nutrients in national statistics is a result of the overall decrease in fertiliser consumption after 1990.
- In recent period slightly positive balance for nitrogen, a slightly negative balance for phosphorus and negative balance for potassium were calculated (Figure 1, 2, 3).
- The negative balance for potassium and the slightly negative for phosphorus are reflected in decreasing contents of these nutrients in soil (Figure 4).

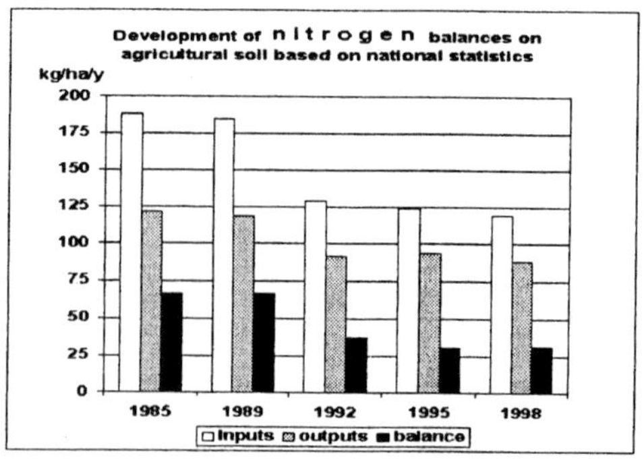

Figure 1. Nitrogen balances (monitoring plots)

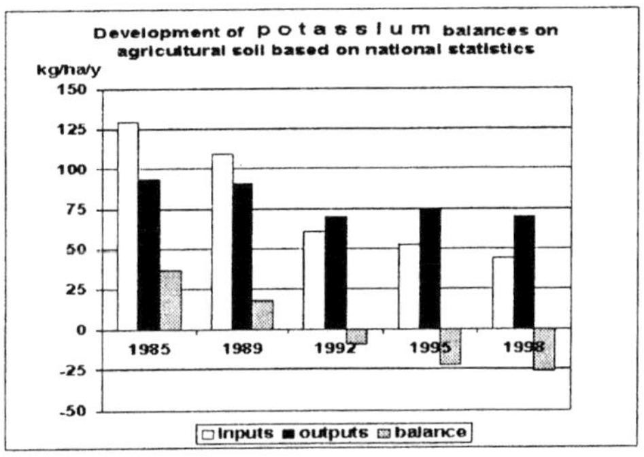

Figure 2. Potassium balances (monitoring plots)

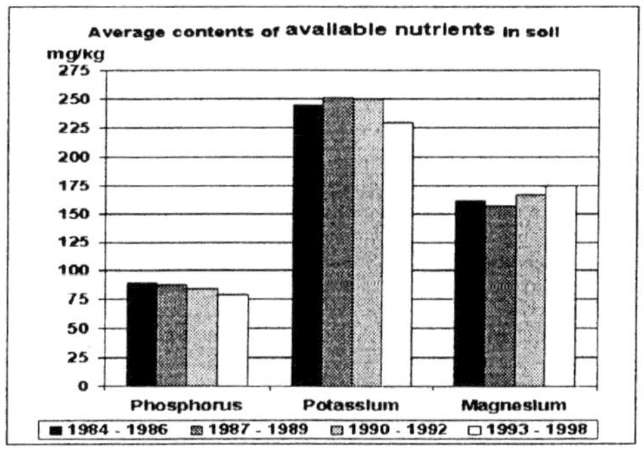

Figure 4. Nutrient contents of nutrient in the soil for different periods

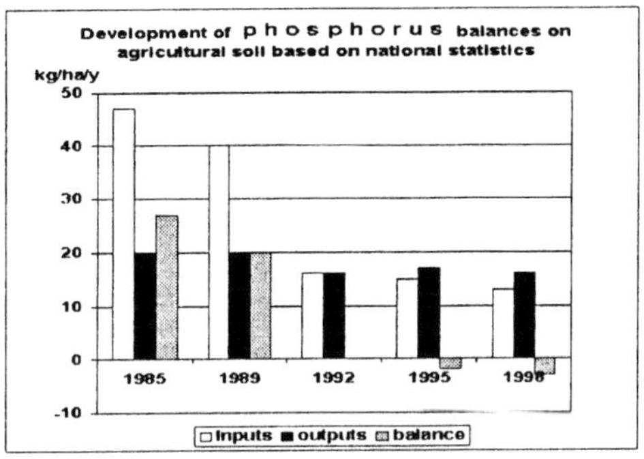

Figure 3. Phosphorus balances (monitoring plots)

Table 1. Field data from monitoring plots: input of nutrients (in kg ha^{-1} $year^{-1}$)

Year	FERTILISERS			MANURE			DEPOSITION			SEEDS+PL. MAT.		
	N	P	K	N	P	K	N	P	K	N	P	K
1992	61,9	15,8	25,8	31,3	6,1	26,3	18,6	1,2	2,2	1,5	0,3	0,6
1993	61,9	18,6	20,8	45,2	7,6	40,7	18,6	1,1	2,2	1,5	0,3	0,6
1994	61,9	15,7	18,4	40,2	6,8	33,1	18,6	1,1	2,2	1,5	0,3	0,6
1995	61,9	16,8	18,0	45,6	7,7	45,7	18,6	1,1	2,2	1,5	0,3	0,6
1996	61,9	13,2	18,6	43,2	7,3	38,8	18,6	1,1	2,1	1,5	0,3	0,6
1997	61,9	18,1	17,8	27,2	4,8	25,0	18,6	1,1	2,3	1,5	0,3	0,6
1998	61,9	15,5	23,2	34,5	6,6	32,8	18,6	1,1	2,3	1,5	0,3	0,6
1999	61,9	10,5	14,1	16,6	2,2	16,3	18,6	1,1	2,3	1,5	0,3	0,6

Table 2. Field data from monitoring plots: output of nutrients (in kg ha^{-1} $year^{-1}$)

Year	MAIN PRODUCT			BY-PRODUCTS			OTHER LOSSES			**BALANCE**		
	N	P	K	N	P	K	N	P	K	N	P	K
1992	-101,0	-20,7	-58,7	-13,2	-2,8	-20,7	-3,0	-0,2	-0,6	-3,8	-0,3	-25,0
1993	-102,8	-18,7	-69,4	-6,6	-1,4	-9,8	-3,0	-0,2	-0,6	14,8	7,3	-15,4
1994	-101,2	-20,3	-60,3	-8,0	-1,6	-12,6	-3,0	-0,2	-0,6	10,0	1,8	-19,3
1995	-85,2	-17,8	-50,8	-8,7	-1,8	-13,9	-3,0	-0,2	-0,6	30,7	6,2	1,3
1996	-92,3	-18,5	-53,4	-9,3	-2,1	-13,5	-3,0	-0,2	-0,6	20,7	1,0	-7,4
1997	-89,5	-17,8	-58,8	-7,3	-1,7	-9,3	-3,0	-0,2	-0,6	9,5	4,6	-23,0
1998	-89,5	-19,0	-52,4	-6,0	-1,3	-8,7	-3,0	-0,2	-0,6	18,1	3,0	-2,9
1999	-81,9	-17,3	-42,3	-5,6	-1,2	-8,6	-3,0	-0,2	-0,6	8,1	-4,5	-18,2

W. J. Horst et al. (Eds.), Plant nutrition – Food security and sustainability of agro-ecosystems. 870–871, 2001 870

Nutrient uptake and crop productivity on Podzolic soils (Alfisols)

G. Y. Elkina

Institute of Biology, Komi Science Center, Ural Division, Russian Academy of Sciences, Kommunisticheskaya, 28, Syktyvkar, 167982, Russia, elkina@ib.komisc.ru

Key words: crop productivity, macro- and micronutrients, nutrient interaction, nutrient uptake, optimization of plant nutrition

Abstract

Various methods to assess crop demand for nutrients were studied. Application of optimal amounts of macro- and micronutrients significantly increased yields of potato, annual grasses, and clover. The effect of nutrient uptake on crop productivity was revealed. Negative influence of exchangeable Al on the crop productivity on podzolic soils (Alfisols) was observed.

Introduction

Podzolic soils of the middle taiga subzone in their natural condition have scarcely favorable properties for crop growth. They are characterized by high acidity, low humus content, and low nutrient content. To improve their fertility, not only acidity and rising of humus content are necessary, but also providing of higher soil nutrient content for the crops is important. Hence, it is necessary to assess crop demand for macro- and micronutrients and to determine relationships between nutrient uptake and crop productivity.

Materials and methods

Field experiments were carried out on a loamy podzolic soil with the following agrochemical properties: humus content 16.3 g kg^{-1}, pH (1.0 *M* KCl) 4.0, sum of exchangeable cations 2.6 cmol kg^{-1} soil, hydrolytic acidity 6.2 cmol kg^{-1} soil, available phosphorus content 103 mg kg^{-1}, available potassium (0.2 *M* HCl) 92 mg kg^{-1}, hydrolyzable nitrogen 27 mg kg^{-1}. Description of methods are given in "Methods for Agrochemical Research" (1975). We have established the following treatments: a balance method, the standard rates currently recommended by the State Agrochemical Service, and a method of optimum nutrition (with and without micronutrients). For the last treatment we applied nutrient concentrations proposed by Rinkis *et al.* (1989) and modified according to the needs of podzolic soils

Results and discussion

There are various methods to assess crop demand for nutrients, however, optimization of crop nutrition with macro- (N, P, K, Ca, Mg, S) and micronutrients (B, Mo, Co, Zn, Cu) allows the highest productivity. Application of inorganic fertilisers and dolomitic meal in accordance to the optimization method increase potato yield from 19.6 t ha^{-1} (control plot) to 56.3 t ha^{-1} (average for 4 years); the fresh matter yield of annual grasses (oat and pea) increased from 7.4 to 29.4 t ha^{-1} (average for 3 years), and that of clover hay from 1.4 to 4.0 t ha^{-1}. B, Mo, Co, Zn, and Cu content in podzolic soils do not reach the level considered as optimal. Therefore, the application of micronutrients has led to an additional increase for potato yield of 9.5 - 13.6 t ha^{-1}, that of annual grasses of 2.9 - 4.9 t ha^{-1}, and that of clover hay of 0.6-0.8 t ha^{-1}. Significant yield increases were also obtained using the balance method to calculate NPK rates: 26.2 t ha^{-1} for potatoes and 16.1 t ha^{-1} for green material were achieved. Application of the standard rates currently recommended by the State Agrochemical Service was less effective.

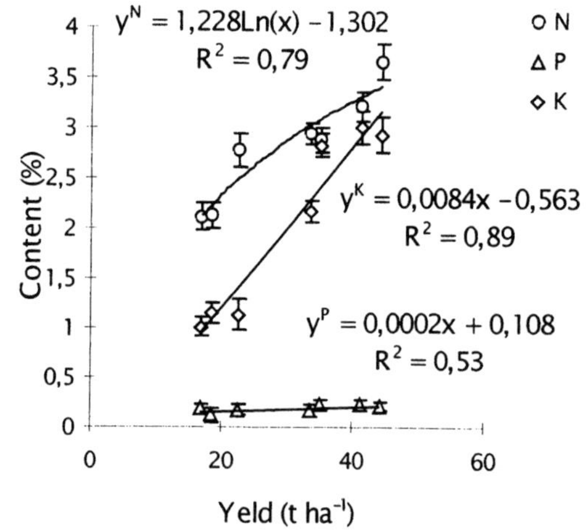

Figure 1. Relation between potato yield and N, P and K content in above ground plant material

Synthesis of organic compounds which forms the crop productivity is related to nutrient uptake. Figures 1 and 2 demonstrate the relation of crop yield and N, P and K content in above ground plant material for potato and oat. The potato productivity correlates with N, P and K uptake (r = 0.73 - 0.77) but is less affected by the content of these nutrients in tubers (r = 0.42 - 0.77). Yields of the vegetative mass of oat and pea also show correlation with N, P, and K content (r = 0.62 - 0.94). Nitrogen content in clover is more stable and less affected by the application of lime and inorganic fertilisers, though the latter increases significantly the crop productivity and the nutrient leaching. The yield – nutrient content curves are normally linear and rarely logarithmic.

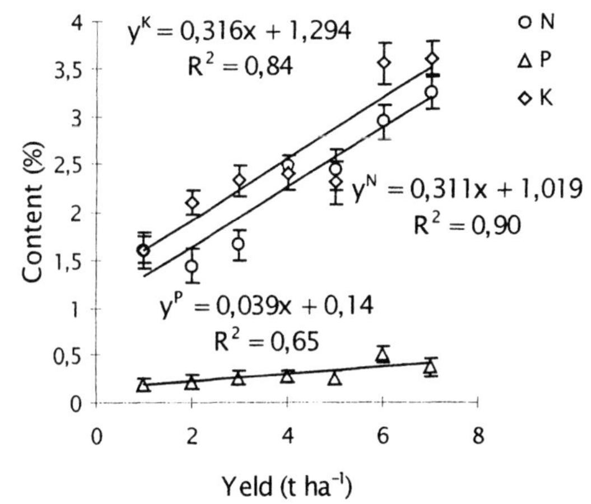

Figure 2. Relation between oat yield and N, P and K content in above ground plant material

Productivity of potatoes and of oat (vegetative mass) benefited from increased uptake of Ca and Mg due to liming and fertiliser application: correlation coefficients were 0.55 - 0.97. Application of dolomitic meal at a rate of 11 t ha^{-1} changed K:Ca:Mg ratio in the potato tops from 1:0.5:0.5 (control plot) to 1:2.1:1.2. The changes were due to a decrease in K uptake. Liming combined with inorganic fertilisers, including K-containing ones, changed the ratio in a smaller degree. Application of ground dolomite did not lead to significant changes in Ca and Mg uptake by the vegetative mass of clover and peas, but caused significant yield increase.

One of the reasons for low productivity of podzolic soils is a high content of exchangeable Al. In our field experiment exchangeable Al (1.0 *M* KCl) content was 1.1 cmol kg^{-1}. Negative correlations were revealed between Al content in plants and yield. Yields of potatoes show negative correlation with Al content in tubers and tops (Fig. 3). Liming and fertilising decreased Al uptake, as well as that of Mn and Fe, and eliminated negative effects of these elements on plants.

In principal nutrients demonstrated a positive correlation with respect to their uptake by potato tops. Synergetic effects were revealed for the following pairs: N-Ca, N-Mg, P-K, P-Ca ($r = 0.72 - 0.73$). Interaction between Al and Mn and macronutrients (except for K) was of antagonistic nature.

For the vegetative mass of oat interdependency with nutrient content ($r = 0.65 - 0.99$) was also shown. With regard to Al and Mn uptake, the strongest antagonism was observed in the pairs Al-K and Al-N, as well as Mn-Ca and Mn-K.

N, P and K contents in pea were positively correlated ($r = 0.73 - 0.95$), but negatively correlated with Al and Mn content. Ca favored N, P and Mg uptake by pea. Clover revealed antagonism between Mg and K. Furthermore antagonism was found between Al, Mn and Fe on the one hand and NPK on the other. Ca, Mg and P impeded Zn and Cu uptake by potato tops. Zn and Cu uptake by oat correlated positively with N and K uptake.

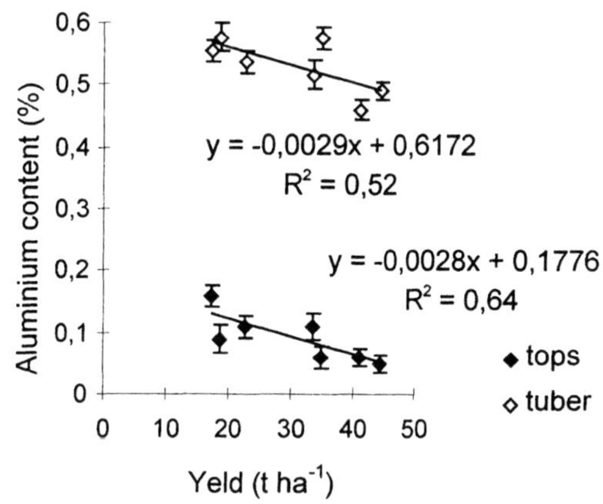

Figure 3. Dependency of potato productivity on Al content in tops and tubers

The study showed different crop strategies with regard to nutrient uptake. Some of the crops demonstrated increases in both the content of a limiting nutrient and productivity. The others showed an increase in productivity following application of lime and fertilisers which was not accompanied by an increase in nutrient contents exceeding that at control plots. Most of the macronutrients favored each other's uptake.

References

Rinkis G Ya, Ramane HK, Paegle GV and Kunitsckaya TA 1989 (In Russian.) Zinatne publ. Riga. 195 p.

Methods for agrochemical research. (In Russian) 1975 Ed. Sokolov AV. Kolos publ. Moscow. 656 p.

W. J. Horst et al. (Eds.), Plant nutrition – Food security and sustainability of agro-ecosystems. 872–873, 2001

Effects of reduced tillage intensity on soil Nmin dynamics following Oilseed rape cultivation

T. Lickfett

Institute of Agricultural Chemistry, Faculty of Agriculture, University of Goettingen, Von-Siebold-Str. 6, D-37075 Goettingen, Germany. tlickfe@gwdg.de

Key words: nitrate leaching, oilseed rape, soil nitrate dynamics, tillage intensity

Abstract

Three tillage systems (mouldboard plough, rototillage, zero tillage and direct seeding) were investigated to find out, if the phenomenon of raising soil N_{min} in autumn/winter after the cultivation of oilseed rape is affected by a different tillage intensity. The two-year field experiments on two loamy soils in Germany showed, that zero tillage and direct seeding of the post-crop winter wheat can contribute to reduce nitrogen losses by delaying mineralisation processes towards next spring, but the effect was uncertain.

Introduction

The mineral N content of the soil (N_{min}) after the cultivation of oilseed is often higher than after cereals and still raises in autumn. This may be caused by a surplus of the N balance (N fertilisation minus N export by the seeds from the field) as well as by an enhanced net nitrogen mineralisation of soil incorporated residues (roots, leaves, stems, pod walls, seeds). As the soil N_{min} content usually exceeds the autumnally N uptake of winter wheat as the traditional subsequent crop, a severe risk of nitrogen losses occurs at the beginning of the leaching period. Therefore the cultivation of oilseed rape is already restricted by regulations in some water protection areas with a high leaching risk.

It has been shown that a balance between N fertilisation and N export by the seeds is accompanied with a drastic reduction of the seed yield, without soil N_{min} in autumn being effectively minimized (Lickfett, 2000).

These findings mainly derived from sites in Germany with a traditional high tillage intensity, which aims at the incorporation of crop residues into the soil and the control of volunteer rape by single or multiple stubble breaking, followed by use of mouldboard plough and rotary harrow for seedbed preparation. Though the acreage of tillage systems with continuously reduced intensity has increased, N_{min} data are rare. Therefore it's the question, whether the nitrate problem after rapeseed cultivation occurs likewise in such tillage systems.

Materials and methods

A monitoring of the $CaCl_2$-extractable soil N_{min} content was performed on two experimental fields near to Goettingen (Tab. 1), Lower Saxony, Germany, where effects of tillage systems differing in intensity and depth (mouldboard plough vs. rototillage vs. zero-tillage) on the productivity of an oilseed rape- winter wheat- winter barley crop rotation are studied since 1994/95. Weather conditions are shown in Fig. 1.

Table 1. Soil properties (0-30cm)

site	soil type	soil texture	pH	C_{org} / N ratio
	Gleyic Luvisol	loam	6.9	14
Weendelsbreite (51.56°N, 9.95°E)	Luvisol	loam	6.8	14

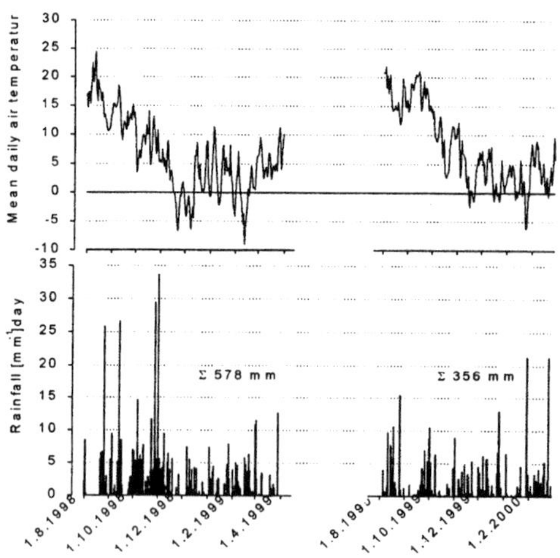

Figure 1. Weather conditions

Agronomical management (N fertilisation, plant protection) was in accordance to the recommendations of the official extension service. In each year, N_{min} investigations run from harvest of oilseed rape in july/august until february/march in the subsequent crop winter wheat, which was sown in early october.

Results and discussion

As expected soil N_{min} content in the ploughed plot (= control) significantly raised within the first eight weeks after harvest of oilseed rape, especially on "Torland"site in

1998, up to 120 kg ha⁻1 (Fig. 2). Due to heavy rainfall (Fig. 1) the overall N_{min} content decreased to about 40 kg ha^{-1}, accompanied by a relative increase in the deeper soil layers until march of subsequent year. Though the N_{min} dynamics in 1999/2000 was less characteristic, downward movement of nitrate occurred anyway. In all cases, N_{min} depletion by N uptake of winter wheat was less than 20 kg ha^{-1} until december (data not shown).

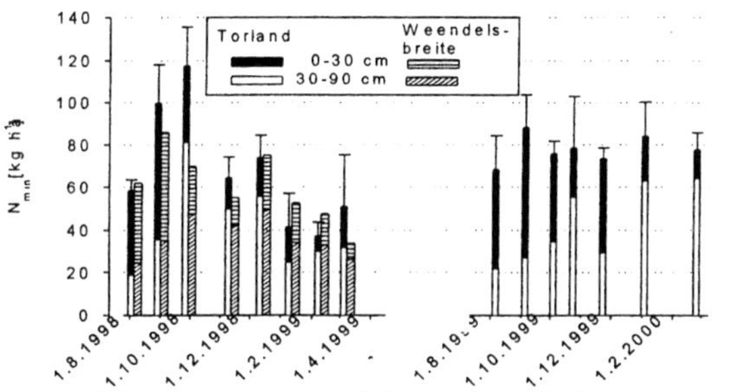

Figure 2. Time course of soil N_{min} in the A_p horizon (soil layer 0-30cm) and in the 30-90cm layer of the „plough"-variant of two sites in autumn/winter after harvest of oilseed rape (error bars represent the std. deviation of the entire 0-90 cm profile).

As compared to the control plot, a lowered N_{min} content in 0-30 as well as 0-90cm soil depth during critical weeks in autumn 1998 was observed only in the zero tillage plots (Fig. 3) with chemical weed control and direct seeding of winter wheat. Raising N_{min} values in february 1999 mark the end of a (desired) delay in mineralisation, giving the chance for a better synchronization of mineralisation processes in soil and N requirement of winter wheat. The shallow mulching of the crop residues in the "rototillage"-plots seems to stimulate mineralisation even more than in the plough-variant. With the aim of low soil nitrate content in periods with low N uptake by the crop, the rototillage treatment seems to be less favourable. Nevertheless, in 1999/2000 even the zero tillage plot showed a similar high osccillation like the rototillage plot, especially in the top soil layer.

Conclusions and outlook

Since the effectiveness of the prevention of N losses in autmun/Winter after harvest of oilseed rape cultivation by a permanently lowered tillage intensity may vary from year to year, more attention should be focused even on uncommon solutions, e.g. an alternative layout of the oilseed rape crop rotation by maximizing the non-tillage-period after harvest of rapeseed by introducing a summer crop (e.g. oats, maize, spring wheat), before winter wheat is grown in the subsequent year. Both on numerous experimental plots as well as on farmers fields this layout has often been successful in keeping soil N_{min} low in autumn. The N uptake by volunteer rape as a catch crop and the reduced N net mineralisation in times without any tillage are important components of that groundwater protecting strategy.

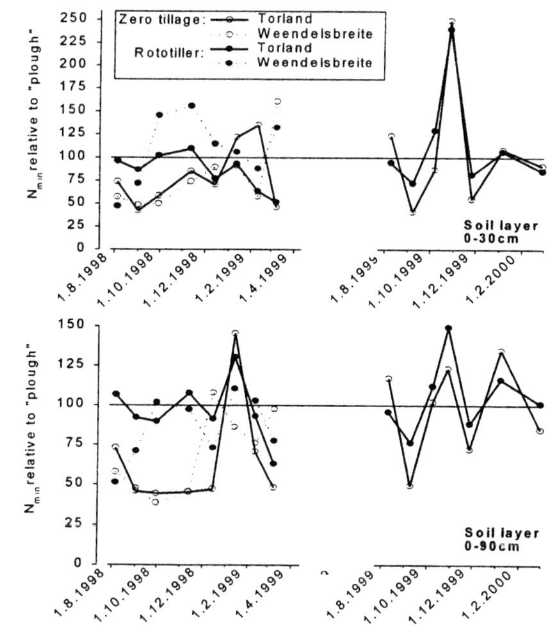

Figure 3. N_{min} dynamics of different soil layers after harvest of oilseed rape cultivation in different tillage systems relative to mouldboard plough.

Acknowledgement

Thanks to Prof. Dr. H.-H. Hoppe, Institute of Plant Pathology and Plant Protection of Goettingen University, for providing his experimental fields and to G. Dormann, R. Hilmer and S. Schubert for technical assistance.

References

Lickfett T 2000 *In* Stickstoffeffizienz landwirtschaftlicher Kulturpflanzen. Eds. C Möllers and H Wurl. pp. 9-30. Erich Schmidt Verlag, Berlin.

W. J. Horst et al. (Eds.), Plant nutrition – Food security and sustainability of agro-ecosystems. 874~875, 2001

Effect of ammonium levels and pH on nitrite accumulation in soils

Q. R. Shen[1)], W. Ran[2)] and Z. H. Cao[2)]

[1)] *College of Natural Resources and Environmental Sciences, Nanjing Agricultural University, Weigang, Nanjing, 210095, China, qrshen@smtp.njau.edu.cn*

[2)] *Institute of Soil Science, Chinese Academy of Science, Beijingdonglu, Nanjing, 210008, China, zhcao@issas.ac.cn*

Key words: ammonium, nitrification, nitrite, pH, soil

Abstract

Oxidation of ammonium-N in soils proceeds through nitrite-N, a form that is usually present in very dilute concentrations in the environment. Accumulation of nitrite-N in ecosystems has been reported. To evaluate the risk of nitrite-N accumulation, we incubated three soils with ammonium and measured nitrite-N accumulation over time. The soils used were a Yellow-brown earth (pH 5.67), a Fluvo-aquic loam (pH 7.89) and a Fluvo-aquic sand (pH 8.20) all from eastern China. No measureable nitrite-N accumulated at any time or at any concentration of added ammonium in the Yellow-brown earth soil. Huge accumulations of nitrite-N occurred in the two calcareous soils. The concentrations of nitrite-N in both alkaline soils were related to applied ammonium-N levels. The effect of pH on nitrite accumulation was demonstrated by using soil slurries of Fluvo-aquic sand under continuous aeration and buffers of different pH. Nitrite-N concentration increased as the soil slurry pH increased. Ammonia oxidizers from the original soil (pH 8.2) adapted to the new medium of low pH (pH 5.35). We suggest that nitrite accumulation in field soils and its subsequent environmental impact should be investigated.

Introduction

Nitrite-N (NO2--N), as an intermediate in nitrification and denitrification, is very toxic to plants and animals. The NO2--N anion rarely accumulates in terrestrial and aquatic ecosystems. However, reports that the NO2--N has accumulated in ecosystems have been noted. For example, in 1994, annual averages of NO2--N concentrations of 13 transects across the Hai-Luan river, in northern China, were significantly high up to 2365 µg of NO2--N L-1 (Editorial Board for China Environment Year Book, 1995). This greatly exceeded the national environment standard (< 60 µg of NO2--N L-1) for protecting surface waters (Xia, 1997). Similar data have been reported by the EBCEY from 1990 to 1998.

Soils are now being heavily fertilised in most agricultural regions of China. For example, the average rate of N fertiliser application was more than 300 kg of N ha-1 season-1 in Shanghai in 1996. It is not uncommon to fertilise vegetable crops at rates of more than 1000 kg of N ha-1 year-1 in suburban areas of Beijing (Cao *et al.*, 1999). The great concern is that NO2--N pollution of waters due to excessive N-fertiliser applications may rise, and subsequently destroy the biodiversity of the aquatic ecosystem. Investigations on occurrence of NO2--N and its environmental impact are still needed.

The objective of this study was to clarify the mechanisms of NO_2^--N accumulation during nitrification. We present evidence that NO_2^--N contents in soils were affected by NH_4^+-N input levels and soil pH.

Materials and methods

Soils

The soils used (Table 1) were surface samples (0 - 20 cm), from continuously cultivated upland fields. Samples of Yellow-brown earths were collected from eastern China subjected to a subtropical humid climate. Samples of Fluvo-aquic loams, and Fluvo-aquic sands were also from eastern China, but subjected to a mid-latitudinal humid continental climate. All samples were air-dried and crushed to pass through a 2-mm screen. The basic properties of the three soils are summarized in Table 1.

Incubation procedure

Subsamples (10 g) of oven-dried soils were amended with 2 mL of 0, 2, 4, 10, 20, 40, or 100 mM $(NH_4)_2SO_4$ solution. The treatments were respectively marked as T0, T1, T2, T3, T4, T5, and T6. Each treatment was replicated three times. After addition of NH_4^+-N, the samples were immediately incubated at 25°C in a humidified constant-temperature chamber. The N transformations were monitored at 12, 24, 48, 96, 192, 384 hrs incubation.

To determine the effect of pH on accumulation of NO_2^--N, 15 g of fresh Fluvo-aquic sand were weighed into a 250 mL Erlenmeyer flask then 50 mL of 3.22 mM $(NH_4)_2SO_4$ and 100 mL of 120 mM phosphate buffer or 200 mM borate buffer were added. The initial pH values were adjusted to 5, 6, 7, 8, 9 or 10 with 2 M HCl or 2 M NaOH. Each treatment was replicated three times. Each flask was then continuously aerated at 25°C under 0.5 MPa. The air was passed through 5 M NaOH before entry into the flasks to remove CO_2. Every day 5 mL of slurry was removed to measure the inorganic N concentration.

Analytical procedures

The contents of NH_4^+-N, NO_2^--N, and nitrate-N (NO_3^--N) in filtrates were measured colorimetrically by the indophenol, Griess-Ilosvay, and cadmium reduction Griess-Ilosvay methods (Mulvaney, 1996), respectively.

Results

Dynamic changes of NO_2^--N in three soils amended with different rates of NH_4^+-N

Within all incubation times and at all levels of NH_4^+-N, we did not detect NO_2^--N accumulation in any sample of Yellow-brown earth. Huge accumulations of NO_2^--N were observed in the Fluvo-aquic loam and Fluvo-aquic sand (Fig. 1).

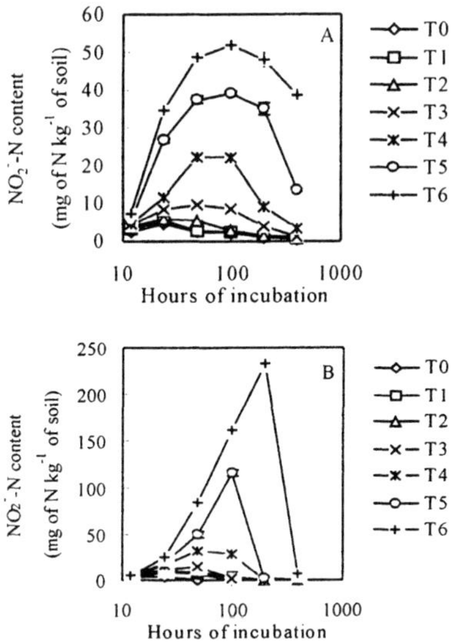

Figure 1. Changes of NO_2^--N in Fluvo-aquic loam (A) and in Fluvo-aquic sand (B) under different rates of NH_4^+-N application.

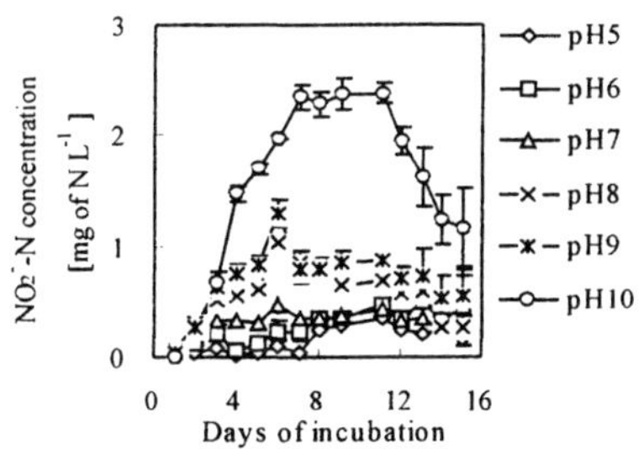

Figure 2. Effect of pH on NO_2^--N accumulation in Fluvo-aquic sand slurries with phosphate (pH 5~8) or borate (pH 9~10) buffers at 25°C under continuous aeration.

Effect of pH on NO2--N accumulation in soil slurries with aeration

During incubation, all soil slurry pH in all treatments fluctuated around their initial values no more than ±0.2 unit (except in treatment pH 5, which was increased to about 5.35 and in pH 10, which was decreased to 9.55). During the first 2 days, the greatest NO_2^--N concentration was found in pH 8 treatment, but after the third day the highest NO_2^- accumulation occurred in the pH 10 treatment. Although no NH_4^+-N was detectable in soil slurries of pH > 8 after 6 days of incubation, the concentrations of NO_2^--N in those slurries were still high (Fig. 2).

Discussion

Our experiments showed that the amount of NO_2^--N that accumulates in soils depends on the NH_4^+-N level and the soil type. Accumulation of NO_2^--N in the two calcarious soils always increased with the increase of NH_4^+-N level. No accumulation of NO_2^--N occurred in the acid Yellow-brown earth. This is in agreement with results of earlier studies (Alexander, 1977). Unfortunately, much more N fertiliser is applied to soils today than any earlier times, but no attention is being given to the potential loss of biodiversity in the aquatic ecosystem caused by toxic nitrite from agricultural sources.

Also, our experiments revealed notable appearances of NO_2^--N in soil slurries buffered with pH 5 and pH 6 solution after 7 days of incubation (Fig. 2). This implies that ammonia oxidizers from the original soil (pH 8.2) can adapt to the acid medium. In contrast, the nitrite oxidizers need time to revive, or need NO_2^--N to induce their activity, which explains accumulations of nitrite that vary with pH in some media.

References

Alexander M 1977 Introduction to Soil microbiology 2nd edition, 251-272. John Wiley & Sons, New York.

Cao ZH and Cai ZC 1999 *In* Chemistry for protection of the environment. Eds. Cao ZH and Pawlowski L. pp. 20-27. Nanjing University Press, Nanjing.

Editorial Board for China Environment Year Book (EBCEYB) 1995 China Environment Year Book. 323-326. China Environment Year Book Press, Beijing.

Mulvaney RL 1996 *In* Methods of Soil Analysis. Part 3, Chemical Methods. Eds. Bigham JM *et al.*. pp. 1123-1184. SSSA, ASA, Madison, Wisconsin, USA.

Xia YH 1997 *In* Practical Handbook for Monitoring Environmental Pollutants. Eds. ST Liu and YH Xia. pp. 310-311. Atomic Energy Press, Beijing.

Table 1. Selected chemical and physical properties of soils

Soils	pH	Organic matter g kg^{-1}	Total N g kg^{-1}	NO_3^--N g kg^{-1}	NH_4^+-N g kg^{-1}	Sand g kg^{-1}	Silt g kg^{-1}	Clay g kg^{-1}
Yellow-brown earth	5.67	14.24	1.05	3.08	7.44	254	400	346
Fluvo-aquic loam	7.89	22.57	1.43	63.16	4.71	294	418	288
Fluvo-aquic sand	8.20	9.50	0.70	16.93	2.54	507	385	108

W. J. Horst et al. (Eds.), Plant nutrition – Food security and sustainability of agro-ecosystems. 876~877, 2001

Spatial variability of available phosphorus and site specific P fertilizer recommendations in a wheat field

M. R. Karaman, S. Ersahin and A. Durak
Gaziosmanpaşa Univ., Agricultural Faculty, Department of Soil Sci., Tokat-Turkey, rkaraman@mail.gop.edu.tr

Key words: phosphorus, site specific, spatial variability, wheat

Abstract

Knowledge of the spatial variability in soil available phosphorus content is essential in determining the local P needs of crops. This study was conducted to determine spatial variability in available P content of topsoils and subsoils on a 8.5 ha alluvial wheat field. Phosphorus values were more variable in topsoils than in subsoils since the topsoils have been frequently mixed with plowing activities and applied P fertiliser. Contour maps for available P contents further indicated that available P deposited differed in the subsoils.

Introduction

Soil physical and chemical characteristics greatly affect the P nutrition of plants. The utilization of P fertilisers by crops generally is low (10-30 %), and fixation of phosphorus significantly affect P accumulation in the soil (Tisdale *et al.*, 1993). The accumulation of P fertiliser in the soil leads to the environmental pollution and economical losses. Also, a great variability in available P values occurs in agricultural areas. Therefore, in large broadcast applications of P, soil analyses are essential for effective P fertiliser recommendation and increased P use efficiency. In the soil fertility analysis, considering local needs of crops for P fertiliser depending on spatial variability of soil available P will make valuable contributions to the beneficial use of field soil (DeCourt *et al.*, 1996).

Materials and methods

This study was conducted to determine spatial variability in available phosphorus content in topsoils (0-30 cm) and subsoils (30-60 cm) on a 8.5 ha alluvial wheat field near Tokat Airport, 25 km north of downtown Tokat, Turkey. Topsoil and subsoil samples were taken on a 25 x 25 m grid. Available P contents (P_2O_5) in the soil samples were determined by the method from Olsen *et al.* (1954). Other analyses were made by routine methods, and statistical values were calculated with the computer program (StatMost, 1995).

Results and discussion

Maximum, minimum, coefficient of variance (C.V.), kurtosis and skewness values revealed that a great spatial variability occurred in available P content, C.E.C. and O.M. content for both topsoil and subsoil (Table 1). Coefficients of variation were 32, 29 and 22% for available phosphorus, C.E.C. and O.M. content for topsoil, respectively. Of the subsoil characteristics, O.M. had the greatest C.V. (63%). All other selected soil properties generally exhibited a medium variability in both topsoil and subsoil. Wilding and Dress (1983) also showed that the soil properties such as soluble P and exchangable cations were commonly more variable than many other soil properties. A three-dimensional map of available phosphorus contents in the topsoil is presented in Fig. 1.

Table 1. Selected physical and chemical properties of the topsoils and subsoils.

Soil properties	Topsoil						Subsoil					
	Max.	Min.	Mean	C.V. %	Kurtosis	Skew.	Max.	Min.	Mean	C.V. %	Kurtosis	Skew.
Sand, %	46.97	21.94	33.46	18.23	2.25	0.134	37.79	7.93	20.94	31.91	2.54	0.340
Silt, %	52.23	30.83	42.00	12.33	2.14	-0.055	56.25	35.24	43.85	9.07	2.99	0.048
Clay, %	31.01	17.79	24.60	10.44	3.65	0.012	49.64	22.05	35.50	14.57	2.98	0.377
$CaCO_3$, %	20.95	7.28	14.42	17.15	2.75	-0.113	22.60	7.32	15.45	19.55	2.99	-0.385
pH	9.50	7.03	8.14	3.21	10.38	0.811	9.90	7.26	8.30	5.23	6.36	1.494
Org. matter, %	5.61	1.91	3.85	21.70	2.25	-0.141	4.01	0.01	1.43	62.93	2.76	0.464
C.E.C, $mol_c\ kg^{-1}$	1.02	0.20	0.41	28.62	8.32	1.577	0.68	0.01	0.37	39.99	3.18	-0.574
Av.P_2O_5, kg ha^{-1}	125.9	27.5	66.0	31.79	2.55	0.250	121.3	25.2	69.7	35.69	2.03	0.080

This study was partly sponsored by The Scientific and Technical Research Council of Turkey, Project No: TARP-1871.

According to boundary values from General Directorate of Rural Services (Ulgen and Yurtsever, 1984), the topsoil was 1% very low, 47% low, 41% medium, and 11% high in available P; the subsoil was 3% very low, 36% low, 38 % medium, and 23 % high (Fig. 2).

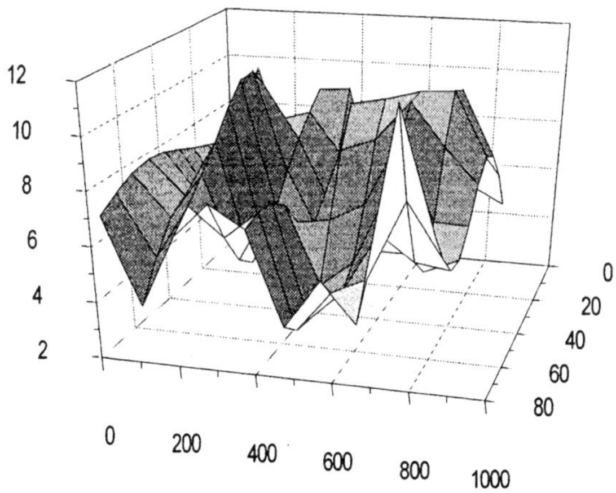

Figure 1. Three-dimensional map for available phosphorus

Three-dimensional maps and contour plots for available P indicated that considerable variability in available P content occurred in the experimental area. Available P contents of topsoil were greater than those of subsoil due to continuous addition of available P to topsoil. Mallarino (1996) and Borges and Mallarino (1997) also emphasized that the problems for soil sampling caused by spatial variability are aggraveted for relatively immobile nutrients such as P and K because of the limited mix of soil and the high residuality of these nutrients. A significant variability occurred in P fertiliser rate recommendations, and depended on the available P content each grid-sampling point. Thus, the classical P fertiliser recommendation program would lead to over fertilisation in 38848 m^2 and underfertilisation in 35206 m^2 of 85000 m^2 total experimental field. Site-specific P fertiliser recommendation which would consider the spatial variability would supply more realistic P fertiliser rates for optimal crop production, and it would benefit both the environment and production economics. But, as a result of economical limitations, one of several interpolation methods can also be used for gridding by estimating values for non-measured locations between sampled points (Wollenhaupt *et al.*, 1994). The results further indicated that available P deposited existed in the subsoil. Thus, deep plowing could be performed every four to five years to mix P from this source with the topsoil and enhance P uptake by the crop.

References

Borges R and Mallarino AP 1997 Soil Sci. Soc. Am. J. 61, 846.

DeCourt H, Darius PL and Baerdemaeker JD 1996 Computers and Electronics in Agriculture 14, 179-196.

Mallarino AP 1996 Soil Sci. Soc. Am. J. 60, 1473.

Olsen SR, Cole CV, Watanable FS and Dean LA 1954 Estimation of available phosphorus in soils by extraction with sodium bicarbonate. U.S. Dept. of Agriculture, Cir. No 939, Washington DC.

StatMost 1995 Dataxiom Software Inc. User's Guide: StatMost. 5th Ed. Dat. Soft. Inc., LA, CA. USA.

Tisdale SL, Nelson WL, Beaton JD and Havlin JL 1993 Soil Fertility and Fertilisers. Macmillan Publishing Company, New York, USA.

Ülgen N and Yurtsever N 1984 General Directorate of Soil-Water Service. Publishing No 47, Fertiliser Guidebook No:8, Ankara, Turkey.

Wilding LP and Dress LR 1983 *In* Application of geostatistics to spatial studies of soil. Eds. BB Trangmar, RS Yost and G Uehara. Advances in Agr. Vol:38.

Wollenhaupt NC, Wolkowski RP and Clayton MK 1994 J. Prod. Agric. 7, 441-448.

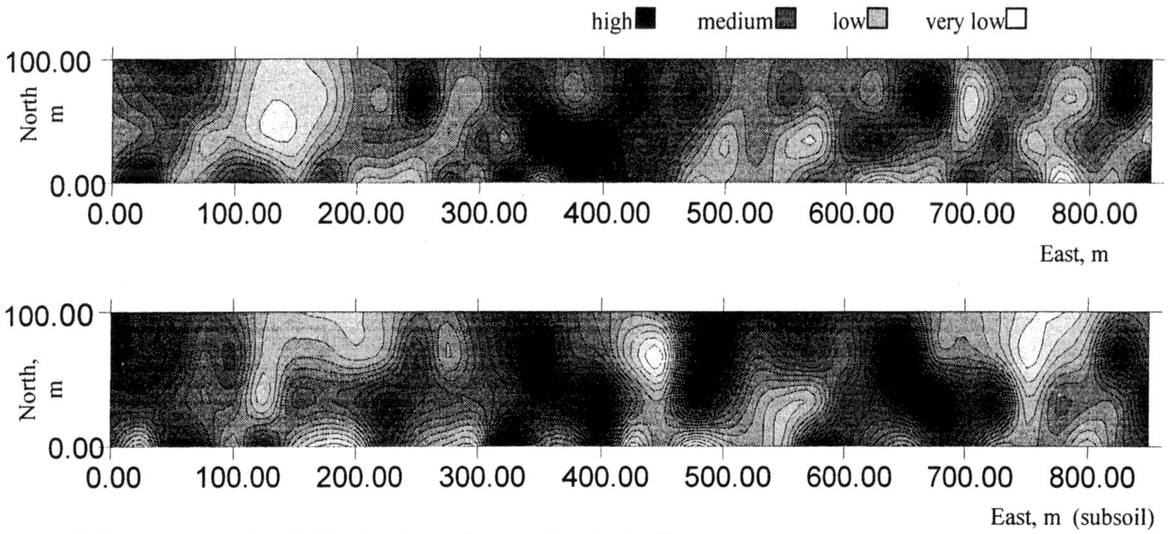

Figure 2. Contour maps of available phosphorus for topsoil and subsoil

W. J. Horst et al. (Eds.), Plant nutrition – Food security and sustainability of agro-ecosystems. 878~879, 2001

Remote sensing of soil properties to support site specific farming

T. Selige and U. Schmidhalter
Chair of Plant Nutrition, Department of Plant Sciences, Technical University of Munich, Am Hochanger 2, Freising, D 85350, Germany, selige@weihenstephan.de

Key words: available water capacity, multispectral remote sensing, nitrogen fertilising, wheat, yield potential

Abstract

Multispectral remote sensing was used to improve the inventory of soil heterogeneity at the field level. Ground measurements of morphological crop parameters were collected from representative soil sites. Spectral information at visible, infrared and thermal wavebands was recorded from the airborne scanner Daedalus AADS 1268. The spectral information was transformed into soil information using bioindicative transfer functions, based on cause and effect relationships of the soil-plant system. This procedure enables the spatial detection and the quantification of soil properties. The available water capacity of the rooting zone accounted predominantly for the heterogeneity of crop stands and yield formation of winter wheat. At specific development stages the crop stand conditions were identified as sensitive bioindicator of the plant available soil water capacity. The thermal emission and its relationship to the transpiration of crops was recognised as most suitable to detect quantitatively soil properties via crop stand conditions of winter wheat. The soil map derived from remote sensing and bioindicative transfer functions explained the pattern of nitrogen uptake and yield formation.

Introduction

Precision farming is the challenging technique for the future agriculture. Site specific decision support systems require detailed information about the natural resources as they vary across a field. The variability and the spatial pattern of crop growth within fields depend on the variability and the spatial pattern of soil properties. Hence, detailed information about the spatial pattern and the functional properties of the soils is required to enable site specific management.

In this study remotely sensed imagery was used to enable the digital mapping of soil resources with high resolution and timely flexibility. Remote sensing technologies are not able to view into the soil profile but records the spectral characteristics of soil-crop surfaces. The conception of this research is capitalizing on these cause-effect relationships of the soil-plant-sensor system. It takes advantage of crop stand characteristics as indicators of soil profile characteristics. The mapping process derives soil quality via the spectral response from crop stand conditions.

Materials and methods

The investigated area was located in Sachsen-Anhalt, Germany. The area is characterised as a sligthly undulated plain at 70 m altitude with an intensive arable farming. With 450 mm mean annual precipitation and 9°C mean annual temperature, the region has a negative water balance during the vegetation period. Chernozem is the predominant soil type.

The remote sensing device used for this investigation was the airborne line-scanner Daedalus AADS 1268 ATM. The scanner records wavebands of visible, near, middle and thermal infrared wavelength at 11 spectral channels. The investigation area was recorded with 5 m geometric resolution. Five fields cultivated with different winter wheat varieties were selected for this paper. Soil properties, plant development and crop stand condition were measured on the ground at representative soil sites. The available water storage capacity (AWC) and the rootability (rz) were derived from soil texture and texture changes within the soil profile. Grain yield and biomass of each soil site were determined. Nitrogen content of straw and grain yield was measured using elementar analyser technique.

Results

Detecting soil quality by remote sensing has to consider that soil is predominantly covered by vegetation and its surface is most of the time not visible. Furthermore, essential soil properties can not be discerned from soil surface appearance. However, it is well known that crop growth depends on soil attributes. Hence it should be feasible to use the crop stand condition as a bioindicator of soil productivity and regulatory functions.

Biomass is one of the important parameters to differentiate crop stand conditions. For regions with negative water balance during the growing season, the site specific availability of soil water is the main limiting soil resource. Apart from groundwater and lateral water, plants can only utilize the available water storage capacity of the root zone (AWCrz).

This soil parameter has been recognized to be the central regulatory attribute of the soil productivity function. AWCrz correlates highly with crop biomass. Additionally the effects of groundwater uptake and lateral water are included in the crop stand condition. This allows an improved derivation of soil productivity via crop stand conditions compared to classical soil core samplings. In this study, the variability of the plant available water storage capacity of the root zone (AWCrz) accounted for 93 % (R^2 = 0,93***) of the variability of winter wheat biomass at the development stage EC 77 (milk ripeness) when the leaves started to become yellow.

The biomass at this development stage indicates also the pattern of the later harvested grain yield. The crop stand at development stage EC 77 accounted for 96 % (R^2 = 0,96***) for the grain yield variability of winter wheat. These results suggest that the crop stand condition can also be used to forecast grain yield and its pattern across fields.

The correlation between plant available water capacity and grain yield underlines the important role of water availability as the main soil resource to be considered in site specific management.

Different wavebands and spectral indices were tested for their useability to differentiate crop stands. The statistical analysis identified the thermal emission as the spectral range with the best linear correlation to the biomass fresh weight. The surface temperature of crop stands differentiates synchronously with the biomass (R^2 = 0.90***) caused by the surface cooling effect of transpiration. Other spectral wavebands and indices were found to be much less sensitive.

Based on the transfer function, soil resource maps of winter wheat fields werecalculated. Classes with width of 50 mm AWC could significantly be delineated. The results were found to be non-sensitive to the five different winter wheat varieties that were used in this study.

The map of AWC was also used to calculate the efficiency of site specific nitrogen fertilising. The relevant data sets of three representative soils are listed in Table 1. The data emphasize the role of water availability for improving site specific N-efficiency.

Low available water capacity limited the yield formation significantly and large amounts of N-surplus remained in the soil. The AWCrz was found to determine N-uptake by grain yield (r^2 = 0.90***). The biomass at EC77 already reflects the spatial heterogeneity of grain yield N-uptake quantitatively (r^2 = 0.94***).

Nitrate leaching to groundwater may occur under that kind of conditions. The efficiency of uniform high N-supply will often be low on fields that are characterised by high variability of available water capacity. The efficiency of N-supply can be reduced dramatically as the data in Table 1 show. This study proved that under conditions of negative water balance during the growth season the map of available water capacity formed the best process-oriented basis to apply site specific management of N-fertilisation. It is also expected from this input map to serve as a basis for decision systems of seed density and crop stand management in general.

Table 1. Data set of three representative soil sites for nitrogen balance and calculation of N supply efficiency

		Sand	Sandy Loam	Loamy Silt
AWCrz	*(mm)*	40	140	220
Grain yield	*(t/ha)*	1.1	4.9	8.3
Grain N-uptake	*(kg/ha)*	35	125	192
N-supply	*(kg/ha)*	180	180	180
N balance	*(kg/ha)*	+145	+ 55	- 12
N-efficiency	*(%)*	19	69	106

Conclusions

The pattern of crop stand conditions at specific "indicative" development stages reflected soilborne heterogeneities within fields. The most indicative stages of plant development to derive soilborne crop stand conditions and site properties were identified for winter wheat. Differences in crop stand conditions were correlated with their spectral response.

The plant available soil water capacity and the rooting depth were identified as the most important soil attribute for yield formation in the investigated region. Hence the relationship of the soil-plant-sensor system soilborne site properties to be derived from multispectral and remotely sensed imagery via scanning crop stand conditions.

The soil map derived from remote sensing and bioindicative transfer functions explained the pattern of grain yield formation and enabled grain yield to be forecast. It provided a process-oriented basis to apply site specific N-supply with high efficiency. This enabled cost reductions in fertilising and an ecological sound management.

The combination of a few ground check plots and multispectral remote sensing provided a flexible and operational technique to detect quantitatively the spatial pattern of soil properties at the field level for extended regions. The presented procedure will support site specific farming with basic informations about the soilborne yield potential. This enables a site specific crop and risk management with targeted resource application of seed, fertiliser and growth regulaton.

Acknowledgement

This research was supported by the Bundesministerium für Bildung und Forschung (BMBF), Bonn, Germany (project number 03397401).

W. J. Horst et al. (Eds.), Plant nutrition – Food security and sustainability of agro-ecosystems. 880~881, 2001

N loss by denitrification from potted ornamental plants

H. Agner and M. K. Schenk
Institute of Plant Nutrition, Department of Horticulture, University of Hannover, Herrenhaeuser Str. 2, D 30419 Hannover, Germany, heidi.agner@mbox.pflern.uni-hannover.de

Key words: denitrification, N loss, N_2O, ornamental plants, water content

Abstract

Denitrification was measured in cultures of Euphorbia pulcherrima 'Sonora Red' and Pelargonium zonale 'Grand Prix' using the acetylene inhibition method in a flow-through-system. P. zonale was grown in peat, E. pulcherrima in peat and compost-peat substrate. Mean N losses by denitrification amounted to 0.05 mg (peat) and 0.11 mg (compost-peat) pot-1 irrigation event-1 for E. pulcherrima and to 0.75 mg pot-1 irrigation event-1 for P. zonale. Addition of 1 mg l-1 glucose to the fertigation solution increased N losses for E. pulcherrima (peat) to 0.77 mg pot-1 irrigation event-1, which is comparable to the mean N loss of P. zonale. This indicates that C is the most limiting factor in the combination E. pulcherrima and peat substrate and that P. zonale probably promotes denitrification by supplying C to the substrate. N loss in P. zonale was highly correlated to water content of the substrate (r2=0.77), while correlation for E. pulcherrima was low in both, peat and compost-peat. Supplying glucose to the peat substrate of E. pulcherrima improved the correlation from r2=0.07 to r2=0.67. Thus, denitrification proved to be determined by water status of the substrate unless it was inhibited by lack of C. An increase in NO3-N supply from 150 to 450 mg l-1 fertigation solution did not affect denitrification.

Introduction

Denitrification requires O_2 deficiency, availability of NO_3 and easily decomposable C, as well as a pH and temperatures favorable to microbial activity. Greenhouse production of ornamental pot plants offer such favorable conditions for denitrification: frequent irrigation and fertilisation induce periods of anaerobiosis and provide nitrate to the substrate. Furthermore pH values of 6 to 6.5 and warm temperature favor microbial activity and organic material derived from plant roots as well as organic components of the substrat may serve as carbon sources to bacteria. However, up to now there is no information available on N losses due to denitrification in potted ornamental plants. Thus, the aim of our research was to estimate N losses due to denitrification in potted ornamental plants.

Materials and methods

The flow-through-chamber-method (Daum & Schenk 1996) was used with and without addition of acetylene (C_2H_2) to the chamber atmosphere to detect total N losses in form of (N_2O + N_2) and N losses in form of N_2O, respectively. Pots were irrigated by "common ebb and flood" fertigation (150 mg NO_3-N l^{-1}), and N_2O emissions were measured in short intervals by gas chromatography (Mosier & Mack 1980). Measurements were taken during a complete culture period of *Euphorbia pulcherrima* 'Sonora Red' and *Pelargonium zonale* 'Grand Prix'. As substrate plain peat was used for both crops and peat blended with 30% compost was additionally used for *E. pulcherrima*.

Results

During the entire measurement period, total N losses were highest in *P. zonale* reaching a mean value of 0.75 mg N per pot and irrigation event. Emissions from pots of *E. pulcherrima* grown in peat were about 1/10 compared to those of *P. zonale.* Mean N losses in compost-peat were twice as high as in peat (Tab.1). However, variation was very high in all treatments.

Table 1. Mean N losses (mg N_2O + N_2 pot^{-1} irrigation $event^{-1}$) by denitrification from *E. pulcherrima* and *P. zonale*

species:	*E. pulcherrima*		*P. zonale*
substrat:	peat	peat-compost	peat
mean	0.050	0.110	0.750
std	0.025	0.063	0.687

Variation between measurements could partly be explained by temperature changes in the greenhouse due to solar radiation. Still, correlation between mean temperature and mean N loss was not significant for *E. pulcherrima* in peat and weak for *E. pulcherrima* in compost-peat (r^2=0.32) as well as for *P. zonale* in peat (r^2=0.22).

A high correlation (r^2=0.77) was found between total N loss and substrate water content (vol. %) at the end of the experiment for *P. zonale* (Fig.1). Correlation between N loss and substrate water content was weak for *E. pulcherrima* in peat (r^2=0.07) and in compost-peat (r^2=0.18) (data not shown). Addition of 1 mg glucose l^{-1} to the fertigation solution in peat substrate increased the correlation to r^2=0.67 (Fig.2). Furthermore, amending glucose increased mean N losses from 0.05 mg to 0.77 mg per pot.

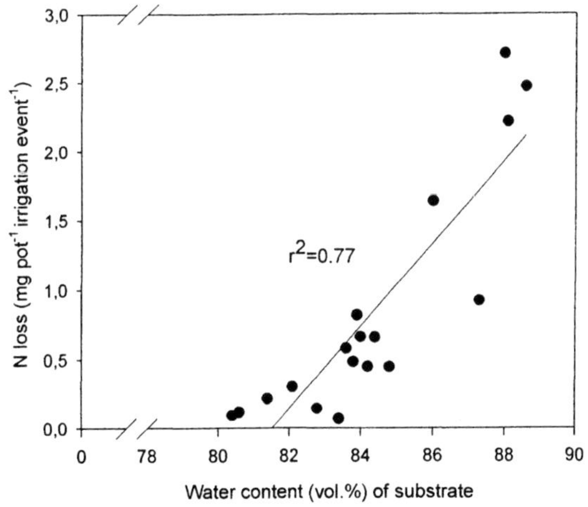

Figure 1. Influence of water content (vol. %) of the substrate on N loss (N_2O + N_2) from pots with *P. zonale* grown in peat

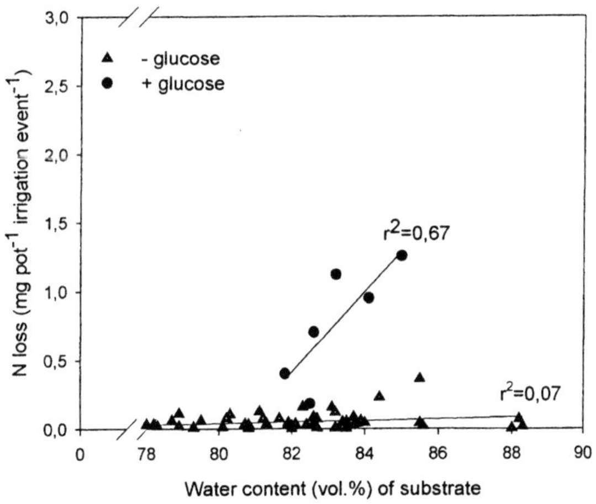

Figure 2. Influence of water content (vol. %) of the substrate on N loss (N_2O + N_2) from pots with *E. pulcherrima* (peat) +/- 1 g glucose l^{-1} fertigation solution

Increasing NO_3 supply to *E. pulcherrima* (peat) from 150 mg to 450 mg per l fertigation solution did not affect emissions of $N_2O + N_2$ (data not shown).

Discussion

High variation of N losses by denitrificaton has often been observed and discussed in literature. Generally, variation is attributed to uneven distribution of C, NO_3 and O_2 depletion (Christensen *et al.* 1990, Parkin 1987). The presented growing system provides homogeneity to an extent which can hardly be expected under field conditions. Nonetheless, variation is high between measurements as well as within replications Tab.1).

Variation in temperature could not justify changes in N loss during the growing period. This was proven by weak or not existing correlation between temperature and N loss for all treatments tested in our study.

Total N loss of *P. zonale* grown in peat showed a high correlation to the water content of the substrate at the end of the experiments (Fig.1). Water content amoung pots varied slightly inspite of uniform irrigation. Its effect on O_2 availability could be responsible for variation in denitrification. When glucose was supplied, N loss of *E. pulcherrima* (peat) also showed a high correlation to the water content of the substrate (r^2=0.67). This indicates that water content of the substrate determines the intensity of denitrification in the presented system as long as C was not limiting.

Without addition of glucose, N loss of *E. pulcherrima* (peat) was lowest of all treatments and only showed a poor correlation to water content (r^2=0.07). The same species planted in compost-peat resulted in about twofold higher N-losses (Tab.1). The correlation between N loss and water content increases to r^2=0.18. This effect can be attributed to the pool of easily decomposable C contained in compost.

Inspite of its share of compost, N loss by *E. pulcherrima* grown in compost-peat was still about seven fold lower than N loss by *P. zonale* grown in plain peat (Tab. 1). This illustrates the effect of plant species on denitrification. Similar observations on the effect of plant species on denitrification have been reported by Mahmood *et al.* (1997), Qian *et al.* (1997) and Haider *et al.* (1987). The favorable effect of plants on denitrification is usually attributed to the supply of C to the rhizosphere. When glucose was used to increase available C in pots grown with *E. pulcherrima* (peat), N loss reached a level comparable to that of *P. zonale*. Thus, *P. zonale* seems to promote denitrification through supply of organic C compounds while denitrification in *E. pulcherrima* is severely limited by C availability.

The lack of effect of increased N supply on denitrification from *E. pulcherrima* (peat) could be due to deficiency of C, as discussed above. This conclusion is supported by the observation that even in cultures of *P. zonale* the amount of N lost by denitrification is no more than 3.5% of the amount taken up ba the plants per pot. Thus, it seems unlikely that the availability of NO_3 was ever limiting to denitrification in system under study.

References

Christensen S, Simkins S and Tiedje JM 1990 Soil Sci. Soc. Am. J. 54, 1608-1613.

Daum D and Schenk MK 1996 Plant and Soil 183, 69-78.

Haider K, Mosier A and Heinemeyer O 1987 Soil Sci. Soc. Am. J. 51, 97-102.

Mahmood T, Ali R, Malik KA and Shamsi SRA 1997 Biol. Fertil. Soils 24, 323-328.

Mosier AR and Mack L 1980 Soil Sci. Soc. Am. J. 44, 1121-1123.

Parkin TB 1987 Soil Sci. Soc. Am. J. 51, 1194-1199.

Qian JH, Doran JW and Walters DT 1997 Soil Biol. Biochem. 29, 1451-146.

W. J. Horst et al., Plant nutrition – Food security and sustainability of agro-ecosystems. 882~883, 2001

Effects of various nitrogen fertilisers on nitrous oxide emissions from arable soils

B. Leick[1)] and C. Engels[2)]

[1)] *Institut für Pflanzenernährung (330), Universität Hohenheim, D-70593 Stuttgart, Germany, leick@uni-hohenheim.de*

[2)] *Abteilung Agrarökologie, Fachgruppe Geowissenschaften der Universität Bayreuth, D-95440 Bayreuth, Germany, christof.engels@uni-bayreuth.de*

Key words: ammonium, arable soils, nitrate, nitrification inhibitor, nitrogen fertiliser, nitrous oxide emission

Abstract

The main cause for increased nitrous oxide (N_2O) emission from agricultural soils is nitrogen fertilisation. In the present investigation the effect of the form of N fertilisers on N_2O emission was studied. Emission of N_2O was measured in a field experiment using the closed chamber method. 60 kg N ha^{-1} were applied on arable soil to oats (*Avena sativa*) in the form of calcium nitrate, ammonium sulphate, ammonium sulphate in combination with the nitrification inhibitor dicyandiamide and calcium ammonium nitrate. N_2O emissions increased within two days after fertilisation. After 6 to 8 days, emissions decreased again to the levels before fertilisation. The rapid decrease of N_2O emissions one week after fertilisation was associated with a decrease in the content of mineral N in the soil. Emissions were generally below 1% of the applied N, presumably because of the fast removal of mineral N from the soil by plant uptake and/or immobilisation. N_2O emission was highest after fertilisation with calcium ammonium nitrate and was nearly prevented by fertilisation with ammonium sulphate in combination with nitrification inhibitor. The present results suggest that the amount of N fertiliser-induced N_2O emission is strongly dependent on the form of N fertiliser and may be effectively reduced by fertilisation of ammonium-N in combination with a nitrification inhibitor.

Introduction

Agricultural soils are a major source of nitrous oxide (N_2O). Nitrous oxide contributes to the greenhouse effect (Bouwman, 1990), and furthermore is involved in the depletion of ozone in the stratosphere (Crutzen, 1981). N_2O is produced through nitrification and denitrification (Granli and Bøckman, 1994). Both processes can occur simultaneously and the rate of the N_2O production depends not only on the availability of ammonium (NH_4^+) and nitrate (NO_3^-) in the soil but also on other factors such as soil moisture content and organic matter content (Shepherd *et al.*, 1991).

The main source for increased N_2O emission from agricultural soils is nitrogen (N) fertilisation. Nitrous oxide emission induced by N fertilisers may vary between 0.01 and 10% of the fertiliser, depending on soil factors like availability of oxygen and dissolved carbon, and agricultural measures like synchronisation of fertiliser application and plant demand and selection of the form of N fertilisation.

The objective of this study was to investigate the immediate effect of the N form on the N_2O emission from arable soils.

Materials and methods

The experiment was carried out on arable soil with oats (*Avena sativa*). Nitrogen was added in the form of calcium nitrate (CN), ammonium sulphate (AS), ammonium sulphate in combination with the nitrification inhibitor dicyandiamide (AS+DCD) and calcium ammonium nitrate (CAN). The total N surface application was 60 kg ha^{-1}.

Nitrous oxide emissions from soil were determined using the closed chamber method. The chamber covered 1 m^2 of the soil and had a volume of 258 l. Opening and closing of the chamber as well as the gas sampling was controlled by PC. Concentration of N_2O in the samples was measured using a gas chromatograph (Hewlett Packard 5890) fitted with a ^{63}Ni electron capture detector (ECD) and a column packed with Porapak Q. Operating temperature of the detector and the column was 280°C and 40°C, respectively.

At the time of N_2O measurement soil samples were taken at a depth of 0-10 cm for determination of water content and mineral N content (NH_4^+-N and NO_3^--N).

Results

N_2O emissions were very low before fertilisation (Fig. 1A). Generally, the emission increased within two days after fertilisation, and decreased again to the level before fertilisation after 6 to 8 days. N_2O emissions increased coincidently with an increase in soil water content after precipitation (4 days after fertilisation).

N_2O emissions were slightly higher from fertilisers containing nitrate (CN, CAN) than from AS. Higher N_2O emissions from NO_3^- containing fertilisers were associated with increased NO_3^- content in soil, whereas NH_4^+ content was increased particularly after fertilisation with AS (Fig 1B; 1C). N_2O emission was nearly completely prevented by application of NH_4^+ in combination with DCD.

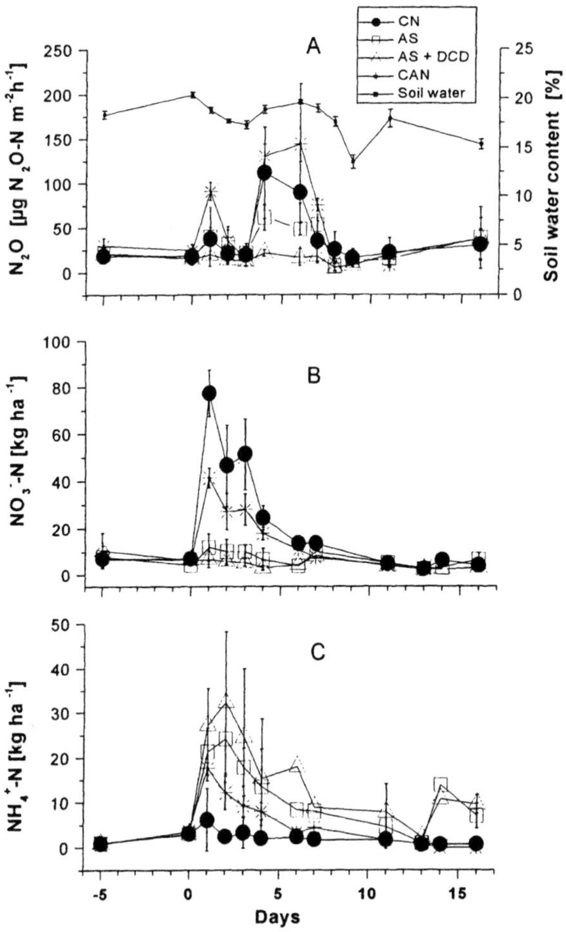

Figure 1: Time course of N_2O fluxes (A), NO_3^- content in the soil (B) and NH_4^+ content in the soil (C) after fertilisation of arable soil (n=4).

Discussion

The rate of N_2O emission followed changes in soil mineral nitrogen contents. Increasing NO_3^- availability in the soil had an effect on the N_2O emission, indicating that denitrification was an important source of nitrous oxide production.

Changes in the N_2O flux corresponded with changes in soil moisture. Kessavalou *et al.* (1998) found that short pulses of N_2O occurring after wetting could be a significant component of annual greenhouse gas fluxes. The soil water content influenced the availability and diffusion of oxygen. Denitrification rate increased with increasing soil water content and anaerobic conditions in the soil, respectively (Weier *et al.*, 1993). In contrast, Peterson (1999) did not find a relationship between increasing soil water content and accumulated N_2O losses.

Amberger (1989) showed that dicyanidiamid is an effective nitrification inhibitor when it is applied in combination with organic fertilisers. Furthermore, the use of nitrification inhibitors in combination with ammonium fertiliser can maintain the NH_4^+-N for a prolonged time in the soil and reduce the potential of N_2O production through denitrification (Terry *et al.*, 1978). In the present experiment we found a nearly complete prevention of the N_2O losses after application of inorganic NH_4^+ fertiliser in combination with DCD. The use of nitrification inhibitors could be an effective possibility to reduce N_2O losses in agricultural systems.

The N_2O losses were between 0.1 % and 0.3 % of the applied nitrogen, i.e. substantially lower than those predicted by the emission factor of 1.25 % used in the IPCC (1997) guidelines. This was probably due to the close synchronisation of fertiliser application and plant demand in the present experiment, which is indicated by the fast decrease in the mineral N content in soil within one week after fertilisation.

The present results show that the amount of nitrogen fertiliser-induced N_2O emission is strongly dependent on the form of the N fertiliser. Emissions may be less than 1% of the applied N when excessive accumulation of mineral N in soil is avoided. Furthermore, the use of ammonium fertiliser in combination with a nitrification inhibitor (DCD) is an effective measure to reduce the N_2O emission immediately after N application.

References

Amberger A 1989 Commun. Soil Sci. Plant Anal. 20, 1933-1955.

Bouwman AF 1990 Ed *In* AF Bouwman. Soil and the greenhouse effect. pp 61-127. Wiley, Chichester.

Crutzen PJ 1981 Ed *In* CC Delwiche: Denitrification, nitrification and atmospheric nitrous oxide. pp 17-45. Wiley, New York.

Granli T and Bøckman OC 1994 Norw. J. Agr.Sci. 12, 1-128.

Intergovernmental Panel on Climate Change 1997 IPCC guidelines for national greenhouse gas inventories. Workbook. Paris, France.

Kessavalou A, Doran JW, Mosier AR and Drijber A 1998 J. Environ. Qual. 27, 1105-1116.

Peterson S 1999 J. Environ. Qual. 28, 1610-1618.

Shepherd MF, Barzetti S and Hastie DR 1991 Atmos. Environ. 25A, 1961-1969.

Terry RE, Nelson DW, Sommers LE and Meyer GJ 1981 Soil Sci. Soc. Am. J. 45, 506-513.

Weier KL, Doran JW, Power JF and Walters DT 1993 Soil Sci. Soc. Am. J. 57, 66-72.

W. J. Horst et al., Plant nutrition – Food security and sustainability of agro-ecosystems. 884~885, 2001

Field emissions of NH_3 and NO_x following urea application to wheat

A. Weber, R. Gutser and U. Schmidhalter

Chair of Plant Nutrition, Department of Plant Science, Technical University of Munich, Am Hochanger 2, Freising-Weihenstephan, D 85350, Germany, aweber@weihenstephan.de

Key words: ammonia, emissions, fertiliser, nitrogen oxides, urea, wheat

Abstract

Within the scope of international efforts for air pollution abatement, emissions of N-containing trace gases from sustainable and effective cultivation become of more importance. The objective of this investigation was to determine actual NH_3 and NO_x emissions from surface applied urea fertiliser in field experiments with modified instrumentation based on the dynamic chamber technique. In 1999 and 2000 urea or CAN (calcium-ammonium-nitrate) were supplied to winter wheat in several applications (N application rate: 80 kg N ha^{-1}). Nitrogen emissions were measured continuously for 10 days after each fertilisation. Under weather conditions favouring losses (high temperature and lack of precipitiation) urea application led to NH_3 emissions up to 4.4 kg NH_3-N ha^{-1} or 5.5 % of the applied nitrogen. The NH_3-emissons of CAN reached a maximum of 1.2 kg NH_3-N ha^{-1} or 1.5 % of the applied nitrogen. NO_x emissions were low (up to 140 g NO_x-N ha^{-1} for both fertilisers). The results from this study suggest that the risk of NH_3 losses is significantly smaller under the semi-humid weather conditions of Southern Germany than has previously been assumed.

Introduction

The potential for gaseous N losses from urea fertiliser following surface application has been recognised for many years. With the increasing use of urea as a source of N, all over the world, it is gaining more attention. NH_3- and NO_x-emissions from applied urea, and thus its fertiliser use efficiency, depend on soil characteristics and environmental factors (Rachhpal-Singh, 1987; Vermoesen *et al.*, 1996). The release of NH_3 by hydrolysis of urea is affected by pH, NH_4^+ concentration in the soil solution, soil moisture, soil temperature and soil texture (Fleisher *et al.*, 1987). The most important process related to the production of NO_x in soils can be nitrification as well as the result of denitrification and chemodenitrification (Williams *et al.*, 1992). Although an increase in NO_x emissions with addition of N has been recorded (Galbally, 1989), the effect of fertiliser type on NO_x losses has not been thoroughly investigated (Maggiotto *et al.*, 2000).

The objective of this study was to determine actual NH_3 and NO_x emissions from surface-applied urea and CAN fertiliser in field experiments under the environmental conditions of Southern Germany.

Materials and methods

Site description

The flux measurements were performed on an experimental area of the Chair of Plant Nutrition, TU-Munich in Southern Bavaria, Germany (800 mm annual precipitation, 7.4 °C air temperature), from 25 March to 23 June 1999 and from 22 March to 14 June 2000. The soil is characterised as a silty loam from loess with pH($CaCl_2$) 6.3.

Mineral N-fertiliser were given to winter wheat as granular urea and CAN (calcium-ammonium-nitrate) in several applications to the soil surface (application rate 80 kg N ha^{-1} to unfertilised plots, respectively).

NH_3 and NO_x volatilisation measurements

NH_3 and NO_x emissions were measured with a modified instrumentation based on the dynamic chamber technique. Ambient air was continuously drawn through chambers, (covering ¼ m² soil surface) with a constant flow rate. Sample air was collected and led to the NH_3/NO_x-analyser (CLD700AL + NH_3-Converter, Fa. EcoPhysics, Gürnten, Switzerland) by magnetic valves. NH_3 and NO_x fluxes were calculated from the difference of NH_3 and NO_x concentration in ambient and sample air and from the airflow through the chamber.

To minimise the influence of the artificial climate inside the chambers (higher temperature, drying out of the soil surface) on the volatilisation processes, the chambers were moved to an undisturbed part of the plot up to four times a day dependent on the weather conditions.

Six chambers were used simultaneously to get comparative measurements with different N fertilisers. For ten days after each fertilisation NH_3 and NO_x fluxes were determined.

Additional data of temperature and precipitation were collected from a meteorological station close to the experimental area.

Results and discussion

In both years, significant differences were found between the NH_3 emissions in the single periods (Table 1). Urea led to higher NH_3 emissions (up to 4.4 kg NH_3-N ha^{-1} or 5.5 % of the applied nitrogen) than CAN (up to 1.2 kg N ha^{-1} or 1.2 %). Maximum losses for both fertiliser types were found, when they were applied on 26 April 2000. For the other application dates the losses were less or even NH_3 deposition was found. These results are in contrast to Schulz *et al.* (1999) establishing NH_3 emissions of 23 % (urea) or 11 % (CAN) of the applied N by a micrometeorological technique in fields of Eastern Germany.

Higher NH_3 losses (up to 50 %) than determined in this study were also found in laboratory experiments and under tropical climate in rice culture (Vermoesen *et al.*, 1996; De Datta *et al.*, 1989; Freney *et al.*, 1993).

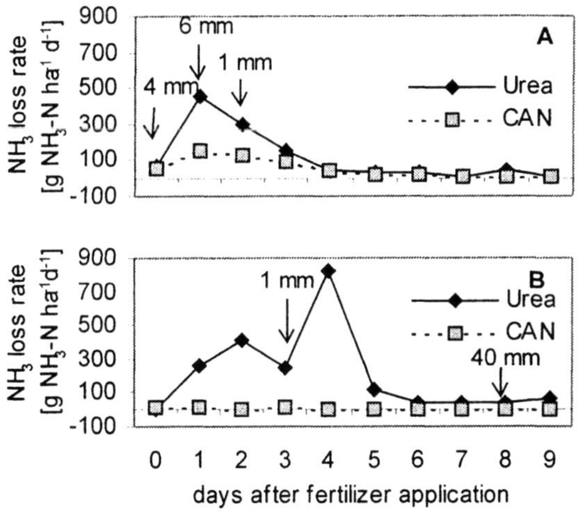

Figure 1. Course of NH_3 emissions as well as occurrence and quantity of rainfall following the fertiliser application in 5 June 2000 (**A**) and 10 May 1999 (**B**).

NH_3-volatilisation was directly influenced by the temperature during the measurement period. In 1999 as well as in 2000 NH_3 losses increased significantly at temperatures higher than 15 °C. Higher temperature lead to enhanced NH_3 losses, because the equilibrium between NH_4^+ and NH_3 in the soil solution shifts to NH_3 and also the velocity of NH_3 volatilisation (Urease activity) increases (Vermoesen *et al.*, 1996).

Thus, this temperature effect is covered by the occurrence, distribution and quantity of rainfall during the first five days of the measurement period. Comparing the course of the NH_3 losses in the measurement periods after 5 June 2000 and 10 May 1999, shows a distinct decrease when rain occurs immediately after the fertilisation (Figure 1A). Late precipitation had no more influence on the total NH_3 losses of the period (Figure 1B). The relation between soil moisture and NH_3 emissions was discussed by Vermoesen *et al.* (1996).

NO_x volatilisation were low (up to 153 g NO_x-N ha^{-1}) for both fertilisers compared to NH_3. A correlation between the environmental conditions and NO_x losses was not found. Urea, as an ammonium-based fertiliser, tended to a smaller release of NO_x than CAN (nitrate-based) in accordance with investigations of Beauchamp (1997) and Maggiotto *et al.* (2000).

The results from this study suggest that the risk of NH_3 and NO_x losses is significantly smaller under the prevailing environmental conditions of Southern Germany than has previously been assumed.

References

Beauchamp EG 1997 Can. J. Soil Sci. 77, 1-57.

De Datta SK, Trevitt ACF, Freney JR, Obcemea WN, Real JG and Simpson JR 1989 Soil Sci. Soc. Am. J. 53, 1275-1281.

Fleisher Z, Kenig A, Ravina I and Hagin J 1987 Plant Soil 103, 205-212.

Freney JR, Keerthisinghe DG, Chaiwanakupt P and Phongpan S 1993 Plant Soil 155/156, 371-373.

Galbally IE 1989 *In* Exchange of Trace Gases between Terrestrial Ecosystems and the Atmosphere. Eds. MO Andreae and DS Schimel. pp. 23-38. John Wiley & Sons, New York.

Maggiotto SR, Webb JA, Wagner-Riddle C and Thurtell GW 2000 J. Environ. Qual. 29, 621-630.

Rachhpal-Singh 1987 Fertiliser Research 13, 277-285.

Schulz D and Dämmgen U 1999 *In* Berichte über Landwirtschaft 77(1). Ed. Bundesminist. pp. 84-89. Ernähr. Landw. Forst. Landwirtschaftsverlag, Münster-Hiltrup.

Vermoesen A, Demeyer P, Hofman G and Van Cleemput O 1996 Trans. 9th Nitr. Workshop, TU Braunschweig.

Williams EJ, Hutchinson GL and Fehsenfeld FC 1992 Global Biogeochem. Cycles 6, 351-358.

Table 1. Cumulative NH_3 and NO_x losses, precipitation and temperature within 10 days following surface application of urea and CAN to wheat at several dates.

Date of Fertilisation	NH_3 Emissions Urea		CAN		NO_x Emissions Urea		CAN		Temperature	Precipitation
	%[1]	kg N ha^{-1}	%	kg N ha^{-1}	%	G N ha^{-1}	%	g N ha^{-1}	°C	mm
1999										
25 March	0.6	0.5	<0.1	<0.1	0.1	51	0,1	115	7.7	11.1
06 April	-0.1	>-0.1	<0.1	>-0.1	0.2	153	0,2	130	7.7	26.1
26 April	0,5	0.4	0.1	0.1	0.0	36	0,1	91	13.3	0.5
10 May	2.6	2.0	0.1	0.1	0.1	98	0,1	100	19.1	49.5
17 June	-0.1	>-0.1	-0.1	>-0.1	0.1	46	0,2	139	12.4	22.1
2000										
22 March	0.6	0.5	<0.1	<0.1	>-0.1	-15	0.1	43	5.7	41.7
10 April	0.2	0.1	0.2	0.2	0	-4	>-0.1	-11	8.0	32.3
26 April	5.5	4.4	1.5	1.2	0.1	67	0.1	92	15.4	4.8
05 June	1.4	1.2	0.7	0.5	0.2	144	0.1	52	18.5	9.5

[1] percent of applied N (80 kg N ha^{-1})

W. J. Horst et al., Plant nutrition – Food security and sustainability of agro-ecosystems. 886~887, 2001

Application of the life cycle assessment methodology to investigate the environmental impact of different nitrogen fertiliser application rates in a crop rotation

F. Brentrup, J. Küsters and H. Kuhlmann
Hydro Agri, Centre for Plant Nutrition, Hanninghof 35, D-48249 Dülmen, [1)] frank.brentrup@hydro.com

Key words: crop production, environmental impact, LCA, Life Cycle Assessment, nitrogen fertilisation

Abstract

In this study the environmental impact of a reduced (Nopt-40%) and an optimum (Nopt) N fertilisation in a crop rotation is investigated using the Life Cycle Assessment (LCA) methodology. This method allows for an inventory and evaluation of all environmental impacts associated with the entire crop production system inclusive of all pre-chain processes. The study is based on a long-term field trial with cereals and maize. The study reveals that the Nopt-40% fertilisation is preferable to the Nopt fertilisation concerning climate change, human toxicity, acidification and eutrophication. On the other hand the Nopt application rate shows lower indicator values for abiotic resource depletion, land use, ecotoxicity and formation of photo-oxidants. It can be concluded that none of the compared N fertiliser regimes is clearly preferable from an environmental point of view. To determine a distinct environmentally optimum N application rate, it is first necessary to identify an environmental indicator that summarises all environmental effects.

Introduction

The intensity of arable farming and in particular the use of mineral nitrogen (N) fertilisers are the focus of a continuous debate on the environmental impacts of agriculture. A reduction in mineral N fertiliser application is seen as an option to reduce these environmental impacts.

In this study the environmental impact of a reduced and an optimum N fertilisation in a crop rotation is investigated. To take account of the differences in yield between the different N treatments all environmental impacts are related to one ton of grain yield and not to the area of one ha.

The application of the Life Cycle Assessment (LCA) methodology allows for an inventory and aggregation of all resources used and emissions released associated with the entire crop production system and all pre-chain processes. LCA was originally developed to analyse industrial products and processes, but has also proven to be applicable to agricultural systems (e.g. Audsley *et al.*, 1997; Cederberg, 1998). However, established LCA methods such as the Eco-indicator 95 (Goedkoop, 1995) often do not cover all environmental impacts relevant for agricultural systems (Brentrup *et al.*, 2000). Therefore, in this study a new LCA approach has been developed and applied to the crop production system.

Materials and methods

This LCA study is based on a long-term field trial, which started in 1989 on 2 sites near Dülmen, north-western Germany. On both sites cereals (winter wheat, winter barley) and maize (corn cob mix) were grown. The straw remained on the field. On each site 2 N treatments were established: 1. optimum N rate (Nopt) and 2. 40% below Nopt (Nopt-40%). The calculation of the N rate for Nopt was according to official German recommendations based on the Nmin method (Wehrmann & Scharpf, 1986). Further information on the field trial is given in Table 1.

Table 1. Information on the field trial

	Nopt-40%	Nopt
total N supply (kg N/ha), incl. Soil NO_3-N in spring		
w-wheat	125	210
w-barley	115	190
maize	110	180
yield (t DM/ha)		
average	7.53	9.05
w-wheat (n=5)	6.87	9.09
w-barley (n=3)	6.13	7.60
maize (n=4)	9.40	10.08

Application of P, K, lime, and crop protection agents was according to common farm practice.

All inputs (i.e. resources used) and outputs (i.e. emissions released) related to this crop production system (Figure 1) in the period of 1994-1999 (i.e. in two 3-years crop rotations) have been compiled for both sites and both treatments. The individual inputs and outputs are aggregated into so-called “environmental effects” based on equivalency principles (e.g. kg CO_2-equiv. for Climate Change, Figure 2). This step is called “Life Cycle Impact Assessment” (LCIA, for more details see Brentrup *et al.*, 2000). For each of the two N levels an average indicator value for each environmental effect considering all years and sites was calculated. The total effect values resulting from the LCIA for each of the two N fertilising levels are defined as their “environmental profile”.

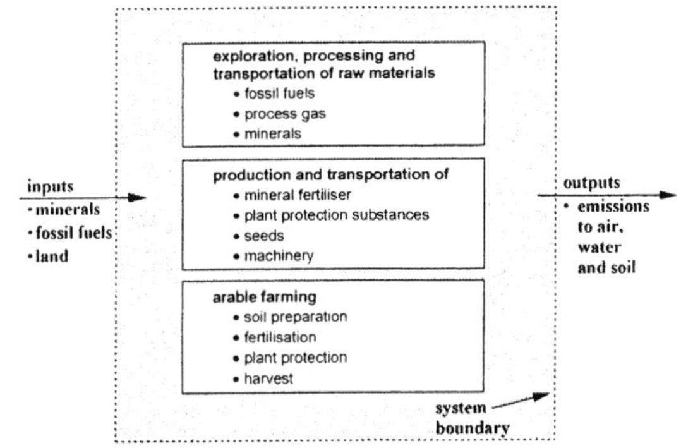

Figure 1. Description of the arable farming system and of the pre-chain processes

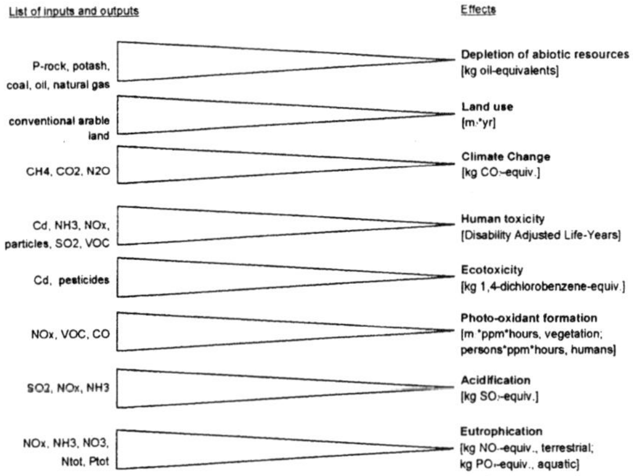

Figure 2. Aggregation of inputs and outputs to environmental effects.

Results and discussion

Table 2 gives the indicator values per ton of grain for Nopt-40% and Nopt for the environmental effects.

Lower indicator values for climate change, human toxicity, acidification and eutrophication are calculated for Nopt-40%. This is due to lower N_2O, NH_3 and NO_3 emissions per ha in the Nopt-40% treatment. By dividing the emission rates per ha by the average yield achieved in each treatment (Tab. 1) the emissions are related to one ton of grain. Related to one ton of grain the Nopt-40% treatment still shows the lowest values for N_2O, NH_3 and NO_3 emissions and consequently for the respective environmental effects.

Related to one ton of grain, the indicator values for abiotic resource depletion and the formation of photo-oxidants are higher for Nopt-40%, although on a per ha basis the main responsible inputs and outputs (fossil fuels, NO_x, VOCs) are lower for Nopt-40%. This is due to lower yields per ha for Nopt-40% compared to Nopt.

Solely yield differences are responsible for higher indicator values for land use and ecotoxicity for Nopt-40%, as the main responsible inputs (land) and outputs (Cd, pesticides) are equal per ha for both treatments. Thus two competing aspects determine the differences in the environmental profiles of Nopt and Nopt-40%: 1. differences in inputs and outputs per ha dependent on the N fertiliser application rate, and 2. differences in the yields achieved per ha.

Table 2. Indicator values for the environmental effects

Effect	Nopt-40%	Nopt
	[unit/ton of grain][§]	
Depletion of abiotic resources	45.25	**43.19**[§§]
Land use	75.46	**61.03**
Climate change	**273.21**	317.12
Human toxicity	$\mathbf{2.4*10^{-4}}$	$2.5*10^{-4}$
Ecotoxicity		
terrestrial	0.21	**0.17**
fresh water	102.51	**94.58**
seawater	54.08	**44.14**
Formation of photo-oxidants		
vegetation	2374.56	**2065.52**
humans	0.15	**0.14**
Acidification	**1.06**	1.31
Eutrophication		
terrestrial	**4.51**	5.85
aquatic	**0.62**	2.15

[§] for units see Figure 2; [§§] bold = lower value

It can be concluded that neither of the N fertiliser regimes is clearly preferable from an environmental point of view.

Assuming the goal of crop production is high yields with minimum environmental impacts, it would be necessary to improve the Nopt treatment in finding ways to reduce the N_2O, NH_3 and NO_3 emissions, because relative to one ton of grain these emissions still increase with increased N fertilisation. The Nopt-40% treatment could be improved by increasing the production efficiency (i.e. yields) without additional N fertilisation.

Due to the competing aspects of increasing inputs and outputs with increasing N rates and decreasing yields with decreasing N rates, no production regime will result in lowest values for all environmental effects.

To determine a distinct environmentally optimum N application rate, first it is necessary to define an environmental indicator that summarises all environmental effects. Therefore it is necessary to develop methods which allow for an integration across all effects to one environmental indicator.

References

Audsley E *et al.* 1997 Harmonisation of Environmental Life Cycle Assessment for Agriculture, Final Report Concerted Action AIR3-CT94-2028, EC DG VI Agriculture, Silsoe.

Brentrup F, Küsters J, Kuhlmann H and Lammel J 2000 Europ. J. Agronomy, (accepted).

Cederberg C 1998. SIK-Rapport Nr 643, Gothenburg

Goedkoop M 1995 The Eco-indicator 95, Final Report, Amersfoort.

Wehrmann J, Scharpf HC 1986 Z. Pflanzenernähr. Bodenk. 149, 428-440.

W. J. Horst et al., Plant nutrition – Food security and sustainability of agro-ecosystems. 888–889, 2001

Growth regulators for cereal and oil crops on the basis of 2,3-dichloroisobutyric acid and chlormequat chloride and residue analyses in the grain of oat

W. Gans and W. Merbach

Institute of Soil Science and Plant Nutrition, Department of Plant Nutrition, University of Halle, Adam-Kuckhoff-Str. 2, D 06108, Halle (Saale), Germany, gans@landw.uni-halle.de

Key words: chlormequat chloride (CCC), determination of residues, 2,3-dichloroisobutyric acid (DCiB), oat, plant growth regulators (PGRs)

Abstract

Field plot experiments were established for studying the influence of growth regulators MLU 208, MLU 2081 [various mixtures of 2,3-dichloroisobutyric acid (DCiB) with chlormequat chloride (CCC)] and CCC on length of shoot axis, susceptibility to lodging and grain yield of oat (*Avena sativa*) and winter oilseed rape (*Brassica napus*). Oat grains were also analysed for residues. With lodging-resistant oat varieties the regulators MLU 208 and MLU 208L caused stem reductions of 10% and yield increases up to 6.5% despite the fact that the untreated controls showed hardly any lodging. CCC had no yield-increasing effect, although a stem reduction of approximately 8% was recorded. Tests with a variety of poor stem stability resulted in stem length reductions of 10%, prevented lodging and increased grain yields by 7-9.5%. Applying MLU 2081 to oilseed rape prior to flowering shortened the stem by 8-11% and contributed to a grain yield increase of about 5-8%. For early (EC 31/32) and late (EC 45) application of MLU 208 and MLU 208L DCiB-residues of about 0.1 mg kg^{-1} were measured. CCC residues increased with increasing dosage to 0.36 mg kg^{-1}, and in the case of later applications up to 0.43 mg kg^{-1}. Less CCC residues were observed in times of high precipitation or intensive irrigation. However, the permitted maximum concentration of 5 mg kg^{-1} CCC residues in the grain was by far not reached.

Introduction

Nitrogen doses to oat stimulate excessive stem elongation which may lead to lodging. The application of plant growth regulators for retarding stem growth is a suitable measure to prevent lodging. Most formulations contained chlormequat (CCC) as the active agent (Linser and Kühn, 1963; Sturm and Jung, 1969). From the toxicological point of view alternative substances such as 2,3-dichloroisobutyric acid (DCiB) are of interest. According to Schilling (1990) an optimum effect can be achieved when DCiB is applied together with CCC. The studies described here had two objectives: Firstly, to identify which preparations are suited for stem reduction and secondly, to determine whether residues of the active agents DCiB and CCC persisted in the harvested material.

Materials and methods

In field plot trials three varieties of oat (*Avena sativa*) were used ('Alfred' and 'Solidor' [relatively resistant to lodging] and 'Samantra' [susceptible to lodging]). The soil is characterized as Gleyic Luvisol. For the oat varieties 'Alfred' and 'Solidor' the field trials were established as a randomized block design with four replicates of 13.5 m^2 plot size, while for 'Samantra' 13.0 m^2 plots were used. Oilseed rape (*Brassica napus*) experiments were made on 18 m^2 plots. Nitrogen fertiliser to oat (100 kg N ha^{-1}) was applied as calcium ammonium nitrate before sowing. For winter oilseed rape a total of 180 kg N ha^{-1} (40/100/40) was applied during the growth period. In Addition all crops received 130 kg P ha^{-1} and 420 kg K ha^{-1}. The preparations MLU 208 (active substances: 25.1 % DCiB and 12.5 % CCC) and MLU 208L (active substances: 27.6 % DCiB and 13.7 % CCC; L indicates a lipid additive for improving the uptake of DCiB) were applied to oates at 2 growth stages (GS; for details s. Tab. 1). In the oilseed rape trials the preparation MLU 2081 (active substances: 25.1 % DCiB, 12.5 % CCC and 5.2 % ethephon) was used at GS 57. In addition to this, in two treatments the fungicide tebuconazole was supplemented in the form of Folicur®. For CCC determination in oat grain a method of Kretschmann (1974) by thin-layer chromatography (TLC) was applied. The determination of DCiB was carried out by a gaschromatographic method with electron capture detector (Gans *et al.*, 2000).

Results and discussion

Table 1 shows the results of the oat experiments with the lodging resistant varieties 'Alfred' and 'Solidor' and the easily lodging variety 'Samantra' (earlier data has also been included). When applied at GS 31/32 in a quantity of 2 to 3 l ha^{-1} MLU 208 and MLU 208L reduced stem length by approx. 10% and simultaneously improved stem stability. Late applications of MLU 208L at GS 45 decreased stem length slightly more, but grain yields increase. Compared to the control was only negligibly. CCC applied at the same time led to shorter stems, but without any yield effects. Although the control treatment showed no major lodging, yield increases tended to be higher with the MLU preparations.

The oat variety 'Samantra' responded with stem length reductions in the same order of magnitude as the above mentioned varieties. Due to strong lodging in the control treatment, the effect of less lodging was much more obvious. Most likely, the increased application rate in case of MLU 208 improved lodging resistance and yield. In principle, the lipid formulation MLU 208L had a similar effect on stem length and stabilization, but compared to the no-lipid preparation yield increases were smaler. Late applications led to stem reductions comparable to the early treatment resulted in a 8% yield increase. Results with oilseed rape and MLU 2081 are in the lower part of Table 1. The addition of the fungicide tebuconazole in the form of Folicur® to the tank mixture produced reductions in stem length and increases in seed yield.

The residue quantities determined after application of split doses of the test formulations MLU 208 and MLU 208 L at different growth stages were always lower than the legal limit of 5.0 mg kg^{-1} (data not shown).

A potential threat by CCC residues in cereal products seems to be unlikely, as milling removes considerable CCC deposited in the glumes, and so only about 10 % of the CCC content in the intact grain is found in the meal.

The results obtained for oat substantiate previous results with CCC/DCiB mixtures for stem shortening and yield increase in wheat and barley. Positive effects of the CCC/DCiB combination have also been obtained when applied in grass seed cropping (Gans et al, 1995) and oil crops (Gans and Schilling, 1997). It can be concluded that DCiB and CCC containing preparations are generally suited as growth regulators for cereal and oilseed crops.

It is possible that DCiB/CCC applications may be appropriate for use in other crops. The active substances might, be used in the production of ornamental plants for stabilizing the stems of cut flowers.

References

Gans W, Beschow H and Merbach W 2000 J. Plant Nutr. Soil Sci. 163, 405–410.

Gans W, Schilling G and Schöberlein W 1995 VDLUFA-Schriftenreihe, Kongreßband 1995, 729-732.

Gans W and Schilling G 1997 40 Jahre Lehr- und Versuchsstation Seehausen und 50 Jahre Landwirtschaftliche Fakultät der Martin-Luther-Universität Halle-Wittenberg, 78 - 85

Kretschmann F, Ackermann H and Krüger S 1974 Die Nahrung 18(5), 543-550

Linser H and Kühn H 1963 Z. Acker-u. Pflanzenbau 117, 129-154

Schilling G 1990 Wachstumsregulatoren,Tag.-Ber. Akad. Landwirtsch.-Wiss.DDR 11-31

Sturm H and Jung J 1969 Med. Rijksf. Landbouww.Gent 34, 520-528

Table 1. Effect of DCiB/CCC-mixtures on stem length and yield of oat and oilseed rape.

Plant	Preparation	Applied amount g ha^{-1}			Time of application	Plant height at harvest		Lodging score at harvest	Harvest grain yield [5]	
		CCC	DCiB	ethephon		cm	rel.		t ha^{-1}	Relative
Oat	Control	-	-	-	-	114	100	6.9	6.23	100
(Alfred,	MLU 208	420	843	-	GS 31/32 [6]	103	90.4 [2]	9	6.63	106.5
Solidor)	MLU 208 L	295	593	-	GS 31/32	102	89.4 [2]	8.1	6.53	104.8
	MLU 208 L	340	682	-	GS 45	101	89.0 [2]	7.6	6.35	102.0
	CCC	736	-	-	GS 45	105	92.1 [2]	7.1	6.25	100.3
Oat	Control	-	-	-	-	117	100	2.0	5.81	100
(Samantra)	MLU 208	560	1124	-	GS 31/32	104	89.3 [2]	6.1	6.36	109.5 [3]
	MLU 208 L	384	771	-	GS 31/32	104	89.0 [2]	7.2	6.22	107.0
	MLU 208 L	340	682	-	GS 45	103	88.0 [2]	7.6	6.27	108.0
Oilseed	Control	-	-	-	-	163	100	9	4.76	100
rape	MLU 2081	570	1144	236	GS 57	150	91.9	9	5.04	105.9
(Lirajet)	MLU 2081 + TC	570	1144	236	GS 57	145	89.0	9	5.13	107.8 [4]
	TC [1]	-	-	-	GS 57	154	94.4	9	4.80	100.8

[1] Folicur® at 1.5 l ha^{-1} (Folicur contains 250 g l^{-1} tebuconazole)
[2] significant in Tukey-test ($\alpha \leq 0.05$) compared to control
[3] significant in Tukey-test ($\alpha \leq 0.05$) compared to control
[4] significant in t-test ($\alpha \leq 0.05$) compared to control
[5] Harvest grain yield (Oat = 86% dry matter, oilseed rape = 91% dry matter)
[6] GS = growth stage

Symposium 11

Nutrient dynamics in natural and agricultural ecosystems – processes and modelling

Convener:

F. O. Beese

Editor:

Heiner Flessa

W. J. Horst et al.(Eds.), Plant nutrition – Food security and sustainability of agro-ecosystems. 892~893, 2001

Nitrogen saturation in forest ecosystems: Progress and remaining mysteries

J. D. Aber
Complex Systems Research Center, Institute for the Study of Earth, Oceans and Space, University of New Hampshire, Durham, NH 03824, USA john.aber@unh.edu

Key words: abiotic immobilisation, forest health, land use impacts, *mycorrhizae*, N cycling, ^{15}N pool dilution, nitrification, nitrate leaching, NPP

Abstract

More than a decade of intensive research has lead to a clearer understanding of the stages of nitrogen saturation in forest ecosystems, and of the role that land use history can play in preconditioning responses to chronic N deposition. The major unknowns surround processes by which nitrogen is retained in soils. Mycorrhizal assimilation and exudation and abiotic N immobilization are the two most likely candidates. Relatively little is known about the mechanisms or capacities of either process.

Advances in our understanding of N saturation

Nihlgard (1985) first directed attention to the potential for excess nitrogen deposition and availability to damage forest ecosystems. Since then, a coordinated Europe-wide program (NITREX, Wright and Van Breeman 1995, Wright and Rasmussen, 1998) and individual studies in the United States (Aber *et al.*, 1998, Fenn *et al.*, 1997) have combined chronic N addition and removal experiments with regional sampling of forest N status to reveal a progressive syndrome of responses to increasing N availability (Aber *et al.*, 1998).

Figure 1 summarizes current theory relative to the course of N saturation. Key aspects of this pattern of response include significant non-linearities in rates of change. The induction of net nitrification in forests where this process was previously absent, or increases in net nitrification are a key indicator of the stage of N saturation.

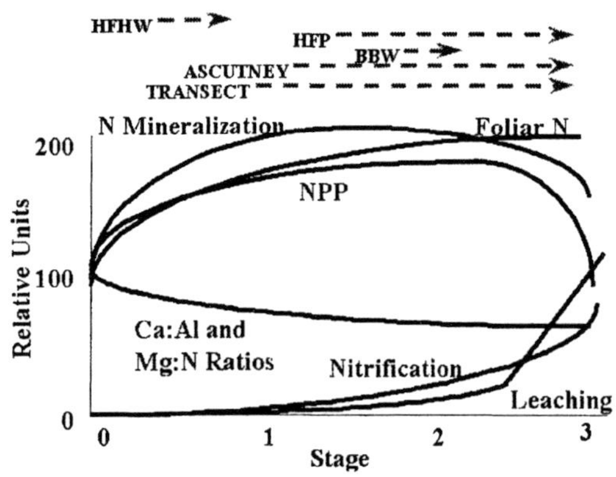

Figure 1. Changes in N cycling and NPP through N saturation. Sites at top of figure have different land use histories (Aber *et al.*, 1998).

The two most important potential negative impacts of excessive N deposition, reduced water quality and decreased tree growth, have been demonstrated both through extensive surveys and intensive experimental treatments. Induction of nitrification and nitrate leaching, or increases in the processes where already present, have been clearly demonstrated, as have reductions in nitrate loss with reductions in N deposition (e.g. Magill *et al.*, 1999, Bredemeier *et al.*, 1998). Reductions in tree growth and/or increases in mortality have been induced in evergreen, needle-leaved species of trees, but not in deciduous broad-leaves (Aber *et al.*, 1998).

A significant role has been hypothesised for previous land use history in the determination of the initial degree of N limitation on ecosystem function, and the amount of N addition required to induce nitrification (Figure 1). In a comparison between responses of two deciduous forests (Magill et al,. 1996), resistance to N saturation was higher in the ecosystem with a strong history of N extraction through harvest, fire and grazing, than a forest with a harvest-only history, even though the former stand receives roughly twice the N deposition.

Remaining mysteries

Under all conditions, forest ecosystems tend to retain a large fraction of added N (an average of nearly 75% in European forests, e.g. Dise and Wright, 1995). Two stands in the U.S. receiving very high and concentrated applications approached 90-100% retention (Magill *et al.*, 1999). This retention provides a profound degree of protection against N saturation. While it is clear that soils are the primary sink for retained N, the mechanisms responsible are very poorly understood.

Recently we summarised the existing literature on N saturation and presented 3 potential processes for N retention. Two of these (Figure 2) involved microbial processing, either by growth of free-living microbes or assimilation and exudation by mycorrhizae. The former was eliminated as a quantitatively important pathway because demands for carbon are high and soil microbes are already carbon limited, and increases in soil CO_2 would be significant, and none have been detected (Aber *et al.*, 1998). One of the greatest remaining mysteries in the biology of soils is the role that mycorrhizae play in element cycling, and, in particular, the carbon costs associated with this role. We need to know much more about nutrient cycling and metabolism in mycorrhizae grown in the field in the symbiotic state.

N Incorp.		C:N Product		Production Respiration		Total Carbon Required	Respired
Microbial Growth							
$\frac{15\ g\ N}{m^2 \cdot yr}$	.	$\frac{8\ g\ C}{g\ N}$	.	$\frac{3\ g\ C}{g\ C}$	=	$\frac{360\ g\ C}{m^2 \cdot yr}$	$\frac{240\ g\ C}{m^2 \cdot yr}$
Mycorrhizal Assimilation							
$\frac{15\ g\ N}{m^2 \cdot yr}$	.	$\frac{6.25\ g\ C}{g\ N}$	.	$\frac{1.25\ g\ C}{g\ C}$	=	$\frac{117.2\ g\ C}{m^2 \cdot yr}$	$\frac{23.4\ g\ C}{m^2 \cdot yr}$

Figure 2. Calculation of carbon costs and CO_2 release for 2 biotic pathways of nitrogen immobilisation(Aber *et al.*, 1998).

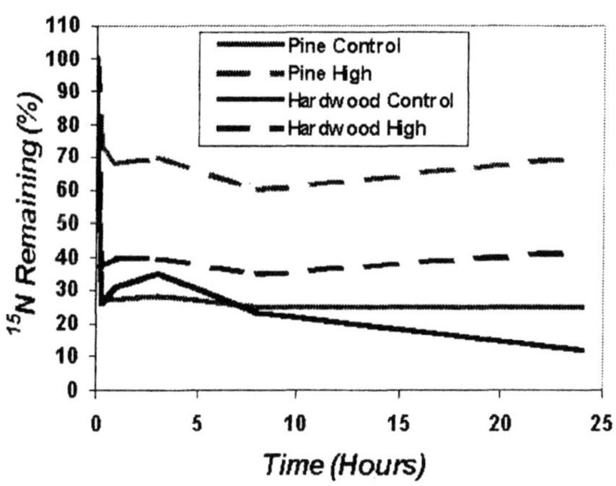

Figure 3. ^{15}N pool dilution data from 4 stands at the Harvard Forest. All sites show significant very rapid N immobilisation usually associated with abiotic processes (Berntson *et al.*, 2000).

The potential role of the third process, abiotic N immobilisation, is more easily addressed. While strictly chemical processes for the incorporation of mineral N into soil organic matter have long been known, the quantitative importance of these mechanisms in field soils has been largely ignored. However, abiotic incorporation of N into foliar litter was demonstrated some time ago by Axelsson and Berg (1988). More recently, ^{15}N pool dilution techniques have been refined and used to demonstrate very high rates of very rapid immobilisation of nitrate (Figure 3, Berntson *et al.*, 1999). A combination of techniques had been used to suggest a very significant role for abiotic N immobilisation in soils from a variety of ecosystem types (Johnson *et al.*, 2000).

Conclusions

We now understand the stages in N saturation. However, we lack the ability to predict with confidence the timing and extent of N saturation in the field because of our fundamental ignorance of the processes by which N is retained in soils. Future research should be focused on the processes of mycorrhizal N assimilation and metabolism, and chemical or non-biological immobilisation in soils.

References

Aber JD, McDowell WH, Nadelhoffer KJ, Magill A, Berntson G, Kamakea M, McNulty SG, Currie W, Rustad L and Fernandez I 1998 BioScience 48, 921-934.

Alexsson G and Berg B 1988 Scandinavian J. Forest Res. 3, 273-279.

Berntson GM and Aber JD 2000 Soil Biol. Biochem. 32, 151-156.

Bredemeier M, Blanck K, Dorhenbusch A, Lamersdorf N, Meyer AC, Murach D, Parth A and Xu YJ 1998 Forest Ecology and Management 101, 281-294.

Dise NB and Wright RF 1995 Forest Ecology and Management 71, 153-162.

Johnson DW, Cheng W and Burke IC 2000 Soil Sci. Soc. Am. J. 64, 1503-1514.

Magill A, Aber J, Berntson G, McDowell W, Nadelhoffer K, Melillo J and Steudler P 1999 Ecosystems ,(submitted).

Magill AH, Downs MR, Nadelhoffer KJ, Hallett RA and Aber JD 1996 Forest Ecology and Management 84, 29-37.

Wright RF and Rasmussen L 1998 Forest Ecology and Management 101, 1-8.

Wright RF and van Breeman N 1995 Forest Ecology and Management 71, 1-6.

W. J. Horst et al.(Eds.), Plant nutrition – Food security and sustainability of agro-ecosystems. 894~895, 2001

Carbon and nitrogen balances in Danish forest and agro ecosystems

L. Rasmussen
Risö National Laboratory, Denmark, Department of Plant Biology and Biogeochemistry, PBK-124, P.O. Box 49, Roskilde, DK-4000, DK : Denmark, lennart.rasmussen@risoe.dk

Key words: carbon, ecosystem, *Fagus silvatica*, fluxes, *Hordeum vulgare*, nitrogen

Abstract

Carbon exchange between the terrestrial biosphere and the atmosphere is one of the key processes that needs to be assessed in relation to the greenhouse effect. At the same time nitrogen plays an important role for turnover processes in the carbon cycle. This study presents data of net ecosystem carbon exchange which confirm that forest and agro ecosystems in Denmark act as carbon sinks, but only to a minor degree compared to the emission of carbon dioxide from fossil fuels. The annual carbon balance ranges from 1-2 t C per ha per year. The study also includes measurements of methane and water mediated carbon fluxes. Their contribution to the carbon balance of the ecosystems are of minor importance. The input of nitrogen to the forest and agroecosystems was 16-25 kg N per ha per year, highest in the forest due to dry deposition. The leaching of nitrate from the forest ecosystem was only a few kg N per ha per year, whereas it amounted to 50-70 kg N per ha per year from agricultural fields. The findings point to the conclusion that afforestation on agricultural fields will improve the groundwater quality with respect to nitrate content, but may also lead to formation of less groundwater due to higher evapotranspiration in forests.

Introduction

The climate is changing due to the greenhouse effect – the future is expected to be hotter and with more vigorous rainstorms (IPCC, 1995). The single most important greenhouse gas is CO_2, and the overall carbon balance plays an important role in understanding and predicting the positive or negative feedback of CO_2 on the future climatic situation. Simultaneously, nitrogen plays an important role in regulating or modifying carbon sequestration. Therefore, it is important to understand and quantify the fluxes of C and N and the net release or uptake of carbon in terrestrial ecosystems (Valentini *et al.*, 2000).

The main objectives of the present study were to quantify and compare the gaseous and water mediated fluxes of C and N compounds in a European beech *(Fagus silvatica* L.) forest and an agro ecosystem with barley *(Hordeum vulgare* L.). The study included measurements of the fluxes of minor gaseous carbon and nitrogen compounds. In total it provided quantification of the net ecosystem fluxes of NO, NO_2, NH_3, N_2O, CO_2, CH_4, H_2O and water mediated inorganic and organic N and C compounds.

Materials and methods

The field site is located at Gyrstinge near Sorø, Zealand, Denmark. The soil is rich in clay and lime, pH 5-8, Mollisol (Soil Taxonomi). The forest is an 80 years old beech stand - average tree diameter at breast height 40 cm, 430 stems per ha – surrounded by agricultural fields. A more detailed description of the site is presented by Pilegaard *et al.* (2001). In 1998 the crop was barley fertilised with 45 kg N as NH_4NO_3 and 60 kg N as manure per ha. Measurements of the fluxes of the gaseous compounds were performed with automated eddy covariance methods from a 57 m high mast in the forest and a 12 m high mast in the field. Atmosphere/canopy fluxes were measured up through the canopy from a 23 m high tower. Atmosphere/soil fluxes were measured by soil chamber techniques. Open field precipitation and throughfall was collected in funnels. Stemflow was sampled with automated tipping bucket systems. Soil water sampling was performed with suction cups (PRENART) placed in a depth of 1 m. CO_2 and H_2O was measured on-line in the field with automated detectors. Water mediated N and C compounds were measured in the laboratory with standard analytical techniques.

Results and discussion

The automated eddy covariance measurements of CO_2 fluxes showed large variations during the day, during the season, during the years and between the forest and agro ecosystems. The net carbon exchange of terrestrial ecosystems is the results of a delicate balance between uptake (photosynthesis) and loss (respiration). Under favourable conditions, the net ecosystem flux was dominated by photosynthesis during daytime, and by respiration at night. In the leafless period of the year respiration was also dominant. The fluxes of carbon from the atmosphere to the forest and agro ecosystems were in the range of 11-12 t C ha^{-1} yr^{-1}. Since the total ecosystem respiration was in the range of 10 t C ha^{-1} yr^{-1}, of which about 60% was soil respiration, the net ecosystem exchange was in the range of 1-2 t C ha^{-1} yr^{-1}. This is a medium value for C sequestration in European ecosystems, the values ranging from 7 t C ha^{-1} yr^{-1} in southern Europe to zero (or even emission) in northern Europe (Valentini *et al.*, 2000). The emission or oxidation of CH_4 only influenced the total C flux of the ecosystems with a few kg C ha^{-1} yr^{-1}. Leaching of C in the form of TOC and HCO_3^- was less than 100 kg C ha^{-1} yr^{-1}.

The inputs of N with rain, throughfall and stemflow to the ecosystems are presented in Fig. 1. In general, the contributions of N deposition from rain and throughfall were in the same range during the year. Stemflow accounted for about 17% and 37% of the total flux of water and nitrogen respectively to the forest floor. The total flux of nitrogen to the forest floor amounted to c. 25 kg N ha^{-1} yr^{-1} and wet deposition to the agricultural fields (and forest) to 16 kg N ha^{-1} yr^{-1}. The difference between the wet deposition and the sum of throughfall and stemflow can be ascribed to dry deposition. However, dry deposition may have been higher than the 9 kg N ha^{-1} yr^{-1}, since nitrogen can be taken up in the canopy, as has been reported for Norway spruce (Hansen, 1996). The presented N deposition values represent medium range values for European ecosystems (Gundersen *et al.*, 1998).

The concentration of NO_3-N in soil water under the root zone was very low in the forest, less than 1 mg L^{-1}. There was no significant difference between samples taken close to the tree trunks and in an equal distance between tree trunks. In the agricultural field the NO_3-N concentration was higher, 10-50 mg L^{-1}. Using a hydrological ET model, the leaching of NO_3 - N from the agricultural field could be calculated to be in the range of 50-70 kg N ha^{-1} yr^{-1}. This is a relatively high leaching rate compared to general findings for agricultural fields in Denmark (Hvelplund and Østergaard, 1979). In the forest leaching was only a few kg N ha^{-1} yr^{-1}, which is the general picture for forests in low-medium N deposition areas in Europe (Gundersen *et al.*, 1998). In the summer period there was almost no leaching from the root zone, since both trees and agricultural crops utilised all available soil water. The emission of NO from the forest floor was less than 1 kg N ha^{-1} yr^{-1}. The emission of N_2O was also very small 0.5-3 kg N ha^{-1} yr^{-1}, highest values being obtained from the agricultural field.

Taking the atmospheric deposition in consideration and the internationally used "critical load" criteria (Gundersen *et al.*, 1998), the critical load for N seems to be exceeded by 5 kg N ha^{-1} yr^{-1} (Beier *et al.*, 2001).

Conclusions

The net carbon ecosystem exchange is relatively modest in both the examined forest and agro ecosystems. However, scaling up the examined area to the total area of Europe, carbon sequestration in forest could amount to 2% to 40% of the CO_2 emission from fossil fuels (Martin *et al.*, 1998). But taking into consideration that a part of the produced biomass from forest and agro ecosystems is burned or decomposed relatively fast, the long-term carbon sequestration in terrestrial ecosystems may be reduced to 2% to 3% of the CO_2 emission (Nabuurs *et al.*, 1999).

Anticipating that dry deposition of N in agricultural fields is relatively low, the atmospheric deposition of N is about 40% higher in forests than in agricultural fields. Comparing forest and agro ecosystems, afforestation will undoubtedly improve the groundwater quality with respect to NO_3^- content, although the amount of formed groundwater might be less due to the higher evapotranspiration in forests.

References

Beier C, Rasmussen L, Pilegaard K, Ambus P, Mikkelsen T, Jensen N O, Kjøller A and Ladekarl U L 2001 Water Air Soil Pollut. (in press).

Gundersen P, Emmett B A, Kjønaas O J, Koopmans C J and Tietema A 1998 For. Ecol. Manage. 101, 37-55.

Hansen K 1996 Atmosph. Envir. 30, 4065-4076.

Hvelplund E K and Østergaard H S 1979 Ugeskrift f. jordbrug 124, 643-646.

IPCC 1995 *In* Climate Change 1995 Ed R T Watson. Cambridge University Press, Cambridge.

Martin P H, Valentini R, Jacques M, Fabbri K, Galati D, Quaratino R, Jarvis P, Jensen N O, Lindroth A, Grelle A, Ceulemans R, Schulze E-D, Grünwald T and Thorgeirsson H 1998 Ambio 27, 582-584.

Nabuurs G J, Dolman A J, Verkaik E, Whitmore A P, Daamen W P and Mohren G M J 1999 RIVM National Institute of Public Health and the Environment, Report no 410200030.

Pilegaard K, Hummelshøj P, Jensen N O and Chen Z 2001. Agricultural and Forest Meteorology 107, 29-41.

Valentini R, Matteucci G, Dolman A J, Schulze E-D, Rebmann C, Moors E J, Granier A, Gross P, Jensen N O, Pilegaard K, Lindroth A, Grelle A, Bernhofer C, Grünwald P, Aubinet M, Ceulemans R, Kowalski A S, Vesala T, Rannik Ü, Berbigier P, Loustau D, Gudmundsson J, Thorgeirsson H, Ibrom A, Morgenstern K, Clement R, Moncrieff J, Montagnini L, Minerbi S and Jarvis P G, 2000 Nature 404, 861-865.

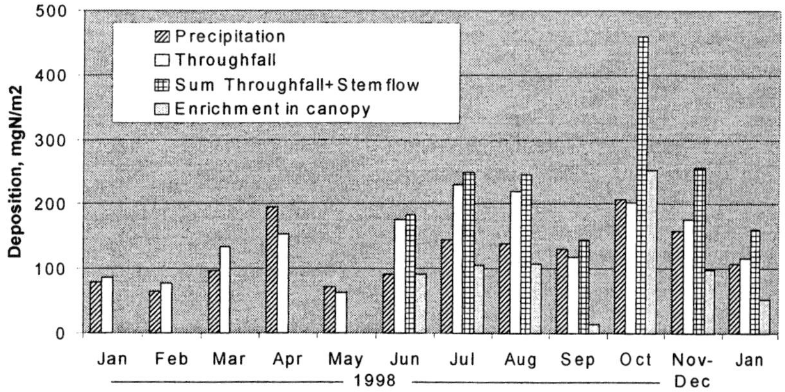

Figure 1. N deposition with precipitation, throughfall and stemflow in a Danish beech forest.

W. J. Horst et al.(Eds.), Plant nutrition – Food security and sustainability of agro-ecosystems. 896~897, 2001

Modelling nutrient cycles in grazed pasture systems of Southern Australia

R. J. Simpson[1)], A. D. Moore[1)], A. E. Richardson[1)], J. R. Donnelly[1)], G. J. Blair[2)] and I. R. Johnson[3)]

[1)] *CSIRO Plant Industry, GPO Box 1600,, Canberra, ACT,, 2601, Australia, richard.simpson@pi.csiro.au*

[2)] *Department of Agronomy and Animal Science, University of New England,, Armidale, NSW 2351, Australia*

[3)] *IMJ Consultants Pty Ltd,, PO Box 1590, Armidale, NSW 2350, Australia*

Key words: modelling, nutrient cycling, pastures, phosphorus availability

Abstract

Effective use of P fertiliser is critical to productive grazing systems in the higher rainfall areas of southern Australia. The efficacy of a three compartment model of inorganic P (Pi) in the soil (soil solution, readily-exchangeable and "fixed" Pi) to predict plant-available Pi was examined for use in a computer-based simulation tool (NutriAce) being developed to assist graziers making fertiliser investment decisions. Predictions of plant-available Pi explained 80% of the variation in soil test values.

Introduction

In southern Australia, regular inputs of P fertiliser are necessary for high production in grassland systems because most soils are deficient in P and N. The fertiliser corrects P-deficiency and promotes legume growth and N fixation. Here we describe progress in modelling P availability in soil for a computer-based model being developed to simulate C, P, N and S cycling in temperate grazing systems. Our ultimate aim is to provide a tool that may be used to assist graziers making fertiliser investment decisions.

Materials and methods

The model.

NutriAce (version 1.1) is an extension of a grazing systems simulation tool, GrassGro (Moore *et al.*, 1997) in which pasture growth is driven using historical daily weather records. Sheep or cattle consume pasture or supplements and the nutrients are partitioned to growth, maintenance and animal products, or are returned to the pasture in excreta. The model distributes excreta non-uniformly to urine patches, animal camps and across the field. Nutrients enter soil organic matter and plant-available or poorly-available inorganic pools as appropriate, and are cycled between the soil pools and the soil decomposer biomass (McCaskill and Blair, 1990). These processes and sorption reactions (for phosphate, sulphate, ammonium) determine the availability of nutrients for plant growth. Pasture may comprise two species which compete for light, nutrients and water. Plant nutrient uptake is based on concepts of de Willigen and van Noordwijk (1994). *Potential* nutrient uptake is determined by the demands for growth and 'luxury' nutrient consumption. Roots grow to a specified depth and are distributed exponentially across soil layers. Root morphology is defined by specifying average diameter and specific root length. Nutrient movement to roots occurs by diffusion and mass flow, but soil conditions and the extent of root development constrain potential uptake to determine *actual* uptake. If the nutrient supply from the soil fails to meet the demands for growth, nutrients are mobilised from older shoot tissues. Plant growth is slowed when the supplies of nutrients from soil, recycling within the plant and N-fixation (legumes) fall short of the 'critical' supply required for maximum growth. Nutrients are lost from the grazing system by volatilization (N), leaching (N, S), and export in animal products (C, P, N, S) or by accumulation in animal camps (C, P, N, S). NutriAce is still under development and has some areas of uncertainty. For this exercise, the model was set up to test prediction of P availability to pasture.

P availability.

Organic P is partitioned between four soil organic matter pools (labile, resistant, humus, decomposer biomass), and a three pool Pi model (soil solution, readily-exchangeable, fixed Barrow and Shaw, 1975) is assumed. The Pi concentration in soil solution equilibrates instantaneously with the concentration in the exchangeable phase according to a phosphate sorption isotherm. Presently, this is defined by a laboratory test (Ozanne and Shaw, 1967) combined with fertiliser and pasture yield data. Fluxes between the sorbed phase and the 'fixed' (sparingly-available) Pi pool are slow and are computed using first-order kinetics modified by soil moisture and temperature. The rate constants are based on data from Barrow (1974).

Field experiment and its simulation.

A P-fertility trial was commenced in 1994 on a yellow podzolic soil near Canberra, Australia, and was grazed continuously by young Merino sheep. Basal fertiliser dressings were applied to ensure that P and N were the only limiting nutrients. Treatments were: no P fertiliser, or fertilised with triple superphosphate (20.7% P) to achieve a target surface soil concentration of 10-12 mg kg^{-1} plant-available Pi (bicarbonate extractable P, 0-10 cm, Olsen and Sommers, 1982) during May-June. Pastures were predominantly annual grasses and subterraneum clover (*Trifolium subterraneum)* with about 15% perennial grass (*Phalaris aquatica*). Clover content ranged typically between 0-20% in unfertilised pasture and 10-40% in fertilised pasture. Brome grasses (*Bromus molliformis, B.*

diandrus) and silver grasses (*Vulpia myuros, V. bromoides*) were the dominant annual grasses in fertilised and unfertilised pastures, respectively.

In the simulations shown here, fertilised pasture (*B. molliformis-T. subterraneum*) and unfertilised pasture (*Vulpia spp.-T. subterraneum*) were grazed with 18 sheep ha^{-1}. Root morphology was assumed to be that of these species grown at low P fertility (Hill *et al.*, 2001)

Results

A reasonable simulation of pasture production was achieved despite restriction to the two main species in each treatment (eg. fertilised pasture Fig. 1a). The pattern of change in the simulated readily-exchangeable Pi of surface soil was similar to that in the plant-available Pi measured in bicarbonate extracts (Fig. 1b). Simulated and measured values from both treatments were related by the equation:

exchangeable Pi (simulated) =

1.87*available Pi (measured) + 6.45 ($R^2 = 0.81$)

The simulation: a) failed to predict a number of very high soil test values measured soon after fertiliser application. It is likely that these particular values were due to inadvertent collection of fertiliser particles in the soil sample and b) predicted higher exchangeable Pi than was measured in winter-early spring, 1996. The reason is unknown, but it is surprising that the soil test values exhibited only a slow increase after the fertiliser was applied in that year.

Discussion

The predictions of plant-available Pi were generally good and explained about 80% of the variation of all soil test values. Some challenges remain before the model can be used on farms. The Pi sorption curve relating solution and readily-exchangeable Pi is presently difficult to define without the benefit of an array of fertiliser x pasture yield information. It is also possible that for longer simulations, it will be necessary to alter the sorption characteristic as the P content of the soil increases (Barrow *et al.*, 1998). Nevertheless, these results indicate that for short simulation periods (5-10 years), a simple 3-pool model of Pi availability can reflect soil Pi transactions adequately.

Acknowledgement

NutriAce development has been funded in part by The Woolmark Company, Land & Water Resources Research & Development Corporation, and Meat & Livestock Australia. RJS is grateful to The Plant Nutrition Trust for a travel grant to attend the IPNC.

References

Barrow NJ and Shaw TC 1975 Soil Sci. 119, 167-177.
Barrow NJ 1974 Soil Sci. 118, 380-386.
Barrow NJ, Bolland MDA and Allen DG 1998 Aust. J. Soil Res. 36, 359-357.
de Willigen P and van Noordwijk M 1994 Soil Sci. 157, 162-170.
Hill J, Simpson RJ, Moore AD and Chapman D 2001 *In* Proc. Int. Plant Nutr. Colloquium, Hannover.
McCaskill MR and Blair GJ 1990 Fert. Res. 22, 161-172.
Moore AD, Donnelly JR and Freer M 1997 Agric. Syst. 55, 535-582.
Olsen SR and Sommers LE 1982 *In* Methods of soil analysis, Part 2, Eds AL Page, RH Miller and DR Keeney. pp 403-430, ASA-SSSA, Madison.
Ozanne PG and Shaw TC 1967 Aust. J. Agric. Res. 18, 601-612.

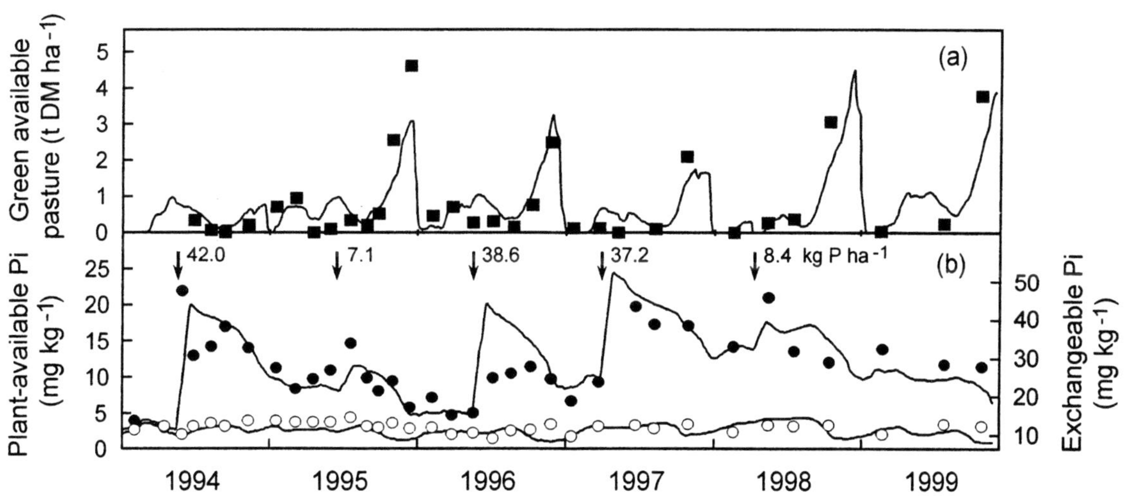

Figure 1. Simulated (lines) and measured amounts of (a) green available pasture (squares) in the fertilised pasture, and (b) plant-available Pi (0-10 cm depth) measured in bicarbonate soil extracts from fertilised pasture (closed circles) and unfertilised pasture (open circles). Values are the mean of 3 replicates. Arrows indicate timing and quantities of P applied.

W. J. Horst et al.(Eds.), Plant nutrition – Food security and sustainability of agro-ecosystems. 898~899, 2001

Modelling soil acidification under pastures - The GRAZPLAN soil acidity model

J. Braschkat, A. D. Moore, R. J. Simpson and P. J. Randall
CSIRO Plant Industry, GPO Box 1600, Canberra ACT 2601

Key words: modelling, pasture, soil acidification

Abstract

Soil acidification increasingly constrains production in agricultural systems in many parts of Australia. The GRAZPLAN soil acidity model has been developed to simulate the impact of grazing enterprises on soil acidification. It enables the user to simulate the acidification process over long periods of time using historical weather data. The soil acidity model is linked to the nutrient cycling model NutriAce, which simulates soil water balance, pasture growth and animal biology for grazing systems in temperate Australia. The paper illustrates the agreement between simulation and field data.

Introduction

In Australia, an estimated 33 million ha of agricultural soils are classed as highly acidic and another 48 million ha as mildly acidic (AACM, 1995). Soil acidification is a process that occurs naturally at low rates. However, during the 20th century the introduction of subterranean clover (*Trifolium subterraneum*) and the use of phosphate fertiliser on grazed pasture in Australia has lead to higher productivity and as a consequence to increased rates of soil acidification (Bromfield *et al.*, 1980). Today, after 50 to 75 years under improved pasture with acidification rates up to 3.4 kmol ha^{-1} $year^{-1}$ (Helyar and Porter, 1989) low soil pH is likely be a major constraint of plant production. Amelioration by liming is an expensive option for the extensive wool production systems that are the main enterprise on large areas. NutriAce (Simpson *et al.*, 2001) is being developed to help farmers to optimise specific management decisions, e.g. lime or fertiliser application rates. NutriAce is an advanced developmental version of the decision support tool named GrassGro (Moore *et al.*, 1997) and is capable of simulating the cycling of C, N, P and S in the grazed pasture systems of temperate Australia. The soil acidity model, which forms part of NutriAce, uses the outputs of the C and nutrient cycles and the water balance model simulated by NutriAce to calculate proton budgets layer by layer down the soil profile. These are then converted to soil pH changes in each soil layer using calculated or, alternatively, measured soil pH buffer capacity. The value of these simulations depends upon how closely they are related to reality and to this end simulated outputs are compared with measured field data. One data set used for testing the simulation is from a field in the catchment of the Pejar dam, near Crookwell, NSW. This site, on a yellow duplex soil formed on granitic parent material, had been under continuous subterranean clover pasture and grazed by sheep for a period of 75 years.

Materials and methods

Soil samples within the sheep grazed field and several virgin sites under native vegetation were taken in October 2000 from 12 different depths. The samples were air-dried and analysed for pH ($CaCl_2$), pH buffer capacity (batch equilibration) and C and N (mass spectrometry). In addition, separate soil cores were taken to determine soil bulk density and water content at wilting point, field capacity and saturation. Using the measured values on soil properties, NutriAce was initialised to simulate a wool growing enterprise with 4.3 sheep ha^{-1} and an annual P fertiliser application rate of 39.4 kg ha^{-1} single superphosphate. Both values are based on the stocking rate and fertiliser history documented in farm records.

Results

Soil pH profiles

Soil pH measured on the virgin sites indicated that the shape of the pH profiles varied considerably.

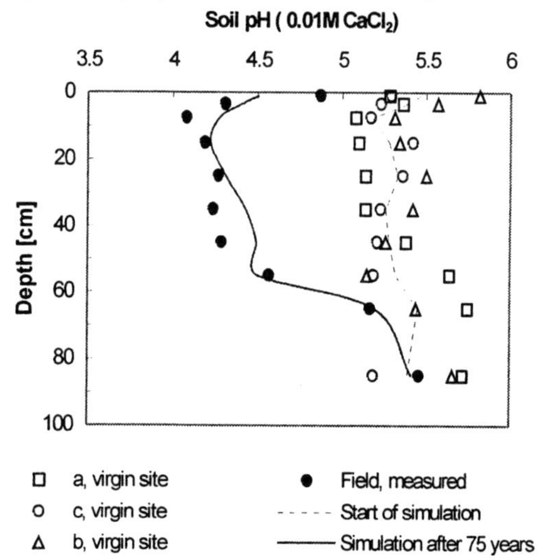

Figure 1. Soil pH under different virgin sites and under grazed pasture after 75 years.

Due to this large variability the average soil pH of the profiles a, b and c was used to initialise the simulation run. Compared to the virgin sites, soil pH under improved pasture had dropped approximately one unit in the soil layers down to a depth of 60cm.

Soil C and N profiles

During the 75 year period under agricultural use the C and N contents of the top 10 cm increased considerably due to the establishment of subterranean pasture with P fertiliser applications resulting in increased nitrogen fixation which in turn stimulates productivity of the grass component of the pasture (Figures 2 and 3).

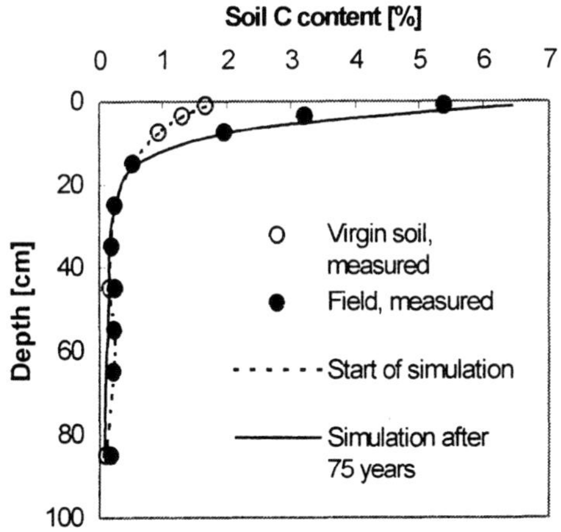

Figure 2. Soil C profiles under virgin site and under grazed pasture after 75 years.

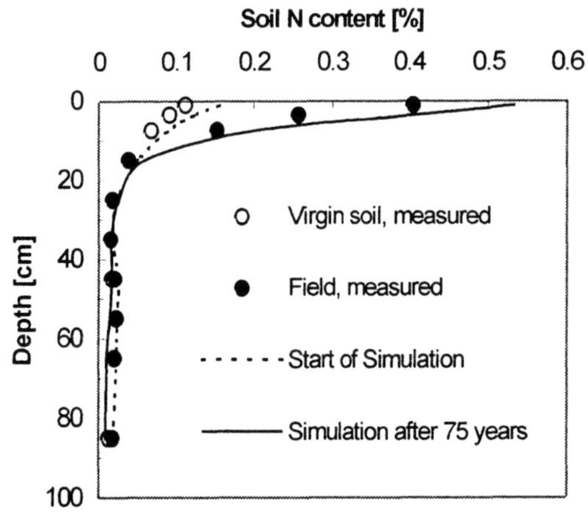

Figure 3. Soil N content under virgin site and under grazed pasture after 75 years.

The simulation captures the build up of organic matter. In fact, accumulation of organic anions in soil organic matter is the largest contributor to the acid budget of this production system (Table 1) comprising 50% of the total acidification. Nitrate leaching is the second most important process followed by losses to sheep camps and the removal of organic anions in products.

Table 1. Total acidification in the pasture simulation and the contribution of different processes to the acid budget.

Source	kmol H^+ ha^{-1} $year^{-1}$	%
Organic matter accumulation	1.68	50
Nitrate leaching	1.20	36
Losses to sheep camps	0.40	12
Product removal	0.04	2
Total	3.32	100

Discussion

Soil acidification under sheep grazed pastures is largely caused by build up of organic matter, nitrate leaching and removal of organic anions (Helyar *et al.*, 1989). NutriAce is capable of simulating these processes and hence is a useful research tool to study soil acidification. The current version of the soil acidity model is still lacking a feedback to plant growth due to low pH and high Al concentrations. This will be important when trying to simulate the production of acid-sensitive species. The pasture simulated in the present example is dominated by acid-tolerant species and in such cases the model can be used to estimate the lime needed to compensate for acidification caused by a grazing enterprise.

Acknowledgement

The work is supported by the Vincent Fairfax Family Foundation and the Land and Water Resources Research and Development Corporation.

References

AACM 1995 LWRRDC Canberra ACT

Bromfield SM, Cumming RW, David DJ and Williams CH 1980 Australian Journal of Experimental Agriculture and Animal Husbandry 28, 181-191.

Helyar KR and Porter WM 1989 *In* Soil Acidity and Plant Growth Ed. AD Robson. pp. 61-101.

Moore AD, Donnelly JR and Freer M 1997 Agricultural Systems 55, 538-582

Simpson RJ, Moore AD, Richardson AE, Donnelly JR, Blair GJ and Johnson IR 2001 Proceedings of the XIV International Plant Nutrition Colloquium, Hannover, Germany.

W. J. Horst et al.(Eds.), Plant nutrition – Food security and sustainability of agro-ecosystems. 900~901, 2001

Simulation study of nitrogen supply in boreal forests using model of soil organic matter dynamics ROMUL

O.G. Chertov[1)*], A.S. Komarov[2)*], M. Nadporozhskaya[1)], S.S. Bykhovets[2)], S.L. Zudin[2)*]

[1)] *Biological Research Institute, St. Petersburg State University, 2 Oranienbaum Rd., 198904 St. Petersburg-Peterhof, Russia; Email: soil@onti1.bio.spbu.ru, chertov@efi.fi*

[2)] *Institute of Physics, Chemistry and Biology of Soils, Russian Acad. of Sci., 142292 Pushchino, Moscow Region, Russia; Email: komarov@efi.fi, byhovets@issp.serpukhov.su, zudin@efi.fi*

* *Present address: European Forest Institute, Torikatu 34, 80 100 Joensuu, Finland*

Key words: boreal forests, simulation model, soil organic matter dynamics, litter fall quality, available nitrogen, forest floor

Abstract

A set of model simulations was performed to quantify the effects of environmental factors and soil properties on available nitrogen pool and dynamics in European boreal forests. The results of simulation of the effect of soil temperature and moisture show a maximum of available nitrogen in times of optimal soil temperature and moisture. The increasing of nitrogen and ash content in forest litter has also positive influence on a production of available nitrogen in forest soils. If a soil has a well-developed surface organic horizon (forest floor in Moder and Raw humus), then the majority of available nitrogen is produced in the organic layer. The results of this simulation supports the important role played by forest floor in the soil nutrition of boreal forest ecosystems.

Introduction

The role of soil conditions on tree growth is generally well known. However, the quantitative experimental study of the nitrogen supply in boreal forest soils meets serious methodological problems. We can mention Wilde's (1958) classic experiments on relative productivity of different soil horizons. The models of soil organic matter (SOM) and nitrogen dynamics are being intensively developed now (Powlson *et al.*, 1996; Smith *et al.*, 1997). An application of the models can elucidate some aspects of plant nutrition in boreal forests.

Materials and methods

The model

A model of SOM dynamics, ROMUL (Chertov *et al.*, 2001), was used to quantify the main patterns of available nitrogen dynamics in forest soils. It is based on a classical concept of "humus type" (Humusform). The rate of litter and SOM mineralisation and humification is dependent on quality of litter, soil temperature and moisture, and some soil parameters in the model. The input parameters are the amount and quality of litter, meteorological data and initial amounts of 6 fractions of soil organic matter and corresponding nitrogen. The model allows the calculation of the dynamics of SOM and nitrogen, including the evaluation of the amount of mineral available nitrogen (bulk sum of ammonium and nitrate N). The model has been used as a soil compartment in forest ecosystem model EFIMOD (Chertov *et al.*, 1999) and is now implemented in the FinFor model by Kellomäki *et al.* (1993).

Simulation

A set of following simulations was performed to quantify the effects of environmental factors and soil properties on the available nitrogen pool and nitrogen dynamics in European boreal forests:

1. A one year simulation with model runs for soil with SOM pool 10 kg m^{-2} and C/N 20 (*a*) at constant soil moisture (15 mass % in mineral horizons) with varying temperature from –3°C to +25°C, and (*b*) at constant soil temperature 15°C with varying soil moisture from 3% to 60% in mineral horizons.
2. A 50-year simulation using climate data from south boreal forests (mean annual temperature +4°C, 650 mm precipitation) on 3 podsolic loamy soils: with Mor (Raw humus, SOM pool 7 kg m^{-2} and C/N 35); Moder (SOM 10 and C/N 20); and Mull (SOM 10.5 and C/N 12). The litter input 0.5 kg m^{-2} $year^{-1}$ for these soils had the quality of conifer, mixed and deciduous forests (C/N 70, 40 and 28, respectively) with mostly above ground input (on forest floor) in Mor and below ground in Mull.

Results

The simulation of the effect of soil temperature displays a non-linear increasing of available nitrogen between –3°C and 10°C only, when the pool of mineral available nitrogen (AN) increases from zero to 3.3 10^{-2} kg m^{-2} $year^{-1}$. After that point the AN pool increases to a maximum 5.2 10^{-2} kg m^{-2} $year^{-1}$ at 25°C linearly. A simulation experiment with different soil moisture also shows a maximum AN in interval of optimal soil moisture (15–40 mass % in mineral layer) where it reaches 5.9 10^{-2} kg m^{-2} $year^{-1}$.

The increasing of nitrogen and ash content in forest litter fall has positive influence on AN pool in forest soils (Tab.1): as soon as litter nitrogen content increases from 0.4 to 1.2 g 100 g^{-1} (from Mor to Mull soil) then the total AN production becomes also 3 times higher.

More interesting is a simulated result on the role of

soil horizons in nitrogen supply. In a soil with a well-developed surface organic horizon (forest floor in Raw humus), the majority of available nitrogen is produced in organic layer even in a case of a significant SOM pool in the mineral topsoil (28% of total SOM in our simulation). In contrast, the mineral horizon plays the principal role in nitrogen supply in Mull soil.

Table 1. Production of available nitrogen pool, kg m^{-2} $year^{-1}$, and contribution of SOM fractions to the pool, %, at the end of 50-year simulation

Soil	Fresh litter	Forest floor	Labile humus	Stable humus	Total
Mor	0,00007	0,00120	0.00017	0,00065	0,00209
	3.3	57.5	8,1	31.1	100.0
Moder	0,00009	0,0075	0.00065	0,00239	0,00388
	2.3	19.3	16.8	61.6	100.0
Mull	0,00016	0,00081	0.00100	0,00445	0,00642
	2.5	12.6	15.6	69.3	100.0

It should be pointed out that there is a clear correlation of AN output with the pattern of litter input. In Raw humus with surface litter (needles and root litter in the organic layer) AN is mostly produced in the forest floor. In Mull soil there is the opposite distribution of litter input with 70% of litter input being root litter in mineral horizons and there being correspondingly a maximum AN production.

Generally, the simulation shows the important role of fast decomposing fractions of SOM (fresh litter, humified forest soil and labile humus in mineral horizon) in the nitrogen supply in forest soils. The contribution of these fractions to the AN pool reaches 70% in Raw humus.

Discussion

The results of this simulation support the existing concepts on the role of environmental factors (temperature and water) and litter quality in nitrogen supply in boreal forests, and quantify the important role of the forest floor on soil nutrition in boreal forest ecosystems.

It is well known that soils with a surface organic layer (Raw humus and peat) dominate in European boreal conifer and mixed forests (Chertov, 1981). The simulation experiment specifies a crucial role of the organic layer in nutrient supply under these conditions (Tab. 1). We conclude that any disturbance of organic layer or its removal will lead to serious damage of the nitrogen supply and will strongly affect forest productivity. In this case the models of SOM dynamics and forest ecosystems can serve as an additional tool to explore the effects of the supply or removal of nutrients in the forest (Wilde, 1958; Sukachev, 1964; Schultze, 2000).

This specific pattern of forest nutrition is also typical on Raw humus soils in other climatic conditions. We performed simulations for Dutch climate (mean annual temperature 9°C, precipitation 800 mm), and found 80% of AN is produced in the forest floor in a dry poor sandy site with a 10 cm Mor layer.

The simulation shows that the role of organic layer and labile humus in the CO_2 evolution is prevailing even in Mull soil: about 50% of carbon dioxide is produced from the forest floor, 30% from labile humus and only 20% from stable humus in Mull soil. In forest Mor the role of easily decomposed SOM fractions is absolutely dominating: about 95% of CO_2 originates from forest floor and labile humus.

The simulation experiment with ROMUL model on SOM and nitrogen dynamics clarifies the prospects of modelling to imitate natural processes of soil functioning and development. It clearly indicates aspects of nitrogen supply that can be rather complicated to measure in the field. Simulations also are a tool for comparative evaluation of the efficiency of nitrogen fertilisation, impact of silvicultural regimes and the effect of atmospheric nitrogen deposition on tree nutrition on different soils and under various climatic conditions.

References

Chertov OG 1981 Ecology of forest lands. Nauka, Leningrad, (in Russian).

Chertov OG, Komarov AS, Tsiplianovsky AV 1999 Forest Ecol. Management 116, 189-206.

Chertov OG, Komarov AS, Nadporozhskaya MA, Bykhovets SA, Zudin SL 2001 Ecol. Modelling 136, (in print).

Kellomäki S, Väisänen H, Strandman H 1993 FinnFor: a model for calculating the response of the boreal forest ecosystem to climate changes. Research Note 6, University of Joensuu, Finland, 1-120.

Powlson DS, Smith P and Smith J (Eds.) 1996 Evaluation of Soil Organic Matter Models. NATO ASI Series, I 38. Springer, Berlin Heidelberg.

Schultze E-D (Ed.) 2000 Carbon and nitrogen cycling in European forest ecosystems. Ecological Studies 142. Springer, Berlin, Heidelberg, New York.

Smith P, Smith JU, Powlson DS, McGill WB, Arah JRM, Chertov OG, Coleman K, Franko U, Frolking S, Jenkinson DS, Jensen LS, Kelly RH, Klein-Gunnewiek H, Komarov AS, Li C, Molina JAE, Mueller T, Parton WJ, Thornley JHM and Whitmore AP 1997 Geoderma 81, 153-225.

Sukachev VN (Ed.) 1964 Fundamentals of forest biogeocoenology. Nauka, Moscow, (in Russian).

Wilde SA 1958 Forest soils. Wiley, New York.

W. J. Horst et al.(Eds.), Plant nutrition – Food security and sustainability of agro-ecosystems. 902~903, 2001

The effect of site specific fertilisation on N_2O emissions and N-leaching - measurements and simulations

J. C. Munch, A. Berkenkamp and U. Sehy

Institute of Soil Ecology, GSF - Research Center for Environment and Health, Ingolstädter Landstrasse 1, Neuherberg, D 85764, Germany, munch@gsf.de

Key words: nitrate leaching, nitrous oxide, N-turnover model, precision agriculture

Abstract

A field study and model simulations were undertaken to evaluating the impact of precision farming measures on N loss through leaching and N2O emission. Sites that were conventionally fertilised were compared with sites that were subjected to site-specific fertilisation. Flux rates of N2O were measured weekly. Simulations of the N-turnover were performed to quantify the rate of N-leaching using the model system Expert-N. In the high yield area, emissions of nitrous oxide from two sites, one under conventional and one under site-specific management, were not significantly different in the period following differentiated fertilisation. However, in the low yield area, a site-specific reduction of fertiliser resulted in a reduction of N2O emissions by 33% compared to the site under conventional fertilisation. Correspondingly, N2O emissions related to plant yield from the same site were significantly lower than from the other sites under investigation. Nitrate leaching rates obtained from model simulations, however, did not differ according to the management practices. Although in total more fertiliser was applied to the field in the precision farming variant, the share of fertiliser-N lost to the atmosphere as N2O or leached as nitrate was not higher than in the variant that was conventionally fertilised.

Introduction

In agricultural systems, nitrogen is frequently the limiting nutrient for crop yield. In order to ensure high yield while minimizing losses of nitrogen to the environment it is desirable to implement measures aiming at the adaptation of fertiliser rates to the need of the growing crop. The aim of the study was to assess the effects of precision farming on agricultural yield and on the losses of nitrogen to the groundwater and the atmosphere.

Materials and methods

Investigations were carried out at the Research Station Scheyern of the FAM Research Network on Agroecosystems, located 40 km north of Munich. The field under investigation is 5.99 ha in area. It was cropped to maize in 1999, and cropped to winter wheat in 1999/2000. Data are presented from measurements between June 1999, when site-specific fertilisation was implemented, and June 2000.

The field was subdivided into two sets of strips: a) strips that were managed according to precision farming measures and b) strips that were conventionally managed. Areas of long term high yield (H) (corresponding to 66% of the total area or 3.95 ha), and areas of long term low yield (L) (corresponding to 34% of the total area or 2.04 ha) were identified from yield maps. In both areas, we monitored one site fertilised according to precision farming practices (P), and one site fertilised according to conventional agricultural practices (C) for N_2O fluxes. N_2O flux rates were measured weekly using closed chambers in five replicates per site.

Soil profile data were obtained from 24 points of a 50 by 50 m grid. The high and the low yield area comprised 15 and 9 grid points, respectively. To each point, the two scenarios (C) conventional management and (P) site-specific management were applied. The sum of nitrate leached was calculated from the amount of drainage water and the nitrate concentration in 90 cm depth and was averaged for the respective yield areas.

For modelling nitrate leaching the model system ExpertN was used (Stenger *et al.*, 1999). This system allows one to choose from several algorithms to simulate water flow, soil heat transport, C and N turnover, plant growth and tillage. Parameterisation and model testing was undertaken in earlier studies with data from the same site (Stenger *et al.*, 1995).

Yield data were obtained from the Chair of Agronomy and Plant Breeding, TU Munich.

Results and discussions

The temporal course of N_2O fluxes is characterised by several emission peaks, which differ in magnitude and duration (data not shown). However, these emission peaks occurred both in the high and in the low yield area, indicating common driving variables.

In order to assess the effects of site-specific fertilisation on cumulated N_2O emissions, the investigation period was considered in total, but also subdivided into three periods following management practices (Tab. 1). In the high yield area, emissions from the site fertilised conventionally (HC) were not significantly different from emissions from the site fertilised site-specifically (HP) (Tab.1). Apparently, due to intensive mineralisation from the soil, soil nitrate contents were not limiting for N_2O emissions irrespective of fertiliser amounts. However, in the low yield area, site LP, receiving less fertiliser due to precision farming measures, emitted 33 % less N_2O than site LC (Tab. 1). This applies to the total investigation period and the period following site-specific fertilisation (Tab. 1).

Table 1. N fertiliser rates and cumulative N losses

	HP	HC	LP	LC
Period A	08.06.99-06.06.00 (complete investigation period)			
$N_2O^{1)}$	8.7^a±1.2	7.9^a±1.13	3.3^b± 0.5	4.7^c±0.83
$NO_3^{2)}$	34.1^a±14.2	30.5^a±12.6	25.5^a±9.7	27.5^a± 10.4
fertiliser$^{3)}$	*290*	*260*	*235*	*260*
Period B	08.06.99-26.09.99 (after differentiated fertilisation (maize))			
$N_2O^{1)}$	4.2^a±0.8	3.8^a±0.4	2.0^b±0.2	3.1^c±0.5
$NO_3^{2)}$	3.8^a±0.8	3.7^a±0.8	3.9^a±1.0	3.2^a± 0.8
fertiliser$^{3)}$	*175*	*150*	*125*	*150*
Period C	27.09.99-26.03.00 (after maize harvest/ sowing of winter wheat until first fertilisation)			
$N_2O^{1)}$	1.3^a±0.4	1.7^a±0.5	0.3^b±0.2	0.4^b±0.1
$NO_3^{2)}$	23.1^a±10.6	20.4^a±9.2	16.5^a±6.7	18.4^a±7.3
fertiliser$^{3)}$	*%*	*%*	*%*	*%*
Period D	27.03.00-06.06.00 (after differentiated fertilisation (winter wheat))			
$N_2O^{1)}$	3.2^a±0.6	2.4^b±0.8	1.0^c±0.2	1.2^c+0.4
$NO_3^{2)}$	7.1^a±3.0	6.4^a±2.7	5.0^a±2.2	5.7^a±2.5
fertiliser$^{3)}$	*115*	*110*	*110*	*110*

$^{1)}$kg N_2O-N ha^{-1} $^{2)}$kg NO_3 -N ha^{-1} $^{3)}$kg N ha^{-1}
HP: high yield, fertilised according to precision farming HC: high yield, conventional, uniform fertilisation
LP: low yield, fertilised according to precision farming
LC: low yield, conventional, uniform fertilisation
data are shown as arithmetic mean with standard deviation
different letters indicate significant differences between sites

One aim of the study was the evaluation of the effect of site-specific fertilisation on the whole field, necessitating an extrapolation of the data obtained for the sites to the area of the overall field. The measured N_2O flux rates from the sites in the high and low yield area were used to calculate flux rates on basis of the share of the respective zones.

As stated above, 66% of the investigated field had been assigned high yield area, and 34% had been assigned low yield area. Accordingly, assuming the overall field were fertilised uniformly according to conventional practices, 260 kg ha^{-1} fertiliser-N were applied to the whole field, resulting in N_2O emissions of 6.80 kg N_2O-N ha^{-1} and nitrate leaching losses of 29.5 kg NO_3-N ha^{-1}. This corresponds to 2.6% of the fertiliser-N lost as N_2O and 11.4% lost as nitrate. If, in contrast, the overall field were fertilised according to precision farming measures, on average 271.3 kg fertiliser-N were applied, resulting in 6.83 kg N_2O-N ha^{-1} emitted and 32.2 kg NO_3-N ha^{-1} leached. This corresponds to 2.5% of the fertiliser-N lost as N_2O and 11.6% lost as nitrate. Thus, if the whole field were fertilised according to precision farming measures, in total more fertiliser were applied. However, the share of fertiliser-N lost to the atmosphere as N_2O and leached as nitrate were not higher than in the conventionally fertilised variant.

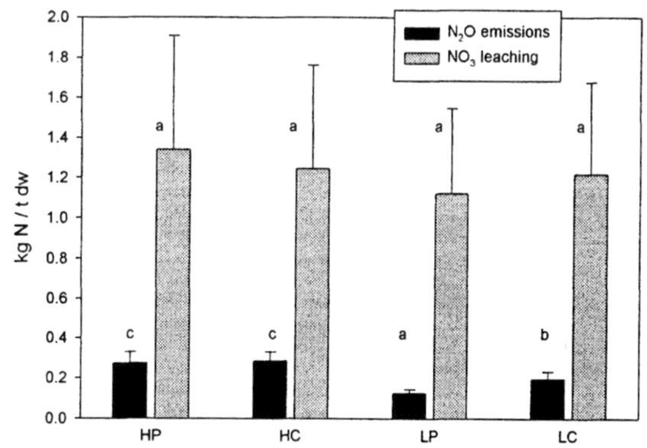

Figure 1. N loss related to yield (periods B and C, tab.1) different letters indicate significant differences between sites

In order to assess the environmental impact of management practices, N losses (N_2O, nitrate) from June 1999 until the fertilisation of winter wheat in March 2000 were related to maize yield (Fig.1). N_2O emissions based on plant yield from site LP were significantly lower than N_2O emissions from sites HP, HC and LC (Fig.1).

Correspondingly, N use efficiency, expressed as kg plant dry weight divided by kg fertiliser N revealed a tendency towards higher efficiency in site LP as compared to the other sites (data not shown). On the contrary, nitrate leaching rates were not significantly different between the treatment variants (Fig.1).

References

Stenger R, Priesack E and Beese F 1995 Plant Soil 171, 323-332.

Stenger R, Priesack E, Barkle G and Sperr C 1999 *In* NZ Land Treatment Collective. Proceedings Technical Session 20. Modelling of Land Treatment. Eds. M Tomer, M Robinson and G Gielen.

W. J. Horst et al.(Eds.), Plant nutrition – Food security and sustainability of agro-ecosystems. 904~905, 2001

Modelling controlled nutrient release from coated fertilisers: Deterministic and statistical model

A. Shaviv

Faculty of Agricultural Engineering, Technion-IIT, Haifa, Israel. agshaviv@tx.technion.ac.il

Key words: controlled release, mechanism, statistical approach

Abstract

Two modelling approaches are presented: a mechanistic model describing the release from membrane coated controlled release fertilisers, CRFs, and a statistical model based on the mechanistic one. The mechanistic model consists of three stages: a lag period, a stage of constant release and a stage of gradual decay of the rate. The model emphasises the importance of the vapour pressure gradient and permeability in affecting the lag period and the role of the concentration gradient and fertiliser permeability through the coating during the linear and decay periods of release. The model predicts the release in terms of the geometrical and chemo-physical parameters: granule radius and coating thickness, vapour and solute permeability, saturation concentration of the fertiliser and its density. The statistical model is based on the mechanistic one stressing the significance of variation of granule radius, coating thickness and nutrient permeability within a population of coated granules. The model is of particular importance for CRF's developers since it emphasises quantitatively the importance of standard deviations of granule radius and coating thickness and the inter-relation between water and nutrient permeability through the coating.

Introduction

The controlled release fertilisers, CRFs, based on coating with hydrophobic polymers are perceived as the ones that provide the best control over release under field conditions (e.g., Trenkel, 1997; Shaviv, 2000). The release from these fertilisers and especially the ones coated with organic polymers or resins is by the diffusion mechanism (Shaviv *et al.*, 2000). The pattern of temporal release from a population of coated fertilisers is generally linear or sigmoidal (Raban *et al.*, 1997; Shaviv, 2000) and it can be well synchronised with the pattern of nutrient uptake by plants (Shaviv, 1996). A mechanistic mathematical model which describes the release from a single coated granule under the diffusion mechanism was recently developed by Shaviv *et al.*, (2000). The release is controlled by: the mass transfer properties of the coating material, properties of the active agent within the granule, and coating thickness and granule radius. The release consists of three stages: lag period; a stage of constant release and finally a phase of gradual decay of the release rate. The pattern of cumulative release from such a granule is thus sigmoidal. Yet the release pattern obtained for a population of coated granules may be significantly different from the one expected for one single granule due to large variation in properties such granule radius, coating thickness or permeability of water or solute (Zaidel, 1996). In this presentation both the mechanistic model and a statistical model based on it are presented stressing the special cases/situations where they may be applicable.

Model development

Mechanistic model for diffusion from a single granule

The conceptual model for the release stages through diffusion from a polymer coated CRFs consists of three stages (Shaviv *et al.*, 2000): i. A lag period during which water penetrates the granule and dissolves part of the internal solid. The driving force for this process is the vapour pressure gradient across the coating. ii. Once a critical volume of saturated solution accumulates inside the granule the second stage of the constant release rate starts. iii. After the solid fertiliser in the core is dissolved the concentration of the internal solution decreases due to the continuing release and the fact that water continues to flow into the granule due to the water potential gradient driving it. Accordingly, the driving force for the nutrient release decreases and the release rate decays. This is the third stage of the release denoted the "decay phase". The conceptual model serves as the basis for expression of fractional release, $g(r, l, t)$, under the diffusion mechanism for a CRF with radius r and coating thickness l:

$$g(r,l,t)=\begin{cases} 0 & t<t' \\ \dfrac{3P_s \mathrm{C_{sat}}}{rl}(t-t') & t'\le t<t^* \\ 1-\dfrac{\mathrm{C_{sat}}}{\rho_s}\exp\left[-\dfrac{3P_s}{rl}(t-t^*)\right] & t\ge t^* \end{cases} \quad (1)$$

where, $t'=\dfrac{\gamma rl}{3P_h \Delta \mathrm{P}}$ is the duration of the lag-period;

$t^* = t' + \left(1-\dfrac{\mathrm{C}_{sat}}{\rho_s}\right)\dfrac{rl\rho_s}{3P_s\mathrm{C}_{sat}}$ is the duration of the linear release period; γ is the total fraction of voids within the granule; P_h and P_s are water and solute permeability coefficients, respectively; ρ_s is density of the fertiliser and

C_{sat} is the saturation concentration.

Statistical Model for release from a population of granules

When a large population of N coated granules is considered it is assumed that granules differ one from the other by two main geometrical factors: granule radius r and coating thickness l. The release from such a system could be described by solving simultaneously N differential equations, which practically does not appear to be an effective approach. Alternatively, a statistical treatment is presented which accounts for the effects of variation in CRF properties (e.g., coating thickness, granule radius, permeability) on the release from a large population of granules. The approach presented below is a modification/elaboration of the description of the release from a population of coated granular drug proposed by Dappert and Thies (1978).

In a real population of coated granules the fractional release at time t, $C(t)$ takes the form (Shaviv *et al.*, 2001):

$$C(t) = \int_{r_{\min}}^{r_{\max}} \int_{l_{\min}}^{l_{\max}} g(r,l,t)\,\psi(r,l)\,\mathrm{d}r\,\mathrm{d}l \tag{2}$$

where the fractional release at time t is described by the function $g(r_i, l_i, t)$ which is basically the same for all individual granules (Eq. 1) and only the radius r_i and thickness l_i differ among granules; $\psi(r,l)$ is a probability density function defined for a definite range of the radii: $r_{\min} \le r \le r_{\max}$ and the coating thickness: $l_{\min} \le l \le l_{\max}$. Assuming that the two variables r and l are independent, the probability density function $\psi(r,l)$ can be expressed as a product of the two probability density functions of each of the variables: $\psi(r,l) = \varphi_r(r)\,\varphi_l(l)$.

Results and discussion

The two approaches were verified against results obtained for release from single granules using Eq. 1 and for release from populations of granules applying the integration in Eq. 2. Parameters and rate constants used for simulation were evaluated experimentally with individual granules and with their population. The statistical model was also used to perform a sensitivity analysis.

The mechanistic model stresses the importance of the product $r \times l$ in determining the rate of release in all three stages. Smaller water permeability induces a larger lag period and lower nutrient permeability reduces the rate of release, while the saturation concentration increases it.

The sensitivity analysis indicated the importance of water permeability in controlling the lag period and solute permeability in governing the rate of linear release and the total duration of the release. This is expected from Eq. 1, but due to variations in $r \times l$ there might be some "smearing" effect, which shortens the lag and the period of linear release. Increasing the mean values of granule radius or coating thickness, which are normally distributed, increases the lag period and the total period of release. The half-life time of the release is inversely proportional in this case to product of the mean values. The variation of radii and coating thickness, within the realistic ranges of this study, affected the release only when the standard deviation became very large, or when the water permeability was reduced without affecting the solute permeability (practically possible). Shifting from a normal distribution to a lognormal one had negligible effect on the release, whereas a polynomial and uniform distribution tended to shorten the lag period and to prolong the release. This effect was amplified when the water permeability was reduced without changing the solute permeability.

The statistical model provides an effective tool for better designing and improving CRF release characteristics. The mechanistic model can also (practically) be applied for populations of granules, which governing factors of release do not vary to an extent that changes the sigmoidal release pattern significantly.

References

Dappert T and Thies C 1978 J. Membrane Sci. 4, 99-113.

Raban S, Zeidel E and Shaviv A 1997 *In* Third Int. Dahlia Greidinger Sym. on Fertilisation and The Environment. Eds. JJ Mortwedt and A Shaviv. pp. 287-295. Technion, Haifa.

Shaviv A 1996 *In* Progress in Nitrogen Cycling Studies. Eds. Van Cleemput *et al.* pp. 285-291. Kluwer Academ Pub. The Netherlands.

Shaviv A 2000. Advan. Agron. 71, 2-49.

Shaviv A, Raban S and Zeidel E 2000 Eur. J. Soil Sci., (submitted).

Shaviv A, Raban S and Zeidel E 2001 Eur. J. Soil Sci., (submitted).

Trenkel ME 1997 IFA, Paris.

Zaidel E 1996 D. Sc. Thesis, Agric. Engn. Technion - IIT.

W. J. Horst et al. (Eds), Plant nutrition – Food security and sustainability of agro-ecosystems. 906~907, 2001

Inventory and budget of nutrients in managed German forest stands - a method for sustainable nutrient management

P. Rademacher, B. Mueller-Using, H. Meesenburg and K. J. Meiwes
Forest Research Institute of Lower Saxony/ Department of Environmental Control, Graetzelstr. 2, Goettingen, D 37079, Germany, peter.rademacher@nfv.gwdg.de

Key words: biomass production, nutrient budget, nutrient export, nutrient storage, tree compartments

Abstract

The biomass production and the nutrient accumulation in trees was measured and calculated in northwest German pine and oak stands for tree compartments and whole stands. Biomass and the nutrient export were modelled for different harvesting procedures. Ca and Mg budgets calculated from fluxes with atmospheric deposition, silicate weathering and seepage output were negative. Ca and Mg export by harvesting is increasing this deficit. For the oak stand the effect of two harvesting intensities are compared with respect to Ca export. The necessity of recompensing measures such as Ca fertilisation is discussed.

Introduction

Considerable atmospheric input of acids and nitrogen has induced an accelerated leaching of nutrients from forest ecosystems in northwest Germany. Thus sustainable forest management today is more difficult than has hitherto been customary. Particularly whole tree harvesting has come under criticism, because the additional export of slash reduces the nutrient pools in the forest ecosystem. These issues have been investigated in a pine and an oak stand in northwest Germany.

Materials and methods

The pine stand (115 years old, II. yield class), which grows on a sandy soil at the northwest German forest site FUHRBERG, has a low to medium supply of nutrients. The soil is a podzol. The oak stand (115 years old, II. yield class) is situated 50 km north at the forest site SELLHORN, the nutrient supply is medium, and the soil is a dystric cambisol. In the pine stand 35 trees were harvested and each tree was separated in below and above ground tree compartments (Table 1). In the oak stand only the above ground compartments were investigated for 6 trees. Data of the roots were extrapolated from Murach and Wiedemann (1988). Biomass compartments with a diameter < 7 cm were weighed. Solid wood sections > 7 cm were determined by volume measurements (Rademacher *et al.*, 1992). The element content as well as the density of the biomass was specified in aliquots.

The increment and the allocation of nutrients over a whole rotation period were simulated for each tree compartment using the growth model BWIN (Nagel, 1999) and the biomass and nutrient model B-BAUM (Rademacher, unpubl.). Additional data for atmospheric input, seepage output and silicate weathering as well as the nutrient pools in organic layers and mineral soil have been derived from neighbouring monitoring sites (Meesenburg *et al.*, 1995; Rademacher *et al.*, 1999; Rademacher *et al.*, 2001).

Results

The actual tree biomass was 170 t ha^{-1} and 230 t ha^{-1} in the 115 years old pine and oak stand, respectively (Table 1). Compared to the pine stand, the oak stand had 40% more biomass, 50% more Mg, 200% more N, P and K and about 300% more Ca.

Table 1. Biomass (BM) and element pools in tree compartments of a pine and an oak stand (115 years old).

PINE-compartments	BM (t/ha)	N (kg/ha)	P (kg/ha)	K (kg/ha)	Ca (kg/ha)	Mg (kg/ha)
needles	2,9	50,0	3,3	12,6	9,2	2,1
branches <7cm	15,1	55,8	3,6	15,4	29,9	5,0
bark of branches >7cm	0,4	2,7	0,2	0,9	4,2	0,3
wood of branches >7cm	5,0	2,6	0,1	1,7	4,9	1,0
bark of stems >7cm	8,7	29,8	2,3	10,1	63,4	3,7
wood of stems >7cm	110,0	52,6	3,0	29,2	79,3	17,7
roots > 0,2cm	22,2	39,7	2,3	19,0	18,7	5,4
roots <0,2cm	4,1	24,1	2,7	7,2	3,4	1,3
total tree	168,3	257,2	17,5	96,0	213,1	36,4
OAK-compartments	BM (t/ha)	N (kg/ha)	P (kg/ha)	K (kg/ha)	Ca (kg/ha)	Mg (kg/ha)
leaves	2,3	66,5	4,4	18,6	12,0	4,0
branches <7cm	24,8	154,8	13,1	45,3	120,2	12,3
bark of branches >7cm	16,6	116,3	7,3	37,7	283,8	11,0
wood of branches >7cm	62,2	63,8	10,0	64,5	28,9	5,7
bark of stems >7cm	13,6	91,9	4,4	20,1	283,2	6,0
wood of stems >7cm	81,6	114,4	11,1	58,9	42,8	6,5
roots > 0,2cm	25,0	58,3	5,6	35,2	72,4	5,0
roots <0,2cm	3,8	38,8	2,9	10,0	7,4	3,6
total tree	229,9	704,8	58,8	290,3	850,7	54,1

One third of the Ca pool in the pine stand was allocated in the wood > 7 cm and another third in the corresponding bark. In the oak stand only 10% of total Ca was allocated in the wood > 7 cm and 67% in the bark. About 50% of the Ca in the oak wood > 7 cm was situated in the branches, in pines this was less than 10%.

During a whole rotation period biomass values of about 450 t ha^{-1} 125 yr^{-1} and 850 t ha^{-1} 185 yr^{-1} were calculated for the pine and the oak stand, respectively, with a rotation period which is 60 years longer in the oak stand (Table 2). In the oak stand the higher biomass production is accompanied by a Ca accumulation of about 2300 kg ha^{-1} 185 yr^{-1}. In the pine stand only 560 kg Ca ha^{-1} 125 yr^{-1} are needed. 70-80% of the biomass and the Ca is allocated in the above ground parts of the trees and 20-30% in the roots.

Table 2. Biomass (BM) and Ca pools in pine and oak stands and possible nutrient export by harvest in a whole rotation period. Low utilisation techniques, which leave up to 80% of the crown wood > 7 cm at the site, are not common in pine stands.

biomass compartment	BM (t/ha)		Ca (kg/ha)	
tree species	PINE	OAK	PINE	OAK
total tree biomass	436	843	559	2316
above ground tree biomass	310	670	438	1861
wood > 7 cm incl. bark	292	614	335	1514
wood > 7 cm incl. bark/ high utilization	248	555	285	1359
wood > 7 cm incl. bark/low utilization	-	376	-	896
wood > 7 cm without bark/high utilization	229	496	158	248

The utilisation of the total above ground wood > 7 cm incl. bark would yield about 300 t ha^{-1} biomass in the pine and 600 t ha^{-1} in the oak stand. This is 67 and 73% of the total tree biomass, respectively. The Ca export by the utilisation of wood > 7 cm would diminish the Ca storage of the ecosystem by about 335 kg ha^{-1} in the pine stand and by about 1500 kg ha^{-1} in the oak stand, approximately 60 and 65% of the stand's storage, respectively. In pine and oak stands, harvest techniques are common which leave a minimum 10-20% of those solid wood types at the site which are not able to be sold or do not yield any margin. Accordingly, the Ca export would decrease between 50 kg ha^{-1} (pine) and 150 kg ha^{-1} (oak). In oak stands forest harvest procedures which do not even use up to 80% of the crown wood > 7 cm are usual. Thus, about 240 t ha^{-1} wood > 7 cm would remain at the site and the corresponding decrease of the Ca export would amount up to 600 kg ha^{-1}. Debarking of the wood at the site would decrease the biomass export by only about 10%, but would reduce the Ca export by about 45% in pine and 80% in oak stands.

Discussion

Fig. 1 shows the fluxes and the pools of Ca at the oak site: during a rotation period of 185 years the atmospheric input of Ca amounts to about 500 kg ha^{-1}, whereby the seepage output of about 1400 kg ha^{-1} is three times higher. The difference between atmospheric input and seepage output can not be balanced by the Ca input by silicate weathering (76 kg ha^{-1}). Due to this negative budget the Ca pools in the organic layer (84 kg ha^{-1}) and the mineral soil (rd. 200 kg ha^{-1}) will be depleted, if the flux rates do not change.

Under these conditions, different intensities of oak wood utilisation may be considered to reduce the Ca deficit. If only stem wood > 7 cm without bark is utilised, the corresponding Ca export is 248 kg ha^{-1}, whereas 1359 kg ha^{-1} is removed from the site if wood > 7 cm with bark is used. To compensate the Ca losses due to harvesting, an application of 0.6 t ha^{-1} lime in case of the utilisation of debarked wood and of 3.5 t ha^{-1} in case of the utilisation of wood and bark would be required. The sustainability of nutrient cycles should be considered within the framework of an economical and ecological examination of forest management practices.

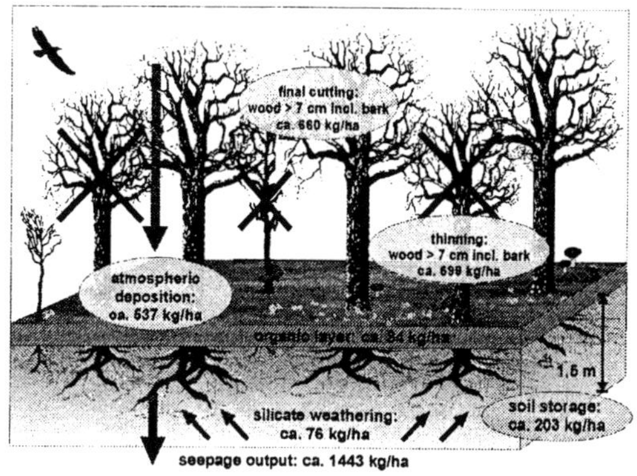

Figure 1. Budget of calcium during a rotation period of 185 years in an oak stand in northwest Germany. Considered fluxes: atmospheric deposition, silicate weathering and seepage output as well as removal of timber leaving 20% of smaller wood on the stand's ground. Actual pools in organic layer and mineral soil 0-1.5 m.

References

Meesenburg H, Meiwes KJ and Rademacher P 1995 Water, Air and Soil Pollution 85, 611-616.

Murach D and Wiedemann H Ber. d. Forschungszentrums Waldökosysteme/ Waldsterben d. Univ. Göttingen 10, 287 p.

Nagel J 1999 Schr. aus d. Forstl. Fak. d. Univ. Göttingen u. d. Nds. Forstl. Vers. Anstalt 28, 128 p.

Rademacher P, Buß B and Müller-Using B 1999 Forst und Holz 54, 330-335.

Rademacher P, Meesenburg H and Müller-Using B 2001 Forstarchiv 71, accept. for publ..

Rademacher P, Ulrich B and Michaelis W 1992 *In* Luftverunreinigungen und Waldschäden am Standort "Postturm", Forstamt Farchau/ Ratzeburg. 92/E/100. Ed W Michaelis and J Bauch. pp 149-186. GKSS-Forschungszentrum, Geesthacht.

W. J. Horst et al. (Eds), Plant nutrition – Food security and sustainability of agro-ecosystems. 908~909, 2001

Long-term nutrient dynamics of a spruce stand fertilised periodically for more than 50 years

M. Mindrup[1)], K. J. Meiwes[1)] and P. K. Khanna[2)]

[1)] *Forest Research Institute of Lower Saxony, Graetzelstr. 2, Goettingen, D 37079, Germany, michael.mindrup@nfv.gwdg.de*

[2)] *Division of Forestry and Forest Products, CSIRO, P.O. Box E4008, Canberra - Kingston, ACT 2604, Australia, Partap.Khanna@ffp.csiro.au*

Key words: atmospheric, calcium, deposition nitrogen, phosphorous, potassium

Abstract

Soil and needle samples were periodically analysed from 1964 - 1994 in a long-term fertiliser experiment in a spruce stand growing on a sandy soil in north Germany. The stand was established in 1929 and received a number of applications of N, P, K, Ca fertilisers. Application of lime and P-fertiliser increased levels of Ca and P in the surface soil and needles, whereas the effects of K and N fertilisers were short-lived. High atmospheric inputs of N increased N concentrations in the needles to levels which were equivalent to those measured one year after application in 1971 of 41 kg N ha^{-1}. These long term observations in a spruce stand suggest that a number of plant and soil factors interact in a complex way to determine the long-term nutrient dynamics in a forest.

Introduction

The long-term nutrient dynamics of a forest is primarily determined by their biogeochemical cycling. During the life of a spruce stand the cycling of nutrients and thus the nutrient status is expected to change due to changes in the stand and environmental conditions. Any evaluation of the long term changes in the nutrition of a spruce stand will depend upon the knowledge of how tightly a nutrient cycle works, and what proportion of the applied nutrients enter that cycle. The availability of nutrients which primarily refers to their supply will determine the cycling pattern (amount and intensity of cycling) of each nutrient. The supply will change when either the inputs (changes in atmospheric N or S inputs or fertiliser inputs) or the decomposition patterns (change in climatic variables or the substrate quality – high decomposable material after N inputs) will change with time. The objective of this study was to address some of these issues by analysing data from a spruce stand which was first fertilised in 1929 and received fertilisers periodically until 1970.

Materials and methods

The experimental site is located in the northwest of Germany in the sandy pleistocene region (longitude 35830 latitude 58730) and belonged to the former forestry district Oerrel (compartment 269/281) (Seibt *et al.*, 1968). Mean annual temperature is 8.3 °C and mean annual rainfall 702 mm.

The spruce stand was established in 1929 with treatments following randomised block design and three replicates. Total amount of nutrients (kg ha^{-1}) applied as fertiliser from 1929 to 1970 were: N (258), P (85), K (217), Ca (1931) and Mg (17). Fertiliser application occurred during the following periods: 1929 - 1938, 1948/49, 1956/58 and 1970; 830 kg Ca ha^{-1} as lime in 1956; 41 kg N ha^{-1}, 7 kg Mg ha^{-1} and 75 kg K ha^{-1} in 1970. The site has sandy soil and moder as humus type. Nutrient status of trees was assessed by analysis of current and three years old needles of the 7th whorl.

Results

Effect on soil properties

Thirty eight years after the last application of lime the pH of the forest floor had increased significantly from 3.0 (control) to 3.7 in the L+F layers and significantly from 2.7 (control) to 3.4 in the H layer. In the 0-5 cm soil the mean pH value on limed plots was significantly higher (3.3) than control (3.0). In deeper soil layers no change in soil pH after liming was observed. The mean Ca content of the forest floor increased on liming from 2.4 to 9.0 mg g^{-1} (L+F layers) and 1.6 to 8.0 mg g^{-1} (H layer). These mean values of control and limed plots were, however, not significantly different. However, proportion of Ca on the exchange sites of the mineral soil increased significantly for the surface 30 cm depth (0-5 cm = 4.0% (control) to 32.6% (limed), 5-10 cm = 1.6% (control) to 8.6% (limed), and 10-30 cm = 1.3% (control) to 4.5% (limed).

P-fertiliser application in 1929 and 1956 affected the P-cycling resulting in a long-term increase in P levels in the soils. In 1994 P content of the forest floor increased to 0.94 mg g^{-1} (0.82 mg g^{-1} in control) in the L+F layers and to 0.77 mg g^{-1} in the H layer (0.57 mg g^{-1} in control). In the 0-5 cm soil total P increased from 140 mg g^{-1} in control plot to 222 mg g^{-1} fertilised. In deeper soil layers the differences were small.

Application of K and N fertilisers did not show any effect on the soil chemical properties.

Effect on tree nutrition

N concentrations of the spruce needles in 1964 and 1966 did not show any effects after application of N fertiliser in 1956. However the values in the years after

1966 lie higher than those measured before, indicating the effect of atmospheric N inputs. Needle N values measured one year after fertiliser application in 1970 were similar (15.5 mg N g^{-1}) to those measured in 1994 on the control site. The effect of N application is usually considered to be of short to medium term because of its storage in the plant which in most cases occurs together with an expansion in the canopy mass and area.

Effects of P fertiliser on needle P concentrations were long lasting, but the values obtained depended on the age of the needles tested. For the period 1964 – 1971, 1-yr old needles from fertilised trees showed significantly (t-test) higher P concentrations (1.6 mg P g^{-1}) than those from control (1.2 mg P g^{-1}). However no significant differences in P concentrations were observed in one year old needles collected from control and fertilised treatments during 1976 – 1994. On the other hand, P content of 3-yr old needles measured in 1994 was significantly higher (1.2 mg P g^{-1}) in fertilised trees than those from control (1.0 mg P g^{-1}).

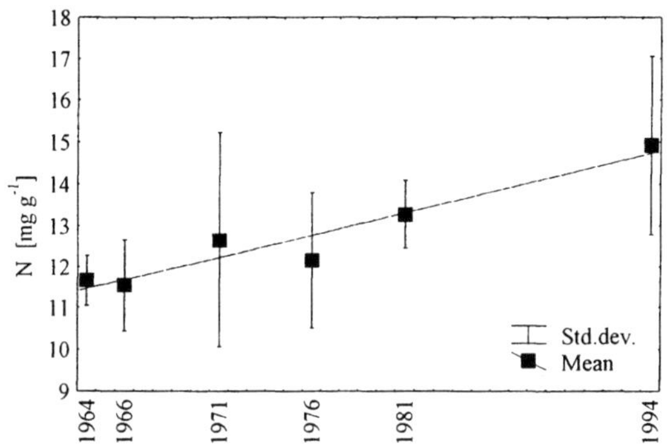

Figure 1. N contents of 1-yr old (7th whorl) needles of spruce in the unfertilised control during 1964 - 1994

Effect of liming on Ca content of 1- and 3-yr old needles was evident until 1994 (control 4.8 resp. 4.4 mg Ca g^{-1}, limed 9.1 resp. 12.9 mg Ca g^{-1}); in 4 out of 7 measurements statistically significant results were obtained. Concentration of K in the needles improved on application of K-fertiliser, but the values were significant from control only in 1966 and 1971 samplings. In 1971 one-yr old needles had 6.1 mg K g^{-1} in the control compared to 8.3 mg K g^{-1} in the fertilised stand.

Longterm changes in the N nutrition of trees

Nitrogen concentration in 1-yr old needles from control trees (Fig 1) has continuously increased from 11.7 mg g^{-1} in 1964 to 14.9 mg g^{-1} in 1994 (highly significant linear regression with $p < 0.001$; $R^2 = 0.39$).

Mean values of N concentration in 1994 were significantly higher than those from 1964, 1966 and 1976 ($p < 0,05$; Tukey's-HSD test for unequal N). Similar results were obtained for 3-yr old needles. These increase in N levels in the needles is related to high atmospheric N inputs. In a nearby located pine stand in Sellhorn an annual atmospheric input of 32 kg N ha^{-1} was reported (Meesenburg *et al.*, 1995).

Conclusions

The application of P fertiliser and lime caused a long-term increase in the concentration of P and Ca in the spruce needles. The increase was evident in the samples collected in 1994, 38 years after the last application of these two elements in 1956. Despite a small amount of P applied compared to the total amount present in this highly acid soil, it increased the concentrations of P in the forest floor. These changes in the P content of needles and the forest floor may be due to high fraction of P uptake from fertiliser, and its translocation and withdrawal from the senescent needles in the canopy. High Ca concentrations in the needles of the spruce were more related to the long term changes in the supply and uptake of Ca as indicated by high exchangeable Ca in the surface horizons of this soil. There was some indication of a small increase in K in the spruce needles on K application but due to its high mobility in the plant and soil systems any changes may only be measured by careful assessment of throughfall and litter fall components of K cycle. These long term observations in a spruce stand suggest that a number of plant and soil factors interact in a complex fashion to determine the long term nutrient dynamics in a forest.

Acknowledgement

We wish to thank Dr. J. B. Reemtsma for providing accurate needle data.

W. J. Horst et al. (Eds), Plant nutrition – Food security and sustainability of agro-ecosystems. 910–911, 2001

Changes in the nitrogen budget of forest ecosystems under sustained reduction of inputs - analysis of state and prognoses based on a case study in a Norway spruce forest at Solling, Germany

N. P. Lamersdorf[1)], M. Bredemeier[2)] and F. Beese[1)]

[1)] *Institute of Soil Science and Forest Nutrition, Göttingen University, Büsgenweg 2, Göttingen, D 37077, Germany, nlamers@gwdg.de*

[2)] *Forest Ecosystems Research Centre, Göttingen University, Büsgenweg 1, Göttingen, D 37077, Germany, mbredem@gwdg.de*

Key words: atmospheric deposition, element budgets, nitrate leaching, nitrogen, Norway spruce

Abstract

Results of a large scale field experiment with a permanent roof construction below the canopy in a 67-year-old Norway spruce (*Picea abies* (L.) Karst.) plantation at Solling, central Germany are presented. Throughfall water was continuously collected, partly de-ionised and re-sprinkled below the roof. Compared to the control plot, input fluxes for H^+; SO_4-S, NH_4-N, and NO_3-N were reduced to 22%, 51%, 14%, and 57%, respectively. Responses of soil water chemistry, fine root growth, and above ground stand parameters indicate a temporal and spatial response sequence (a strong and quick decrease of soil solution concentration of SO_4^{2-}, NO_3^-, and Al^{3+} in the upper soil layers, an improvement of the fine root growth, and a decrease of the nitrogen needle content, but only temporary changes in tree growth parameters). No significant change of the soil solution pH was observed. Annual flux budget for sulphate remained negative (net output), due to the ongoing dissolution of aluminium-sulphates, formerly accumulated in deeper soil horizons. Mean nitrogen budgets remained positive (net accumulation) but strongly decreased during the observation period. N-output fluxes exhibit an exponential increase with increasing throughfall input rates of inorganic nitrogen compounds.

Introduction

Clean air legislation in many European countries led to rapid and drastic changes in atmospheric deposition of sulphur and acidity in most forest ecosystems of Central Europe. However, for inorganic nitrogen compounds (NH_4^+, NO_3^-) no trend can be observed so far (Meesenburg *et al.*, 1997). Under current ambient deposition (30-40 kg ha^{-1} a^{-1}), leaching of nitrate is often observed, although the biggest part of the N-input is stored within forest ecosystems. In order to study the effects of strongly reduced atmospheric inputs, an ecosystem-scale field experiment with a roof cover below the canopy was established in a typical Norway spruce plantation on a strongly acidified soil at Solling, central Germany. The roof construction facilitated manipulations of throughfall chemistry, quantity and temporal distribution. Results of "clean rain" experiments are reported.

Materials and methods

Roof plots are located at 500 m a.s.l., in a currently (2001) 67-year-old Norway spruce (*Picea abies* L. Karst.) plantation. Average annual temperature is 6.4°C, average annual precipitation is 1090 mm. Soil type is a spodic dystric cambisol (FAO), strongly acidified ($pH_{(KCl)}$ 3.1 - 4.2, base saturation 5 %). Humus layer is a mor type ($pH_{(KCl)}$ < 3.0, C/N ratio = 25). Transparent roofs (3 in total, each 300 m^2 + 1 ambient control plot) were constructed in 1990 underneath the forest canopy about 3.5 m above the forest floor. The roof plot *D1* was used for the "clean rain" treatment and plot *D2* served as the roof control plot (no further manipulation than re-sprinkling of collected water). "Clean rain" application started in autumn of 1991, about two years after a pre-experimental phase with no manipulations. Compared to plot *D2*, deposition rates (mean of 1992-1997) for *D1* were reduced to the following input [kg ha^{-1} a^{-1} (% of *D2*)]: H^+ = 0.15 (22%), SO_4-S = 15.6 (51%), NH_4-N = 2.1 (14%) NO_3-N = 8.3 (57%). For further methodological details see Bredemeier *et al.*, (1998), Lamersdorf *et al.*, (1998), Xu *et al.*, (1998).

Results

Soon after beginning the experiment SO_4^{2-}, NO_3^-, and Al^{3+} concentrations decreased significantly in the upper soil horizons, but soil solution pH remained well within the aluminium buffer range (pH < 4.2). Flux budget for nitrogen remained positive (net accumulation) but decreased from +13.6 (mean of 1990-1991) to +2.8 (mean of 1992-1997) kg ha^{-1} a^{-1}. For sulphate, however, the output fluxes below the rooting zone (1 m soil depth) are strongly related to the seepage water fluxes, and budgets decreased due to the relatively wet years in the mid 1990's from -3.1 (1990-1991) to -28.2 (1992-1997) kg ha^{-1} a^{-1}. The net loss is attributed to the dissolution of aluminum-sulphates formerly accumulated in deeper soil horizons. Fine root growth improved, more fine root tips and more intensive deep rooting occurred and the fine root biomass of plot *D1* increased from about 60% in 1990 to more than 100% since 1994, relative to the control plot *D2* (see also Murach and Parth, 1999). Nitrogen needle content at *D1* decreased significantly ($p < 0.05$) in four of seven observation years (1993, 1995-1997, mean = 12.9 mg g^{-1})

below the reference values of plot *D2* (mean = 14.2 mg g^{-1}). Concerning tree growth parameters, there was no consistent response during the total observation period (1990-1998). The diameter increment improved significantly only in 1993 and 1995 and height increment showed no significant differences, compared to the control plot (see also Dohrenbusch *et al.*, 1999).

Discussion

Fig. 1 shows that the annual seepage output of nitrogen (in fact 100% nitrate at the site) is strongly related to the input of inorganic nitrogen compounds (at Solling about 50% nitrate and 50% ammonium). Including all experimental plots (3 roof plots + 1 ambient control) and a total of 8 years of observation, results indicate an exponential increase of the nitrate output with increasing input rates (throughfall). However, input rates of *D1* were artificially reduced to such low rates (mean of 1992-1997 = 10.4 kg ha^{-1} a^{-1}) that a further reduction of the input would not significantly influence the output rates. According to our data there is a threshold of about 15 kg ha^{-1} a^{-1} of atmospheric throughfall input below which no further reduction of the output appears. We therefore conclude that atmospheric nitrogen deposition should be at least reduced to rates of about 15 kg ha^{-1} a^{-1} to prevent the system output from a permanent nitrate contamination. On the other hand data indicate that internal nitrogen consuming or nitrate producing processes could contribute irregularly and significantly to the net annual nitrogen flux budgets. Such processes could be climate-induced nitrification pulses or an increase of the net plant N-assimilation or microbial immobilization rates.

Acknowledgement

The Solling roof project was funded by the German Federal Ministry of Education and Research (BMBF) and by the EU, DG XII.

References

Bredemeier M, Blanck K, Dohrenbusch A, Lamersdorf N, Meyer AC, Murach D, Parth A and Xu Y-J 1989 Forest Ecology and Management 101, 281-293.

Dohrenbusch A, Jaehne S and Meyer A-C 1999 AFZ/DerWald. 60-62.

Meiwes K-J, Schulze A and Rademacher P 1997 Arbeitshefte Boden. 77-95. Nds. Landesamt f. Bodenforschung.

Lamersdorf NP *et al.* 1998 Forest Ecology and Management 101, 95-109.

Meesenburg H, Meiwes K-J, Schulze A, and Rademacher P 1997 Arbeitshefte Boden. pp 77-95. Nds. Landesamt f. Bodenforschung.

Murach D and Parth A 1999 AFZ/DerWald. 58-59.

Xu Y-J, Blanck K, Bredemeier M and Lamersdorf NP 1998 Forest Ecology and Management 101, 295-306.

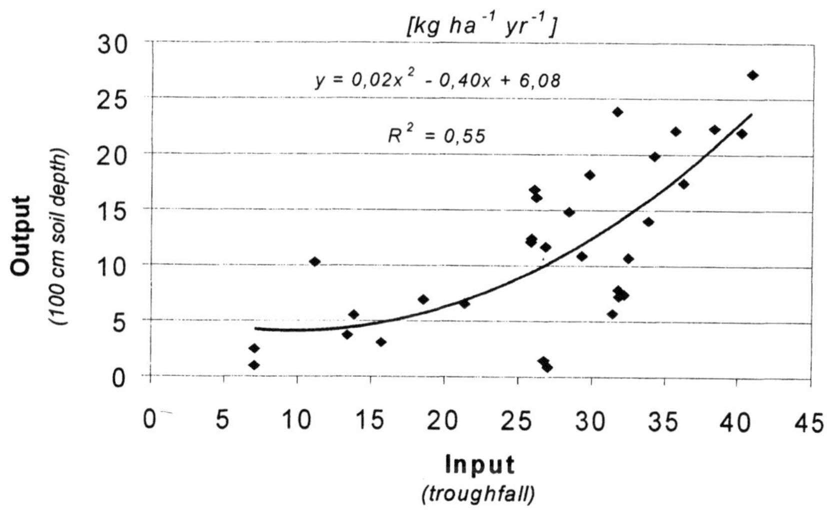

Figure 1. Input/output relation for annual flux rates of inorganic N-compounds ($NH_4^+ + NO_3^-$), 1990-1997, Solling roof project, all plots (rates for *D1* in 1992: Input = 10.6, Output = 22.8 kg ha^{-1} a^{-1}, not included in regression calculation).

W. J. Horst et al. (Eds), Plant nutrition – Food security and sustainability of agro-ecosystems. 912~913, 2001

Inverse estimation of parameters in a soil nitrogen turnover model for a beech forest

E. Priesack[1)], S. Gayler[1)], R. Brumme[2)], N. Bartsch[3)] and T. Vor[3)]

[1)] *Institute of Soil Ecology, GSF National Research Center for Environment and Health, Neuherberg, Oberschleißheim, D 85758, Germany, priesack@gsf.de*

[2)] *Institute of Soil Science and Forest Nutrition, Georg-August-University of Göttingen, Büsgenweg 2, Göttingen, D 37077, Germany, rbrumme@gwdg.de*

[3)] *Institute of Silviculture, Georg-August-University Göttingen, Büsgenweg 1, Göttingen, D 37077, Germany, nbartsc@gwdg.de*

Key words: beech forest, mineralisation, nitrification, nitrous oxide emission, nitrogen model, parameter estimation, soil nitrogen turnover

Abstract

Anthropogenic deposition of acids and nitrogen influences decomposition in contrary ways. Acid deposition reduces and nitrogen deposition may increase decomposition. The net effect on the N transformation rates is still questionable. To simulate N turnover in forest ecosystems (both above and belowground) rates of plant litter decomposition including C- and N-mineralisation, nitrification and denitrification have to be determined. Also the vertical gradient of these rates from OL, OF, OH, to the mineral soil horizons including their dependance on temperature and moisture has to be known. Therefore laboratory soil column experiments with ^{15}N were conducted to quantify nitrogen and carbon transformation rates. Field studies in a forest gap and a mature stand were performed to quantify net-N-mineralisation and nitrification by the use of the "close-top tube" method. Additionally, element losses were quantified by input-output studies and the emissions of CO_2 and N_2O were quantified by the closed-chamber method. These experimental datasets are used to calibrate corresponding N turnover submodels of the model system Expert-N, in order to simulate N availability in soils of a beech forest stand. This is achieved by a new inverse modelling technique, which enables one to estimate soil depth dependent parameters. Using additional, independent data sets the comparison of measured and simulated mineral N contents demonstrates the model's ability to describe seasonal patterns of soil respiration, N_2O emission and soil mineral N pool sizes.

Introduction

Because of their important role in providing information about expected possible long term changes in terrestrial ecosystems due to human impacts, carbon (C) and nitrogen (N) turnover processes are studied most intensively. Particularly rates of N mineralisation and soil respiration need to be determined to predict changes in soil C and N storage. We analysed laboratory and field experiments which were conducted to study C and N transformations in the forest soils of Solling. From the laboratory experiments, parameters of a C and N turnover model are estimated. This includes maximal rates for mineralisation and nitrification. These estimated parameters are then used to simulate soil N transport and transformation and to compare simulation results with data on ammonium and nitrate concentrations from the field experiments. It is the aim to see if the modelling C and N turnover processes can provide means to predict mineralisation, immobilisation and turnover in forest soils in a realistic way.

Material and methods

Laboratory experiment

Undisturbed soil cores (15 cm diameter) of the upper organic layer, 0-10 cm, and 10-20 cm soil depth were used to study the vertical distribution of C and N transformation rates in the soil profile of a beech stand in the Solling. The columns were installed in a microcosm system (Hantschel *et al.*, 1994, Loftfield *et al.*, 1997) and automatically irrigated with 4 mm d^{-1} of artificial throughfall. CO_2 emissions were measured automatically. Soil solutions were sampled and analysed to calculate net-nitrogen-nitrification and leaching losses of deposited ammonium and nitrate. The nitrogen in the irrigation solution was applied at a rate of 41 kg ha^{-1} yr^{-1} as NH_4NO_3, where ammonium and nitrate were labelled in separate experiments with 50 atom% ^{15}N to study the behavior of deposited ammonium and nitrate (Brumme *et al.*, 2002).

Field experiment

Four gaps of 30 m diameter were cut in a beech stand in the Solling in 1989. Two gaps were limed with 3 t ha^{-1} of fine dolomite. Precipitation, litter fall, photosynthetically active radiation, air and soil temperature were monitored. Deposition was determined above the forest, under the canopy, in the gap and in the stand (stem flow). From 1990-1998 soil solution was sampled every month with ceramic cups of the centres of the unlimed and limed gaps and of the stand at 10 cm and 80 cm depths (Vor, 1999).

Parameter estimation procedure

The estimation of solute transport and transformation parameters often is a highly nonlinear optimisation problem and graphs of the involved objective functions most often have a complex geometry containing many local minima. Therefore besides the usually applied Levenberg-Marquardt algorithm (Press *et al.*, 1992) an alternative approach based on an annealing-simplex method (Press *et al.*, 1992) was proposed (Pan and Wu, 1998). Both methods were implemented into the Expert-N program package, which provides routines to simulate water flow, N-turnover and N-transport in agroecosystems (Stenger *et al.*, 1999).

To accelerate and stabilise the estimation procedure both methods were combined with a multilevel approach (Chavent *et al.*, 1994; Knabner and Igler, 1997). This method consists of three different steps: first start the optimisation with the least possible number of degrees of freedom (if possible n=1); second interpolate the optimisation result for a parametrisation with one or more degrees of freedom added and use this as next start value; third repeat the second step as long as the optimisation result is not heavily disturbed by oscillations. This multi-level method may directly use information on soil horizons. Starting with average parameter values constant throughout the whole soil profile, assuming a homogeneous soil, and continuing by dividing the profile successively to approximate the given division into soil horizons often helps to reach an adequate parameter estimation. This is especially true for cases of high numbers of degrees of freedom and highly nonlinear objective functions.

The optimisation methods were used to fit nitrification and denitrification rates from data of a laboratory soil column experiment. Comparison of results obtained by the Levenberg-Marquardt and by the Simulated Annealing method showed only minor differences below the range of the numerical simulation error for the examples considered (Priesack and Achatz, 1999).

References

Brumme R, Priess J, Steinmetz G, Wang CP and Raubuch M 2002 *In* Human Impact on N and C Cycle in Beech Forest Ecosystems. Ed. Brumme R. Springer Verlag, Heidelberg, Berlin, New York. (in press).

Chavent G, Zhang J and Chardaire-Riviere C 1994 *In* Inverse Problems in Engineering Mechanics. Eds. Bui HD, Tanaka M, Bonnet M, Maigre H, Luzzato E and Reynier M. pp. 265-272. Balkema.

Hantschel R, Flessa H and Beese F 1994 Soil Sci. Soc. Am. J. 58, 401-404.

Knabner P and Igler B 1997 Preprint. Universität Erlangen-Nürnberg.

Loftfield N, Flessa H, Augustin J and Beese F 1997 J. Env. Qual. 26, 560-564.

Pan L and Wu L 1998 Water Resour. Res. 34, 2261-2269.

Press WH, Teukolski SA, Vetterling WT and Flannery BP 1992 Numerical recipies in FORTRAN, 2nd ed. Cambridge Univ. Press.

Priesack E and Achatz S 1999 *In* Modelling of Transport Processes in Soils. Eds. Feyen J and Wiyo K. pp. 641-649. Wageningen Pers, Leuven.

Stenger R, Priesack E, Barkle G and Sperr C 1999 *In* NZ Land Treatment Collective Proceedings Technocal Session 20: Modelling of Land Treatment Systems. Eds. Toner M, Robinson M and Gielen G. pp. 19-28. New Plymouth.

Vor T 1999 Berichte des Forschungszentrums Waldökosysteme, Reihe A 163, 1-168.

W. J. Horst et al. (Eds), Plant nutrition – Food security and sustainability of agro-ecosystems. 914~915, 2001

Ecosystem effects after ameliorative liming of a catchment at the Harz mountains, Germany

H. Meesenburg[1)], K. J. Meiwes[1)], M. Wagner[1)] and J. Prenzel[2)]
[1)] *Niedersächsische Forstliche Versuchsanstalt, Grätzelstr. 2, D-37079 Göttingen*
[2)] *Institute of Soil Science and Forest Nutrition, Büsgenweg 2, D-37077 Göttingen*

Key words: liming, organic layer, restoration, simulation, soil solution, streamwater, tree nutrition

Abstract

The effects of the ameliorative liming of a whole catchment on soil chemistry, soil solution, streamwater, and nutrition of the trees were studied at Steile Bramke, Harz mountains, Germany. Nine years after liming most of the added Ca and Mg are stored in the organic layer. Only in the top mineral soil substantial higher pools of Ca and Mg can be observed. In soil solution improved conditions for tree roots occurred within the main rooting zone. Streamwater has been nearly unaffected until present. The nutrition of the trees with Mg, Ca and P has been improved significantly. Long-term simulations suggest a duration of the liming effects of about 150 years.

Introduction

A major part of forest soils in Central Europe are highly acidified due to acid deposition and long term exploitation of forests. Liming is a frequently used measure against adverse effects of acidification (Meiwes, 1995). Usually, liming is conducted in forest practice with dosages of 2 to 5 to/ha to compensate for the actual acidic inputs. However, if a restoration of soils is aspired to, the lime requirement is much higher. Liming may enhance nitrification, inducing the risk of increased nitrogen leaching from the soil. Hence, liming measures have to be planned carefully to minimise negative impacts on ecosystems.

Study sites

The Steile Bramke catchment (38 ha) is situated in the Harz mountains, Germany, at an elevation of 510 to 670 m a.s.l. adjacent to the control catchments Lange Bramke and Dicke Bramke. The catchments are stocked with spruce plantations. Cumulative deposition of acids is very high due to regional mining industry during the last millenium and long range transport during the last century. During the last two decades, deposition of acids decreased at Lange Bramke and other sites in northern Germany (Meesenburg *et al.*, 1995). However, Critical Loads of acidity are actually exceeded by far at the study sites (Eberl *et al.*, 1999)

Methods

An ameliorative dosage (16 t ha^{-1}) of dolomitic lime was applied in 1989 to the Steile Bramke catchment. The lime dosage was calculated according to the base neutralising capacity of the upper mineral soil (0-40 cm) up to pH 5.0. Soil sampling was conducted at four experimental plots before liming in 1988 and in 1998. For sampling of soil solution at the plots, tension free lysimeter plates were installed under the organic layer and ceramic suction cups in 20, 40, and 80 cm depth of the mineral soil. Streamwater was monitored weekly since 1987 at the catchment outlets of Steile, Dicke, and Lange Bramke. Needle samples of the 1st and 3rd needle year from the 7th whirl were taken in 1988, 1993, 1997 at the experimental plots and at the Dicke Bramke catchment from five trees each. Analytical methods for soil samples, soil solution, streamwater, and foliar analyses are described by König and Fortmann (1996).

Results and discussion

Soil chemistry

Nine years after liming 78% of the lime has been dissolved. 36% and 45% of the added Ca and Mg has been stored in the organic layer, respectively, mainly adsorbed on ion exchange sites (Table 1). The effective ion exchange capacity has been increased about twofold in this period. Only 3% and 8% of the added Ca and Mg has been stored in the mineral soil, respectively. Below 5 cm depth, the base saturation remained lower than 20%.

Table 1. Pools of Ca and Mg in kg ha^{-1} in soils of Steile Bramke in 1988 and 1998

compartment	1988		1998	
	Ca	Mg	Ca	Mg
lime dosage	3982	1745		
undissolved lime			860	379
organic layer	224	40	2536	1202
0-5 cm	16	6	66	60
5-10 cm	12	5	40	36
10-20 cm	24	12	58	47
20-30 cm	13	7	30	23
30-50 cm	15	5	22	10

Soil solution

In soil solution acid neutralising capacity (ANC), pH, and concentrations of Ca^{2+} and Mg^{2+} increased after liming significantly. The Ca/Al ratio was raised to a level which is regarded as non toxic for tree roots. However, ANC remained negative below 20 cm depth. An increase of NO_3^- contents in soil solution could only be observed under the organic layer and in 20 cm depth.

Streamwater

Ca^{2+} and Mg^{2+} concentrations in streamwater decreased from 1989 until the present. Since the same trends have been observed for adjacent unlimed catchments, a liming effect can be excluded. Decreasing trends occured also for NO_3^- in streamwater of all investigated catchments. For other solutes, also no liming effects could be observed. For the given site conditions of the catchment even a liming dosage of 16 t ha^{-1} is not sufficient to alter the chemistry of streamwater significantly and to reduce the risk for streamwater biocoenosis or drinking water supply due to acidification.

Tree nutrition

The nutrition of the trees in the catchment has been improved for Ca and Mg. The Mg nutrition has been elevated to a range which can be rated as sufficient to optimal, whereas severe Mg deficiency was observed before liming (Fig. 1). A deterioration of the K nutrition after liming due to the Ca-K antagonism (Zech, 1970) was not found at Steile Bramke. Despite no P being applied during liming, P content of the foliage also increased significantly. Better conditions for P mineralisation might be attributed to this unexpected result.

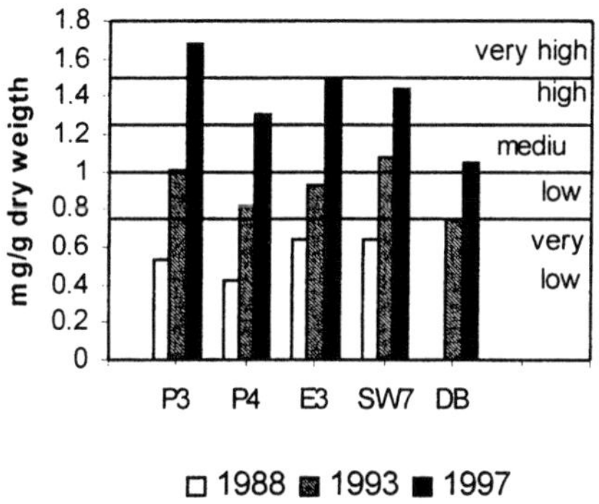

Figure 1. Mg content of 1st year spruce needles from intensive measuring plots at Steile Bramke (P3, P4, E3, SW7) and Dicke Bramke (DB) in 1988, 1993 and 1997 (limits of supply ranges after Arbeitskreis Standortkartierung, 1996).

Simulation of long-term effects of liming

For the estimation of the long-term effects of the liming measure, the Batch Equilibrium Model (BEM) has been applied to the Steile Bramke catchment. BEM simulates the composition of soil solution and soil solid phases with a system of joint chemical equilibrium reactions (Prenzel, 1991; Prenzel and Meiwes, 1994). A cation exchange system, gibbsite, jurbanite, and alunite have been defined as solid phases for the simulation. One organic and two mineral soil layers (0 – 10 cm and 10 – 50 cm depth) have been defined. Contents of Ca and Mg in soil solution can only be modelled if an elevated ion exchange capacity of the organic layer is used. Thus, ion exchange capacity can be interpreted as an effective parameter for all processes incorporating and releasing cations in the organic layer. The simulation shows an increased base content of the organic layer for about 100 years after liming. For the mineral soil, a duration of liming effects of about 150 years was estimated. A base saturation of 20% will be achieved only in the top 10 cm of the mineral soil (Fig. 2). Model results show that ion exchange in the organic layer is a key process for the depth effect of liming measures.

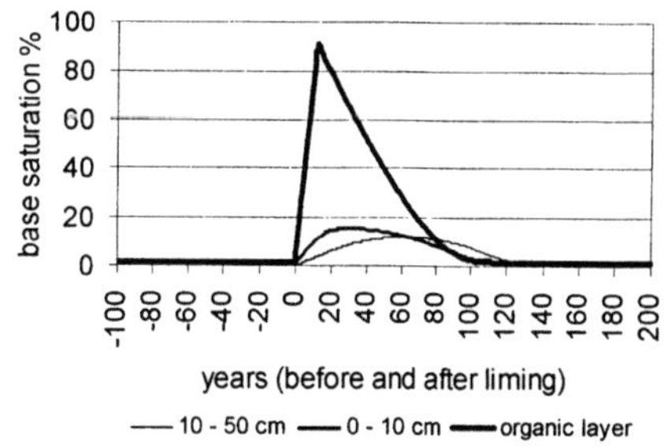

Figure 2. Simulated time series of base saturation for three soil compartments at Steile Bramke

Conclusions

The liming measure has substantially improved the soil conditions and the nutrition of the forest stand at Steile Bramke 9 years after liming. Negative effects such as enhanced NO_3^- leaching could not be observed. According to the model results, the intended base saturation of 20% and a pH of 5.0 within the main rooting zone will not be achieved because the organic layer has not accounted for the lime requirement. Thus, the properties of the organic layer have to be taken into account when liming programs are planned. The risk of negative effects through high dosage liming measures are estimated as low for comparable site conditions.

Acknowledgement

The study was supported by the European Community Project No. 93.60.DL.0120.

References

Arbeitskreis Standortkartierung 1996 ,Arbeitskreis Standortkartierung in der Arbeitsgemeinschaft Forsteinrichtung, Eching bei München, IHW-Verl.

Eberl C, Hooge H, Meesenburg H, Müller-Using B, Rademacher P, Stüber V and Wachter H 1999 Mitt. Deutsche Bodenkundl. Ges. 90, 279-306.

König N and Fortmann H 1996 Ber. Forschungszentrum Waldökosysteme B46-49.

Meesenburg H, Meiwes KJ and Rademacher P 1995 Water Air Soil Poll. 85, 611-616.

Meiwes KJ 1995 Water Air Soil Poll. 85, 143-152.

Prenzel J 1991 Introduction to BEM (Batch Equilibrium Model). Ber. Forschungszentrum Waldökosysteme B28.

Prenzel J and Meiwes KJ 1994 J. Environm. Qual. 23, 1212-1217.

Zech W 1970 Forstw. Cbl. 89, 1-9.

W. J. Horst et al. (Eds), Plant nutrition – Food security and sustainability of agro-ecosystems. 916~917, 2001

Contribution of root turnover to nutrient cycling in beech forests

K.-H. Wu[1)], G. Jentschke and D. L. Godbold[2)]
[1)] *Institute of Forest Botany, University of Göttingen, Büsgenweg 2, Göttingen, D 37077, Germany.*
[2)] *School of Agricultural and Forest Sciences, University of Wales, Bangor, Bangor LL572UW, UK. d.l.godbold@bangor.ac.uk.*

Key words: aluminium, *Fagus sylvatica*, nutrient cycling, root turnover

Abstract

In two mature beech (*Fagus sylvatica*) forests growing on an acid brown earth or a terra fusca rendzina, the contribution of turnover to ecosystem nutrient cycling was investigated. The estimate of the amount of elements returned to the soils via turnover was dependent upon the calculation method used. For most elements with the exception of Fe and Al, the input via root l was about one third of the total input (leaf + root litter). For Al, higher inputs via root turnover were found in deep soil layers.

Introduction

Fine root turnover is a major component in nutrient cycling in forests. An accurate measurement of root production and turnover is central to the determination of nutrient return via fine roots. Nutrient return to the soil is often calculated from the average element contents in fine roots over a season multiplied by the annual root turnover. However, the element concentrations in fine roots and root turnover show significant fluctuations over a season and at different soil depths. In the present work return of elements to the soil have been estimated using both conventional methods and by summing fluctuations in fine root element storage.

Materials and methods

Sites

Both study sites are mature beech (*Fagus sylvatica*) forests. The site Göttinger Wald was an approximately 130 years-old stand growing on a terra fusca rendzina. The Solling B1 site was approximately 150 years-old stand growing on an acid brow earth and the organic layer is a typical mor type humus.

Fine root sampling

Root samples were taken every 4 weeks in growing season between April and October in 1995 and every 8 weeks in 1996. At each site 4 plots, 3.5 m × 3.5 m, were divided into 49 subplots. At each sampling time, cores were taken from 3 subplots resulting in 12 cores per site. Coring locations were determined by random co-ordinates. Subplots were only sampled once. The core samples were taken, from Solling B1, to a depth 40 cm below the upper part of the mineral soil and from Göttinger Wald only to a depth of 20 cm because of the stoniness of the soil. The samples then were sliced into humus (OF+OH), 0-5 cm, 5-20 cm and 20-40 cm levels from the Solling stand, and 0-5 cm and 5-20 cm levels from the Göttinger Wald.

Element analysis

Fine roots (< 2mm) were sorted into live and dead roots, dried to constant weight and wet-ashed at 180°C under pressure in Teflon vessels. P, S, K, Ca, Mg, Fe and Al were determined by AES-ICP (Spectroflame ICP M, Spectro), N using an automated C/N analyser (Perkin Elmer).

Calculation of fine root turnover and element inputs

Fine root turnover was calculated using the maximum-minimum method (McClaugherty *et al.*, 1982). Elements inputs into the soil were estimated either as the product of annual fine root turnover and mean element contents (T), or by changes in fine root element storage (S) (Wu, 2001). Fine root element storage was estimated at each sampling date, and calculated for the different soils depths.

Results

The different methods used to calculate element inputs results for a number of elements in strongly different values (Tab. 1). For Göttingen Wald, the values calculated were similar for all elements except Fe and Al which were lower when calculated by changes in storage. However for Solling, all elements except Fe and Al were considerably higher when calculated by changes in storage.

The mean element inputs calculated from changes in element storage for the years 1995-1996 are shown in Tab. 2. Input of N, Fe and Al was higher at the Solling site compared to the Göttinger Wald site. A manyfold higher Ca input due to fine root turnover was determined at the Göttinger Wald site compared to the Solling site. At Göttinger Wald for all elements except Fe and Al, the input via root litter was ca 28% of the input via leaf litter. At Solling, root litter input accounted for only 10 and 17% of Ca and K input, respectively. At both sites, 90% of the Al return to the soil was via root turnover.

For N the input was independent of soil depth (Tab. 3). However, for Ca at Göttinger Wald and Al at Solling a higher input was found at the deeper soil depths.

Table 1. Element inputs using different methods of calculation.

Element	Element input (S as % of T)	
	Göttinger Wald	Solling
N	105	163
P	82	155
S	97	148
K	104	136
Ca	83	129
Mg	80	128
Fe	65	98
Al	66	108

Table 2. Absolute element inputs due to root turnover and as a percentage of total litter input (roots + leaves).

Element	Solling		Göttinger Wald	
	Root input kg ha^{-1} a^{-1}	% of root + leaf	Root input kg ha^{-1} a^{-1}	% of root + leaf
N	37	36	20	27
P	2.3	32	0.9	23
S	3.0	48	1.9	28
K	4.0	17	4.4	20
Ca	3.4	10	20	31
Mg	1.2	39	1.6	27
Fe	8.2	78	4.6	71
Al	20	95	12.2	90

Table 3. Element inputs due to root turnover at different soil depths

Site	Soil depth	Root input kg ha^{-1} a^{-1}		
		N	Ca	Al
Göttinger Wald	0-5 cm	8.0	7.6	4.8
	5-20 cm	12.0	12.4	7.6
Solling	Humus	9.4	1.4	1.8
	0-5 cm	7.7	1.1	2.2
	5-20 cm	10.7	0.6	4.2
	20-40 cm	9.5	0.9	10.8

Discussion

The results show that the calculation method used may affect the estimates of element inputs to soils. Calculations using the product of annual fine root turnover and mean element contents appear to underestimate the input of a number of elements in Solling. The return via root litter is on average about one third of total input via biomass (roots + leaves), and is thus a considerable ecosystem input. For Solling, the return to the soil via root litter is similar to the atmospheric N deposition (Meersenburg *et al.*, 1995). For elements which are poorly transported to leaves, such as Al, the root litter inputs dominate. The high inputs of Al via root turnover in the deeper soil layers may have significant effects on the amount of organically complexed Al found at these soil depths.

References

McClaugherty CA, Aber JD, Melillo JM 1982 Ecology 63, 1481-1490.

Meesenburg H, Meiwes KJ, Rademacher P 1995 Water Air Soil Poll. 85, 611-616.

Wu K-H 2001 PhD Thesis, University of Göttingen, Germany.

W. J. Horst et al. (Eds), Plant nutrition – Food security and sustainability of agro-ecosystems. 918~919, 2001

Nitrate reductase activity in fine roots of norway spruce - A biochemical marker for nitrogen conditions in the soil?

M. Genenger[1)], E. Frossard[2)], S. Zimmermann[1)] and I. Brunner[1)]

[1)] *Swiss Federal Research Institute WSL, Zürcherstrasse 111, Birmensdorf, CH 8903, Switzerland, genenger@wsl.ch*

[2)] *Institute of Plant Science, Research Station Eschikon, ETH Zürich, Eschikon 33, Lindau, CH-8315, Switzerland, frossard@ipw.agrl.ethz.ch*

Key words: fine roots, N saturation, nitrate reductase activity, *Picea abies*, spruce, wood ash

Abstract

The roots of trees are integrating organs over temporal variations reporting on various properties of the surrounding rhizosphere. In a field experiment the reactions of Norway spruce fine roots to wood ash and liquid fertilisation was investigated for 2 years. The liquid complete fertilisation contained 70-100 kg N and was meant, as with the wood ash (mainly Ca, K, Mg, P), to improve the nutritional balance of the trees. The nitrate reductase activity (NRA) of the fine roots was enhanced by the liquid fertilisation and through the ash treatment. A direct correlation between nitrate concentrations in the soil solution and the NRA in the fine roots could not be proven, but the NRA was significantly correlated with the additional N input on the liquid fertilised plots. The elevated NRA with the wood ash treatment could be caused by a change in the nitrate availability but also by other direct consequences of an upshift of the soil pH. The NRA as an indicator for N status on the ecosystem level and in the long term view is confirmed as a valuable additional parameter.

Introduction

World-wide the anthropogenic production of reactive N is rising at a high rate (Galloway, 1998). Especially N limited ecosystems like forests are subjected to strong nutritional imbalances and mainly still unknown changes (Rennenberg and Gessler, 1999). A N-saturated forest shows an increased nitrate mobility in the soil in combination with a saturated N status of the plants (Aber et al., 1998). The activity of the nitrate reductase, as an inducible and uptake controlling enzyme (Tischner, 2000) in fine roots, could be a good marker for the N status of the plant.

In the so called HARWA project the influence of a complete liquid fertilisation and fertilisation with wood ash on various parameters in a spruce forest was tested. Wood ash recycling should be evaluated as one method of sustainable forest management. Root NRA has been investigated in connection with soil solution properties.

Materials and methods

The experiment was conducted in a 70 year old spruce (*Picea abies* (L.) Karst) forest on the Swiss Plateau, 450 m a.s.l. Wood-ash (A), 4 t ha^{-1} yr^{-1} were applied by hand in May '98 and July '99, respectively. A complete liquid fertilisation (LF) modified after Ingestad and Lund (1986), was applied at 70 kg N ha^{-1} yr^{-1} (1998) and 100 kg N ha^{-1} yr^{-1} (1999). The fertilisation was irrigated during the growing season (May to September). In the control plots (C) no treatment was applied. Fine root (< 2 mm) samples were taken in May, June, August, and October 1998 and April, June, August, and October 1999. In each plot the soil solution was collected regularly at the soil surface by the aid of lysimeters and in 10 cm depth with suction-cups.

In vivo NRA of the roots was determined with a method modified after Yandow and Klein (1986). The collected soil solutions have been analysed for nitrate and ammonium concentrations. Data represent the mean of three plots for one treatment.

Results

Table 1. Statistical analyses on treatment effects on the NRA in fine roots, differences (±SE) and probabilities, not significant ns, $p<0.05$ *, $p<0.01$ **, $p<0.001$ ***

Time	Difference LF-C			Difference A-C		
1998	Mean	SE.	p^a	Mean	SE	p^a
May	-0.92	± 5.8	ns	0.38	± 8.2	ns
June	0.70	± 5.5	ns	14.49	±10.2	ns
August	0.27	± 7.5	ns	18.33	±10.1	ns
October	21.49	± 6.7	**	55.13	±17.1	**
1999						
April	14.09	± 5.2	*	46.19	±14.5	**
June	23.01	± 9.8	*	56.34	±13.2	***
August	45.61	±17.4	*	61.59	±20.6	**
October	40.32	±10.8	**	65.84	±17.2	**

a = one-way ANOVA, treatment effect on the NRA with nested data-set from each month (DATADESK)

The nitrate concentrations in the soil solution were enhanced up to nearly ten-fold by the fertilisation treatment (not shown). After the irrigation period the nitrate concentration dropped back to the natural level. In the ash plots a nitrate peak after each ash application was observed. In the control plots a seasonal aspect of the nitrate in the soil solution was reported. In particular in the summer of 1998 (dry year) the nitrate concentrations in the soil solution were higher than in the winter.

The first ash treatment was followed by an increase in the NRA of Norway spruce fine roots (Table 1). Furthermore, the liquid fertilisation resulted in a higher NRA from the end of the first year on. The effects on the NRA through the ash treatment were higher (up to 100% increase) than the effects deriving from the fertilisation (up to 65% increase). The NRA of the roots from the nitrogen fertilised plots was correlated with the cumulative amount of N applied (Fig. 1).

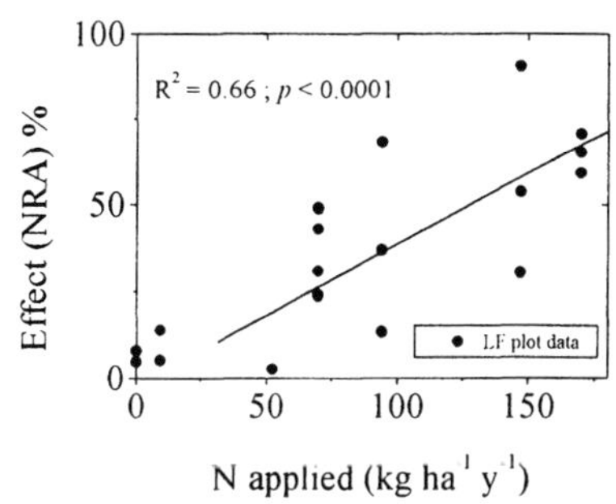

Figure 1. Effect of LF treatment on the NRA in the fine roots correlated to the N applied, cumulative for both years.

Discussion

Although the NRA reacted positively to the LF treatment, a direct correlation between the nitrate concentrations in the soil solution and the NRA in the fine roots could not be drawn. The *in vivo* NRA resembles the nitrate reduction capacity of the measured tissue (Bauer, 1997; Högberg et al., 1986). Greenhouse experiments have shown, that the NRA in the roots of conifers reacts slowly to changes in the nitrate concentration (Genenger, unpublished data; Min et al., 1998) compared to broadleaf trees. Therefore, the reactions to N inputs were reflected more in a long term range. This hypothesis is supported in the present study by a significant correlation between the cumulative N input and the NRA of Norway spruce fine roots. Similarly, Bauer (1997) could correlate the NRA data of twigs of birch and beech to the measured N deposition.

As expected from other experiments (Kahl et al., 1996; Meiwes, 1995) the ash treatments were followed by nitrate peaks. However, the increase of NRA was not connected to the nitrate mobilisation in the soil solution. One possible explanation might be the elevated pH as it was observed in our study but also in former studies (Clarholm, 1998; Eriksson, 1998; Kahl *et al.*, 1996). That has a positive effect on the nitrate availability. Consequently plants use a higher amount of easy to access nitrate, which is metabolised through the nitrate reductase. Another explanation is a high spatial heterogeneity in the soil N availability, which is reported after an ash treatment (Grogan et al., 2000). This heterogeneity might have caused the NRA of some root samples to react strongly while others showed a not induced NRA. The wood ash powder with a high reactivity (Eriksson, 1998) might also have caused a stress reaction in the fine roots with elevated enzyme activities.

In a study along an N-deposition gradient in Europe Högberg *et al.* (1998) tested the abilities of the *in vivo* NRA of Norway spruce fine roots to determine the different N properties in the soil along that transect. The results of the present study confirm the differences in the NRA of spruce fine roots deriving from higher N input. The use of NRA of fine roots as an indicator is limited through influences of other environmental factors on this enzyme. However, as an additional marker for the N-status of the soil and plant it is valuable.

Acknowledgement

We thank S. Brodbeck for skillful work in the laboratory and field. Furthermore we are grateful to W. Landolt as head of the HARWA project and his group, to the laboratory team of the Soil Ecology group and the central laboratory. The project was funded by the Swiss Federal Office for Education and Science (BBW) and the Swiss Agency for the Environment, Forests and Landscape (BUWAL) as part of the COST E6 Eurosilva Project: Forest tree physiology research.

References

Aber J, McDowell W, Nadelhoffer K, Magill A *et al.* 1998 BioScience (48), 921-934.

Bauer G 1997 PhD Thesis. University Bayreuth, Bayreuth.

Clarholm M 1998 Scand. J. For. Res. (2), 65-75.

Eriksson HM 1998 Scand. J. For. Res. (2), 43-55.

Galloway JN 1998 Env. Poll. (1), 15-24.

Grogan P, Bruns TD and Chapin FS 2000 Oecologia (122), 537-544.

Högberg P, Granström A, Johansson T and Lundmark-Thelin A 1986 Can. J. For. Res. (16), 1165-1169.

Högberg P, Högbom L and Schinkel H 1998 Tree Physiol. (18), 823-828.

Ingestad T and Lund AB 1986 Scand. J. For. Res. (1), 439-453.

Kahl JS, Fernandez IJ, Rustad LE and Peckenham J 1996 J. Environ. Qual. (25), 220-227.

Meiwes KJ 1995 Water Air Soil Pollut. (85), 143-152.

Min X, Siddiqi MY, Guy R, Glass ADM and Kronzucker HJ 1998 Plant Cell Environ. (21), 1039-1046.

Rennenberg H and Gessler A 1999 Water Air Soil Pollut. (116), 47-64.

Tischner R 2000 Plant Cell Environ. (23), 1005-1024.

Yandow TS and Klein M 1986 Plant Physiol. (81), 723-725.

W. J. Horst et al. (Eds), Plant nutrition – Food security and sustainability of agro-ecosystems. 920–921, 2001

Fine roots of trees: Indicators for the element availability in forest soils and for forest soil conditions

I. Brunner, S. Zimmermann and M. Genenger
Swiss Federal Research Institute WSL, Zürcherstrasse 111, Birmensdorf, CH 8903, Switzerland, brunner@wsl.ch

Key words: element concentrations, fine roots, liquid fertilisation, Norway spruce, wood ash application

Abstract

Using element concentrations in fine root of Norway spruce (*Picea abies*), changes of the element availability in the forest soils after the treatments "control", "wood ash", "liquid fertiliser", and "water" could be detected. After the ash and liquid fertiliser treatments, K, Mg, and P concentrations in fine roots were increased, whereas elevated Ca concentrations were detected only after the ash, and elevated N concentrations only after the liquid fertiliser treatment. An enhancement of the soil pH due to the ash treatment led to decreased concentrations of Mn, Fe, and Al in the fine roots. The concentrations of the heavy metal Zn were not affected by the treatments.

Introduction

Leaves or needles of forest trees are widely used as indicators for the nutritional status of forest soils (Linder, 1995). However, Al and heavy metals accumulate in the fine roots of trees and are only poorly transported to the needles or leaves. Therefore, fine roots have the potential 1) to indicate the nutritional status of forest soils, and 2) to indicate plant availability of Al and heavy metals which is dependent on the soil pH (Persson *et al.*, 1996; Bakker, 1999).

Materials and methods

In a Norway spruce stand on the Swiss Plateau near Zurich a fertilisation experiment was conducted in 1998 using either wood ash (mainly Ca, Mg, K, P) or an optimal liquid fertilisation (mainly N, P, K, Mg, Ca; Ingestad and Lund, 1986). Two further treatments were the application of water alone and a control. In total, four subplots were established per treatment. The liquid fertiliser treatment was applied every night during one vegetation period (in total 70 kg N ha^{-1} yr^{-1}), whereas the ash treatment was applied once in the spring (4 t ha^{-1}).

Top soils were collected before (mean pH 3.2) and after the treatments in the springs on a 5 x 5 m grid, in total 110 sampling points, with the aid of a soil coring cylinder. Soils were sieved, the soil pH measured with $CaCl_2$, and the fine roots washed, further processed, and their element concentrations measured with an ICP-AES or a C/N analyser.

Results

After one year, an enhancement of the concentrations of K, Ca, Mg, and P in Norway spruce fine roots was observed after the ash treatment (Fig. 1). Within the the liquid fertiliser treatment, elevated K, Mg, N, and P concentrations were obvious (Fig. 1). The increase of K, Mg and P were for both treatments in a similar range. Due to the elevation of the soil pH by 0.5 within the ash treatment, Mn, Fe and Al concentrations in fine roots were decreased (Fig. 1). Elevated concentrations of the heavy metals (Cd, Cu, Zn) in fine roots compared to the other treatments, however, were not recorded, although the ash contained relatively high amounts of Cu and Zn (Cd and Cu were only measured after the treatments, data not shown).

Discussion

The fertilisation treatments in the present study increased the concentrations of nutrient elements such as K, Mg, or P in fine roots. Thus, the fine roots element concentrations reflected the nutrient conditions in the soil (compare also Feger *et al.*, 1998). On the other hand, an increase of the soil pH due to the ash treatment lowered the concentrations of elements such as Al or Mn. Fine root element concentrations, therefore, are additionally useful to indicate shifts in soil pH, which can occur after soil acidification processes (compare also Persson *et al.*, 1996; Bakker, 1999). Since fine roots accumulate Al and heavy metals, they are more suitable indicators than tree needles or leaves, since they not only reflect the nutritional status of the forest soils but also soil conditions, relevant to plants such as soil acidification or heavy metal deposition.

References

Bakker MR 1999 For. Ecol. Manage. 122, 7-16.

Feger KH, Raspe S and Zoettl HW 1998 *In* Oekosystemforschung im Schwarzwald. Ed S Raspe, KH Feger and HW Zoettl. pp 394-409. Ecomed, Landsberg.

Ingestad T and Lund A-B 1986 Scand. J. For. Res., 439-453.

Linder S 1995 Ecol. Bull. 44, 178-190.

Persson H, Majdi H and Clemensson-Lindell A 1995 Ecol. Bull. 44, 158-167.

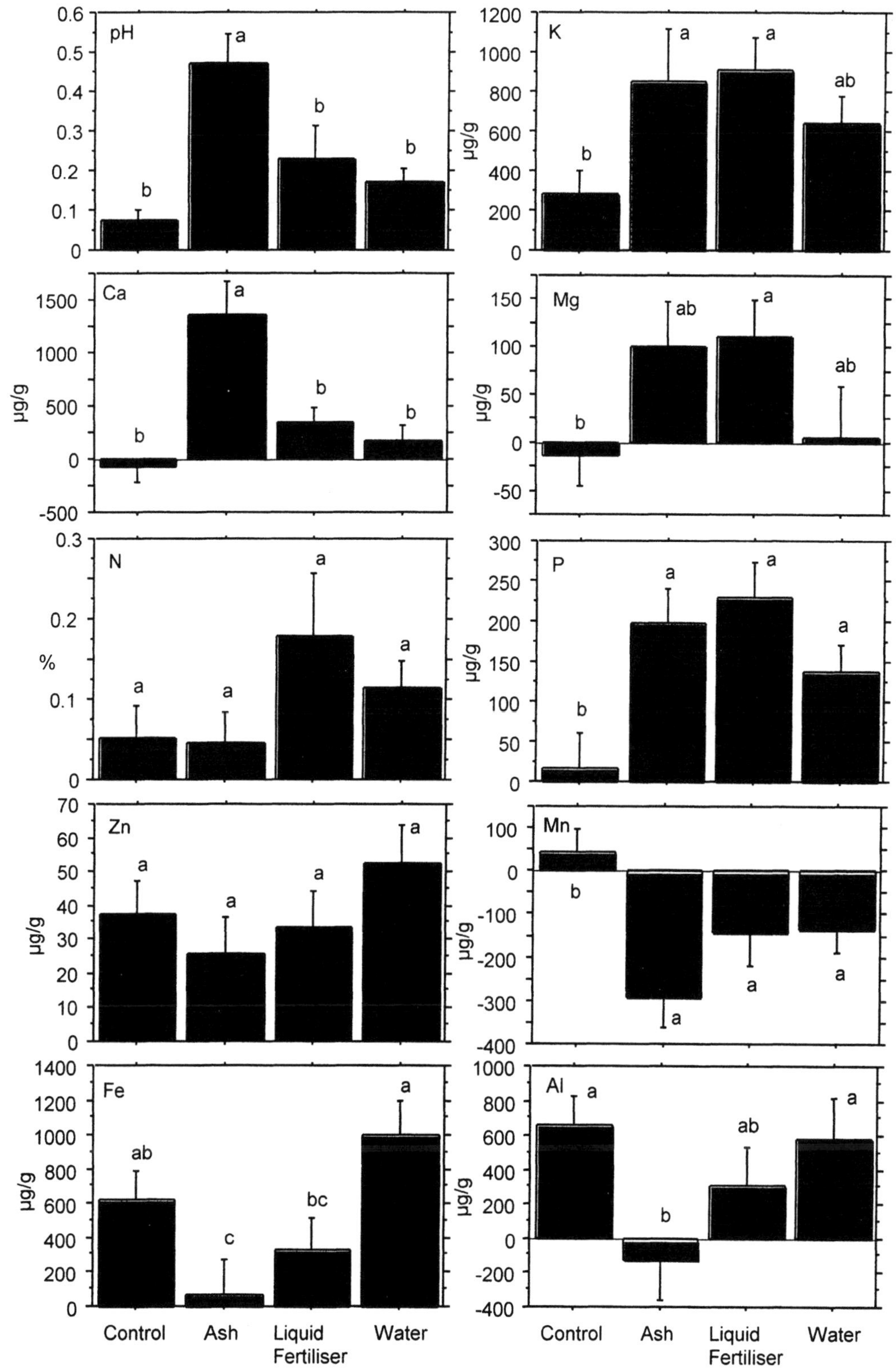

Figure 1. Differences of the soil pH and the fine root element concentrations before and after the treatments. Different letters indicate a significant difference (p=0.05).

W. J. Horst et al. (Eds), Plant nutrition – Food security and sustainability of agro-ecosystems. 922–923, 2001

Parameterizing net N mineralisation from 'old humus' in sandy arable soils of a drinking water catchment (Fuhrberger Feld)

S. Heumann[1)], J. Könecke[2)], J Böttcher[1)] and G. Springob[1)]

[1)] Institute of Soil Science, University of Hannover, Herrenhaeuser Str. 2, Hannover, D 30419, Germany, heumann@ifbk.uni-hannover.de

[2)] Stadtwerke Hannover AG, Am Wasserwerk 2, Burgwedel, D 30938, Germany, Jutta.Koenecke@enercity.de

Key words: incubation experiments, landuse, net N mineralisation, 'old humus', parameterisation, sandy arable soils

Abstract

N mineralisation from 'old humus' is of special importance in drinking water catchments with arable soils which have high organic N contents. This is true for many soils in the catchment Fuhrberger Feld close to Hannover due to peculiarities in historical landuse. To parameterise N mineralisation, and thereby to quantify the contribution of 'old humus' to N mineralisation in the field, we conducted long term aerobic incubation experiments - both in the lab (at 35°C) and under field conditions. Soils with distinct, but widely differing cultivation history were used. Cumulative net N mineralisation curves were parameterised using the two-fractions first-order approach in order to be able to subtract N released from recent residues. The resulting net N mineralisation from slowly decomposable 'old humus' (per g of org. N) spanned an extremely wide range and was strongly influenced by former landuse (forest and grassland). Simulated N mineralisation using field temperature records fitted well to measured net mineralisation data from field incubation of undisturbed soil columns (r^2=0,65), thus showing that the derived parameter values are reasonable. It is further remarkable that the studied sandy soils revealed only approx. 50% of the mineralisation rates from 'old humus' (per g of org. N) when compared to reference loess soils. However, as organic N contents are very high in a number of these soils, net amounts of N resulting from this pool are considerable - especially during winter when there is only little plant uptake.

Introduction

Groundwater-contaminating N losses from agricultural soils include N from current agricultural practices (fertiliser N) as well as N being mineralised from organic materials. Among these different fractions N mineralisation from 'old humus' (soil organic matter excepting recent residues) is of special importance in drinking water catchments, because it cannot be easily controlled by simple agricultural management practices. In addition in the catchment *Fuhrberger Feld* (300 km^2) many soils provide relatively high contents of organic N due to peculiarities in historical landuse. Furthermore landuse changes such as conversion of grassland into arable land have been reported to result in dramatic losses of organic matter (Monreal and Janzen, 1993) combined with enhanced N mineralisation (Strebel *et al.*, 1988).

In order to parameterise N mineralisation and thereby to quantify the contribution 'old humus' to net N mineralisation in the field we conducted long-term incubation experiments - both in the lab and under field conditions - using soils with distinct, but widely differing cultivation history.

Materials and methods

About 130 sandy arable top soils from the Fuhrberg region and 10 reference loess soils (west and south of Hannover) were sampled after harvest of 1998, 1999 and 2000. The sandy soils were mainly Podzols, Gleyic Podzols and Gleysols. The loess areas are dominated by Luvisols. By means of landuse maps, fields were catagorised by historical landuse (old arable land, former grassland or forest) and the approximate time of first cropping.

Field-moist soil was incubated aerobically in the lab at 35°C for 200 - 300 days. Mineralised N (NO_3- and NH_4-N) was leached with 0,02M $CaCl_2$ up to 8 times during incubation. The simultaneous two-fraction first-order kinetic equation (e.g. Nuske and Richter, 1981) was fitted to curves of cumulative N mineralisation. For comparisons, average N mineralisation rates of slowly decomposable ‚old‘ organic matter were calculated using the derived parameter values.

Field mineralisation was measured in undisturbed soil columns (20 cm in diameter and 30 cm in depth) over winter. Six columns per field were taken immediately after harvest of different field crops in 1999 and 2000 in almost 30 fields. The columns were protected from rainfall in order to avoid leaching of mineral N and incubated for about 4 months under similar temperature conditions as in the field.

Results and conclusions

The resulting net N mineralisation of 'old humus' (at 35°C, per unit of organic N) spanned an extremely wide range and showed clear influences of former landuse like grassland or forest (Table 1). Rates in former forest soils also seemed to be influenced by time of first cropping, whereas there were no such trends in former grassland soils. The sandy soils revealed - on an average – approx. 50% of the mineralisation rates when compared to reference loess soils (Table 1). Rate coefficients themselves were also usually a lot smaller than in loess soils. A larger part of the organic material probably has to be considered as relatively inert material. Therefore it is strongly recommended to adjust mineralisation parameters to the properties of these sandy soils.

N mineralisation in undisturbed soil columns (winter 1999/2000) was simulated for the incubation period using the obtained parameter values, field temperature records and temperature functions for the rate coefficients (Nordmeyer and Richter, 1985). Results fitted well to measured N mineralisation (r^2=0,65) showing that the derived parameter values are reasonable.

Net amounts of N resulting from mineralisation of 'old humus' during only 4 winter months ranged from 3 to about 20 kg N ha^{-1} (results from the simulations).

Therefore, 'old humus' can indeed contribute considerably to net N mineralisation, since leaching of only 23 kg N ha^{-1} during winter would be sufficient on these sites to reach the EU limit for nitrate in drinking water (50 mg L^{-1}) in the uppermost groundwater.

Table 1. Average mineralisation rates of old organic matter per % total soil N of sandy arable soils with different landuse history (old cropland on sandy soils = 100) and relation to reference loess soils.

Sandy arable soils in the catchment *Fuhrberger Feld*					Reference loess soils
Old cropland	Sites with different landuse in the past				Old cropland
	Cropped approx. since	Former grassland	Former forest	Former wet grassland	
100	1780				244
	1901	163	129		
	1961		95	105	
	1975		51	156	
	1987	154		127	
	1991			154	
	1995	105		112	

W. J. Horst et al. (Eds), Plant nutrition – Food security and sustainability of agro-ecosystems. 924~925, 2001

A study of nutrition and toxic elements in the Carpathian Mountain forests using foliar analysis of white fir (*Abies alba* Mill.), Norway spruce (*Picea abies* Karst.), beech (*Fagus sylvatica* L.) and humus.

B. Mankovská
Forest Research Institute, T. G. Masaryka str. 22, 960 92 Zvolen, Slovakia, mankov@fris.sk

Key words: air pollution, chemical foliar and humus analysis, white fir, Norway spruce, beech, Carpathian Mountains

Abstract

Concentrations of Al, B, Ca, Cu, Fe, K, Mg, Mn, N, Na, P, S and Zn in the foliage of tree species (*A. alba*, *P. abies* and *F. sylvatica*) and in the humus from 25 localities of Carpathian Mountain forests (Czech Republic, Poland, Romania, Slovakia and Ukraine) and concentrations of As, Be, Cd, Co, Cr, Hg, Li, Ni, Pb, Se, Sr, and V in *A. alba* are described. Additionally, the composition of individual particles in stomata of leaves were analysed and the suitability of foliage of forest tree species as bioindicators of environmental contamination with heavy metals is discussed.

Introduction

The Carpathian forests represent unique reservoirs of many endemic, rare, and unusual plant and animal species in Central Europe. White fir, spruce and beech are well distributed in the Carpathian Mountains (Elvingson, 1998). Chemical foliar analysis is a widely used diagnostic and monitoring method in environmental studies (Mankovská, 1996). The aim of this paper is to present actual data of the first survey of the chemical composition of foliage of three forest tree species and humus.

Materials and methods

Research sites were established in Slovakia, the Czech Republic, Poland, Ukraine and Romania in December 1995-1998. Altogether 1101 trees of predominant or dominant class in the second age category (21-40 years old) were sampled. Samples of foliage were analysed unwashed. The accuracy of the analytical data published in this paper was verified and tested. Observations of foliage surfaces were performed with a Scanning Electron Microscope JEOL 840 A and by using an EDX analyser LINK 10000. Particles deposited on surface and in stomata were assessed according to their morphology and EDX spectra (Mankovská, 1996). The data were evaluated by common statistical methods (Student test, PCA).

Results

Foliar analyses of *A. alba*, *P. abies* and *F. sylvatica* and humus analysis (Tab. 1) show that the greatest load by air pollutants is in the Slovak part of Carpathian Mountains, namely in the surroundings of industrial plants where also the highest values of studied elements were found. Statistically significant differences between concentrations of all study elements in the foliage of forest tree species were found between all studied countries. The equilibrium of S and N in foliage is a precondition of their normal growth (Fig. 1). Foliage of forest tree species from contaminated regions can be considered a monitor of the accumulation of polluting elements (Mankovská, 1996). Metal concentrations in needles were evaluated based on pollution of forest environment with heavy metals and threshold concentrations of the studied metals for plants. The highest values of the studied elements were found in the Slovak portion of the Carpathians in the vicinity of industrial sources. The highest values of Al, Ba, Cd, Cr, F, K, Mg were found near to the Thermal Power Plant ENO Zemianske Kostolany and the Aluminium Plant in Ziar nad Hronom. The highest concentrations of Cu, Fe, Hg, Zn were determined in the localities influenced by the Non-ferrous Metallurgic Plant in Krompachy and the Mercury Processing Plant in Rudnany. High concentrations of As, Fe, Mn, S, and V occurred in the vicinity of the Ferrous Metallurgic Plants in Podbrezová and Košice. Above normal concentrations of Be, Cr, Na, Ni and Pb in fir needles reflect regional pollution of the environment with these elements. Elevated concentrations of Co, Se and Li, in the absence of the Slovak sources of these elements, indicate effect of long-range transport from the heavily polluted areas in Poland and the Czech Republic. Liming of the pollution-weakened forest stands was a probable reason for increased concentrations of Ca in fir and spruce needles. This study proves that foliage of forest tree species can be used as bioindicators of environmental contamination with heavy metals. Stomata of observed foliage in the vicinity of non ferrous plants contained aluminium as a part of minerals, ashes (Al-Si) and particles Al_2O_3; iron as a part of particles of minerals, ashes and Fe_2O_3 (vicinity of metallurgy complex and thermal power plant); potassium as a part of mineral and biological particles and Mn, Na, V, Ni, Zn, Cr, As, Ba and Cu as part of industrial particles.

References

Elvingson P 1998 Acid News 4, 14.

Mankovská B 1996 Geochemical atlas of Slovakia -forest biomass, Geologic Servis Bratislava, Slovakia, ISBN 80-85314-51-7.

Stefan K, Furst A, Hacker R, Bartels U 1997 Forest Foliar Condition in Europe, EC-UN/ECE-FBVA Brussels Geneva Vienna, ISBN 3-901347-05-4.

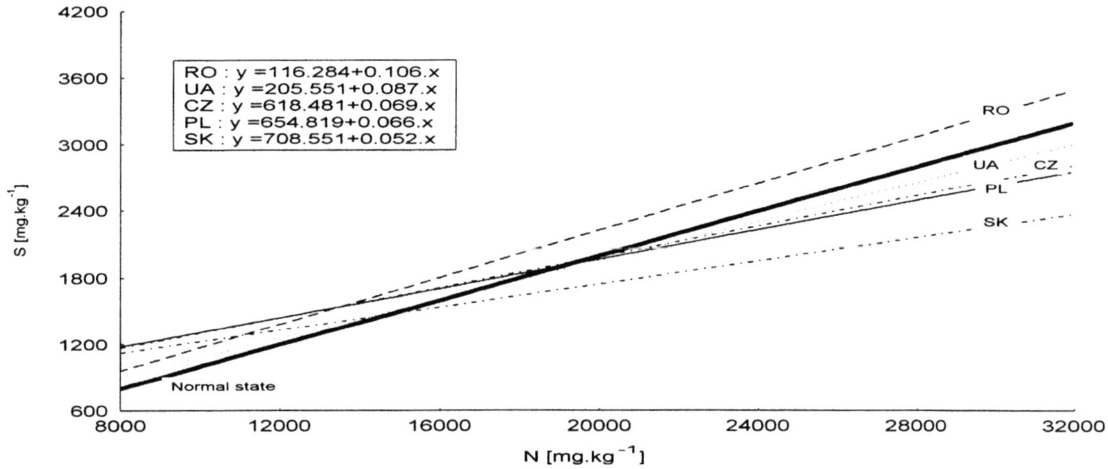

Figure 1. Relationship between the concentration of N and S in two years old needles of *A. alba*.

Table 1. Concentration of elements in the foliage of *F. sylvatica*, *P. abies*, *A. alba* and humus layer (arit. means in mg kg^{-1})

Element	CZ	PL	RO	SK	U	CZ	PL	RO	SK	U
			F.sylvatika					*P.abies*		
Al	172	138	151	94	139	98	69	95	163	229
B	23	25	22	28	21	19	19	18	13	17
Ca	10546	7552	10393	10402	9564	2754	2891	5893	5561	6267
Cu	11.0	13.1	13.0	11.3	11.6	5.4	6.5	9.7	7.0	8.7
Fe	297	163	593	372	1046	214	113	1666	159	1429
K	10425	11549	8637	9843	8495	5897	6127	9343	7236	7547
Mg	1968	1648	1965	2244	2299	1080	772	1289	1274	1292
Mn	862	1774	1743	457	1093	215	647	1000	501	800
N	26240	28230	23123	13942	26620	12800	12744	15167	12510	16575
Na	89	68	120	87	99	43	101	143	73	57
P	1369	1452	1796	1267	1409	1416	1174	1646	1360	1561
S	2408	2510	2541	1626	2584	1446	1477	1947	1352	1583
Zn	34	30	41	32	54	31	16	63	43	67
			A.alba					humus layer		
Al	194	163	348	229	208	1164	4349	7598	4775	1169
B	17	17	16	18	15	5.5	10.6	10.8	14.4	11.8
Ca	7651	8209	11895	9262	7473	4413	3850	10633	16142	10548
Cu	4.3	5.0	4.4	6.9	4.6	17.6	21	27.4	20.0	13.2
Fe	116	215	123	92	177	2177	4666	7618	4046	1282
K	4346	6453	5079	5586	5262	1183	1584	2252	1650	1294
Mg	952	1519	1506	1488	1092	512	936	2384	4056	966
Mn	966	1174	897	1622	534	210	700	1518	959	1192
N	11030	12457	12306	15557	17233	17200	19047	14952	17788	17500
Na	52	60	75	48	33	85	102	125	103	64
P	947	1464	1596	980	1085	1201	1105	1134	923	1042
S	1430	1371	1426	1607	1675	2590	2746	1760	1705	1513
Zn	39	39	40	30	34	49	95	79	105	51

Note: Literature limit values in mg.kg^{-1}: 50-150 Al; 10-60 B; 4000-8000 Ca; 2-3 Cu; 200-300 Fe; 5000-1000 K; 1000-1500 Mg; 200-1000 Mn; 13000-25000 N; 20-80 Na; 700-1400 P; 1000-1500 S; 20-60 Zn (Mankovská, 1996; Stefan *et al.*,1997)

W. J. Horst et al. (Eds), Plant nutrition – Food security and sustainability of agro-ecosystems. 926~927, 2001

Nutrient concentrations in leaves of peach trees subjected to acid rain

Klimenko O. E. and Klimenko N. I.
Steppe Department, Nikitski botanical garden, Pitomnik, Gvardeyskoje Simferopol Crimea, 97513, Ukraine, mb085@scp.post.crimea.ua

Key words: air pollution, mineral nutrition, peach tree

Abstract

Acid rain may damage fruit trees and affect the productivity of peach production. It is difficult to evaluate the degree of damage induced by acid rain because of non-specific symptoms. We investigated the effect of simulated acid rain on the mineral nutrition of peach trees. The acid rain applied varied in composition and pH (pH 2, 3, 4 and 5). The changes of the nutrient concentrations in leaves were used as an indicator for tree damaging and to assess the resistance of different peach varieties to acid deposition.

Introduction

The pollution of air by sulfur and nitrogen oxides is a global ecological problem. As a result there is acid rain and nitrogen deposition which may lead to an increase or decrease of ecosystem productivity. Fruit trees are sensitive to air pollution. However, there is little information on the effect of acid deposition on peach trees.

Materials and methods

We studied the effect of simulated acid rain (SAR) (solutions with sulfate or nitrate differing in pH) on mineral composition of peach leaves of the two varieties 'Favorita Morettini' (early ripening) and 'Doktorskiy' (middle ripening) in a field experiment. These varieties had a different resistance to the SAR. The SAR had pH values of 2, 3, 4 and 5. Frequency of SAR application was monthly for a period of 5 months from blossom till leaf-fall. Control trees were not treated with SAR. Analysis of leaves was done after wet combustion (Semenjuk, 1984)

Results

The effects of simulated acid deposition on nutrient concentrations in leaves are shown in Table 1. The SAR based on HNO_3 did not cause considerable changes in the leave nutrient concentrations of both varieties. By contrast, the application of SAR based on H_2SO_4 considerably affected the nutrient contents of the leaves (compared to the control).

The leaves of the variety 'Favorita Morettini' treated with SAR based on H_2SO_4 showed the following changes: Nitrogen concentrations were significantly increased at pH values of the SAR of ≤ 3; a significant reduction of the leaf Ca concentration started at pH ≤ 4; the Mg concentration increased with decreasing pH of SAR (except for the pH 5) and the concentrations of P, K and S were hardly or not affected by SAR.

The following results were found for the variety 'Doktorskiy' treated with SAR based on H_2SO_4: The leaf N concentration was increased at pH 4, but a significantly decreased N concentration was found at pH ≤ 3; the SAR-induced decrease of the Ca concentration was smaller than for the variety 'Favorita Morettini'; there was a significant decrease of the leaf K concentrations for all SAR(H_2SO_4)-treatments and the concentrations of P and S were hardly or not affected by SAR.

We found an accumulation of S in falling leaves for the SAR of pH 2 for both varieties.

Discussion

Our results show that peach varieties may differ in their response to acid precipitation. The changes in nutrient concentrations in leaves may be used as an indicator for regions with high deposition rates and for evaluating the different resistance of peach varieties to acid deposition.

References

Semenjuk GM 1984 Mineral nutrition diagnosis of fruit cultures. Shtiintsa, Kishineu.

Table 1. Composition of simulated acid rain (SAR) and nutrient concentrations (%) in leaves of peach trees (varieties 'Favorita Morettini' and 'Doktorskiy'), August 1997.

SAR	pH	N	P	K	Ca	Mg	S	
							August	September
				'Favorita Morettini'				
H_2SO_4	6.0 (control)	2.57	0.45	1.34	1.61	0.50	0.23	0.38
	5.0	2.80	0.57	1.22	1.79	0.28	0.18	0.28
	4.0	2.46	0.58	1.38	1.22	0.83	0.24	0.37
	3.0	3.08	0.55	1.39	1.39	0.67	0.30	0.33
	2.0	3.72	0.53	1.35	1.18	0.88	0.19	0.46
HNO_3	5.0	2.61	0.67	1.57	1.57	0.42	0.21	0.31
	4.0	2.48	0.64	1.58	1.43	0.45	0.24	0.36
	3.0	2.64	0.59	1.37	1.22	0.46	0.24	0.32
	2.0	2.23	0.57	1.31	1.50	0.45	0.20	0.37
		(0.19)*	(0.03)*	(0.03)*	(0.22)*	(0.10)*	(0.01)*	(0.02)*
				'Doktorskiy'				
H_2SO_4	6,0 (control)	2.81	0.33	1.87	1.50	0.66	0.27	0.30
	5.0	2.87	0.34	1.39	1.29	0.50	0.32	0.35
	4.0	3.34	0.37	1.34	1.57	0.59	0.29	0.37
	3.0	2.30	0.33	1.33	1.36	0.64	0.30	0.31
	2.0	2.48	0.43	1.47	1.36	0.42	0.31	0.58
HNO_3	5.0	2.66	0.35	1.34	1.57	0.56	0.26	0.25
	4.0	2.67	0.25	1.35	1.57	0.67	0.25	0.28
	3.0	2.45	0.32	1.36	1.50	0.63	0.28	0.26
	2.0	2.76	0.37	1.37	1.36	0.56	0.25	0.27
		(0.27)*	(0.05)*	(0.03)*	(0.20)*	(0.09)*	(0.02)*	(0.01)*

* Minimal significant difference

W. J. Horst et al. (Eds), Plant nutrition – Food security and sustainability of agro-ecosystems. 928~929, 2001

Nutritional effects on leaf area index and growth of a young *Eucalyptus nitens* plantation

P. Smethurst[1)], C. Baillie[2)] and M. Cherry[3)]

[1)] *Cooperative Researech Centre for Sustainable Production Forestry and CSIRO Forestry and Forest Products, GPO Box 252-12, Hobart, Tasmania, 7001, Australia, Philip.Smethurst@ffp.csiro.au*

[2)] *Soil and Water Science Department. Faculty of Agriculture, University of Alexandria, GPO Box 252-12, Hobart, Tasmania, 7001, Australia, Craig.Baillie@ffp.csiro.au*

[3)] *Cooperative Researech Centre for Sustainable Production Forestry and CSIRO Forestry and Forest Products, GPO Box 252-12, Hobart, Tasmania, 7001, Australia, Maria.Cherry@ffp.csiro.au*

Key words: *Eucalyptus nitens*, fertilisers, forests, growth, LAI, leaf area index, plantations

Abstract

Natural and fertiliser-induced variation in leaf area index (LAI) of a 6-year-old *Eucalyptus nitens* plantation was related to growth (annual increment in stem biomass). Heavy rates of N and P fertilisation increased LAI and growth. Values of LAI ranged from moderate (3.9) to quite high (9.4), with little temporal variation during the study. A quadratic function fitted to the positive relationship between LAI and growth explained 59% of the variation.

Introduction

Leaf area index (LAI) is an important determinant of a forest plantation's capacity to fix carbon and grow. The use of LAI estimates in forest research has increased greatly during the past decade. Forest managers are also interested in using LAI estimates to monitor and predict plantation growth, but we lack quantitative links between LAI and growth for many species, including *Eucalyptus nitens*, a frost-tolerant eucalypt widely grown in Tasmania, Australia. Our aims were to (a) determine the effects of fertilisation on LAI and growth of a *Eucalyptus nitens* plantation in Tasmania, and (b) quantify the relationship between LAI and growth during the 7th year.

Materials and methods

Seedlings were planted at a density of 1270 trees ha^{-1} on cultivated mounds of a clay-loam soil derived from mudstone. The site had a cool-temperate climate

Frosts were common in winter, summer droughts were rare and mild, and mean annual rainfall was 1444 mm. Six fertiliser treatments (Table 1) were arranged in a randomised block design with 5 replicates. During the first 6 years, cumulative rates of fertiliser application ranged from 0 to 1600 kg Nitrogen (N) ha^{-1} and 0 to 700 kg Phosphorus (P) ha^{-1}. Experimental units contained 70 trees, and all measurements were restricted to the central 30 trees.

Leaf area index (LAI) was estimated by interception of diffuse radiation in all plots at ages 5.7 and 7.0 years. A subset of plots was also measured for LAI at 6.2 years. Stem volume growth was estimated from measurements of height and over-bark diameter (at 1.3 m height) at 5.8 and 6.8 years.

Results and discussion

By 5.8 years, fertilisation had significantly increased LAI and stem volume in the treatments that had received the 3 highest rates of fertilisation, and there was no significant benefit to growth of applying more than 600 kg N and 300 kg P ha^{-1}, even when applied in many split doses (Fig. 1). The range of LAI produced was very large, i.e. 3.9 to 9.4 for individual plots, which covered the range of moderately full canopies to others that were unusually dense.

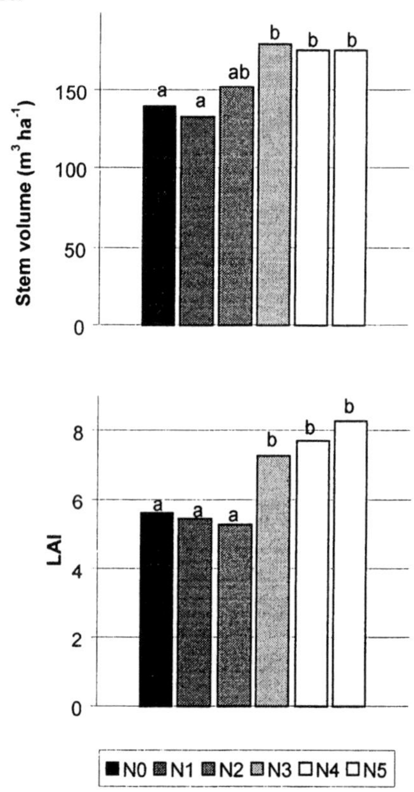

Figure 1. Stem volume and LAI at 5.8 years. Differentletters indicate a significant difference.

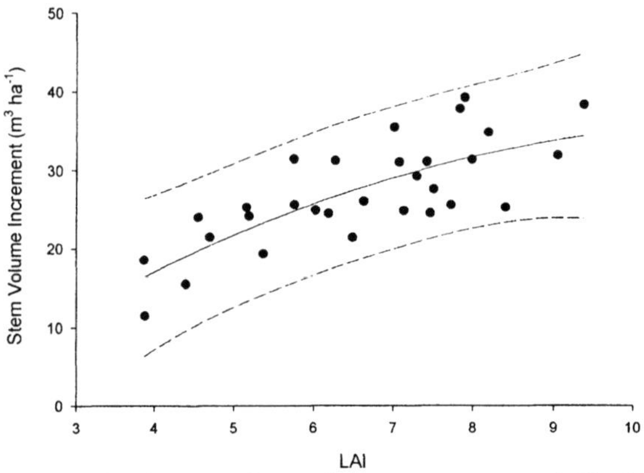

Figure 2. Stem volume increment versus LAI for individual plots showing the fitted curve and the 95% prediction interval ($y = -8.14 + 7.63x - 0.331\ x^2$, $R^2 = 0.59$).

The LAI of all treatments changed little between 5.7 and 6.2 years, as indicated by an $R^2 = 0.95$ (n = 16) for a linear relationship between the two measurement dates for individual plots. The relationship between LAI at 5.7 and 7.0 years was less strong ($R^2 = 0.78$, n = 30), but there was an average decrease in LAI of only 0.6 between 5.7 and 7.0 years. There was a significant but rather imprecise relationship between LAI at 5.7 years and growth of individual plots between 5.8 and 6.8 years ($R^2 = 0.59$) (Fig. 2).

These results indicate that fertilisation of this eucalypt plantation significantly increased LAI and consequent stem growth, but LAI alone was a rather imprecise predictor of current or subsequent growth. These data provide the first available relationship between LAI and stem volume growth of *E. nitens* over such a wide range of LAI, but this relationship should be used with caution in other circumstances until further work determines the extent of variation in the relationship that might be due to age and environment. Future research is expected to improve LAI as a predictor of growth by developing a process-based understanding of the interactions between light interception, climatic factors, and the availability of water and nutrients.

Acknowledgement

We appreciated assistance by Sandra Hetherington, Norske Skog, in managing the site, and Robin Cromer, Charlie Turnbull and Ann LaSala of the CRC and CSIRO for assistance in designing and establishing the experiment.

Table 1. Nitrogen and P fertiliser applications (kg elemental ha^{-1}). Fertilisers were broadcast over the entire plot area as ammonium sulphate (years 0.2 to 3.2), urea (year 6.0), and triple superphosphate.

Treatment	0.2 years N:P	1.2 years N:P	2.2 years N:P	3.2 years N:P	6.0 years N:P	Total N:P
N0	0:0	0:0	0:0	0:0	0:0	0:0
N1	75:37.5	0:0	75:37.5	0:0	400:100	550:175
N2	150:75	0:0	150:75	0:0	0:100	300:250
N3	300:150	0:0	300:150	0:0	0:0	600:300
N4	600:300	0:0	600:300	0:0	400:100	1600:700
N5[a)]	300:150	300:150	300:150	300:150	0:100	1200:700

[a)] Annual rates in the N5 treatment during the first 4 years were split into four doses per year.

W. J. Horst et al. (Eds), Plant nutrition – Food security and sustainability of agro-ecosystems. 930~931, 2001

Possible role of the soil in the Sissoo forest (*Dalbergia sissoo*, Roxb.) decline in the Nepal terai

S. P. Sah[1)], C. K. Sharma[1)] F. Sehested[2)]

[1)] *Department of Biological and Environmental Sciences, Kathmandu, University, Dhulikhel, Kavre, Nepal.*
[2)] *Danish Forestry Extension, Koteshwor, Kathmandu, Nepal*

Kew words: *Dalbergia sissoo*, forest decline, sissoo forest, soil, water logging

Abstract

This paper is concerned with the study of the recent sissoo (*Dalbergia sissoo*, Roxb.) decline in the plain land (locally called "terai") of Nepal. The Sissoo tree species, due to its fast growing nature, quality timber, easy propagation and drought resistance has been the most favourable plantation species for the private as well as government sector for the last 3 decades in the region. There are a variety of symptoms for sissoo decline. The dying of the crown is most prevalent. This paper makes a comparative study of 30 different sissoo forest stands (both natural and plantation). The various aspects of soil characteristics were studied and it has been found that soil is not the sole responsible factor involved in sissoo decline, because sissoo has been found to be dying on all types of soils.

Introduction

The accelerated rate of deforestation in the past in Nepal has resulted in severe forest degradation in the country. Fortunately, plantations of fast growing tree species present in the country have become very popular to overcome this problem. One of these fast growing tree species is sissoo (*Dalbergia sissoo*, Roxb.). Unfortunately, for the last 3-4 years, this tree species has been dying in plantation forests. Similar decline phenomenon has been reported from the adjoining border area of India in Bihar and Utter Pradesh (Sharma *et al.*, 2000) as well as from other Indian sub-continent nations. The objective of this paper is to compare the stand characteristics of different sissoo stands in both natural and plantation sites.

Methods

The 30 research sites (both natural and plantation sites) werer selected in the lowland of Nepal from Central Development Region (Hetauda) to the Eastern Development Region (Jhapa district) for the present study on the basis of the reconnaissance survey. Both healthy and diseased plantation sites have been selected for this purpose. All these selected sites have been assessed for their soil characteristics.

Results and discussion

The stands selected in this study have a wide range of variations such as young (4 years old) to older plantation (28 years old), natural to plantation sites, healthy to high mortality stands etc. From the study of all these stands, the following inferences have been drawn (see Table 1):

- Out of a total 30 stands, the majority (29 stands) have were dying. The decline was from low mortality to high mortality. Only the Sissoodhari stand, which was natural, was fully healthy. Another natural site (Surunga) also showed a low decline trend. Out of 30 stands, 5 stands (Tilkane, Jatuwa-SC, Belgachhi, Dhanusha-1, and Lakhanpur) had a very high mortality.
- No relationship between pH of soil and the stand mortality was obtained. The pH of different habitat soils ranged from highly acidic (4.5) to highly alkaline (8.0). The stands were dying on all types of soils.
- Soil texture of the top soil in the majority of the sites was heavily-textured and sub-soils loosely-textured. No definite trend existed between the soil texture and the stand mortality, however Sissoo dieback was much more prominent in the stands on stiff clay and water-logged soils.
- Water logging has been found to intensify the decline. However, there was no correlation between the stand mortality and the water drainage status of the sites.

References

Sharma M K, Aryal R M and Pokhariyal TC 2000 *Dalbergia sissoo* in India. Paper presented at "Regional Seminar on Sissoo Dieback, Kathmandu, Nepal, pp. 16.

Table 1. Declining status of different plots and their soil characteristics

(SC = Sandy-Clay, SL = Sandy-loam, CL=Clayey-loam, SCL=Sandy-clay-loam, SiL=Silty-loam, WHC = Water Holding Capacity).

Forest sites	Soil pH		WHC (%)		Soil-texture		Soil density (g.cm3)	Soil porosity (%)	Soil Colour	
	Top-	Sub-soil	Top-	Sub soil	Top-	Sub-soil			Top soil	Sub soil
(A). HEALTHY STANDS										
1.Sissoodhari, Hetauda	7.54	7.37	60	40	Loamy	SL	1.43	46	2.5Y4/2	10Y3/2
B) STANDS WITH LOW MORTALITY										
1. Surunga:	4.38	5.54	70	33	Sandy	Dandy	1.34	49	10YR4/2	10YR6/2
2. Chuchche Khola	5.02	5.91	50	50	Clayey	SL	1.45	45	10YR6/2	2.5Y7/3
3. Sauraha	4.97	5.23	47	40	Clayey	SiL	1.63	39	10YR4/2	10YR5/4
4. Baghamara	7.38	7.58	70	45	SC	CL	-	40	10YR4/2	10YR6/2
5. Pushapalpur	5.78	6.16	30	35	SL	SL	1.56	41	2.5Y6/2	2.5Y6/2
6. Lalbhitti	5.46	6.02	50	40	SL	SL	1.42	46	2.5Y6/2	2.5Y6/2
7. Jatuwa	5.47	5.61	50	46	SL	Clayey	1.55	42	2.5Y6/3	2.5Y6/2
8. Janpath-East	5.33	6.50	60	50	Clayey	SL	1.30	51	2.5Y6/3	2.5Y7/3
9. Ghinaha Ghat	5.22	5.22	60	50	SiL	SiL	1.60	40	2.5Y6/3	2.5Y6/3
10. Duhabi	7.47	7.84	50	50	Clayey	SCL	1.36	49	2.5Y6/2	2.5Y6/3
11. Aurabari	5.10	6.48	80	67	Clayey	SL	1.47	45	2.5Y6/3	2.5Y5/2
C) STANDS WITH MODERATE MORTALITY										
1. Churia Mai	6.46	5.32	50	40	SL	SL		39	7.5YR2.5/2	10YR4/4
2. Ram Nagar:	6.39	6.56	40	30	SL	SL	1.43	46	2.5Y5/2	2.5Y6/2
3. Dhanusha-2	6.29	6.61	45	40	Clayey	SL	1.34	49	2.5Y5/2	2.5Y7/3
4. Janpath (west)	7.8	7.97	60	70	Clayey	Clayey	1.45	45	2.5Y6/2	2.5Y6/2
5. Rani Sikiyahi:	6.01	5.37	60	60	Clayey	Clayey	1.44	48	2.5Y7/2	2.5Y7/3
6. Katahari:	5.61	6.76	60	60	Clayey	CL	1.44	46	2.5Y5/2	2.5Y6/2
7. Pokariya:	7.27	5.54	60	60	CL	SL	0.94	65	2.5Y5/3	2.5Y5/3
8. Mrigauli	5.73	6.40	50	60	SL	Sandy	1.42	49	2.5Y6/2	2.5Y6/2
9. Dandi Ghat:	5.91	6.55	50	56	SL	Sandy	1.36	49	2.5Y5/2	2.5Y6/2
10. Bhokra	5.11	5.25	50	67	SL	Clayey	1.55	42	2.5Y3/2	2.5Y6/2
11. Satashi Dham	5.40	5.66	60	60	SL	SC	1.25	53	2.5Y5/2	2.5Y6/2
12. Top Gachhi:	6.71	5.50	50	46	Clayey	Sandy	1.38	48	2.5Y6/2	2.5Y5/2
13. Dharampur	6.17	6.41	50	50	SC	SL	1.48	44	2.5Y5/2	2.5Y7/2
D). STANDS WITH HIGH MORTALITY										
1. Tlkane:	6.01	5.45	50	55	SL	SL	1.40	47	7.5YR2.5/1	7.5YR2.5/1
2. Dhanusha-1	6.23	6.32	50	50	Clayey	CL	1.55	42	2.5Y7/3	2.5Y7/3
3. Jatuwa-SC	5.11	5.70	60	70	CL	C	1.51	43	2.5Y6/3	2.5Y6/3
4. Belgachhi	7.10	7.50	40	30	SC	Sandy	1.49	44	2.5Y6/2	2.5Y6/2
5. Lakhanpur	4.99	5.27	40	40	Sandy	Sandy	1.16	56	2.5Y6/3	2.5Y6/3

2.5Y4/2 = dark grayish brown; 2.5Y3/2 = very dark grayish brown; 10YR5/4 = yellowish brown
2.5Y6/3 = light yellowish brown; 2.5Y6/2 = light brownish grey; 10YR6/2 = light brownish gray
2.5Y7/2 = light grey; 2.5Y5/2 = grey brown; 10YR4/4 = Brown
2.5Y5/3 = light olive brown; 2.5Y7/3 = pale yellow; 10YR4/2 = Dark gray brown
2.5Y7/2 = light gray; 2.5Y7/4 = pale yellow; 10YR3/2 = very dark grayish brown
2.5Y6/4 = light yellowish brown; 7.5YR2.5/2 = very dark brown; 7.5YR2.5/1 = Black

W. J. Horst et al. (Eds), Plant nutrition – Food security and sustainability of agro-ecosystems. 932~933, 2001

Relations between the use of N-fertilisers and the resulting N-balances on sandy and marshy soils in Lower Saxony/Germany – A farm study

K. Möller and E. Przemeck

Institute of Agricultural Chemistry, Georg-August-University, Göttingen, Von-Siebold-Str. 6, D 37075 Göttingen, Germany, eprzeme@gwdg.de

Key words: nitrogen-balance, nitrogen-fertiliser reduction, nitrogen use of fertilisers

Abstract

The nitrogen balance is seen as an important environmental indicator for a sustainable agricultural crop production. It is expressed as N-input minus N-output on the farm land. The results of a research project (6 years investigation period) in Lower Saxony/Germany point out that the N-balance on farm lands (sandy and marshy soils) significantly correlates with the applied amount of N fertilisers. It decreases clearly if farmers reduce the use of N-fertilisers to under the economical level. On average, the reduction of the use of N-fertilisers resulted in 50-80% lower N-surpluses on the farm land, but this decrease was connected with yield losses of up to 10-20% for cereals.

Introduction

Generally, in central Europe with its favourable climatic conditions economic reasons are responsible for the high intensity of field farming, which is linked with a high nitrogen use in the field. If the N-amount is not based on a realistic yield expectation, nitrogen pools will increase in the soils, which can be responsible for the pollution of water, air etc.. The calculation of N-balances can help to find out whether the N-amount was in a correct relation to the yield. In order to investigate whether the intensity of crop production, especially in respect to the use of N-fertilisers, has consequences for the N-balances in the field and for the farm productivity, a research-project was implemented on specific farms. 26 farms in two different pedo-climatic regions of Lower Saxony were involved in an intensive analysis of medium-term effects of a reduced use of N-fertilisers.

Materials and methods

For 6 years, 26 farms participated in the project in two different pedo-climatic areas of Lower Saxony, Germany (Tab 1): in the more continental area of Lüchow-Dannenberg [LD] (predominantly sandy podzols) and in the more humid area of Aurich-Leer [AL] (predominantly young marshy gleysols). The farm types varied from commercial farms to mixed farms to forage-growing farms with milk production. In each region 9 farms ('quota farms') were allowed to buy 60% of the 1991 calculated amount of fertiliser N recommended by the official extension service, including organic manure. Farmers were free to portion this quota to their crops and to change their crop pattern as they wished. Four farms of each region were allowed to buy the full budget of N-fertilisers according to fertiliser recommendations ('reference farms'). Field specific N-balances of all arable land and of the grassland were calculated from 1992 to 1997 on each farm. They were based on notes of field records for the N-input and yield and N content determinations for the N-output (details see Möller, 1999). To calculate the grassland yield, data on silage yield, animal grazing, milk production a.o. were used according to Ehlerding (2000).

Table 1. The participating farms (1992-97)

region		LD	AL
soils		sandy	marshy
climate		more dry	more humid
number of farms	reference	4	4
	N-quota	9	9
farm-land (ha)	reference	782	476
	N-quota	760	644
average yield (t ha^{-1})	reference	6.2 rye	8.2 wheat
	N-quota	5.5 rye	7.7 wheat

Results

In deviation from the originally 1991 fixed N-budget, the average of the actually applied N in the quota farms was 24% (LD) and 36% (AL) less compared to the reference farms (Fig. 1). The N-output from the fields with the harvested materials was in both regions in the reference farms only 15-20 kg N ha^{-1} higher than in the quota farms. But the N balances resulted in a surplus of about 70 kg N ha^{-1} in both reference farms groups and of 16 (AL) and 41 (LD) kg N ha^{-1} in the quota farms.

If the N-balances were related to individual fields, their wide variability could be recognised (Tab. 2). The mean values covered largely the average balance of the total field areas (Fig. 1). The balances on the sandy soil (LD) were equalised by the reduced N-amount in quota farms in a minor degree than on the much more fertile marshy soils (AL), where less yield losses (Fig. 1) were found in case of a reduced use of N-fertilisers.

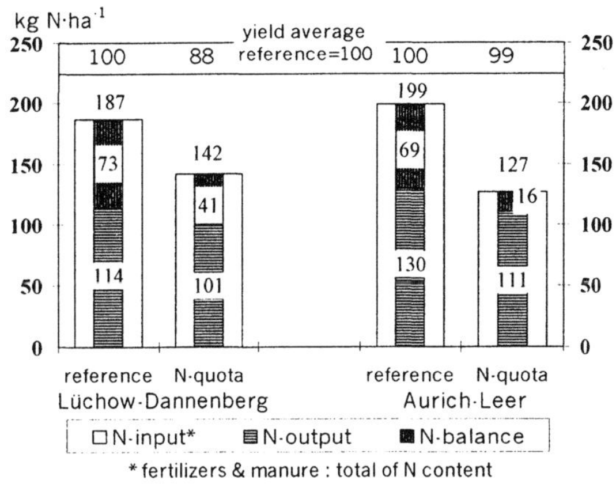

Figure 1. Comparison of yield and N-balances in average per hectare on the total farm land (1992-97).

Moreover, in both regions strong deviations from the average of the balances occurred (Tab. 2), which indicates on the one hand a high N-output from some fields of the quota farms with negative balance values of about 100 kg N ha^{-1} and on the other hand an (intolerable) surplus of sometimes over 100 kg N ha^{-1}. Exceptional high N-balances in quota farms were noticed only on fields with organic manure application. But basically, the N-balance values strongly depended on the intensity of the use of N-fertilisers, which is demonstrated in Fig. 2 for LD and in Fig. 3 for AL, where each individual field is marked.

Table 2. Mean and variability of N-balances (1992-97)

farm groups (arable land)	fields n	lowest kg N ha^{-1}	average kg N ha^{-1}	highest kg N ha^{-1}
LD reference	145	-19	+62	+145
LD quota	165	-99	+44	+226
AL reference	32	+17	+71	+146
AL quota	108	-107	+4	+111

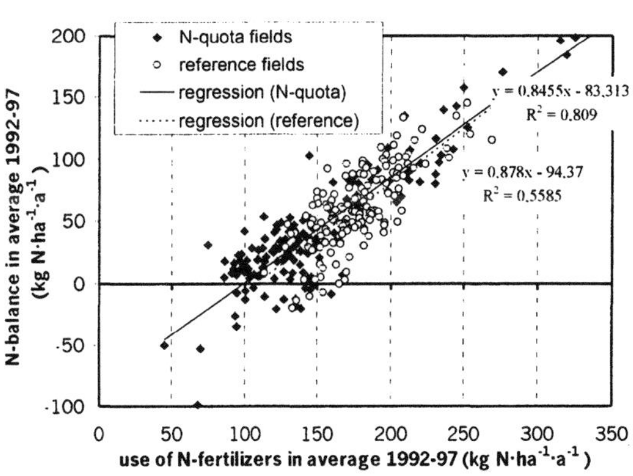

Figure 2. Relations between use of N-fertilisers and N-balance on individual fields (arable land) in average of 6 years in the region with sandy soils, LD (1992-97)

On the sandy soils in LD, the relations between applied N-fertilisers and the resulting N-balance were close in quota and reference farm groups. However on the fertile marshy soils in AL the dependence between the N-input and the resulting N-balance was less close, which can be explained by the higher N-pool in these soils, which can amount to 20 t N ha^{-1} (total) and more. The results show that the increase of the regression line is nearly the same in both regions, despite different soil and climatic conditions.

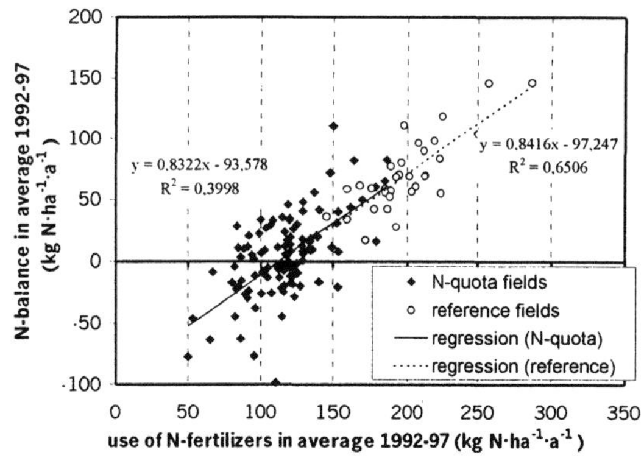

Figure 3. Relations between use of N-fertilisers and N-balance on individual fields (arable land) in average of 6 years in the region with marshy soils, AL (1992-97)

Discussion

Under real farm conditions there is a strong relation between the intensity of N-fertilisation and the resulting N-balances (field balance) in plant production. Therefore the simple calculation of N-balances proves to be suitable for a field-specific examination of concepts for using N-fertilisers in the farm. To explain striking deviations of balance results, the prevailing pedo-climatic conditions have to be taken into account. On the sandy soil in LD e.g. water stress can be a decisive reason for an unexpected low N-output from the field with a balance surplus. A 40% reduction of the N farm-budget can cause the long-term effect of a N-exhaustion with the consequence of considerable and in time increasing losses of the potential yield, especially on less fertile soils. This also has to be taken into account.

Acknowledgement

Promoted by the Niedersächsisches Ministerium für Ernährung, Landwirtschaft und Forsten and the European Community and co-ordinated by the Forschungs- und Studienzentrum Landwirtschaft und Umwelt, Universität Göttingen.

References

Ehlerding M 2000 ifwP, Bonn, Diss. Göttingen.
Möller K 1999 Hainholz, Diss. Göttingen.

W. J. Horst et al. (Eds), Plant nutrition – Food security and sustainability of agro-ecosystems. 934~935, 2001

Effects of subsurface drainage design on nitrate leaching and crop yield

A. El-Sadek and J. Feyen

K.U.Leuven, Institute for Land and Water Management, Vital Decosterstraat 102, B-3000 Leuven, Belgium

Key words: controlled drainage, conventional drainage, hydrology, water quality

Abstract

Some of the highest losses of nitrate to surface waters come from drained agricultural land. This article studies the effect of subsurface drainage density on nitrate losses and crop yield under Belgian farming conditions, using the nitrogen version of DRAINMOD (Brevé *et al.*, 1997). DRAINMOD (Skaggs, 1997) was used to simulate the performance of the drainage system of the Hooibeekhoeve experiment, situated in the sandy region of the Kempen (Belgium), and this for a 14-year (1985-1998) period. In the analysis a continuous cropping with maize was assumed. Daily NO_3-N losses were predicted for a range of drain spacings and depths, two drainage strategies (conventional and controlled) and three fertiliser application dressings (225, 275, 325 kg N ha^{-1}). Losses of N in subsurface drainage occur almost entirely in the NO_3-N form. Losses of organic and mineral N in the form of NO_3-N in surface runoff are negligible. Hydrologic results indicated that increasing drain spacing or decreasing drain depth reduces drainage discharge while it increases runoff. The use of controlled drainage reduces subsurface drainage and increases runoff. Results also revealed that increasing the drain spacing or decreasing the drain depth reduces nitrate-nitrogen (NO_3-N) drainage losses and net mineralisation, while increasing denitrification and runoff losses. Controlled drainage caused a predicted reduction in drainage losses and an increase in denitrification and runoff losses. The optimal combination of drain density and management is one that maximises profits and minimises environmental impacts. Simulated results indicated that NO_3-N losses to the environment could be substantially reduced by reducing the drainage density below the level required for maximum profits based on grain sales. The study concluded that if the environmental objective is of equal or greater importance than profits, drainage systems can be designed and managed to reduce NO_3-N losses while still providing an acceptable profit.

Introduction

Phosphorus is a major limiting factor for crop production on many tropical and subtropical soils. Given the limited access of small farmers to fertiliser P, it is desirable to identify and incorporate plant species into cropping systems which can mobilise P from soil-P pools that are unavailable to less P-efficient food crops. Some leguminous cover crops are reported to possess a remarkable capacity to mobilise soil P through root exudates, for instance Cajanus cajan (Ae *et al.*, 1990) and especially Lupinus albus (Gardner *et al.*,1983). In response to low P supply Lupinus albus develops cluster roots (Gardner *et al.*, 1981) which are major sites of H^+ and citrate excretion (Marschner *et al.*, 1986, Dinkelaker *et al.*,1989). The release of organic anions into the rhizosphere enhances the mobilisation of sparingly soluble soil P (Li *et al.*, 1997). One of the objectives of the present work was the development of a technique which allowed the quantification of organic anion exudation of plants grown in nutrient solution and in soil in situ.

Materials and methods

In this study, the DRAINMOD model was calibrated using historic field data collected for a 14-year period (1985-1998) on the experimental site of the Hooibeekhoeve, situated in the community of Geel, the northeastern sandy region, the Kempen, of Belgium. During the period of experimentation the field was cropped with maize. Only organic manure was applied as fertiliser. Missing data, required to run the model, such as the soil hydraulic parameters (van Genuchten and Nielsen, 1985), were either measured on undisturbed soil cores or reconstructed by using the pedo-transfer functions of Vereecken (1988), as indicated by Ducheyne and Feyen (1999). DRAINMOD-N was applied to evaluate long-term effects of several drainage and related water table control systems on nitrogen-nitrate losses. In the scenario-analysis it was assumed that the field, for which DRAINMOD-N was calibrated, was equipped with a subsurface drainage system consisting of parallel, 10 cm diameter, corrugated plastic drains. The drainage designs evaluated consisted of four drain depths (0.75, 1.00, 1.25, and 1.50 m), four drain spacings (25, 50, 100, and 300 m), and three fertiliser strategies (225, 275, and 325 kg N ha^{-1}). The management treatments included conventional and controlled drainage. The maize production practices used in the simulations are characteristic for the sandy region of the Kempen.

Results

For the experimental field in the Hooibeekhoeve, Geel, effects of drainage system management, drain spacing and depth on subsurface drainage and surface runoff are shown in Fig. 1. The average annual rainfall of the simulation period (14-year) is 86.75 cm. Simulation results indicate that increasing the drain spacing reduces subsurface drainage while it increases surface runoff and ET. Increasing the drain depth increases drainage and decreases ET and runoff. Furthermore, controlled drainage reduces subsurface drainage and increases surface runoff, as compared to conventional drainage. The magnitude of these changes increases with the intensity of controlled drainage.

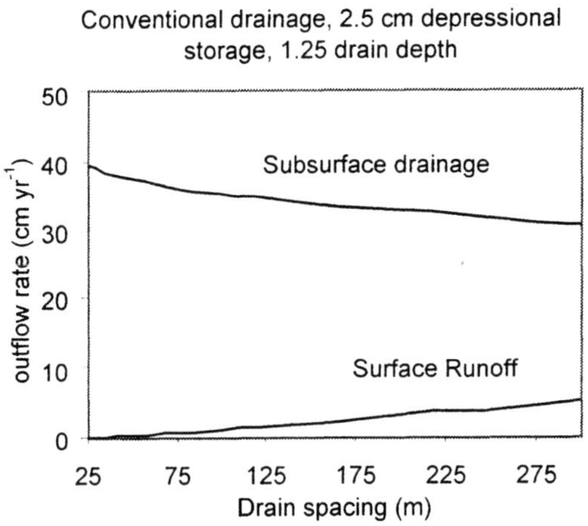

Figure 1. Predicted average annual subsurface drainage and surface runoff as affected by system management, drain depth and spacing.

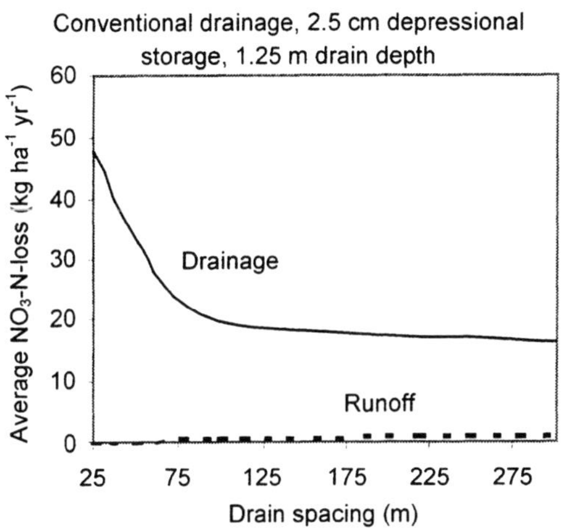

Figure 2. Predicted average annual NO_3-N losses in drainage and runoff as affected by system management, drain depth and spacing. Results are for a 275 kg N ha^{-1} fertiliser strategy.

Effects of drainage system management, drain spacing and depth on the nitrogen budget components (drainage losses and runoff losses) are shown in Fig. 2. Simulation results reveal that increasing the drain spacing reduces NO_3-N drainage losses and net mineralisation, but it increases NO_3-N runoff losses and denitrification. On the other hand, increasing the drain depth increases drainage losses and net mineralisation, and it decreases runoff losses and denitrification. Furthermore, increasing fertiliser application increases drainage losses, runoff losses and all nitrogen components. The use of controlled drainage decreases the subsurface drainage intensity. Hence it reduces drainage losses and increases denitrification and runoff losses, as compared to the use of conventional drainage. The effect of controlled drainage on net mineralisation is not clear. Results of the simulation analysis show that total NO_3-N losses (subsurface drainage plus surface runoff) can be substantially reduced with controlled drainage during the winter season. This can be explained by the climatological conditions in Belgium where drainage is usually highest during the winter and early spring months.

Discussion

The optimal combination of drainage design and management is one that maximises profit and minimises environmental impact. Results of the scenario-analysis indicate that NO_3-N losses to the environment could be substantially reduced by reducing the drainage density below the level required for maximum profits from grain sales. That is, if the environmental objective is equal or of greater importance than profits from the agriculture crops, the drainage systems can be designed and managed to reduce NO_3-N losses while still providing an acceptable profit. The study revealed that for the given climate-crop-soil combination there are different scenarios of drain depth and spacing which result in a similar reduction of the NO_3-N loss to surface waters. The foregoing enables the decision maker to adjust the drainage system design according to the real field situation and the need to control water quality. The analysis presented in this paper further demonstrates the applicability of DRAINMOD-N for quantifying effects of drainage design and management on agricultural profits and NO_3-N losses to the environment for a specific combination of climate, crop and soil.

References

Brevé MA, Skaggs RW, Parsons JE and Gilliam JW 1997 Trans. ASAE, 40 (4), 1067-1075.

Ducheyne S and Feyen J 1999 Int. Workshop on Modelling of transport processes in soils at various scales in space and time. Leuven, Belgium, 24-26 Nov.: 457-466.

Skaggs RW 1997 North Carolina State University, Raleigh, North Carolina, USA.

van Genuchten MTh and Nielsen DR 1985 Annales Geophysicae, 3(5): 615-628.

Vereecken H 1988 PhD-Thesis No. 171, Faculty of Agricultural and Applied Biological Sciences, K.U.Leuven, Belgium, 254 p.

W. J. Horst et al. (Eds), Plant nutrition – Food security and sustainability of agro-ecosystems. 936~937, 2001

Measured and simulated nitrate leaching in vegetable culture

K. R. Végh[1)] and I. Cserni[2)]

[1)] *Research Institute for Soil Science and Agricultural Chemistry of HAS, Herman O. 15. Budapest,H-1022, Hungary, krvegh@rissac.hu*

[2)] *Kecskemét College, Faculty of Agriculture, Erdei F. tér 1-3. Kecskemét, H-6000 Hungary, cserni@kfk.hu*

Key words: intensive vegetable cultivation, sandy soil, *Foenuculum vulgare Mill. Convar. Azoricum Mill. Thell.*, lysimeter, simulation of water and nitrogen dynamics

Abstract

Excessive use of chemical fertilisers can lead to pollution of groundwater with nitrogen, especially if associated with high rainfall or irrigation, shallow rooted crops and sandy soils. Nitrate leaching in a high production fennel culture was studied in sandy soil using lysimeters. The amount of drainage water and its nitrogen content were measured several times during the growing period. Simulation of water and nitrogen dynamics in the soil was tested by comparing measured drainage and N-losses to those predicted by SOIL and SOILN models. In the simulation the change of the parameter values characterising the growth, depth distribution and N uptake efficiency of root systems resulted in reduced nitrate leaching from the soil by increasing the capacity of the fennel plants to acquire water and nitrogen from the soil.

Introduction

Vegetable crops which can produce high yield during short growing period have a high daily demand for nitrogen and water and require a buffer amount of available nutrients and water in their root zone (Cserni, 1996). However, production of vegetable crops with low nutrient acquisition efficiency often involves heavy inputs of fertiliser and irrigation especially in light texture soils, consequently with high risk of nitrate leaching to the groundwater (Németh, 1996). In order to develop crop and site specific management practices more information is required on the dynamics of soil moisture and nitrogen in intensive vegetable production. The goal of our work was to study environmental consequences of intensive fennel production in the region and test the ability of the SOIL/SOILN models (Jansson *et al.*, 1991, Johnsson *et al.*, 1987) to predict nitrate leaching from the rooting zone and interactions with the water and nitrogen use of the crop.

Materials and methods

Experiments were carried out in the lysimeter station of Kecskemét College. Lysimeters (60 cm height, 60 cm diameter) were filled with sandy soil (with 94.7% sand, 2.66% clay, 0.64% organic matter content, pH=7.8) from the local, traditionally vegetable growing area. Topsoil (0-25 cm) and subsoil (25-60 cm) horizons were carefully layered in the lysimeters. After 5 years of settle, soil physical properties were determined before starting fennel cultivation in 1998. In 1999 soil was fertilised on 5 July (sowing day) and on 4 August with NH_4NO_3 solution at a rate of 60 - 60 kg N ha^{-1}. Drip irrigation was applied to prevent rapid drying out of the uppermost sand layer during early growth and to supply sufficient water for fast biomass accumulation in the exponential growth phase.

The SOIL water and heat model was used to simulate soil moisture dynamics and to provide driving variables for the soil nitrogen and crop growth model SOILN. Daily meteorological data (driving variables for SOIL model) were measured in the experimental site. Leaf area, fresh and dry matter accumulation of leaves, stems, bulbs and roots and their nutrient contents were sampled on eight occasions. The measured and derived growth parameters were used in the parameter optimization of the growth model. The SOIL and SOILN model outputs were fitted on the amount and nitrate content of drainage water which was measured several times during the growing period (from early July until late October), and vegetable yield. The model calibrated with plant, soil water, nitrate leaching data was then used to simulate N dynamics in the soil.

Results and discussion

Fennel growth was quite slow in July compared to the leaf area and biomass accumulation during the following three months (Table 1.).

Table 1. Leaf area (LA) and dry mass (DM) production of one fennel plant

Days After sowing	LA cm^2	Shoot DM g	Root DM g
35	55	0.93	0.23
42	153	1.85	0.69
49	376	4.41	0.76
58	600	8.74	1.62
89	1634	35.60	6.27
104	3676	50.52	10.6

In the Kecskemét region the upper limit of required irrigation water in soils with good water regime for vegetables is about 300 mm in the growth period (Szalóki, 1989). For optimum plant water supply the frequency of irrigation has to be increased for sandy soils. In 1999 precipitation was unusually high and accumulated water input (precipitation: 360 mm + irrigation: 200 mm) exceeded total evapotranspiration (340 mm) during the growing season. As a consequence of high rainfalls, deep percolation of soil water was sufficient in July and August, when both the above and belowground biomass of the seedlings were relatively low (Table 1 and 2.).

Table 2. Measured (Meas.) and simulated (Sim.) amount of drainage water and nitrate leaching under fennel cultivation on days of collection

Days After Sowing	Water mm Meas.	Sim.	Nitrate $g.m^{-2}$ Meas.	Sim.
1	10.5	14.4	0.30	0.38
4	9.0	11.4	0.24	0.19
8	16.5	15.7	0.55	0.64
20	16.8	18.5	4.80	4.33
35	10.7	12.1	3.80	3.72
53	16.7	16.7	3.55	3.31

The large quantity of water infiltrated to the soil shifted the accumulation zone of nitrate dissolved from recent applied fertiliser downwards to the almost saturated soil. Because of the low water and nitrogen demand of the seedlings, nitrogen accumulated in the 25-60 cm layer by the end of the second week after the N application, and the total amount of N applied in fertiliser was leached by the end of August.

Rooting depth (estimated from taproot length data) was about 10, 40 and 50 cm on the 20th, 45th and 60th days of growth. By the change of the parameter values characterising the growth, depth distribution and N uptake efficiency of root system, simulated nitrate leaching could be minimised.

These results of the simulation support our understanding of nitrate leaching in conditions of intensive vegetable production in sandy soil.

A genotype with rapid root proliferation and higher acquisition efficiency would be more suitable for intensive cropping in this environment.

Acknowledgement

This work was supported by the Hungarian Scientific Research Foundation (OTKA) T 023348.

References

Cserni I 1996 Kée KFK Bulletin 102-110.

Jansson P-E 1991 Swedish University of Agricultural Sciences, Division of Hydrotechnics (Uppsala). Report 165, 72 p

Johnsson H, Bergström L, Jansson P-E and Paustian K 1987 Agric Ecosys Environ 18, 333-356,

Németh T 1996 Organic matter and nitrogen content of soils (In Hungarian) MTA TAKI, Budapest,

Szalóki S 1989 In Handbooc of irrigation practice (in Hungarian), Ed G Szalay. pp100-155. Mezgazda Press, Budapest.

W. J. Horst et al. (Eds), Plant nutrition – Food security and sustainability of agro-ecosystems. 938~939, 2001

Nitrogen balances derived from long-term experiments with vegetable crops

J. Rühlmann[1)] and B. Geyer[2)]

[1)] *Institute for Vegetable and Ornamental Crops Grossbeeren/Erfurt e.V., Theodor-Echtermeyer-Weg 1, Grossbeeren, D 14979, Germany, Ruhlmann@igzev.de*

[2)] *Institute of Horticultural Crop Science, Humboldt University Berlin, Lentzeallee 75, Berlin, D 14195, Germany, bernd.geyer@agrar.hu-berlin.de*

Key words: long-term experiments, nitrogen balance, vegetable

Abstract

Our approach to calculate nitrogen balances should show which quantities of the nitrogen supply were recovered in the plant-soil-system. Two time scales were compared - the cultivation period and the decadal time scale of long-term experiments. The nitrogen uptake of the vegetable crops within the cultivation period differed between 53% and 72% of the nitrogen supply depending on vegetable species. The recovery rate at the end of cultivation period was about 80% of the nitrogen supply - independent of vegetable species. In contrast, the mean annual recovery rate in the plant-soil-system of all tested treatments was only 50% of the mean annual supply in the long-term experiments.

Introduction

Nitrogen balances are an essential instrument to quantify nitrogen losses from the plant–soil–system. In the past, calculations of nitrogen balances often have excluded the net nitrogen mineralisation and they were related only to the cultivation period. Therefore, our approach should show which quantities of the nitrogen supply (including the net nitrogen mineralisation) were recovered in the plant-soil-system both after one cultivation period and in the long-term mean of investigations within long-term experiments.

Materials and methods

The base of our investigations were two long-term experiments with vegetable crop rotations in sites situated in Grossbeeren, south of Berlin (Richter, 1999). The first experiment was setup in 1972 on three different soils – a diluvial sandy soil, an alluvial loam, and a loess loam. Factors investigated in this experiment were the nitrogen fertilisation (three levels) and manure application (four kinds). This experiment was used to calculate long-term nitrogen balances. The second experiment was setup in 1986 on a sandy soil to investigate the influence of 4 levels of nitrogen fertilisation (0, 50%, 100%, and 150% of the expected nitrogen uptake of plants) and 3 levels of farmyard manure application (0, 30, and 60 t ha^{-1}) on the proportion of nitrogen mineralised from soil on the nitrogen uptake of plants. This experiment was used to calculate the nitrogen balances related to the cultivation period.

The balances for the cultivation period and the long-term nitrogen balances were calculated in different ways.

The nitrogen supply for the cultivation period was defined as the sum of mineral soil nitrogen (0-60 cm) at the begin of cultivation period, added fertiliser nitrogen, and nitrogen derived from atmospheric deposition and net nitrogen mineralisation (NNM). This NNM was calculated using the model CANDY (Franko and Rühlmann, 1991). The nitrogen recovery was estimated by summarising the nitrogen uptake by plants and the mineral soil nitrogen (0-60 cm) at the end of cultivation period.

The nitrogen input of the long-term experiments was defined as the mean of all nitrogen quantities added annually by fertilisers, manure, and crop residues. The nitrogen recovery included the nitrogen taken up by plants and the measured changes of total nitrogen content of soils.

Results

The relation between nitrogen supply and nitrogen uptake by plants showed differences depending on plant species. The tested vegetable species were divided into two groups (Fig. 1).

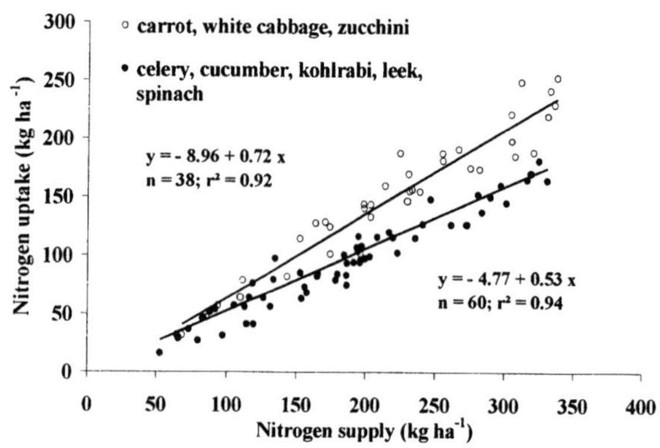

Figure 1. Nitrogen uptake by plants in dependence on nitrogen supply.

The first group (carrot, white cabbage, zucchini) took up about 72% of nitrogen supply whereas this proportion was nearly 20% lower in the second group (celery, cucumber, kohlrabi, leek, spinach). The nitrogen recovered by uptake and residual mineral nitrogen in soil at harvest was about 81% of nitrogen supply and independent on vegetable species (Fig. 2).

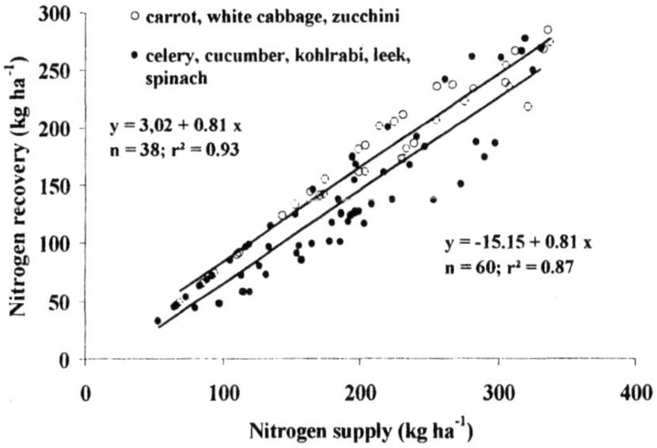

Figure 2. Nitrogen recovery in plant and soil depending on nitrogen supply.

The estimated mean nitrogen recovery in the long-term experiments of about 50% in the mean of all included treatments was clearly lower than in the balances related to the cultivation period.

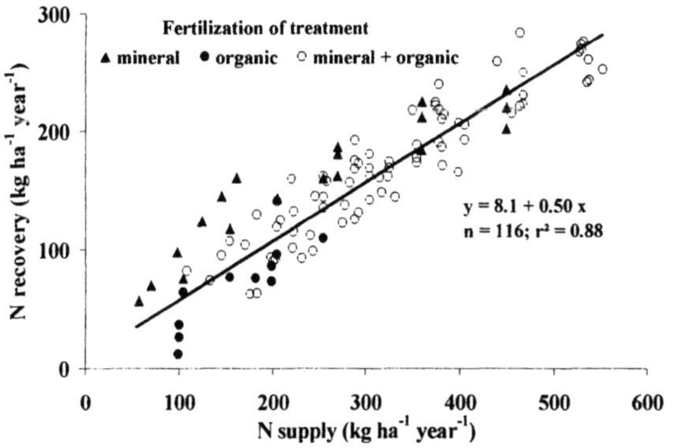

Figure 3. Nitrogen recovery in plant and soil depending on nitrogen supply.

Figure 3 shows that treatments fertilised exclusively with mineral nitrogen were characterised by an above average recovery, whereas that of the organic treatments was below average.

Discussion

The results shown in Figure 1 indicate that the tested vegetable species were different in their nitrogen uptake efficiency. Generally, this may be caused by differences in the length of vegetation period and in rooting depth. The vegetable species of the first group (carrot, white cabbage, zucchini) were characterised by a long vegetation period and by a root system with allowed them to use water and nutrients up to depth of 60 – 90 cm. In contrast, the cucumber, kohlrabi, and spinach as representative of the second group had a much shorter vegetation period and a smaller rooting depth. Nevertheless, the presence of leek and celery within the second group with a cultivation period and rooting depth very similar to the first group shows that other processes such as total and/or active root surface may influence the nitrogen uptake efficiency.

The difference in the nitrogen uptake between the two groups (Fig. 1) and the similarity in their nitrogen recovery (Fig. 2) were clearly caused by different values of the residual mineral nitrogen in soil at harvest (data not shown). The higher residual values of the second group may be related to the risk of higher nitrogen losses after the cultivation period. Because the mean nitrogen recovery related to the cultivation period was about 80% for both groups it may assumed that the dimension of mean nitrogen losses during this period amounted 20%.

In contrast, the mean annual nitrogen recovery in the plant-soil-system of all tested treatments of only 50% of the mean annual supply (Fig. 3) indicates that the mean annual nitrogen losses from this system may also amount 50%. No differences were estimated depending on soil (diluvial sandy soil, alluvial loam, loess loam; data not shown). Therefore it was assumed that the dimension of losses may be independent of the kind of the dominating process of loss (leaching vs. denitrification). The above average nitrogen recovery in the mineral fertilised treatments may be interpreted as a good synchronisation between the demand of plants and the supply within these treatments. It was most pronounced in the range of nitrogen supply up to about 300 kg ha^{-1}. Higher supply led to a lower recovery due to the limited capacity of nitrogen uptake by plants. On the other hand, long-term application of manure led to significantly higher soil organic matter contents (data not shown) which were related to an increase of nitrogen mineralisation – also and especially outside of the cultivation period. Therefore the below average recovery of organic fertilised treatments may be an expression of a lower correspondence between the demand of plants and the supply within the organic fertilised treatments.

References

Franko U and Rühlmann J 1991 Gartenbauwiss. 56, 174-179.

Richter K 1999 *In* Ökologische Hefte der Landwirtschaftlich-Gärtnerischen Fakultät. Humboldt University, 228 pp. Berlin.

Rühlmann J 1996 Acta Horticulturae 428, 95-104.

W. J. Horst et al. (Eds), Plant nutrition – Food security and sustainability of agro-ecosystems. 940–941, 2001

Mineralisation of newly accumulated nitrogen

R. Nieder[1)], H. P. Dauck[1)] and D. K. Benbi[2)]
[1)] *Institute of Geography and Geoecology, Braunschweig Technical University, Langer Kamp 19c, 38106 Braunschweig, Germany*
[2)] *Department of Soils, Punjab Agricultural University, Ludhiana, India*

Key words: long-term N accumulation, N mineralisation potential, soil organic matter "equilibria"

Abstract

Since the 1950s, large amounts of nutrients (N, P and K) have been accumulated in German arable soils due to a surplus in the balance between the amounts of nutrients applied and those removed in crop harvest. In western Germany (the former FRG), a significant part (about 30-100 kg N ha^{-1} yr^{-1}) of the surplus N (more than 100 kg N ha^{-1} yr^{-1}) has been incorporated into newly formed soil organic matter after deepened tillage since about 1970. The buffering capacity of the soils for reactive C, N, and P compounds will reach its limits in the near future (around the year 2005), after organic matter "equilibria" will have been re-established in most soils. Long-term incubation experiments with loess soils (luvisol A_p material) show that the actual N mineralisation capacity is significantly higher than in the 1970s.

Introduction

In western Germany, deepening of plough layers (from < 25 to > 35 cm) has been a feature in intensively cultivated areas. This practice causes a long-term build-up of soil organic matter (SOM). After dilution of SOM in the A_p due to mixing of underlying C-poor subsoil material, C-, N-, S- and P mineralisation potentials are decreased. Subsequently, SOM contents and mineralisation potentials present before the onset of deeper ploughing are expected to become re-established within several decades. Accumulation rates on 120 plots from 16 farms in Lower Saxony yielded up to 1.0 t N and 16 t C ha^{-1} (1967-1998) in loess soils and 2.4 t N and 25.9 t C ha^{-1} (1972-1998) in sandy soils.

Materials and methods

The soils used in this study (Table 1) are from the southern Lower Saxony loess area. The main crops are cereals (winter wheat and winter barley) and sugar beet. In 1985 and 2000, field moist samples (aliquots equivalent to 20 g dry matter) were incubated in triplicate at 35° C for a total of 155 (1985) or 354 (2000) days, respectively (method according to Stanford and Smith, 1972). All calculations are based on a soil bulk density of 1.5 kg L^{-1} and a 35 cm plough layer. For the 2000 experiment, mineralization parameters were estimated from the cumulative amounts of mineralised NO_3^--N and NH_4^+-N via non-linear regression, using an exponential model with two first-order kinetics reactions:

$$N(t) = N_1 \cdot \{1-\exp(-k_1 \cdot t)\} + N_2 \cdot \{1-\exp(-k_2 \cdot t)\}.$$

N_1 is the nitrogen mobilised from a more available fraction, N_2 that from a more resistant fraction of the potentially mineralisable N_0, k_1 and k_2 are the respective rate coefficients.

Results

The results of the experiment conducted in 1985 (Figure 1a) show a significant difference in the courses of N mineralisation between soils with "equilibrium" N contents and recently deepened A_p horizons with diluted SOM contents. This was confirmed by additional experiments in the 1980s (Nieder, 1990). The double-exponential model yielded a good fit for the courses of N mineralisation in the experiment conducted in 2000 (Figs. 1b-1d). The N_1 fraction varies widely among the A_p deepening phases (Table 2). This may be mostly due to recent management effects (inputs of crop residues and plant roots). The resistant N_2 fraction is strongly affected by tillage. However, the 1967-1970 and 1972-1975 deepened plough layers exhibit a similar N_2 fraction and thus seem to have reached an "equilibrium" in their mineralisation potentials. This potential has increased significantly during the last three decades due to the recent N accumulation. According to Richter *et al.* (1982), the N_2 fraction was about 600 to 800 kg N ha^{-1} 30 cm^{-1} in the 1970s. In contrast, in the soils of the last tillage deepening phase (1980-1982), the mineralisation potential at present is only about 70% of that in the soils with earlier tillage deepening.

References

Nieder R 1990 Kali-Briefe (Büntehof) 20 (2), 197-203
Richter J, Nuske, A, Habenicht W and Bauer J 1982 Plant and Soil 68, 379-388.
Stanford G and Smith SJ 1972 Soil Sci. Soc. Am. Proc. 36, 464-472.

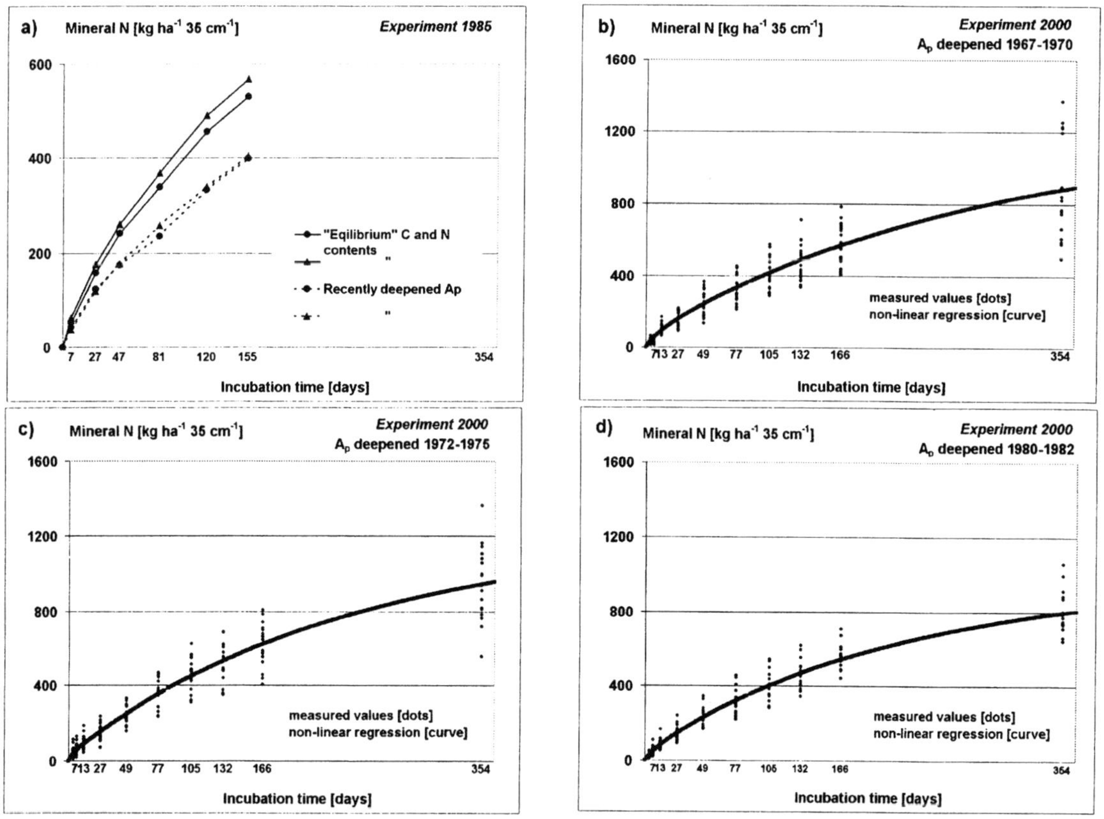

Figure 1. Cumulative N mineralisation (35° C) in loess soils with „equilibrium“ C and N contents and recently deepened A_p incubated in 1985 (Fig. 1a) as well as in soils with deepened tillage 1967-1970 (Fig. 1b), 1972-1975 (Fig. 1c) and 1980-1982 (Fig. 1d) incubated in 2000

Table 1. Properties of arable soils (luvisols from loess) with different A_p-deepening phases (incubation experiments 1985 and 2000)

Experiment	A_p-deepening phase	Number of Farms	Number of Soils	C_{org} (%)	Total N (%)	Clay (%)	Silt (%)	Sand (%)
1985	1967-1970	1	2	1.01	0.11	14.4	n.d.	n.d.
	1980-1982	1	2	0.85	0.09	12.1	n.d.	n.d.
2000	1967-1970	2	6	1.06	0.12	16.2	79.6	4.2
	1972-1975	2	6	1.15	0.12	15.9	80.7	3.4
	1980-1982	2	6	1.01	0.11	14.8	79.5	5.7

Table 2. Optimized parameters for potentially mineralisable N (kg ha^{-1} 35 cm^{-1}) and rate coefficients k (d^{-1}) of arable soils from incubation experiment 2000

	N_1	k_1	N_2	k_2
A_p deepened 1967-1970	63	0.086	1153	0.0035
A_p deepened 1972-1975	20	0.331	1162	0.0042
A_p deepened 1980-1982	43	0.158	790	0.0062

W. J. Horst et al. (Eds), Plant nutrition – Food security and sustainability of agro-ecosystems. 942~943, 2001

Recycling of N between soil and plants during the growing season

M. A. Jimenez[1], R. Gutser[2], H. Schmid[2] and J. C. Munch[1]

[1] *GSF- Research Center for Environment and Health, Institute of Soil Ecology, P.O. Box 1129, Neuherberg, D-85758, Germany, jimenez@gsf.de; munch@gsf.de*

[2] *Chair of Plant Nutrition, Technical University of Munich, Freising-Weihenstephan, D 85350, Germany, gutser@weihenstephan.de*

Key words: ^{15}N, N rhizodeposition, root biomass production, soil microbial biomass

Abstract

In spite of the known below-ground biomass production of plant roots that concurrently introduce significant amounts of carbon and nitrogen into the soil, the effects of these inputs on N cycling in the soil-plant system are seldom considered. Two field data sets were used to estimate N rhizodeposition by winter wheat plants: (1) a N-turnover experiment to determine the N-fluxes derived from ^{15}N-labelled clover-residues incorporated into the plough layer of defined plots; (2) a root production experiment to assess the above (shoot) and below ground (gross and net root) biomass production of winter wheat in different fields but near the ^{15}N plots. Based on the depth distribution of winter wheat net root biomass (root production experiment) and on soil organic ^{15}N depth distribution (^{15}N-turnover experiment), the root N input into soil was estimated to be 282 kg ha^{-1}. This was equivalent to 54% of total net N-assimilation of winter wheat. These results give substantial evidence for a N-loop between soil and growing plants, whereby a part of the net mineralised N taken up by the plants is continuously returned into soil by their roots.

Introduction

The proportion of assimilated N recovered below ground at plant maturity ranges from 14% to 57% in annual plants (Jensen, 1996; McNeill *et al.*, 1996; Janzen, 1990). Rhizodeposition represents a labile fraction of soil N (Jensen, 1996). Due to the difficulties using ^{15}N labeling techniques in the field, the N rhizodeposition was quantified in pot experiments, where rooting depth and volume were most likely to be restricted as compared to conditions in the field. Thus, we estimated N rhizodeposition under field conditions using complementary data of ^{15}N fluxes in the plant-soil system and of root dry mass production.

Materials and methods

The experiments were carried out at the FAM Research Station Scheyern (Bavaria, Germany). In the N turnover experiment ^{15}N labelled clover residues were incorporated into the plough layer in Oct. 23, 1996. The ^{15}N signature and N turnover in different soil pools were followed until harvest of the successive crop (winter wheat). The net and gross root production (root length and volume) was measured between 1995 and 1997 at different winter wheat fields adjacent to the ^{15}N plots.

Results

Between spring 1997 and the harvest of winter wheat, a significant increase of organic ^{15}N and concurrent decrease of mineral ^{15}N was found in the subsoil (Fig. 1). During the first 130 days after straw incorporation, a rapid 50% decrease of initial soil organic ^{15}N levels was observed (Fig. 2). This stabilised after day 138, whereby ^{15}N contents remained constant until the harvest of winter wheat at day 286. Within 14 days following incorporation of clover residues, maximum ^{15}N contents in soil microbial biomass were observed, which decayed exponentially with a turnover rate of 0.0024 day^{-1} (half-life of 289 days, Fig. 2). A pronounced decrease in root length and root dry mass was found with increasing soil depth (Table 1). The gross root production in the upper 30 cm between tillering and early yellow ripeness of wheat plants was 2.6 fold higher than the average values of net root production during this period.

Based on depth distribution of root dry mass and of ^{15}N in the soil, the gross N-input into the soil (0-90 cm) by winter wheat roots in the N-turnover experiment was estimated at 282 kg ha^{-1} (Table 1), equivalent to 54% of total net N-assimilation of winter wheat.

Discussion

Our results emphasise that the ^{15}N which was taken up by winter wheat returned partly into the soil as rhizo-deposits, thus inducing a net increase of organic ^{15}N in subsoil within the growing period. The stabilisation of organic ^{15}N in the plough layer was the net effect resulting from N mineralisation and N rhizodeposition.

References

Janzen H H 1990 Soil Biol. Biochem. 22, 1155-1160.

Jensen E S 1996 Soil Biol. Biochem. 28, 65-71.

McNeill A M, Zhu C and Fillery I R P 1996 Aust. J. Agric. Res. 48, 295-304.

Table 1. N-inputs by winter wheat roots into the soil in the field experiment deduced from the depth distribution of root dry mass and ^{15}N-signature in organic N.

Depth	root dry mass [a]	Soil organic ^{15}N		Soil organic N [d]
		total	derived from roots	derived from roots
cm	% of total	% of added ^{15}N		kg N ha^{-1}
0-20	48	41.1	15.0 [b]	161
20-50	30	9.5	9.5 [c]	103
50-90	22	1.8	1.8 [c]	19

a) according to consistent results observed in the root production experiment between 1995 and 1997 (not shown)
b) assuming that ^{15}N deposition by winter wheat roots was proportional to the root dry mass in each depth
c) assuming that all ^{15}N in the subsoil was derived from winter wheat roots
d) assuming that mean ^{15}N signature in the roots was identical to that measured in the shoots of winter wheat plants

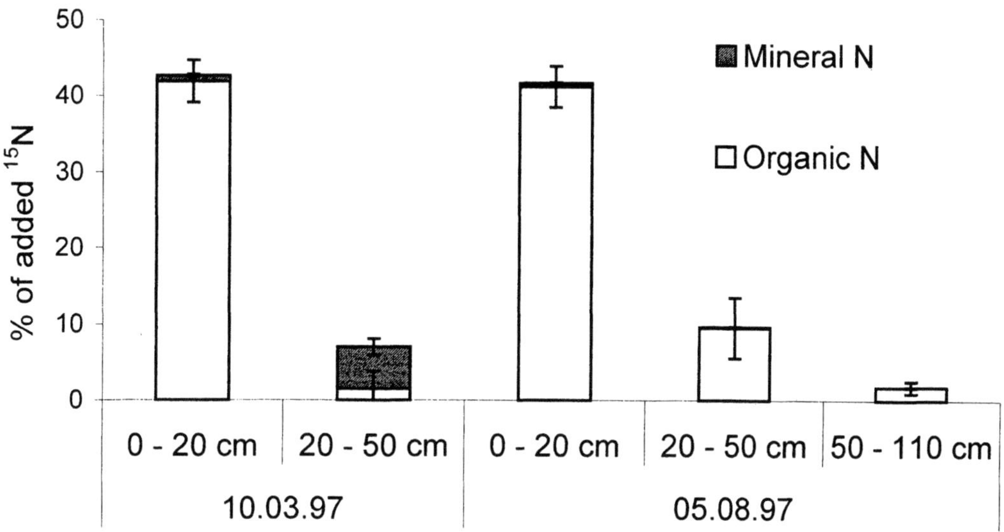

Figure 1. Depth distribution of clover-derived ^{15}N in soil.

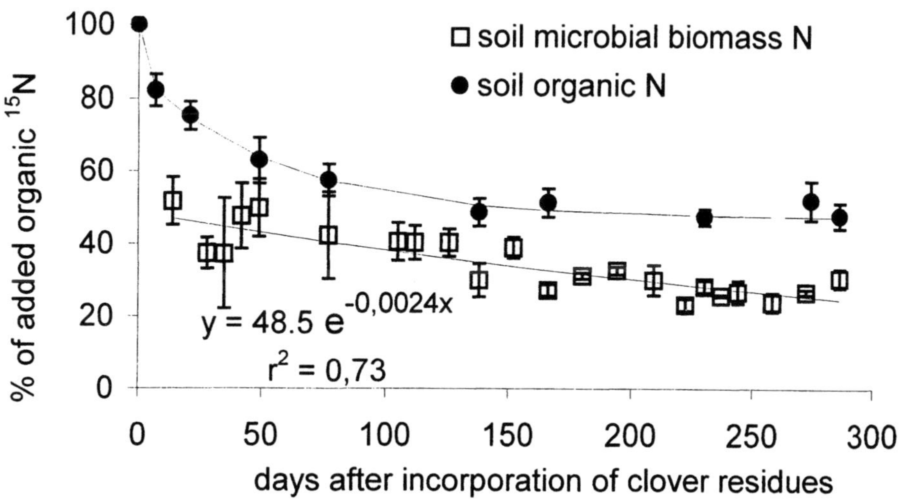

Figure 2. Measured values (points) and fitted temporal course (lines) of clover derived ^{15}N in soil organic matter and soil microbial biomass (0-20 cm depth).

W. J. Horst et al. (Eds), Plant nutrition – Food security and sustainability of agro-ecosystems. 944~945, 2001

Sulphate and nitrogen net mineralisation in coarse-textured soils in western Australia

G. C. Anderson and I. R. P. Fillery

CSIRO, Plant Industry Private Bag No 5 Wembley WA 6913 Australia. g.anderson@ccmar.csiro.au

Key words: mineralisation, nitrogen, sulphate, sulphur

Abstract

The replenishment of soil inorganic S and N from soil organic sources by the process of mineralisation is a major process that determines the capacity of the soil to supply S and N to plants. Rates of net S mineralisation in coarse-textured soils in Western Australia ranged from –4 to 14 kg S ha^{-1} over the growing season (May to November), while rates of net N mineralisation ranged from 29 to 130 kg N ha^{-1} over the same period. The low rates of net S and N mineralisation show that the application of S and N fertiliser is required to avoid S and N deficiency in cereal and canola crops. Net S immobilisation occurred when the level of soil S was greater than 5 kg S ha^{-1}, and net S mineralisation occurred when the level of soil S was less than 3 kg S ha^{-1}. When the level of soil S ranged between 3 and 5 kg S ha^{-1}, mineralised S was related to soil organic C, labile C and total S content. Hence, mineralised S appeared to occur by both biochemical and biological S mineralisation. In contrast, N mineralisation was independent of the level of inorganic soil N content and related to the level of soil organic C, labile C and total N content.

Introduction

Sulphur and nitrogen (N) in soil exist predominantly in organic forms. Organic sulphur and nitrogen must be mineralised to inorganic sulphate (S) and inorganic N before these nutrients can be absorbed by plants. The Mediterranean climate of Western Australia is characterised by intense rainfall following the sowing of crops in May-June. This rainfall often results in significant leaching events that deplete the levels of soil inorganic S and N. Hence, the replenishment of the inorganic S and N from organic sources by the process of mineralisation is a major process that determines the capacity of the soil to supply S and N to growing plants.

The aims of this paper are to determine the rates of S and N mineralisation for a range of soil types within the agricultural zone of Western Australia and to define relationships with various soil measurements.

Materials and methods

Soil samples were collected at monthly intervals over the growing season (25 May to 30 November 1999) from ten locations throughout the agricultural region of Western Australia. The soils' chemical composition is presented in Table 1.

Measurements of net S and N mineralisation were undertaken from the border plots of field fertiliser trial sites. The border plots were chosen as the sampling area to avoid areas that had received fertiliser applications. Soil samples were collected under wheat at sites 1, 4, 6, and 7, under canola at sites 9 and 10 and from areas without crop at sites 2, 3, 5 and 8.

Net S and N mineralised over the growing season were determined by measuring the accumulation of inorganic S and N in soil contained in PVC tubes (80 mm diameter by 120 mm long) and S and N that was retained in anion exchange resin traps that were sealed into the lower end of each soil core. A detailed description of the technique to measure net N mineralised is given in Anderson *et al.* (1998). This technique was modified to include the measurements of net S mineralised. Soil S content was measured by the procedure described by Watkinson and Kear (1994). Sulphate was extracted from anion exchange resin by shaking 5 g of wet resin with 100 mL of 1M HCl for 1 hr at 25 °C on an end over end shaker. The HCl solution was filtered and then measured for S using an ICP-AES.

Table 1. Soil chemical composition

Site number	C (%)	Labile C[A] (%)	Total S (μg g^{-1})	Total N (%)
1	0.58	0.162	54	0.05
2	0.54	0.180	50	0.04
3	0.69	0.280	80	0.05
4	0.73	0.149	98	0.06
5	1.02	0.204	94	0.07
6	0.66	0.233	70	0.06
7	1.61	0.317	134	0.13
8	2.19	0.640	147	0.15
9	2.51	1.303	411	0.18
10	1.11	0.375	48	0.07

[A] Blair (1995)

The intact cores collected at each site were transported and incubated at a common site located in the central wheat belt of Western Australia (Wongan Hills).

Net S and N mineralised were calculated as the change in soil S and N status over the incubation period plus the amount of S and N retained in the resin trap.

Results and discussion

The low rates of net S and N mineralisation measured in soils collected from the agricultural region of Western Australian indicate that application of S and N fertilisers is required to avoid S and N deficiency in cereal and canola crops, grown in the region.

Table 2. Net S and N mineralisation (kg ha^{-1})

Site number	Net S min. (kg ha^{-1})	Net N min. (kg ha^{-1})
1	5 (2)	23 (6)
2	-16 (2)	26 (6)
3	5 (1)	42 (11)
4	10 (1)	38 (5)
5	-6 (4)	32 (8)
6	7 (1)	57 (8)
7	3 (2)	49 (8)
8	4 (6)	158 (9)
9	10 (4)	170 (19)
10	1 (2)	60 (3)

Values in the parentheses are the standard deviations

The amounts of net S mineralised ranged from –4.1 to 14.4 kg S ha^{-1} (-6.8–12.3 μg S g^{-1}) over the period May to November (Table 2). These low rates of net S mineralisation were related to the low soil total S content (Table 1). Net S mineralised amounts were equivalent to 9.7% of the total S content of the soil. Eriksen (1995) observed net S mineralisation rates of 3.3 to 6.7 μg S g^{-1} yr^{-1} in Danish soils. These amounts of S mineralisation were equivalent to 1.7-3.1% of total organic S and were insufficient to maintain soil S supply to prevent S deficiency in rye grass, barley or canola.

In terms of daily rates, net S mineralisation ranged from -0.1 to 0.1 μg S g^{-1} soil during August and September. In contrast, higher rates (0.1-0.5 μg S g^{-1} d^{-1}) have been observed when the soil was incubated under ideal moisture and temperature conditions (Williams 1967). The higher rate of mineralisation observed are related to the greater (125–1100 μg S g^{-1} soil) total S content of the soils used in these studies.

The amounts of net N mineralised ranged from 29 to 130 kg N ha^{-1} over the period May to November (Table 2). These amounts of net N mineralisation were equivalent to 5.6% of the soils total N content. In terms of daily rates, net N mineralisation ranged from 0.1 to 0.9 μg N g^{-1} d^{-1} during September. Similar rates (0.1-1.0 μg N g^{-1} d^{-1}) have been measured under field conditions (Norton, 2000). Williams (1967) observed higher rates (0.2–2.1 μg N g^{-1} d^{-1}) when the soil was incubated under laboratory conditions.

There was greater release of N than S during the process of mineralisation but the amounts of S and N mineralised were not correlated. The ratio of N:S mineralised (4-43) was greater than the N:S ratio of the soil (6–10). Sakadevan *et al.* (1993) also observed that the amounts of S and N mineralised were much lower than those estimated by multiplying the amounts of respired C by either the C:S or C:N ratio of the soil organic matter. Sakadevan *et al.* (1993) argued that unlike mineralised C, the S and N mineralised are not lost from the soil and therefore remain susceptible to immobilisation. Nevertheless, N mineralisation was observed to be related to the C:N ratio with approximately 2.2 times more C mineralised compared to the amount of N mineralised. In contrast, S mineralised was not related to the C:S ratio due to the occurrence of biochemical mineralisation.

The level of soil inorganic S in the 0-10 cm soil layer during the experimental period was less than 3 kg S ha^{-1} at sites 1, 3, 4, 6; between 3-5 kg S ha^{-1} at sites 7, 8, 9 and 10 and greater than 5 kg S ha^{-1} at sites 2 and 5. Net S immobilisation occurred when the level of soil inorganic S was greater than 5 kg S ha^{-1} and net S mineralisation occurred when the level of soil inorganic S was less than 5 kg S ha^{-1}. Over all soils this effect of end product promotion/inhibition on the process of net S mineralisation suggests that biochemical mineralisation was more important than biological mineralisation. Nevertheless, net S mineralisation was related to the level of organic C (r^2=0.7), labile C (r^2=0.9) and total S (r^2=0.9) in soils when soil inorganic S was between 3-5 kg S/ha. This suggests that biological mineralisation was dominant process in soils with intermediate quantities of inorganic S.

There was a large variation in the level of inorganic N (5-40 kg N ha^{-1}) over the growing season. Nevertheless, N mineralised was observed to be independent of the level of inorganic soil nitrogen. As a result N mineralised was related to the level of organic C (r^2=0.8), labile C (r^2=0.7) and total N content (r^2=0.8). This suggests that net N mineralisation was mainly occurring by the process of biological mineralisation.

Acknowledgement

GRDC and CSBP Future Farm provided funding for the project. The Plant Nutrition Trust (Sam Tisdale Award) provided financial assistances to Dr Anderson to attend the colloquium.

References

Anderson GC, Fillery IRP, Dolling PJ and Asseng S 1998 Aust. J. Agric. Res. 49, 329-343.

Blair GJ, Lefroy RDB and Lisle L 1995 Aust. J. Agric. Res. 46, 1459-1466.

Eriksen J, Mortensen JV, Nielsen JD and Nielsen NE 1995 Agriculture Ecosystems and Environment 56, 43-51.

Norton JM 2000 *In* Handbook of Soil Science, Ed ME Summer. pp. C148-C181. CRC Press, Florida USA.

Sakadevan K, Hedley MJ and Mackay AD 1993 J. Soil Sci. 44, 73-83.

Williams CH 1967 Plant Soil 26, 205-223.

Watkinson JH and Kear MJ 1994 Comm. Soil Sci. Plant Anal. 25, 1015-1033.

W. J. Horst et al. (Eds), Plant nutrition – Food security and sustainability of agro-ecosystems. 946~947, 2001

Nitrogen mineralisation kinetics in soils: An assessment of some modelling approaches

D. K. Benbi[1] and J. Richter[2]

1) Department of Soils, Punjab Agricultural University, Ludhiana, 141004, India. dbenbi@pau.chd.nic.in

2) Institute for Geography and Geoecology, TU Braunschweig, Langer Kamp 19c, D 38106 Braunschweig, Germany.

Key words: mineralisation-immobilisation turnover, microbial biomass, model comparison, N mineralisation, N mineralisation kinetics, N mineralisation models, N mineralisation potential, soil organic matter

Abstract

A number of modelling approaches have been used to describe N mineralisation kinetics in soils. In this paper we present a comparative evaluation of some modelling approaches to predict net N mineralisation in soils. The models range in complexity from zero order to multi-compartment first order. The following conclusions emerged: (i) the single pool first-order kinetic model is inadequate to describe net N mineralisation in soils as it provides the worst fit to the data and systematically deviates from the measured cumulative N mineralisation; (ii) In disturbed soil samples, a minimum of two pools of mineralisable N are considered essential to significantly contribute towards N mineralisation. .

Introduction

Kinetic models ranging in complexity from zero order to multi-compartment first order have been used to describe nitrogen (N) mineralisation in soils. Stanford and Smith (1972) hypothesised that the rate of mineralisation is proportional to the amount of potentially mineralisable N in soil that constitutes a single, distinct, homogenous fraction of soil organic N. Several authors (Molina *et al.*, 1980, Nuske and Richter, 1981) have proposed that more than one fraction of soil organic N may be directly mineralised each with its specific rate of decomposition. The single and multi-fraction first order kinetic models are based on the assumption that the N mineralisation potential of soil is a definable quantity, whereas many researchers have suggested that, with the incubation methods, it may not be possible to define the N mineralisation potential of soils as the slowly mineralisable N component may in fact follow zero-order kinetics (Bonde and Rosswall, 1987). They advocated the use of a mixed first and zero-order kinetic (FOZO) model.

This paper presents a comparative evaluation of the modelling approaches adopted for predicting net N mineralisation in soils.

Materials and methods

We made a critical assessment of single fraction first order kinetic model (FOSC), two-fraction first order kinetic model (FOTC), modified single fraction first order kinetic model (MFO), and mixed first order plus zero order model (FOZO) (Table 1) on previously published data from Griffin and Laine (1983), Lindeman and Cardenas (1984), Chae and Tabatabai (1986) and Richter *et al.* (1989). The models were assessed with respect to parameter estimates and t ratios (parameter estimate/ standard error) associated with the mean parameter estimates and fit to the experimental data as indicated by the residual sum of squares (RSS). Statistical significance of differences in RSS between any two models were assessed by F-test.

Table 1. Integral form of kinetic models used to predict net N mineralisation in soils

Model	Equation
First-order single compartment (FOSC)	$N_m = N_0(1-e^{-kt})$
Modified first-order (MFO):	$N_m = N_0 - (N_0 - N_e)e^{-kt}$
First-order two compartment (FOTC) :	$N_m = N_1(1-e^{-k_1 t}) + N_2(1-e^{-k_2 t})$
First-order plus zero-order (FOZO):	$N_m = N_1(1-e^{-k_1 t}) + Kt$

N_m= cumulative N mineralised in time t; k, k_1, k_2, K are rate coefficients; N_1 and N_2 represent decomposable and recalcitrant organic N fractions, respectively; and N_e is N mineralised in first seven days

Results

The parameter estimates produced by each model varied considerably. For example, the N mineralisation potential (N_0) in different soils estimated by the FOSC model ranged from 52-313 mg kg^{-1} soil as compared to 71-497 mg kg^{-1} by the FOTC model. The mineralisation rate constant (k) for the FOSC model ranged from 0.016 to 0.662 per week. The rate constants for decomposable and recalcitrant organic N fractions estimated by the FOTC model ranged from 0.408-5.110 and 0.015-0.152 per week, respectively. Not all the models could be fitted to each data set. For example, the FOZO model could not be fitted to fine loamy (Harps) soil and it yielded non-significant (p=0.05) parameter (N_1, k_1) estimates in fine silty (Downs) soil from the studies of Chae and Tabatabai (1986). A realistic solution was not possible for FOTC model in both these soils. In arable and ploughed grassland soils (Richter *et al.*, 1989), the FOTC model yielded inconsistent parameter estimates for the easily decomposable pool

(N_1, k_1). This was probably due to the nature of the samples (from deepend plough layer) and the relatively short incubation period (82 days). To the other three soils/ treatments from the studies of Richter *et al.* (1989), the FOTC model provided good fit to the data. The MFO model produced a significant (p=0.05) negative N_e value in Harps soil, which has no meaning. Whereas in Downs soil the t ratios for parameter estimates with the model were non-significant. There was no consistent trend in t ratios associated with parameter estimates for various models perhaps because of variable number of parameters estimated with different models.

The goodness-of-fit as indicated by the RSS showed that FOSC model provided the worst fit to the data followed by FOZO model. There were systematic differences between the measured and the FOSC fitted curves. The fitted curve under-estimated the measured values at the beginning and ending of the incubation period, and over-estimated at intermediate times. With the exception of soil from ploughed grasslands (air-dry treatment, Richter *et al.*, 1989), in all the other soils the fit with FOZO model was poorer compared to MFO model. The FOTC model offered the best fit to the data and it invariably produced the minimum residuals. However, the differences in residuals between MFO and FOTC model were significant (p=0.05) only in 3 out of 11 cases. Physically, the MFO model may also be interpreted as a two compartment model with mineralisation potentials of N_e and N_0. But N_e is defined as the dead microbial biomass created due to air-drying of soils prior to incubation whereas the N_1 fraction in the FOTC model is defined as easily decomposable plant material. In incubation studies using field moist soils without a prominent initial flush of mineralised N, the model may produce physically unrealistic N_e values (e.g. Harps soils; Chae and Tabatabai, 1986)

Discussion

We have discussed different modelling approaches to describe N mineralisation kinetics in soils. A number of other equations have been used to describe N mineralisation during incubations, some of which undoubtedly produced a good fit to the experimental data. We regard fitting data to the curves described by two compartment models as useful for the purpose of obtaining parameters needed for modelling N dynamics in soils or to allow data to be tabulated in comparable forms. However, in view of the complexity of the system and interdependence of a number of biotic and abiotic factors, one should not try to infer the mechanism of the reaction from a good fit between experimental points and a particular mathematical form alone. Though the partitioning of organic matter into two or more fractions each with a distinct N_o and k is gross simplification of the turnover process, it is easier to get parameter estimates this way and is, therefore, in accordance with the principle of parsimony. Simulation models of nitrogen turnover in the soil-crop system which employ three or more organic nitrogen pools are difficult to parameterise and these do not give better results than the simpler models (de Willigen, 1991).

Large differences in N mineralisation potentials and the rate of mineral N production could originate from differences in time of sampling, probably due to seasonal variation in microbial biomass and the amount of undecomposed crop residues in the soil. Microbial biomass has a central role in organic matter transformations. Microorganisms act both as a source (mineralisation) and a sink (immobilisation) for N during microbial biomass turnover. Therefore, models for N mineralisation kinetics should, ideally, include a component for microbial biomass. This would lead to the second order rather than the first order kinetics as the N mineralisation rate would depend on the concentration of both the substrate as well as the microbial biomass.

In conclusion it may be stated that (i) the single pool first-order kinetic model is inadequate to describe net N mineralisation in soils as it provides the worst fit to the data and systematically deviates from the measured cumulative N mineralisation. (ii) In disturbed soil samples, a minimum of two pools of mineralisable N are considered essential to significantly contribute towards N mineralisation, irrespective of the fact whether the soil samples were dried and subsequently rewetted or kept field moist prior to incubation.

References

Bonde TA, and Rosswall T 1987 Soil Sci. Soc. Am. J. 15, 1508-1514.

Chae YM and Tabatabai MA 1986 J. Environ. Qual. 15, 193-198.

de Willigen P 1991 Fertiliser Res. 27, 141-149.

Griffin GF and Laine AF 1983 Agron. J. 75, 124-129.

Lindemann WC and Cardenas M 1984 Soil Sci. Soc. Am. J. 50, 323-326.

Molina JAE, Clapp CE and Larson WE 1980 Soil Sci. Soc. Am. J. 28, 778-781.

Nuske A and Richter J 1981 Plant Soil 59, 237-247.

Richter GM, Hoffmann A, Nieder R and Richter J 1989 Soil Use and Management 5, 169-173.

Stanford G and Smith SJ 1972 Soil Sci. Soc. Am. Proceedings 36, 465-472.

W. J. Horst et al. (Eds), Plant nutrition – Food security and sustainability of agro-ecosystems. 948~949, 2001

Validation of the EPIC model for nitrogen and phosphorus uptake by intercropped maize and cowpea on Alumi Haplic Acrisol

I. de Barros[1)], T. Gaiser[2)] and V. Römheld[1)]
[1)] *Institute of Plant Nutrition (330), University of Hohenheim, Stuttgart, 70593, Germany, inbarros@uni-hohenheim.de*
[2)] *Institute of Soil Sciences (310), University of Hohenheim, Stuttgart, 70593, Germany, tgaiser@uni-hohenheim.de*

Key words: cowpea, maize, modelling, nitrogen, phosphorus

Abstract

Simulation models can be used for predicting several aspects of agricultural activities such as irrigation, conservation practices and crop management and their impact on yield, nutrient availability and environment. Most part of the research on crop models has been carried out for temperate or subtropical conditions. In order to evaluate the accuracy of the EPIC 8120 model to tropical semi-arid conditions on a highly acid soil, the results of a three year field experiment measuring biomass production and nutrient uptake of 3 years field experiment for intercropping maize and cowpea were compared with the results of simulations. EPIC overemphasised the effects of water shortage and aluminium stress on the production and nutrient uptake by the crops. This indicates the need to calibrate and adjust the model for those conditions in order to be able to consider the interaction of nutrient and water stress and the dynamics of pH as changed by tillage and lime.

Introduction

The EPIC model (Williams at al., 1984) was originally developed to simulate effects of climate, soil, conservation practices and crop management on soil erosion and its impact on long term crop yield and requirement of fertiliser. In the current version of EPIC, only two nutrients are considered. Simulated nitrogen dynamics include loss of nitrate, organic nitrogen transport by sediment, denitrification, immobilisation, mineralisation, biological nitrogen fixation, crop uptake and fertilisation. Simulated phosphorus processes include runoff of soluble phosphorus, sediment transport of mineral and organic phosphorus, mineralisation, sorption-desorption process, crop uptake and fertilisation (Jones *et al.*, 1984). The objective of this study is to evaluate the capability of the EPIC model to simulate the biomass production and nutrient uptake by maize and cowpea on a highly acid soil under the semi-arid conditions of northeastern Brazil.

Materials and methods

In order to test the simulation results of uncalibrated EPIC 8120, an experiment was carried out on a Latossolo Amarelo Álico (Alumi-Haplic Acrisol) (pH ($CaCl_2$): 3,7; aluminium saturation of 65%) in the Picos region (Brazil) in 1998, 1999 and 2000. Cowpea was intercropped with maize. The variety of maize was BR 5039 "São Vicente" and that of cowpea was "EPACE 10". The final stand was 4,5 plants of maize and 9 plants of cowpea per square meter. The statistical layout was a RCB design (Randomised Complete Blocks) with 4 replications. Each plot had an effective area of 7 x 5 meters.

The following treatments were used: 1) Complete fertilisation (nitrogen - 30 kg ha^{-1}, phosphorus - 18 kg ha^{-1}, potassium- 33 kg ha^{-1} and dolomitic lime – 3 t ha^{-1}); 2) Complete fertilisation without nitrogen; 3) Complete fertilisation without phosphorus; 4) Complete fertilisation without lime; 5) Control (without fertiliser application). Cowpea and maize samples were collected at harvest. From 40 randomised holes per plot (80 plants for cowpea and 40 for maize) in order to determine dry matter production. All above ground parts of the plants were harvested and separated into grains and crop residues. After drying N and P content were analysed. The data of the field experiments were compared with the results of the simulations by EPIC model version 8120.

Results

The results of the three years experiment show that phosphorus is the most nutrient constraint for production of maize and cowpea (Table 1).

As the results of the experiment show a strong difference between some treatments each year on biomass production and nutrient uptake, it is clear that an interaction between water and nutrient availability affects biomass production and uptake by the crops more than each factor alone.

The results of the simulations of biomass production and nutrient uptake by the crops shows differences between the years. However, in each year, no differences were found in the treatments with lime as well as in the treatments without lime application. Only a difference between these two groups was found (Tables 1)

In the first year, lime was incorporated only in the upper 5 cm, whereas a better incorporation, up to about 20 cm deep, was carried out at the beginning of the second year by disk bedder during the land preparation of the field experiment. The simulations of the three years of cultivation did not show changes in the soil profile on the parameters associated with acidification (pH, Al saturation) (data not shown).

Discussion

In the simulation process by EPIC, several stress factors for root and above ground biomass are calculated each day and the most limiting of them is applied to a potential root and biomass development rate, which represents the maximal rate of a crop under ideal conditions. However, interactions between the stress factors are not explicitly considered. Under the conditions of the field experiment, only water stress was considered by the model as a reducing factor for the above ground biomass development, due to the low precipitation, the high percolation in the sandy soil and reduced potential root development due to aluminium stress. The nutrient stress factors (N and P) did not play any role for the simulated biomass development. This might be the reason why the simulated biomass for complete fertilisation, complete fertilisation without N and complete fertilisation without P did not show any difference.

As the model did not change the chemical parameters associated with soil acidity by tillage in the second year, the model considered that just the first 5 cm of the soil were amended in the treatments with lime application over the whole three years period. This poor amendment could explain the small differences found in the results of the simulations between the treatments with and without liming.

As the aluminium saturation in that soil is very high (65%), and as the model considered that just 5 cm of soil in the treatments with lime application were amended, a very small fraction of the soil profile was considered to be favourable for root development and nutrient acquisition, affecting directly the above ground biomass development.

As the biomass production and N uptake in the simulations for complete fertilisation were lower than the results of the field experiment, it seems that the model overemphasises the effects of water and aluminium stress on the crop development, biomass production and N uptake. It seems to be necessary to calibrate and adjust the model in order to predict with acceptable accuracy the production and nutrient uptake by the maize and cowpea in soils under semi-arid tropical conditions

References

Jones CA, Cole CV, Sharpley AN and Williams JR 1984 Soil Sci. Soc. Am. J. 48, 800-805.

Williams JR, Jones CA and Dyke PT 1984 Trans. of Am. Soc. Agric. Eng. 27, 129-44.

Table 1. Biomass production, N and P uptake by maize and cowpea on the chapada in a field experiment at Picos, Noertheastern Brazil in comparison with simulations by EPIC 8120

		Biomass of maize and cowpea (kg/ha)				Nutrient uptake by maize and cowpea (kg/ha)			
		Maize		Cowpea		Nitrogen		Phosphorus	
Year	Treatment	OBS	SIM	OBS	SIM	OBS	SIM	OBS	SIM
1998	Control	501	820	1117	1080	39	54	2.12	6.96
	Complete (Cpt)	1933	1080	2706	1390	86	69	6.51	8.59
	Cpt –N.	1570	1080	2979	1390	84	69	6.31	8.59
	Cpt –P	993	1080	1783	1390	60	69	3.23	8.59
	Cpt -Lime	1409	820	2558	1080	80	54	6.18	6.96
1999	Control	237	840	425	990	18	53	0.33	7.00
	Complete (Cpt)	1146	1050	2163	1260	81	67	5.73	8.62
	Cpt –N.	1024	1050	1365	1260	48	67	3.37	8.62
	Cpt –P.	916	1050	637	1260	31	67	1.60	8.62
	Cpt -Lime	1207	840	1200	990	53	53	4.00	7.00
2000	Control	1202	790	498	1020	33	52	4.32	6.82
	Complete(Cpt)	1654	1220	1425	1330	76	69	12.51	8.69
	Cpt –N.	1666	1220	1280	1330	60	69	9.00	8.69
	Cpt –P.	1258	1220	624	1330	38	69	4.33	8.69
	Cpt -Lime	1661	790	1091	1020	55	52	8.52	6.82

OBS = Observed values; SIM = Simulated values

W. J. Horst et al. (Eds), Plant nutrition – Food security and sustainability of agro-ecosystems. 950–951, 2001

Influence of rain distribution on yield and management practices for a winter wheat cultivation

J. Bobert and U. Schmidhalter

Department of Plant Sciences, Chair of Plant Nutrition, TU München, Am Hochanger 2, Freising, D 85350, Germany, bobert@toax.de

Key words: nitrogen efficiency, precision agriculture, water balance

Abstract

For the development of management strategies for precision agriculture spatial site properties and temporal aspects have to be taken into account. This study shows the temporal influence of precipitation distribution on yield and nitrogen fertilisation efficiency from sites in a region with a negative seasonal water balance.

Introduction

Water is one of the most important factors for the crop growth. Precision agriculture strategies are mainly adjusted to the spatial variability of soil properties such as water storage capacity, texture, and relief. In regions with a negative seasonal water balance and fluctuating precipitation amounts and distribution, management strategies have to take into account this temporal variation as well as the spatial variation. This article describes a method to assess the temporal and spatial influences on yield pattern, and the effects of uniform fertiliser applications.

Materials and methods

The study was carried out in a field in Wulfen, Germany. Wulfen is situated in the Eastern part of Germany 100 km southeast from the highland ranges of the Harz. The average precipitation is about 450 mm per year; fluctuating from 250 to 620 mm. 12 monitoring sites were selected within a field with very heterogeneous loess soils and textures ranging from sandy to clayey loam. The relief consists of gentle slopes from North to South. At the monitoring sites the volumetric water content was measured with capacitance probes. To determine the plant development on the monitoring sites, biomass, growth stages and yields were monitored. Long-term weather data covering a forty-year period were obtained from a nearby weather station. The soil water balance, weather and crop data were used as input for the modelling package DSSAT (Decision Support System for Agrotechnology Transfer, IBSNAT, 1999). A bio-indicative model (Selige *et al.*, 2001) was used to classify zones with width of 50 mm available water storage capacity of the root zone AWCrz to transfer the point data into area. The N-efficiency was determined using an empirical approach.

Results

Figure 1 shows the spatial distribution of grain yield and nitrogen efficiency of the farm standard nitrogen application rate (180 kg N ha^{-1} yr^{-1}). Average grain yield was 4100 kg ha^{-1} (ranging from 1100 to 6500 kg ha^{-1}). The average N-efficiency was 58%, fluctuating between 12 and 90%.

Figure 2 shows the simulated grain yield distribution (as index) at the monitoring site number five for a thirty-year period from 1969 to 1999. Average grain yield over the period was 4100 kg ha^{-1}, ranging from 1300 to 6600 kg ha^{-1}. Efficiency of the standard N-application fluctuated between 19 to 92 % with an average of 58 %. These results clearly show that the development of precision agriculture strategies in regions with a negative seasonal water balance has to take into account the temporal variation of precipitation as well as the spatial heterogeneity of soil properties. The fluctuation of yield and N-fertilisation efficiency in the given example indicates the need for management approaches containing risk assessment strategies for yield expectations and nitrogen fertilisation.

References

IBSNAT 1999 DSSAT35, A Decision Support System for Agrotechnology Transfer, University of Hawaii, Honolulu, Hawaii.

Selige T and Schmidhalter U 2001 Site specific soil resource mapping using remote sensing. Proceedings of the 3rd European Conference for Precision Agriculture, Montpellier, France.

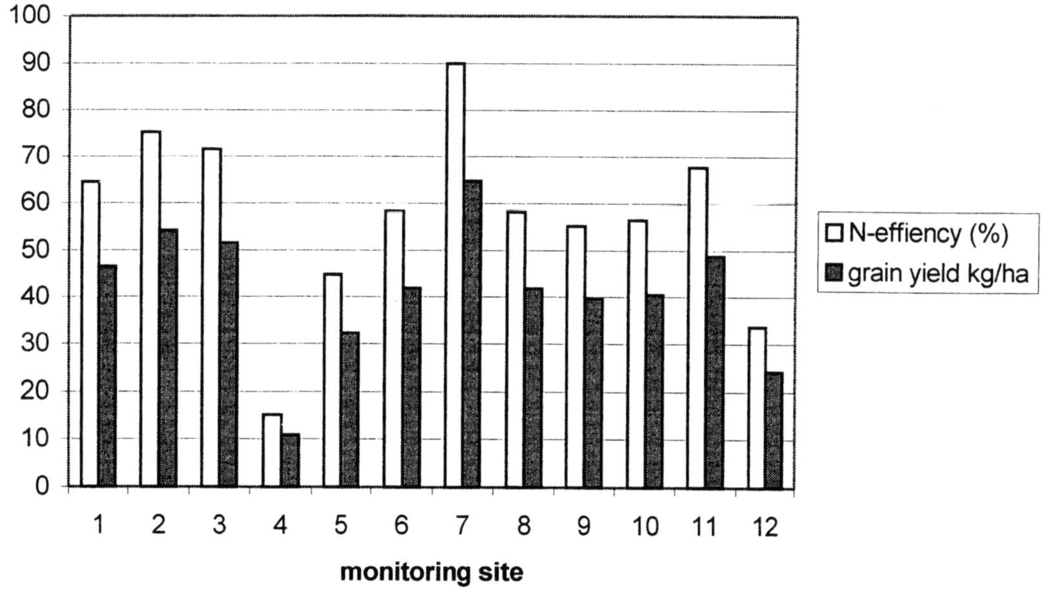

Figure 1. Spatial yield distribution and N-efficiency on 12 monitoring sites in Wulfen

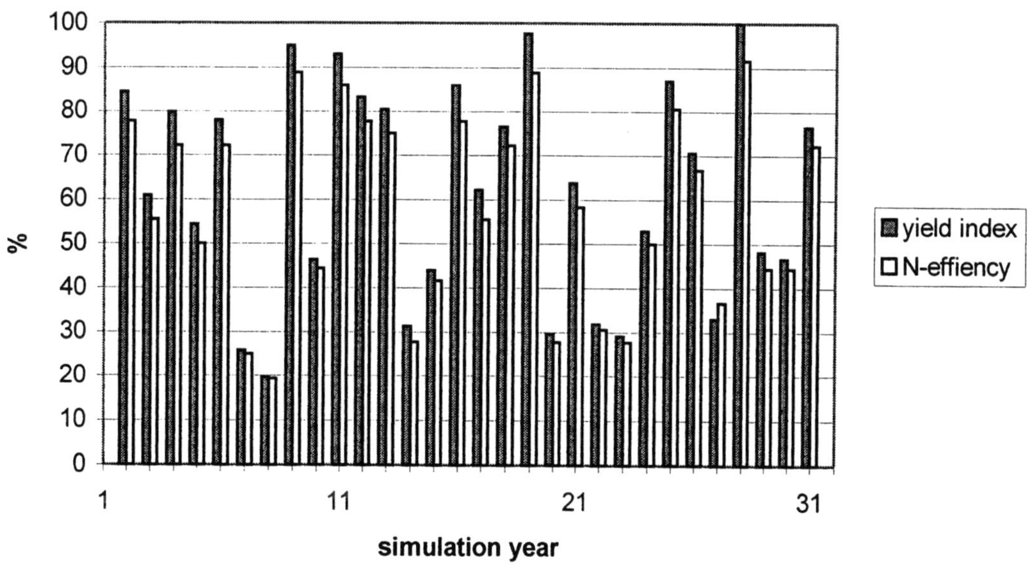

Figure 2. Timely yield distribution and N-efficiency on monitoring site five in Wulfen

W. J. Horst et al. (Eds), Plant nutrition – Food security and sustainability of agro-ecosystems. 952~953, 2001

Long-term trial on soil formation at Halle/Saale

H. Beschow and W. Merbach

Institute of Soil Science and Plant Nutrition, Department of Plant Nutrition, Martin-Luther-University of Halle/Wittenberg, A.-Kuckhoff-Str. 17b, Halle/Saale, D 06108, Germany, beschow@landw.uni-halle.de

Key words: soil development, soil organic matter, loess

Abstract

In 1948, a model soil formation trial was arranged in Halle/Saale by Schmalfuß. Loess was used as substrate. The soil formation was studied using different rates of P and N mineral fertilisation. The trial was subdivided into two parts, N treatments (0, 100, 200 and 400 kg N ha^{-1}) with constant P application and P treatments (0, 9, 17 and 35 kg P ha^{-1}) with constant N application. In the first 30 years, yields increased strongest in the highest fertiliser treatment compared with the no-N or no-P fertiliser treatment. Afterwards yields remained nearly constant at the different fertilisation treatments. With higher levels of fertiliser application, humus accumulation and biological activity increased faster than the C_{org} content in the topsoil. Soil organic matter improved the nutrient availability. It is assumed that decalcification and a decrease in the C/N ratio will continue. The lower C_{org} and N_{org} contents and the lower cation exchange capacity in comparison with naturally developed soils indicate a still ongoing soil formation process.

Introduction

Long-term experiments are useful tool for studying C and N dynamics in soils. The influence of environmental conditions, fertilisation, vegetation and management on the steady state (formation and decomposition balance) of the C pools in the soil has been extensively investigated (Körschens *et al.*, 1998). Quantitative and qualitative changes of soil organic matter (SOM) were mainly studied in 'old' soils. The process of formation and stabilisation of SOM during soil development from SOM free parent material to the steady state of the SOM pool is little understood. The trial on soil development in Halle/Saale can help clarify these questions, because the experiment was started with substrate which had a very low SOM content. It began 50 years ago and is still continuing (Schmalfuß, 1960, 1965, 1966; Schilling, 1995; Beschow *et al.*, 1999). The results could be important for the cultivation of the overburden of open cast mining operations.

Materials and methods

Experimental design

Glazed clay cylinders, 1m length and 40 cm diameter, were embedded in the ground and covered with a wire cage. The cylinders are open at the bottom and stand on a layer of gravel. Loess (15% clay, 80% silt, 5% sand, 178.6 mg C (100 g)$^{-1}$, 19 mg N (100 g)$^{-1}$, 1.27 mg P (100 g)$^{-1}$ (Neubauer method), 17% $CaCO_3$) obtained from an open cast mine near Etzdorf (not far from Halle/Saale) was filled into the cylinders. Annual precipitation was 467 mm between 1965 and 1989, and annual mean air temperature was 9°C.

The trial was subdivided into two parts, N treatments (0 (N0), 100 (N1), 200 (N2) and 400 kg N ha^{-1} (N4)) with constant P application (P4) and P treatments (0 (P0), 9 (P1), 17 (P2) and 35 kg P ha^{-1} (P4)) with constant N application (N4). Potassium application was constant. For 50 years plants have been cultivated in an irregular sequence (maize (17 times), ryegrass (13), spring wheat (5), alfalfa (5), sugar beet, spring barley, broad beans (each 3), other plants (1-2)). No organic manure was applied. Each treatment was replicated 6 times. Above-ground matter and beet roots were removed, dry matter was determined annually.

Soil parameter

The following parameters were determined in soil samples: C_{org}, N_{org}, N_{min}, lime content, microbial biomass, P fractionation, cation exchange capacity, pH.

Alkaline extract of soil

2% HCl (60 ml) was added to 5g soil (< 2mm) for 4 hours. The soil was filtered, washed and then shaken three times with 0,5% NaOH for 5h. The extracts were centrifuged and united (250 ml). Light absorption at 530 nm of the alkaline extract was measured to characterise the degree of SOM accumulation (content of brown coloured, alkaline soluble humic substances). Extracts were diluted 1:1 with 0.5% NaOH.

Results

In Table 1 an overview of mean shoot dry matter per year ([Σ shoot dry matter] per trial period) and important soil parameters of the topsoil and the alkaline extract in treatments without N (N0P4) fertilisation and in the highest fertilised treatments (N4P4) are given.

In the highest fertilised treatment (N4P4), the mean shoot dry matter per year and the C_{org} and N_{org} contents in the topsoil were higher than those of the treatments without N fertilisation (N0P4). The relative increase of N_{min} was always higher than that of N_{org} in the topsoil. In the treatment N0P4 phosphorus was enriched in the fractions of water soluble, labile and microbial P (lightly soluble P) in the highest degree (32,8% of total P). Light absorption of the alkaline extract increased more than its $C_{org,alk}$ content.

In Table 2 the contents of microbial carbon (C_{mic}) and

lime as well as the potential cation exchange capacity (CAC_{pot}) of the topsoil are shown.

Table 2. Microbial carbon (C_{mic}), Cation exchange capacity (CAC_{pot}) and lime content of the topsoil (0-20 cm)

Treatment	C_{mic} [mg (100g soil)$^{-1}$]	CAC_{pot} [mmol (100g soil)$^{-1}$]		Lime [%]	
	1996	1979	1996	1979	1996
N0P4	36,5	n.d.	13,5	17,0	17,9
N4P4	60,7	12,2	14,1	16,9	15,0

Discussion

An evaluation after 48 trial years illustrates that large yield differences per year between treatments without N or P fertilisation and completely fertilised treatments had occurred only in the initial 30 years.
Afterwards, no further increase was observed and the differences between the treatments remained nearly constant. When comparing now the highest N and P fertilisation treatment (N4P4) with the treatment without N fertilisation (N0P4) the mean shoot dry matter per year is 2.5fold but the C_{org} content of the topsoil only 1.2fold higher. The modification in SOM formation may originated in the altered shoot/root ratio due to nutrient deficiencies (Beschow et al, 2000). Not only the quantity but also the quality of SOM from the Halle trial has changed. This is supported by a further reduction of the C/N ratio, a faster increase in alkaline soluble SOM accumulation compared to the C_{org} development in the topsoil and a high level of microbial activity. SOM formation also promoted the nutrient supply (high increase of N_{min}, storage of P as soluble, labile and microbial bound P).

Further new formation and stabilisation of SOM and continuation of soil development can be expected because of increased decalcification and further narrowing of the C/N ratio during the trial period and lower C_{org} and N_{org} contents and lower cation exchange capacity in comparison to the naturally developed soils (derived from loess) of the experimental field in Etzdorf.

References

Beschow H, Gransee A, Merbach W 2000 J.Plant Nutr. and Soil Sci. 163, 359-365.

Körschens M, Weigel A, Schulz E 1998 J. Plant Nutr. Soil Sci. 161, 409-424.

Schilling G 1995 Mitteilgn. Dtsch. Bodenk. Gesell. 78, 145-152.

Schmalfuß K 1960 Z. Pflanzenern. u. Bodenk. 90, 50-58.

Schmalfuß K 1965 A. Thaer-Archiv 9, 3-7.

Schmalfuß K 1966 Sitzungsber. Sächsische Akad. Der Wiss, Leipzig, 107, 1-13.

Table 1. Mean shoot dry mass per year and soil parameter of the topsoil (0-10 cm). In parenthesis relative data. 1958=100. n.d.=not determined

Treatment	Year	Mean shoot dry mass per year [g pot^{-1}]	C_{org}	N_{org}	Total soil N_{min} [mg (100 g)$^{-1}$]	P_{total}	$P_{lightly soluble}$	Alkaline extract $C_{org.alk}$ [mg (100 g)$^{-1}$]	E (530nm)
N0P4	1958	97	770 (100)	49 (100)	0,3 (100)	n.d.	n.d.	189 (100)	0,08 (100)
	1968	71	1200	83	0,4	n.d.	n.d.	212	n.d.
	1979	80	1360	86	2,5	83	34	358	0,66
	1996	77	1380 (179)	107 (218)	4,7 (1570)	56	38	400 (212)	0,70 (875)
N4P4	1958	150	970 (100)	65 (100)	0,8 (100)	n.d.	n.d.	255 (100)	0,10 (100)
	1968	157	1480	113	0,7	n.d.	n.d.	330	n.d.
	1979	212	1580	115	2,7	107	12	347	n.d.
	1996	201	1720 (177)	159 (245)	4,2 (525)	69	15	445 (175)	0,72 (720)

W. J. Horst et al. (Eds), Plant nutrition – Food security and sustainability of agro-ecosystems. 954~955, 2001

Model type selection for the examination of SOM dynamics and soil assessment as CO_2 source/sink based on EuroSOMNET database

O. D. Sirotenko[1)], L. K. Shevtsova[2)] and V. A. Romanenkov[2)]

[1)] All-Russian Research Institute of Agricultural Meteorology, Lenina Ul.,82, Obninsk Kaluga Region, 249030, Russia, cxm@obninsk.ru

[2)] All-Russian Institute for Fertilisers and Agricultural Soil Science, Pryanishnikova st.,31, Moscow, 127550, Russia, viua@online.ru

Key words: carbon dioxide, databases, long-term field experiments, mathematical models, model evaluation, model integration, simulation, soil organic matter

Abstract

The study of long-term sustainability of arable soils and their role as CO_2 source/sink requires an adequate selection of mathematical models based on available results of field experimentation and monitoring data to identify parameters. The evaluation of the reliability and shortcomings of some model types was based on the SOM dynamics in Russian long-term (duration more than 20 years) field experiments included in the European Soil Organic Matter Network. Performance of non-linear empirical static models proposed by the authors and process-oriented models of SOM dynamics with different level of complexity are discussed. The simulated results of SOM dynamics over 30 years and CO_2 emission for different initial SOM levels, typical crop rotations, fertilisation rates and selected climate change scenarios are compared. Simulation of soil processes for different field treatments for continuous crops and crop rotations made it possible to identify the most important management factors which influence long-term soil sustainability on a regional and district level and to reveal the potential role of different management scenarios maintaining SOM levels.

Introduction

Complex process-based models (for example, CENTURY) are suitable for the simulation of soil organic matter (SOM) dynamics and CO_2 emission.

For predictive purposes, there is usually a need for site specific calibration of the models parameters, which is the main obstacle to the confident use of models on a large scale for non-uniform territories (European part of Russia in our example).

We have attempted to minimise this problem by integrating long-term information from Russian field experiments suitable for use with the Climate-Soil-Yield (CSY) simulation system (Sirotenko *et al.*, 1995).

Materials and methods

The CSY system was developed for estimating the impact of the global greenhouse effect on Russian agriculture and contains 3 modules:

- Weather-Yield dynamic simulation models of agroecosystem productivity
- Stochastic models permitting the generation of an annual series of daily meteoelements corresponding to the specified climate norms
- A database including hydrometeorological information over the 100 year period, scenarios of possible climate change and data about physical and agrochemical soil properties for the whole territory of Russia.

Construction of a SOM database for Russian Federation field experiments within the framework of the European Soil Organic Matter Network Project and the Global Change and Terrestrial Ecosystems (GCTE) programme in an international standard format made it possible to revise accepted views about the role of specified agricultural management options in SOM transformation. At present, the Russian EuroSOMNET database includes 24 long-term experiments which provide the basis for CSY system evaluation (Shevtsova *et al.*, 2000). The datasets from the soddy-podzolic soil zone (podzoluvisols) were selected for the integration.

A special soil carbon and nitrogen submodel was designed to develop the original version of the CSY system. The simulation of SOM dynamics is based on the following differential equations:

$$\frac{dx_1}{dt} = hm_R x_2 - (m_H + s_H)x_1,$$

$$\frac{dx_2}{dt} = k_0 \frac{P_0 x_3}{\alpha_N + x_3} - m_R x_2 + f_m,$$

$$f_N + V + \gamma_1 m_H x_1 + \gamma_2 m_R x_2 = \gamma_3 (1-k_0) \frac{P_0 x_3}{\alpha_N + x_3} + K_N x_3,$$

where x_1, x_2 are old (O) and young (Y) OM

x_3- is the effective N mineral content

m_H and m_R are the rate constants for decomposition

f_m is the C input to soil via mineral fertilisers

h is the humification coefficient

f_N is the N input to soil via mineral fertilisers

V is the N input from atmosphere (including nitrogen fixation)

γ_1, γ_2 and γ_3 are the N rates in O and Y organic material and plant biomass, respectively

K_N is the factor of mineral N losses and s_H is the factor of SOM losses with leaching. P_0 is the potential productivity, k_0 denotes the rate of crop-residues and above-ground biomass

α_N is a constant (P_0 , k_0 , α_N are crop specific constants).

m_H and m_R in the submodel can be calculated as

$$m_H = m_H^0 r_c \left(\frac{x_3}{x_0} \right)^a , \quad m_R = m_R^0 r_c \left(\frac{x_3}{x_0} \right)^b ,$$

where m_H , m_H^0 , a , b and x_0 are constants and r_c is a rate modifying parameter for taking the impact of climate into account.

The construction of the EuroSOMNET database gave the possibility to examine and calibrate a static Model of Humus Balance (Shevtsova *et al.*, 1996). Based on the information available for podzoluvisols zone of Russia, the model was run to calculate required FYM rates to reach specific SOM levels, optimise field structure and provide high crop yields.

Non-linear regression equations were identified which combined the long-term SOM database for more than 500 treatments (soil data and management options, including EuroSOMNET data) and climatic information.

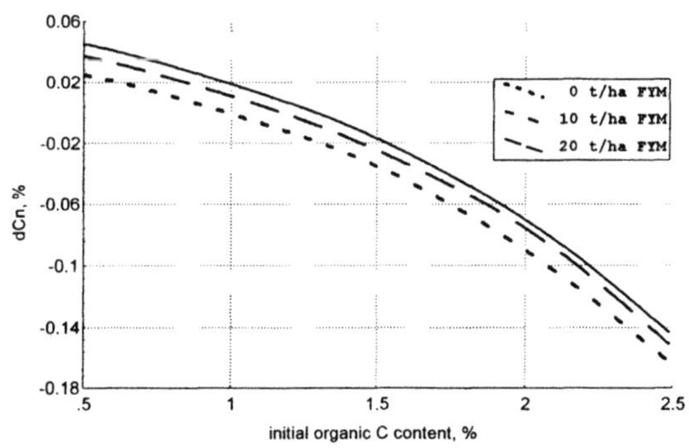

Figure 1. Annual change of soil organic matter in heavy loam soils as affected by initial C level and different farmyard manure (FYM) rates. Climatic conditions of Moscow Region.

Results and discussion

Model simulation with the combined long-term SOM database which included EuroSOMNET data and climatic information calculated with the CSY system have revealed a regular effect of climatic conditions on SOM dynamics. Among climatic variables affecting humus change, the most important ones were precipitation during the vegetation period, potential evapotranspiration and bioclimatic potential. Figure 1 illustrates simulation results for the climatic conditions of the Moscow Region.

Of all options examined, the most important way to prevent SOM decrease in the long-term for intensive agricultural system in the Moscow region is the increase of percentage of grass in crop rotation at the expense of row crops. FYM application also shows some potential of increasing soil sustainability, while the influence of mineral N fertilisation rates on SOM dynamics is extremely low. Thus the change of crop rotation is a substantial factor for decreasing CO_2 emission from arable land in the Moscow Region but single agricultural management changes are not able to prevent C losses as CO_2 as a result of prevailing mineralisation of SOM.

The dynamic submodel parameters were calibrated for 24 treatments in a 9-year experiment "Snegiry" of Dokuchaev Soil Institute (Moscow Region). Preliminary results are quite satisfying: the measured SOM changes were reasonably well reproduced by the model ($R^2 > 0.8$) for the plough layer.

The next step is the integration of the submodel into the CSY system and the identification of additional parameters for the main crops of the podzoluvisols zone. Maps of r_c spatial distribution and an estimation of h-soil clay content relationship are considered as the last stage of the modelling.

The simulation results of SOM dynamics over 30 years and CO_2 emission for different initial SOM levels, typical crop rotations, fertiliser rates and selected climate change scenarios were used to derive the relationship between land management practices and changes in soil carbon stock. The results suggest: a) based on a combined static and dynamic simulation it is possible to optimise utilisation of agricultural resources available for farmers and agricultural cooperatives (fertilisers, ameliorants, crops) taking into consideration an upper sound SOM limit and expected yield; and b) annual SOM dynamics for the same fertilisation rates and crop rotation, soil texture and initial SOM level in different climatic conditions were opposite for different regions of podzolic soil zone of Russia, revealing a gradient of annual SOM change.

References

Sirotenko O D, Abashina E V and Pavlova V N 1995 Meteorologia i gidrologia 4, 107-114 (in Russian).

Shevtsova L K, Romanenkov V A and Volodarskaya I V 1996 *In* Soil Organic Matter Network (SOMNET): 1996. Model and Experimental Metadata Eds P Smith, J U Smith, D S Powlson. pp 165-166. Wallingford, UK.

Shevtsova L K, Romanenkov VA, Rodionova V N, Volodarskaya I V 2000 *In* Information technologies, information measuring systems and appliances in study processes in agriculture. Ed. Aleinikov A.F. pp137-140. Novosibirsk (in Russian).

W. J. Horst et al. (Eds), Plant nutrition – Food security and sustainability of agro-ecosystems. 956~957, 2001

Removal of N and P from eutrophic pond water by using plant bed filter ditches planted with crops and flowers

K. Abe and Y. Ozaki

Department of Soils and Fertilisers, National Agriculture Research Center, Tsukuba, Ibaraki 305-8666, Japan, abekaoru@narc.affrc.go.jp

Key words: eutrophic pond water, particulate N and P, N and P removal, amenity, organic matter

Abstract

We evaluated the effectiveness of plant bed filter ditches in treating eutrophic pond water containing particulate N and P and the effect of adding organic matter (rice straw and dried papyrus stems) to the bed filter to make N treatment more efficient in winter. The experimental ditch was filled with zeolite (particle size: 6-8 mm) and planted with crops and flowers. A plant-free ditch effectively removed particulate P, but the PO_4-P dissolved out from particulate P accumulated in the bed filter. Adding plants to ditches removed PO_4-P successfully. Plant uptake and filtration by bed filters played an important role in removing P in ditches containing plants. A plant-free ditch completely removed particulate N and removed NO_3-N to some extent. Ditches containing plants removed NO_3-N and particulate N well in summer. The N balance showed N was removed mainly by denitrification and plant absorption. Adding organic matter to ditches improved N removal efficiency in winter and early spring.

Introduction

Recently, eutrophication in agricultural ponds has begun adversely affecting rice production and residential living environments in Japan (Fujita *et al.*, 1986, Hidaka, 1990). We designed plant bed filter ditches in which both terrestrial and aquatic plants are used to remove nitrogen (N) and phosphorus (P) from domestic wastewater (Abe *et al.*, 1997, 1999, Abe and Ozaki, 1998). In this study, we evaluated the effectiveness of plant bed filter ditches in the treatment of eutrophic pond water containing particulate N and P. The effect of adding organic matter to ditches to promote denitrification and increase the lower removal efficiency in winter and early spring was analysed.

Materials and methods

(1) Plant bed filter ditch design for pond water treatment

A plant bed filter ditch filled with zeolite as bed filter material (particle size: 6-8 mm) enables both terrestrial and aquatic plants to be cultivated when zeolite depth is adjusted to a plant species' saturation tolerance. The bed filter surface was about 0.1 m higher than the water level for terrestrial species and the same as the water level for aquatic species. The experimental ditch was a concrete ditch, 0.4 m wide, 0.3 m deep, and 11.2 m long. Water was 0.2 m deep.

(2) Operational conditions

The planted ditch contained impatiens (*Impatiens sultani* Hook. f.), peppermint (*Menta piperita* L.), African marigold (*Tages erecta* L.), papyrus (*Cyperus papyrus* L.), and kenaf (*Hibiscus cannabinus* L.) in summer and autumn and English daisy (*Bellis perennis* L.), peppermint, hanana (*Brassica campestris* L. var.) and Italian ryegrass (*Lolium multiflorum* Lam.) in winter and spring.

The planted/organic matter ditch contained the same plants as those listed above and organic matter within the bed filter in the midstream. Nets packed with 1.5 kg of dried rice straw and 1.0 kg of papyrus stems (cut into roughly 2 cm fragments) were buried in mid-October, 1997 and removed at the end of May, 1998 (Fig. 2). The control ditch did not contain plants (plant-free ditch).

Pond water was pumped and supplied continuously to ditches at a loading rate of about 200 Lm^{-2} (ditch area) d^{-1}. Retention time was about 20 hours.

Results and discussion

N and P concentrations in ditch effluents and ditch water

T-N and T-P concentrations of effluent from ditches containing plants decreased to less than those recorded in plant-free ditch during the experimental period. The concentration of the planted ditch effluent decreased to less than 0.3 mg L^{-1} for N in summer and 0.02 mg L^{-1} for P after September 1997. The minimum nutrient concentration for water bloom formation has been reported to be 0.3 mg L^{-1} for N or 0.02 mg L^{-1} for P (Kojima *et al.*, 1977).

Concentrations of N and P compounds in ditches from inlet to outlet determined September 2, 1997 are shown in Figure 1. In the plant-free ditch, particulate N (T-N minus TDN) and P (T-P minus TDP) were effectively removed, unlike dissolved N and P. TDP (total dissolved P) and PO_4-P concentrations increased considerably from inlet to outlet in the plant-free ditch, suggesting that particulate P was dissolved in the bed filter under reduced conditions or by microbial decomposition. The planted ditch removed NO_3-N, PO_4-P and particulate N and P well (Fig. 1). NO_3-N not removed by filtration through the zeolite bed filter and PO_4-P dissolved from particulate P were absorbed by ditch plants.

Effect of addition of plants and organic matter to the ditches on N and P removal efficiency

Nitrogen removal rates in the planted ditch and the planted/organic matter ditch generally exceeded

0.6 g $m^{-1}d^{-1}$ in summer (Fig. 2). The addition of plants to the ditch increased the N removal efficiency by 30% in summer and autumn (August to October, 1997 and June to October, 1998). The N removal rate decreased in winter and spring. Adding plants to ditches increased the N removal efficiency by 15% in winter and spring (October, 1997 to May, 1998).

Burying organic matter in the ditch (October, 1998 to May, 1999) raised N removal rate by 0.1 - 0.2 g m^{-2} d^{-1}, improving N removal efficiency by 10% (Fig.2).

The phosphorus load was much lower than the N load which was 20 times (summer) - 70 times (winter) as high as the P load. However, the amount of N uptake by the plants in the ditches was 10 - 20 times as large as the P uptake. Thus almost all of influent P was removed in the ditches containing plants though excess N was absorbed only partially and remained in ditch effluent, especially in winter. P removal efficiency of the plant-free ditch declined 9 months after the beginning of the experiment (Fig. 2), because particulate P accumulating in the ditch continued to be dissolved during this period.

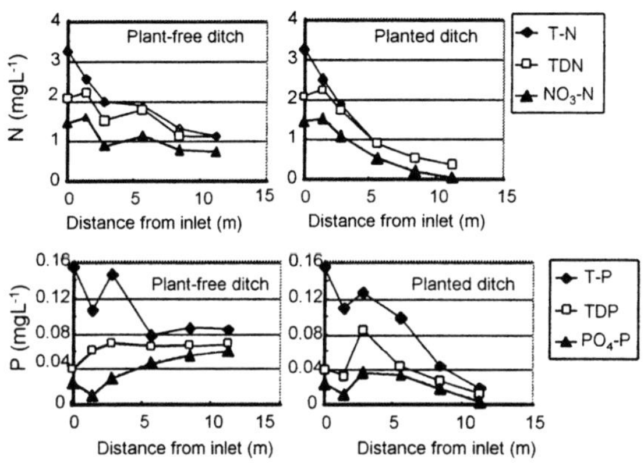

Figure 1. T-N, TDN, NO_3-N, T-P, TDP, and PO_4-P concentrations in ditches from inlet to outlet on September 2, 1997.

N and P balance

In the plant-free ditch, 13.7% of N removed from pond water remained in the bed filter and depositor of the ditch, while 94% of P removed from pond water remained in the plant-free ditch. These data indicated that P was removed mainly by filtration and precipitation in the plant-free ditch, while N was assumed to be removed mainly by denitrification using organic compounds in the suspended solids accumulating in the bed filter as hydrogen donors.

In the ditches containing plants (the planted ditch and the planted/organic matter ditch), plant absorption accounted for 23.0% - 24.8% of N and 44.5% - 50.2% of P removed from the pond water. These data indicated that P was removed mainly by filtration and plant absorption, while N was removed mainly by denitrification and plant absorption.

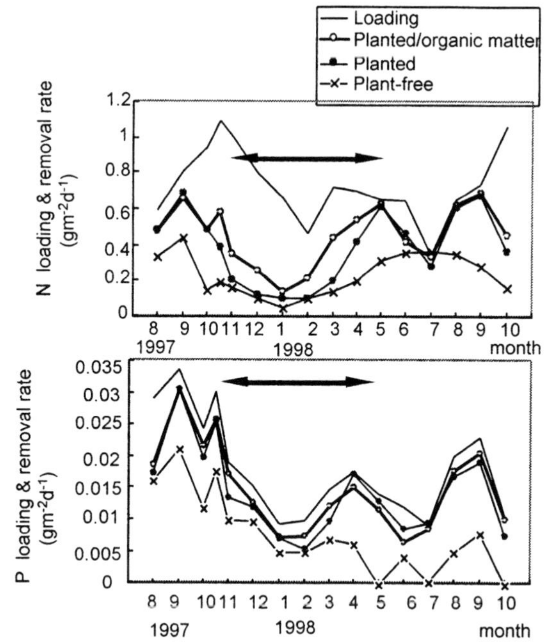

Figure 2. Changes in T-N and T-P removal rates of the ditches (monthly average).

References

Abe K, Ozaki Y and Kihou N 1997 Soil Sci. Plant Nutr. 43, 35-43.

Abe K, and Ozaki Y 1998 Soil Sci. Plant Nutr. 44, 599-607.

Abe K and Ozaki Y 1999 Soil Sci. Plant Nutr., 45, 409-417.

Fujita T, Gomi T and Kotani M 1986 Jour. JSIDRE 54, 623-630.

Hidaka S 1990 Bull. Sait. Agric. Exp. Stn., 44.

Kojima S, Sudo R, Sakurai T and Matsumoto T 1997 *In* Techniques for N and P removal and prevention of eutrophication, pp77-86, IPC (in Japanese).

W. J. Horst et al. (Eds), Plant nutrition – Food security and sustainability of agro-ecosystems. 958–959, 2001

Nitrous oxide and dinitrogen losses from flooded fen peat with and without reed canarygrass (*Phalaris arundinacea* L.) and common reed (*Phragmites australis* (Cav.) Trin. ex Steud.) plants

J. Augustin[1)], U. Münchmeyer[2)] and R. Russow[2)]

[1)] *Institute of Primary Production and Microbial Ecology, Centre for Agricultural Landscape and Land Use Research, Eberswalder Str. 84, Müncheberg, D 15379, Germany, jaug@zalf.de*

[2)] *Department of Soil Sciences, UFZ Centre for Environmental Research Leipzig-Halle, Theodor-Lieser-Strasse 4, Halle, D 06120, Germany, muenchmeyer@bdfl.ufz.de, russow@bdfl.ugz.de*

Key words: common reed, denitrification, fen peatland, ^{15}N gas flux method, reed canarygrass, reflooding

Abstract

There is little information about the effect of wetland plants on N turnover and N gas emission (N_2O, N_2) in recently reflooded minerotrophic mires (fens). We carried out lab model experiments with two typical wetland plants, reed canary grass (*Phalaris arundinacea* L.) and common reed (*Phragmites australis* (Cav.) Trin. ex Steud.) on fen peat substrate using the ^{15}N gas flux method. The reflooded fen peat substrate and the tested plants influenced the amount of gaseous N losses, its composition, and the way of its emission from the peat substrate to the atmosphere in a differentiated manner. Irrespective of plants the denitrification losses were between 49.9 to 74.5 % of the applied N. Whereas common reed narrowed the ratio between N_2 and N_2O up to 11:1, red canary grass caused a strong expansion of this ratio up to 343:1. The denitrification N gas liberation through the shoots of reed canary grass was not significant. In case of common reed approx. 20% of the fertiliser derived N_2 and approx. 10% of the fertiliser derived N_2O were exhaled by the plants.

Introduction

Minerotrophic mires called fens (accumulated N amounts of 5 to 120 t ha^{-1}) are widespread organic wetlands in the young Pleistocene landscapes of Northern Germany ($1{\cdot}10^4$ km^2). Most of these mires have been drained or otherwise altered resulting in strong soil degradation and rapid mobilisation of the high organic C and N content. During the last years dramatic shifts in agriculture of Northeast German plains have opened chances for peatland restoration, e. g. by reflooding. However, there is a considerable uncertainty regarding the effect of these measures on the N transformation and emission processes. This applies especially to the effect of aquatic macrophytes on N turnover and N loss processes in reflooded fen soils. Investigations in similar wetlands led to inconsistent results. Sometimes typical wetland plants stimulated both nitrification and denitrification (Bodelier *et al.*, 1997, Chen *et al.*, 1998). On the other hand, the intense N uptake of competitive plant species often led to a decrease of gaseous N losses from wetlands (e. g. Dean and Biesboer, 1985). We examined the effect of two common wetland plants like reed canary grass (*Phalaris arundinacea* L.) and (*Phragmites australis* (Cav.) Trin. ex Steud.) on the N transformation processes and on N gas emission (N_2O, N_2) of reflooded fen peat substrates..

Materials and methods

The investigations were conducted as lab model experiments by means of ^{15}N gas flux technique (Münchmeyer *et al.*, 1999). Plant growth took place in compartmented root cylinders (height 40 cm, diameter 8 cm). A strongly degraded fen substrate was used (Senitz-Welse-Lowland in the Northeast of Berlin, C_t 22%, N_t 1.67%, bulk density 0.42 g cm^{-3}, pH 6.2). A first experiment was done with reed canary grass (water level was held 10 cm below soil surface) and a second experiment with common reed (water level 2 cm below soil surface). Planted and unplanted pots were supplied with 300 mg N as ^{15}N labelled KNO_3 (^{15}N abundance 47% at% $^{15}N_{exc}$), 90 or 120 days after starting the experiments by means of suction cups. During the experiments the ^{15}N abundance in the gases N_2O and N_2 were determined with an isotope mass spectrometer (Russow *et al.*, 1996) and the N_2O concentration with a photo acoustic gas monitor or a GC (detector: ECD). With the aid of the compartmented root cylinders it was also possible to distinguish between N gas evolution through the plants (short term gas analysis in a closed shoot headspace) and gas evolution from the fen substrate (long term gas analysis by a dynamic system in the root compartment). ^{15}N abundance in soil, water, and plant samples were analysed by emission spectrometry (Münchmeyer *et al.*, 1999).

Results and discussion

Especially in the cylinders without plants the recovery rates of the applied ^{15}N fertiliser were surprisingly small (Table 1). The reason for this could be associated with methodical problems while recording N_2 emissions. Caused by its small solubility in water, most of the N_2 produced may have been trapped in the soil in the form of gas bubbles which were probably released at the finishing of the measurements (Chen *et al.*, 1998). The reflooded peat substrate and the tested plants influenced the amount of gaseous N losses, its composition, and the way of its emission from the peat substrate to the atmosphere in a differentiated manner. The presence of the plants generally caused an acceleration of the N gas emission from the fen substrate (data not shown). Whereas common reed narrowed the ratio between N_2 and N_2O up to 11:1, red canary grass caused a strong expansion of this ratio up to 343:1. The denitrification N gas liberation through the shoots of reed canary grass was not significant. In case of common reed approx. 20% of the fertiliser-derived N_2 and approx. 10% of the fertiliser-derived N_2O were exhaled by the plants (table 1). Similarly, high portions of the denitrification losses were also observed by Stengel *et al.* (1987) in an experimental system with common reed. Up to now, however, it is unknown whether these plant species-specific effects on N gas losses are based on differences in the internal gas transport systems, in oxygen release to the anaerobic root environment, in the competitiveness for soil N acquisition, and/or in the promotion of the microbial N turnover processes by root exudates (Chen *et al.*, 1998, Münchmeyer *et al.*, 1999, Ottosen *et al.*, 2000). Our findings indicated that besides substrate quality the characteristics of the plants must be considered in order to be able to estimate the possible ecological effects of a changed land use (e. g. reflooding) on the N transformation and loss processes correctly.

References

Bodelier P L E, Wijlhuizen A G, Blom C W P M and Lannbroek H J 1997 Plant and Soil 190, 91-103.

Chen D L, Chalk P M and Freney J R 1998 Nutrient Cycling in Agroecosystems 51, 269-279.

Münchmeyer U, Augustin J, Russow W and Merbach W 1999 *In* Stoffumsatz im wurzelnahen Raum. Ed W Merbach. pp 159-166. B G Teubner Verlagsgesellschaft, Stuttgart.

Ottosen L D M, Risgard-Petersen N and Nielsen L P 2000 Aquatic Microbial Ecology 19, 81-91.

Russow W, Stevens R J and Laughlin R J 1996 Isotopes Environ. Health Stud. 32, 289-297.

Stengel E, Carduck W and Jebsen C 1987 *In* Aquatic plants for water treatment and resource recovery. Ed K R Reddy and W H Smith. pp 543-550. Magnolia Publishing Inc., Orlando, Florida.

Table 1. Accumulated gaseous N losses and distribution of the fertiliser derived ^{15}N in the plant-fen peat-atmosphere-system. All data as % of amount of ^{15}N applied.

Fraction	**Phalaris arundinacea**		**Phragmites australis**	
	unplanted	planted	unplanted	planted
soil	**7.7**	**6.2**	**2.4**	**3.7**
plants	-	**7.0**	-	**19.5**
gaseous losses	**49.9**	**68.9**	**74.5**	**73.3**
thereof:				
N_2 emission soil	49.6	64.5	72.4	55.4
N_2 emission shoots	-	4.2	-	11.6
N_2O emission soil	1.2	0.2	2.2	5.7
N_2Oemission shoots	-	0.001	-	0.6
relationship N_2O-N/N_2-N	1 : 41	1 : 343	1 : 33	1 : 11
other fractions[1]	**11.9**	**0.3**	**0.3**	**0.3**
^{15}N-recovery	**57.7**	**88.7**	**77.2**	**96.8**

[1] soil solution from suction cups and water around soil columns

W. J. Horst et al. (Eds), Plant nutrition – Food security and sustainability of agro-ecosystems. 960–961, 2001

Production, consumption and emission of nitrous oxide in profiles of Stagnic Gleysols under changing soil moisture conditions

M. Trimborn and W. Werner

Institute of Agricultural Chemistry, University Bonn, Karlrobert-Kreiten-Str. 13 D 53115 Bonn, Germany, m.trimborn@uni-bonn.de

Key words: denitrification, subsoil, nitrous oxide, nitrogen-15, N_2:N_2O ratio, N_2O consumption

Abstract

N_2O profile and N_2O emission were studied on undisturbed soil columns of two different Stagnic Gleysol (Pseudogleye) under changing soil moisture conditions. The N_2O emissions principally showed the known relations to soil moisture, but particularly high emissions were only observed during periods of changing soil moisture. Thereby a strong relation could be found between the site of denitrification and dense layer in soil profile. While determining total denitrification ($N_2 + N_2O$) by adding $^{15}N\text{-}NO_3^-$ to subsoil, high N_2O production was found there during a period of alternating soil moisture. In the same time, consumption of ^{15}N-labelled N_2O in soil profile could be clearly quantified. The adaptation and composition of microbial communities may be a crucial factor in this process. The experiment pointed out the significance of subsoil denitrification, but N_2O produced in this process does not necessarily result in increased emissions.

Introduction

It is known that denitrification occurs not only in topsoil but also in deeper soil layers where it contributes to reduction of nitrate leached from the rooting zone of agricultural fields. In the past, subsoil denitrification was estimated by modelling (measuring the denitrification potential) or balances (nitrate or ^{15}N-mass balance). Investigations of soil cores (with or without added carbon) only give a rough view into real processes since they are not influenced by surrounding soil layers (Funk *et al.*, 1995; Castle *et al.*, 1998; Clough *et al.*, 1999). Till now there are no direct measurements in different depths of undisturbed soil profiles and so little is know about localisation of N_2 and N_2O production or consumption of N_2O on its way to soil surface. The enlightenment of these processes, which may play an important role in hydromorphic soils, was the aim of the investigation presented here.

Materials and methods

Two Stagnic Gleysols (S1, S2) with different waterlogging conditions were investigated. The nitrogen turnover and soil moisture regimes of this soils were described by previous field experiments. S1 has a dense layer (> 40 cm depth) that prevents infiltration of water and leads to frequent waterlogging or dry conditions in the topsoil. The subsoil is poorly rooted and shows low microbial activity. S2 contains a layer (> 70 cm) that is often saturated by laterally infiltrating water. Its moisture conditions in the topsoil are more balanced compared to S1.

Four undisturbed soil columns were taken from both soils with PVC-tubes (75 cm, Ø 20 cm). During 140 days definite moisture conditions (see Fig. 1) were adjusted by drainage or irrigation. In the end of the wet period subsoil was ^{15}N-labelled in 53 cm depth by injection of a KNO_3-solution (95 Atom-% ^{15}N). Soil atmosphere was sampled from artificial pores (20 cm, Ø 0,4 cm) drilled into the columns in six different depths. Gaseous emissions from soil surface were measured by sealing the top of the PVC-tubes. Gas samples were analysed by GC-ECD or IRMS. Denitrification was calculated by flux and soil content of ^{15}N-labelled N_2 and N_2O as described by Arah (1992).

Results

Immediately after the first of three irrigations during the period of rewetting, N_2O emission increased (Fig. 1), but max. N_2O emission was observed 12 days later when field capacity was exceeded.

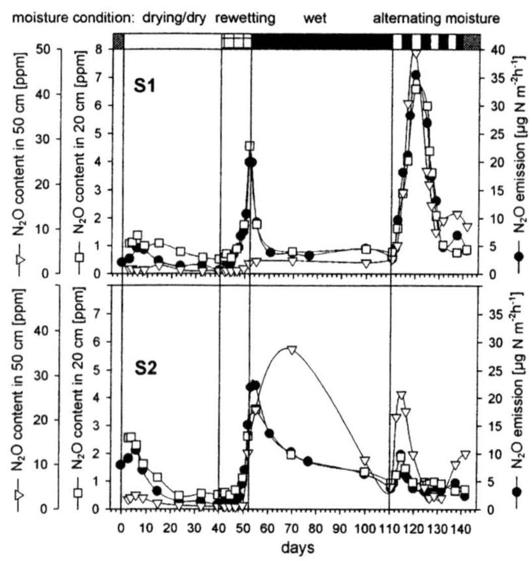

Figure 1. N_2O emission and N_2O content in 20 and 50 cm depth under different soil moisture conditions.

There was only a narrow emission peak although in the following wet period soil moisture was kept at field capacity. From N_2O concentrations in the soil profile it can be concluded that at soil S1 (dense layer near topsoil) N_2O production occurs primarily in topsoil, whereas at soil S2 (often saturated in subsoil) N_2O is also produced in subsoil. During the wet period denitrification in topsoil was comparatively low at S1 and S2. But at S2 high N_2O concentrations in subsoil could be observed over a long time which were not related to higher emissions.

Table 1. Cumulated emissions of N_2O from topsoil and subsoil and total denitrification in subsoil during period of changing soil moisture (30 days) [mg N m^{-2}].

	S1	S2	t-test[1]
topsoil N_2O-emission	2.89	2.47	n.s.
subsoil N_2O-emission	8.08	0.85	*
share of subsoil	74.5 %	26.3 %	**
subsoil denitrification	93.6	30.5	*
N_2:N_2O ratio	11.7	39.0	*

[1]: n.s.: not significant; *: = 5 %; ** = 1 %

After the wet period soil was alternately dried and irrigated so that soil moisture in topsoil fluctuated around field capacity (83-103 %). Hereupon specially S1 reacted with high N_2O emission whereas S2 showed only a small increase. An explanation for this was possible by ^{15}N-labelling of subsoil. In S1 75 % of the emitted N_2O was produced in subsoil whereas in S2 the share of subsoil was only 25 % (Tab. 1). The N_2:N_2O ratio of ^{15}N-labelled gas emitted from S2 was nearly 4 times higher than from S1, although there was no significant difference in subsoil atmosphere at 50 cm depth between S1 and S2 (Fig. 2).

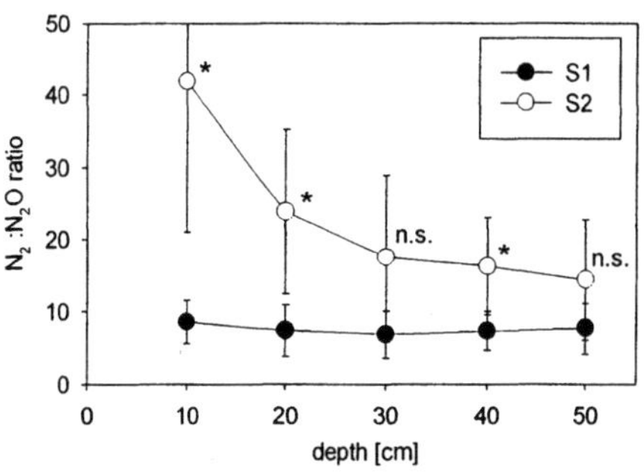

Figure 2. Variation in N_2:N_2O ratio of gas coming form denitrification in subsoil (50 cm). Error bars are standard deviation (n=4). Symbols are explained in Tab.1.

In soil profile of S2 the N_2:N_2O ratio raised to an order of magnitude that corresponded to the N_2:N_2O ratio of emissions. Hence a consumption of N_2O could be concluded and we calculated that only a third of N_2O produced in subsoil was emitted at soil surface.

Discussion

The amounts of N_2O emitted during the periods of constant moisture are consistent with the known relation between soil moisture and denitrification. But even the emissions during the wet period were unimportant compared to emission peaks occurring at changing soil moisture conditions. The soil type and composition of microbial communities influence both the amount and distribution of denitrification in soil profile (Funk *et al.*, 1996; Menyailo *et al.*, 1999). Due to the dense layer the topsoil of S1 dried faster and also rewetted faster under equal irrigation. Hence denitrifiers at S1 react faster than at S2. Our field experiments showed higher amounts of microbial biomass in subsoil of S2 which might be a reason for the high amounts of N_2O found there during the wet period. The consumption of N_2O that could be demonstrated for the period of alternating moisture, explains why high amounts of N_2O did not result in adequate N_2O emissions. Microbial communities show great differences in their ability to consume N_2O (Chèneby *et al.*, 1998). So it can be concluded that microorganisms in topsoil of S2 are adapted to the consumption of N_2O from subsoil, especially as the subsoil of S2 is often saturated and favourable for denitrification.

The experiment pointed out the significance of subsoil denitrification of hydromorphic soils. But N_2O produced in this process does not necessarily result in increased emissions. In field studies this may be the case for N_2O emissions during freeze-thaw cycles. N_2O produced in subsoil accumulates beneath the ice layer and will be released while thawing (Burton and Beauchamp, 1994). These N_2O emissions occurring within a few days can exceed the emission of the rest of the year. It is unlikely that N_2O production in subsoil is particularly pronounced even during freezing periods. It is more likely that also in other periods – especially during periods of seepage – N_2O is produced in subsoil but is consumed on its way to the soil surface. During freezing periods this consumption is inhibited so that subsoil production will be observable by adequate increased N_2O emissions.

References

Arah JRM 1992 Soil Sci. Soc. Am. J. 56, 795-800.

Burton DL and Beauchamp EG 1994 Soil Sci. Soc. Am. J. 58, 115-122.

Castle K, Arah JRM and Vinten AJA 1998 Biol. Fertil. Soils 28, 12-18.

Chèneby D, Hartmann A, Hénault C, Topp E and Germon JC 1998 Biol. Fertil. Soils 28, 19-26.

Clough TJ, Jarvis SC, Dixon ER, Stevens RJ, Laughlin RJ and Hatch DJ 1999 Soil Biol. Biochem. 31, 31-41.

Funk R, Maidl F-X and Fischbeck G 1996 Z. Pflanzenernähr. Bodenk. 159, 207-213.

Menyailo OV and Huwe B 1999 J. Plant Nutr. Soil Sci. 162, 533-538.

W. J. Horst et al. (Eds), Plant nutrition – Food security and sustainability of agro-ecosystems. 962~963, 2001

Effect of pH on denitrification losses from different arable soils

B. W. Hütsch[1)], S. Zhang[2)], K. Feng[2)], F. Yan[1)] and S. Schubert[1)]

[1)] *Institute of Plant Nutrition, Justus Liebig University, Interdisciplinary Research Center (IFZ), Heinrich-Buff-Ring 26-32, Giessen, D 35392, Germany, birgit.w.huetsch@ernaehrung.uni-giessen.de*

[2)] *Department of Agronomy, University of Yangzhou, Sunong Road 12, Yangzhou, 225009, China*

Key words: acidification, arable soils, denitrification, liming, nitrous oxide, soil pH

Abstract

Denitrification of nitrate can result in remarkable gaseous nitrogen losses, the emission of N_2O being particularly important from an environmental point of view, as it contributes to the greenhouse effect and the destruction of the ozone layer. The influence of soil pH on denitrification and on the proportion of released N_2O and N_2 is yet unclear, and therefore a study was conducted comparing a loamy and a sandy arable soil. After stepwise acidification of the soils and addition of nitrate (40 mg NO_3-N kg^{-1}), soil slurries were anaerobically incubated and N_2O emission rates determined, without and with acetylene inhibition. Maximum denitrification losses were observed with pH 5.2 and 5.9 in the sandy and the loamy soil, respectively, consisting mainly of N_2O (between 68% and 85%). Under less acidic and particularly under alkaline conditions gaseous N losses decreased, combined with a reduction in the N_2O proportion to approx. 16%. Interestingly, the highest N_2O emissions occurred well below the native soil pH. Thus, regular liming of these arable soils is benefical, as with higher pH less total denitrification losses can be expected with a much smaller proportion of the detrimental N_2O.

Introduction

Under anaerobic conditions denitrification of nitrate can result in remarkable gaseous nitrogen losses, emitted either as NO, N_2O or N_2. Attention has been focused particularly on N_2O, as this is an important greenhouse gas and contributes to the destruction of the stratospheric ozone layer. However, total denitrification emissions should be also considered, as they contribute significantly to the loss of valuable nitrogen fertiliser. Soil pH has an important influence on all biological processes, including N transformations, and there have been several studies to assess the effect of acidity on denitrification in terms of amounts and the composition of gaseous products. However, the obtained results are not unequivocal. Simek and Hopkins (1999) found that the optimum pH value for potential denitrification was between 7 and 8, and that the reduction of pH in arable soils led to reduced denitrification. In investigations on a grassland soil, where different pH levels had been established for many decades, N_2O production increased with increasing pH (Ellis *et al.*, 1998). However, if the soil pH of a native, acid plot from the same grassland site was increased, it caused a reduction in N_2O fluxes (Yamulki *et al.*, 1997). With an increase in pH less N_2O and more N_2 were evolved (Weier and Gilliam, 1986; Stevens *et al.*, 1998). The proportion of N_2O appeared to be negatively related to total N_2O + N_2 losses (Schenk and Daum, 1997).

In order to characterise the effect of varying soil pH on denitrification and on the proportion of released N_2O and N_2 in more detail, an incubation experiment was conducted using two arable topsoils with different texture.

Materials and methods

In autumn 2000 soil samples were collected from the plough layer of a Luvic Phaeozem derived from loess (silty loam: 21% clay, 67% silt; pH 7.7), and an Eutric Cambisol (silty sand: 7% clay, 36% silt; pH 6.4). The soils were slightly air-dried at ambient temperatures (5-10°C), sieved (≤ 5 mm mesh width), and stored at 5°C prior to the incubation experiments.

Moist soils equivalent to 30 g on an oven-dry basis (105°C drying) were weighed into 250 ml Erlenmeyer flasks and adjusted to different pH values (3 replicates per treatment) by adding certain amounts of NaCl or HCl, dissolved in 1.5 ml of deionized water. After an aerobic preincubation at 25°C for 10 days, 40 mg NO_3^--N kg^{-1} soil were added with 50 ml solution of KNO_3 attaining a soil slurry. The gas atmosphere was exchanged by N_2 and the flasks were anaerobically incubated on a rotary shaker (120 rpm) at 25°C for 24 h. After renewing the N_2 atmosphere, gas samples from the headspace were collected immediately afterwards (time 0) and after 45 and 90 minutes. In some experiments 10 vol% of the headspace were exchanged by acetylene prior to gas sampling in order to estimate N_2 release. Soil pH was determined in the Erlenmeyer flasks (soil: solution = 1:2.5), and the concentration of N_2O was measured with a gas chromatograph equipped with an electron capture detector. N_2O emission rates were calculated and are expressed as µg N_2O-N kg^{-1} dry soil h^{-1} ± standard error (s.e.).

Results and discussion

In the loamy soil by far the smallest N_2O emission was observed at pH 7.7, which was the native soil pH (Tab. 1). With pH decrease to 5.9 a remarkable increase in N_2O emission rates occurred, which decreased at more acid soil conditions. From the measurements with acetylene, which inhibits the reduction from N_2O to N_2 and thus gives the total denitrification losses (NO losses are negligible under the incubation conditions), N_2 emission rates between 191 and 253 µg N_2-N kg^{-1} h^{-1} were calculated with no significant differences between the 4 pH levels. However, the N_2 proportion of the total denitrification losses (N_2O + N_2) varied considerably, being 84%, 38%, 15%, and 37% with pH 7.7, 6.9, 5.9, and 5.0, respectively.

Table 1. N_2O emission rates (t_{45}-t_0) of the loamy soil as affected by different pH values, determined without and with acetylene inhibition (means of 3 replicates ± s.e.).

Soil pH	N_2O emission rate (μg N kg^{-1} h^{-1})	
	without acetylene	with acetylene
7.7	36.2 ± 1.5	227.6 ± 12.2
6.9	420.5 ± 18.2	674.0 ± 31.1
5.9	1339.0 ± 59.5	1584.5 ± 148.3
5.0	349.2 ± 20.2	552.7 ± 24.9

Comparable results were obtained with investigations of the sandy soil, having the highest N_2O emission rate at pH 5.2 and much smaller values at pH 6.4 (native pH) and 7.6 (Tab. 2). N_2 emission rates lay between 79 and 0 μg N_2-N kg^{-1} h^{-1}. Apart from pH 4.5, which had the smallest rate, no significant treatment differences occurred. The proportion of N_2 decreased with decreasing pH from 83% at pH 7.6 to 60%, 32%, and 0% at pH 6.4, 5.2, and 4.5, respectively.

Table 2. N_2O emission rates (t_{45}-t_0) of the sandy soil as affected by different pH values, determined without and with acetylene inhibition (means of 3 replicates ± s.e.)

Soil pH	N_2O emission rate (μg N kg^{-1} h^{-1})	
	without acetylene	with acetylene
7.6	10.6 ± 0.5	62.4 ± 8.0
6.4	33.5 ± 3.1	84.4 ± 4.4
5.2	164.2 ± 2.7	242.8 ± 21.6
4.5	77.5 ± 4.0	75.5 ± 2.1

The denitrification potential of the sandy soil was at least 2 to 3 times smaller than that of the loamy soil.

It is generally recognised that the rate of denitrification increases with pH from acidic conditions (pH 3 to 5) to neutral or slightly alkaline conditions (Granli and Bockman, 1994). Also in the experiments of Weier and Gilliam (1986), Stevens *et al.* (1998), and Simek and Hopkins (1999) pH increases from acid to alkaline favored denitrification losses. However, in our own investigations decreases of the prevailing soil pH were induced, causing an increase in denitrification losses followed by a decrease at pH values of 5.0 and below. The maximum emissions resulted mainly from N_2O (between 68% and 85%), and for this nitrogen gas the effect of soil pH is more complex. Stevens *et al.* (1998) reported the highest N_2O flux with pH 6.5, and the minimum fluxes at pH 6.0 and 8.0; Weier and Gilliam (1986) found maximum N_2O being evolved at pH $\leq$ 5.8. This is in agreement with our own investigations, where maximum N_2O emissions occurred at pH 5.2 and 5.9 in the sandy and loamy soil, respectively. Interestingly, these maximum rates were found well below the native soil pH (pH 6.4 and 7.7, respectively), which had been adjusted by regular liming of the arable soils. From an environmental point of view this is quite beneficial, as with higher pH less total denitrification losses occurred with a much smaller proportion of the detrimental N_2O. Thus, soil acidification should be avoided by liming, particularly if a site is prone to frequent anaerobic conditions favouring denitrification. However, further experiments are needed to investigate interactions with available organic carbon and shifts in soil moisture content.

References

Ellis S, Howe MT, Goulding KWT, Mugglestone MA and Dendooven L 1998 Soil Biol. Biochem. 30, 359-367.

Granli T and Bockman OC 1994 Norwegian J. Agric. Sci., Supplement No. 12.

Schenk MK and Daum D 1997 *In* Plant Nutrition – For Sustainable Food Production and Environment. Eds. T Ando *et al.* pp. 567-568. Kluwer Academic Publishers. Printed in Japan.

Simek M and Hopkins DW 1999 Biol. Fertil. Soils 30, 41-47.

Stevens RJ, Laughlin RJ and Malone JP 1998 Soil Biol. Biochem. 30, 1119-1126.

Weier KL and Gilliam JW 1986 Soil Sci. Soc. Am. J. 50, 1202-1205.

Yamulki S, Harrison RM, Goulding KWT and Webster CP 1997 Soil Biol. Biochem. 29, 1199-1208.

Symposium 12

Plant nutrition and sustainable development

Convener:

Andreas Bürkert
Volker Römheld

Editors:

Andreas Bürkert
Volker Römheld

W. J. Horst et al. (Eds), Plant nutrition – Food security and sustainability of agro-ecosystems. 966~967, 2001

Challenges for sustaining productivity gains and environmental quality in intensive grain production systems of Asia and the United States

A. Dobermann and K. G. Cassman

Department of Agronomy and Horticulture, University of Nebraska, Lincoln, NE, 68583-0915, United States, adobermann2@unl.edu

Key words: crop intensification, maize, nutrient use efficiency, rice, site-specific nutrient management

Abstract

Are present nutrient management recommendations for the world's major cereal cropping systems adequate to sustain the productivity gains required to meet food demand while also assuring acceptable standards of environmental quality? Current management guidelines and the scientific basis for them were evaluated in large-scale, mechanized maize (*Zea mays* L.)-based cropping systems of the USA and more labour-intensive, small-scale irrigated rice (*Oryza sativa* L.) production systems in Asia. The principal challenges in both systems are similar: (1) there is no evidence for significant increases in the genetic yield potential of rice or maize in both systems during the past 30 years, (2) farm yields are presently about 50 to 60% of the attainable yield potential, and (3) nutrient management guidelines rely on mostly static, empirical approaches. Research must seek to develop nutrient management approaches tailored to the technologies and spatial scales relevant to each system that optimise profit, preserve soil quality, and protect natural resources in systems that consistently produce at levels of 70 to 80% of the attainable yield. Significant advances in soil chemistry, crop physiology, plant nutrition, molecular biology, and information technology must be applied to this effort. Future field-oriented plant nutrition research must be of a more strategic, interdisciplinary, and quantitative nature.

Introduction

Double- and triple-crop continuous rice and rice-wheat systems occupy 40 million ha land area in south and southeast Asia. Rice yields increased by 2.5% yr^{-1} from 1967 to 1984, but yield growth rates dropped to 1.2% from 1984 to 1996. Current average yields of irrigated rice are equivalent to 60-65% of the climatic yield potential and the average plant recovery efficiency of fertiliser N is only 30% (Dobermann *et al.*, 2001). Rainfed and irrigated maize, grown either in rotation with soybean (*Glycine max L.*) or as a continuous monocrop, are the predominate cropping systems in North America. During the past 30 years, average maize yields have increased steadily at about 100 kg ha^{-1} yr^{-1}. However, maize yields are only about 50-60% of the attainable yield potential (Duvick and Cassman, 1999), and only 30-40% of the applied fertiliser-N is typically taken up by the plant. Water pollution has become a serious issue. The ecological intensification of these important cereal production systems is justified by concerns about food security, the availability of adequate land and water resources, and environmental quality (Cassman, 1999). In the following we review some of the key issues related to the crops' germplasm and the management of soil fertility and plant nutrients.

Yield response and plant nutrient requirements

In Asia, fertiliser recommendations for rice are based on empirical yield response functions that are extended on a district or regional scale. Commercial soil testing is rarely used. Most of the existing soil tests are of limited use because they fail to predict the indigenous nutrient supply under field conditions. Local calibration research has led to a wide range of fertiliser recommendations for maize in the USA. Farmers routinely use commercial soil testing to determine field-specific application needs. Algorithms for estimating N rates typically include a yield goal and credits for crop rotation and organic sources of N input. Recommendations for managing P and K follow sufficiency, build up and maintenance, or replenishment of crop removal concepts (Hergert *et al.*, 1997). Critical soil test levels for P and K are derived from response functions using relative yields, assuming that such levels are independent from seasonal fluctuation in the yield potential. Many concepts currently in use were developed when average yields were in the 5-6 Mg ha^{-1} range for maize and 3-4 Mg ha^{-1} for rice. However, to meet global demands yields must reach 12-13 Mg ha^{-1} for maize and 7-8 Mg ha^{-1} for rice in 30 years.

Current empirical approaches to nutrient management may not be adequate to meet future goals for several reasons. Empirical calibration requires investment in repeated multi-location field testing. Likewise, present soil tests are 'static' and may not assess soil nutrient supply rates during peak times of crop demand at elevated yield levels or plant-induced mobilisation of soil nutrients. Existing algorithms assume linearity between crop yield and nutrient input requirements yet crop nutrient requirements are not linearly associated with yield level (Witt *et al.*, 1999). The ability to predict soil nutrient supply and plant accumulation in absolute terms rather than as a chemically extracted soil fraction or a relative yield response to applied inputs will become a key to fine-tuning nutrient management. Modelling the relation between nutrient supply, uptake and yield provides a more robust solution to understanding variation in crop response as a function of climatic yield potential and macronutrient interactions (Janssen *et al.*, 1998, Witt *et al.*, 1999).

Site-specific nutrient management

Can a generic basis for site-specific management be developed for both small-scale and large-scale agriculture? In Asian rice systems, a new approach for field-specific nutrient management has been tested, which included: (1) field-specific estimation of the indigenous nutrient supply, (2) estimation of balanced crop nutrient requirements for an economically efficient yield target and (3) real-time adjustment of N doses based on monitoring of the plant N status (Dobermann and White, 1999). On-farm research allowed developing generic models as well as local options for adjustment according to differences in the environment and cropping practices. In field testing at 179 sites, average rice yields increased by 7-11%, N losses from fertiliser were reduced by 30-40% and profits increased by \$46 ha^{-1}. An advantage of this concept is that updates are easier to implement because there is less reliance on soil tests, and because nutrient uptake and yield are estimated in absolute terms.

Precision farming in the USA has mostly focused on variable application rates of N, P, K, or lime using different local fertiliser algorithms for each of those in combination with soil samples collected from grids or soil management zones within a field. Many of the case studies conducted have failed to demonstrate consistent and significant agronomic, economic, or environmental benefits (Pierce and Nowak, 1999). Reasons for this failure include: (i) insufficient characterisation of spatial variation in indigenous nutrient supply and yield goals, (ii) use of fertiliser algorithms that are not suited for site-specific management, and (iii) insufficient post-emergence adjustment of N to account for seasonal variation in yield potential. More attention should be paid to understanding of spatial cause-effect relationships that can be generalised and extrapolated.

Germplasm improvement and nutrient use efficiency

Improved germplasm with greater yielding ability contributes to greater nutrient use efficiency due to higher growth rates and increased recovery of nutrients from soil and applied fertiliser. In both USA maize and Asian rice systems, germplasm improvement during the past 30 years has resulted in greater yield stability as a result of substantial increases in tolerance to abiotic and biotic stresses (Duvick and Cassman, 1999, Peng *et al.*, 1999). Less certain is whether there has been a significant increase in yield potential and whether breakthroughs in complex traits such as yield potential or specific traits that confer greater nutrient use efficiency can be achieved. In rice, research is ongoing to develop new plant types with a 25% larger yield potential, C4-photosynthesis, or increased biological N_2 fixation. These efforts will not have measurable impact in the near future and we are not aware of comparable efforts on maize. Biotechnology and molecular breeding may facilitate the development of crop genotypes which have a root architecture more congruent with the distribution of nutrients in the soil profile. This might result in improvement in nutrient use efficiency in different tillage systems and soil types. Little is known about practical nutrient management of varieties and hybrids with specific, higher value end uses.

Future field research

Research that focuses on empirical relationships often fails to account for rapid advances in the basic sciences, improved germplasm, yield levels, cropping systems, information technologies and changes in the soil resource base itself. There is also an imbalance between research at the field level vs. research at meso-spatial scales to obtain a quantitative understanding of crop response to environmental variation. More high-quality, strategic multi-site research should be conducted at locations and scales that are relevant for crop production and using agroecological zones concepts and tools such as crop models. Linkages between production research and globally oriented environmental research must be improved to obtain a better understanding of the net global warming potential of intensive agricultural systems and how it can be manipulated through better plant nutrition. Yield level, amount and decomposition of crop residues, soil organic matter and soil N dynamics determine soil C sequestration and greenhouse gas emissions and can be manipulated through plant nutrition.

Conclusions

Despite significant progress in technologies and fundamental scientific knowledge, the basic concepts for applying fertilisers have changed little during the past 30 to 40 years. The commonly used approaches to research and management may be insufficient for harnessing the potential breakthroughs from crop genetic improvement, and manipulating the net global warming potential in intensive cereal systems. Future strategies for site-specific nutrient management must seek to achieve more dynamic management of spatially variable resources and be based on quantitative understanding of seminal relationships between yield and nutrient uptake and the congruence between nutrient supply and crop demand. More resources should be devoted to strategic field research that follows systems approaches.

References

Cassman KG 1999 Proc. Natl. Acad. Sci. 96, 5952-5959.

Dobermann A and White PF 1999 Nutr. Cycl. Agroecosyst. 53, 1-18.

Duvick DN and Cassman KG 1999 Crop Sci. 39, 1622-1630.

Hergert GW, Pan WL, Huggins DR, Grove JH and Peck TR 1997 *In* The site-specific management for agricultural systems. pp. 283-300. ASA, CSSA, SSSA, Madison, WI.

Janssen BH 1998 Field Crop Res. 56, 197-201.

Peng S, Cassman KG, Virmani SS, Sheehy JE and Khush GS 1999 Crop Sci. 39, 1552-1559.

Pierce FJ and Nowak P 1999 Adv. Agronomy 67, 1-85.

Witt C, Dobermann A, Abdulrachman S, Gines HC, Wang GH, Nagarajan R, Satawathananont S, Son TT, Tan PS, Tiem LV, Simbahan GC and Olk DC 1999 Field Crop Res. 63, 113-138.

W. J. Horst et al. (Eds), Plant nutrition – Food security and sustainability of agro-ecosystems. 968~969, 2001

Does the farming system affect the nitrogen fertiliser value of animal manure?

M. Langmeier[1], A. Oberson[1], M. Kreuzer[2], P. Mäder[3], D. Dubois[4] and E. Frossard[1]

[1] *Institute of Plant Sciences, Swiss Federal Institute of Technology (ETH), Eschikon 33, P.O. 185, Lindau, 8315, Switzerland, astrid.oberson@ipw.agrl.ethz.ch*

[2] *Institute of Animal Science, Swiss Federal Institute of Technology (ETH), ETH Centre, Zurich, 8092, Switzerland*

[3] *Research Institute of Organic Agriculture, Ackerstrasse, Frick, 5070, Switzerland*

[4] *Swiss Federal Research Station for Agroecology and Agriculture (FAL), Reckenholzstrasse 191, Zurich, 8046, Switzerland*

Key words: ^{15}N, farming systems, N fertiliser value

Abstract

Increases in soil microbial biomass and activity often observed as a consequence of organic farming raise the question whether these changes affect the fertiliser efficiency of animal manure. The hypothesis that on a long-term organically managed soil (ORG) a greater portion of nitrogen (N) applied as cattle manure is mineralised and taken up by plants than on a conventionally managed soil that had received exclusively mineral fertilisers (MIN) was tested in a pot experiment. Soil taken from ORG and MIN systems of a field experiment were mixed with ^{15}N labelled cattle manure or mineral N and sown with italian ryegrass (*Lolium multiflorum*). The N uptake of the ryegrass was significantly higher on the ORG than the MIN soil in all fertiliser treatments and in the non-fertilised control. The utilisation of the applied fertilisers was similar in ORG and MIN. The major part of the increased N uptake on ORG was due to a significantly higher N supply from the soil in ORG compared to MIN. The increased capacity of the ORG soil to supply N to plants became more important at later cuts when the significance of fertiliser N had declined. It is concluded that long-term organic farming led to an increase of the capacity of the soil to sustain plant N nutrition.

Introduction

Long-term organic farming has been shown to increase soil microbial biomass and activity (Fliessbach and Mäder, 2000). This raised the question whether these changes in soil biology affect the fertiliser efficiency of applied animal manure. In a pot experiment, the hypothesis was tested that on a long-term organically managed soil (ORG) a greater portion of nitrogen (N) applied as cattle manure is mineralised and taken up by plants than on a conventionally managed soil that had received exclusively mineral fertilisers (MIN).

Materials and methods

The ORG and MIN soils were taken from a long-term field experiment conducted near Basel (Switzerland) since 1978 (Besson and Niggli, 1991). The ORG and MIN systems differed in the amount and form of applied fertilisers as well as the plant protection strategy. Crop rotation and soil tillage were the same during 19 years before samples were taken in 1997 (Fliessbach and Mäder, 2000). The ORG and MIN soils were mixed with ^{15}N labelled cattle manure and sown with *Lolium multiflorum* in a pot experiment. Total N uptake and N derived from fertiliser (Ndff) were determined in six cuts of the ryegrass (19, 36, 64, 96, 134, 162 days after sowing) and the real coefficient of utilisation of fertiliser N (RCUN) calculated (Ndff/applied N, in %; Jensen *et al.*, 1999). The ^{15}N labelled manure was obtained by feeding ^{15}N labelled hay to a cow and separately collecting faeces and urine (Powell and Wu, 1999; Kröber *et al.*, 2000). Thus the two manure components could be tested separately. Fertiliser treatments were faeces only (56 mg N_{tot}/kg soil) or an equal amount of faeces mixed with urine (slurry, 108 mg N_{tot}/kg soil). Controls were ^{15}N labelled NH_4NO_3 (101 mg N/kg soil; mineral N) and a treatment receiving no fresh N (non-fert). All the N was applied in one portion at the onset of the pot experiment. Other macro- and micronutrients were supplied at equal amounts to all treatments and re-added after different cuts to avoid limitations.

Results and discussion

Total N uptake of ryegrass was significantly higher on ORG than on MIN soil (Fig. 1) throughout the pot experiment in all fertiliser treatments and in non-fert. This difference was due to a higher supply of N from ORG compared to MIN soil (Ndfsoil, Fig. 1, 2).

Utilisation of applied N was lowest for faeces and highest for mineral N (Tab. 1), and was higher in this pot experiment than under field conditions (Jensen *et al.*, 1999). Utilisation of the applied manure was only slightly affected by the farming system history of the soils. Faeces N utilisation was slightly higher on ORG than on MIN and was enhanced by additional presence of urine N on both soils (Tab. 1). A greater immobilisation of urine N on ORG than on MIN was suggested by a lower RCU of urine N on ORG than MIN (Tab. 1). This compensated for the higher utilisation of faeces N resulting in a similar utilisation of total slurry N on both soils.

Soil microbial activity and biomass C and N were higher in ORG than MIN soils, both absolutely as well as in relation to the total and light fraction soil organic matter

C and N (Fliessbach and Mäder, 2000). These soil properties clearly had a lower impact on plant N uptake from fresh fertilisers than on the capacity of the soil to supply N (Fig. 1). Differences in soil N contribution between MIN and ORG increased with time (Fig. 2). About 81% of the total N uptake from the mineral N fertilisers occurred with the first cut, while for faeces and slurry N, these portions were around 47 and 67% (Fig. 3). Thus, the increased capacity of the ORG soil to supply N to plants became more important at later cuts when the significance of fertiliser N had declined. However, the total N uptake by the first cut when fresh fertiliser N was available was higher than the sum of N taken up during all subsequent cuts (Tab. 2) because yields of *Lolium multiflorum* markedly decreased after the first cut in all treatments and became low after the second cut (Fig. 2, Fig. 3).

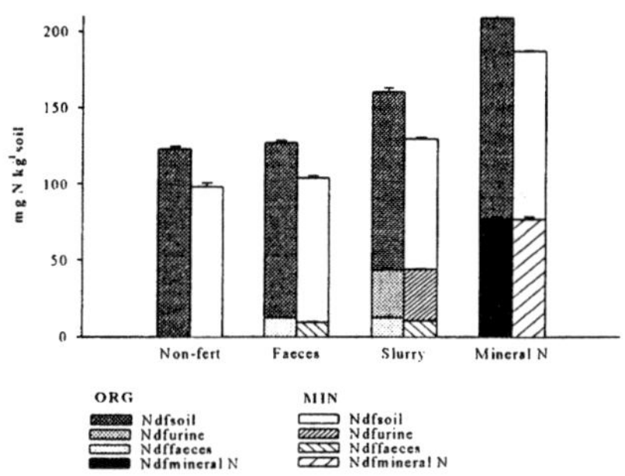

Figure 1. Total N uptake from fertiliser and soil N on soils of different farming systems. Mean and standard error of four replicates.

Table 1. Real coefficient of utilisation of N (RCUN) of manure components and mineral N during six cuts of *Lolium multiflorum* grown on the ORG and MIN soil.

RCUN	ORG	MIN	
	— % —		
Faeces	19	17	**
Slurry	41	41	ns
$Faeces_{slurry}$	22	18	***
$Urine_{slurry}$	62	66	*
Mineral N	76	76	ns

Values are means of four replicates. Differences between soils were significant at* P = 0.05 ** P = 0.01 *** P = 0.001

Table 2. Total N uptake by *Lolium multiflorum* of the first cut and of the total of the following five cuts in the different fertiliser treatments on the ORG and MIN soil.

	ORG		*MIN*	
Cut	*1.*	*2.-6.*	*1.*	*2.-6.*
Treatment	*mg N kg^{-1} soil*			
Non-fert	*72±0.7*	*51±1.1*	*63±2.8*	*35±0.7*
Faeces	67±0.9	59±1.2	60±0.8	44±0.6
Slurry	94±1.4	66±1.6	79±1.2	50±0.9
Mineral N	144±2.0	65±1.0	139±1.4	48±0.6

Means and standard errors of 4 replicates.

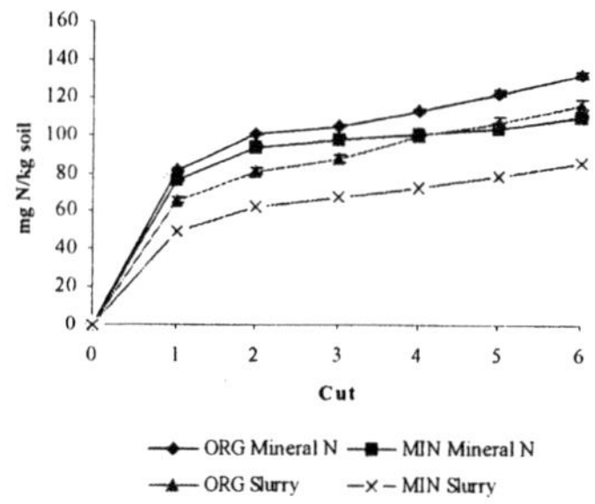

Figure. 2. Cumulative N uptake from the soil (Ndfs) by *Lolium multiflorum* in selected fertiliser treatments applied on the ORG and MIN soils. Mean and standard error of four replicates.

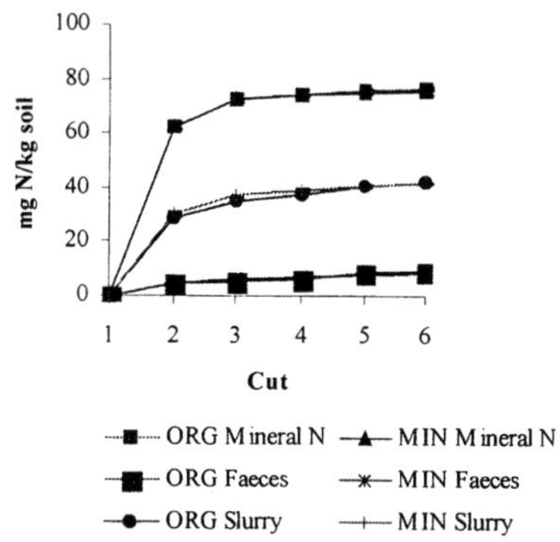

Figure 3. Cumulative fertiliser N uptake (Ndff) by *Lolium multiflorum* in different fertiliser treatments applied on the ORG and MIN soils. Mean and standard error of four replicates

Conclusions

Short term N fertiliser efficiency of manure was only slightly altered through long-term application of organic fertilisers in organically managed soils. However, the capacity of the soil to sustain plant N nutrition was enhanced by 19 years of organic farming practices compared to a long-term minerally fertilised soil. Quantitative and temporal patterns of this different soil N supply should be evaluated under field conditions to determine (i) whether the N release is synchronised with the demand of growing annual crops and (ii) to which extent soil N release can overcome N limitations. In addition, the potential risk of N losses in different farming systems has to be assessed.

References

Besson JM and Niggli U 1991 Schweiz. Landw. Forsch. 31, 79-109.

Fliessbach A and Mäder P 2000 Soil Biol. Biochem. 32, 757-768.

Jensen B, Soerensen P, Thomsen IK, Jensen ES and Christensen BT 1999 Soil Sci. Soc. Am. J. 63, 416-423.

Kröber TF, Külling DR, Menzi H, Sutter F and Kreuzer M 2000 J. Dairy Sci. 83, 2941-2951.

Powell JM and Wu Z 1999 Agron. J. 91, 814-818.

W. J. Horst et al. (Eds), Plant nutrition – Food security and sustainability of agro-ecosystems. 970~971, 2001

The role of N nutrition on lowland rice yields along an agroecological gradient in West Africa

M. Becker[1)], M. C. S. Wopereis[2)] and D. E. Johnson[3)]

[1)] *Institute for Agricultural Chemistry, Karlrobert Kreitenstr. 13, Bonn, D-53115, mathias.becker@uni-bonn.de*
[2)] *West Africa Rice Development Association, B.P. 2551, Bouake 01, Côte d'Ivoire*
[3)] *University of Greenwich, Long Ashton Experimental Station, Bristol, UK*

Key words: Nitrogen, N use efficiency, ORYZAS, *Oryza sativa*, weeds, yield gap

Abstract

Experiments in over 250 farmers' lowland rice fields of West Africa determined yield gaps and responsible management factors. Simulated potential yields increase from 6.8 Mg ha^{-1} in the forest zone to nearly 10 Mg ha^{-1} in the Sahel. Actual farmers' yields rarely exceed 1.8 Mg ha^{-1} in rainfed and 3.5 Mg ha in irrigated fields. Between 30 and 60% of the observed yield gap was related to N deficiency. Particularly water management, early weed control and timly N application increased N use efficiency with the largest impact in irrigated fields of the savanna zone.

Introduction

Rice in West Africa is grown all along the agroecological gradient between the humid forest zone to the Sahara desert margin. In this region, currently about 3 million ha are planted to lowland rice and the area is rapidly expanding (WARDA, 1996). Main lowland production systems comprise rainfed (~65%) and irrigated lowland rice (~20%) and in both, yields are far below world average. Nitrogen deficiency is wide-spread and speculated to be the main factor contributing to low yield levels (Windmeijer and Andriessen, 1993). Relating rice yield to quantitative information on N availability and agronomic practices will improve priority setting and technology targeting. Applicable technical options are likely to vary with the production system and agroecological zone.

Materials and methods

We determined spatial and temporal variability of N-related yield gaps in more than 257 rainfed and irrigated lowland rice fields across West Africa, from the high rainfall monomodal forest zone in the south (Becker and Johnson, 1999) over the savanna belt to the Sahel in the north (Wopereis *et al.*, 1999). Rice grain yield and N uptake were determined under farmers' management as well as in super-imposed researcher-managed subplots (no N input control treatment, farmers' N application plus early hand weeding at 28 days after seeding/transplanting and split-application of N at 60-120 kg ha-1, depending on site and cropping season with 50% given at panicle initiation stage). Yield gaps were calculated as differences between actual farmers' and researchers' rice yield and the potential simulated yields (ORYZA-S crop growth model; Dingkuhn and Sow, 1997). Observed yield variability was attributed to agronomic practices using multiple regression analysis.

Results and discussion

Simulated potential yield increased from 6.8 Mg ha^{-1} in the forest, over 9.1 Mg ha^{-1} in the savanna to 9.8 Mg ha^{-1} in the Sahel following increasing solar radiation (gross photosynthesis) and diurnal temperature amplitude (respiration losses). Accordingly, potential crop N requirements increased from 129 in the forest, 180 in the savanna, to 206 kg N ha^{-1} in the Sahel. Irrespective of the large potential, yields in farmers' fields averaged only about 1.7 Mg ha^{-1} in the rainfed (forest and savanna) and 3.5 Mg ha^{-1} in the irrigated fields (all environments). Accordingly, the yield gap was large and increased in irrigated fields from 3.3 Mg ha^{-1} in the forest to nearly 6 Mg ha^{-1} in the Sahel (Figure 1).

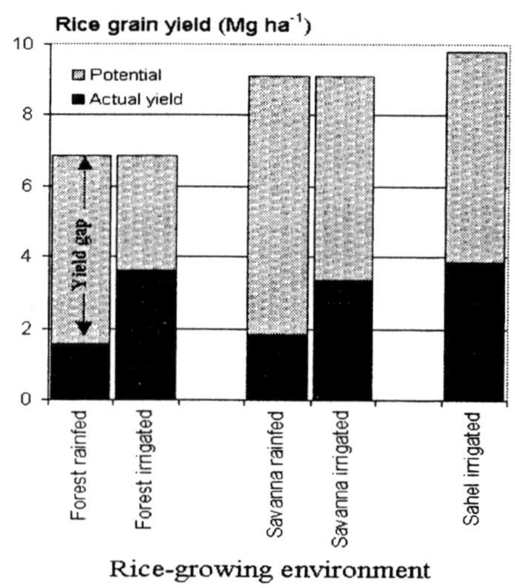

Figure 1. Yield gap between observed farmers' yield and simulated potential yield in West Africa

With a soil N supply of 46-78 kg N ha^{-1} (based on rice N uptake from no N input control plots) and mean fertiliser application rates in irrigated fields of 54-115 kg N ha^{-1}, considerably higher yield levels would have been expected. The agronomic efficiency of applied mineral N-NUE was generally very low (1-5 in rainfed and 5-12 in irrigated fields). An inefficient use of N appears thus to be a key factor associated with the large observed yield gaps. A cross-regional linear multiple regression analysis with environmental descriptors and agronomic crop management attributes indicated that despite a large diversity in environments, production strategies and agronomic management practices, only water control, timely weeding and N were significantly related to yield (r^2_{yield} = 0.65***) and NUE (r^2_{NUE} = 0.59***). Improved water management increased mean yields by 1.8 Mg ha^{-1} and mean NUE from 7.8 to 11.4 7 kg kg^{-1} N). Early weeding alone was associated with yield increases of 0.2 Mg ha^{-1} (P<0.003) and topdressing of half the N rate at panicle initiation stage with 0.5 Mg ha^{-1} (P<0.01).

A comparative analysis of farmers'- and researcher-managed plots is presented in Table 1. Improved weed control significantly increased yield and NUE only in the rainfed forest (by 0.4 Mg ha^{-1} or 6 kg kg^{-1} N) and the irrigated sahelian systems (by 1.0 Mg ha^{-1} or 8 kg kg^{-1} N). A timely application of farmers N rate increased yields more in irrigated (1.1 - 1.8 Mg ha^{-1}) than in rainfed systems (0.2 - 0.6 Mg ha^{-1}), with largest effects in the environments with the highest production potential (savanna and Sahel). It may be concluded that improved agronomic management, without higher N application rates, can close 11-17% of the yield gap in rainfed and 40-60% of the yield gap in irrigated systems. In relative terms, N management effects were largest in the forest (leaving 28% of unexplained yield gap), while absolute gains were largest in the Sahel (Figure 2).

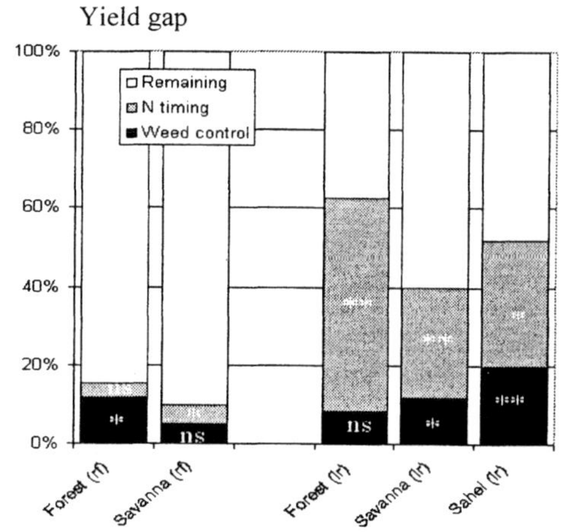

Figure 2. Differential management effects (timing of weeding and N application) on yield gaps in rainfed and irrigated lowland rice systems of West Africa.

References

Becker M and Johnson DE 1999 Field Crop Res. 60, 201-208.

Dingkuhn M and Sow A 1997 *In* Irrigated rice in the Sahel. Eds. KM Miezan, MCS Wopereis, M Dingkuhn, J Deckers and TF Randolph. pp. 361-379. WARDA, Bouaké, Côte d'Ivoire.

WARDA-West Africa Rice Development Association 1996 Rice trends in sub-Saharan Africa. WARDA, Bouaké, Côte d'Ivoire.

Windmeijer P and Andriesse W 1993 Inland valleys in West Africa. IRLI Publication No 53, Wageningen, The Netherlands

Wopereis MCS, Donavan C, Nebie B, Guindo D and N'Diaye MK 1999 Field Crops Res. 61, 125-145.

Table 1. Effect of the timing of weed control and N application on the agronomic efficiency of fertiliser N (NUE) in rainfed (n=102) and irrigated (n=157) farmers' fields along an agro-ecological gradient in West Africa. Values followed by the same letter are not significanly different by DMRT at 5%

	-----Rainfed rice------		------------ Irrigated rice ------------		
	Forest	Savanna	Forest	Savanna	Sahel
Farmer's N input (kg ha^{-1})	30	23	54	67	115
Researcher's N input (kg ha^{-1})	30	30	60	60-90 [a]	90-120 [a]
NUE (kg grain kg^{-1} N)					
Farmer's management	4 b	3 b	12 b	5 c	11 c
Improved weed control	10 a	5 b	15 b	10 bc	18 b
Timely nitrogen split	13 a	9 a	27 a	19 a	23 a

[a] lower rate in wet season, higher rate in dry season rice, following extension recommendations

W. J. Horst et al. (Eds), Plant nutrition – Food security and sustainability of agro-ecosystems. 972–973, 2001

Causes of legume-rotation effects in increasing cereal yields across the Sudanian, Sahelian and Guinean zone of West Africa

A. Bürkert[1)], M. Bagayoko[2)], S. Alvey[3)] and A. Bationo[4)]

[1)] *Institute of Crop Science, University of Kassel, Steinstr. 19, D-37213 Witzenhausen, Germany, buerkert@wiz.uni-kassel.de*

[2)] *Institut d'Economie Rurale (IER), B.P. 258, Bamako, Mali, Minamba.Bagayoko@ier.ml*

[3)] *Department of Environmental Sciences, University of California Riverside, Riverside, CA 92521, USA, smalv@mail.ucr.edu*

[4)] *TSBF-UNESCO, United Nations Complex, Gigiri, P.O. Box 30592, Nairobi, Kenya, a.bationo@cgiar.org*

Key words: acid phosphatase, mineral nitrogen, mycorrhizae, nematodes, pH, phosphorus mobilisation, rhizosphere

Abstract

On-farm experiments and pot trials were conducted on eight West African soils to explore the mechanisms governing the often reported legume rotation-induced cereal growth increases in this region. Crops comprised pearl millet (*Pennisetum glaucum* L.), sorghum (*Sorghum bicolor* Moench), maize (*Zea mays* L.), cowpea (*Vigna unguiculata* Walp.) and groundnut (*Arachis hypogaea* L.). In groundnut trials the observed 26 to 85% increases in total dry matter (TDM) of rotation cereals (RC) compared with continuous cereals (CC) in the 4th year appeared to be triggered by site- and crop-specific early season differences in nematode infestation (up to 6-fold lower in RC than in CC), enhanced Nmin and a 7% increase in mycorrhizal (AM) infection. In cowpea trials yield effects on millet and differences in nematode numbers, N_{min} and AM were much smaller. Rhizosphere studies indicated effects on pH and acid phosphatase activity as secondary causes for the observed growth differences between RC and CC. In the study region legume-rotation effects on cereals seemed to depend on the capability of the legume to suppress nematodes and to enhance early N and P availability for the subsequent cereal.

Introduction

Legume rotations have been advocated as a low input strategy to increase cereal yields on acid sandy soils of sub-Saharan West Africa that are notoriously low in phosphorus (P) and nitrogen (N). However, this view is based on a very limited number of comparative yield trials providing site-specific results (Bationo *et al.*, 1998; Nicou, 1978; Stoop and van Stavern, 1981). To critically assess the merits of legume rotations for different soils and crops of the region, information is needed about the mechanisms governing cereal yield increases.

Materials and methods

Field experiments with pearl millet, sorghum, maize, cowpea and groundnut were conducted from 1995 to 1999 under rainfed conditions at eight sites in Niger, Burkina Faso and Togo (Bagayoko *et al.*, 2000b). The sites provided a gradient of annual rainfall from 510 to 1300 mm with a corresponding range in topsoil (0-0.2m) pH from 4.1 to 5.6, in cation exchange capacity (CEC) from 0.8 to 3.3 $cmol_c\ kg^{-1}$, in mineral N (Nmin) from 5 to 18 mg kg^{-1}, in organic carbon from 1.5 to 6.5 g kg^{-1} and in clay content from 20 to 160 g kg^{-1}. The field experiments were used to record data of crop yields, root infection with arbuscular mycorrhizae (AM) and nematode infestation. Additionally, pot and root box experiments with CC and RC soil from two of the sites (Fada and Gaya) were set up to investigate soil-root interactions under controlled conditions.

Results and discussion

Legume-induced increases in cereal total dry matter (TDM) as recorded in the field experiments were site- and crop-specific, relatively consistent over years, but tended to grow over time. In the 4th trial year groundnut rotations led to average TDM increases in RC between 26 and 85% compared with CC. Respective increases in cowpea systems varied between –3 and 26% (Table 1).

Early season N_{min} levels in RC of cowpea sites tended to be higher than in CC but such differences with up to 114% at Fada (Bagayoko *et al.*, 2000b) and 89% at Koukombo (Alvey at al., unpublished) were only significant at groundnut sites. AM infection of millet roots was significantly higher in RC than in CC (Table 2) but differences declined over the growing season. The same was found for sorghum (data not shown). Regardless of the amount of applied P in the multi-factorial field experiment, nematode numbers were up to six-times lower in RC than in CC at Fada but no rotation effects on nematode infestation were noted at Gaya (Figure 1).

Under controlled conditions legume-rotation effects on cereal growth were even larger than in the field and similar effects on nematodes and AM infection were found (data not shown). Bagayoko *et al.* (2000a) reported from the field trials at Goberi that the pH at the rhizoplane of millet was 1.9 units higher than in the bulk soil at 45 days after sowing (DAS) and 1.2 units at 75 DAS. Respective differences were 0.6 units at 50 DAS and 0.4 units at 80 DAS at Gaya. At all sites, pH levels followed the pattern rhizoplane>rhizosphere>bulk soil. For sorghum at Kouaré, pH increases at the rhizoplane reached a maximum of 0.3

units, much smaller than in millet, and were not statistically significant (data not shown). Rotation-induced pH increases in the respective bulk soils were noted under controlled conditions at Fada (Figure 2) and to a lesser degree also at Koukombo and Kaboli (data not shown). These differences are still poorly understood, as are those in the rhizosphere, and may be due to particular properties of decomposing groundnut crop residues (Yan and Schubert, 2000).

Table 1. Legume-rotation effects on cereal total dry matter (TDM) in sub-Saharan West Africa. Data are means across P and N levels after three cropping cycles.

Site	Rain mm yr^{-1}	System	TDM kg ha^{-1}
Bani	510	Continuous millet	3080
		Millet after cowpea	3420 **
Sadoré	560	Continuous millet	3500
		Millet after cowpea	4310 ***
Kara	590	Continuous millet	4430
		Millet after cowpea	5580 ***
Goberi	600	Continuous millet	4220
		Millet after cowpea	4800 ***
Gaya	800	Continuous millet	2600
		Millet after cowpea	2510 ns
Fada	850	Continuous sorghum	2680
		Sorghum after cowpea	3630 ***
Kouk.	1100	Continuous maize	4990
		Maize after groundnut	6280 ***
Kaboli	1300	Continuous maize	3440
		Maize after groundnut	6350 ***

, * significant at P<0.01 and P<0.001; ns=not significant

Table 2. Root infection by mycorrhizae (AM) in millet (0-0.3m soil depth) as affected by cropping systems in a field experiment at Sadoré, Goberi and Gaya in 1997.

Cropping system	Sadoré 35 DAS[a]	Goberi 45 DAS	Gaya 50 DAS
	AM infection (% of roots)		
Continuous millet	23a[b]	27a	11a
Millet after cowpea	32b	48b	31b

[a] days after sowing; [b] significantly different at P < 0.001

Acid phosphatase activity at the root surface of sorghum was lower in rotation soil indicating improved P nutrition of seedlings and a subsequently smaller exudation of phosphatase from the root. In the rhizosphere, of rotation sorghum, however, an up to 74% higher phosphatase activity was noted (Figure 2). This was most likely the consequence of microorganisms that may have substantially contributed to the consistent increase of the measured P fractions in rotation soil compared with continuous soil (Alvey *et al.*, 2001).

The combination of (i) improved P availability through changes in soil pH, earlier AM infection, increased rhizosphere phosphatase activity and decreased parasitic nematode populations and (ii) higher levels of early season N_{min} likely explain the observed growth differences between RC and CC on the acid nutrient poor soils of sub-Saharan West Africa. Site-specific differences of such rotation effects likely depend on the legume component used (groundnut effects>cowpea effects), the ability of the cereal crop to cope with low nutrient availability (millet>sorghum>maize) and nematode infestation levels.

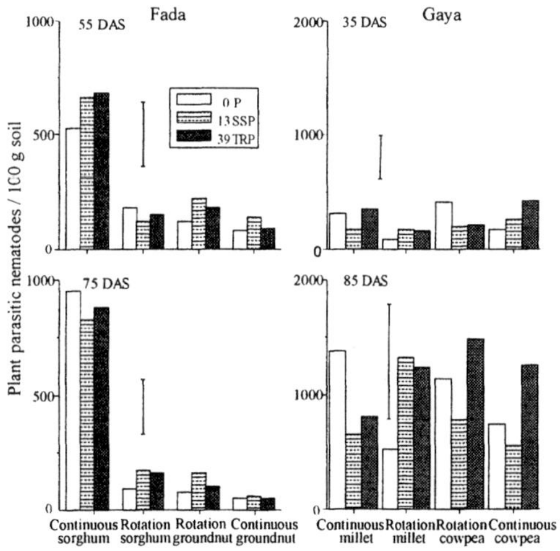

Figure 1. Effects of cropping system on nematodes without P, with annually 13 kg P ha^{-1} SSP and with 39 kg P once as Tahoua rockphosphate (TRP) at Fada and Gaya in 1998.

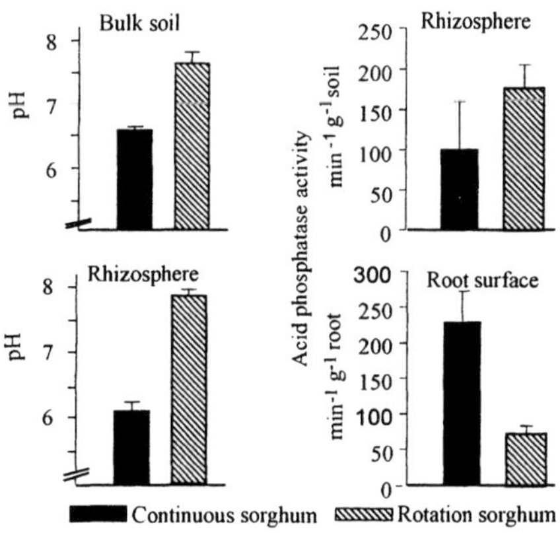

Figure 2. Effects of cropping system on pH and on acid phosphatase activity at the root of sorghum at Fada 37 days after sowing (after Alvey *et al.*, 2001).

References

Alvey S, Bagayoko M, Neumann G and Buerkert A 2001 Plant Soil (in press).

Bagayoko M, Alvey S, Buerkert A and Neumann G 2000a Plant Soil 225, 117-127.

Bagayoko M, Buerkert A, Lung G, Bationo A and Römheld V 2000b Plant Soil 218, 103-116.

Bationo A, Lompo F and Koala S 1998 Agric. Ecosyst. Environ. 71, 19-35.

Nicou R 1978 Agron. Trop. 33, 51-61.

Stoop WA and van Stavern JP 1981 *In* Biological Nitrogen Fixation Technology for Tropical Agriculture. Eds. PC Graham and SC Harris. pp. 653-657. CIAT, Cali, Colombia.

Yan F and Schubert S 2000 Plant Soil 220, 279-287.

W. J. Horst et al. (Eds), Plant nutrition – Food security and sustainability of agro-ecosystems. 974~975, 2001

Integrated approach for improved P nutrition of plants in tropical acid soils

G. Keerthisinghe[1)], F. Zapata[1)], P. Chalk[1)] and P. Hocking[2)]

[1)] *Joint FAO/IAEA Division of Nuclear Techniques in Food and Agriculture, International Atomic Energy Agency, Wagramer Strasse 5, Vienna, A 1400, Austria, G.Keerthisinghe@iaea.org*

[2)] *CSIRO Plant Industry, GPO Box 1600, Canberra, ACT 2601, Australia, P.Hocking@pi.csiro.au*

Key words: organic acids, phosphorus availability, rock phosphates, root exudates, soil acidity

Abstract

Phosphorus deficiency is a major constraint to crop production in the humid and sub-humid tropics, as most of the soils in these regions are inherently low in total P and possess a high phosphate sorption capacity resulting in extremely low availability of P. Thus, to provide an adequate supply of P to crops, high P inputs are needed, which is not an economically viable option for many farmers in developing countries. In this context, the use of crop species capable of secreting substantial quantities of organic acids from their roots is an effective strategy to mobilise P bound to Fe and Al in acid soils. Moreover, the organic acid-secreting crops increase the use efficiency of cheaper forms of P fertilisers such as unprocessed rock phosphates. The use of acid-tolerant, P-efficient genotypes in conjunction with the application of rock phosphate is proposed as a cost-effective strategy to develop more sustainable production systems on acid soils in the tropics, which is one of the main objectives in a newly initiated FAO/IAEA project on tropical acid soils.

Introduction

Phosphorus deficiency is a major constraint to crop production on tropical acid soils such as Oxisols and Ultisols which cover about 40% of the world's tropical area. Most of these soils have high phosphate sorption capacities, and thus only about 10-20% of applied P is taken up by plants in the year of application. Consequently, substantial P fertiliser inputs are required to sustain adequate food and fibre production on these soils. Unfortunately, these fertilisers have to be imported by most developing tropical countries using valuable foreign currency, and their supply to resource-poor farmers in rural areas is very restricted. Thus, the development of a strategy to enable crop plants to access sparingly soluble soil P reserves and to use cheaper forms of P fertiliser such as rock phosphate is highly desirable. In view of this, the Joint FAO/IAEA Division of Nuclear Techniques in Food and Agriculture has initiated a co-ordinated research project on the "Development of management practices for sustainable crop production systems on tropical acid soils." A major objective of this project is to identify P-efficient genotypes that can access fixed P in tropical acid soils through the exudation of organic acids and to integrate these plants into existing cropping systems.

Materials and methods

Phosphorus status of tropical acid soils

A total of 51 soil samples from 15 countries participating in the FAO/IAEA co-ordinated research project on "The use of nuclear and related techniques for evaluating the agronomic effectiveness of P fertilisers, in particular rock phosphates" were collected and analysed for their P status using a range of techniques including Bray-P, Olsen-P, Resin-P and Fe-strip-P. A ^{32}P exchangeable kinetic method was used to assess the dynamics of soil P, that is istopically exchangeable-P and the P-fixation capacity (Montange and Zapata, 2001). Representative data from 16 Ultisols and Oxisols collected from 7 countries are shown in Table 1.

Organic acid exudation from plant roots and uptake of P

The ability of P-deficient plants to exude substantial quantities of organic acids was demonstrated using white lupin (*Lupinus albus*) as an example (Keerthisinghe *et al.*, 1998). The effects of exuded organic acids on P uptake by different crop species were investigated using the L-value technique described by Hocking *et al.* (1997).

Results

The total P content of all the soils was extremely low, and 14 soils showed a total P content of less than 500 mg kg^{-1} (Table 1).

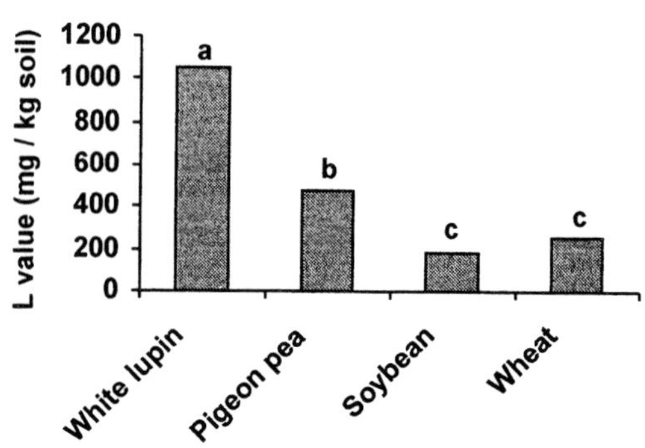

Figure 1. L-values for crop species grown in an Oxisol. Histograms with the same letter do not differ at P = 0.05 (Hocking *et al.*, 1997).

Most of these soils were also low in available P as measured with the Bray 2-P extractant (Table 1), with other P extraction techniques showing a similar trend. All soils showed a high P-fixation capacity; nine soils fixed more than 90% of the added P within one minute.

Table 1. Phosphorus status of selected Oxisols and Ultisols

Country	Location	Soil taxo nomy	Total P	Bray 2-P	P fixation capacity
			(mg kg^{-1})		(%)
Brazil	Piracicaba 1	Oxisol	171	18.0	90.0
Brazil	Piracicaba 2	Oxisol	165	19.1	98.3
China	Kunming	Ultisol	281	16.5	84.0
China	Huazhou	Ultisol	310	46.9	86.0
China	Yiyang	Ultisol	572	49.2	95.5
China	Jinhua	Ultisol	388	25.1	99.6
China	Yingtan	Ultisol	235	2.4	99.8
Chile	Metrenco	Ultisol	659	3.3	98.2
Ghana	Tikobo	Ultisol	138	12.9	63.0
Ghana	Ankasa	Oxisol	116	13.4	84.0
Ghana	Abenia	Oxisol	127	13.1	81.0
Ghana	Ayinasi	Oxisol	118	19.8	66.0
Ghana	Boi	Oxisol	226	21.2	84.0
Kenya	Kakamega	Ultisol	458	13.4	98.8
Malaysi a	Rengam	Ultisol	81	6.8	81.0
Vene zuela	El Pao	Ultisol	110	2.8	94.0

Hocking *et al.* (1997) demonstrated the ability of different crop species to access fixed soil P from an Oxisol using the L-value technique. The L-values for selected species (Fig. 1) clearly show that crops such as white lupin and pigeon pea (*Cajanus cajan*) accessed soil P that was unavailable to the other crops. The ability of these crops to acquire fixed soil P was attributed to exudation of organic acids from their roots.

Discussion

Low P availability and a high P fixation capacity are characteristic of weathered acid soils in the tropics. Soluble P fertilisers applied to these soils react readily with sesquioxides to form sparingly-soluble P forms. This results in a very low P concentration in the soil solution which is a major constraint to P uptake by plants. Although the problem can be partly rectified by the application of large quantities of P fertiliser, this is not a viable option to farmers in developing countries. Consequently, it is important to consider other options such as the use of organic-acid secreting crops in crop rotations or in multiple cropping systems to solubilise fixed P in tropical soils.

There is considerable evidence that indicates the importance of organic acid exudation from roots in the acquisition of fixed soil P. As mentioned above, some plants such as white lupin and pigeon pea have the ability to access fixed P and this is attributed to the exudation of organic acids into the rhizosphere. Under P-limiting conditions, white lupin exudes large quantities of citrate (Dinkelaker *et al.*, 1989; Keerthisinghe *et al.*, 1998), and pigeon pea responds by increased exudation of malonic and piscidic acids (Otani *et al.*, 1996). These organic acids increase the availability of P in acid soils, mainly by chelation of Al and Fe bound to P and by suppressing re-adsorption and precipitation of inorganic P (Gerke *et al.*, 1994). This is important as mobilisation of P due to organic acid secretion occurs in the rhizosphere, so that the solubilised P is directly available to plants. The use of P-efficient species as cover crops or in rotations and in multiple cropping systems could contribute to the P nutrition of less P-efficient crops by converting fixed P into more plant-available forms. Access to only a small proportion of the fixed inorganic P in many tropical soils using organic acid secreting plants could significantly reduce P fertiliser inputs as well as increasing the use-efficiency of P fertiliser, resulting in lower P inputs and more sustainable agricultural systems. Moreover, the inclusion of organic acid secreting crops may enable the use of cheaper, locally available forms of P fertiliser such as unprocessed rock phosphates. Preliminary studies conducted on a high P fixing Andosol in Chile have shown that P-efficient wheat (*Triticum aestivum* L.) and rapeseed (*Brassica napus* L.) cultivars can be grown profitably in combination with the application of locally available rock phosphate. Therefore this strategy has been proposed to promote more sustainable agricultural production in the region (Montenegro and Zapata, 2001; Pino *et al.*, 2001). However, the integrated approach of using P-efficient genotypes in combination with the application of rock phosphate needs to be further tested to determine whether it can increase and sustain agricultural production on a range of tropical acid soils.

References

Dinkelaker B, Römheld V and Marschner H 1989 Plant Cell Environ. 12, 285-292.

Gerke J, Römer W and Jungk A 1994 Z. Pflanzenernaehr. Bodenk. 157, 289-294.

Hocking PJ, Keerthisinghe G, Smith FW and Randall PJ 1997 *In* Proceedings 18th International Plant Nutrition Colloquium Eds. T Ando *et al.* pp. 305-308. Kluwer Academic Publishers, Dordrecht.

Keerthisinghe G, Hocking PJ, Ryan PR and Delhaize E 1998 Plant Cell Environ. 21, 467-478.

Montange D and Zapata F 2001 IAEA-TECDOC (In Press).

Montenegro A and Zapata F 2001 Nutrient Cycling Agroecos. (In press).

Otani T, Ae N and Tanaka H 1996 Soil Sci. Plant Nutr. 42, 553-560.

Pino I, Zapata F, Navia M, Luzio W and Parada AM 2001 IAEA-TECDOC (In Press).

W. J. Horst et al. (Eds), Plant nutrition – Food security and sustainability of agro-ecosystems. 976~977, 2001

Root turnover and nutrient cycling in native and introduced pastures in tropical savannas

I. M. Rao, C. Plazas and J. Ricaurte
Centro Internacional de Agricultura Tropical (CIAT), A. A. 6713, Cali, Colombia, i.rao@cgiar.org

Key words: carbon, nitrogen, nutrient cycling, phosphorus, root biomass, root length, root turnover

Abstract

The study was carried out at the Carimagua Research Station of CORPOICA-CIAT in the Eastern plains (Llanos) of Colombia. Measurements of root production and turnover were made on two introduced (9-year-old) pastures (grass alone - *Brachiaria dictyoneura* CIAT 6133; and grass + legume pasture - *B. dictyoneura* + *Arachis pintoi* CIAT 17434) compared to a native pasture. Annual root production (biomass and length) was significantly greater in the introduced pastures compared with the native pasture. Although root biomass and root length turnover rates were similar among native and introduced pastures, the greater amounts of live roots in the introduced pastures contributed to strongly superior root turnover and N and P cycling from the introduced pastures compared to that of the native pasture.

Introduction

Root production, turnover and decomposition are key processes in the carbon and nutrient dynamics in agroecosystems (Aerts *et al.*, 1992; Long *et al.*, 1989; Jackson *et al.*, 1997; Smucker, 1990; Vogt *et al.*, 1998). Root production very often exceeds shoot production and thereby determines to a large extent the input of organic matter and nutrients into the soil. A common approach to measuring root turnover is sequential sampling of root mass and estimating root production by adding increments of standing stocks of living and dead roots (Vogt *et al.*, 1998). Assuming steady state conditions for the living root system, the rate of root death in a soil equals root production. Previous research showed that the average live root biomass of the introduced pastures was significantly larger than that of native pastures and this strongly superior root production in introduced pastures resulted in a substantial storage of carbon in the soil (Fisher *et al.*, 1994; Rao, 1998). But the extent of root turnover in introduced pastures compared to native pastures remains unknown. The aim of this study was: (i) to determine differences in root turnover among native and introduced pastures; and (ii) to assess the impact of root turnover on nutrient cycling, particularly of N and P.

Materials and methods

The study was carried out at the Carimagua Research Station of CORPOICA-CIAT in the Eastern plains (Llanos) of Colombia. Measurements of root production and turnover were made on two introduced (9-year-old) pastures (grass alone – *Brachiaria dictyoneura* CIAT 6133; and grass + legume pasture – *B. dictyoneura* + *Arachis. Pintoi* CIAT 17434) and compared to a native pasture. Both introduced pastures were managed with rotational grazing system (a 4-paddock system with 7/21 days for both wet and dry seasons).

Root sampling involved collection of replicate soil cores from 0 to 80 cm depth (Rao, 1998), washing with a hydroneumatic elutriation system (Gilson's Inc., Benzonia, Michigan, USA) and sorting of live and dead roots from the soil matrix. Sampling occurred in all three pastures with monthly intervals between May 1995 to May 1996. Root production (P) per sampling interval was calculated depending on the direction of change in standing stocks of living (L) and dead (D) roots: a) if $dL > 0$ and $dD > 0$, then $P = dL + dD$; b) if $dL > 0$ and $dD < 0$ then $P = dL$; c) if $dL < 0$ and $dD > 0$ then $P = dL + dD$ or $P = 0$ (in case of a negative P); if $dL < 0$ and $dD < 0$ then $P = 0$ (Aerts *et al.*, 1992). Annual estimates were calculated by summing the estimates of all sampling intervals within the year. Assuming steady state conditions for the root system on an annual basis, root biomass (dry) turnover rate (yr^{-1}) was calculated by dividing mean root biomass production by average living root biomass. Root length turnover rate (yr^{-1}) was calculated by dividing mean root length production by average living root length.

Results and discussion

Analysis of variance showed significant effects of sampling date ($P < 0.05$) and pasture type (native *versus* introduced) on living root biomass and root length. Significant seasonal fluctuations were observed for both living root biomass and length. The larger rainfall during June-July and October-November caused significant amounts of root turnover during the rainy season. Annual root production (biomass and length) was significantly larger in the introduced pastures compared with the native pasture (Tables 1 and 2). Root biomass and root length turnover rates were similar among native and introduced pastures. However, the increased amounts of live roots in the introduced pastures could contribute to the much superior root turnover from the introduced pastures compared to that of the native pasture (Table 1 and 2).

The processes of production, death and decomposition of roots are well known to be simultaneous and counteracting (Vogt *et al.*, 1998). Observed standing live or dead roots are the end products of these processes. Inferring root production from changes in standing live and dead roots alone does not account for processes of production and turnover. A certain degree of error must therefore be expected with estimations of primary production of roots based on repeated measurements of live and dead roots. Two major factors contribute to this error: (1) dead roots are difficult to sort out from total root samples; and (2) some of the dead root material is decomposed between samplings.

Table 1. Root biomass production, decomposition and rate of turnover for native and introduced pastures on a clay loam Oxisol at Carimagua, Colombia.

Pasture	Total Annual Root Biomass Production kg ha^{-1}	Total Annual Root Biomass Decomposition kg ha^{-1}	Root Biomass Turnover Yr^{-1}
Bd	6117	5249	2.29
Bd + Ap	5661	5595	2.23
Savanna	2408	1845	2.24

Bd = *Brachiaria dictyoneura* CIAT 6133
Ap = *Arachis pintoi* CIAT 17434

Table 2. Root length, decomposition and rate of turnover for native and introduced pastures on a clay loam Oxisol at Carimagua.

Pasture	Total Annual Root Length Production km m^{-2}	Total Annual Root Length Decomposition km m^{-2}	Root Length Turnover yr^{-1}
Bd	40.7	20.5	2.61
Bd + Ap	24.8	12.4	1.80
Savanna	17.5	12.1	2.40
$LSD_{0.05}$	2.3		

Bd = *Brachiaria dictyoneura* CIAT 6133
Ap = *Arachis pintoi* CIAT 17434

Studies on net primary productivity of natural grass ecosystems in the tropics have shown that these ecosystems are far more productive than had previously been estimated (Long *et al.*, 1989). Accounting for the turnover of shoot and root material at only 0-20 cm soil depth resulted in net primary productivities that were up to to five times higher than those obtained through standard procedures recommended by the International Biological Programme. Below-ground estimations ranged from 610 to5680 kg/ha. Annual root turnovers, estimated as production divided by mean biomass, were between 1.7 and 4.0. Results from the present work that extended to 0 to 80 cm soil depth indicated that the extent of root turnover could be around 2.3 in natural and introduced tropical pastures while net production of roots may be 2 to 3 times larger in introduced pastures.

The larger amounts of standing live roots of introduced pastures could contribute to a more spontaneous cycling of nutrients and sequestration of soil carbon upon root death and decomposition (Rao, 1998). In contrast, a lower contribution to nutrient cycling and carbon sequestration is expected from NS pasture because of its low below-ground root biomass. Changes through time in sizes of nutrient pools in roots indicated that the amount of N cycled through roots of introduced legume-based pastures could be as much as 34 kg ha^{-1} y^{-1} compared to 29 kg ha^{-1} y^{-1} in grass alone and 16 kg ha^{-1} y^{-1} in native savanna pastures. Cycling of P via root turnover from introduced pastures could be five to six times larger than those of native savanna pasture: 7.2, 5.8 and 1.2 kg ha^{-1} y^{-1} in grass alone, in grass + legume and in native savanna pastures, respectively.

Agroecosystems that incorporate the beneficial features of introduced pastures, such as extensive root growth and turnover and permanent above-ground cover, may thus contribute to reduced nutrient losses by storing large quantities of nutrients in the below-ground biomass, augmenting soil organic matter and protecting the soil. Such beneficial features of introduced pastures might usefully be incorporated into the design of agropastoral systems appropriate for the South American savannas, where the potential for nutrient loss rates through leaching and erosion can be significant.

References

Aerts R, Bakker C and De Caluwe H 1992 Biogeochemistry 15, 175-190.

Fisher MJ, Rao IM, Ayarza MA, Lascano CE, Sanz JI, Thomas RJ and Vera RR 1994 Nature 371, 236-238.

Jackson RB, Mooney H and Schulze ED 1997 Proc. Natl. Acad. Sci. USA 94, 7362-7366.

Long SP, Garcia Moya G, Imbamba SK, Kamnalrut A, Piedade MTF, Scurlock JMO, Shen YK and Hall DO 1989 Plant Soil 115, 155-166.

Rao IM 1998 *In* Root Demographics and their Efficiencies in Sustainable Agriculture, Grasslands and Forest Ecosystems. Ed. JE Box. pp. 19-41. Kluwer Academic Publishers, Dordrecht.

Smucker AJM 1990 Remote Sensing Reviews 5, 237-248.

Vogt KA, Vogt DJ and Bloomfield J 1998 *In* Root Demographics and their Efficiencies in Sustainable Agriculture, Grasslands and Forest Ecosystems. Ed. JE Box. pp. 687-720. Kluwer Academic Publishers, Dordrecht.

W. J. Horst et al. (Eds), Plant nutrition – Food security and sustainability of agro-ecosystems. 978~979, 2001

Contribution of above- and below-ground interactions to intercropping

F. S. Zhang[1], L. Li[1] and J. H. Sun[2]

[1] The Key Laboratory in Plant Nutrition, MOA, Department of Plant Nutrition, China Agricultural University, Beijing, 100094, China, zhangfs@mail.cau.edu.cn

[2] Institute of Soil and Fertiliser, Gansu Academy of Agricultural Sciences, Lanzhou, 730070, China

Key words: above-ground, below-ground, border row effect, intercropping, interspecific interaction, nutrient uptake

Abstract

Wheat/maize and wheat/soybean intercropping are the common cropping systems for high yield in the irrigated area of northwest part of China. A root barrier technique was employed to determine the relative contribution of above- and below-ground interaction to yield and nutrient uptake in the intercropping systems. Land equivalent ratios (LER) of grain yield were 1.21-1.58 and 1.23-1.26 for wheat/maize intercropping and wheat/soybean intercropping, respectively. Grain yields of wheat located at border row increased by 74% and 53% over those at inner rows when intercropped with maize and soybean, respectively. Correspondingly, N uptake by the wheat at the border row increased by 109% and 53%. There were 98% and 54% increases of P uptake by wheat located at border row compared with those at inner rows. Potassium uptake showed a similar trend as for N and P uptake for the two intercropping systems. Out of the yield increase, 47 and 30% was attributed to above-ground interactions, and 27 and 23% attributed to below-ground interactions between two species in wheat/maize and wheat/soybean intercropping, respectively. The contribution of above-ground and below-ground interactions to the increase of N uptake by wheat were 50 and 59%, respectively, for wheat/maize intercropping and 23 and 20% respectively, for wheat/soybean intercropping. Similarly, contribution of above-ground and below-ground to the increase of P uptake were 56 and 42%, respectively, for wheat/maize intercropping, and 26 and 28%, respectively, for wheat/soybean intercropping.

Introduction

Intercropping has been practiced in many parts of the world (Francis, 1986). It is an environmentally friendly and sustainable approach to increase crop production in agricultural systems (Stuelpnagel, 1993 Exner, 1999). Wheat/maize and wheat/soybean intercropping are common cropping systems for high yield in irrigated fields of north-west China. Yield benefits from intercropping become more and more recognized, especially in China. However, research on mechanisms underlying the yield improvement by intercropping is limited. The objective of this study was to determine the relative contribution of above-ground and below-ground interactions between intercropped species to the intercropping advantage of grain yields of wheat/maize and wheat/soybean intercropping systems.

Materials and methods

The root barrier technique was employed to determine the relative contribution of above-ground and below-ground interactions to yield advantage and N, P and K uptake in a wheat/maize and wheat/soybean intercropping system. The experiments had two treatments: a) plastic sheet barrier and b) no barrier for roots. The barriers were inserted between two intercropped species with 0.7 m depth and 1 m length in micro-plots at two field sites. The study was conducted in 1997 at two sites. Baiyun site was (38°37'N, 102°40'E) located 15 km north of Wuwei city, Gansu province at an altitude of 1504 m above the sea level. Annual mean temperature was 7.7°C. Cumulative temperatures above 0°C and 10°C were 3646°C and 3149°C, respectively. The frost-free period was 170-180 days. Total solar radiation was 5988 MJ m^{-2} $year^{-1}$. Annual precipitation was 150 mm, and potential evaporation was 2021 mm. Jingtan site was (37° 05' N, 104° 40' E) located in Jinyuan county, Gansu province, at an altitude of 1645 m above sea level. Annual mean temperature was 6.6 °C and accumulated temperatures above 0 °C and 10 °C were 3208 °C and 2622 °C, respectively. The frost-free period was 160-170 days. Total solar radiation was 6162 MJ m^{-2} $year^{-1}$. Annual precipitation was 200-250 mm, and potential evaporation was 2369 mm. The soils at both sites were classified as an Aridisol.

Results

Land equivalent ratios

Land equivalent ratios (LER), an index of showing intercropping advantage, of grain yield were 1.21-1.58 and 1.23-1.26 for wheat/maize intercropping and wheat/soybean intercropping, respectively. There were significant yield advantages for both intercropping systems.

Border row effects of intercropped wheat

Grain yields of wheat located at the border row increased by 74% and 53% compared with the inner rows when intercropped with maize and soybean, respectively (Table 1). Correspondingly, N uptake by the wheat at the border row increased by 109% and 53% (Table 2). There were 98% and 54% increases of P uptake by wheat located at border row compared with those at inner rows (Fig. 1). Potassium uptake showed a similar trend as for N and P uptake for the two intercropping systems.

Table 1. Contribution of above-ground and below-ground to border row advantage in grain yield

Intercropping system	Yield and border effect	Above- and below-ground[1)]	Above-ground only[2)]	Below-ground only[3)]
Wheat/ maize	Border row yield (g/m of row)	128	107	
	Inner row yield (g/m of row)	74	73	
	Border row over inner row (%)	74	47	27
Wheat/ soybean	Border row yield (g/m of row)	134	118	
	Inner row yield (g/m of row)	87	91	
	Border row over inner row (%)	53	30	23

1) No root barrier was inserted between roots of both species.
2) Root barrier was inserted between intercropped species.
3) The difference between 1) and 2).

Table 2. Contribution of above-ground and under-ground to border row advantage of N uptake

Intercropping system	Yield and border effect	Above- and below-ground[1)]	Above-ground only[2)]	Below-ground only[3)]
Wheat/ maize	Border row yield (g/m of row)	3.72	2.86	
	Inner row yield (g/m of row)	1.78	1.90	
	Border row over inner row (%)	109	50	59
Wheat/ soybean	Border row yield (g/m of row)	4.13	3.40	
	Inner row yield (g/m of row)	2.91	2.77	
	Border row over inner row (%)	42	23	19

1), 2) and 3) are the same as Table 1.

Relative contribution of above-ground and below-ground interactions to border row effect

Out of the yield increase, 47 and 30% was attributed to above-ground interaction, and 27 and 23% attributed to below-ground interaction between two species in wheat/maize and wheat/soybean intercropping, respectively. The contributions of above-ground and below-ground interaction to the increase of N uptake by wheat were 50 and 59%, respectively, for wheat/maize intercropping and 23 and 19% respectively, for wheat/soybean intercropping (Table 1). Similarly, contribution of above-ground and below-ground to the increase of P uptake were 56 and 42%, respectively, for wheat/maize intercropping, and 26 and 28%, respectively, for wheat/soybean intercropping (Fig. 1). Clearly, the Contributions of above-ground and below-ground to nitrogen, phosphorus and potassium uptake advantage of intercropped wheat were equally important in the intercropping systems.

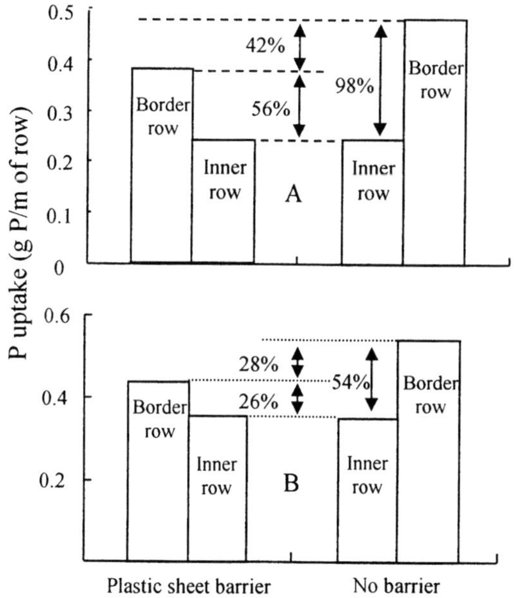

Figure 1. Contribution of above-ground and below-ground to the border effect on P uptake by the intercropped wheat with maize (A.) and soybean (B.) Interspecific below-ground interactions were minimized with the plastic sheet root barrier for roots. No root barrier allowed interspecific below-ground interaction and above-ground interactions.

Conclusions

Significant yield advantages of wheat/maize and wheat/soybean intercropping were observed, with 1.21 – 1.58 and 1.23 – 1.26 of land equivalent ratios (LER). For wheat intercropped with maize and soybean, yields of the border rows were higher than that of the inner rows by 74% and 53%, respectively. Nutrient acquisition by the plants in the border row were also higher than that in inner rows. Out of the yield increase in wheat, more than 1/3 of the increase came from below-ground interactions. The relative contribution of below-ground interactions to increase in N and P acquisition were about 1/2 for wheat intercropped with maize and soybean.

Acknowledgement

We are grateful to the Major State Basic Research Development Program of the People's Republic of China (G1999011707) and the National Natural Science Foundation of China (39790100 and 30070450).

References

Exner DN, Davidson DG, Ghaffarzadeh M and Cruse RM 1999 Am. J. Altern. Agric. 14, 69-77.

Francis CA 1986 *In* Multiple Cropping Systems. Ed CA Francis. Macmillan, New York.

Stuelpnagel R 1993 Proceedings of the International Crop Science Congress. Ames, Iowa.

W. J. Horst et al.(Eds.), Plant nutrition – Food security and sustainability of agro-ecosystems. 980~981, 2001

Organic matter and its relation to maize crops on acid soils of Colombia

L. A. Rojas[1)], J. E. Baquero[2)], M. Ramírez[1)], F. Rodriquez[1)] and G. Roveda[1)]

[1)] Colombian Corporation of Agricultural Research, CORPOICA A.A. 240142 Bogotá , Colombia, lrojas@corpoica.org.co

[2)] Colombian Corporation of Agricultural Research, CORPOICA A.A. , A.A. 3129, Villavicencio-Meta, Colombia, jbaquero@corpoica.org.co

Key words: aluminium, lime, maize cultivars, organic matter, phosphorous, soil acidity, *Zea mays*

Abstract

To examine genotype by environment interactions with respect to the adaptation of maize (*Zea mays* L.) cultivars on acid soils of the tropics characterised by aluminium toxicity and low availability of nitrogen and phosphorus, a long term field experiment was carried out on an Oxisol of the Eastern plains of Colombia. Treatments applied were as follows: two sources of organic matter (OM; chicken manure and cowpea shoots at 5 t ha^{-1} each), two phosphorus rates (25 and 50 kg P ha^{-1}), two levels of limestone (0 and 1.5 t ha^{-1}) and three maize varieties with differential tolerance to soil acidity. The experimental results showed that grain yields remained fairly stable with time at all treatment levels. The application of OM combined with lime and P produced the highest yield increment compared to the unamended control. The maize cultivar Sikuani by far out-yielded the other cultivars in the unamended control treatment. Particularly the soil acidity-sensitive cultivar ICA had very low yields in the check treatment. On the other hand, cultivar ICA had the largest grain-yield response to lime, P and OM applications followed by the cultivars Clavito and Sikuani. Application of lime produced the greatest yield increase in cultivar ICA. Application of OM further significantly enhanced maize yields, whereas the effect of Cowpea green manure was rather small. Mean grain yields over the five seasons at the highest level of soil amendments (combination of lime, P and OM) were 4843, 3879, and 4741 kg ha^{-1} for ICA, Clavito, and Sikuani, respectively. With combined lime/organic matter application the available P and the exchangeable aluminium in the soil changed from 6 to 32 mg kg^{-1} and from 2.1 to 0.2 $cmol_c$ kg^{-1}, respectively, whereas the Al, Ca, and Mg saturation percentages changed from 73 to 6%, 10 to 35%, and 3.7 to 9%, respectively.

Introduction

The Eastern Plains (Llanos Orientales) of Colombia comprise an area of about 26 million hectares. The most suitable area for agriculture has an extension of about 4.6 million hectares. Maize has been a suitable crop for both, large and small farmers of this region because it can be used by humans and animals at a low level of mechanisation and industrial processing (Valencia, 1992). However, the main consequences of low soil fertility, such as clay minerals with low cation exchange capacity (CEC), high acidity, high Al, low organic matter (OM) content, and low availability of plant nutrients, particularly of phosphorus (P; Salinas, 1984), require the application of lime and mineral fertilisers thereby increasing the production costs for maize in that region. The introduction of soil acidity-tolerant maize cultivars like ICA-Sikuani V-110 (Sikuani) has allowed to reduce lime applications, providing a partial solution to soil acidity as a restriction to crop growth. However, this approach did not solve the P fixation problem. It is also unclear whether the long-term use of soil acidity-tolerant and/or P-efficient cultivars will enhance soil acidification or allow sustainable crop production on these soils. The main objective of this research was to evaluate agronomic practices, such as the application of lime, mineral P and OM on soil Al and P availability in a long term experiment with different maize cultivars.

Materials and methods

The field experiments were carried out on a Tropeptic Haplorthox soil from the piedmont of the Eastern acid plains of Colombia, at 320 m.a.s.l., with a mean temperature of 27° C and a relative humidity of 70-90%. The soil characteristics are shown in Table 1. Treatments were as follows: three organic matter rates (no OM, chicken manure and cowpea shoots at 5 t ha^{-1} each, incorporated into the soil), two levels of P (25 and 50 kg P ha^{-1} applied as calcium superphosphate), two limestone levels (0 and 1.5 t ha^{-1}) and three maize varieties with a different Al resistance (Sikuani, an improved soil acidity-tolerant cultivar), Clavito (a local cultivar) and ICA V-109 (an acid-soil sensitive cultivar, ICA). The experimental design was factorial with randomised complete blocks of 36 treatments and 4 replications totalling 144 plots. A basic fertilisation was performed by adding N, K, Ca, Mg, Zn and B (at levels of 100, 70, 60, 30, 1.3 and 0.3 kg ha^{-1}) to the soil. Maize was grown during five consecutive semesters, from 97a season through 99a season. In each year soil chemical parameters were determined at the anthesis stage of the crop and grain yield was measured at harvest.

Results and discussion

The results did not show any effect of the varieties on soil characteristics during the five semesters of maize cropping (Table 1). It was observed that individual treatments of lime and organic matter (especially chicken manure) decreased the exchangeable Al in the soil diminishing therefore both, soil acidity and Al saturation percentage. The latter parameter decreased from 72 to 4.2%. The combination of lime and chicken manure led to a decrease of exchangeable Al to undetectable levels.

Table 1. Soil chemical parameters of the experimental soil over time.

Time	pH	OM %	Al	Ca	Mg	K	Na	CEC
					cmol$_c$ kg$^{-1$}			
I.T.[1)]	4.5	1.8	2.1	0.4	0.15	0.09	0.13	2.90
A.T.[2)]	5.2	1.9	0.2	1.1	0.29	0.17	0.14	3.15

	Al	Ca	Mg	K	P	sand	silt	Clay
		Saturation %		mg kg^{-1}			%	
I.T.[1)]	73	10.0	3.7	2.2	6.0	50	30	20
A.T.[2)]	6	35.0	9.0	5.4	32.0			

[1)] I.T.: Initial time at the onset of the experiment
[2)] A.T.: After five growing seasons and combined treatment applications

Lime and organic matter increased the exchangeable Ca and Ca saturation percentage from 14 to 36% and of Mg from 3.7 to 9%. According to ICA (1992), the adequate mean values of these parameters in the soil should be 30-50% for Ca and 15-25% for Mg. CEC increased from 2.90 to 3.15 cmol$_c$ kg^{-1}.This may still be classified as very low. However, the colloidal complex of the soil became saturated mainly with Ca and Mg instead of toxic Al. Lime (1.5 t ha^{-1}) and P application (50 kg ha^{-1}) did not increase plant available soil P. Phosphorus availability was increased with either chicken manure (with or without lime) or cowpea with lime: soil P concentration increased from 6 to 32 mg kg^{-1} after five semesters of treatment application.

Possibly due to their fast mineralisation, the organic materials used in the present study did not lead to increases in soil organic matter, however, the decrease in Al and the increase of exchangeable bases and P led to an overall increase in soil fertility.

Mean grain yields of the three varieties across all treatments were in the order Sikuani > ICA > Clavito with values of 3586, 3220 and 2645 kg ha^{-1}, respectively. Cultivar Sikuani by far out-yielded the other cultivars in the unamended control ('check') treatment (data not shown). Particularly the soil acidity-sensitive cultivar ICA had very low yields in the check treatment (Fig. 1). On the other hand, cultivar ICA had the largest grain-yield response to lime, P and OM applications followed by the cultivars Clavito and Sikuani.

Application of lime produced the greatest yield increase in cultivar ICA. Application of OM further significantly enhanced maize yields, whereas the effect of cowpea green manure was rather small. Mean grain yields over the five seasons at full soil correction (combination of lime, P and OM applications) were 4843, 3879, and 4741 kg ha^{-1} for ICA, Clavito, and Sikuani, respectively. This clearly reflects the superiority of the improved *versus* the local cultivar (Clavito) and confirms that breeding for soil acidity-tolerance did not lead to a loss in productivity at the high input level (Sikuani)

Overall, the results clearly show that the breeding and use of soil acidity-tolerant cultivars in combination with moderate rates of lime, P and organic matter application represent an important component of the sustainable use of the acid soils of the Eastern plains of Colombia for maize production. Further research is needed to better understand the complex interactions between genotype, organic matter, and climate on plant nutrient uptake and soil chemical and physical parameters.

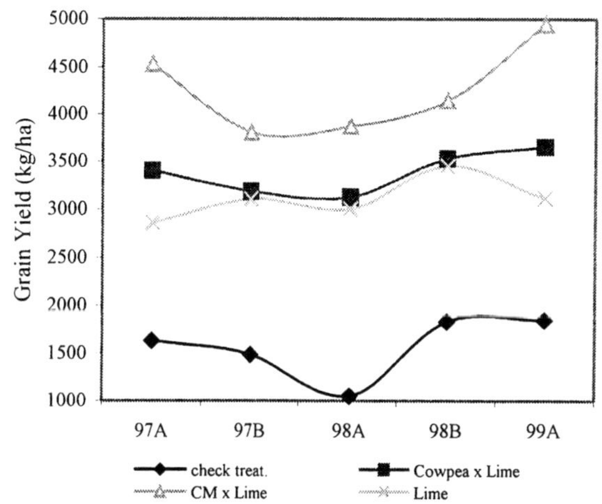

Figure 1. Effect of organic matter and lime applications on the grain yield of soil acidity-sensitive maize cultivar ICA V-109.

References

ICA 1992 Technical Assistance Manual No 25. 64pp.

Salinas J 1984 Fertilidad de suelos - Diagnostico y Control. pág 397-418. Colombia.

Valencia RA 1992 Revista ICA Colombia. 27, 255-268.

Interaction effects of organic materials and lime on grain yield and nutrient aquisition of three maize varieties grown in an Oxisol of the Colombian eastern plains

J. E. Baquerol and L. A. Rojas

Colombian Corporation of Agricultural Research, A.A. 3129, Villavicencio-Meta, Colombia, jhbernal@villavicencio.cetcol.net.co

Key words: Colombian eastern plains, lime, maize, organic matter, oxisols.

Abstract

An experiment was conducted in order to study the effects of some interactions resulting from the addition of organic materials (Cowpea residues and chicken manure) and two levels of lime, (0 and 1500 kg ha^{-1} dolomite lime) on the grain yield and nutrient acquisition of three maize genotypes with contrasting genetic background. The experiment design was a randomised block with factorial arrangement and four replications. Grain yield increased in treatments that received either 5t ha^{-1} cowpea, residues 5 t ha^{-1} chicken manure, or 1500 kg ha^{-1} dolomite lime. Yield improvement was higher in the Al sensitive variety, (ICA-V 109) than in the Al tolerant variety, (Sikuani-V110). Treatments that received only chicken manure had grain yield slightly higher than treatments with 1500 kg ha^{-1} of lime. There was a positive effect on grain yield and nutrient acquisition in treatments both, organic materials and dolomite lime. Leaf concentration of N, P, Ca, Mg and K were higher with the incorporation of cowpea residues, chicken manure and lime. The Al tolerant variety Sikuani V-110 had higher leaf concentrations of Al, N, P, Mn and Fe. The results indicated that the varieties which are more efficient in nutrient uptake under aluminum stress had a good chance to grow well in acid soils low in nutrients. Furthermore, the results also confirmed the importance to preserve and/or increase the soil organic matter in agricultural production systems which is particularly recommended for the acid soils of the Eastern Plains of Colombia.

Introduction

Soil acidity, high Al saturation and low soil fertility are the major soil constraints for agricultural production in the savannas of the Eastern Plains of Colombia (Blamey *et al.*, 1990). Limestone incorporation and genotype tolerance to Al are some of the strategies to overcome soil acidity (Valencia and Leal, 1996). The use of organic matter is another way to diminish the toxic effect of Al in plants and also to increase the availability and uptake of mineral nutrients in plants (Hargrove and Thomas, 1981; Rodriguez, 1998).

Materials and methods

The study was conducted in a fine, kaolinic soil, isohypertermic tropeptic haplorthox of the Eastern Plains of Colombia. Two sources of organic matter, chicken manure and cowpea residues, at two rates, (0 and 5 t ha^{-1}), two lime levels, (0 and 1.5 t ha^{-1}) and three maize varieties were evaluated in a randomised complete block design with factorial arrangement of treatments. Maize genotypes with contrasting genetic background were use in this study: Sikuani V-110 (Al tolerant variety), Clavito (traditional maize cultivar) and ICA V 109 (Al susceptible variety) Grain yield was obtained by harvesting three main rows from each plot. For analysis of nutrient concentrations ear leaves were collected at initial silk stage.

Results

A positive interaction was observed for grain yield when 5 t ha^{-1} of chicken manure or cowpea residues were incorporated into the soil together with 1.5 t ha^{-1} of dolomite lime (Tab. 1). The grain yield increase was more pronounced in the Al susceptible variety ICA V 109 than in the Al tolerant genotype and the traditional cultivar. The Al susceptible variety ICA V 109 had the highest grain yield (4481 kg ha^{-1}) with applications of 5 t ha^{-1} of chicken manure and 1.5 t ha^{-1} of dolomite lime. The increases of grain yield of maize varieties were higher with the incorporation of 5 t ha^{-1} of chicken manure than with the application of 1500 kg ha^{-1} of dolomite lime (Tab. 1).

Genotypical differences were found in Al and P concentrations of maize leaves. The Al tolerant cultivar Sikuani V 110 had five times higher leaf Al concentrations than the Al susceptible variety ICA V 109 and the traditional cultivar Clavito (Tab. 2). Leaf P concentration were higher in maize grown with 5 t ha^{-1} of chicken manure. No significant increase was found in leaf P concentrations with the addition of lime or cowpea residues (Tab 3). Leaf P concentrations in all varieties ranged between 3.2 mg g^{-1} and 6.1 mg g^{-1}. Similar as with P, the Al tolerant variety had higher N, Mn and Fe than the Al susceptible and the traditional cultivar.

Discussion

The results from this study indicate that the Oxisols of the Eastern Plains of Colombia could be used successfully for agriculture production if the soil chemical constraints have been resolved. The positive interaction between dolomite lime and chicken manure increased the grain yield of the maize genotypes and improved the uptake of mineral nutrients by the plants. Similar results have also been found by Ahmad and Tan (1986) and Estrada et al. (1995).

The Al tolerant genotypes Sikuani V 110 is highly efficient in uptake and utilization of mineral nutrients under Al stress. Therefore, this genotype had a great advantage in growth and production in acid soils low in mineral nutrients like the Oxisol of the Eastern Plains. The high leaf Al concentration in the Al tolerant genotype indicates that genotypic differences among species in the ability to tolerate high amounts of toxic Al in the soil should be chosen (Taylor 1987).

The results also confirm the importance to preserve and/or increase the soil organic matter content of agricultural production systems which is highly recommend-ed for the acid soils of the Eastern Plains of Colombia.

References

Ahmad F and Tan KH 1986 Soil Sci. Soc. Am. J. 50, 656-661.

Blamey F, Wheeler D, Edwards D and Christie R 1990 J. Plant Nutr. 13, 713-728.

Estrada G, Diaz AL and Figueiredo AF 1995 Suelos Ecuatoriales 25, 57-60. Bogotá Colombia.

Hargrove WL and Thomas GW 1981 *In* Chemistry in the Soil Environment, American Society of Agronomy and Soil Science of America. pp. 151-165. Wisconsin USA.

Rodríguez HF, Lozano Y and Rojas LA 1998 Trabajo de Grado, Departamento de Química, Universidad Nacional de Colombia: Santa fe de Bogotá D.C.

Taylor GJ 1987 J. Plant Nutr. 10, 1213-1222.

Valencia R and Leal D 1996 Suelos Ecuatoriales 26, 35-44.

Table 1. Grain yields of three maize genotypes (kg ha^{-1}) as affected by application of organic matter and lime.

Variety	Not Organic matter		Cowpea residues		Chicken manure	
	Without lime	1.5 t ha^{-1} lime	Without lime	1.5 t ha^{-1} lime	Without lime	1.5 t ha^{-1} lime
Sikuani	3723	3796	3820	4027	4122	4333
Clavito	1734	2541	2748	2669	3103	3331
ICA V109	2201	2988	3262	3402	3762	4481

C.V= 7.8. r^2= 0..91. L.S.D.=106

Table 2. Concentration of Al in leaves (mg g^{-1}) as affected by application of organic matter and lime.

Variety	Not Organic matter		Cowpea residues		Chicken manure	
	Without lime	1.5 t ha^{-1} lime	Without lime	1.5 t ha^{-1} lime	Without lime	1.5 t ha^{-1} lime
Sikuani	500	580	770	400	420	690
Clavito	70	30	70	120	30	30
ICA V109	80	40	60	20	80	80

C.V= 177.6. r^2= 0..57. L.S.D.=230

Table 3. Concentration of P in leaves (mg g^{-1}) as affected by application of organic matter and lime.

Variety	Not Organic matter		Cowpea residues		Chicken manure	
	Without lime	1.5 t ha^{-1} lime	Without lime	1.5 t ha^{-1} lime	Without lime	1.5 t ha^{-1} lime
Sikuani	4.6	4.4	4.4	4.6	5.6	6.1
Clavito	3.2	3.6	3.5	3.7	5.0	4.0
ICA V109	4.0	4.2	3.9	4.3	5.2	5.4

C.V= 13.4. r^2= 0.74. L.S.D.=0.30

W. J. Horst et al.(Eds.), Plant nutrition – Food security and sustainability of agro-ecosystems. 984~985, 2001

Three years performance of a tolerant and a susceptible maize cultivar on non-amended and amended acid soil

C. The[1)], H. Calba[2)], W. J. Horst[3)] and C. Zonkeng[1)]

[1)] *Institute of Agricultural Research for Development (IRAD), P.O. Box 2067 Yaoundé, Cameroon. THE@ICCNET.Cm*

[2)] *CIRAD, TA 40/01 Avenue d'Agropolis 34 398 Montpellier, Cedex 5, France. Henri.calba@cirad.Fr*

[3)] *Institute of Plant Nutrition, Department of Horticulture, University of Hannover, Herrenhäuser Str. 2, D-????? Hannover, Germany, Horst@mbox.pflern.uni-hannover.de*

Key words: aluminium toxicity, acid-tolerant cultivar, *Cassia septabilis*, chicken manure, lime, phoshorus

Abstract

A long term agronomic experiment was conducted from 1996 to 2000 at Ebolowa, Cameroon. On the acid soil with high Al supply maize grain yield of the soil acidity-tolerant cultivar ATR-SR-Y was 61 % higher compared to the sensitive cultivar Tuxpeno sequia. The annual application of 60 kg P ha^{-1} for three consecutive years did not significantly increase the grain yield of the soil acidity-tolerant cultivar. Except in the first year lime addition resulted in a significant increase of grain yield of the tolerant (82%) and particularly the susceptible cultivar (208%). This corresponded to a significant decrease in exchangeable Al and to a significant increase in pH and exchangeable Ca and Mg contents of the soil. The application of chicken manure or green manure (*Cassia septabilis* leaves) significantly increased maize grain yield. These increases in yield were partly attributed to an increase of exchangeable Ca and available P of the soil with chicken manure and to a decrease of exchangeable Al with the application of *Cassia septabilis* leaves.

Introduction

Maize (*Zea mays* L.) is grown on approximately eight million hectares of acidic soils (Brewbaker, 1985; Pandey and Gardner, 1992). On these soils maize yield is reduced due to Al or Mn toxicities, and because of Ca, Mg, P and Mo deficiencies (Aldrich et al; 1975, Granados et al; 1993).

In tropical Africa, acid soils cover 29% of the continent (Eswaran *et al.*, 1997). To meet Sub Saharan African food security by the year 2050, Jacques du Guerny (1997) estimated that food production should be increased seven-fold as compared to the 1995 level. This implies the development of strategies for advanced, resource-friendly, sustainable, and economic production systems on acid soils including the introduction of improved acid soil-tolerant germplasm, the amelioration of soil acidity using phosphorus, lime and/or organic amendments.

A long term agronomic experiment was conducted from 1996 to 2000 both in Ebolowa, Cameroon and in the Eastern Acid plains of Colombia with the objective of assessing the relative advantages of the use of acid soil-tolerant maize cultivars compared to the use of fertilisers and organic manure to correct soil acidity.

Material and methods

The ATR-SR-Y maize cultivar, acid soil-tolerant and developed by maize breeders of Cameroon, and the sensitive cultivar Tuxpeno sequia from Mexico were used.

Soil amendments per year were as follows: phosphorus at O and 60 kg ha^{-1}, lime at 0 and 2 t ha^{-1}, chicken manure at 0 and 4 t ha^{-1} and green manure (*Cassia septabilis* leaves) at 0 and 4 ton ha^{-1}. Treatments were arranged as a randomised complete block design with six replications and a plot size of 6 x 6 m, consisting of eight rows plot^{-1}, each 6 m long. Plant density was 5,3333 plants ha^{-1}. In all plots additional fertiliser application occurred as follows: 100 kg ha^{-1} N, 24 kg ha^{-1} P_2O_5 and 14 kg ha^{-1} K_2O according to local recommendations.

Soil data were collected at the beginning of the experiment and after three consecutive years of soil amendments. Soil samples were analysed at CIRAD, Montpellier, France.

Results and discussion

At the beginning of the experiment, soil chemical analysis revealed that the pH (H_2O) ranged from 4.43-4.65 and exchangeable Al ranged from 1.57 µmol kg^{-1} to 2.47 µmol kg^{-1}. The Mg content varied from 0.09 µmol kg^{-1} to 0.38 µmol kg^{-1}.

After three consecutive years of soil amendments and maize cultivation, none of the soil characteristics had changed significantly as compared to the control plots.

On acid soil (control plots) the acid soil-tolerant cultivar exhibited a 61% higher grain yield than the susceptible cultivar. Soil amendments generally significantly increased maize grain yields by 5% and 42% for the tolerant and the susceptible cultivar, respectively.

Phosphorus application did not change pH or Mg. However, exchangeable Al and H^+ tended to increase, and the available (Olsen) P content of the soil increased in plots planted with the Al-sensitive but P-efficient cultivar Tuxpeno sequia.

Except for the first year of application (1997), the lime effect was highly significant and increased with the duration of the experiment. Lime application alone led to a 82% and 208% yield increased compared to the unlimed control for the tolerant cv ATP-SR-Y and the susceptible cv Tuxpeno sequia, respectively. The combination of chicken manure with lime gave an additional 50% and

40% more grain yield compared to the lime application alone for the tolerant and the susceptible maize cultivar, respectively. These results were partially explained by the decrease in Al exchangeable in the soil and the increase in Ca, Mg and pH.

Chicken manure application alone produced 73% and 15% more grain yield than the control for both the tolerant and susceptible cultivar, respectively. These yield increases were attributed to the increased availability of soil Ca and P as the pH of the soil was not affected.

The effect of green manure on grain yields was positive and significant except for the 1997 cropping season and the tolerant cultivar. *Cassia septabilis* leaf application was more beneficial to the susceptible cultivar (86% increase in grain yield compared to a 27% yield increase for the tolerant cultivar). The positive effect of green manure application could be related primarily to a decrease in exchangeable Al.

Lime application resulted in a reduction of exchangeable Al allowing a more efficient uptake of N and P absorption particularly for the sensitive cultivar. This observation is in agreement with the findings of Raij and Quaggio (1997). Plots planted with the susceptible cultivar had higher levels of available P than control plots. This is in agreement with results by EMBRAPA, Brasil (personal communication) stating that Tuxpeno sequia is Al-sensitive but P-efficient. Application of chicken manure resulted in a decrease in exchangeable Al which confirms the findings of Keltjens (1997). The lack of a significant corresponding pH increase was also observed by Zekeng (1992). The slow rate of pH change compared to lime suggested that acid soil correction with manure might take a longer time than the three years investigated in this study.

The results clearly show that for soil acidity-sensitive maize cultivars organic amendments readily available to small farmers can at least partly substitute for lime application. However, at this site in Cameroon, lime application was necessary and highest yields were only achieved when lime and chicken manure were combined. The use of soil acidity-tolerant cultivars can greatly reduce the need for lime application and thus contribute to the overall sustainability of cropping systems.

Acknowledgement

The financial support by EU-INCO is highly appreciated.

References

Aldrich SR, Scott WO and Leng ER 1975 Modern corn production. A & L Publication Champaign, IL.

Eswaran H, Reich P and Beigroth F 1997 *In* Plant soil interactions at low pH. Eds. Moniz AC *et al.* pp. 159-164. Brazilian Soil Science Society, Campinas, Brasil.

Granados G, Pandey S and Cebalos H Crop Sci. 33, 936-940.

Keltjens WG 1997 *In* Plant soil interactions at low pH. Eds. Moniz AC *et al.* pp. 109-117. Brazilian Soil Science Society, Campinas, Brasil.

Raij, B and Quaggio JA 1997 *In* Plant soil interactions at low pH. Eds. Moniz AC *et al.* pp. 205-214. Brasilian Soil Science Society, Campinas, Brasil.

Zekeng P 1992 The effect of teperature and soil acidity on the decomposition rate of manure. 112 P. Thesis for the degree of Ph.D. Michigan State University.

Table 1. Grain yield of two maize cultivars in the third year and soil characteristics after three consecutive years of soil amendment

Treatments	Soil acidity-tolerant cv ATR-SR-Y							Soil acidity-sensitive cultivar Tuxpeno sequia						
	Grain yield $kgha^{-1}$	pH H_2O	Exchangeable cations Al	H	Ca	Mg	Avail. P	Grain yield kg ha^{-1}	pH H_2O	Exchangeable cations Al exch.	H^+	Ca	Mg	Avail. P
Control	2918	4.54	2.32	0.16	0.42	0.13	22.00	1282	4.51	2.25	0.17	0.64	0.14	17.66
Phosphorus	2895	4.45	2.47	0.19	0.43	0.14	21.20	2420	4.47	2.45	0.19	0.54	0.12	27.16
Lime	4720	4.77	1.30	0.11	1.21	0.73	18.52	5410	5.38	0.70	0.05	1.49	0.84	21.25
Chicken manure	5245	4.61	1.61	0.18	0.78	0.21	31.39	5080	4.54	2.20	0.20	0.72	0.16	29.45
Cassia sept.	3164	4.75	1.96	0.24	0.57	0.18	22.15	3333	4.90	1.76	0.19	0.61	0.18	22.50

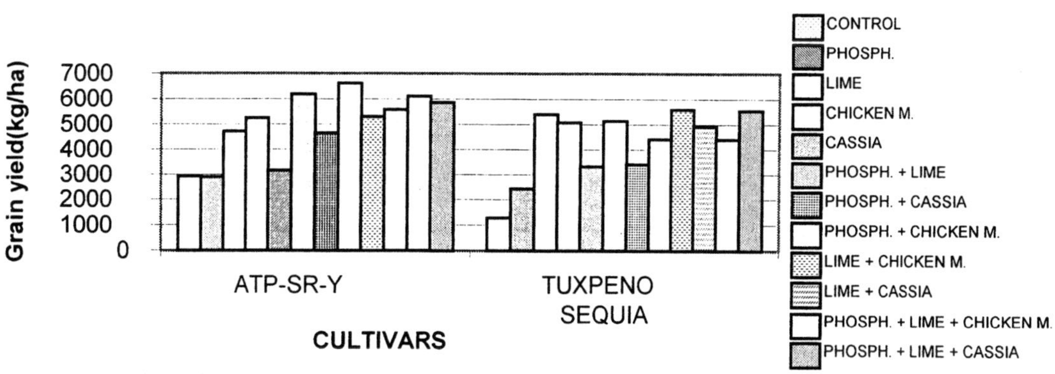

Figure 1. Grain yield of two maize cultivars as affected by phosphorus, lime, chicken manure, *Cassia septabilis* leaves application after 3 consecutive years of soil amendments

W. J. Horst et al.(Eds.), Plant nutrition – Food security and sustainability of agro-ecosystems. 986~987, 2001

The effect of bio-compost application on crop yield and nitrogen dynamic in the soil

R. Schulz[1)], H. Al-Najar[1)], J. Breuer[2] and V. Römheld[1)]

[1)] *Institute of Plant Nutrition (330), University of Hohenheim, 70593 Stuttgart, Germany, Schulzru@.uni-hohenheim.de*

[2)] *Institute of Agricultural Chemistry (710), University of Hohenheim, 70593 Stuttgart, Germany*

Key words: bio-compost, nitrogen mineralisation, nitrogen immobilisation, nitrogen dynamic

Abstract

Since 1997 bio-compost with and without optimised mineral N fertilisation is applied to maize, winter wheat and summer barley (crop rotation) in a randomised field experiment on a Luvisol in south-west Germany. The aim of this experiment is to evaluate the effect of bio-compost application on crop yield and N-net mineralisation in soil. In the bio-compost application treatment (100 kg total N ha^{-1} a^{-1}, that is about 7.5 t DM) with optimised mineral N fertilisation yield of summer barley (1999) and maize (2000) was higher than in the optimised mineral N treatment without bio-compost application. Similar results were obtained with the higher bio-compost application rate (400 kg total N ha^{-1} a^{-1}, that is about 30 t DM) without additional mineral N fertilisation. The yield increase can be attributed to an improvement of soil structure in the Luvisol. During the vegetation period of summer barley net-N-mineralisation on the field plots of the bio-compost treatments with additional mineral N fertilisation was lower than in plots without additional mineral N application. The higher Nmin values after harvest (Sept. 1999) in bio-compost treatments with additional mineral N-fertilisation can be partially attributed to the remineralisation of immobilised N at the end of the growth period.

Introduction

The aim of utilising municipal and industrial organic wastes in agriculture is to foster more sustainable recycling systems in which waste products are reused in agriculture as secondary organic fertilisers (Schulz and Römheld, 1997). Compost application can improve soil structure and nutrient supply to plants and thus reduce the input of mineral fertilisers. On the other hand repeated applications of high amounts of compost can result in an accumulation of organically bound N in the soil and thus in an increased risk of nitrate leaching after mineralisation and nitrification.

Materials and methods

A field experiment was started in 1997 as randomised block design (4 replications) on a loess-derived Luvisol in south-west Germany (mean annual temperature of 8.5°C; 685 mm rainfall) with the following treatments: (1) control treatment without N fertilisation, (2) optimised mineral fertiliser application (110 kg N ha^{-1} as NH_4NO_3), (3) 100 kg total N ha^{-1} as bio-compost with additional 'optimised' mineral N fertilisation (125 kg N ha^{-1}), (4) 400 kg total N ha^{-1} as bio-compost with additional optimised mineral N (100 kg N ha^{-1}), (5) 100 kg total N ha^{-1} as bio-compost without additional mineral N, (6) 400 kg total N ha^{-1} as bio-compost without additional mineral N.

The bio-compost applied was fresh and sieved (20 mm) with C:N ratio ranging from 14 to 24. Optimisation of mineral N fertilisation depended on the results of Nmin analyses and nitrate tests in plants (Wollring and Wehrmann, 1990; Geyer and Marschner, 1990). Net-N-mineralisation was calculated according to the following equation: N mineralised = N uptake by the plant + Nmin (after harvest) - (N min before cultivation + N fertiliser application) (Mengel *et al.*, 1999). Crop rotation was maize (1997), winter wheat (1998), summer barley (1999) and maize (2000).

Results and discussion

Compared with control, the yield of barley (1999) and maize (2000) increased by mineral fertiliser and bio-compost application (Tab. 1), which was mainly attributed to an improved N supply. The most striking effect, however, was the yield increment of up to 57% by bio-compost application at optimised mineral N fertilisation in comparison with optimised mineral N fertilisation without bio-compost application. This yield increase appeared to be mainly caused by improved soil structure and not by an enhanced nutritional status of the plants (Al-Najar *et al.* 2000).

As shown in the N balance for summer barley (Tab. 2) the bio-compost treatments with high application rates had a high N surplus of 300 - 450 kg N ha^{-1}.

During the growth period of barley the apparent net-N-mineralisation increased with the amount of bio-compost applied to the soil. But in both compost treatments with additional mineral N fertilisation net-N-mineralisation was distinctly lower (50 kg N ha^{-1}) than that without additional mineral N (Tab. 2). In treatment (2) with mineral N fertilisation (without compost) N was immobilised. As there were no indications of N losses during the growth period, it can be concluded from these results that the broadcast applied mineral fertiliser N was preferentially immobilised by microorganisms in the upper soil layer. Similar results were found in field experiments with ^{15}N labelled fertiliser application by Peschke and Mollenhauer (1998) and by Nieder *et al.* (1993), who found an immobilisation of fertiliser N up to 35 kg N ha^{-1} during the growth period of winter wheat. As shown by Kwast *et al.* (1994) immobilisation of mineral fertiliser N is decreased by the application of manure.

N immobilisation increased with N supply at beginning of the growing period (Engels and Kuhlmann, 1992). The immobilised N was partly remineralised at the end of the growth period (Peschke and Mollenhauer, 1998) and thus increased the Nmin content after harvest. In September 1999 Nmin data in the compost treatments with additional optimised mineral fertilisation were higher (20-30 kg N ha^{-1}) than those of the corresponding bio-compost treatments without additional mineral N fertilisation (Table 2).

The decrease of Nmin content in the upper soil layer (0-30cm) and the increase in deeper layers (60-90) during the winter season (Sept. 1999 to April 2000) indicate nitrate leaching of mineralised N in the bio-compost treatments. The increase of the total Nmin content in the soil shows that the mineralisation potential and thus the risk of nitrate leaching increases with the number of repeated applications of high amounts of bio-compost.

References

Al-Najar H, Schulz, R, Breuer J and Römheld V 2000 VDLUFA Schriftenreihe 53, (in press).

Engels T and Kuhlmann H 1993 Z. Pflanzenernähr. Bodenk. 156, 149-154.

Geyer B and Marschner H 1990 Z. Pflanzenernähr. Bodenk. 153, 341-348.

Kwast A, Klasink A and Bischoff R 1994 VDLUFA Schriftenreihe 38, 159-162.

Mengel K, Schneider B and Kosegarten H 1999 J. Plant Nutr. Soil Sci. 162, 139-148.

Nieder R, Kersebaum KChr, Widmer P and Richter J 1993 Z. Pflanzenernähr. Bodenk. 156, 293-300.

Peschke H and Mollenhauer S 1998 Z. Pflanzenernähr. Bodenkd. 161, 9-15.

Wollring J and Wehrmann J 1990 Z. Pflanzenernähr. Bodenk. 153, 47-53.

Schulz R and Römheld V 1997 J. Soil Sci. Plant Nutr. 43, 1051-1056.

Table 1. Yield of summer barley (1999) and maize (2000) (BC=Bio-Compost)

Treatment	summer barley (1999) dry matter (t ha^{-1}) grains	straw	maize (2000) dry matter (t ha^{-1})
(1) Control (without fertilisation)	1.37 ± 0.09[1]	1.37 ± 0.11	11.6 ± 1.2
(2) Mineral N fertilisation (optimised)	3.10 ± 0.11	2.30 ± 0.14	14.5 ± 1.4
(3) BC (100 kg N ha^{-1}a^{-1}) + optimised mineral N fertilisation	3.70 ± 0.12	2.52 ± 0.15	16.4 ± 0.8
(4) BC (400 kg N ha^{-1}a^{-1}) + optimised mineral N fertilisation	4.88 ± 0.21	3.05 ± 0.17	17.6 ± 1.1
(5) BC (100 kg N ha^{-1}a^{-1})	2.38 ± 0.10	1.62 ± 0.08	14.6 ± 1.4
(6) BC (400 kg N ha^{-1}a^{-1})	4.32 ± 0.20	2.54 ± 0.15	17.9 ± 1.5

[1] Numbers indicate one standard deviation (4 replications)

Table 2. Nitrogen balance (kg N ha^{-1}) of summer barley (1999), N-net mineralisation and Nmin. For a description of treatments see Table 1.

Treat-ment	N-Input: Nmin March 1999 0-90 cm (0-30)	Mineral fertiliser	Bio-compost	Total	N-Output: N-uptake (grain + shoot)	N-Surplus	Net-N-minerali-zation	Nmin Sep. 1999 0-90 cm (0-30)	Nmin April 2000 0-90 cm (0-30)
(1)	47 (12)	-	-	47	34 ± 3[1]	13	46 ± 11	59 (30)	64 (27)
(2)	53 (15)	110	-	163	88 ± 6	75	-23 ± 14	52 (25)	71 (23)
(3)	56 (14)	125	100	281	95 ± 6	186	8 ± 3	94 (45)	79 (28)
(4)	72 (18)	100	400	572	119 ± 8	453	60 ± 10	113 (60)	154 (43)
(5)	52 (14)	-	100	152	43 ± 4	109	56 ± 15	65 (35)	74 (27)
(6)	64 (17)	-	400	464	81 ± 5	383	107 ± 21	90 (50)	98 (34)

[1] Numbers indicate one-standard deviation (4 replications)

W. J. Horst et al.(Eds.), Plant nutrition – Food security and sustainability of agro-ecosystems. 988~989, 2001

Agricultural application of municipal solid waste compost in the Gaza Strip: *in situ* measurement of nutrient and heavy metal leaching

A. Kaschl[1)], E. Neumann[1)], Y. Chen[2)] and V. Römheld[1)]

[1)] *Institute of Plant Nutrition (330), University of Hohenheim, Stuttgart, 70593, Germany, plantnut@uni-hohenheim.de*
[2)] *Department of Soil and Water, Faculty of Agricultural, Food and Environmental Quality Science, Hebrew University of Jerusalem, P.O. Box 12, Rehovot, Israel, yonachen@agri.huji.ac.il*

Key words: copper, chloride, groundwater quality, monitoring boxes, nitrate, potassium, preferential flow

Abstract

The vertical displacement of nutrients and heavy metals after application of municipal solid waste (MSW) compost was studied on a field plot in the Gaza Strip. Resin-filled monitoring boxes were installed under the root zone during the rainy season. Leaching rates between replicates varied considerably, indicating the importance of preferential flowpaths in this sandy soil. Copper was the only heavy metal which eluted in higher quantities from the topsoil due to the addition of MSW compost to the soil. Among the macronutrients an indication of higher leaching rates were only found for potassium, whereas nitrate concentrations in the boxes remained unaffected by compost addition. No imminent risk of groundwater contamination was discernible.

Introduction

In the Gaza Strip overpopulation and adequate waste disposal is an urgent problem. The local production of MSW compost and its agricultural use would constitute a sensible way of practising sustainable agriculture at the same time as it reduces the amount of generated waste to deposit, helps to improve soil quality and decreases the necessary amounts of mineral fertilisers. However, due to strong rainfalls concentrated in the winter months (300-350 mm on average), the abundance of coarse, sandy soils with little organic matter and a very shallow groundwater depth (1-2m), increased leaching of pollutants from the MSW compost might endanger groundwater quality. To explore these phenomena leaching rates under field conditions were determined in this study.

Materials and methods

A long-term field experiment was established in 1996 in Gaza to investigate the agricultural application of MSW compost. Up to 65 t dry weight ha^{-1} of MSW compost were applied yearly in September. Due to the lack of source separation the heavy metal levels of the compost were very high, exceeding the limits adopted by many European countries (Amlinger, 1996; Kaschl *et al.*, 1998). The soil of the field plot was a typical calcareous sandy soil with high pH (90% sand, 1.5% organic matter, pH_{H2O}=8.5, 6.1% $CaCO_3$). In the winter season 98/99 and 99/2000 'monitoring boxes' filled with a cation/anion resin mixture (Merck Ion exchanger Amberlite IRN-150, Darmstadt, Germany) were installed in the control plots receiving no fertilisation and in the plots receiving the highest compost charges. Twelve boxes (replicates) were installed in each of the two treatments. The monitoring boxes were placed just under the root zone with the soil above the boxes left undisturbed (Fig. 1). After the recovery of the boxes at the end of the rainy season, the resin material inside the boxes was divided into layers α, β and γ (from top to bottom). Nitrate (NO_3^-) and chloride (Cl^-) were measured colorimetrically and with a silver electrode, respectively, after extraction with 1M Na_2SO_4 from the resin. Cations were quantified by ICP-OES in a 2M HNO_3 extracting solution.

Results and discussion

Cl^- is a conservative tracer of little interaction with soil components and can therefore be used to monitor water fluxes. Large differences between individual boxes suggested a significant spatial variability between leaching rates (Fig. 2), with the bulk of the vertical transport taking place along high-velocity flowpaths ('fingers') in the soil. No indications of upward flow as a result of high theoretical evaporation rates were found. For the 99/2000 growing season, 287 kg Cl^- ha^{-1} a^{-1} (average of twelve repetitions) were leached from the root zone in the control plots, whereas 388 kg Cl^- ha^{-1} a^{-1} leached in the compost-amended plots (both α and β-layers combined). These values are high but can be explained by the proximity to the seashore and the consequent atmospheric Cl^- input via dry and wet deposition from the sea. Moderate losses of NO_3^- were recorded, with no significant difference between compost-amended plots (74 kg ha^{-1} a^{-1}) and control plots (66 kg ha^{-1} a^{-1}). On the other hand, higher leaching of K on compost-amended plots was suggested by the results of the 99/2000 season, during which average amounts of rainfall were recorded in the study area (314mm), as opposed to the extremely dry season of 98/99 (161mm). Leaching rates for K were 2.8 kg ha^{-1} a^{-1} in compost-amended and 0.6 kg ha^{-1} a^{-1} in control plots in 99/2000. Among the studied heavy metals (Cu, Zn, Cd, Ni, Cr, Pb, Hg) only Cu showed increased leaching rates in the compost treatments. Maximum fluxes of up to 150 g Cu ha^{-1} a^{-1} were measured, but the average flux values for the winter season 99/2000 amounted to 27.2 g Cu ha^{-1} a^{-1} (Fig. 3). In the control plots no leaching of Cu could be measured. As suggested by simulations under laboratory conditions, Cu is strongly bound to dissolved organic matter from the compost and can as such be displaced with the water flow. The average leaching rates of Cu, however, were not high, so that groundwater quality should not be adversely affected in the short term.

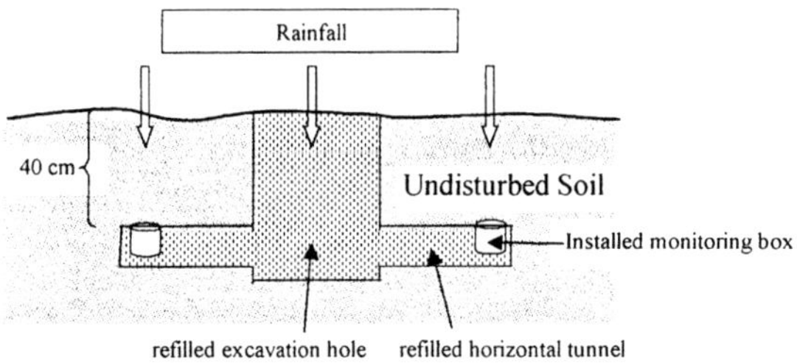

Figure 1. Schematic representation illustrating the positioning of the monitoring boxes in the field.

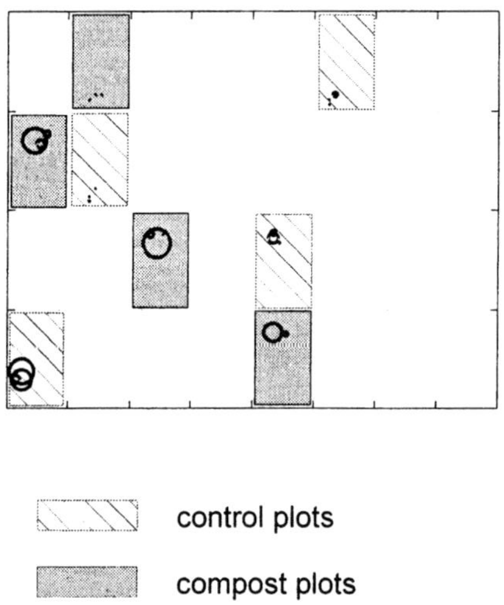

Figure 2. A drawing of the randomized field experiment with the location of the monitored plots highlighted according to the legend. Each circle represents one of the 24 installed monitoring boxes, with the center (not shown) of each circle indicating the location of the box in the field and the size of the circle (radius) representing the amount of Cl^- measured in the two top layers (α and β) of the box at that location. Circles with very small radii (equivalent to low amounts of Cl^- in the box) appear as dots.

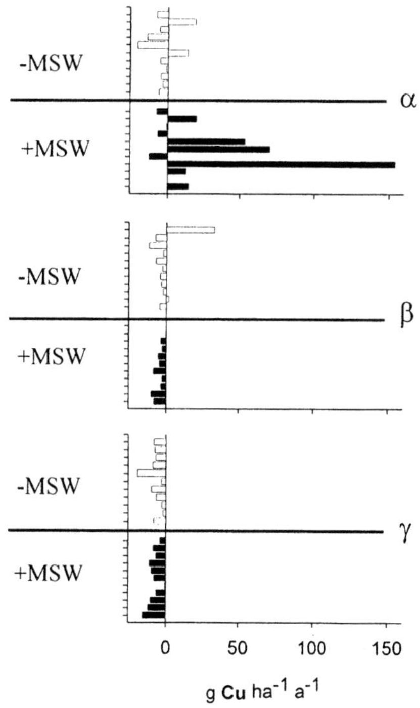

Figure 3. Cu analyzed in the layers α (top), β (middle), γ (bottom) in the monitoring boxes (99/2000). The white columns represent individual boxes from the control plot, while solid black columns show individual boxes from the compost-amended plot. Negative values result from subtracting the Cu content of resin from a blank not installed in the field.

Acknowledgement

The authors would like to thank W. A. Bischoff for providing the know-how and assistance to install the monitoring boxes, the members of EPRI Institute, Gaza City for active participation in the field work and the DFG for financing this project.

References

Amlinger F 1996 *In* The Science of Composting. Eds. M de Bertoldi, P Sequi, B Lemmes and T Papi. pp. 314-328. Blackie Academic & Professional, London.

Kaschl A, El Nahhal Y, Abu Mourad T, Tubail K, Safi J, Hadar Y, Chen Y, and Römheld V 1998 *In* VDLUFA Schriftenreihe 49. pp. 251-254. VDLUFA-Verlag, Darmstadt.

W. J. Horst et al.(Eds.), Plant nutrition – Food security and sustainability of agro-ecosystems. 990–991, 2001

Chemical properties of different composts consisting of sewage sludge, cement kiln dust and plant residues

A. Zayed, S. Mabrouk and M. Abdou
Soil and Water Science Department, Faculty of Agriculture, Suez Canal University, Ismailia, Egypt

Key words: cement kiln dust (CKD), fertiliser value, heavy metals, plant residues (PR), sewage sludge (SS)

Abstract

Sewage sludge (SS), cement kiln dust (CKD), and plant residues (PR) were composted at different ratios under different conditions. Increasing CKD application decreased organic carbon content and C/N ratio of the final compost. NPK contents differed only slightly between the different compost types. CKD application only slightly affected total Cd, Cu, Zn and Ni concentrations in the final compost. Concentrations were lower than maximum permissible concentration for agricultural use.

Introduction

Large amounts of sewage sludge (SS) and cement kiln dust (CKD, by-pass dust released during clinker burning) are produced in urban Egyptian regions. Disposal of SS by soil application has been practiced as a resourse utilisation especially for N and P (Burnham *et al.*, 1992). However, potential risks of SS application should be considered, especially the presence of pathogens and the contamination with heavy metals. Now there is growing realization that turning the daily produced wastes into some useful resources by composting and nutrient stabilization is necessary. By this way, reduction of pathogens is expected during composting. CKD is used to reduce the excessive amount of bioavailable P as well as to stabilize heavy metals (Logan, 1992). Our work aimed to assess the chemical properties of different compost consisting of SS, CKD and PR in various ratios. Additionally, the effects of aeration conditions during composting were studied.

Materials and methods

Sewage sludge, plant residues and cement kiln dust were composted at volume ratios of 3:2:0, 5:4:1 and 2:2:1 (s. Table 1). These mixtures were exposed to the following aeration conditions during composting: (i) pile turned twice weekly, (ii) pile suction 0.5, (iii) pile suction 1.0, (iv) pile blowing 0.5 and (v) pile blowing 1.0 $m^3\ min^{-1}\ ton^{-1}$.

Electrical conductivity in compost waste-water and pH of the compost (compost-water suspension ratio: 1:2.5) were determined according to standard methods recommended by the U.S. salinity laboratory (Richards, 1954).Total concentrations of N, P, K, Zn, Pb, Ni, Cu and Cd in the final compost were determined according to APHA (1995).

Results and discussion

Organic carbon signifcantly decreased during composting. There was a further decrease of organic carbon in the final compost resulting from CKD application (not shown). Only slight differences in organic C contents resulted from the different aeration conditions. It was found that the organic C degradation increased in the following order:

1. Exposition to blowing with 1.0 $m^3\ min^{-1}\ ton^{-1}$: 9.36 %,
2. Exposition to blowing with 0.50 $m^3\ min^{-1}\ ton^{-1}$: 9.96 %,
3. Exposition to suction with 0.50 or 1.0 $m^3\ min^{-1}\ ton^{-1}$: 10.13%,
4. Turning twice weekly: 10.48 %.

Amending SS and PR with CKD decreased the C/N ratio of the final compost, however, additional effects were caused by the different aeration treatments (Table 1).

Application of CKD considerably increased the EC of the compost waste-water, EC values being 3.72, 4.77 and 6.69 dSm^{-1} for the treatment which had received 0, 10 and 20% CKD, respectively. Similar CKD effects were observed regarding the pH in the final compost.

In agreement with other reports (Dam Kofoed, 1983) compost composition and compost treatment affected the NPK contents of the final composts only slightly (Table 1).

Treatments had no consistent effects on the heavy metal concentration of the final composts. Compared with legislative regulations in Germany (Sauerbeck and Leschber, 1992), concentrations of most heavy metal were much lower than maximum permissible concentrations in sewage sludge for agricultural use, however Cd concentration are near the critical limit.

References

American Public Health Association (APHA) 1995 Standard Methods for the Examination of Water and Wastewater, 19th ed. Washington, DC.

Burnham JC, Hatfield N and Bennett GF 1992 *In* Innovation and uses for Lime. Eds. DD Walker *et al.* pp. 128-141. ASTM STP 1135. ASTM., Philadelphia.

Dam Kofoed A. 1983 *In* Utilization of Sewage Sludge on Land: Rates of Application and Long-Term Effects of Metals. pp. 2-21. Proceedings of a seminar held at Uppsala.

Logan TJ 1992 Chemistry and bioavailability of metals and nutrients in CKD-stabilized sewage sludge. *In* Proc. The Environ. Technol. Expo. East Chicago. pp.

493-502. Cahners Publ. Co. Des Plaines.

Richards LA 1954 Diagnosis and Improvement of Saline and Alkali Soils. U.S.D.A. Handbook No. 60.

Sauerbeck DR and Leschber R 1992 *In* Effects of Organic Contaminates in Sewage Sludge on Soil Fertility, Plants and Animals. pp. 1-13. Commission of the European Community, Luxemberg.

Table 1. Final characteristics of composts produced from sewage sludge, cement kiln dust and plant residues in different ratios and exposed to different aeration condition during composting.

Treatments (see Tab. 2)	pH	EC (dSm^{-1})	C/N ratio	N	P %	K	Zn	Cu	Ni ($mg\ kg^{-1}$)	Cd
1	7.17	2.64	9.44	1.13	0.414	0.82	90.76	33.16	30.66	6.30
2	7.18	3.81	7.47	1.35	0.430	0.85	93.60	39.18	36.00	7.68
3	7.12	4.63	7.57	1.02	0.432	0.86	115.30	42.18	50.00	8.76
4	6.74	3.62	7.81	1.13	0.414	0.91	91.90	37.20	39.34	8.76
5	6.82	3.88	8.11	1.21	0.346	0.85	108.48	35.60	39.34	9.81
6	7.44	3.77	6.26	1.27	0.486	0.87	102.74	37.66	36.66	10.00
7	7.37	4.49	7.55	1.05	0.368	0.88	109.60	37.92	32.66	10.20
8	7.40	4.91	7.38	1.05	0.409	0.73	35.03	35.70	28.00	10.00
9	7.32	6.04	5.90	1.28	0.351	0.81	55.92	37.42	30.66	10.20
10	7.42	4.66	5.31	1.15	0.409	0.83	91.30	36.94	28.00	10.35
11	7.61	4.18	6.85	0.94	0.376	0.72	93.04	32.94	50.00	10.05
12	7.59	7.32	5.41	1.22	0.430	0.88	105.04	32.46	34.66	10.07
13	7.88	8.29	6.10	1.06	0.328	0.76	94.78	31.22	28.66	9.57
14	7.75	8.42	6.78	0.94	0.307	0.79	59.36	31.52	22.00	9.51
15	7.65	5.24	7.48	0.79	0.382	0.80	45.68	27.18	26.66	8.85

Table 2. Description of treatments:

Treatment	Volume ratios of the components			**Aeration**
	Sewage sludge	Plant residues	Cement dust	
1	3	2	0	Pile turned twice weekly
2	3	2	0	Pile suction 0.5 $m^3\ min^{-1}\ ton^{-1}$
3	3	2	0	Pile suction 1.0 $m^3\ min^{-1}\ ton^{-1}$
4	3	2	0	Pile blowing 0.5 $m^3\ min^{-1}\ ton^{-1}$
5	3	2	0	Pile blowing 1.0 $m^3\ min^{-1}\ ton^{-1}$
6	5	4	1	Pile turned twice weekly
7	5	4	1	Pile suction 0.5 $m^3\ min^{-1}\ ton^{-1}$
8	5	4	1	Pile suction 1.0 $m^3\ min^{-1}\ ton^{-1}$
9	5	4	1	Pile blowing 0.5 $m^3\ min^{-1}\ ton^{-1}$
10	5	4	1	Pile blowing 1.0 $m^3\ min^{-1}\ ton^{-1}$
11	2	2	1	Pile turned twice weekly
12	2	2	1	Pile suction 0.5 $m^3\ min^{-1}\ ton^{-1}$
13	2	2	1	Pile suction 1.0 $m^3\ min^{-1}\ ton^{-1}$
14	2	2	1	Pile blowing 0.5 $m^3\ min^{-1}\ ton^{-1}$
15	2	2	1	Pile blowing 1.0 $m^3\ min^{-1}\ ton^{-1}$

W. J. Horst et al.(Eds.), Plant nutrition – Food security and sustainability of agro-ecosystems. 992~993, 2001

Utilisation of boiler ash from biomass heating plants as fertiliser substitut

G. Uckert, R. Mette and B. Sattelmacher
Institute of Plant Nutrition and Soil Science, Kiel University, Olshausenstr. 40, Kiel, 24118, Germany, guckert@plantnutrition.uni-kiel.de

Key words: *Avena sativa* L., biomass, boiler ash, combustion, fertiliser substitute, heavy metals, hedgerows, nutrient cycle, oat, recycling

Abstract

The utilisation of ash which comes from recently established biomass heating plants in Schleswig Holstein is suggested as a fertiliser substitute for arable land. In a case study the quality of different ash fractions of a combined heat and power (CHP) plant was investigated, which is using untreated wood chips from hedgerow biomass. Pot experiments with oats were set up in two different soils - a nutrient poor sandy soil with a pH of 4.7 and a sandy loam soil with pH 5.6. Two different ash treatments were used - 2 t/ha and 6 t/ha - representing an adequate supply of potassium, phosphorus and lime, respectively. Yield of oat in poor sandy soil was significantly increased (100 %) in the 2 t ash treatment. The increase in sandy loam soil was about 60 % of the unfertilised control. The ash treatment of 6 t/ha (approximately 2000 kg CaO-equivalents) on the sandy soil increased pH by 0.7 and on the sandy loam soil by 0.5 units relatively to the unlimed control.

Introduction

Hedgerows as typical nature conservation elements in Schleswig Holstein's agricultural ecosystems have to be pruned every 10 - 15 years. Their biomass can be used as fuelwood from the field. The wood chips made out of the shrubs were burned in one of the recently established biomass heating plants. The combustion of this kind of fuelwood leaves 1 - 2 % of ash-residues. The residues of hedgerow biomass contains high amounts of mineral nutrients (Uckert 1997) similar to short rotation coppice (Ericsson 1992). These amounts differ significantly from nutrient concentrations analysed in unmanaged forest ecosystems (Ellenberg 1986). Thus, it seems to be promising to use the ash residue as fertiliser substitute.

To run a total heating period of a 5 MW combined heat and power (CHP) plant untreated wood chips from about 70 km hedgerow biomass have to be required. Combustion of this biomass leaves about 60-100 t of ash residues. Depending on the combustion technology and ash fractions (boiler or filter ash) even the ash of untreated wood might contain heavy metals in concentrations which could increase the soil content of these elements. If ash residues are considered as fertiliser substitute in agriculture, analysis of the composition of ash have to be done. Additionally effects on plant and soil have to be investigated.

Materials and methods

A pot experiment with oat was conducted on two different soil types. Soils were a: a nutrient poor sandy soil with 44 mg kg^{-1} P, 34 mg kg^{-1} K (both VdLUFA DL-method) at a pH of 4.7 and b: a sandy loam, low in extractable nutrients (92 mg kg^{-1} P, 117 mg kg^{-1} K; pH 5.6). Plants were grown in 'Mitscherlich' pots containing 4 kg of air dried soil. All pots received 0.5 g N as ammonium nitrate. Two different ash treatments - 2 t/ha and 6 t/ha – were tested against zero control and mineral nutrient (K, P, Mg, CaO) supply equivalent to the nutrient content of 2 tons ash. For total amounts of nutrients see *Table 1*. The irrigation was done by fibre glass wicks with additional flow in a closed system. Treatments were harvested after 4 month. In oat plants yield and total plant biomass, nutrient and heavy metal concentrations in grain and straw were determined. In soils nutrient and heavy metal contents were measured. For analysis, plant, soil and ash samples were treated with 32 per cent concentrated HNO_3 for 2 1/2 hours at 80 °C, which extracted total amounts in plants and even the less mobile fractions in the soils and ash residues. Extractions were mesured with an Inductive Coupled Emission Spectrometer (ICP).

Table 1. Nutrient and heavy metal content supplied in the 2 t/ha ash treatment

2 t ash/ha	Nutrients				
	P	K	CaO	Mg	
kg/ha	60	160	720	60	
g/pot	0.19	0.5	5.28	0.19	
	Heavy metals				
	Pb	Cd	Cr	Cu	Zn
g/ha	12	8	30	200	800
mg/pot	0.04	0.03	0.09	0.36	2.52

Results and discussion

The ash fertiliser treatments influenced the yield of oat significantly (see *Figure 1*). On the poor sandy soil the grain yield in the 2 t ha^{-1} ash treatment was more than 100 % higher than the untreated 0-Control (without P, K, Mg). A three time higher amount of ash (a 6 t ha^{-1} equivalent) increased the yield again by about 40 %. This had almost the same effect as the mineral control with readily available nutrients. On the sandy loam soil a similar but reduced response was expressed as shown in following *Figure 1*. The smaller difference in grain yield between the ash treatment and the 0-control in the sandy loam soil depends on its higher buffering capacity in comparison to the poor sandy soil.

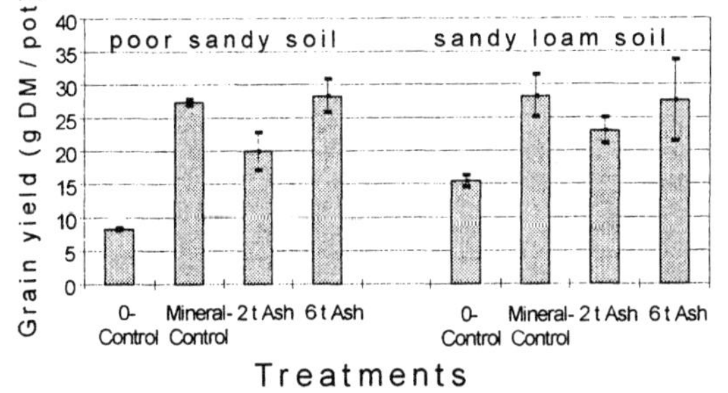

Figure 1. Effect of boiler ash at two application rates on grain yield of oat in a pot experiment in comparision with an untreated and mineral fertiliser control

With respect to the liming effects of CaO and MgO ash-treatments of 6 t/ha (ca. 2000 kg CaO-equivalents) increased pH on the sandy soil by 0.7 and on the sandy loam soil by 0.5 relative to the unlimed control at harvest.

According to Obernberger (1997a) analysis on ICP Emission Spectrometer of different ash fractions of a biomass heating plant indicates low concentrations of Cd, Pb and Zn in the boiler ash and concentrations up to 100 times higher in the electro filter ash (*Table 2*).

Table 2: Heavy metal concentration of different ash fractions of an untreated wood burning biomass heating plant in Schleswig-Holstein

Ash-fractions:	Pb	Cd	Cr mg kg^{-1}	Cu	Zn
Boiler-ash	6.3	0.2	9.9	71.8	32.5
Cyclon-ash	47.1	20.9	11.7	87.1	895.9
E-filter-ash	209.4	41.9	12.0	114.4	4195.2

Conclusions

High amounts of macro nutrients in ash residues of combusted hedgerow wooden chips compared to ash from forest residues (Obernberger *et al.* 1997) indicate a profitable nutrient uptake from fertilised arable land, as indicated through ecosystem analysis in the hedgerow field crop interface (Mette and Sattelmacher 1994). Additionally this high values results in a higher proportion of nutrient rich bark of young stems which where pruned every 10-15 years (Uckert *et al.*, 1997). Negligible concentrations of most relevant heavy metals (Cd and Pb) introduced through deposits and soil specifics in addition with concentrations during combustion process indicate the utilisation of seperated boiler ash as a fertiliser substitute. Therefore this ash fraction should be integrated in future nutrient cycles on a local scale. On the other hand filter ash residues, even from untreated hedgerow shrubs contain high amounts of heavy metals and have to be disposed.

First fertiliser experiments had shown increasing effects on yield, soil nutrient content and soil pH. Further investigations will show the impacts on crop growth of different species and the possible heavy metal contamination of plants caused by different ash qualities.

References

Ellenberg H, Mayer R and Schauermann J 1986 Ökosystemforschung: Ergebnisse des Sollingprojekts 1966 – 1986, Eugen Ulmer Stuttgart.

Ericsson T, Rytter L and Linder S 1992 *In* Ecophysiology of Short Rotation Forest Crops Ed C P Mitchell. pp 35-65 Elsevier Applied Science, London and New York.

Mette R and Sattelmacher B 1994 *In* Agroforestry and Land Use Change in Industrialized Nations Eds E Welte Szabolcs R F Huettl. Proceedings of the 7th International Symposium of CIEC Berlin, pp 275-284.

Obernberger I Biedermann F Widmann W and Riedel R 1997 Biomass and Bioenergy, No. 12:3, pp 211-224.

Uckert G Mette R and Sattelmacher B 1997 VDLUFA-Schriftenreihe 46, Kongressband 1997, pp 583-586.

Heavy metal pollution in a rhodic ferralsol of Cuba

O. Muñiz, J. Molina, S. Quicute and J. Estevez

Estación Experimental "La Renée". Instituto de Suelos. Carretera Bejucal – Quivicán, Km 33½, Quivicán. Habana. Cuba. Email: larenee@ceniai.inf.cu.

Key words: critical levels, heavy metals, soil pollution

Abstract

Because of their fertility, productivity and extension, Rhodic Ferralsol are the most important agricultural soils in Cuba and receive, year after year, large quantities of fertilisers and pesticides. Nevertheless, there are no commonly accepted threshold values for heavy metals on this tropical soil type. In this report, different analytical procedures for the determination and evaluation of Cd, Ni, Pb, Zn and Cu concentrations on Rhodic Ferralsol were compared to the results of data obtained from the test crop *Sorghum vulgaris* grown with increasing quantities of these heavy metals. The results suggest that total and DTPA available concentrations of heavy metals are valuable parameters to judge upon the heavy metal pollution of Rhodic Ferralsol.

Introduction

Due to the application of mineral fertilisers, of organic amendments such as manure and rock phosphates and of pesticides, increasing quantities of heavy metals (HM) are applied to Cuban soils with a potential danger to the environment. In view of this it is surprising that there are still no commonly accepted threshold values for heavy metals on Rhodic Ferralsols the predominant soil type of the region. The objective of this study therefore was to contribute to the establishment of such thresholds based on a better understanding of the plant availability of HM on Rhodic Ferralsols.

Materials and methods

A typical Rhodic Ferralsol from Quivican, province of Havana was used for this study. This soil is characterised by a pH in water of 6.90, 26 mg kg^{-1} of organic matter (Walkley and Black method), a cation exchange capacity of 16 $cmol_c.kg^{-1}$, 26 mg kg^{-1} of P and 41 mg kg^{-1} K (0.5 *M* H_2SO_4 extraction).

Plant experiments (one for each HM studied) were carried out in pots containing one kg of soil using *Sorghum vulgaris* (6 plants per pot) as a test crop. The soil had been amended with increasing quantities of the HM in solution (Table 1). Each treatment had 4 repetitions.

The plants were harvested 45 days after germination. Recorded were total dry matter of the shoot and the concentration and total uptake of HM.

Table 1. HM rates applied to the soil (mg kg^{-1} of soil).

Cd = 0 - 2 - 4 - 12 and 24 mg $Cd.kg^{-1}$
Pb = 0 - 37.5 - 75 - 150 and 300 mg $Pb.kg^{-1}$
Ni = 0 - 100 - 200 - 400 and 800 mg $Ni.kg^{-1}$
Zn = 0 - 200 - 400 - 800 and 1600 mg $Zn.kg^{-1}$
Cu = 0 - 50 - 100 - 200 and 400 mg $Cu.kg^{-1}$

The following analytical methods to determine HM availability in the soil were used: (1) total concentration: to 1 g of sample, in Pt crucible, add 5 ml of HF and let overnight, add 5 ml of HF, 5 ml of HNO_3 and 5 ml $HClO_4$. Heat almost to dryness (white fumes). Add 5 ml of concentrated HCl and make volume to 50 ml with H_2O.; (2) extraction with DTPA (Lindsay and Norvell, 1978); (3) extraction with EDTA in N ammonium acetate at a 1:5 soil:extractant ratio; (4) extraction with N ammonium acetate (NH_4Ac), at a 1:5 soil:extractant ratio; (5) extraction with ammonium nitrate (NH_4NO_3) at a 2:5 soil:extractant ratio.

Cd was determined by polarography and Ni, Pb, Zn and Cu by atomic absorption spectrophotometry. Results were statistically evaluated by linear correlation and ANOVA.

Results and discussion

With recovery rates between 85 and 115% total concentrations of HM in the soil determined by Method 1 closely mimicked the applied HM rates. The HM extraction capacity of the other four methods for the five elements under study declined in the order DTPA > EDTA > NH_4Ac > NH_4NO_3.

Good linear correlations (r values between 0.99 and 0.92) were obtained between soil total content (Method 1), DTPA extractant (Method 2) and EDTA extractant (Method 3). Nevertheless, for low HM concentrations correlations were poor with NH_4Ac and NH_4NO_3 extractants (Methods 4 and 5). This was most likely due to the low detection limits of the available equipment.

Methods 1, 2 and 3 also lead to significant correlations with the HM concentration and uptake in sorghum (Table 2). Therefore, all three methods would appear suitable for the determination of Ni, Cd, Pb, Zn and Cu in Rhodic Ferralsols. Nevertheless, the DTPA extractant proved to me more convenient than EDTA because it is more sensitive and is also more widely used in Latinamerica allowing a better comparison of results.

Finally, preliminary potentially toxic HM contents (Table 3) were obtained considering as evaluation criteria, the soil's HM content that caused significant negative effect on sorghum yield (total dry matter production) or

HM phytotoxic values according to the international literature (Kabata-Pendias and Pendias, 1992; Alloway, 1995). The critical Ni values obtained were much higher than those in the literature likely because Rhodic Ferralsols are able to accumulate high amounts of Ni. The 1:1 clays and Al and Fe hydroxides of these soils were found to have a very high fixation capacity for this element making it thereby unavailable to plants (Muñiz and Haydée Irigoyen, 1989).

The results suggest that Cd, Ni, Pb, Zn and Cu total and DTPA available concentrations are valuable parameters to judge upon the HM pollution of Rhodic Ferralsol. However, further studies are required considering different soil conditions and crops HM uptake ability.

References

Alloway BJ 1995 2nd edition, Blackie Acad. and Prof. United Kingdom.

Kabata-Pendias A and Pendias H 1992 2nd edition, CRS Press. Boca Raton.

Lindsay WL and Norvell WA 1978 Soil Sci. Soc. Amer. J. 42, 421-428.

Muñiz O and Irigoyen H 1989 Informe en archivo del Instituto de Suelos. La Habana.

Table 2. Linear correlation (r) between heavy metal (HM) concentrations in the soil determined according to different methods and sorghum HM plant concentration (C) and total HM uptake (U).

No	Analytical method	Ni		Cd		Pb		Zn		Cu	
		C	U	C	U	C	U	C	U	C	U
1	TotalContent	0.88	0.87	0.98	0.97	0.98	0.96	0.99	0.88	0.77	0.80
2	DTPA	0.83	0.83	0.98	0.96	0.98	0.96	0.95	0.95	0.70	0.83
3	EDTA	0.78	0.78	0.95	0.93	0.98	0.96	0.99	0.86	0.73	0.83

Table 3. Potential toxicity threshold for HM concentrations in a Rhodic Ferralsol (mg kg^{-1}).

Analytical method	Ni	Cd	Pb	Zn	Cu
Total concentration	1100	6	350	600	600
DTPA extractable	35	1	60	80	110

W. J. Horst et al.(Eds.), Plant nutrition – Food security and sustainability of agro-ecosystems. 996~997, 2001

Physicochemical processes effecting ^{137}Cs and ^{90}Sr migration in soils and uptake by plants

E. N. Rovdan and A. M. Abramets

Institute for Problems of Natural Resources Use and Ecology, National Academy of Sciences of Belarus, Staroborisovsky Tr. 10, Minsk, 220114, Belarus, elenarov@yahoo.com

Key words: bentonite, *Beta vulgaris* L., ^{137}Cs, diffusion, distribution coefficient, electrolyte, kaolin, migration, peat, soddy-podsolic soil, ^{90}Sr, subsurface brine water, transfer factor

Abstract

Soil-plant transfer of radionuclides involves plant physiological and soil physicochemical aspects. The objectives of the study was to investigate the impact of electrolytes on the mobility and bioavailability of radio nuclides in various natural growth media. For this purpose diffusive mobility and sorption activity of 137Cs and 90Sr were studied in media such as peat. soddy-podsolic sand, quartz sand, bentonite and kaolin. Furthermore, field plots on a contaminated soil were installed in the Gomel region in the south part of Belarus and planted with red beet (Beta vulgaris L.) as a root crop to study the soil-plant transfer of 137Cs and 90Sr. as affected by application of subsurface brine water. The results indicated that diffusion of radio nuclides in soil solution and adsorption at the soil matrix are the main processes limiting uptake of this radio nuclides by plants. Soil-based counter-measures for radioactive contaminated soils included addition of chemical reagents to inundate the soil-plant transfer pathway with innocuous elements or to increase the binding potential of the soil. Both resulted in a reduction of the soil-plant transfer factors for radionuclides. To predict the residence half times of radio nuclides in the soil layer with the growing roots of the vegetation and to evaluate the possible vertical migration across the soil profile and the plant uptake a mathematical model has been developed. This model will help also to assess possible countermeasures against radio nuclide uptake by plants in various ecosystems such as the application of subsurface brine water at different rates.

Introduction

The transport of ions to the root surface is governed by the combination of three different mechanisms: diffusion, mass flow driven by transpiration and interception as a consequence of soil displacement due to root growth. For various elements in a soil one of these transport mechanisms can play a dominant role. For uptake of radionuclides by plant roots diffusion of these elements in the soil-root interface is one of the process limiting factor (Nerpin *et al.*, 1975; Nye *et al.*, 1980). A soil-based countermeasure for radioactive contaminated soils could be the addition of inert materials which inundate the pathway to root surface of radionuclides with innocuous elements or increase the binding of these elements to the soil. Both will reduce the transfer to plants. The objective of the presented study was to investigate the effect of subsurface brine water to reduce the transfer of radionuclides (^{137}Cs and ^{90}Sr) from a sodpodsolic sandy soil to red beet (*Beta vulgaris* L.) as a root crop.

Materials and methods

Adsorption activity and mobility by diffusion of both radionuclides ^{137}Cs and ^{90}Sr were investigated in growth media such as peat soil, quartz sand, bentonite, kaolin and soddy-podsolic sandy soil. For this purpose experimental plots (1 m^2) were installed in the region of Gomel (Belarus) with contaminated soil The soil had the following characteristics: Soddy-podsolic with sandy texture, pH_{KCl} 6.3, 1.6 % organic matter, exchangeable K 21.5 mg/100g and P 15.7 mg/100g, radioactivity of ^{137}Cs 3.8 10^{-2}k Bq m^{-2} and of ^{90}Sr 111 k Bq m^{-2}. As crop plant red beet (*Beta vulgaris* L.) was chosen. The used subsurface brine water was from the region of Belarus Polesje and contained various macro and trace elements including physiological relevant ones for plants, such as (g l^{-1}): NH_4^+ (50); K (6) Ca (40); Mg (4); Na (40); Mn, (0.03); Br (3); Sr (3).

Results and discussion

Soil-plant transfer of radionuclides involves plant physiological and soil chemical aspects. The flow of a radionuclide from the soil to the root depends on its bioavailability in a distinct soil and on the specific capacity of a plant to take up and to accumulate this particular element. This transfer of radionuclides is assessed by the soil-plant transfer factors, defined as the ratio of concentrations of the radionuclides in the plant and the soil. This transfer integrates the action of two processes: a plant physiological and a soil chemical process. The transfer coefficient is strongly correlated to the distribution coefficient of radionuclides in soils (Sheppard, 1985).

The results have shown that the distribution coefficient for ^{90}Sr was low for all growth media studied. In comparison, the distribution coefficient and consequently the effective diffusion coefficient of ^{137}Cs were still lower. For the sorption capacity the following sequences were obtained: peat > bentonite > kaoline > soddy-podsolic soil > quartz sand for ^{137}Cs and peat > kaoline > bentonite > soddy-podsolic soil> quartz sand for ^{90}Sr.

The diffusion of elements in the rhizosphere is the main process limiting uptake of radionuclides from the soil by roots. Effective diffusion coefficient (D_{ef}) are for ^{137}Cs in peat = 0.9 x 10^{-13} soddy-podsolic soil 1.6 x 10^{-13}, quartz sand 1.8 x 10^{-13}, bentonite 2.9 x 10^{-13} and kaoline 3.2 x 10^{-14} $m^2 s^{-1}$ and for ^{90}Sr in peat 3.2 x 10^{-12}, soddy podsolic soil 6.7 x 10^{-12}, quartz sand 1.5 x 10^{-11}, bentonite 1.1 x 10^{-11} and kaoline 7.0 x 10^{-12} $m^2 s^{-1}$.

The presence of electrolytes in the soil water, such as HCl, CsCl, SrCl, $CaCl_2$, KCl and KOH decreased the sorption of ^{137}Cs at all growth media with exception of bentonite. The greatest effect was achieved by application of non-radioactive CsCl. As similar picture was observed with ^{90}Sr. The lowest values for the distribution coefficients were got with application of non-radioactive $SrCl_2$ into the soil solution of all studied growth media. Thus, the presence of electrolytes in the soil solution will strongly decrease the distribution coefficient for radionuclides and, as a result of this enhance their mobility by diffusion.

Behavior of radionuclides in soils and their distribution and function in plants follows in general the same principles as of stable isotopes as analogs of the radionuclides. However, the stable isotopes are present at much higher concentrations than radionuclides. Therefore, the features of soil sorption processes are stipulated by extremely low concentrations of radionuclides in soils. The migration of these ultra low concentrations of radionuclides in soils and their uptake by roots depend on their dilution by macro concentration of stable isotopes of the same element.

The results of the field experiment indicated that ^{137}Cs has a lower biological activity compared with ^{90}Sr. The transfer factor for ^{137}Cs and ^{90}Sr from soddy-podsolic sandy soil to red beet as a root crop was 0.00007 and 0.0009, respectively. The application of subsurface brine water to the soil reduced the accumulation of ^{137}Cs by 25-30% and of ^{90}Sr by 80-90%. But the addition of this brine water not exceeding 50 ml/m^2 does not decrease the uptake of radionuclides by red beet. The mechanism behind the possible effect of subsurface brine water is the decrease of radionuclide uptake by plants due to an increase of its stable isotopes in the soil.

The extent of various electrolytes on the distribution coefficient value for ^{137}Cs in peat is declining in the following sequence: Cs Cl >> HCl > KCl > Ca Cl_2. In particular, stable Cs as CsCl has a strong effect. Its presence in the soil solution at an amount of only 1 x 10^{-3} mol L^{-1} varies the distribution coefficient (Kd) almost 50-fold and enhances the diffusion coefficient by 2 orders. The comparable sequence for ^{90}Sr is
$HCl>SrCl_2>CaCl_2>KCl>KOH$.

For manipulation of the mobility of radionuclides in soils it is necessary to consider the kinetics of radionuclide desorption into the soil solution.

The total quantity of radionuclides in a soil exploited by plant root systems is related to the balance between the deposit and disappearance from plough layer due to removal by harvest and leaching in deeper soil layers. However, only distinct fraction of the radioactivity remaining in the soil might be available for plant uptake. This fraction depends on the physico-chemical characteristics of a soil and may change with time due to ageing as well as to modification of the soil properties such as those induced by the application of countermeasures.

A mathematical model has been developed which describes the vertical migration of radionuclides. This model requires information on the actual radionuclide transport process in the soil (e.g. sorption properties, diffusion, water infiltration). The model was aimed at the prediction of the residence half times of radionuclides in the soil layer containing the roots of the vegetation. Further, the vertical migration across the soil profile and the radionuclide transfer from the soil profile to plants should be evaluated with this model (Rovdan, 1997).

References

Nerpin SV and Chudnovsky AF 1975 Energy- and mass-exchange in the plant-soil-air system. Gidrometeoizdat, Leningrad.

Nye PH and Tinker PB 1980 *In* Solute movement in the soil- root system. Eds Usiarov OG. Kolos, Moscow.

Rovdan EN 1997 Forecast and control of radioecolodical condition of peat-bog soils. Ph.D Thesis, Institute for Problems of Natural Ressources use, Minsk.

Sheppard MJ 1985 Health Phys. 49, 106-111.

W. J. Horst et al.(Eds.), Plant nutrition – Food security and sustainability of agro-ecosystems. 998~999, 2001

Control of Al/Fe activities in soils of the humid tropics by a balanced fertiliser management

K. Sommer
Agrikulturchemisches Institut der Universität, Karlrobert Kreiten-Str. 13, 176, 53115 Bonn, Germany

Key words: Al activity, fertilisation, liming, oxisols, pH adjustment, phosphate availability

Abstract

Soils of the humid tropics are predominantly acid and nutrient-deficient. The acidity leads to aluminium (Al) and iron (Fe) being major constituents on the soil's sorption complex, which is activated by the application of fertilisers with monovalant anions such as chlorides or nitrates. As a consequence, Al and Fe can substantially contribute to the fixation of phosphorus (P) and molybdenum (Mo). In temperate zones activities of Al or Fe in soils are controlled by liming. In addition to its prohibitive costs for most field crops, liming of soils in the humid tropics raises a number of problems. After the exchange from the sorption complex by calcium (Ca) and a raise in pH, Al and Fe in the soil are precipitated as hydroxides or oxides. Because of high rainfall and continuous leaching, the pH-increase after liming is only temporary and soon thereafter a re-exchange of Ca from the sorption complex of soils takes place by activated Al and Fe. Based on these problems a concept of fertilisation for soils in the humid tropics was developed in a series of pot experiments. To keep the intensity of the bonding of SO_4^{2-} to Al high, care was taken to not raise the soils' pH (measured in 0.05 *M* K_2SO_4 or in 0.1 *M* KCl) by liming above 5.0 or 4.3, respectively. Thus the effect of liming was largely limited to neutralise the exchangeable H^+ acidity on the sorption complex of the soil. The acidity derived from exchangeable Al and its hydration was neutralised by adding sulphate fertilisers such as $(NH_4)_2SO_4$, K_2SO_4, $MgSO_4$ and $CaSO_4$.

Introduction

The acidity of many tropical soils is based on the Al and Fe ions at their sorption complex, which are activated by the application of fertilisers containing chlorides or nitrates. This leads to a high P and Mo fixation potential on the one hand and to species-dependent Al toxicity on the other. In tropical soils the application of lime to control the activities of Al or Fe is often not recommendable given its high cost, a rapid decline of the pH with heavy rainfall and subsequent leaching and a re-exchange of Ca from the sorption complex by activated Al and Fe. In plantations of bananas, sugar cane and coffee, these processes are often enhanced by the decomposition of organic matter after the application of mineral fertilisers.

Materials and methods

A series of experiments, at the beginning as screening tests, with four replications each was conducted in pots filled with 1.6 kg of an oxisol from the Buenos Aires area in the south of Costa Rica, Central America, in 1976/77. The Al content of this soil was about 45 % Al_2O_3 and the soil's pH values were 5.3 in H_2O, 4.3 in 0.1 M KCl and 5.0 in 0.05 M K_2SO_4. Adequate amounts of mineral fertilisers were mixed into the soil as phosphates, sulphates, chlorides or nitrates in different combinations. At the beginning, the test crop chosen was tomato (*Lycopersicum esculentum* L.). The responses to different fertiliser treatments were classified according to the intensity of growth of the plants. All treatments with extreme intensities in growth, very good or very poor, were analysed for nutrient uptake and their response according to soil acidity.

Results and discussion

Good growth according to the yields of dry matter and the uptake of nutrients was obtained in treatments with an "open nutrient system" (Table 1). This means that by the sorption of fertilisers to the soil a balanced neutralisation of active bonds of aluminium by sulphate or phosphate took place. This can be followed by the pH-values analysed in the soils. They were equal in H_2O and 0.05 M K_2SO_4, while the pH-value analysed in 0.1 M KCl was lower. Because of the more intensive uptake of cations in comparison to anions the exchange of nutrients and their uptake by the plants is easy. At this low range of pH below 5.0, analysed in 0.05 M K_2SO_4, the strength of bonding between aluminium and sulphate was high enough to prevent the hydration and protonisation of Al, accompanied by strong fixations of P and Mo. In the soil of this treatment the pH-value in 0.1 M KCl was below that in water and in 0.05 M K_2SO_4. Furthermore after the uptake of cations by the plants in high surplus in comparison to phosphate and especially sulphate-activated bonds of aluminium can return to the free sorption complex of the soil.

From the beginning, poor growth occurred in treatments with "closed nutrient systems" (Table 1). By addition of Ca as $Ca(OH)_2$, aluminium exchanged from the sorption complex of the soil will form strong bonds with phosphate. In spite of the fertilisation of high amounts of phosphate the plants suffered phosphate deficiency. Under these conditions adequate anions with equal bonding intensities to Al as phosphate for the exchange of phosphate by the plants are missing. At a higher range of pH the bond intensity of sulphate in comparison to phosphate is not high enough to form $AlOHSO_4$ as a stable complex. Therefore, plants face difficulties to take up phosphate.

In the "open sulphate / phosphate nutrient system" even under the worst soil conditions the hydration of Al is prevented by the formation of $AlOHSO_4$ provided that soil pH is kept below 5.0 in 0.05 M K_2SO_4 or 4.0 in 0.1 M KCl.

Since 1970 this fertilisation concept was improved and tested in different research projects of the Bong Mining Company, Liberia, West Africa, the University of Bonn, Germany, and the University of San Jose, Costa Rica, Central America. From 1972 to 1979, up to the beginning of the civil war, this concept was also used in agricultural projects of the Bong Mining Company, Liberia, West Africa. After clearing the bush the land was used for intensive cultivation of field crops, vegetables and lowland rice, the selection of varieties and seed production for local farmers and for raising of cattle on intensive pastures.

References

Kamprath EJ 1972 *In* Soils of the humid tropics. National Academy of Sciences, Washington, D.C.

Mehlich A 1964 Soil Sci. Soc. Am. Proc. 28, 492-496.

Shainberg ME, Miller WP, Farina MPW, Pavan MA and Fey MV 1989 Use of gypsum on soils: a review; Advances in Soil Science, Springer-Verlag New York, Berlin, Heidelberg.

Summer ME 1993 Adv. Agron. 51, 1-15.

Table 1. Testing of nutrient systems in pot experiments with tomatoes (*Lycopersicum esculentum* L) as test plants

Nutrient systems	Dry matter		Nutrient contents in %					pH of soil after harvest		
	g/pot	S_x	N	P	K	Ca	Mg	H_2O	KCl	K_2SO_4
Control					No fertilisers applied					
	0,14	---	2,03	----	----	----	----	5.3	4.4	4.9
"Open" nutrient	Fertilisers applied: $(NH_4)_2SO_4$; KH_2PO_4 ; K_2SO_4 ; $CaSO_4$; $MgSO_4$									
system	7.76	$\pm$ 0.16	1.94	0.12	3.6	1.0	0.50	4.7	4.4	4.7
"Closed" nutrient	Fertilisers applied: NH_4NO_3 ; KH_2PO_4 ; KCl ; $Ca(OH)_2$; $MgCl_2$									
system	5.67	$\pm$ 0.19	1.84	0.13	4.0	1.1	0.45	4.6	4.5	4.9

Fertilisers applied per pot: N = 125 mg; P_2O_5 = 300 mg; K_2O = 500 mg; CaO = 639 mg; MgO = 150 mg

W. J. Horst et al. (Eds), Plant nutrition – Food security and sustainability of agro-ecosystems. 1000–1001, 2001

Optimum rice yield under short-term cover crop fallow systems

R. K. Akanvou[1)] M. Becker[2)] M. J. Kropff[3)] L. Bastiaans[3)] and G. Dea[1)]

[1)] *Centre National de Recherche Agronomique (CNRA), B.P. 856 Korhogo, Côte d'Ivoire, kgo.cnra@aviso.ci*
[2)] *Institute of Agricultural Chemistry (ACI), University of Bonn, Karlrobert-kreiten-Str. 13, 53127, Bonn, Germany*
[3)] *Crop and Weed Ecology Group, Wageningen University, Haarweg 333, lammert.bastianns@cwe.dpw.wau.nl*

Key words: legume cover crops, *Oryza sativa*, residual effects, savannah zone, yield loss

Abstract

In the tropical savannahs of West Africa, nutrient-deficiency (N and P) and weed infestation of agricultural lands contribute to low yields in food crop production. The existence of a pronounced dry period between cropping seasons makes intercropping cover legume species with upland rice the only feasible option for short fallow improvement. A factorial experiment with two cover crops (*Cajanus cajan* and *Stylosanthes hamata*), by two rice cultivars (Wab450-24-3-2-P18-HB [V4] and Wab56-50) and four legume relay-sowing dates (0, 14, 28, 56 days after rice sowing; DARS) was used to assess rice yield and the N contribution of cover crops to the systems. Grain yield losses varied between 1 % and 97 % in V4, and between 1 % and 100 % in Wab56-50 in relay with C. cajan, whereas with S. hamata, yield decreases were between 1 % and 52 in V4 and between 8 % and 61 in Wab56-50. Residual effects of legume residues were not linearly related to the amount of fallow N returned to the soil. Optimum rice yield depended on the combination of rice cultivar, legume species and legume sowing date.

Introduction

Introducing legume cover crops into rice cropping systems for short fallow improvement offers alternatives to farmers with limited resources to increase yield. For sustainable rice production, cover crops may have an important role in supplying N and contributing to the control of weeds and pests (Nye and Greenland, 1960; Buckles, 1999). Becker and Johnson (1998) showed that upland rice yield increased on average by 30 % following dry season cover crops. For socio-economic reasons, relay cropping with cover legumes is sometimes advocated. In this situation, yield losses may occur because of interspecific competition for light, water and nutrients. The extent of competition-induced yield losses in the companion food crop and the amount of accumulated legume biomass depend on the competitive ability of the component species and on the date of legume seeding (Kropff and Van Laar, 1993). In each system, N accumulation by the legume is important to enhance soil fertility and expected to be proportional to biomass accumulation.

The objectives of this study were to evaluate effects of relative sowing dates for two contrasting cover crop legumes in two rice cultivars with different competitive ability. Nitrogen accumulation and residual effects of cover legumes on subsequent upland rice crop were investigated.

Materials and methods

Experiments were conducted from 1997 to 1999 at the West Africa Rice Development Association (WARDA) research station at Bouaké, Côte d'Ivoire. The site is located in the savannah zone with a cumulative annual rainfall of 1000 mm. Two rice cultivars were intercropped with two cover crops using an additive experimental design in randomised complete blocks with three replicates. All treatment combinations are presented in Table 1. Each year, the legumes remained in the field during the dry season after rice harvest. With the onset of the rainy season, residues were cut and incorporated. Nitrogen content was determined by colorimetric analysis. Rice grain yield and biomass of each crop were determined on a 3 m^2 harvest area. Data on rice grain yield and final biomass were fitted using a logistic function to the relative sowing time of the legume (DARS) as an explanatory variable.

Results and discussion

Advancing the establishment date of legumes reduced yield and shoot biomass of upland rice. Total above ground dry weight of rice increased significantly ($P<0.001$) by delaying legume sowing from 0 DARS to 56 DARS. Introducing the legumes between 0 and 28 DARS caused the strongest reduction. Rice grain yield followed the same trend as the reduction in total shoot biomass, except at a high competitive stress when the relative reductions in grain yield surpassed the reduction in total shoot biomass. Reduction in panicle number or ~~a~~ in the number of spikelets per panicle caused reductions in grain yield.

When intercropped with *C. cajan*, grain yield of V4 decreased by 97 % at 0 DARS, 7 % at 28 DARS but only 1 % at 56 DARS. With *S. hamata* rice grain yield decreased by 52 % at 0 DARS, 1.5 % at 28 DARS and remained unaffected at 56 DARS. Likewise, reduction in grain yield of Wab56-50 was higher with *C. cajan* (100 % at 0 DARS, 37 % at 28 DARS and no effects at 56 DARS) than with *S. hamata* (61 % at 0 DAS and 8 % at 28 DARS).

Biomass of *C. cajan* intercropped with rice decreased logistically resulting in reductions of 53 % in V4 and 31 % in Wab56-50 at 28 DARS. At the same time, reduction in dry biomass of *S. hamata* was 82 % and 79 % when intercropped with V4 and Wab56-50, respectively. At 56

DARS, more than 90 % reduction was observed in each legume species. Likely, it was the combined effects of radiation unavailability and short growing period due to delayed sowing that reduced the biomass of the undersown legume.

For evaluation of the rice cover legume intercropping system both the amount of legume biomass and the current loss in rice grain yield are important. It is obvious that the choice of component crops determines the potential of this intercropping system (Figure 1). Adjusting the relative planting time to the benefit of the cover crops allows more biomass accumulation, but this occurs at the cost of rice grain yield. *C. cajan* at simultaneous planting accumulated about 12 t ha^{-1} biomass by completely outcompeting rice, whereas *S. hamata* produced about 4 t ha^{-1} at maximum, thereby causing a rice yield reduction of 50%. At later introduction times, the differences between V4 and Wab56-50 were apparent. The optimum date for intercropping *C. cajan* in upland rice considering a possible loss in rice yield of up to 5% was 30 to 35 DARS. *S. hamata* is only favourable with a competitive rice cultivar. It could be introduced two to three weeks after rice sowing.

Table 1. Treatments combinations

Factors	Levels	Description
Rice cultivars	2	Wab56-50, V4
Legume species	2	*C. cajan, S. hamata*
Legumes Introduction date[1]		
1997	4	0, 28, 56, 84 DARS [2]
1998	4	0, 14, 28, 56 DARS
Treatment combinations		
Rice monoculture	2	
Rices * Legumes * Dates	16	2 x 2 x 4
Legumes * Dates	8	2 x 4

[1] First sowing date in 1997 was June 18th, and July 2nd in 1998
[2] Days after rice sowing

Similar to legume biomass, N accumulation in leaves declined with later establishment dates of the cover crops. In the *C. cajan* fallow, the highest amount of incorporated residue N was 100 kg ha^{-1} at 0 DARS, followed by 52 kg N ha^{-1} at 14 DARS, 35 kg N ha^{-1} at 28 DARS and 2.5 kg N ha^{-1} at 56 DARS. In *S. hamata* incorporated N amounted to 25, 18, 7 and 1 kg ha^{-1}, respectively. Rice grain yields following a *C. cajan* fallow were significantly higher (P=0.001) than those following *S. hamata*. Legume establishment date in the previous year had a significant effect (P=0.0036) on grain yield. The residual effect of the *C. cajan* fallow on the subsequent rice yield depends on the amount of available N added to the system (Akanvou *et al.* 2000, Becker and Johnson, 1998). Unexpectedly, early-established fallow did not produce the highest increase in grain yield. The relative increase in grain yield of subsequent rice (relative to rice after rice without fallow crop) following *C. cajan* was 6.7 % at 0 DARS, 23 % at 14 DARS, 34 % at 28 DARS and about 4 % at 56 DARS.

The results indicate that apart from the amount of fallow N returned to the soil, its availability to the next rice crop is important for the short-term gain of this production system.

It was demonstrated that grain yield in rice cover legume intercropping systems was determined by a combination of factors which comprised rice cultivar, legume species and sowing dates of the legumes. Additionally, the choice of suitable combinations of species should consider residual effects of the legumes in either short or long rotation periods.

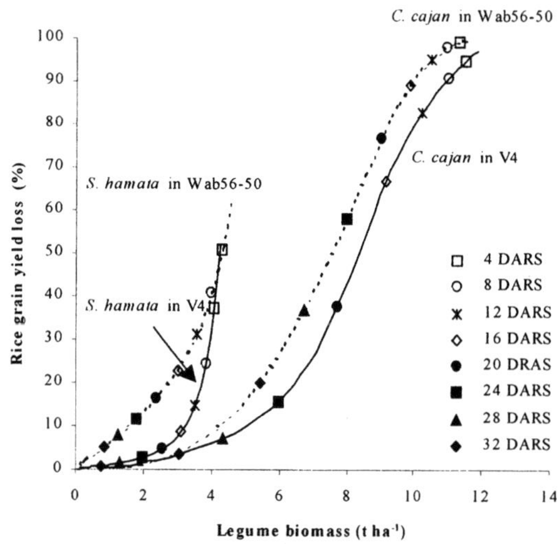

Figure 1. Relationships between rice yield loss and accumulated legume biomass. Symbols on the lines indicate the respective time of legume sowing.

References

Akanvou R, Becker M, Chano M, Johnson DE, Gbaka-Tcheche H and Touré A 2000 Biol. Fertil. Soils, 31, 501-507.

Becker M and Johnson DE 1998 Biol. Fertil. Soils 27, 358-367.

Buckles D and Triomphe B 1999 Agroforestery Systems 47 (1/3), 93-122.

Kropff MJ and Van Laar HH 1993 Cab International Publishers, UK., pp. 267.

Nye PH and Greenland DJ 1960 Technical Communication n° 51. Commonwealth Agricultural Bureaux, Farnham Royal, England.

W. J. Horst et al. (Eds), Plant nutrition – Food security and sustainability of agro-ecosystems. 1002~1003, 2001

Dynamics, balance and recycling of nitrogen in lowland rice-based cropping system

R. K. Shrestha[1] and J. K. Ladha[2]

[1] *Soil Science Division, Nepal Agricultural Research Council, Khumaltar, Lalitpur, Nepal, shrestha@infoclub.com.np*

[2] *International Rice Research Institute, P. O. Box 3127, Makati Central Post Office, 1271 Makati City, Philippines, j.k.ladha@cgiar.org*

Key words: balance, dynamics, nitrogen, recycling, transition crop

Abstract

In intensive rainfed lowland, a large amount of residual mineral-N is lost through leaching and denitrification at flooding of rice (*Oryza sativa* L.). The conservation and recycling of this N are essential to maintain groundwater quality and system sustainability. Experiments conducted in rice-sweet pepper (*Capsicum annuum* L.) in farmers' fields at four sites aimed to (i) quantify the N release from dry-to-wet transition crops residues in combination with mineral N, and (ii) estimate the soil N balance. Significant build up of soil NH_4-N (24 kg ha^{-1}) from transition crop residues such as indigo (*Indigofera tinctoria* L.) alone or mixed with mungbean (*Vigna radiata* L.) was observed at 15 d after incorporation (DAI), and declined thereafter. With corn (*Zea mays* L.), NH_4-N build up was observed at a later stage (60 DAI). Nitrogen release from corn residues was lower compared to indigo and mungbean. It was improved when mixed with mineral N. Nitrate-N peaked in the upper soil layer before flooding followed by its leaching and disappearance later. Crop residues incorporated plot maintained low NO_3 throughout the soil profile. An increasing ability of the transition crop to capture soil N was observed with an increase in initial soil mineral N (r = 0.96**). This resulted in an increase in N uptake (r = 0.84*), indicating recycling of residual soil N. Nitrogen balances indicated that integration of a transition crop reduced N losses by 33 to 72%. Nitrogen released from crop residue incorporation could supply N equivalent to 87 kg ha^{-1} to rice. The results suggest that a transition crop alone cannot completely reduce the N losses, and therefore strategies for reducing N fertiliser rates to match N demand of dry season crops are needed.

Introduction

Rainfed lowland rice encounters a complex and unpredictable environment. Rainfed lowlands in Ilocos Norte, Philippines, have a system of wet season (WS) rice followed by dry season (DS) high value crops: sweet pepper (*Casicum annuum* L.), tomato (*Lycopersicum esculentum* L.), garlic (*Allium sativum* L.), mungbean, corn, and tobacco (*Nicotiana tabacum* L.). The transition from DS aerobic to WS anaerobic soil conditions occurs for approximately 2½ months. Farmers usually apply high N fertiliser to the DS crops. Large amount of mineral N is accumulated (101-627 kg ha^{-1} in 100-cm soil profile) (Shrestha and Ladha, 1998) resulting large losses (549 kg ha^{-1}) in rice-sweet pepper rotation (Tripathi *et al.*, 1997). The study had the following objectives: (i) to quantify the amounts of N released from transition crops' residues, and (ii) to estimate the soil mineral N balance.

Materials and methods

The 4 sites with the sweet pepper-rice cropping system in Ilocos Norte, Philippines were selected out of 7 sites used for other study based on NO_3 levels in groundwater. The soils were Inceptisol with organic C 5.4-8.5g kg^{-1}, Kjeldahl N 0.7-1.2g kg^{-1} in upper 25 cm depth and pH 7-8.

Amount of fertiliser-N applied to the sweet pepper farmers in sites 1, 3, 5 and 7 were 600, 698, 467, and 475 kg ha^{-1}, respectively. At site 1, the transition crops grown were indigo, indigo+mungbean and corn in addition to fallow. During the WS, rice received either no fertiliser-N, applied as prilled (PU) or tablet urea (TU) alone at 87 kg N ha^{-1} or in combination ($^1/_3$ fertiliser- + $^2/_3$ residue-N) or equal amount of residues-N. At other sites, the transition treatments had indigo and fallow, and WS treatments include no fertiliser-N, fertiliser-N and residue-N equivalent to 87 kg N ha^{-1}. A randomised complete block design with 3 replications was used.

Soil samples were collected from 4 depths at 25-cm intervals and extracted with KCl. Bulk density, particle density and water-filled pore spaces were determined by gravimetric water content. Nitrate-N in KCl extracts was determined by Cd reduction (Dorich and Nelson, 1984) and NH_4-N by steam distillation method (Bremner, 1965).

Result and discussion

The majority of the extractable NH_4-N measured during rice remained in the 0- to 25-cm layer. The incorporation of transition crop residues affected NH_4-N levels mostly at 15 DAT. Ammonium-N release from indigo/mungbean residue (24 kg ha^{-1}) was comparable with indigo (Table 1). The NH_4-N release from indigo/mungbean residue was about three times higher than that of corn. A lower release of N by corn was most likely due to its high C:N ratio of 36 compared with 14 to 15 for indigo and mungbean. The remaining N from corn residue is likely to be mineralised during the succeeding years (Jensen, 1992). Residue quality had a significant effect on NH_4-N release, but the effects tend to disappear when combined with fertiliser-N (Table 1).

Table 1. Effect of transition crop residues and fertiliser-N on NH_4-N (0-25 cm) dynamics under rice at site 1.

	Treatment transition	WS rice[a]	Ammonium-N (kg ha^{-1}) 15 DAT	29 DAT	60 DAT	90
1	Fallow	0 N	4.1 e	6.5	3.6 b	1.5
2	Fallow	PU	7.0 de	3.1	7.3 a	1.0
3	Fallow	TU	7.0 de	5.3	5.3 ab	1.1
4	Indigo	I R	11.9 cd	5.7	6.2 ab	1.3
5	Indigo	I R+ PU	18.8 ab	7.4	7.5 a	1.8
6	Indigo	I R+ TU	6.3 e	5.6	6.0 ab	1.7
7	Indigo+Mungo	IM R	24.1 a	4.8	3.1 b	1.9
8	Indigo+Mungo	IM R+ PU	18.7 b	9.5	5.9 ab	1.8
9	Indigo+Mungo	IM R+ TU	7.0 de	5.8	4.6 ab	2.8
10	Corn	C R	7.1 de	9.5	7.7 a	1.2
11	Corn	C R+ PU	13.5 bc	5.1	3.4 b	1.1
12	Corn	C R+ TU	5.1 e	9.0	4.7 ab	1.1
		F test	***	ns	*	ns
		CV (%)	28	56	36	37

[a] 0 N, no nitrogen; PU, prilled urea; TU, tablet urea; IR, indigo residue; IMR, indigo mungo residue; CR, corn residue

Soil NO_3 during the rice crop declined at upper depths increasing at lower depths (Fig. 1). The decline or loss of NO_3-N in the no-residue-incorporated plots was higher than in the incorporated. At 7 d before trans-planting rice, soil NO_3-N was low in 0-25 cm depth, due to catch crop N uptake, but a peak was observed at 25- to 50-cm depth. At 29 DAT, substantial NO_3-N moved down to 75- to 100-cm depths. At 60 DAT, NO_3 started to disappear at lower depths and was minimal at 90 DAT indicating a large loss of NO_3-N, particularly in the farmers' practice.

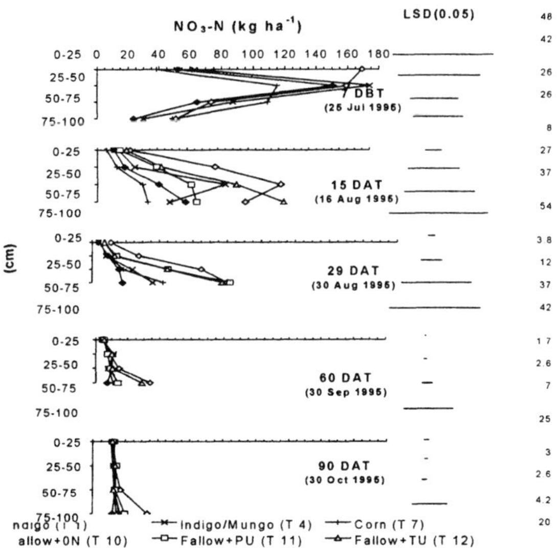

Figure 1. Soil NO_3-N during rice crop as affected by time and transition crops.

As indigo residue resulted in low levels of NO_3 throughout the profile, it may indicate that indigo was able to capture N. Mineral N balance indicated a N loss of 208 (corn) -516 kg (fallow) ha^{-1} (100cm soil depth) (Table 2). The efficiency of transition crops in decreasing N loss was in the order of corn > indigo+mungbean > indigo. At sites 3 and 5, losses were similar to site 1. At site 7, the loss in the indigo plot was negligible. The reduction in N loss due to integration of a catch crops ranged from 33 to 72% of the N lost in the fallow. The inability of the transition crops to reduce N loss at sites 1, 3, and 5 compared to 7 is due to high build-up of soil mineral N, which was beyond the capacity of the transition crop to capture soil N. This indicates, the strategy of the N transition crop alone is not effective enough to reduce N loss. There is an urgent need to explore other avenues, especially reducing the N fertiliser rates to better match their demand for N to DS crops.

Table 2. Apparent mineral N balance (kg ha^{-1}) for 0-100 cm soil depth.

Treatment DTW transition	Wet season	Soil mineral N Max. (A)	Min. (B)	Crop N uptake Transition crop (C)	Rice (D)	Apparent loss of soil mineral-N (A-B-C-D)
		Kg ha^{-1}				
		Site 1				
Fallow	0 N	602	43	-	43 f	516
Fallow	PU	602	44	-	78 b	480
Fallow	TU	602	55	-	91 a	456
Indigo	I R	604	51	151 a	71 cde	338
Indigo	I R+ PU	604	51	151 a	66 cde	343
Indigo	I R+ TU	604	53	151 a	70 cde	337
Indigo + Mungo	IM R	539	40	118 b	65 de	318
Indigo + Mungo	IM R+ PU	539	55	118 b	72 cde	303
Indigo + Mungo	IM R+ TU	539	50	118 b	74 bc	299
Corn	C R	494	49	177 a	61 e	208
Corn	C R+ PU	494	43	177 a	64 de	211
Corn	C R+ TU	494	43	177 a	66 cde	209
CV (%)				*11*	*8*	
		Site 3				
Fallow	0 N	908	216	-	95 b	597
Indigo	I R	744	156	205	131 a	252
CV (%)					*4*	
		Site 5				
Fallow	0 N	506	43	-	38 c	425
Fallow	PU	506	76	-	80 a	350
Indigo	I R	514	54	126	57 b	277
CV (%)					*64*	
		Site 7				
Fallow	0 N	313	58	-	47 b	208
Fallow	PU	313	78	-	80 a	155
ndigo	I R	260	46	131	58 b	25
CV (%)					*10*	

References

Bremner JM 1965 *In* Methods of soil analysis. Part 2. Eds. CA Black *et al.* pp. 1149-1179. ASA Madison, WI.

Dorich RA and Nelson DW 1984 Soil Sci. Soc. Am. J. 48, 72-75.

Jensen ES 1992 J. Soil Sci. 43, 335-345.

Shrestha RK and Ladha JK 1998 Soil Sci. Soc. Am. J. 62, 1610-1619.

Tripathi BP, Ladha JK, Timsina J and Pascua SR 1997 Soil Sci. Soc. Am. J. 61, 812-821.

Effects of cover crops species on *Passiflora edulis* nutrition

L. R. M. de Andrade, A. M. de Carvalho, I. de C. Mendes, L. Vivaldi, C. T. Karia and N. T. V. Junqueira
Embrapa Cerrados, BR 020, Km 18, Planaltina, DF, Zip Code 70301-970, Brazil, leide@cpac.embrapa.br

Key words: Cerrados region, soil enzymes, soil nematodes, soil quality

Abstract

This work was carried out to evaluate the impact of cover crops species in the nutrient cycling, soil quality, and passion fruit (*Passiflora edulis*; PF)–productivity. The treatments consisted of four leguminous species: forage *Arachis pintoi*, *Crotalaria spectabilis*, *Canavalia ensiformis*, *Cajanus cajan*, cultivated in the interrow area, plus a control treatment (spontaneous species). In the first year, there was a significant effect of the cover crop on PF fruit production and on phosphorus (P), sulphur (S) and copper (Cu) concentration in leaves. The highest fruit productions were observed in the *A. pintoi* and the control treatments. The highest concentration of P was found in PF plants cultivated with *C. ensiformis*. Nevertheless, neither colonisation of plant roots by vesicular-arbuscular mycorrhizal (AM) fungi nor soil phosphatase activity, nor soil P content could explain the differences in the tissue P concentration. The cover crop and PF plants increased the number of AM spores in the soil. *C. spectabilis* negatively affected the S content in PF plants tissue. The fruit production was significantly correlated with the content of S in tissue, but not with the soil sulfatase acitivity. In the control treatment, PF plants accumulated less Cu than with cover crops species. Biochemical parameters such as soil phosphatase and arylsulfatase activities were not significantly different among treatments, which is possibly related to soil fertility. In PF rows, the root-knot nematode *Meloidogyne* was found to be mainly associated with the *C. spectabilis* treatment. However, no negative effects on PF fruit production were associated with the presence of this plant-parasite.

Introduction

The interrows of passion fruit (*Passiflora edulis* f. *flavocarpa* Deg.; PF)–orchards located in the Cerrados region of Brazil are usually kept without weeds, that can compete for water and nutrients with the main crop. Over years, it is common to observe the physical, chemical and biological degradation of the soils in these rows due to the erosion and leaching of nutrients caused by strong rains, machinery traffic, heavy use of pesticides, and organic matter degradation. Such negative effects can cause reductions in PF production and quality. The establishment of cover crops in the interrows is an alternative to overcome these problems. Cover crops can protect the soil, reduce weeds and pest proliferation and avoid soil nutrients leaching. The root system of cover crops can also alter the soil biological properties with positive effects on nutrient cycling, soil quality and plant productivity.

This work was carried out to evaluate the impact of different species on soil quality and on PF nutrition and production.

Materials and methods

The field experiment was established over a 2-year period beginning in January 1999. The soil was a clay-loam red yellow Oxisol, whose characteristics are described in Table 2. The PF seedlings were spaced 3 m (between plants) x 2,5 m (between rows). Four leguminous species: *Arachis pintoi*, *Crotalaria spectabilis*, *Canavalia ensiformis* L., and *Cajanus cajan* were cultivated in the interrow strips of the orchard. Each treatment was replicated three times in a randomised block design. The seeds of the leguminous species were hand-planted in rows 0.50 cm apart, except for forage groundnut that was planted vegetatively in furrows 5-7 cm deep. *A. pintoi* is a perennial species, whereas the other legumes were sowed annually, in January, and harvested at their maximum flowering stage, about 180 days after emergence. After cutting cover crops were left as mulch on the soil surface. At the 10th month the concentration of major and minor plant nutrients in leaves of PF was determined. Fruit production was measured regularly over a 15 months period.

In the second year, immediately before cutting the cover crops, colonisation with and spores of vesicular-arbuscular mycorrhizae (AM) as well as nematodes populations were assessed in roots and soil samples collected from 0 to 20 cm depth. For the determination of soil chemical properties (pH_{H2O}, P, S, B, exchangeable cations) and organic matter content soil samples were collected from 0 to 5 cm and from 5 to 20 cm depth within crop rows and in interrow areas. Soil subsamples collected from 0 to 5 cm depth in the cover crops area were used for determination of soil enzymes related to P (acid phosphatase) and S (arylsulfatase) cycling.

Data were analysed using the SAS software package (SAS Institute, 1996, Cary, NC). Duncan´s Test was used to test for treatment differences. Pearson's correlation coefficient was used to measure the response of variables to the treatments.

Results and discussion

There was a significant cumulative effect ($P < 0.05$) of the cover crops on PF fruit production over the 15 months (Table 1). Highest fruit productions were obtained in the *A. pintoi* and the control treatment. Since *A. pintoi* had an initially slow establishment, the reduction in fruit production in the others cover crop treatments is likely due to the competition during the vegetative development of PF.

The preliminary evaluation showed a significant effect ($P < 0.05$) of cover crops on sulphur (S), phosphorus (P) and copper (Cu) concentration in PF leaves. With *C. spectabilis*, PF plants had the lowest S tissue concentration and this was significantly correlated with the fruit production ($r = 0.0167$). But sulfatase activity in soil was not related with the S content in the tissues of any specie. In the control treatments, PF accumulated less Cu than in cover crop treatments.

The benefits of a species used as green manure for crop production partly depend on nutrient transfers from their tissue to other plants. These transfer mechanisms include mulch decomposition, leaching of nutrients from tissue in the soil, underground root decomposition, root exudation and direct transfer via mycorrhizae. Two years after the onset of the experiment, soil analyses indicated that cover crops induced slight chemical (pH, nutrient availability) changes in the soil (data not shown). Enzyme activity in soil is a function of microbial and root activity, which are in turn regulated, among other factors, by substrate quality (nutrient availability). The activities of soil phosphatase (on average 275 μ *p*-NP g^{-1} soil) and arylsulfatase (on average 14 μ *p*-NP g^{-1} soil) were not significantly different among treatments ($P > 0.05$), which is possibly related to S and P levels in soil.

Mycorrizal fungi in association with plant roots may also improve P uptake from soils. The highest concentration of P was found in plants with *C. ensiformis* as a cover crop, and the lowest in the control treatment. However, soil P, soil phosphatase activity and mycorrhizal associations with roots were only weakly correlated with differences in the P concentration of PF leaves ($r = 0.937$, 0.871 and 0.463, respectively). Overall, after two years, the number of AM spores in the whole area increased more than eight times. Root colonisation with AM was, on average, 57 % in cover crops roots and 70 % in PF roots.

More then 11 genera of nematodes were found in the soil and root samples, including phytoparasite and saprophyte species. The population distribution pattern of those with higher incidence in the orchard is shown in Figure 1. *Meloidogyne* is one of the most harmful nematode genus for PF in Brazil (Sharma *et al.*, 1999). The highest incidence of this root-not nematode was observed in *C. ensiformis* (91,5%, with 84 individuals per 50 g soil). In PF rows, the genus *Meloidogyne* was found mainly associated with *C. spectabilis* (45,7% of incidence, with 14 individuals per 50 g soil). However, no negative effects of this parasite on PF fruit production were observed in our study.

References

Sharma RD, Junqueira NTV and Gomes AC 1999 *In* Comunicado Técnico. Ed. Embrapa Cerrados. pp. 1-4. Embrapa Cerrados, Planaltina, DF.

Table 1. Effect of cover crop species on the nutrient content in passion fruit leaves.

Cover crop	Fruit production	Nutrient concentration in leaves										
		N	P	K	Ca	Mg	S	B	Cu	Fe	Mn	Zn
	t ha^{-1}	g kg^{-1}						mg kg^{-1}				
A. pintoi	17.6 a	39.5	2.4 cb	16.3	14.1	2.97	3.57 a	23.9	5.52 a	128	54.4	33.0
C. spectabilis	13,0 b	38.3	2.6 ba	19.0	13.8	2.97	3.10 b	22.0	5.35 a	155	49.5	33.3
C. ensiformis	14,3 b	38.7	2.7 a	18.5	14.0	3.25	3.40 a	22.9	5.75 a	124	42.7	35.0
C. cajan	14,4 b	38.4	2.6 ba	19.3	13.1	2.80	3.55 a	23.5	5.67 a	118	44.7	35.9
Control	16,1 ba	39.4	2.3 c	20.7	14.7	3.10	3.67 a	24.0	4.25 b	122	49.5	32.4
CV (%)	10.3	7.6	5.0	17.2	13.5	10.3	7.4	8.0	9.9	16.9	18.0	6.3

Means with the same letter, in the same column, are not significantly different.

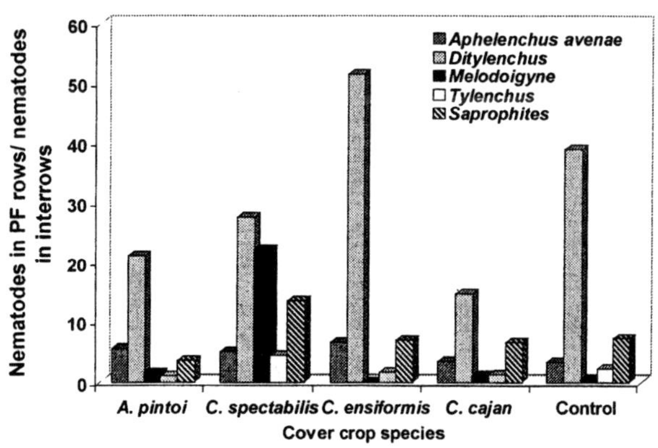

Figure 1. Relationship between nematode density in PF rows and in interrow areas.

W. J. Horst et al. (Eds), Plant nutrition – Food security and sustainability of agro-ecosystems. 1006~1007, 2001

Cultivation of legumes and grass/legume mixtures on arable land as a sustainable management

W. Grzebisz[1)], J. Kryszak[2)], W. Szczepaniak[1)] and R. Gaj[1)]

[1)] *Agricultural University, Department of Agricultural Chemistry, grenata@au.poznan.pl*

[2)] *Department of Grassland Sciences, street Wojska Polskiego 71F, 60-625 Poznań, Poland*

Key words: grass/legume mixtures, green manure, legumes, organic N, wheat grain yield

Abstract

Including into crop rotation leguminous plants is probably one of the ways to decrease the negative environmental impacts of excessive amounts of mineral nitrogen N applied by farmers. The agronomical value of the organic N of legumes or their mixtures with grasses as a potential N source for subsequntly cultivated winter wheat was assessed in field experiments. The potential nutritional value of organic N was affected both by biodegradability, of plant residue which could be related to botanical composition of a given sward and structure of plant organs ploughed under and their N content. The obtained yields of winter wheat grain ranged from 5.1 to 6.2 t/ha, i.e. satisfactory both for farmers and environmentalists'.

Introduction

Concern for negative environmental impacts of excessive amounts of mineral N fertiliser used by farmers in the high-input agricultural technologies have grown dramatically in the last two decades (Jordan and Weller, 1996; Voet *et al.*, 1996). Agricultural systems to protect the environment are under broad investigation. Legumes, which agronomical value has been recognized for centuries, may recommend the most "environmentally friendly" farming system (Badaruddin and Meyer, 1990; Danso and Papystylianou, 1992).

The objective of this study was to explore organic N of legumes or their mixtures with grasses as a potential N source for subsequent grown cereals.

Methods

An assessment of N fertilising value of two year old grass/legume mixtures was performed on the basis of two series of experiments conducted in years 1987-1990 and 1997 - 2000. In the first step of the experimental procedure the 4th cut and remaining plant residues of the two-old sward were ploughed down. In the second step, the main plots were divided into two sub-plots and were sown by winter wheat. One part of each sub-plot was fertilised with 100 kg N/ha and the other one was not fertilised with mineral N.

Results and discussion

The N-degradability balance was calculated on the basis of data on roots (+ sward) biomass and N amounts incorporated into the soil after ley plowing at the end of September. In the first experiment the positive N balances were found both for red clover and its mixture with italian ryegrass (Tab.1). Yields of plant residues decreased in the order clover (8.3 t ha^{-1}), clover/ryegrass mixture (7.1 t ha^{-1}) and ryegrass (6.7 t ha $^{-1}$). Nitrogen concentration in clover tops, treated as green manure, ranged from 2.8% to 4.9% and in roots from 1.3% to 2.3%. For ryegrass these values were in the range 1,8 % - 2.7% and 1.1% - 1.6%, respectively. Yields of winter wheat grain were high and significantly responded to the net N supply from organic N resources. It was found, using the quadratic regression model (data not shown), that the net N balance about + 80 kg N ha^{-1} secured to harvest the highest grain yield of winter wheat, without fertiliser N application.

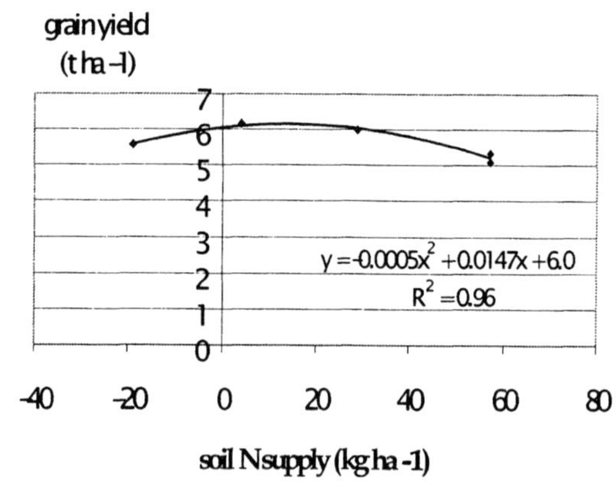

Figure 1. Relationship between total N supply and grain yield

In the second experiment the biomass of sward residues was high and accordingly to the sward botanical compostion, varied from 9.6 to 12 t ha^{-1} (Tab.2). Nitrogen concentration in the above-ground sward ranged from 2.7% to 3.3% and in roots (+ stubble) from 1.55% to 1.77%. As is shown in Table 1, the highest values of the net N supply were found for clover/grass mixtures. Winter wheat significantly responded to N supply from organic sources (Fig.1). The highest yields of grain were harvested on treatments where the net N balance was slightly positive (+ 15 kg N ha^{-1}).

In this studied case, the net N balance amounting to + 15 kg N ha^{-1} was sufficient to obtain the highest yield of wheat grain.

Conclusions

The results indicate the cheapest way to attain sustainable management of arable soils. The net N supply value, calculating on the basis of the threshold N value of 1.8% in plant residues, might be used as an useful indicator of organic nitrogen contribution of legumes and their mixtures with grasses to subsequent cereals. Therefore, legumes or legume/grass mixtures grown on arable soils fulfill both the environmentalists requirements for the *low-input* N system and also farmers requirements for the *high-output* farming practice.

References

Badaruddin M and Meyer DW 1990 Crop Sci. 30, 819-825.

Danso SKA and Papastylianou I 1992 J. Agric. Sci. Cambrigde 119, 13-18.

Jordan TE and Weller DE 1996 BioScience 46, 655-664.

Lewis O 1986 Plants and nitrogen. Studies in Biology no. 166. p. 104. E. Arnold, London.

Voet van der E, Kleijn R and Haes de H Environ. Conservation 23, 120-132.

Table 1. Potential supply of N by different types of sward and grain yield of winter wheat

Type of sward	kg N ha^{-1}			Grain yield
	needs*	supply**	balance	t ha^{-1}
Red clover	151.2	234.5	+ 83.3	5.8
Italian ryegrass	115.2	115.9	+0.7	5.2
Red clover/ryegrass mixture	127.8	158.1	+ 30.3	5.6

* calculated with respect to plant residues biomass and threshold N value =1,8% (Lewis, 1985); ** N incorporated into soil with plant residues

Table 2. Potential supply of N by different types of sward and grain yield of winter wheat

Type of sward	kg N ha^{-1}			Grain yield
	needs*	supply**	balance	t ha^{-1}
[1]Rc + F + T	164	221	+ 57	5.33
Rc + IR + C	144	201	+ 57	5.09
Rc + Wc + Pr + Ir	167	196	+ 29	5.96
Aa + F + Ir	200	104	+ 4	6.15
Aa + Ir + C	221	202	- 19	5.56

[1]Rc – Red clover; F – Festulolium; T – Timothy grass; Ir – Italian ryegrass; Wc – White clover; Aa-Alfalfa; Pr - Perennial ryegrass; C-Cocsfoot

*, ** see Table 1

W. J. Horst et al. (Eds), Plant nutrition – Food security and sustainability of agro-ecosystems. 1008~1009, 2001

Productivity and yield quality of fodder galega (*Galega orientalis Lam.*) - Grass mixed swards

A. M. Adamovich

Department of Crop Production, Faculty of Agriculture, Latvia University of Agriculture, Liela Str. 2, Jelgava, LV - 3001, Latvia, alexadam@cs.llu.lv

Key words: fodder galega, grasses, mixtures, productivity, yield quality

Abstract

Fodder galega (*Galega orientalis* Lam.) is a promising, perennial and fast - growing forage legume, which is useful for the production of diverse grass forage. Field trials (1997 to 2000) were established with the aim of studying continuous green forage production from fodder galega-grass swards in the stage of intensive growth. Thirty mixed binary- and multi-component swards were developed on a stagnic Luvisol and fertilised with either 0 kg N or 90 kg N ha^{-1}. The swards were composed of fodder galega cv. 'Gale' and seven grass species. Swards were cut three to four times during the growing season. It is concluded that fodder galega in mixtures with grasses of various growth patterns provides continuous green forage production during the whole summer season. The botanical composition and frequency of cutting affected the productivity of a sward (7.5 to 11.2 t ha^{-1} DM), metabolizable energy (8.32 – 13.41 MJ kg^{-1} DM) as well as NDF and ADF contents. The swards receiving no fertiliser N were more productive and of high quality.

Introduction

Fodder galega (*Galega orientalis* Lam.) is an early-maturing perennial forage legume high in productivity and the capacity to fix atmospheric nitrogen (Raig, 1994). It originates from the mountain regions of the Caucasus. Experiments on fodder galega show that this long-lived legume survives in pure stands of 15 to 19 years and provides annual dry matter (DM) yields of 9.56 to 11.0 t ha^{-1} (Adamovich, 1998; Driķis, 1995). Compared with other legumes, pure fodder galega stands are not thinning out between the years of stand use, providing stable yields of green feed and seeds. The use of the N-fixing symbiotic potential of fodder galega grown in mixtures with grasses contributes to the production of ecologically safe forage and animal products.

Materials and methods

Field experiments were conducted during 1997 to 2000 in a stagnic Luvisol (FAO, 1988) with pH_{KCl} 6.9, containing mobile P 80 mg kg^{-1} (DL – method) and K 99 mg kg^{-1}, humus content 22 to 25 g kg^{-1} of soil. Binary- and multi-component seed mixtures were composed of fodder galega 'Gale' and grass species: meadow foxtail (*Alopecurus pratensis*), cocksfoot (*Dactylis glomerata*), meadow fescue (*Festuca pratensis*), red fescue (*Festuca rubra*), timothy (*Phleum pratense*), perennial ryegrass (*Lolium perenne*), smooth meadow grass (*Poa pratensis*). Field trials on the productivity of fodder galega-grass swards were conducted, in which 7 binary- and 22 multi-component mixtures were included. The total seeding density of each seed mixture was 1000 geminating seeds per m^2. The ratio of fodder galega / grass seeds in mixtures was as follows: in 7 binary mixtures 400:600, in 14 three-component mixtures 400:300:300, in 5 four-component mixtures 400:200:200:200, in 1 five-component mixture 400:150:150:150:150 and in 2 six-component mixtures 400:120:120:120:120:120. The plots were fertilised as follows: P 40 and K 150 kg ha^{-1} and two N fertiliser treatments N0, N90 (2 x 45 kg ha^{-1}) as amonium nitrate. Stands were sown in early May 1997. Swards in binary - and multi-component mixtures were cut three to four times yearly. The botanical composition of the sward was determined at each cut. Each mixture was analysed for crude protein (CP) content by a modified Kjeldahl method. The content of crude fibre (CF), neutral detergent fibre (NDF) and acid detergent fibre (ADF) (analysed by van Soest, 1991) were determined for cut 1 and cut 2 in all mixtures in fodder galega bud stage. The difference between metabolisable energy (ME) was calculated based on the chemical composition of dry matter (DM), using digestabilty coefficients.

Results

Due to a slow growth pattern of pure fodder galega swards in the seeding year, the highest green fodder and DM yields were reached only in the second and third experimental year (Adamovich, 1998; Driķis 1995).

Inclusion of a grass component in a mixture resulted in yield increase by 27 to 35% in the first production year. In 3 production years of pure galega stands the following average yields of DM and CP were attained in the bud stage: 9.12 t ha^{-1} DM and 1.84 t ha^{-1} CP by 3-fold cutting during growing season, and 5.01 t ha^{-1} DM and 1.31 t ha^{-1} CP at four-cut treatments, respectively.

Significant changes in the productivity of galega-grass swards were observed in all the treatments depending on supply of N fertiliser and frequency of cutting. (Figs. 1 and 2.).

Receiving no fertiliser N, there were developed highly productive binary- and three-component galega-grass swards providing the following average yields of 9.82 to 10.0 t ha^{-1} DM and 1.82 to 1.96 t ha^{-1} CP at three-cut treatments.

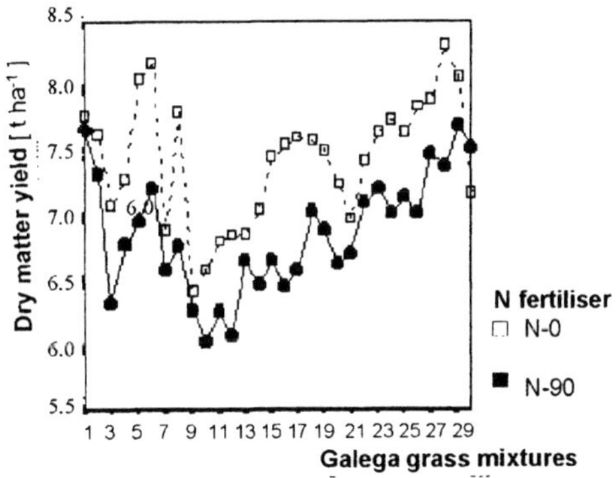

Figure 1. Productivity of fodder galega-grass swards depending on the level of N supply.

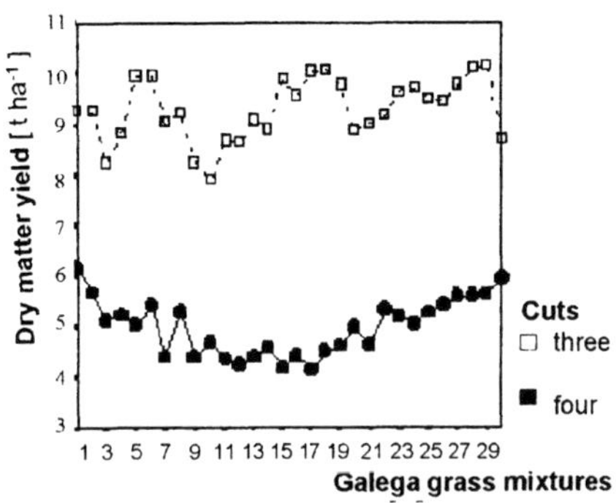

Figure 2. Productivity of fodder galega-grass swards depending on frequency of cutting

Highly productive binary swards were developed when growing fodder galega in association with *Alopecurus pratensis, Festuca rubra, Phleum pratense* and *Poa pratensis*. Most productive fodder galega-grass mixtures were composed of three components providing the following average yields: 10.01 t ha^{-1} DM and 1.86 t ha^{-1} CP. The following three-component mixed swards: *Galega orientalis* 40% + *Lolium perenne* 30% + *Phleum pratense* 30%; *Galega orientalis* 40% + *Lolium perenne* 30% + *Festuca rubra* 30%; *Galega orientalis* 40% + *Phleum pratense* 30% + *Poa pratensis* 30%; *Galega orientalis.* 40% + *Phleum pratense* 30% + *Festuca rubra* 30% provided the highest yields of DM with 11.06, 10.87, 10.86 and 10.51 t ha^{-1} respectively. The crop yield level in a sward was not significantly affected by the increase of the number of the components from 4 to 6 in a sward compared with binary- and three-component mixed swards, but it ensured the stability of yields between the production years.

Contrary to the opinion of Driķis (1995) that the application of N fertiliser in pure galega fields provided significant (1.90 - 2.41 t ha^{-1} DM) yield increase, our investigations show that the application of N fertiliser in binary- and multi-component fodder galega grass swards was detrimental to the total DM yields.

Split application of N fertiliser (2 x 45 kg ha^{-1}) negatively affected the proportion of galega in a sward resulting in decrease of DM yields by 1.7 t ha^{-1} at three-cut treatments compared with unfertilised plots. Frequently, 4-cut treatments had a negative effect on the productivity of galega-grass mixtures. The total yield of DM decreased by 4.05 t ha^{-1} or 45% in all experimental plots at 4-cut treatments.

In branching and bud stages fodder galega excelled with high crude protein contents of 295 ± 18 g kg^{-1} DM. In mixed swards the CP content in fodder galega declined to 276 ± 19 g kg^{-1} DM in bud stage; during early flowering, the average CP content was 206 ± 31 g kg^{-1} DM. This could be explained with a great proportion of plant leaves accounting for 549 ± 57 g kg^{-1} DM in fodder galega yield. The atmospheric nitrogen fixation fully met the demands necessary for the development of fodder galega and contributed to the growth of associated grasses in a sward in treatments receiving no mineral fertiliser N. The average content of metabolisable energy was 12.9 MJ kg^{-1} DM of fodder galega: in mixed galega-grass stands it was 11.9 ± 1,7 MJ kg^{-1} DM.

Literature findings (Driķis, 1995; Moller and Hostrup, 1996) indicate that there is a rapid increase of CF content including NDF and ADF fractions when cutting fodder galega in the early stage of flowering and later. Our studies showed that in mixtures with grasses, the average NDF content did not exceed 382 ± 24 g kg^{-1} DM, compared with a NDF content of 422 ± 38 g kg^{-1} DM in pure galega stands during early flowering.

This could be explained with a different proportion of plant leaves and their position in the canopy structure in galega-grass mixtures compared with pure galega swards. The slower growth pattern of associated grasses in a mixed sward resulted in slower maturation of grass leaves compared with galega, thus contributing to forage production of higher quality.

The results suggest that fodder galega in pure stands or in mixtures with grasses of various growth patterns were productive with high quality and persisted for long time periods. The highest productivity was obtained in three-component mixtures. Galega in pure or mixed stands fixed sufficient amounts of N to eliminate the need for additional mineral N application.

Mixed fodder galega-grass swards provided the production of high-quality grass, hay, and silage. The proper choice of mixed legume-grass swards enabled continuous production of grass during the summer period.

References

Adamovich A 1998 46-51. Jelgava, LV.

Drikis J 1995 Jelgava, LV.

Moller E and Hostrup SB 1996 Acta Agriculturae Scandinavica 46, 97-104.

Raig H 1994 5-24. Saku.

Van Soest PJ, Robertson JB and Lewis BA 1991 Dairy Science 74, 3583-3597.

W. J. Horst et al. (Eds), Plant nutrition – Food security and sustainability of agro-ecosystems. 1010~1011, 2001

Root growth and nitrogen utilization of a leek crop and an undersown catch crop

K. L. Nielsen and K. Thorup-Kristensen

Department of Horticulture, Danish Institute of Agricultural Sciences, Kirstinebjergvej 10, DK-5792 Aarslev, Denmark, kai.nielsen@agrsci.dk

Key words: *Allium porrum*, *Cichorium intybus*, cover crops, nitrate leaching, root growth, soil nitrate depletion

Abstract

After its late harvest a leek (*Allium porrum*) crop will normally leave large quantities of nitrogen (N) in the soil which may be subject to leaching. A possible method to reduce N leaching losses is to establish a catch crop before the harvest of the leek crop. It was the aim of this study to investigate the effects of incorporating a catch crop in a leek crop with minimal competition and to monitor how root growth affected N availability across soil depths at different distances between catch and leek crop rows. To reduce competition between both crops leek was planted at 0.5 and 0.75 m row distances and chicory (*Cichorium intybus*) as a catch crop at an interrow spacing of 0.75 m. Minirhizotron glass tubes were inserted in the middle of the interrow spaces (with or without catch crop established in mid July) and in the leek row. Leek roots grew to a depth of about 0.4 m by mid October with a high root density directly below the leek row but did not extend growth to the centre of the 0.75 m interrow. At the same time the catch crop had extended its roots to a depth of at least 1.25 m and was exploring the full width of the 0.75 m interrow with some roots even below the leek rows. The results show that it is possible to establish a catch crop in leek thereby depleting soil N and reduce the risk of N leaching without jeopardising crop yield.

Introduction

To reduce N leaching catch crops can be established in autumn to take up excess N left over by major crops. After the harvest of leek normally large quantities of N remain in the soil whereby the crop is usually harvested too late to allow the successful establishment of a catch crop thereby preventing N leaching. A possible alternative would be to establish the catch crop before the harvest of leek. To avoid negative effects on leek yield it is important to minimise competition with the catch crop.

Nitrate moves rather freely with the flow of water in the soil, and when present in deeper soil layers it will have to be transported from there by plant roots to become available for the succeeding crop. Although catch crops have been studied intensively, only a few studies have dealt with roots (Lainé et al,. 1993; Thorup-Kristensen, 1993, 2001; Voss *et al.*, 1998). Of the parameters compared, rooting depth was found to be closely related to N depletion (Thorup-Kristensen, 1993). Neither plant dry matter production and N content nor root growth in the upper soil layers was closely related to subsoil nitrate depletion (Thorup-Kristensen, 2001). Deep rooted catch crops will have access to a larger soil volume and can therefore deplete parts of the soil volume not available to more shallow rooted crops.

This study aimed at investigating the effects of incorporating a catch crop in leek with minimal competition and to monitor the effects of root growth on available soil N at various depths and distances between crop rows.

Materials and methods

The experiment was conducted in the fall of 1997 at the Research Centre Aarslev, Denmark (10° 27' E, 55° 18' N) on a typic Agrudalf soil. Leek had been planted by end May 1997.

To reduce competition between both crops leek had been planted in May 1997 at alternating rows spaced 0.5 and 0.75 m and chicory (*Cichorium intybus*) as a catch crop at the 0.75 m interrow spacing in mid July. Minirhizotrons glass tubes (of 70 mm outer diameter and 1.5 m length) were inserted in each plot by the end of May, prior to the planting of leek. They were placed in the centre of the leek row (reference row, position 0 m), between leek rows (position 0.25 m) and in the centre of the catch crop row (position 0.375 m). The minirhizotrons were installed at an angle of 30° from vertical, reaching a depth of approximately 1.25 m. Rooting depth was monitored at 14 day intervals and at 7 day intervals between mid August and late October. Root intensity (defined as the number of root intersections per cm along a 40 mm vertical and a 40 mm horizontal line painted on the minirhizotron tubes) was measured on the tubes down to 1.25 m by mid October. Similarly N_{min} was measured mid October at four depth intervals (0 – 0.25m, 0.25 – 0.50m, 0.50 – 0.75m and 0.75 – 1m). Temperature sums (day °C) were calculated as the sum of daily average temperatures.

Results and discussion

Significant differences in root depth penetration were found between leek (0.197 mm [day °C]$^{-1}$ equivalent to 2.9 mm day^{-1}) and the catch crop (1.31 mm [day °C]$^{-1}$ equivalent to 23.8 mm day^{-1}). This led to significant differences in rooting depth over time with leek reaching a maximum depth of about 0.4 m before harvest and the catch crop exceeding the depth of the minirhizotrons by late September (Figure 1). The presence of the catch crop did not lead to reductions in yield (number and dry weight of leeks m^{-2}) and quality (number of usable leek m^{-2}) of the leek crop (data not shown).

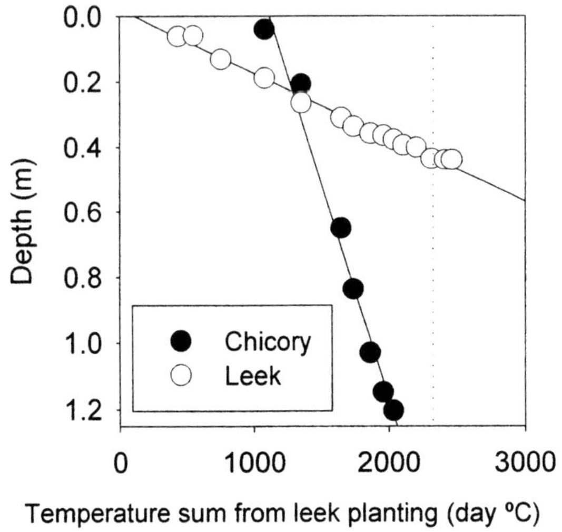

Figure 1. Rooting depth of leek (*Allium porrum*) and Chicory (*Chicorium intybus*) in a typic Agrudalf soil. Time is shown as cumulative temperature from May 25 (planting of leek). The vertical dotted line indicates mid October when root intensity was measured. Linear regressions for depth penetration of leek. Open symbols indicate the rooting depth (m) of leek [1.97×10^{-4} × temperature sum from planting – 0.0221; r^2 = 0.989] and and closed symbols that of chicory [1.31×10^{-3} × temperature sum from sowing – 0.462; r^2 = 0.979].

Mid October leek roots grew to a depth of about 0.4 m with high root intensity below the leek row (0 m from the reference row; Figure 2A), but did not extend its roots to the centre of the 0.75 m interrow (0.375 m from the reference row; Figure 2A, without chicory). At the same time the catch crop had reached a rooting depth of more than 1.25 m.

Mid October the nitrate content under the leek row (0 m from the reference row, Figure 2B, without chicory) was still very high, with 4 – 6 mg kg^{-1} particularly below 0.5 m, whereas the catch crop depleted the full width of the 0.75 m center row for nitrate (0 – 2 ppm) and also reduced the nitrate concentration under the leek row (2 – 4 ppm, Figure 2B, 0 m from the reference row, with chicory).

Our data confirm the results of Thorup-Kristensen (1993, 2001), that rooting depth and root intensity are closely related to nitrogen depletion. The results allow the conclusion that it is possible to establish an efficient catch crop in leek to deplete the soil of residual N thereby reducing the risk of leaching losses.

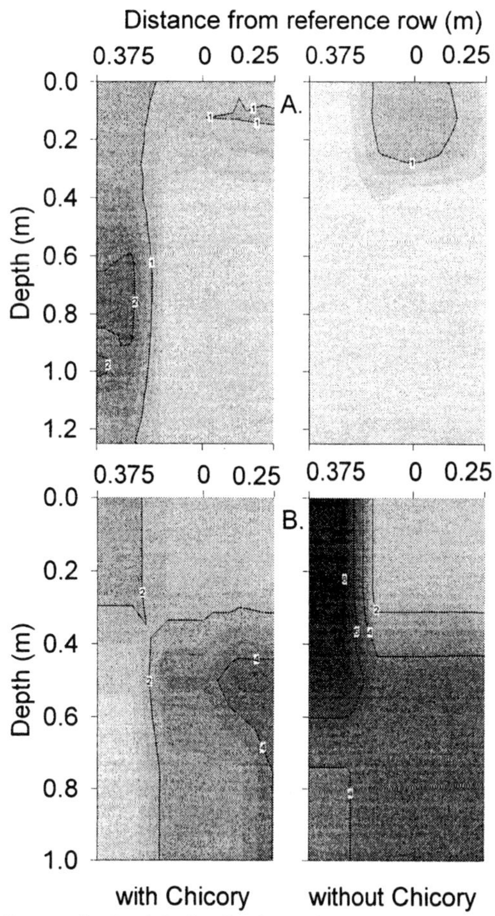

Figure 2. Spatial distribution of (A) root intensity (root intersections cm^{-1} line on the minirhizotrons) and (B) soil nitrate-N (μg nitrate g^{-1} soil dry matter) of leek mid October with (left) and without (right) chicory as a catch crop. Measurements were taken under a leek row (0 m from the reference row) between leek rows (0.25 m from the reference row) and under the catch crop (0.375 m from the reference row).

References

Lainé P, Ourry A, Macduff J, Boucard J and Salette J 1993 Physiol. Plant. 88, 85-92.

Thorup-Kristensen K 1993 Acta Agric. Scand., Sect. B, Soil Plant Sci. 43, 58-64

Thorup-Kristensen K 2001 Plant Soil., (in press).

Vos J, van der Putten PEL, Hussein MH, van Dam AM and Leffelaar PA 1998 Plant Soil 201, 149-155.

W. J. Horst et al. (Eds), Plant nutrition – Food security and sustainability of agro-ecosystems. 1012~1013, 2001

Competition for nutrients by key species of grassland systems on the tablelands of southeast Australia.

J. Hill[1)], R. J. Simpson[1)], A. D. Moore[1)] and D. Chapman[2)]

[1)] *CSIRO Plant Industry, PO Box 1600, Canberra, ACT, 2601, Australia, jacqui.hill@pi.csiro.au*

[2)] *Department of Animal Production, Institute of Land and Food Resources, University of Melbourne, Victoria, Australia, 3010, Australia, d.chapman@landfood.unimelb.edu.au*

Key words: *Bromus molliformis*, critical phosphorus, root morphology, *Vulpia* spp.

Abstract

Grassland systems of southeastern Australia are typically found on infertile soils, and fertiliser inputs are generally required to maintain productivity. Increased soil fertility can lead to changes in the botanical composition of such grasslands. This study addresses the emerging issue of changing botanical composition of pastures as a consequence of increased fertiliser use, by focussing on the P requirements of *Bromus molliformis* and a mixture of *Vulpia* spp. (*V. myuros* and *V. bromoides*) which are the dominant annual grasses in fertilised and unfertilised pastures, respectively. The response of these species to phosphorus (P) application was investigated. Significant differences were found between the species in their requirements for P and in their ability to acquire P from the soil. These differences were related to differences in root morphology.

Introduction

There is increasing pressure in the grazing industries of southern Australia to intensify production by higher stocking rates. These industries are predominantly located on soils deficient in N and P, and at many locations also deficient in S, K and/or micronutrients (Lewis and Sale, 1993). High stocking rates are only possible if P-fertilisers are applied which promote clover growth and N fixation. Thus, P-fertiliser also drives the N cycle of these grassland systems.

Significant botanical changes can accompany increased fertiliser use. For pastures on the Southern Tablelands of New South Wales we have observed a shift from *Vulpia* spp. towards *Bromus* spp. (*B. molliformis* and *B. diandrus*) where fertiliser is applied regularly. This shift has also been observed by Dowling *et al.* (1996). In these cases the change was beneficial, that is from weedy to more productive species. However, there is concern that, in the longer term, native perennial grasses may be displaced irreversibly by annual species (Moore, 1970).

In this study we investigate the growth of annual grasses (*B. molliformis* and *Vulpia* spp.) in response to the application of P. The purpose of the work is to understand how the nutrient requirements and growth response of these plants may contribute to changes in the botanical composition of pasture when P fertiliser is applied.

Materials and methods

Seed of *B. molliformis* and *Vulpia* spp. was collected from a field near Canberra, Australia. The plants were grown in a steam sterilised yellow podzolic soil (Northcote, 1979) taken from the A horizon (2-10 cm). pH was 4.8 in 0.01 M $CaCl_2$ and extractable P 10 ppm in 0.5 M $NaHCO_3$ (Colwell, 1963).

Application of P occurred shortly after germination using KH_2PO_4 solution that was sufficient to wet the first 2-3cm of soil at rates of 0, 4.5, 10.5, 21, 42 and 84 mg P pot^{-1} (6 replicates). Nitrogen in excess of plant requirements was applied as KNO_3 and other essential nutrients were added regularly to ensure they were in adequate supply.

All treatments were harvested when the shoots in the highest nutrient treatment reached approximately 6g dry weight. Roots were washed clean of soil particles and a sub-sample was stored in 50% ethanol. Plant material was dried at 70°C, weighed and digested with concentrated H_2SO_4 and 30% H_2O_2 at 320°C, and total P concentration was measured. Root length and diameter were measured from scanned images using the image analysis software RHIZO™ 4.0B (Régents Instruments, Quebec).

Results

Application of P increased the yield of both *B. molliformis* and *Vulpia* spp. The critical external requirement for P (the amount of P required to produce 90% of maximum shoot DM) was 13 mg pot^{-1} for *Vulpia* spp. It was significantly less ($P < 0.05$) than that of *B. molliformis* (25 mg pot^{-1}) (Fig. 1). In an analogous experiment with N application, there was no significant difference in the critical external N requirement between *B. molliformis* and *Vulpia* spp.

At zero P, the relative growth rate (RGR) of *Vulpia* spp. was significantly larger ($P < 0.05$) than that of *B. molliformis* (Fig. 1). The growth rate of both species increased with P application. However, at the higher P levels, the RGR of *B. molliformis* was significantly larger than that of *Vulpia* spp.

The critical internal concentration of P (the shoot P concentration at which 90% maximum shoot DM were obtained) of the two species did not differ significantly (*B. molliformis* 0.20% and *Vulpia* spp. 0.17%).

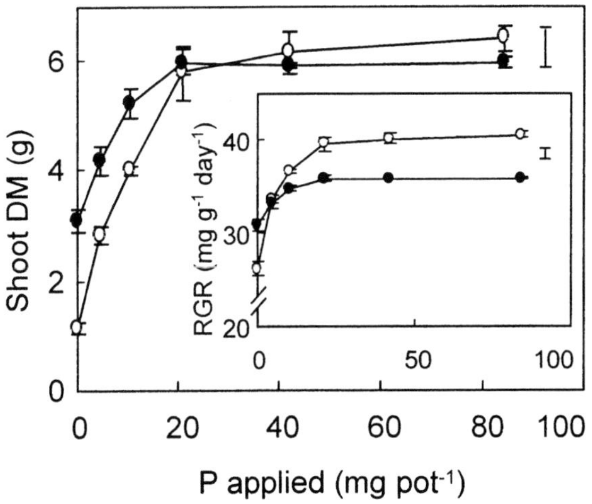

Figure 1. Effect of P application on shoot dry weight and relative growth rate of B. molliformis (o) and Vulpia spp. (●). The vertical bars represent LSD values ($P < 0.05$).

Table 1. Effect of P application on total root length, specific root length (SRL) and average root diameter of *Vulpia* spp. and *B. molliformis*. Standard errors of the mean are given in parenthesis.

	Total root length ($m\ pot^{-1}$)	SRL ($m\ g^{-1}$)	Average diameter (mm)
Vulpia spp.			
No added P	819 (93)	562 (32)	0.16 (0.003)
Above critical P	855 (39)	536 (27)	0.18 (0.003)
B. molliformis			
No added P	268 (11)	436 (22)	0.20 (0.003)
Above critical P	454 (26)	288 (15)	0.23 (0.004)

Root morphology differed between *B. molliformis* and *Vulpia* spp. The total root length and SRL were significantly larger ($P < 0.05$) for *Vulpia* spp. than *B. molliformis* (Table 1) in all treatments. *Vulpia* spp. did not respond in either total root length or SRL to increased P. The total root length of *B. molliformis* increased with increasing P and the SRL decreased, which has also been observed in other species (Bolan *et al.*, 1987). The average diameter of roots increased significantly ($P < 0.05$) with higher levels of P application but was always significantly ($P < 0.05$) larger for *B. molliformis* than for *Vulpia* spp. (Table 1).

Discussion

Vulpia spp. had a low external requirement for P and exhibited a lower response in RGR than *B. molliformis*. Both characteristics are typical of plants adapted to low soil P (Chapin *et al.*, 1993) and they indicate why pastures dominated by *Vulpia* spp. are considered to be poorly responsive to P (Rossiter, 1964). In comparison, *B. molliformis* had a high critical external requirement for P and a large response in RGR to increasing P application. *B. molliformis* has been reported to have both a greater response to P and a broader response range, characteristic of plants adapted to high soil P (Chapin *et al.*, 1993).

The results indicate different strategies in soil exploitation between both species. When nutrients are scarce, root surface area is most important for nutrient uptake (Nye, 1973). The high total root length of *Vulpia* spp. increased both the absorbing surface of the roots and the volume of soil explored compared with *B. molliformis*. The greater SRL of *Vulpia* spp. than *B. molliformis* appears largely the result of thinner roots and is important for P acquisition because carbon (C) invested in root length contributes more to surface area than C invested in root diameter (assuming a constant tissue density). These factors allow *Vulpia* spp. to reach maximum production at lower P levels than *B. molliformis*.

The results are consistent with the shift in botanical composition associated with P fertiliser use in pastures, and they help to understand why *Vulpia* spp. pastures are less productive than *Bromus* spp. pastures. Because the responses to N were similar for the two species, we conclude that the main nutritional influence over the shift in botanical composition of these species is their different abilities to acquire P and to respond to P application.

References

Bolan NS, Robson AD and Barrow NJ 1987 Plant Soil 104, 294-298.

Chapin FS III, Autumn K and Pugnaire F 1993 Am. Nat. 142, 78-92.

Colwell JD 1963 Aust. J. Exp. Agric. Anim. Husb. 3, 190-197.

Dowling PM, Kemp DR, Michalk DL, Klein TA and Millar GD 1996 Rangel. J. 18, 309-326.

Lewis DC and Sale PWG 1993 *In* Pasture management technology for the 21st century. Eds. DR Kemp and DL Michalk. pp. 38-50. CSIRO, Australia.

Moore RM 1970 *In* Australian grasslands. Ed. RM Moore. pp. 169-190. Australian National University Press, Canberra.

Northcote KH 1979 A factual key for the recognition of Australian soils, 4th edn. Rellim, Adelaide.

Nye PH 1973 J. Exp. Bot. 24, 783-786.

Rossiter RC 1964 Aust. J. Agric. Res. 15, 61-76.

W. J. Horst et al. (Eds), Plant nutrition – Food security and sustainability of agro-ecosystems. 1014~1015, 2001

Water and nutrient balance under slash-and-burn agriculture in the Eastern Amazon, Brasil - The role of a deep rooting fallow vegetation

R. Sommer[1)], T. D. A. de Sá[2)], K. Vielhauer[1)], P. L. G. Vlek[1)] and H. Fölster[3)]

[1)] ZEF - Center of Development Research, University of Bonn, Walter-Flex-Str. 3, Bonn, D 53113, Germany, r.sommer@uni-bonn.de

[2)] EMBRAPA-CPATU Belém, Dr. Eneas Pinheiro s/n, Belém, 66095-100, Brazil

[3)] Institute of Soil Science and Forest Nutrition, University of Göttingen, Büsgenweg 2, Göttingen, 37075, Germany

Key words: leaching, nutrient recycling, secondary forest, shifting cultivation, subsoil nutrient accumulation

Abstract

A 4-year-old fallow vegetation was able to noticeably deplete the soil water store down to 6 m depth. The annual uptake from 0.9 to 6 m soil depth reached 400 mm. An equivalent uptake of dissolved nutrients from this layer is likely. Thus, the fallow vegetation might pump up nutrients, which are out of reach for the shallow roots-system of the crops of a preceding cultivation phase. This feature, however, limits the scope of prolonged cropping and any agriculture that reduces the vitality of the fallow vegetation.

Introduction

In the Bragantina region of the Eastern Amazon, shifting cultivation predominates, which is mostly practiced by smallholders with properties smaller than a hundred hectare. After land preparation by slash-and-burn, maize (*Zea mays* L.), beans (*Vigna unguiculata* L.) and cassava (*Manihot esculanta* CRANTZ) are sequentially cultivated for a period of 1.5 to 2 years. In the subsequent fallow period a woody secondary vegetation can re-establish. Nowadays, however, shifting cultivation is under change. Demographic pressure on land leads to an intensified cultivation with shortened fallow periods (3-7 years) and a move to semi-permanent crops like passion fruit (*Passiflora edulis*). This development endangers the essential function of the fallow period to restore soil fertility. As a long-term consequence soil fertility deteriorates and land degrades creating a need for encroaching on primary forest areas.

Recent studies of the deeper subsoil using pits or auger revealed a deep-extending root system for primary forests in South-Pará (Nepstad *et al.*, 1991) as well as for the fallow vegetation in the Bragantina region (Sommer *et al.*, 2000). In the south of Pará, Nepstad *et al.* (1994) showed that deep roots are necessary for a sufficient water supply of the evergreen forest in the dry season. This might also be the case for the evergreen secondary vegetation in the Bragantina region. To examine this assumption, we monitored the deep soil water depletion under a fallow vegetation during the dry season in 1997.

The feature of the fallow vegetation to establish a deep root system might also be important regarding the nutrient dynamics of shifting cultivation. If the root system was important for deep-soil water uptake, an equivalent nutrient uptake might be likely. This would substantially influence the nutrient efficiency of this land-use system and diminish leaching losses. Therefore, leaching was quantified under a slash-burned site during a cultivation phase.

Materials and methods

To study the cultivation-induced leaching, a 7-year-old fallow vegetation was selected. This was slashed and burned in December 1996 and a sequence of maize, cowpea (*Vigna unguiculata*) and cassava was planted in the following 1.5 years. Concentrations of dissolved nutrients were determined in samples of soil solution taken biweekly at 0.9 m, 1.8 m and 3 m soil depth using suction-cup lysimeters (each time 6 repetitions). The nutrient concentrations were combined with water fluxes derived from a soil-water model applied on the cultivation site.

To determine deep soil water uptake, the annual dynamics of the soil water pressure head at different depth in the soil profile were recorded with tensiometers under a 4-year-old fallow vegetation starting in early April 1997. Based on this data set the soil water movement was modelled using laboratory soil-water retention curves, pedo-transfer functions and the soil water model Hydrus-1D (Van Genuchten, 1987).

Results

Due to El Niño effects the dry season of 1997 was exceptionally pronounced, with only 148 mm rainfall between 28th of August and the onset of the rainy season on 8th January 1998. In 1997 total annual precipitation was 2104mm and in 1998 2545mm. According to results of the soil water model, the transpiration of the fallow vegetation amounted to 154 mm within the dry season (Table 1). As much as 74 % of this transpiration water was extracted from the soil profile below 0.9 m depth. For the whole year, 400 mm and 427 mm of water originated from the soil layer of 0.9 to 6 m depth in 1997 and 1998, respectively. Maize, cowpea and cassava with their shallower root system did not deplete the soil water storage below 3 m depth and water from 0.9 to 3 m contributed less than 10 % to the annual water uptake.

Table 1. Soil water extraction under a 4-year old fallow vegetation from 22/8/1997 to 8/1/1998 and for the years 1997 and 1998 considering different soil layers. Percentages refer to the extraction from 0 to 6 m

Soil depth	22/8/97-8/1/98		--- 1997 ---		--- 1998 ---	
	(m)	(%)	(mm)	(%)	(mm)	(%)
0 - 0.9 m	40	26	730	65	853	67
0.9 - 3 m	41	27	199	18	223	18
3 - 6 m	73	47	202	18	204	16
Sum	154		1131		1280	

Considerable amounts of nutrients were leached below the rooting zone of these crops. At 0.9 m depth, these amounted to 77kg nitrate-N, 1.3 kg P, 25 kg K, 149 kg Ca, 34 kg Mg and 7 kg S per hectare over the two-year cropping period. However, during further percolation, nitrate, K, Ca and Mg were strongly reduced, so that losses at 3 m decreased to about 9 kg nitrate-N, 13 kg K, 45 kg Ca and 10 kg Mg per hectare over the two years (Table 2).

Table 2. Amounts of leached nutrients during the cropping period 1997-1998 considering reference soil depths of 0.9, 1.8 and 3 m

Soil depth	Nitrate-N	K	Ca	Mg
	[kg ha^{-1} 2a^{-1}]			
0.9 m	77	25	149	34
1.8 m	28	5	87	21
3 m	9	13	45	10

The decrease with depth in the amounts of dissolved nutrient was apparently related to the cation and anion exchange capacity. The effective cation exchange capacity (ECEC) of the soil at 0.9 m depth was around 1 cmol$_c$kg^{-1} and decreased to around 0.5 cmol$_c$kg^{-1} at 3 m. Anurugsa (1998) studied the anion exchange capacity (AEC) of soil of the study region at 30-50 cm depth. According to his sequential batch experiments AEC reached 0.08 cmol$_c$kg^{-1} at a soil-pH adjusted to 6.5 which increased to 0.4 cmol$_c$kg^{-1} at a pH of 3.1. Even if the ion exchange capacity of the deeper soil is rather low, the thickness of the profile allows that large amounts of exchangeable ions can be retained.

Discussion

About two-third of the annually transpired water of the fallow vegetation was taken up from the upper 0.9 m soil-layer. Still deep soil layers significantly contributed to the water supply of the vegetation. This might help to explain why most of the fallow species are able to maintain an evergreen canopy during the dry season.

The deeper soil layers, however, appeared to be out of the reach for the roots of maize, cowpea, cassava and newly introduced cash crops like passion fruit. Assuming an equivalent uptake of dissolved nutrients during water extraction, a nutrient-pumping mechanism of the deep-rooting fallow vegetation seems likely. The retention of nutrients beyond the time of cropping, therefore, provides the opportunity for the regrowing fallow vegetation to take up and to recycle these nutrients. It is likely that due to a decreasing ionic strength of the percolating soil water after fallowing of the crop land, retained ions are released and might be exported if not recycled by the fallow vegetation.

Our results on the water balance of the 4-year-old fallow vegetation show that even young secondary vegetation is able to extract deep soil water, a characteristic, so far only known for primary forest. If properly integrated into a shifting cultivation system such vegetation therefore provides a safety net against nutrient leaching. Intensified cultivation by semi-permanent or permanent crops suppresses the rootstock of the fallow vegetation and, therefore delays or even prevents its re-establishment. Hence, the nutrient efficiency of these systems is strongly reduced in the long run endangering agricultural productivity and ecosystem functioning as a whole.

References

Anurugsa B 1998 101. Ph.D. Thesis. University of Göttingen.

Nepstad DC *et al.* 1994 Nature 372, 666-669.

Nepstad DC, Uhl C and Serrão EAS 1991 Ambio 20, 248-255.

Sommer R, Denich M and Vlek PLG 2000 Plant Soil 219, 231-241.

Van Genuchten MT 1987 Research Report No. 121. U.S. Salinity Laboratory, USDA, ARS, Riverside. CA.

W. J. Horst et al. (Eds), Plant nutrition – Food security and sustainability of agro-ecosystems. 1016~1017, 2001

Effects of short-duration fallows on P transformations in Oxisols of Western Kenya

E. Bünemann[1)], A. Oberson[1)], P. C. Smithson[2)], B. Jama[2)] and E. Frossard[1)]

[1)] *Institute of Plant Sciences, Swiss Federal Institute of Technology (ETH), Eschikon 33, Lindau, CH 8315, Switzerland, else.buenemann@ipw.agrl.ethz.ch*

[2)] *International Centre for Research in Agroforestry (ICRAF), P.O. Box 30667, Nairobi, Kenya, p.smithson@cgiar.org*

Key words: *Crotalaria grahamiana*, improved fallows, legumes, microbial biomass, microbial P, organic P, Oxisol, P dynamics, tropical soils

Abstract

The introduction of a short-term improved legume fallow into small farmers' maize monocropping in Western Kenya doubled annual maize yields. Organic and microbial P were significantly increased in maize-fallow rotations while inorganic P remained unchanged. Microbial P was limited by C and N rather than by P availability. In spite of very low levels of available inorganic P, the microbial biomass was able to rapidly take up and release P when supplied with easily available sources of C and N. These data provide evidence for the importance of microbially mediated P transformations in tropical soils.

Introduction

Intensified land-use without sufficient compensatory inputs of inorganic or organic nutrients has depleted the soils in the densely populated areas of Western Kenya over the last decades. Unless subsidized, inorganic fertilisation is not an economic option for small farmers in the region. The introduction of an improved fallow into the cropping system takes advantage of the bimodal rainfall pattern with two growing seasons per year, whereby the short rainy season's maize crop often produces little and can be substituted by 6-8 months of fallow. Processes behind observed increases in soil fertility due to improved fallows with legumes such as *Crotalaria grahamiana* are not well understood. As plant growth on these soils is primarily limited by P availability, our main objective was to compare P transformations in soils under maize-fallow rotations and maize monocropping.

Materials and methods

Selected for this study was a three-year-old field experiment on an Oxisol of Western Kenya (pH 5.0, 39% clay, 37 % sand) comparing maize-monocropping (M), maize-crotalaria rotation (C) and maize-natural fallow rotation (NF) at two levels of inorganic P fertilisation (–P: 0 and +P: 50 kg P ha^{-1} a^{-1}, applied as TSP). Soil samples (0-15 cm) were collected at the end of the third fallow period before incorporation of the fallow biomass.

Different parameters of inorganic and organic P as well as total C and N contents were assessed. Nutrients held in the microbial biomass were determined with the fumigation-extraction methods described by Vance *et al.* (1987) for C and N and by Kouno *et al.* (1995) for P. They are reported as amounts rendered extractable by chloroform or hexanol fumigation without the use of conversion factors.

The effect of C and N additions on microbial P was studied in a lab incubation experiment. Glucose was chosen as a model C substrate and added at 2.5 mg C g^{-1} soil with or without N (supplied as NH_4NO_3 at a C:N ratio of 10).

Results

Without addition of inorganic P, the *Crotalaria* fallow doubled annual maize yields as compared to maize monocropping (Table 1). With P fertilisation, yields increased substantially in all cropping systems. Even with added P, grain yield was significantly higher in the *Crotalaria* rotation than in maize monocropping or maize-natural fallow rotation. With or without P, the amount of residues left in the field and incorporated into the soil decreased in the order C>NF>M.

All parameters of inorganic P (Table 2) showed a significant increase with P fertilisation without being affected by the different cropping systems. In the unfertilised soils, plant availability of P_i as indicated by P_{resin} and $_{mod}$Olsen-P was very low. Organic P, total C and N content (Table 3) and C, N and P held in the microbial biomass (Table 4) were significantly influenced by the cropping systems without being affected by P fertilisation, being highest in the *Crotalaria* rotation and lowest in the maize monocropping.

As the absence of a microbial response to inorganic P application indicated possible limitations by other nutrients, the effect of C and N additions on microbial P was studied. While the addition of N alone did not affect microbial P, C supplied as glucose increased microbial P 2.7-fold (data not shown). The combined addition of glucose-C and inorganic N increased microbial P 5.2- to 8.6-fold within two days (Figure 1). Maximum relative changes in microbial P seemed to depend on P availability rather than on the initial size of the microbial biomass. Maximum microbial immobilisation was rather transient, decreasing after 2-4 days.

Discussion

The data show a significant increase of organically and microbially bound P a year after the last incorporation of fallow biomass while inorganic P was not affected by the different cropping systems. Lacking changes in inorganic P pools were also reported by Maroko *et al.* (1999) who studied the effects of 17 months of *Sesbania* fallow on an Oxisol. In their study, the increase in microbial P was significant soon after incorporation of fallow biomass while 18 months later only the weight and P content of the light fraction of soil organic matter were still elevated. The repeated presence of 6-8 months of fallow appeared to have a longer lasting effect on organic and microbial P than a single fallow of longer duration.

In contrast to maize, fallow biomass responded poorly to P fertilisation. Soil microbial biomass values did not differ between –P/+P plots of the same cropping system, suggesting a strong link between plant residue inputs and microbial biomass. Smith (1993) stated that in most ecosystems C is the most limiting element for soil microbial biomass. Additions of C and N to –/+ P soils showed that microbial P was limited by the availability of C > N >> P. In spite of very low levels of available inorganic P, the microbial biomass was able to rapidly take up and release P when supplied with easily available sources of C and N.

In conclusion, our data provide evidence for the importance of microbially mediated P transformations in tropical soils.

Table 1. Annual production[1] (t dry matter ha^{-1})

cropping system	maize grain	total biomass	residues returned
M–P	1.2b[2]	6.3b	1.6c
C–P	2.5a	13.5a	5.9°
NF–P	1.7ab	7.9b	3.9b
M+P	3.5b	11.8b	2.5c
C+P	5.7a	20.8a	7.0a
NF+P	3.0b	10.6b	4.3b

1. Averages of two maize-maize or maize-fallow rotations, respectively
2. Within the same P treatment and column, mean values followed by the same letter are not statistically different at p=0.05

Table 2. Different parameters of inorganic P as affected by P fertilisation[1]

	P_{resin}	$_{mod}$Olsen-P_i	Bray-P_i	H_2SO_4-P_i
	- - - - -	mg kg^{-1}	- - - - -	
–P	1.7b	1.3b	4.9b	81b
+P	6.7a	4.1a	11.1a	124a

1. Means of the three cropping systems

Table 3. Contents of organic P, total C and N as affected by cropping systems[2]

cropping system	P_o Mg kg^{-1}	C_{tot}	N_{tot}
		- - - - g kg^{-1}	- - - -
M	264b	22b	1.6b
C	286a	25a	2.0a
NF	272ab	24a	1.8ab

2. Means of –P/+P treatments

Table 4. Microbial P, C and N as affected by cropping systems[3]

Cropping system	P_{hex}	C_{chl}	N_{chl}
	- - - -	mg kg^{-1}	- - - -
M	3.5c	98c	35c
C	6.4a	147a	42a
NF	5.3b	128b	40b

3. Means of –P/+P treatments

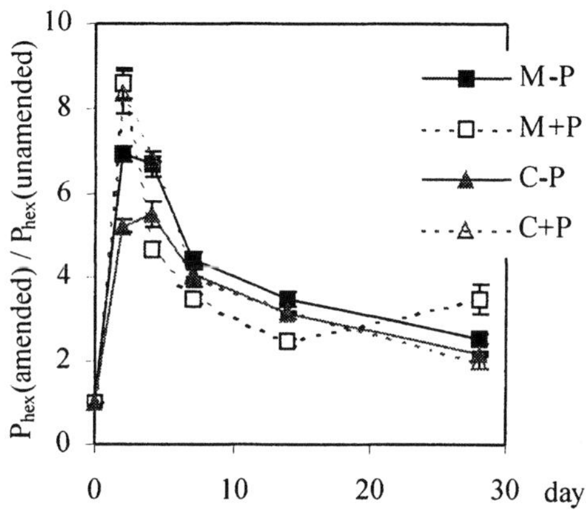

Figure 1. Changes in microbial P (relative to unamended soils) after addition of 2.5 mg glucose-C and 0.25 mg N g^{-1}

References

Kouno K, Tuchiya Y and Ando T 1995 Soil Biol. Biochem. 27, 1353-1357.

Maroko JB, Buresh RJ and Smithson PC 1999 Soil Sci. Soc. Am. J. 63, 320-326.

Smith JL 1993 *In* Soil biology: effects on soil quality. Eds. JL Hatfield and BA Stewart. pp. 91-120. Lewis, Boca Raton.

Vance ED, Brookes PC and Jenkinson DS 1987 Soil Biol. Biochem. 19, 703-707.

W. J. Horst et al. (Eds), Plant nutrition – Food security and sustainability of agro-ecosystems. 1018~1019, 2001

Phosphorus fractionation in oxisols under agroforestry and conventional coffee systems in Brazil

I. M. Cardoso, B. H. Janssen, O. Oenema and Th. Kuyper

Department of Environmental Science, Wageningen University, P.O. BOX 8005, Wageningen, 6700 EC, Netherlands, irene.cardoso@bodvru.benp.wau.nl

Key words: agroforestry, Brazil, coffee, oxisols, phosphorus fractionation

Abstract

Phosphorus is a primary limiting nutrient for crop production in highly weathered tropical soils. We hypothesised that the distribution of P among the various pools is influenced by land use and soil management. The objective of this paper is to compare the soil P pools at different depths in agroforestry and conventional coffee cultivation systems. The study was carried out in Brazil with Oxisols as the dominant soil type. Soils were collected from four coffee (~*Coffea arabica*~) fields, two with an agroforestry system and two with a conventional system. Per field three profiles were sampled at depths of 2-3, 10-15 and 40-60 cm. A simplified sequential P fractionation was carried out, using resin, $NaHCO_3$, NaOH and HCl as extractants. P contents were higher in the conventional systems than in the agroforestry systems, but the ratio of organic P to total P was at all depths higher in the agroforestry fields than in the conventional fields. It may indicate that the influence of agroforestry systems on the dynamics of soil phosphorus lies in the conversion of inorganic P into organic P and may increase in this way the amount of P that is within the reach of agricultural crops.

Introduction

Phosphorus is the major nutrient in relatively short supply in most natural ecosystems and the primary limiting nutrient for crop production in highly weathered tropical soils (Nye and Bertheux, 1957). The deficiency is mainly caused by strong sorption of phosphate to Al- and Fe-(hydr)oxides, which turns large proportions of total P into a form that is sparingly available to plants (Fontes and Weed, 1996). With the increasing cost and scarcity of soluble P, especially for resource-poor farmers (Ae *et al.*, 1990) it is necessary to accelerate P cycling by improving the availability of this P pools. In tropical environments, organic P (Po) may play a substantial role in nutrient cycling (Beck and Sanchez, 1994). It is hypothesised that the distribution of P among the various pools is influenced by land use and soil management, and that in agroforestry systems a part of the unavailable inorganic P is made available to agricultural crops by modifying the P dynamics. The objective of this paper is to compare the soil P pools at different depths in agroforestry and conventional coffee cultivation systems.

Materials and methods

The study was carried out in south-east Brazil, in the Atlantic Coastal Rainforest domain, with a tropical highland climate and Oxisols as the dominant soil type. Soils were collected from coffee (*Coffea arabica*) fields of four farms, two with an agroforestry system and two with a conventional system. The coffee was 15 and 19 years old in the agroforestry farms, and 15-20 and 20-24 years in the conventional farms. Per field three profiles, approximately 10 m apart between the coffee rows, were sampled at depths of 2-3 cm, 10-15 cm and 40-60 cm. A simplified sequential P fractionation was carried out, using resin, $NaHCO_3$, NaOH, diluted and concentrated HCl as extractants (Heddley *et al.*, 1982; Tiessen and Moir, 1993; Beck and Sanchez, 1994).

Results

Total P ranged from 370 to 830 mg kg^{-1} (Table 1). Concentrated HCl extracted the largest portion of P, followed by NaOH (on average 73 and 23%, respectively). The resin, diluted HCl and $NaHCO_3$ fractions were on average 0.5 and 0.8 % and 2.4%, respectively. Organic P in concentrated HCl was very low. Therefore, concentrated HCl fraction is considered as inorganic P. The fractions of $NaHCO_3$-Po and NaOH-Po were on average 1.4% and 15%, respectively, of the total P. On average around 26% of the total P (range 13 - 40%) is considered labile. In this study we consider Resin-Pi, $NaHCO_3$-Pi and –Po, and NaOH-Pi and -Po as labile P, similar to other authors. In general, labile P decreases with increasing depth. There were no systematic differences in the ratio of labile to total P between the agroforestry and conventional systems. The organic P represents on average 62% (range 38 – 76%) of the labile P. Organic P is relatively high in 10-15 cm layer. P contents were higher in the conventional systems than in the agroforestry systems, and also higher in the old fields than in the younger fields. The probable reason is the higher addition of fertiliser P. The agroforestry systems have a higher percentage of organic P when compared per soil layer than the conventional systems.

Discussion

Our results show that the largest P fraction is in the concentrated HCl fraction, which is considered unavailable for plants. Our results show that a low percentage of the total P consists of labile P and that it is not clearly related to cultivation system or age. The higher percentage of labile P close to the surface can be due to fertilisation.

Of the labile pool, a higher percentage of P is organic

than inorganic. This shows the important role of Po in plant nutrition in highly weathered soils. The lower percentage of organic P in 2-3 cm (hence, more inorganic P) is probably due to chemical fertilisation. In 10-15 cm the fertilisation effect is less than in 2-3 cm (less inorganic P) and the roots activity is higher than in 40-60 cm and this can explain the higher percentage of organic P at 10-15 cm than at 40-60 cm. The lower percentage of organic P in conventional systems may also be due to fertiliser use.

Organic P is relatively higher in agroforestry systems than in the conventional systems. The related increase in P turnover rate brings more P within the reach of agricultural crops. The general hypothesis that part of the unavailable inorganic P is made labile to agricultural crops in agroforestry systems is not supported but the higher percentage of organic P in agroforestry systems suggests that agroforestry systems influence the dynamics of phosphorus in the soils. The influence of agroforestry systems on the dynamics of soil phosphorus would than lie in the conversion of inorganic P into organic P. Because the rate of turnover is higher for organic P than for inorganic P, agroforestry in this way would increase the amount of P that is within the reach of agricultural crops.

References

Ae N, Arihara J, Okada K, Yoshihara J and Johansen C 1990 Science 248, 477-480.

Beck MA and Sanchez PA 1994 Soil Sci. 34, 1424-1431.

Fontes MPF and Weed SB 1996 Geoderma 72, 37-51.

Hedley MJ, Stewart JW and Chauhan BS 1982 Soil Sci. Soc. Am. J. 46, 970-976.

Nye PH and Bertheux MH 1957 J. Agr. Sci. 49, 141-159.

Tiessen H and Moir JO 1993 *In* Soil sampling and methods of analysis. Ed. Carter MR. pp. 75-86. CRC Press, Boca Ratton.

Table 1. Average phosphorus concentration in several P pools, labile P, ratio of organic to labile (Org/lab) and total P, in medium-aged and old agroforestry coffee systems (AM and AO)[a], medium-aged and old conventional coffee systems (CM and CO)[b], at three depths (2-3cm, 10-15cm and 40-60cm).

Systems	Depths (cm)	Resine Pi[c]	$NaHCO_3$ Pi	$NaHCO_3$ Po[d]	NaOH Pi	NaOH Po	1M HCl Pi	Conc. HCl[e] Pi	Labile	Org/lab	Total
					mg/kg soil						
AM	2-3	2.62	5.16	8.47	47.89	94.99	2.43	282.81	159.13	0.65	444.37
	10-15	0.73	0.81	6.99	28.21	88.50	1.80	254.10	125.24	0.76	381.14
	40-60	0.73	0.74	4.81	41.69	117.17	1.67	246.35	165.14	0.74	413.16
AO	2-3	6.03	4.72	8.67	29.47	67.65	5.97	439.82	116.54	0.65	562.33
	10-15	1.69	2.86	7.40	20.81	48.66	2.02	414.07	81.42	0.69	497.51
	40-60	1.46	2.92	6.82	24.27	46.26	4.29	445.63	81.73	0.65	531.64
CM	2-3	5.72	12.70	11.89	55.61	68.75	11.84	300.03	154.67	0.52	466.54
	10-15	1.95	8.21	7.31	24.24	51.42	6.05	334.11	93.13	0.63	433.29
	40-60	0.00	6.81	2.24	13.66	24.88	5.42	317.93	47.59	0.57	370.94
CO	2-3	6.38	8.76	9.78	90.21	129.49	1.59	534.74	244.61	0.57	780.94
	10-15	2.71	3.11	6.89	54.30	119.52	0.26	572.76	186.53	0.68	759.55
	40-60	2.49	6.50	2.03	116.77	75.33	1.02	624.51	203.12	0.38	828.66

[a]AM, 15 years old; AO, 19 years old
[b]CM, 15-20 years old, CO, 20-24 years old
[c]Pi, inorganic P; [d]Po, organic P; [e]Conc. HCl, concentrated HCl.

W. J. Horst et al. (Eds), Plant nutrition – Food security and sustainability of agro-ecosystems. 1020~1021, 2001

Fuzzy set analysis and canonical correspondence analysis of soil arthropods (*Coleoptera, Isopoda*) in organic and conventional agroecosystems

E. Hadjicharalambous[1)], K. L. Kalburtji[2)] and A. P. Mamolos[2)]
[1)] *Greek Biotope/Wetland Centre, 14th km Thessaloniki - Michaniona, GR 570 01, Thermi, Greece, Tel.: +3031 998621; Fax: +3031 998643, Helena@ekby.gr*
[2)] *Aristotle University of Thessaloniki, Faculty of Agriculture, Laboratory of Ecology & Environmental Protection 540 06 Thessaloniki Greece, Kalbourt@agro.auth.gr; Mamolos@agro.auth.gr*

Key words: agroecosystems, canonical correspondence analysis, *Colleoptera*, fuzzy clustering, *Isopoda*

Abstract

Fuzzy set theory and canonical correspondence analysis (CCA) are often used for classification and ordination techniques in studies of various arthropod communities. The objective of the present study was to investigate the relationships and trends of the two soil arthropod communities (Coleoptera and Isopoda) in organic and conventional agroecosystem. Coleoptera and Isopoda were sampled by pitfall traps in the following fields: a) two organic vineyards and two conventional b) two organic olive-groves and two conventional, and c) one organic rotation system that involved maize and one conventional maize field. According to species composition in fields, three main fuzzy groups were found, based not on the farming system (either organic or conventional) but on the cultivated plant species (olive-groves, vineyards and maize crop). The CCA grouped together all olive-groves in relation to soil organic matter, all vineyards in relation to soil P and all maize in relation to soil N. In a maize field N is most probable to be the soil limiting element due to the high demand of maize for N. Concerning Coleoptera and Isopoda, CCA placed the rare and unique taxa at the edges of the diagram, while it grouped together most Coleoptera and Isopoda in relation to soil organic matter and Ca.

Introduction

The interaction between agriculture and biodiversity has been reported in several publications (e.g. Baldock *et al.* 1993, Pfiffner and Luca 2000). The objective of the present study was to evaluate the effects of organic farming on biodiversity and this was done by focusing on certain soil arthropods taxa in organic and conventional agroecosystems.

Materials and methods

Sampling was carried out by pitfall traps (Southwood 1978) left for a whole month (October 1999) in the various fields. The different farming systems were practised at least for the last 10 years. Fuzzy set theory (Equihua 1990) was used to obtain both ordination and classification of the two soil arthropod assemblages, while the relationship between soil arthropod taxa [data transformed to ln(x+1)], soil physicochemical parameters and sampling sites were analysed by canonical correspondence analysis (CCA-program CANOCO, ter Braak, 1988).

Results

Four Isopoda families and 21 Coleoptera families were found in the studied fields. Some of the Isopoda were not found in all fields. As for Coleoptera, the families Carabidae, Staphylinidae, Anthicidae and Chrysomelidae were found in all studied fields, although some of their species were not found in every field.

Fuzzy analysis grouped the sampling sites according to the assemblages of Isopoda and Coleoptera in three clusters (Fig. 1). CANOCO sites analysis placed together the fields from each plant species, just like fuzzy analysis did, in relation to certain soil variables (Fig. 2a). As to Isopoda and Coleoptera species (Fig. 2b), CANOCO placed the rare and unique taxa at the edges of the diagram, whereas the dominant taxa of each field were grouped together in relation to specific soil variables (Fig. 2b).

Discussion

Isopod species found in the fields are characteristic inhabitants of agroecosystems and mostly prefer humid habitats (Sfenthourakis 1994). None of them is Greek endemic or protected by national or international legislation.

Coleoptera species found in all studied fields are pest predators (e.g. Carabids and Staphylinids) or detritivores - fungivores (e.g. Anthicids) or herbivores (e.g. Chrysomelids). The dominant species (Carabids), as well as, the representatives of the most diverse family (Staphylinids) were reported pest predators by Pfiffner and Luka (2000) in northwest of Switzerland.

No species seemed to be treatment specialist. This is also supported by the results of fuzzy analysis that grouped together the fields of the same crop, irrespective of the farming system.

The CANOCO analysis grouped the fields according to the cultivated plant species due to their soil physicochemical similarities, whereas it grouped together the dominant taxa in each field in relation to certain soil variables. According to Sacchi and Testard (1971) soils of a high N and Ca content show high biodiversity since these elements are of great importance for many invertebrates

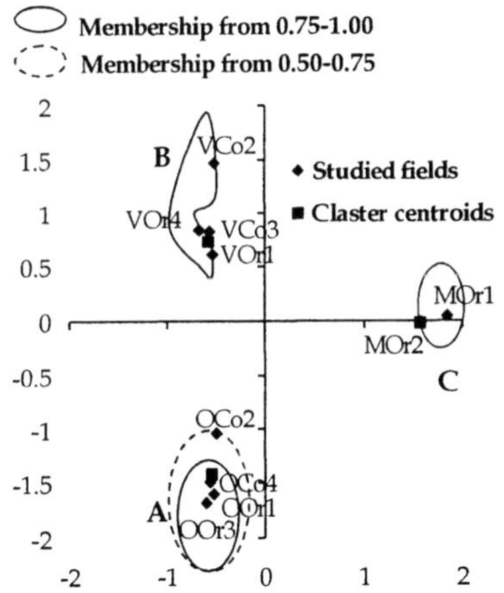

Figure 1. Classification of the studied agroecosystems [conventional (Co) and organic (Or) fields with olive trees (O), vines (V) and maize (M)] with the FUZZY method based on Coleoptera and Isopoda taxa found in them.

(e.g., Dipolopoda, Isopoda, insects, etc.). Therefore, this could be the reason why most of Isopods and Coleoptera taxa are strongly related to the above mentioned soil variables.

References

Baldock D, Beaufoy G, Bennet D and Clark J 1993 Institute for European Environmental Policy, London.

Equihua M 1990 J. Ecol. 78, 519-534.

Pfiffner L and Luka H 2000 Agric. Ecosyst. Environ. 78, 215-222.

Sacchi CF and Testard F 1971 Organismes et milieu. pp 31-275.

Sfenthourakis S 1994 PhD. Thesis. University of Athens.

Southwood TRE 1978 2nd edn. Chapman and Hall, London.

Ter Braak CJF 1988 GLW Wageningen.

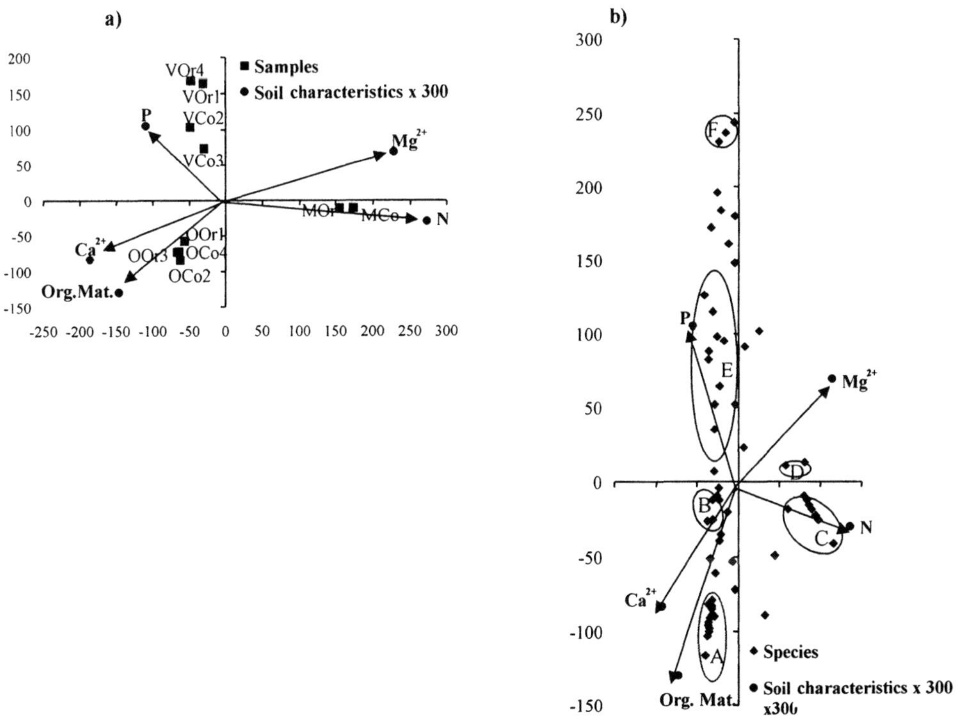

Figure 2. a) CCA ordination diagram with soil physicochemical variables and studied agroecosystems [conventional (Co) and organic (Or) fields with olive trees (O), vines (V) and maize (M)]. b) CCA ordination diagram with soil physicochemical variables and Coleoptera and Isopoda taxa in 10 studied agroecosystems.

W. J. Horst et al. (Eds), Plant nutrition – Food security and sustainability of agro-ecosystems. 1022~1023, 2001

Soil arthropod diversity in relation to weed diversity in organic and conventional agroecosystems

E. Hadjicharalambous[1)], K. L. Kalburtji[2)] and A. P. Mamolos[2)]

[1)] *Greek Biotope/Wetland Centre, 14th km Thessaloniki - Michaniona, GR 570 01, Thermi, Greece.*
[2)] *Aristotle University of Thessaloniki, Faculty of Agriculture, Laboratories of Ecology & Environmental Protection, GR 540 06, Thessaloniki, Greece. Helena@ekby.gr; Kalbourt@agro.auth.gr; Mamolos@agro.auth.gr*

Key words: coleoptera, isopoda, maize, olive-groves, shannon index, soil arthropods, vineyards, weeds

Abstract

There are indications that plant diversity can influence soil arthropod abundance and diversity in terrestrial ecosystems. This study was conducted to investigate the relationships between weed and soil arthropod diversity in organic and conventional agroecosystems. The fields involved in this study were: a) two organic and two conventional vineyards, b) two organic and two conventional olive-groves, and c) one organic rotation system with maize (*Zea mays* L.) and one conventional maize field. Weed sampling was carried out within a randomised complete block design with five replications, where soil arthropods, were sampled in pitfall traps. For both, weeds and soil arthropods, Shannon's diversity index was calculated in all studied fields. Weed diversity indices were lowest in conventional vineyards, conventional maize and in organic maize. The highest diversity indices for soil arthropods were found in organic vineyards and the lowest in conventional maize. Coleoptera and Isopoda had higher diversity indexes in conventional than in organic maize. In all fields diversity indexes were significantly lower for Isopoda than for Coleoptera. A positive trend was found between diversity indices for weeds and arthropods. Generally, diversity indexes for weeds showed higher values in organic fields than in conventional ones.

Introduction

Intensive agricultural production, particularly high input levels of agrochemicals, impoverished crop rotations, field enlargements and field margin removals, has created undesirable environmental and social side-effects (Kromp and Steinberger, 1992). This study therefore focuses on the possible effects of different farming systems on biodiversity as well as on weed species and soil arthropod interactions in organic and conventional agroecosystems, with special reference to Isopoda and Coleoptera communities.

Materials and methods

Soil arthropod fauna and weed flora were studied in organic and conventional agroecosystems of northern Greece which consisted of: a) two organic and two conventional vineyards (*Vitis vinifera* L.), b) two organic and two conventional olive groves (*Olea europaea* L.), and c) one organic rotation system with maize compared to a conventional maize field.

For weeds sampling each field was divided in to five equal rectangles and one sample was obtained from each rectangle in October 1999 using a 30 x 30-cm quadrate. Soil arthropods were sampled during the entire month of October 1999 with pitfall traps (Southwood, 1978). For both, weeds and soil arthropods, Shannon's diversity index H′ [$H' = -\Sigma p_i \ln(p_i)$] (Krebs, 1978) was calculated..

Results

Eighteen weed species (*Alyssum saxatile* L., *Amaranthus blitoides* S. Watson, *Amaranthus retroflexus* L., *Chenopodium album* L., *Cichorium intybus* L., *Convolvulus arvensis* L., *Cynodon dactylon* (L.) Pers., *Cyperus rotundus* L., *Echinochloa crus-gali* (L.) P. Beauv., *Medicago sativa* L., *Paspalum distichum* L., *Portulaca oleracea* L., *Settaria viridis* (L.) Beauv., *Sinapis arvensis* L., *Solanum nigrum* L., *Sorghum halepense* L., *Tribolus terrestis* L. and *Vicia* spp.) and 21 soil arthropod orders (3 Arachnidia, 2 Chilopoda, 2 Diplopoda, 1 Isopoda and 13 Insecta) were found in the studied fields. The highest number of soil arthropod orders (21) was recorded in the organic olive groves while the lowest number (12) was came from conventional maize.

Shannon's diversity indices (H′) for weeds and soil arthropods are shown in Figure 1. In conventional vineyards and in conventional and organic maize weed diversity index had the lower values. The highest diversity indices for soil arthropods were found in the organic vineyards and the lowest in conventional maize. Generally, a positive trend between diversity indexes for weeds and arthropods was found, which is expressed by positive r-values (Fig. 1).

Discussion

The number of soil arthropod orders found in the studied agroecosystems is almost the same as the numbers previously reported from disturbed maquis and phrygana ecosystems of the Greek islands (Magioris, 1991; Marmari, 1991) but lower than those found in Greek pine forests (Radea, 1989). Generally the numbers of soil arthropod taxa determined in the conventional fields are similar to the numbers found in degraded natural ecosystems. Moreover, according to Magioris (1991) biodiversity in a gap (open areas) was much lower than the biodiversity in any other natural habitat.

Generally, for weeds diversity indices were higher values in the organic fields than in conventional ones whereas we could not find this for soil arthropods. With respect to the relationship between plant and arthropod diversity Di Castri and Vitali-Di Castri (1981) and Brown and Southwood (1987) reported that vegetation cover and quality of litter rather than plant diversity are the most important factors affecting soil arthropod diversity.

References

Brown VK and Southwood TRE 1987 *In* 26th Symposium of the British Ecological Society. pp 315-338. Blackweel, Oxford.

Di Castri F and Vitali-Di Castri V *In* Mediterranean type shrub-lands. Eds. Di Castri F, Goodall DW and Specht RL. pp 445-478. Elsevier Scientific Publication Company.

Krebs JK 1978 2nd Edition. Harper and Row, Publishers, New York.

Kromp B and Steinberger KH 1992 Agric. Ecosyst. Environ. 40, 71-93.

Magioris SN 1991 Ph.D Thesis. University of Athens. p. 209.

Marmari A 1991 PhD Thesis. University of Athens. 221 p.

Radea K 1989 PhD Thesis. University of Athens. 256 p.

Southwood TRE 1978 2nd edn. Chapman and Hall, London.

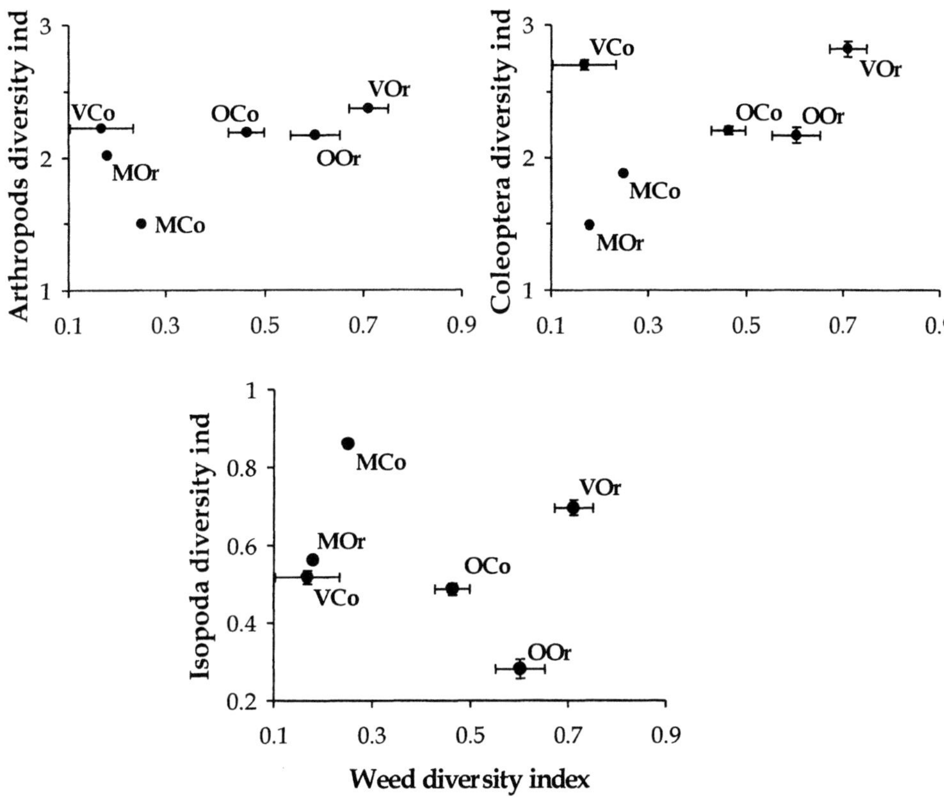

Figure 1. Relationships between mean (± one standard error); n=2 for olive-groves (O) and vineyards (V) and n=1 for maize (M)] diversity indexes of a) weeds and Arthropods b) weeds and Colleoptera c) weeds and Isopoda in organic (Or) and conventional (Co) agroecosystems.

W. J. Horst et al. (Eds), Plant nutrition – Food security and sustainability of agro-ecosystems. 1024~1025, 2001

Plastic film mulching cultivation: a new technology for resource saving water N fertiliser and reduced environmental pollution

L. Wu[1)], Z. Zhu[2)], Y. Liang[3)] and F. Zhang[4)]

[1)] *College of Natural Resources and Environmental Sciences, Zhejiang Univ., 268 Kaixuan Road, Hangzhou 310029, Zhejiang, P.R. China. lhwu@mail.hz.zj.cn*

[2)] *College of Agriculture and Biotechnology, Zhejiang Univ., Hangzhou 310029, P. R. China*

[3)] *College of Natural Resources and Environmental Sciences, Nanjing Agricultural Univ., Nanjing 210095, P.R.China*

[4)] *College of Natural Resources and Environmental Sciences, China Agricultural Univ., Beijing 100094, P.R..China*

Key words: cultivation, film, mulching, nitrogen use efficiency, *Oryza sativa* L.

Abstract

Plastic film mulching cultivation is a promising alternative technology for the conventional continuous flooded cultivation system of rice (*Oryza sativa* L.). The objective in this study was to establish a better rice film mulching cultivation system and to improve the nitrogen use efficiency in this system. Two field experiments were conducted, in a mountainous area in Fenghua district using one season rice, Japonica cultivar 9454 in 1998 and another in a plain area of West Lake district using Indica cultivar Jiazao 935, 1997 planted in late spring. The nitrogen N use efficiency for highest yield treatment as proposed in 1997 and 1998 was checked in 1999 and 2000. For the Indica cultivar Jiazao 935 maximum yield of 6855 kg ha^{-1} were obtained using the plastic film mulching at the half farmer's fertiliser rate. Grain yields with plastic film mulching increased mainly due to an increased panicle number per hectare. This corresponded to higher harvest and nutrient (N, P) index. The average N use efficiency for this higher yield treatment was about 43.0%, compared to 30.3% for the flooded-cultivation with farmer's fertiliser rate in 1999 and 2000. For the Japonica cultivar 9454 plastic film mulching was a better cultivation technology coupled with direct dibble seeding for high yield and N use efficiency. Higher grain yield of 8842 kg ha^{-1} was obtained in the mulching plot with direct seeding in 1998. The N use efficiency with film mulching (48.6%) was also higher than in the flooded-cultivation system with direct seeding in 1999.

Introduction

Rice production is often strongly limited by irrigation water scarcity. In temperate areas low temperature can adversely affect rice during early growth stages. Excessive use of chemicals such as fertilisers, pesticides and herbicides do not only increase costs of production, but also cause serious environmental pollution. Under these circumstances, growing upland and lowland rice using plastic film mulch under non-flooded conditions becomes one of the crop management options to alleviate the above problems and consequently increase grain yield in some areas of China (Xiong, 1996; He, 1997; Liang *et al.*, 1999; Wu *et al.*, 1999). The aim of the present study was to establish a better rice plastic film mulching cultivation system, and to compare N use efficiency under film mulching with continuous flooded rice system in plain and mountainous areas.

Materials and methods

The field experiment 1 with Japonica cultivar 9454 was conducted in a mountainous area at Fenghua district of Zhejiang Province in 1997, where rice grain yield in a single crop system was greatly limited by water scarcity. Treatments included: 1. CF 1 (continuous flooded, direct dibble seeding on 15 June); 2. CF 2 (continuous flooded, sowing on 15 June, transplanting on 10 July); 3. Mulching 1, direct dibble seeding (sowing on 15 June) and 4. Mulching 2, transplanting (sowing on 15 June, transplanting on 10 July). The experiment was conducted as a randomised complete block design with three replicates. Rice were sown or transplanted, according to treatments, at spacing of 25 × 25 cm with three seeds or seedlings per hill (0.48×10^6 seedlings ha^{-1}). Each plot received same fertiliser rate of 150 kg N ha^{-1}, 22 kg P ha^{-1} and 187 kg K ha^{-1}.

The field experiment 2 with Indica cultivar Jiazao 935 was conducted in a plain of West Lake district of Zhejiang Province in 1998, where low temperature was a major constraint to early grain yield of rice in a rice-rice-fallow rotation system. Treatments in a randomised complete block design with three replications were: 1. CF 1, using farmer's flooded rice cultivation method, applying 225 kg N ha^{-1} and 13 kg P ha^{-1}, seedlings were broadcasted at the rate of 1.65×10^6 seedlings ha^{-1}; 2. CF 2, using the same method as CF 1, but plant density was 1.73×10^6 seedlings ha^{-1}. 3. Mulching 1, applying half the farmer's N, and P fertiliser rate. Seedlings were transplanted at a space of 30×30cm (0.33×10^6 seedlings ha^{-1}) with three seedlings per hill; 4. Mulching 2, using the same method as mulching 1, but plant density was 20×20cm (0.75×10^6 seedlings ha^{-1}); 5. Mulching 3, applying a third of the farmer's fertiliser rate. Seedlings were transplanted at a space of 30×30cm; 6. Mulching 4, using the same method as mulching 5, but plant density was 20×20cm. K fertiliser supply for each plot was 56 kg K ha^{-1}. Germinated rice seeds were sown on 26 March and seedlings were broadcasted for CF and transplanted for mulching plots on 21 April.

The N use efficiency for highest yield treatment as

proposed in 1997 and 1998 was checked in 1999 and 2000.

Results

The grain yields for Japonica cultivar 9454 with plastic film mulching were greater than for the flooding treatments (CF 1 and 2), independent whether direct dibble seeded or transplanted (Tab. 1). The highest yield of 8842 kg ha^{-1} was obtained with mulching and direct dibble seeding treatment (Mulching 1) due to more panicle number per hectare and less unfilled grains. Direct dibble seeding (CF 1 and Mulching 1) resulted in more panicles than transplanting in the flooding or mulching cultivation system. The N use efficiency of mulching 1 was 48.6%, compared to 35.7% in the flooded-cultivation system with direct seeding (CF 1) in 1999.

Table 1. Yield components and grain yield of japonica cultivar 9454 in the plastic film mulching cultivation experiment at Fenghua district of Zhejiang Province in 1997

Treat.	Panicle No ha^{-1} ($\times10^6$)	Grain No panicle^{-1}	Unfilled grain (%)	Grain yield (kgha^{-1})
CF1	4.204 ab	108.7a	20.7 b	7673 b
CF2	3.093 c	103.1 a	20.2 b	6062 d
Mulching1	4.630 a	113.2a	16.9 a	8842 a
Mulching2	3.889 b	107.4a	19.5 b	7315 c

Within a column, means followed by different letters are significantly different at the P<0.05 by the LSD test.

Table 2. Yield components and grain yield of Indica cultivar Jiazao 935 in the experiment with plastic film mulching cultivation at West Lake district of Zhejiang Province in 1998

Treat.	Panicle No ha^{-1} ($\times10^6$)	Grain No panicle^{-1}	Unfilled grain (%)	Grain yield (kg ha^{-1})
CK1	4.58 a	73.9 c	24.4 a	5835 cd
CK2	4.25 abc	87.0 b	28.9 a	5983 bc
Mulching1	3.77 bc	104.1 a	27.0 a	6133 b
Mulching2	4.39 ab	89.3 b	22.6 a	6855 a
Mulching3	3.61 c	102.0 a	28.8 a	5475 e
Mulching4	4.30 abc	87.3 b	23.4 a	5745 d

Grain yields for Indica cultivar Jiazao 935 in mulching systems with transplanting seedlings at spacing of 20x20cm were higher than with transplanting seedling at spacing of 30x30cm (Tab. 2). The treatment with mulch and half of the farmer's rates of fertiliser (Mulching 2) had the highest yield of 6855 kg ha^{-1} due to more panicles per hectare. The treatments mulching 2 and 4 with a 1/3 - 1/2 total applied fertiliser rate produced panicle numbers comparable to CF 1 and 2. This indicates that the panicle number increased with mulching. The average of N use efficiency in mulching 2 was 43.0%, which was higher than 30.3% of the flooded-cultivation system (CF2) with broadcasted rice in 1999 and 2000.

The mulching treatment with the highest grain yield (mulching 2) was compared with CF in flag leaf area (FLA), harvest index (HI), N harvest index (HIN), and P harvest index (HIP). Greater HI in mulching 2 was related to higher FLA, HIN and HIP (Tab. 3). No significant differences in K harvest index (HIK) were found between mulched and flooded plots.

Table 3. Flag leaf area, harvest index, and N, P harvest index of Indica cultivar Jiazao 935 in the experiment with plastic film mulching cultivation at West Lake district of Zhejiang Province in 1998

Treat.	Flag leaf area (cm^2)	Harvest index (%)	N harvest index (%)	P harvest index (%)
CF1	41.9 c	60.2 ab	75.0 a	80.2 a
CF2	42.5 bc	57.7 bc	69.3 b	77.6 b
Mulching1	51.3 a	61.0 ab	72.7 ab	84.3 ab
Mulching2	46.8 abc	62.8 a	74.3 a	87.9 a
Mulching3	47.4 ab	60.3 ab	70.1 b	84.4 ab
Mulching4	43.3 bc	53.9 c	61.8 c	80.2 ab

Discussion

To meet the requirement of a high panicle number for high yield under film mulching, planting density should not be too low. About 0.75×10^6 seedlings ha^{-1} is recommended to early rice with the Indica cultivar in the plastic film mulching cultivation. A lower demand for N-fertiliser under film mulching will minimise the environmental pollution due to nitrate leaching and gaseous N losses. For the Japonica cultivar in a single cropping system, plastic film mulching cultivation was better coupled with direct dibble seeding for high yield and resource efficiency.

Thus, plastic film mulching is an alternative technology to the conventional rice cultivation system (Peng *et al.*, 1999). It does not only save about 75% of irrigation water, but also about 50% N-fertiliser, 80-100% herbicides, 30% pesticide and 40% seeds, compared to rice system with continuous flooding (Liang *et al.*, 1999; Wu *et al.*, 1999). Up to now, the used plastic film has to be properly recycled, but in near future a biodegradable film will be used.

Acknowledgement

This research was supported by NKBRSF, PR China, No. G1999011707 and Natural Science Foundation of China, No. 39970146. Many thanks are due to Dr. V. Roemheld for his suggestions on the manuscript.

Reference

He D 1997 China Rice 3, 21.

Liang Y, Hu F, Yang M, Zhu X, Wang G and Wang Y 1999 Scientia Agricultura Sinica 32, 26-32

Peng S, Shen K, Liu J, Luo, X and Wu L 1999 IRRN 24, 9-10

Wu L, Zhu Z, Liang Y, Shi W and Zhang L 1999 J. Zhejiang Agr. University 25, 41-42.

Xiong F 1996. Comm. Agr. Sci. Technol. 10, 8.

W. J. Horst et al. (Eds), Plant nutrition – Food security and sustainability of agro-ecosystems. 1026~1027, 2001

A holistic approach towards sustainable land use management in the Hedgerow - Field crop interface

R. Mette, G. Uckert and B. Sattelmacher

Institute of Plant Nutrition and Soil Science, University of Kiel, Olshausenstr. 40, Kiel, D 24118, Germany, rmette@plantnutrition.uni-kiel.de

Key words: biofuel, boiler ash, N-surplus, nutrient cycling, renewable energy

Abstract

Hedgerows as traditional landscape elements of agricultural ecosystems in Northern Germany were investigated for their above and below ground interactions in the hedgerow - field crop interface. Biomass accumulation of the hedgerow vegetation tended to increase with the management intensity of the farming system. The annual biomass accumulation rate was considerably higher than in natural forest stands or naturally managed ecosystems but was quite comparable with silvo-arable systems or other biofuel alternatives grown in Northern Germany. Following the higher growth rates, total nutrient contents of the hedgerow biomass were higher than those of unfertilised ecosystems. This was based on the fact that yields of arable crops significantly decreased in the proximity of the hedgerows depending on exposition, distance to and age of the hedgerow vegetation. Reduced nutrient extraction by arable crops caused by reduced yield capacity led - in conjunction with continuous high fertiliser input - to nutrient accumulation in the hedgerow –field crop interface. As a consequence the increased nutrient and biomass accumulation of the hedgerow vegetation indicated a favourable nutrient supply out of the hedgerow - field crop interface enabled through intensive root exploration of the hedgerow shrubs in arable land. Burning these hedgerows in modern heating/power plants contributes to an advance in the development of regional carbon, energy and nutrient cycles when - at least- the remaining ash fraction is taken as a fertiliser substitute back to the fields.

Introduction

In the high yielding farming areas of Northern Germany, hedgerows are characteristic landscape elements separating different fields and properties. Typically put onto earth walls, these hedgerows were already introduced as so-called living fences in the 18th century. Similar to other agro-forestry systems, farmers were pruning these hedgerows every 8–10 years using the shrubs as fuelwood or for other purposes.

Despite of their ecological benefits the large existing hedgerow network was considerably reduced between 1960 and 1990 as a consequence of intensification and consolidation of farming systems. Today the remaining hedgerow network of about 46,000 km length is protected by law, since it is considered to be an eminent nature conservation element. To ensure the biodiversity of these hedgerows this special law implies regular pruning of the shrubs every 10–15 years. But up to now farmers were burning the shrubs after labour intensive cutting on the field, because there are no economically attractive alternative uses. Our investigations were focussed on nutrient cycling in the hedgerow – field crop interface as it influences the biomass accumulation of hedgerows and their potential as a future biofuel source.

Materials and methods

Yield and biomass of different arable crops in the hedgerow – field crop interface were determined in 1 m, 3 m, 5 m and 10 m distance to hedgerows respectively. Nutrient inputs through fertilisers and leaf defoliation were quantified in the same way. Nutrient uptake by crops were analysed in different growth stages. For the description of interactions between both ecosystems studies on water household (matrix potential) and root ecology (root length densities etc.) of both arable crops and the hedgerow vegetation were carried out (for details see Mette, 1994). With respect to the high biodiversity of the hedgerows, investigations on hedgerow biomass and nutrient contents were focussed on two hedgerow transects of about one hundred metre length each (for details see Uckert *et al.*, 1997).

Results and discussion

Yield and biomass data from different crops and years (Fig. 1) indicated an yield reduction with closer distance to the hedgerows. The extent of yield reduction was most pronounced to growth duration of the hedgerow vegetation after pruning and the exposition of crops to the hedgerows. Respectively yields in south exposition were less reduced than in north exposition to the hedgerows (for details see Mette, 1994). There also was a strong difference between C_3 (winter rye *Secale cereale* L.) and C_4 plants (corn, *Zea mays* L.) in northern exposition.

As a consequence of reduced yields and biomass production in the hedgerow-field crop interface, nutrient extraction of arable crops decreased leading - with respect to non-reduced fertiliser input - to high nutrient surpluses (Table 1). The recurrent autumn leaf defoliation of the hedgerow vegetation further contributed to this gradient (Table 1). In the long run the accumulation of organic matter from leaf defoliation has to be considered as a additional and continuous source of mineralisation.

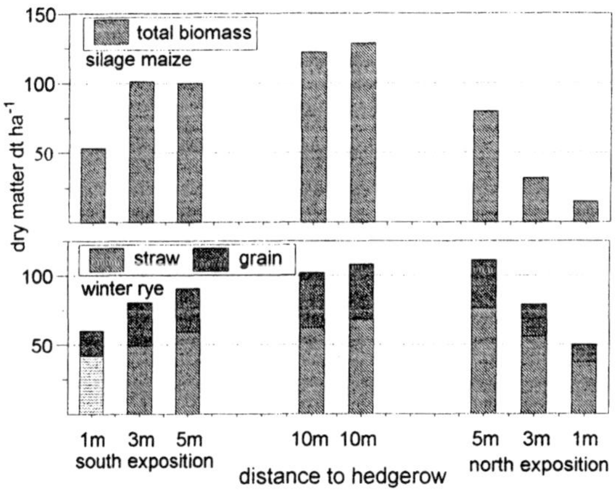

Figure 1. Yield and biomass production as affected by exposition and distance to hedgerows

Table 1. Nitrogen balance (kg N) in north exposition of hedgerows (arable crop: silage maize)

Deposition			18	
Leaf Defoliation	69	28	11	5
Min/org. fertiliser			317	
Nutrient uptake by crops	24	50	156	224
N-surplus	382	315	192	118
Distance to hedgerows	1m	3m	5m	10m

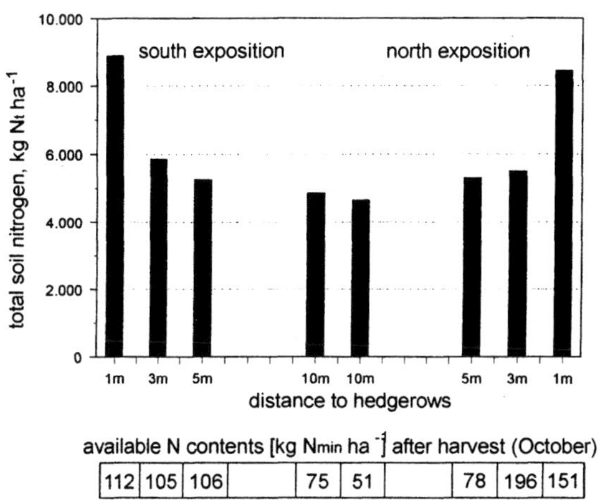

Figure 2. Total soil nitrogen (N) in the hedgerow-field crop interface and available N after harvest

Considering the above mentioned figures, total soil N (N_t) increased with leaf defoliation and reduced nutrient extraction at the border areas of the hedgerows. Over time this may lead to an increase of the available N content after harvest reaching up 200 kg N_{min} ha^{-1} at 3 m distance from the hedgerow (Fig. 2). These amounts of N may be lost during the winter season. It was shown that available N correlated with mean and shallow rooting densities of arable crops growing in the shaded area in northern exposition (Mette and Sattelmacher, 1994). On the other hand intensive root exploration of the hedgerow shrubs into arable land was found up to 10 m distance. This indicated a lateral nutrient uptake from the fertilised arable land contributing to accelerated growth and eutrophication of the hedgerows. This was previously documented through detailed biomass quantifications on different hedgerow segments and depended on the hedgerow species, its spatial distribution and the time of growth after pruning (Uckert *et al.*, 1997). Total biomass accumulation of two typical hedgerows reached up to 6 tons dry matter per 100 m length (Table 2). This is equivalent to a yearly accumulation rate between 5.13 and 7.82 kg m^{-1} dry matter. In addition, high N accumulation in the hedgerow dry matter (Table 2) and high mineral concentrations in the ash fraction after combustion (Uckert *et al.*, unpublished) indicate the luxury consumption of fertiliser inputs of hedgerows growing–close to intensively managed fields. Compared to other renewable energy sources in Northern Germany such as short rotation coppice (*Salix spp.*) or *Miscanthus* - the hedgerow vegetation seem to be an attractive alternative as biofuel (Mette *et al.*, 2000).

Table 2. Dry weigth and nitrogen accumulation of two different hedgerow segments

	Dry weigth, kg 100 m^{-1}		N-accumulation, kg N ha^{-1} yr^{-1}	
Hedgerow	A1	A3	A1	A3
	5912	6060	24.1	37.7

Hedgerow A1: 19 years of growth after pruning
Hedgerow A3: 13 years of growth after pruning

Conclusions

Hedgerows as eminent nature conservation elements of farming systems in Northern Germany are heavily affected by the intensive crop production on neighbouring fields. Their high nutrient status and the increase in biomass production is governed by intensive root exploration of fertilised arable land. With respect to the basic necessity to replace fossil fuels the hedgerow biomass could be used as a renewable biofuel and energy source in future regional circular flow economies. Additional use of the ash residues as fertiliser substitutes would help to further close the cycles and help to fulfil the specific law of nature conservation to conserve and regularly prune these hedgerows as an important component of the typical agricultural systems in this region.

References

Mette R 1994 PhD. Thesis. Ulrich E. Grauer Verlag, Stuttgart. p. 126.

Mette R and Sattelmacher B 1994 *In* Proceedings of the 7th International symposium of CIEC. Eds. E Welte, I Szabolcs and R F Hüttl. pp. 275-284.

Mette R, Uckert G and Sattelmacher B 2000 *In* 1st World Conference and Exhibition on Biomass for Energy and Industry. James and James, (in press).

Uckert G, Mette R and Sattelmacher B 1997 VDLUFA Schriftenreihe 46, Kongressband, pp. 583-586.

W. J. Horst et al. (Eds), Plant nutrition – Food security and sustainability of agro-ecosystems. 1028~1029, 2001

Integration of plant nutrition and soil fertility in international and interdisciplinary teaching activities on sustainable land use and natural resource management

T. Müller[1)], Q. Gausset[2)], P. Van der Keur[3)], O. Mertz[4)], P. Oksen[3)] and A. Aalbæk[6)]

[1)] *Department of Soil Biology, University of Kassel, Nordbahnhofstr. 1a, Witzenhausen, D-37213, Germany, tmuller@wiz.uni-kassel.de*

[2)] *Institute of Anthropology, University of Copenhagen, Frederiksholms Kanal 4, Copenhagen, DK-1220, Denmark, quentin.gausset@anthro.ku.dk*

[3)] *Department of Geography and International Development Studies, Roskilde University Centre, Building 19.1, Post Box 260, Roskilde, DK-4000, Denmark, pvdk@ruc.dk*

[4)] *Institute of Geography, University of Copenhagen, Øster Voldgade 10, Copenhagen, DK-1350, Denmark, om@geogr.ku.dk*

[5)] *Department of Economics and Natural Resources, The Royal Veterinary and Agricultural University, Rolighedsvej 23, Frederiksberg, DK-1958, Denmark, aaa@kvl.dk*

Key words: educational programme, sustainable land use

Abstract

Issues on Plant Nutrition and Soil Fertility were integrated in an ongoing interdisciplinary post-graduate programme on sustainable land use and natural resource management. The programme is cooperating with similar educational activities in Southeast Asia and southern Africa. An interdisciplinary basic course is compulsory for all participants. This course integrates teaching on general issues (such as basic indicators of soil fertility, soil sampling and analysis, soil degradataion), methodological exercises (such as field soil analysis) and three weeks of field work in a tropical country. Further courses have to be selected by the students according to their specialisation such as “Integrated Plant Nutrient and Soil Fertility Management in the Tropics” or “Soil, Water and Plant Nutrients in the Tropics”.

The Danish University Consortium on Sustainable Land Use and Natural Resource Management

The Danish University Consortium on Sustainable Land Use and Natural Resource Management (DUCED-SLUSE) comprises three Danish universities: The Royal Veterinary and Agricultural University (KVL), University of Copenhagen (KU) and Roskilde University Centre (RUC)). It facilitates education and research on a wide range of topics pertaining to sustainable land use and natural resource management. One of the objectives is the improvement and development of relevant university curricula in Denmark

Six assistant or associate professors are employed by the programme at the three Danish Universities. They have backgrounds in agriculture (such as soil fertility and plant nutrition), forestry, physical geography (particularly land use systems and hydrology), human geography and anthropology. Additional university teachers from different disciplines are collaborating with the programme.

The curriculum

To fulfil the above-mentioned objective, a one-year educational programme has been developed at the Danish universities. It integrates students and university teachers from very different disciplines: Agriculture, Horticulture, Forestry, Biology, Chemistry, Physical and Human Geography, Anthropology, International Development Studies, Politics, Economy and Public Administration.

The DUCED-SLUSE-curriculum is fully integrated in the first two semesters of common post-graduate programmes at the three Danish universities. It is the aim of this curriculum to prepare the students for an interdisciplinary cooperation across their specific subjects rather than to educate multi-disciplinary developing workers without specific knowledge in any subject.

The interdisciplinary joint basic course

In the first post-graduate semester, an interdisciplinary basic course is compulsory for all students taking part in the DUCED-SLUSE-programme.

The aims of this basic course are

- to enhance students sensitivity to interdisciplinary approaches in development work in the context of sustainable land use and natural resource management
- to teach the students how to contribute with their specific background (such as in plant nutrition and soil fertility) to interdisciplinary field work and what they can expect from experts with a different background
- to teach the students socio-economic and natural scientific field methods
- to give the students field work experience in a tropical country. This includes close cooperation with local students, government agencies and farmers.

The basic course begins with a three days internship. The course also integrates a three weeks field stay in one of the partner countries.

During the internship socio-economic and natural scientific methods are taught. The latter includes a general introduction to plant nutrition and to basic parameters relevant for the estimation of soil fertility. In the following exercise, the students are trained in soil sampling techniques and field measurements of soil pH, electrical conductivity, soil colour, estimation of soil texture and the measurement of plant available nutrients with an N-P-K-pH soil test kit developed by one of the collaborating universities in Thailand (KASETSART university, Bangkok). The introduction and the exercise are open for all students even for those with a socio-economic background. Although the latter may not use these methods during fieldwork, they may get an impression about their importance if talking about soil fertility. Furthermore, it opens up their mind for natural science input in development programmes.

Another important aspect of the internship is the introduction to the field work and the formation of students groups. Each students group receives a collection of basic data and information on an area to be investigated during the project. Basic data and information were collected beforehand during two to three weeks of field work by local and Danish university teachers. The data collection contains also a broad problem formulation for each group. The final problems identified by the students have an interdisciplinary character and focus on subjects within the field of sustainable land use and natural resource management. Plant nutrition and soil fertility are important aspect in many of the projects. The students of each group are selected with respect to interdisciplinary group formation and with respect to the problem formulation.

After the internship, the students have about six weeks to prepare a field project. At the same time weekly lectures with subsequent plenum or group discussions introduce general subjects: "Land Degradation", "The Concept of Sustainability", "Agricultural Intensification", "Environmental Management", "Intercultural Communication" and "Logical Framework Approach". Plant nutrition and soil fertility are integrated parts of many of the lectures and discussions.

The central part of the basic course is a three weeks field stay in Malaysia, Thailand or southern Africa. The Danish students groups are merged with local students groups which have gone through a similar educational programme and which have prepared field projects on the same problem formulations. E-mail and video-link contact between the Danish and the local students was established before the meeting in the countries of fieldwork, but communication proved to be difficult until the groups met in person.

At the beginning of the field stay, Danish and local students integrate their projects. During the following period of field work, local and Danish students live together with their university teachers in base and field camps. Often, field camps are situated in villages or longhouses. This facilitates the close contact to the local people. Investigations of soil fertility and of issues within plant nutrition are an important part of the field work. Field laboratories are established at base and/or field camps. The field work is interrupted by a mid-term evaluation. Finally, local and Danish students jointly present their results and write a debriefing report.

After returning to Denmark, the Danish students have about six weeks to prepare reports on their field work. Each group has to defend its project report in front of a Danish expert panel consisting of two university teachers and an external examiner. In some cases, university teachers of the collaborating countries are members of the expert panels.

Further educational activities

Parallel to the basic course in the first post-graduate semester and in the following second semester, the students may choose between other specific and/or interdisciplinary courses at their own universities and at the other universities of the DUCED-SLUSE consortium. Some courses have been developed with special focus on the DUCED-SLUSE programme. Examples are "Environmental Impact Assessment" and an "Introduction to Natural Resource Management in Southeast Asia". Courses with emphasise on plant nutrition and/or soil fertility are "Environment, Natural Resources and Agriculture", "Tropical Crop Production", "Integrated Plant Nutrient and Soil Fertility Management in the Tropics", "Soil, Water and Plant Nutrients in the Tropics" and "Soil Pollution in the Tropics".

Cooperation

Similar programmes have already been established in Malaysia (MUCED-SLUSE) and Thailand (TUCED-SLUSE). The local students collaborating during the field stay take part in these programmes. In Southern Africa, a similar programme is under preparation.

Ph.D. courses in collaboration with universities in cooperating countries are currently under development.

Conclusion

DUCED-SLUSE facilitates the integration of Plant Nutrition and Soil Fertility in an interdisciplinary post-graduate programme on Sustainable Land Use and Natural Resource Management. Students with a background in Plant Nutrition and Soil Fertility improved their theoretical knowledge and practiced a wide range of field methods under realistic working conditions in a tropical country. Furthermore, they experienced their role in an interdisciplinary project and they learned how other disciplines can contribute to it. Students without a background in Plant Nutrition and Soil Fertility, especially those from socio-economic sciences, experienced the importance of these subjects for investigations on Sustainable Land Use and Natural Resource Management.

Author index

Appendix of manuscripts

W. J. Horst et al.(Eds.), Plant nutrition – Food security and sustainability of agro-ecosystems. 1040~1041, 2001

Analysis of michorrizae populations in maize (*Zea mays*) cultivated in acid soil under different agronomic treatments

A. M. Serralde and M. Ramírez

Programa Nacional de Recursos Biofisicos. Corporación Colombiana de Investigación Agropecuaria. Centro de Investigación Tibaitatá, Km 14 vía Mosquera. Colombia. mramirez@corpoica.org.co

Key words: acid soil, arbuscular michorrizae, *cowpea*, chicken manure, DNA extraction, PCR amplification, specific primers, universal primers, *Zea mays*.

Abstract

Analysis of native populations of arbuscular michorrizae in symbiotic association whith two maize cultivars (Sikuani V-109 and ICA V-110) cultivated in acid soil using chicken manure as organic source and *cowpea* as treatments. The soil and root samples for each treatment and control were taken three months after germination and evaluated for number of spores and colonisation percentage. The spores where extracted from 10 g by sucrose gradient centrifugation and counted. The root were evaluated microscopically after stained using Trypan Blue. The percentage of colonisation was between 83 and 90%, and the amount of spores per gram of soil was always under 100. The spores isolated from the soil were washed with sterile water for DNA extraction directly in PCR tubes for DNA amplification and evaluated to identify the genus presents in each sample. The knowledge of the ecology of MA fungi will let us propose its use as bio-fertiliser for the conditions evaluated.

Introduction

The symbiotic relation between MA fungi and roots of almost all plants is beneficial, the fungi colonise the cortex of root to obtain carbon from their plant host, while assisting the plant whith the uptake of phosphate and other mineral nutrients from the soil (Harrison & Buuren, 1995). This benefits are seen in many agriculturally important crop species like maize for increasing their productivity. In Los Llanos Orientales of Colombia, the soil has a high content of Aluminium interchange, acidity and low fertility, (CORPOICA, 1998). In order to improve the crop productivity in this region the use of MA fungi as biofertiliser has been proposed, however, the knowledge of native MA fungi is scare.

One of the major limitation for the evaluation of native arbuscular michorrizae is that the morphological characters used for taxonomy are based on the spores produced in soil. This structures might vary, because they are subjected to alterations during spores ontogeny. For this reason the methods using specific DNA primers (Simon et al, 1992; Gardes et al, 1993) for amplification by PCR (Polymerase Chain Reaction) are the most promising for ecological evaluation. (Clapp et al, 1995).

Materials and methods

The soil and root samples were taken randomly (4 repetitions per treatment) three month after germination in La Libertad (Centro de Investigación - CORPOICA, Colombia). The extraction and quantification of spores was done by a combination of the methods propose by Gerdemann and Nicholson (1963) and Jenkins (1964). The method used for the evaluation of colonisation on roots (Phillips and Hayman, 1970) was modified.

The DNA evaluation was done using a variation of the technique from Clapp et al, 1995, that includes a nested PCR. For pre-amplification by PCR the universal primers SS1492 (Simon et al, 1993) and NS21 (Clapp, et al, 1995) were used. The genus specific primers VAGIGA, VAGLO, VAACAU were used in conjunction with VANS1 (Simon et al, 1993). The results of the amplification were examined by gel electrophoresis using 1.5% agarose.

Results and discussion

There are native spores although in low number and there is a high colonisation, over 80% on maize root under the conditions evaluated. However, the differences between treatments are not significant (Table 1) and does not indicate any effect of the treatment on the symbiotic establishment.

Table 1. Total spores per gram

Organic source	Cultivar	No. Spores g^{-1}*	% colonization. *
None	ICA V-110	74	86.11
Cowpea	ICA V-110	74	94.00
Chicken manure	ICA V-110	63	94.69
None	Sikuani	52	87.77
Cowpea	Sikuani	67	92.42
Chicken manure	Sikuani	48	83.28

* Average 4 repetitions

The methodology used for the morphological evaluation of DNA using the combination of universal and specific primers help us to identify the genus that confirm the population. The amplification using one total spore and the specific primers (VAGIGA and VANS1), show differences in the size of the amplified fragment (Figure 1). This differences permit us to evaluate the spores extracted from the soil for each case and to define the composition of the native community. The next step was to evaluate the MA fungi directly from the colonised roots, using the same combination of primers, in order to establish preferential associations under the evaluated conditions.

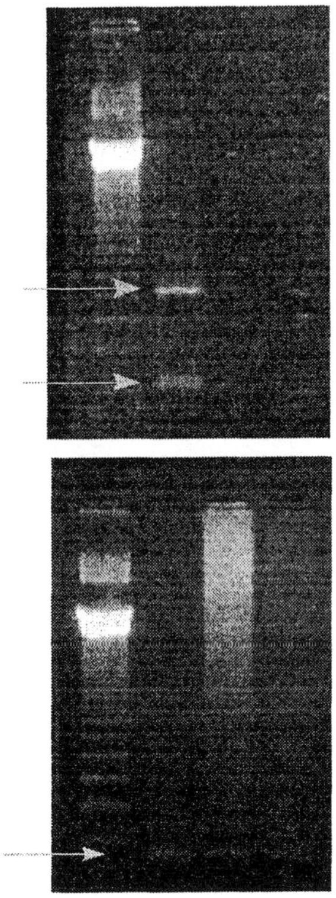

Figure 1. [Genus-specific PCR products generated after amplification of 1 spore (lines 2, 3 and 4) using VANS1 and VAGIGA for spores identified as *Gigaspora* (left) and *Entrophospora* (right). Line 1, size marker (DNA ladder 123bp)]

References

Clapp J, Young J, Merryweather J and Fitter A 1995 New Phytol. 130, 259-265.

Corpoica 1998 Frutos de la Investigación Corpouica cinco años. p.22, 42.

Gardes M, White T, Fortin J, Bruns T and Taylor J 1991 Can. J. Bot. 69, 180-190.

Harrison M and van Buuren M 1995 Nature 378, 626-629.

Simon L, Lévesque R and Lalonde M 1993 Appl. Environ. Microbiol. 59, 4211-4215.

W. J. Horst et al.(Eds.), Plant nutrition – Food security and sustainability of agro-ecosystems. 1042~1043, 2001

The effect of phosphate and lime application on groundnuts and bean yield grown on tropofluvent soils in east Kalimantan

Subroto

Regional Development Research Center GOR 27 September, Jl. Sambaliung 1 Kampus Gn. Kelua Samarinda Kaltim, Indonesia. p3w_unmul@smd.mega.net.id

Key words: phosphate, lime, tropofluvents, nuts production.

Abstract

Tropofluvents soils of river plains in East Kalimantan differ in their soil chemical characteristics.
P application decreased exchangeable Al^{3+} and H^+ significantly. The application of 2.5 t Calcite ha^{-1} increased soil pH to 7.4 and resulted in a 100% CEC. Groundnut yield increased to 2.4 t ha^{-1} due to application of 3.5 t Calcite ha^{-1}. Application of 5 t dolomite ha^{-1} increased string bean yield to 13.19 t ha^{-1}.

Introduction

Application of P fertilizer and liming to maintane soil fertility has to be seen as an important effort to increase food production on humid tropical soil.

Material and methods

The research was carried out on an alluvial river plain of Mahakam starting in 1996.

The first step of the research was a soil classification based on USDA soil taxonomy (Soil Survey 1994). Furthermore a physiographic description was done using a method developed by FAO (1990).

Field trials using a complete randomized block design were conducted with groundnuts and string beans. Triple Super Phosphate (TSP) and 2 types of lime (calcite and dolomite) were applied at increasing rates (see Tab. 3 and 4). A soil analysis was done at the start of the experiment and at harvest with focus on CEC, pH, exchangeable cations, and base saturation. The Duncan Multiple Range test at the 5% level was applied for statistical evaluation of the results obtained.

Results

According to the USDA soil taxonomy (Soil Survey 1994) the soils on the alluvial river plains of Mahakam can be characterized as Tropofluvents, developed from recent riverine alluvium of the Mahakam, which is periodically flooded. Therefore the soil formation process is always interrupted by new material deposited by flooding during the rainy season.

The soil texture varies from fine to coarse, and the permeability is low. The structure is blocky and firm when moist and sticky to plastic when wet.

According to the soil fertility criteria of FAO (1990) these Tropofluvents on alluvial Mahakam river plains in East Kalimantan can be characterized as less fertile. This is indicated by a low pH (< 5.5) and an effective CEC of < 150 mmol kg^{-1} soil, a potential CEC of < 370 mmol kg^{-1} soil, an available P < 150 ppm, and N < 0.2 % (Tab. 1). These results are due to the poor soil parent material as bivalent cations are leached out by high rainfall.

From results shown in Table 2 it can be concluded that P fertilizing decreased exchangeable Al^{3+} and H^+ in the soil. According to Tisdale and Nelson (1966) and Sanchez (1976) application of P fertilizer will cause an increase of soil acidity. This is because the reaction of Al and HPO_4^{2-} or $H_2PO_4^-$, which will increase H^+ concentration in the soil solution.

Table 1. Soil characteristics of tropofluvent soils of river plains in East Kalimantan

	range
pH H_2O	4.5 – 6.0
pH KCl	3.5 – 4.5
EC Ca^{2+} [mmol kg^{-1} soil]	5 – 110
EC Mg^{2+} [mmol kg^{-1} soil]	3 – 25
EC Na^+ [mmol kg^{-1} soil]	0.1 – 0.2
EC K^+ [mmol kg^{-1} soil]	5 – 10
EC H^+ [mmol kg^{-1} soil]	42.5 – 225
EC Al^{3+} [mmol kg^{-1} soil]	0.3 – 42.5
Potential CEC	107 – 374
Effective CEC	43 – 65
Potential BS	17.6 – 38
Effective BS	30.0 – 100
% C	0.3 – 3.0
% Organic Matter	0.6 – 4.1
% N	0.03 – 0.2
Total P [ppm]	400 - 950
Available P [ppm]	3.7 – 150
Total K [ppm]	200 - 1000
Available K [ppm]	56 - 143

P fertilization increased yield of groundnut, mung bean and string bean significantly (data not shown).

Liming using calcite and dolomite changed the chemical soil properties (increase of pH, exchangeable Ca and Mg) and consequently reduced the concentation of exchangeable Al and H in the soil (Tab. 3). This will stimulate nutrient availability in the soils as discussed by Sanchez (1976), Tisdale and Nelson (1966) and Sys

(1977).

Liming increased the yield of groundnuts and sting beans (Tab. 4). The application of 0.5 to 4.0 t Calcite per ha led to conciderably higher groundnut yields compared to the regional and national level.

Table 2. Effects of P application on soil parameters of tropofluvent soils of river plains in East Kalimantan (B = before P application; A = after P application).

	pH		Al^{3+}	H^+
	H_2O	KCl	mmol kg^{-1}	mmol kg^{-1}
B	5.23	4.12	46.0	42.5
A	5.11	3.85	8.1	7.8

Conclusion

Tropofluvents soils on Mahakam riverine alluvial plains clearly responsed to the application of lime as could be shown via higher groundnut and sting bean yields.

The application of calcite and dolomite also had a strong effect to the soil pH, CEC and exchangeable Ca and Mg, resulting in a decreasing Al and H concentration.

Refrence

FAO 1990 Guide Lines Soil Description, Soil Resources, Management and Conservation Service, Land and Water Development Division, Rome, Italy.

Sanchez PA 1976 Properties and Management of Soil in The Tropic. John Wiley and Sons, New York.

Soil Survey Staff 1994 Keys to Soil Taxonomy. USDA – SCS. Sixth Edition. Washington D.C.

Sys C 1977 Tropical Soils. State University of Ghent ITC. For Post Graduate Soil Scientist. Ghent.

Tisdale SL and Nelson WL 1966 Soil Fertility and Fertilizer. The Macmillan Company Collier – Macmillan Limited, London.

Table 3. The effect of liming ($CaCO_3$) on soil characteristics of tropofluvent soils of river plains in East Kalimantan

Dosage t ha^{-1}	pH KCl	pH H_2O	Exchangeable cation Ca	Mg	Na	K	Al	H	C E C (mmol kg^{-1})	Base saturation (%)
0.0	4.9	5.9	11.0	2.0	0.0	0.3	0.04	1.5	159	84
0.5	6.1	7.0	11.1	2.0	0.0	0.2	0.02	1.0	153	87
1.0	6.3	7.4	11.4	2.1	0.0	0.3	0.00	0.0	143	97
1.5	6.4	7.4	11.7	2.5	0.0	0.2	0.00	0.0	150	96
2.0	6.1	7.2	11.2	2.6	0.0	0.4	0.00	0.0	157	90
2.5	6.6	7.4	29.6	5.6	0.0	0.6	0.00	0.0	368	97
3.0	6.1	7.1	18.5	3.7	0.0	0.6	0.00	0.0	222	100
3.5	6.4	7.5	20.1	3.9	0.0	0.6	0.00	0.0	266	93
4.0	6.6	7.5	20.5	4.0	0.0	0.7	0.00	0.0	269	94

Table 4. Groundnuts and sting beans response to calcite and dolomite addition on a tropofluvent soil in East Kalimantan.

Lime	Dosage (t ha^{-1})	Plants	Yield (t ha^{-1})	regional	national
Calcite	0.5	Groundnut	1.9	0.74	1.01
	2.5		2.1		
	3.5		2.4		
	4.0		2.3		
	4.5		2.2		
Dolomite	3.0	Beans	11.52	5.2	15.0
	5.0		13.19		